Global Climate Change
and Plant Stress Management

Global Climate Change and Plant Stress Management

Edited by

Mohammad Wahid Ansari
Zakir Husain Delhi College (Delhi University),
New Delhi, India

Anil Kumar Singh
ICAR-National Institute for Plant Biotechnology
New Delhi, India

Narendra Tuteja
ICGEB,
New Delhi, India

WILEY

This edition first published 2023

Registered Offices
John Wiley & Sons, Inc., 111 River Street, Hoboken, NJ 07030, USA
John Wiley & Sons Ltd, The Atrium, Southern Gate, Chichester, West Sussex, PO19 8SQ, UK

For details of our global editorial offices, customer services, and more information about Wiley products visit us at www.wiley.com.

Wiley also publishes its books in a variety of electronic formats and by print-on-demand. Some content that appears in standard print versions of this book may not be available in other formats.

Library of Congress Cataloging-in-Publication Data
Names: Ansari, Mohammad Wahid, editor. | Singh, Anil Kumar (Principal scientist of plant biotechnology), editor. | Tuteja, Narendra, editor.
Title: Global climate change and plant stress management / edited by Mohammad Wahid Ansari, Anil Kumar Singh, Narendra Tuteja.
Description: Chichester, West Sussex, UK ; Hoboken, New Jersey : Wiley, 2023. | Includes index.
Identifiers: LCCN 2023000299 (print) | LCCN 2023000300 (ebook) | ISBN 9781119858522 (hardback) | ISBN 9781119858539 (adobe pdf) | ISBN 9781119858546 (epub)
Subjects: MESH: Plants–genetics | Plants–metabolism | Stress, Physiological–genetics | Adaptation, Physiological | Carbon Dioxide–physiology | Climate Change
Classification: LCC QK981.3 (print) | LCC QK981.3 (ebook) | NLM QK 981.3 | DDC 572.8/2–dc23/eng/20230518
LC record available at https://lccn.loc.gov/2023000299
LC ebook record available at https://lccn.loc.gov/2023000300

Cover Design: Wiley
Cover Image: © Salvideo/E+/Getty Images

Set in 9.5/12.5pt STIXTwoText by Straive, Pondicherry, India

Prof. Jitendra Paul Khurana
(30 October 1954 to 27 October 2021)

Prof. Jitendra Paul Khurana was an Indian botanist known for his contributions to the fields of plant molecular biology. He obtained a PhD in Botany from the University of Delhi in 1982 and did postdoctoral work at the prestigious Smithsonian Institution, Washington, DC (1985–1986) and the Michigan State University, USA (1986–1988). He was a visiting professor at the USDA, Beltsville, between 1996 and 1998, and at the Waksman Institute of Microbiology, Rutgers University, USA, in 2001. He was a founder faculty member of the Department of Plant Molecular Biology, University of Delhi South Campus. He was a J.C. Bose National Fellow of SERB at the University of Delhi South Campus. He was the Vice President of the Indian National Science Academy (INSA) for International Affairs. Prof. Khurana was Pro-Vice Chancellor (Interim), University of Delhi, and Director, University of Delhi South Campus, for over three years (2016–2019); he also had additional charge as Dean (Colleges).

Prof. Khurana's work on Arabidopsis *mutants led to the identification of a novel blue light receptor, phytotropin 1, which primarily controls phototropism and leaf orientation to capture maximal solar energy for photosynthesis. Recently, his group has demonstrated the role of other blue light receptors, cryptochrome 1 (CRY1), in controlling plant height in mustard and CRY2 in regulating flowering time in both mustard and rice. He played a key role in sequencing of rice, tomato, and wheat genomes as part of the International Consortia. Using in-house expertise, they provided evidence for bZIP and F-box protein-coding genes in regulating light, hormone, and stress signaling leading to panicle and seed development. OsbZIP62 serves as Flowering Locus D (FD), preferentially expressing in the shoot apical meristem, and interacts with the mobile flowering signal "florigen" (FT) to regulate the transition to flowering and panicle development in rice. His recent work stressed on the role of the bZIP and F-box proteins in abiotic stress responses and the interplay of light and environmental stress in plant development. His work is documented in over 200 publications. Professor Khurana was the elected Fellow of all the National Science Academies (INSA, IASc, NASI, and NAAS) and the World Academy of Sciences (TWAS), Trieste, Italy. His other honors and awards include the Tata Innovation Fellowship (2010–2013) by the DBT, "J.C. Bose National Fellowship" by the DST-SERB (2013 onward), Birbal Sahni Medal by the Indian Botanical Society (2011), "Goyal Prize" in Life Sciences (2017) by the Goyal Foundation, Shri Om Prakash Bhasin Award in Biotechnology (2017), and Jawaharlal Nehru Birth Centenary Visiting Fellowship (2019) by INSA, to name a few.*

This book is dedicated to the memory of Prof. Jitendra P. Khurana as a token of our appreciation and respect for him and his achievements.

Contents

List of Contributors

Ganesh Kumar Agrawal
Department of Education, Global Research Arch for Developing Education (GRADE) Academy Pvt. Ltd.
Birgunj, Nepal

Department of Biotechnology, Research Laboratory for Biotechnology and Biochemistry (RLABB)
Kathmandu, Nepal

Naser A. Anjum
Department of Botany
Aligarh Muslim University
Aligarh, India

Mohammad Wahid Ansari
Department of Botany
Zakir Husain Delhi College
University of Delhi
New Delhi, India

Rajeev Nayan Bahuguna
Center for Advanced Studies on Climate Change
Dr. Rajendra Prasad Central Agricultural University
Samastipur, Bihar, India

Agriculture Biotechnology
National Agri-Food Biotechnology Institute
Sector 81, SAS Nagar
Mohali, India

Gurdeep Bains
Department of Plant Physiology
Govind Ballabh Pant University of Agriculture & Technology
Pantnagar, Uttarakhand, India

Shivendra Bajaj
Federation of Seed Industry of India
New Delhi, India

Alma Balestrazzi
Department of Biology and Biotechnology
University of Pavia
Pavia, Italy

Ruchi Bansal
Division of Plant Physiology
ICAR-Indian Agricultural Research Institute
New Delhi, India

Sahana Basu
Department of Life Science
Central University of South Bihar
Gaya, Bihar, India

Jyotsna Bharti
Plant Biology and Biotechnology
Nutritional Improvement of Crops Group
International Centre for Genetic Engineering and Biotechnology (ICGEB)
New Delhi, India

Abhishek Bohra
Centre for Crop & Food Innovation
State Agricultural Biotechnology Centre
Food Futures Institute
Murdoch University, Murdoch
Western Australia, Australia

Isorchand Chongtham
Department of Molecular Biotechnology and Health Sciences
Molecular Biotechnology Center
University of Turin, Turin, Italy

Sudeshna Das
Center for Advanced Studies on Climate Change
Dr. Rajendra Prasad Central Agricultural University
Samastipur, Bihar, India

Karabi Datta
Department of Botany
University of Calcutta
Kolkata, India

Swapan K. Datta
Department of Botany
University of Calcutta
Kolkata, India

Sharma Deepika
Department of Botany
Zakir Husain Delhi College
University of Delhi
New Delhi, India

Laboratory of Soil Biology and Microbial Ecology
Department of Environmental studies
University of Delhi
New Delhi, India

H. K. Dikshit
Department of Genetics
ICAR-Indian Agricultural Research Institute
New Delhi, India

Conrado Dueñas
Department of Biology and Biotechnology
University of Pavia
Pavia, Italy

Murugesh Eswaran
Plant Biology and Biotechnology, Nutritional Improvement of Crops Group
International Centre for Genetic Engineering and Biotechnology (ICGEB)
New Delhi, India

Shraddha Shridhar Gaonkar
Department of Biology and Biotechnology
University of Pavia
Pavia, Italy

Vikrant Goswami
Laboratory of Soil Biology and Microbial Ecology
Department of Environmental studies
University of Delhi
New Delhi, India

Adriano Griffo
Department of Biology and Biotechnology
University of Pavia
Pavia, Italy

Ramwant Gupta
Department of Biology
University of Guyana
Georgetown, South America

Present Address: Department of Botany
Deen Dayal Upadhyaya Gorakhpur University
Gorakhpur, UP, India

Ratnesh K. Jha
Center for Advanced Studies on Climate Change
Dr. Rajendra Prasad Central Agricultural University
Samastipur, Bihar, India

Rohit Joshi
Division of Biotechnology
CSIR-Institute of Himalayan Bioresource Technology
Palampur, Himachal Pradesh, India

Academy of Scientific and Innovative Research (AcSIR)
CSIR-HRDC Campus
Ghaziabad, Uttar Pradesh, India

Rashmi Kaul
Plant Biology and Biotechnology
Nutritional Improvement of Crops Group
International Centre for Genetic Engineering and Biotechnology (ICGEB)
New Delhi, India

Tanushri Kaul
Plant Biology and Biotechnology
Nutritional Improvement of Crops Group
International Centre for Genetic Engineering and Biotechnology (ICGEB)
New Delhi, India

Nafees A. Khan
Department of Botany
Aligarh Muslim University
Aligarh, India

SM Paul Khurana
Amity Institute of Biotechnology
Amity University Haryana
Gurugram, Haryana, India

David Kothamasi
Laboratory of Soil Biology and Microbial Ecology
Department of Environmental studies
University of Delhi
New Delhi, India

Strathclyde Centre for Environmental Law and Governance
University of Strathclyde Glasgow
United Kingdom

Gautam Kumar
Department of Life Science
Central University of South Bihar
Gaya, Bihar, India

Narendra Kumar
Department of Botany
Guru Ghasidas Vishwavidyalaya
a Central University
Bilaspur-495009(C. G.), India

Sudarshna Kumari
Department of Plant Physiology
Govind Ballabh Pant University of Agriculture & Technology
Pantnagar, Uttarakhand, India

Ratna Kumria
Federation of Seed Industry of India
New Delhi, India

Sayanta Kundu
Center for Advanced Studies on Climate Change
Dr. Rajendra Prasad Central Agricultural University
Samastipur, Bihar, India

Anca Macovei
Department of Biology and Biotechnology
University of Pavia
Pavia, Italy

Shuvobrata Majumder
Department of Botany
University of Calcutta
Kolkata, India

Mamun Mandal
Laboratory of Applied Stress Biology
Department of Botany
University of Gour Banga
Malda, West Bengal, India

Asim Masood
Department of Botany
Aligarh Muslim University
Aligarh, India

Piyush Mathur
Department of Botany
Microbiology Laboratory
University of North Bengal
Darjeeling, West Bengal, India

Jyoti Maurya
National Institute of Plant Genome Research (NIPGR)
Jawaharlal Nehru University
New Delhi, India

Moaed Al Meselmani
School of Biosciences
Grantham Centre
The University of Sheffield
Sheffield, England

Francesca Messina
Department of Biology and Biotechnology
University of Pavia
Pavia, Italy

Khaled Fathy Abdel Motelb
Plant Biology and Biotechnology
Nutritional Improvement of Crops Group
International Centre for Genetic Engineering and Biotechnology (ICGEB)
New Delhi, India

Mamta Nehra
Plant Biology and Biotechnology
Nutritional Improvement of Crops Group
International Centre for Genetic Engineering and Biotechnology (ICGEB)
New Delhi, India

Andrea Pagano
Department of Biology and Biotechnology
University of Pavia
Pavia, Italy

Paola Pagano
Department of Biology and Biotechnology
University of Pavia
Pavia, Italy

Hadi Pirasteh-Anosheh
Department of Agronomy Research
National Salinity Research Center
Agricultural Research
Education and Extension Organization (AREEO)
Yazd, Iran

Natural Resources Department
Fars Agricultural and Natural Resources Research
and Education Center
AREEO, Shiraz, Iran

Manoj Prasad
National Institute of Plant Genome Research (NIPGR)
Jawaharlal Nehru University
New Delhi, India

Ravinesh Rohit Prasad
Department of Geography
Fiji National University
Lautoka, Fiji Islands

Swati Priya
Department of Botany
Kurukshetra University
Kurukshetra, Haryana, India

Nandula Raghuram
Centre for Sustainable Nitrogen and Nutrient Management
University School of Biotechnology
Guru Gobind Singh Indraprastha University
New Delhi, India

Randeep Rakwal
Department of Education, Global Research Arch
for Developing Education (GRADE)
Academy Pvt. Ltd.
Birgunj, Nepal

Department of Biotechnology, Research Laboratory
for Biotechnology and Biochemistry (RLABB)
Kathmandu, Nepal

Department of Health and Sport Science, Faculty of
Health and Sport Sciences
University of Tsukuba, Tsukuba, Japan

Faisal Rasheed
Department of Botany
Aligarh Muslim University
Aligarh, India

Anamika Roy
Laboratory of Applied Stress Biology
Department of Botany
University of Gour Banga
Malda, West Bengal, India

Swarnendu Roy
Department of Botany
Plant Biochemistry Laboratory
University of North Bengal
Darjeeling, West Bengal, India

Bhuvnesh Sareen
Division of Biotechnology
CSIR-Institute of Himalayan Bioresource Technology
Palampur, Himachal Pradesh, India

Abhijit Sarkar
Laboratory of Applied Stress Biology
Department of Botany
University of Gour Banga
Malda, West Bengal, India

Samir Sharma
Department of Biochemistry
University of Lucknow
Lucknow, India

Zahid Hameed Siddiqui
Department of Biology
Faculty of Science, University of Tabuk
Tabuk, Saudi Arabia

Genomic and Biotechnology Unit
Department of Biology
Faculty of Science, University of Tabuk
Tabuk, Saudi Arabia

Roshan Kumar Singh
National Institute of Plant Genome Research (NIPGR)
Jawaharlal Nehru University
New Delhi, India

Satish K. Singh
Department of Plant Breeding and Genetics
Dr. Rajendra Prasad Central Agricultural University
Samastipur, Bihar, India

Sonia Khan Sony
Plant Biology and Biotechnology
Nutritional Improvement of Crops Group
International Centre for Genetic Engineering
and Biotechnology (ICGEB)
New Delhi, India

Sudhir K. Sopory
Department of Plant Molecular Biology
International Centre for Genetic Engineering and Biotechnology
New Delhi, India

Rewaj Subba
Department of Botany
Microbiology Laboratory
University of North Bengal
Darjeeling, West Bengal, India

Arul Prakash Thangaraj
Plant Biology and Biotechnology
Nutritional Improvement of Crops Group
International Centre for Genetic Engineering and Biotechnology (ICGEB)
New Delhi, India

Palaniswamy Thangavel
Department of Environmental Science
Periyar University
Salem, India

Baishnab C. Tripathy
Department of Biotechnology
Sharda University
Greater Noida, India

Manas K. Tripathy
Division of Plant and Microbial Biotechnology
Institute of Life Sciences
Bhubaneswar, Odisha, India

Narendra Tuteja
Plant Molecular Biology Group
International Centre for Genetic Engineering and Biotechnology
New Delhi, India

Rajeev K. Varshney
Centre for Crop & Food Innovation
State Agricultural Biotechnology Centre
Food Futures Institute
Murdoch University
Murdoch, Western Australia, Australia

Rachana Verma
Plant Biology and Biotechnology
Nutritional Improvement of Crops Group
International Centre for Genetic Engineering and Biotechnology (ICGEB)
New Delhi, India

Ratnum K. Wattal
Department of Botany
Zakir Husain Delhi College
University of Delhi
New Delhi, India

Priya Yadav
Department of Botany
Zakir Husain Delhi College
University of Delhi
New Delhi, India

Foreword

I am overjoyed to write about this book, *Global Climate Change and Plant Stress Management*, which symbolizes a comprehensive and current exchange on the newest insights into the improvement of crops under climate change and plant stress management. At the present time, the global climate change (see, e.g. The Discovery of Global Warming by Spencer R, Weart, 2008, Harvard University Press) and the population increase are two important restrictions before us, and dealing with these crucial issues is of paramount importance in the field of agriculture. This book deals with a subject of enormous significance not only for plant scientists but also for farmers worldwide. Research on exploring diverse aspects of an easy, money-making, and ecologically oriented practice of pre-soaking seeds in salt solutions (what one calls "halo priming") seems desirable, as it might aid in sustaining agricultural production in our changing environment. The recent trend in climate change involving high salinity, increased temperature, draught, and heavy metal toxicity, as well as negative effects of bacterial, viral, and fungal diseases and insect infestation have gloomy effects on agriculture productivity. To add to this, predicted increase in CO_2, with ocean acidification, is expected to cause a drastic decline in global agriculture productivity. All of this will have a considerable harmful impact on our ecosystem. Thus, this particular volume, which deals with various aspects of plant stress physiology, together with plant stress responses, and physiological and molecular mechanism of plant tolerance to environmental stresses, is particularly welcome. It goes a long way toward finding ways to overcome the gloomy predictions before us.

On a positive note, the potential function of several important genes and, thus, the proteins that they code for, as well as a range of signaling molecules, such as plant hormones, that regulate plant growth and developmental processes, is now available. The above was possible because of detailed studies on responses of tolerant and susceptible agriculturally important crops to climate change from both physiological and biotechnological points of view. On the other hand, developing climate-smart varieties through mutation breeding involving modification of a single gene rather than altering the whole genome is an attractive goal. Recent development in science relies on 'Omics' tools, such as genomics, transcriptomics, epigenomics, proteomics, metabolomics, and phenomics, and this is indeed being actively pursued at many institutions around the world. In addition, CRISPR/Cas techniques provide precision and rapidity to breeding programs to develop smart and nutrition crops in changing environment, which might be a key solution for ensuring food security.

The compilation of a comprehensive volume on this very important and challenging topic has been achieved in the current book entitled *Global Climate Change and Plant Stress Management*, edited by Mohammad Wahid Ansari, Anil Kumar Singh, and Narendra Tuteja; it is both commanding and timely. This book also emphasizes the effect of climate change studies on plant metabolism and adaptive characteristics; it is an up-to-date compilation for the benefit of researchers and academicians. In this book, authors introduce and classify climate change conditions as well as various stress components and then present a detailed discussion related to their effects on plant development, controlling factors of their biome, as well as the *behavior* of plants under climate change conditions and the associated adverse effects. This book also covers the new emerging technical concepts of stress management, which is an advanced concept to sustain agricultural productivity under recent climatic scenarios. Further, this book provides instant access to comprehensive, cutting-edge data, making it possible for plant scientists and others to utilize this ever-growing wealth of information. I strongly believe that this book provides a great deal of global implications not only for food security but also for the socioeconomic condition of communities affected by climate change worldwide. In addition, the knowledge presented in this book is expected to be of great benefit to the farmers, who can understand and exploit the useful crops as per the nature of the climate and benefit from it for public and private investment. The current insightful book is expected to provide key information, in an excellent manner, to students, postdoctoral fellows, plant scientists, and policymakers on what actions to take on *plant stress management* under the expected climate change. I am quite confident that this book will be read, understood, and exploited extensively.

I heartily appreciate the efforts of all the contributing 80 authors from 12 different countries – Australia, England, Fiji Island, Italy, India, Iran, Japan, Kingdom of Saudi Arabia, Nepal, South America, and the United Kingdom, and all the outstanding editors for this well-timed and enlightening publication on the important topic of the effects of global climate change on plants and what to do to alleviate its negative impact.

Govindjee Govindjee
(E-mail: gov@illinois.edu)
Professor Emeritus Plant Biology, Biochemistry and Biophysics,
University of Illinois at Urbana-Champaign
Urbana, IL, USA
15 November 2022

Preface

The existence of living organisms depends on the food synthesized by mainly green plants by capturing energy from sunlight through the process of photosynthesis. At present, a global challenge is to sustain crop productivity in a changing environment to meet the demand of increasing population. However, the current reports of the Intergovernmental Panel on Climate Change have made clear that the urgency to take action on global climate change and agrifood production is not well understood so far. There is an urgent need for agrifood systems to be more versatile to the existing and upcoming impacts of global climate change, which could be achieved through learning from superior practices, encouraging transformative adjustment strategies, plans, and its subsequent actions. A growing tendency toward climate change for the past few decades has badly hit global crop production on the large scale. It imposes environmental variations that include high salinity, very high and low temperatures, draughts, heavy metal toxicity and nutrient loss, the growth of bacteria, viruses, fungi, different pests, and parasites, harmful insect invasions, and increased CO_2 and ocean acidification. This will have a considerably harmful impact on beneficial microbes, plant productivity, restoration efforts, and ecosystem health. Global warming is expected to elicit harsh weather trends, long-lasting droughts, floods and waterlogging, storms, and increased disease incidence, which cause altered growth, impaired photosynthesis, and reduced physiological responses in plants that limit agrifood production.

Global sustainable farming systems are at risk owing to rising and co-occurring temperatures, droughts, and salinity stresses. According to a new report from the World Meteorological Organization (WMO), the impact of water stress and drought hazards, including withering droughts and overwhelming floods, is thrashing African communities and ecosystems. The strategies to deal with increased CO_2 concentrations and global warming and enhance plant tolerance to abiotic and biotic stresses are important targets for sustainable agricultural production. Recent advances in science, such as CRISPR-associated (Cas) protein-based genome editing (CRISPR-Cas) and "Omics" tools such as genomics, transcriptomics, epigenomics, proteomics, metabolomics, and phenomics, have enhanced precision and rapidity in the progress of plant molecular breeding programs to develop nutrient-enriched and stress-tolerant plant variety, which might be the key players in ensuring food security. Additionally, it will contribute a significant amount of potential for developing more resilient and climate-smart crops to respond to the rising threat of climate change and its undesirable effects on agrifood.

In the present book, *Global Climate Change and Plant Stress Management*, we present a collection of 27 chapters by 80 experts from 12 different countries – Australia, England, Fiji Island, Italy, India, Iran, Japan, Kingdom of Saudi Arabia, Nepal, South America, and the United Kingdom. This book offers a current overview of recent developments in sustainable agriculture production in a changing environment. This book aims to accentuate issues of global climate change and food insecurity for billions of people, assuming they will face drastic hunger in the upcoming period. It emphasizes all concerns about carbon and nutrient cycles, global warming, and environmental stresses originating under a scenario of global climate change and thereby badly affecting basic agriculture production, on which the common world's poor depend. This book also presents the potential ways of exploring, investigating, and adopting novel techniques and tools, methodologies, and scientific inventions to realize climate's outcomes on food security. The knowledge convened herein might be inspiring to farmers who may respond to beneficial crops as per the foretold climate. This perceptive will result in good dealings for both scientists indicted for predicting global climate threats and policymakers responsible for influential decisions in the field. The present book is specifically appropriate for environmental and biological science students engaged in interdisciplinary research, research scholars, young scientists, and faculty members. We thank the late Prof. R.C. Pant and Dr. Alok Shukla who helped us during the initial phase of this work.

Editors
Mohammad Wahid Ansari, ZHDC, University of Delhi, India
Anil Kumar Singh, ICAR-NIPB, IARI, Pusa, New Delhi, India
Narendra Tuteja (Superannuated), PMB, ICGEB, New Delhi, India

Author Biographies

Dr. Mohammad Wahid Ansari is currently an Assistant Professor in the Department of Botany, Zakir Husain Delhi College (University of Delhi), India. He has an extensive research and educational background in the field of plant molecular physiology. His special interest lies in plant hormone homeostasis and cross-talk to improve abiotic stress tolerance in plants. Dr. Ansari has published over 75 scientific papers in peer-reviewed international journals with an overall Impact Factor above 234 and citations more than 3401, h-index of 26, and i-10 index of 50. He has contributed 17 book chapters and has edited four books. As a PI, he has completed research project(s), DST/SERB Government of India, and guided PhD student(s). He is a recipient of Young Scientist Fellow-DST (ICGEB), Post-Doctoral Fellow-DBT (ICGEB), Post-Doctoral Fellow-DST (GBPUAT), and Senior Research Fellow-UPCAR (GBPUAT). He is awarded with DBT-CTEP Travel Award, Best Teacher Award (ATDS), and Best research paper award, Government of Uttarakhand. He, as convener/co-convener/coordinator/organizing secretary, has organized 10 international and national conferences/webinars and in-house workshops and has presented paper orally at INPPO, Italy. He has been a member of the Departmental Research Committee (DRC) of the University of Delhi and the Science-Setu program of NII and DBT, Government of India. He is an academic editor of *PLoS ONE* journal and is a member of *Plant Signaling and Behaviors* journal.

Dr. Anil Kumar Singh is currently the Principal Scientist at the ICAR-National Institute for Plant Biotechnology, New Delhi, India. He has been working in the field of plant molecular biology and biotechnology for more than two decades. His group has characterized genomes and transcriptomes of several important organisms, including crop plants and commercially important microbes, and developed gene resources for crop improvement. Dr Singh has published more than 80 articles in peer-reviewed international journals with cumulative Impact Factor >250, >2500 citations, and h-index 28. He has also authored 15 book chapters and delivered invited/keynote talks at >35 national and international conferences in India and abroad. He is serving as editor of various reputed journals, such as *Frontiers in Plant Science*, *PLoS ONE*, *BMC Research Notes*, *Phyton-International Journal of Experimental Botany*, and has guest edited special issues in *Antioxidants*, *Genes*, *Tree Physiology*, and *Physiologia Plantarum*. For his excellent publication record and contribution to plant molecular biology research, he has been conferred membership in the National Academy of Sciences, India (NASI) and Plant Tissue Culture Association-India.

Narendra Tuteja

An elected fellow of numerous National & International academies, Prof. (Dr.) Narendra Tuteja worked as Group Leader at International Centre for Genetic Engineering & Biotechnology (ICGEB) and Director at Amity Institute of Microbial Technology, NOIDA, India. He has made significant contributions to crop improvement under adverse conditions, reporting the first helicase from plant and human cells and demonstrating new roles of Ku autoantigen, nucleolin and eIF4A as DNA helicases. Furthermore, he discovered novel functions of helicases, G-proteins, CBL-CIPK and LecRLK in plant stress tolerance, and PLC and MAP-kinase as effectors for $G\alpha$ and $G\beta$ G-proteins. Dr. Tuteja also reported several high salinity stress tolerant genes from plants and fungi and developed salt/drought tolerant plants. Dr. Tuteja is recipient of many prestigious awards and also featured consecutively in the "World Ranking of Top 2% Scientists" prepared by Stanford University, USA and Elsevier EV.

Citation: 37,786; h-index: 80; i.10-index: 290.

Part 1

Views and Visions

1

Boosting Resilience of Global Crop Production Through Sustainable Stress Management

Rajeev K. Varshney and Abhishek Bohra

Centre for Crop & Food Innovation, State Agricultural Biotechnology Centre, Food Futures Institute, Murdoch University, Murdoch, Western Australia, Australia

Climate change is a global phenomenon with unequivocal evidence of its impact on life worldwide. The average temperature has increased at a rate of 1.7 °C per century worldwide since 1970. The Earth's surface temperature was recorded to be the warmest in the last decade (https://www.un.org/en/climatechange). According to the IPCC report, limiting the global temperature rise to 1.5 °C or even 2 °C calls for drastic and immediate measures for large-scale reductions in greenhouse gas emissions (IPCC 2021). The agriculture sector that provides livelihoods for 1.1 billion people remains one of the major contributors to greenhouse gas emissions from the aspect of land-based emissions and removals (Lamb et al. 2021).

To feed 10 billion people in 2050, cutting-edge technologies and resource-efficient strategies are needed for producing 56% more food than today without expanding the agricultural landscape (Ranganathan et al. 2018). The climate change induced rise/unpredictability in weather extremes, rainfall patterns, and pest-pathogen dynamics severely impacts global efforts to achieve food and nutritional security targets (Varshney et al. 2021a). A paradigm shift is inevitable in crop breeding strategies, which earlier focused on developing high-yielding cultivars that relied heavily on the supply of fossil energy inputs.

Genetic improvement of agricultural crops through breeding practices and advanced genomic tools is one of the most promising approaches to bring sustainable gains in productivity and resilience of the crop production systems (Bohra et al. 2020). Modern genomic resources have greatly expanded our capacity and knowledge to harness the enormous genetic and breeding potential archived in the global as well as local germplasm repositories (Bohra et al. 2022; Varshney et al. 2021b). Recent examples of large-scale genome sequencing in rice (Wang et al. 2018) and chickpea (Varshney et al. 2021c) reveal the high-resolution genetic relationships and beneficial genes or haplotypes for designing future germplasm management and breeding strategies aimed at climate change adaptation. Also, the availability of pangenomes in different crop species highlights the significance of structural variations in environmental adaptation including biotic and abiotic stress response (Della Coletta et al. 2021). The concurrent developments in plant stress phenotyping have unlocked doors for the implementation of machine learning and artificial intelligence tools for building cutting-edge predictive models based on large-scale phenotyping datasets alone or in combination with sequence information.

The precise information about key genetic loci revealed through the latest deep sequencing efforts in different crops paves the way for their rapid and targeted manipulation by genome engineering or gene editing tools. The easily customizable DNA binding specificity of the clustered regularly interspaced short palindromic repeat (CRISPR)/CRISPR-associated protein 9 (Cas9) has stimulated greater interest in this class of programmable sequence-specific nucleases for controlled genetic manipulation of the target sequence (Zsögön et al. 2017). The remarkable efficiency of the CRISPR-Cas9 system for precise genetic manipulation has opened novel routes to create ideal crop cultivars in order to broaden the array of food crops equipped with adaptation features.

Sustainable stress management of crops demands efficient utilization of the germplasm resources and appropriate modifications in cultivation techniques. Deployment of new crop cultivars that perform better in adverse environments forms the straightforward approach to cope with risks associated with the climate change. For example, the Drought Tolerant Maize for Africa (DTMA) initiative has facilitated the release of more than 200 drought-tolerant maize varieties benefitting over 43 million smallholder farmers in sub-Saharan Africa (Katengeza and Holden 2021). For breeding programs aimed at improving crop stress response, focused identification of germplasm strategy (FIGS)

Global Climate Change and Plant Stress Management, First Edition. Edited by Mohammad Wahid Ansari, Anil Kumar Singh, and Narendra Tuteja.

has been extremely useful to develop germplasms carrying specific adaptive traits, as exemplified by the identification of germplasm resources for abiotic stress in soybean (Haupt and Schmid 2020). The increased tolerance levels of new lines resulting from transfer of the gene(s) and QTL(s) controlling stress tolerance traits have been evident in several crops (Varshney et al. 2021b). Like traditional multiline approach, the availability of a series of isogenic lines carrying different resistant genes by genomic breeding strategy provides scope for the deployment of spatial and temporal rotation for disease management (Wing et al. 2018). Functional genomics research in different crops has leveraged our understanding of the morphological, physiological, and biochemical changes in crop plants that occur in response to stress. The enriched knowledge of the mechanisms involved in plant stress responses would help enhance stress management strategies to enable crops to withstand harsh climates. Novel genetic technologies such as TBF1-cassette demonstrate the potential of regulating translation of defense proteins to trigger an immune response without any deterioration in plant performance (Xu et al. 2017). Besides these genetic approaches and exogenous application of phytohormones, phytoprotectants also enhance crop responses to adverse environments. Adoption of such genetic and non-genetic approaches would support stress management strategies that in turn would reduce the indiscriminate use of chemicals as control measures.

The increased uniformity of crop varieties and species on farms has led to a growing realization that diversification of crop production and broadening crop diversity mitigate the risk associated with farming systems relying on a limited number of crop varieties and species (FAO 2010). Introducing more crop and species diversity into production systems would boost the resilience of agricultural systems. Cultivar mixtures, intercropping, crop rotation, and cover crop systems are among the major crop diversification strategies that have shown positive impacts on both crop production and biodiversity (Vernooy 2022). Improved soil fertility, reduced disease and pest damage, enhanced water use efficiency, and reduced risk of complete crop failure are the other outcomes of crop diversification. This has also stimulated efforts on the development and commercialization of farmers' varieties/landraces and underutilized species. Farmers' varieties and landraces represent the evolving reservoirs of genetic variability with high adaptation to the local environmental conditions.

Crop introductions, reintroductions, and domestication of new wild species have also contributed to diversify the farming systems and enhance domestic food production, thus transforming underutilized or minor crops into potential major crops. This is exemplified by the successful introduction of quinoa from Andean region to different countries, which in turn led to the expansion of the number of countries growing quinoa from 8 in 1980 to 75 in 2014 (Bazile et al. 2016). Reintroduction aims to protect endangered species in their natural habitats. Domestication of new wild species or manipulating domestication genes directly into wild plants has received greater attention in recent years (Li et al. 2018). Gene editing tools, owing to their ability to induce targeted genetic modifications, have demonstrated potential for rapid domestication of wild plants in several plant species including tomato (Rodríguez-Leal et al. 2017; Zsögön et al. 2018), ground cherry (Lemmon et al. 2018), and more recently rice (Yu et al. 2021). As demonstrated recently in *Vigna stipulacea* (Takahashi et al. 2019), traditional breeding approaches, including trait introgression and mutagenesis, have been applied for rapid domestication to develop future crops suitable for low-input agriculture.

Efficient seed production and distribution systems would be crucial to deliver the genetic gains from research settings to real-life climates (Varshney et al. 2021c). Besides formal seed systems that are highly regulated, informal seed systems play an important role to facilitate seed production and enable farmers' access to seed particularly in developing countries. For instance, in Africa, more than 90% of the smallholder farmers' requirements for seeds are met by the informal sector that often provides seeds of variable quality (McGuire and Sperling 2016). Policy support from governments in the form of state-sponsored subsidies can contribute significantly to improve the yield and resilience of crop production systems. This is exemplified by a national input subsidy program covering fertilizer and seed in Malawi that boosted maize productivity, resulting in a 53% surplus in 2007, which earlier had recorded a 43% national food deficit in 2005 (Denning et al. 2009). A dynamic seed system that ensures quick deployment and timely replacement of the latest cultivars will be key to impart climate adaptation to farming systems.

In summary, recent advances in plant biology have provided researchers with the tools and knowledge to develop a steady stream of crop technologies suitable for future cultivation requirements. Equally important will be the demonstration of the potential of these new crop technologies in farmers' fields. The factors that remain crucial to maximize the potential of new crop technologies are the adoption of better crop production techniques, the use of digital agriculture tools, minimizing post-harvest losses, and improving market access for farmers (Varshney et al. 2021d). International collaborations engaging multidisciplinary teams that bring together researchers with diverse skills would be essentially required to develop solutions to climate change issues that increasingly challenge the world agriculture.

References

Bazile, D., Jacobsen, S.E., and Verniau, A. (2016). The global expansion of quinoa: trends and limits. *Front. Plant Sci.* 7: 622.

Bohra, A., Jha, U.C., Godwin, I., and Varshney, R.K. (2020). Genomic interventions for sustainable agriculture. *Plant Biotechnol. J.* 18: 2388–2405.

Bohra, A., Kilian, B., Sivasankar, S. et al. (2022). Reap the crop wild relatives for breeding future crops. *Trends Biotechnol.* 40: 412–431.

Della Coletta, R., Qiu, Y., Ou, S. et al. (2021). How the pan-genome is changing crop genomics and improvement. *Genome Biol.* 22: 3.

Denning, G., Kabambe, P., Sanchez, P. et al. (2009). Input subsidies to improve smallholder maize productivity in Malawi: toward an African green revolution. *PLoS Biol.* 7: e1000023.

FAO (2010). *The Second Report on the State of the World's Plant Genetic Resources for Food and Agriculture*. Rome: FAO.

Haupt, M. and Schmid, K. (2020). Combining focused identification of germplasm and core collection strategies to identify genebank accessions for central European soybean breeding. *Plant Cell Environ.* 43: 1421–1436.

IPCC (2021). Climate change widespread, rapid, and intensifying. https://www.ipcc.ch/2021/08/09/ar6-wg1-20210809-pr/ (accessed 29 June 2022).

Katengeza, S.P. and Holden, S.T. (2021). Productivity impact of drought tolerant maize varieties under rainfall stress in Malawi: a continuous treatment approach. *Agric. Econ.* 52: 157–171.

Lamb, W.F., Wiedmann, T., Pongratz, J. et al. (2021). A review of trends and drivers of greenhouse gas emissions by sector from 1990 to 2018. *Environ. Res. Lett.* 16: 073005.

Lemmon, Z.H., Reem, N.T., Dalrymple, J. et al. (2018). Rapid improvement of domestication traits in an orphan crop by genome editing. *Nat. Plants* 4: 766–770.

Li, T., Yang, X., Yu, Y. et al. (2018). Domestication of wild tomato is accelerated by genome editing. *Nat. Biotechnol.* 36: 1160–1163.

McGuire, S. and Sperling, L. (2016). Seed systems smallholder farmers use. *Food Sec.* 8: 179–195.

Ranganathan, J., Waite, R., Searchinger, T., and Hanson, C. (2018). *How to Sustainably Feed 10 Billion People by 2050, in 21 Charts*. Washington, DC: World Resources Institute.

Rodríguez-Leal, D., Lemmon, Z.H., Man, J. et al. (2017). Engineering quantitative trait variation for crop improvement by genome editing. *Cell* 171: 470–480.e8.

Takahashi, Y., Sakai, H., Yoshitsu, Y. et al. (2019). Domesticating *Vigna stipulacea*: a potential legume crop with broad resistance to biotic stresses. *Front. Plant Sci.* 10: 1607.

Varshney, R.K., Barmukh, R., Roorkiwal, M. et al. (2021a). Breeding custom-designed crops for improved drought adaptation. *Adv. Genet.* 2: e202100017.

Varshney, R.K., Bohra, A., Yu, J. et al. (2021b). Designing future crops: genomics-assisted breeding comes of age. *Trends Plant Sci.* 26: 631–649.

Varshney, R.K., Roorkiwal, M., Sun, S. et al. (2021c). A chickpea genetic variation map based on the sequencing of 3,366 genomes. *Nature* 599: 622–627.

Varshney, R.K., Bohra, A., Roorkiwal, M. et al. (2021d). Rapid delivery systems for a food-secure future. *Nat. Biotechnol.* 39: 1179–1181.

Vernooy, R. (2022). Does crop diversification lead to climate-related resilience? Improving the theory through insights on practice. *Agroecol. Sustain. Food Syst.* 46: 877–901.

Wang, W., Mauleon, R., Hu, Z. et al. (2018). Genomic variation in 3,010 diverse accessions of Asian cultivated rice. *Nature* 557: 43–49.

Wing, R.A., Purugganan, M.D., and Zhang, Q. (2018). The rice genome revolution: from an ancient grain to Green Super Rice. *Nat. Rev. Genet.* 19: 505–517.

Xu, G., Yuan, M., Ai, C. et al. (2017). uORF-mediated translation allows engineered plant disease resistance without fitness costs. *Nature* 545 (7655): 491–494.

Yu, H., Lin, T., Meng, X. et al. (2021). A route to *de novo* domestication of wild allotetraploid rice. *Cell* 184: 1156–1170.e14.

Zsögön, A., Cermak, T., Voytas, D., and Peres, L.E. (2017). Genome editing as a tool to achieve the crop ideotype and de novo domestication of wild relatives: case study in tomato. *Plant Sci.* 256: 120–130.

Zsögön, A., Čermák, T., Naves, E.R. et al. (2018). *De novo* domestication of wild tomato using genome editing. *Nat. Biotechnol.* 36: 1211–1216.

2

Sustaining Food Security Under Changing Stress Environment

Sudhir K. Sopory

Department of Plant Molecular Biology, International Centre for Genetic Engineering and Biotechnology, New Delhi, India

The impact of unpredictable variations in temperature and precipitation across the globe, due to changes in climate conditions, will be felt on the growth and survival of plants in general and domesticated crops in particular. The stressful environment will drastically decrease food production and thus create a deficiency in food supply for the ever-growing population, especially in the developing world. On the one hand, one needs to take steps to mitigate any further increase in temperature by reducing greenhouse gases and increasing vegetative biomass. On the other hand, one needs to adopt new scientific methods and strategies to address the issue of food security under changing climate and imposing stress conditions. It is more than ever felt that biotech research, using omics-based data, also referred to as genomics-assisted breeding, can, in addition to the classical methods of plant improvement, offer sustainable solutions toward managing and mitigating the negative impact of abiotic and biotic stresses. Moving forward in those directions, I feel we need to further intensify research in the following areas and also initiate new programs.

1) There is a need to do extensive phenotyping of available germplasm and wild relatives of crop plants to identify those that can tolerate and adapt to certain stress conditions. As is already being done in some laboratories, we have to move from greenhouse-based automatic phenotyping platforms to drone-based and AI-based technology in field conditions (see, for example, Guo et al. 2021). These studies in addition to genotyping the selected accessions for desirable traits will benefit the plant breeder.
2) In addition to taking the path of moving from phenotype to genotype, one can also move from genotypes to phenotypes. Functional genomics has been adopted in many plants to assign function to gene(s) by overexpressing or repressing the gene via plant transformation. Obtaining mutation in a gene is also a very powerful way to assign function(s). This work needs to be undertaken in more crop plants as well as in underutilized crops like millets. However, it is now also possible, with the large whole genomics data combined with neural network tools to do in silico prediction and move from genotype to phenotype, as has been demonstrated recently in Arabidopsis by Raimondi et al. (2022). One needs to see if this will be possible and applicable to crop plants.
3) Since environmental conditions are variable and are changing due to global climate changes, it is very important to build the right kind of futuristic climate models at the regional and local levels, as different varieties of crops have adapted to different regions and soil conditions. Daniel Bebber and colleagues have recently modeled the production of maize, wheat, soybean, rice, and other temperate and tropical crops under future climate scenarios over the twenty-first century (Chaloner et al. 2021). "One of the things that's missing [from existing crop models] is the biological component – the pests, pathogens," says Bebber. "One of our long-term aims is to start building in a pathogen component ... so we have a better appreciation of what the future might look like." I feel more such studies need to be undertaken for specific regions and areas to predict the impact and accordingly find the right solutions.
4) It is time to do field-level molecular studies like transcriptomics, proteomics, and metabolomics on plants growing under the natural ecosystem and under various stress conditions that plants will face during their life cycle. Such studies will also reflect on plant soil, plant microbe, and plant pathogen interactions. For example, in a study on Sorghum, the authors found large changes and variations that probably could have been identified in plants growing in laboratory conditions (Varoquaux et al. 2019).

Global Climate Change and Plant Stress Management, First Edition. Edited by Mohammad Wahid Ansari, Anil Kumar Singh, and Narendra Tuteja.

No wonder that the 2022 year resolution of EMBL for bench scientists is to step out of the lab to study how life really works. It says molecular biologists study the world at the smallest scales... chromosomes, sub-cellular structures, proteins, metabolites which deprives us of the fullest picture. Thus, they are launching a program called "Molecules to Ecosystem," and I feel such programs will be very useful to understand genome functions under natural and stressful environment. What is now being labelled as plant ecological genomics is now being studied in plants that grow under extreme environmental conditions to fish out the "genetic gold mine to boost crop resilience" (Eshel et al. 2021).

5) It is important to monitor molecular changes during the entire life of a plant at all organizational levels (see Munne-Bosch 2022). In most of the studies on the impact of climate change and stress in plants, more often only the effect is studied on above-ground parts. The effect of climate change on below-ground parts of the plants. Liu et al. (2022) synthesized 88 published studies to explain the mismatch between above- and below-ground plant responses to climate warming. One has to realize that there is shoot/root continuum and continuous signaling between different parts. Chen et al. (2021) studied what they call "root economics spectrum." If roots survive in all possibility, the plant will survive or will revive under stress or during recovery. It is thus important to have an integrated approach to build molecular networks and signature related to a trait.
6) Lastly, with all the big data that is going to emerge from various sequencing, omics, biochemistry, physiology etc., and with smart climate models, it is possible for molecular computational analysts using AI and ML tools to generate virtual crops for future or planning breeding in silico. This can guide molecular breeders to generate the required characters in a crop for specific regions and environment.

I am of the opinion that to sustain food security under changing stress environment it is essential to generate qualitative data on plant responses to stress in field conditions and use this data with proper models toward in silico breeding, which should help in developing climate-resilient crops (Marshall-Colon et al. 2017). This in fact is going to be a challenging and exciting area (readers may refer to the special journal: *In silico Plants* from Wiley).

I am happy that the editors have compiled important and relevant chapters that highlight the present scenario and biotechnological approaches toward precision breeding for developing crops for the future.

References

Chaloner, T.M., Gurr, S.J., and Bebber, D.P. (2021). Plant pathogen infection risk tracks global crop yield under climate change. *Nat. Clim. Chang.* 11: 710. https://doi.org/10.1038/S41588-021-01104-8.

Chen, W., Wu, Y., Fritschi, F.B., and Juenger, T.E. (2021). The genetic basis of the root economics spectrum in a perennial grass. *Proc. Natl. Acad. Sci. U.S.A.* 118: e2107541118. https://doi.org/10.1073/pnas.2107541118.

Eshel, G., Araus, V., Undurraga, S. et al. (2021). Plant ecological genomics at the limits of life in the Atacama Desert. *Proc. Natl. Acad. Sci. U.S.A.* 118 (46): e2101177118. https://doi.org/10.1073/pnas.2101177118.

Guo, W., Carroll, M., Singh, A. et al. (2021). UAS-based plant phenotyping for research and breeding application. *Plant Phenomics* 2021: 9840192. https://doi.org/10.34133/2021/9840192.

Liu, H., Wang, H., Li, N. et al. (2022). Phenological mismatches between above- and belowground plant responses to climate warming. *Nat. Clim. Chang.* 12: 97. https://doi.org/10.1038/s41558-01-01244x.

Marshall-Colon, A., Long, S.P., Allen, D.K. et al. (2017). Crops *in silico*: generating virtual crops using and integrative and multi-scale modelling platform. *Front. Plant Sci.* 8: 786. https://doi.org/10.3389/fpls.2017.00786.

Munne-Bosch, S. (2022). Spatiotemporal limitations in plant biology research. *Trends Plant Sci.* 27 (4): 346–354. https://doi.org/10.1016/j.tplants.2021.10.005.

Raimondi, D., Corso, M., Fariselli, P., and Moreau, Y. (2022). From genotype to phenotype in *Arabidopsis thaliana*: *in silico* genome interpretation predicts 288 phenotypes from sequencing data. *Nucleic Acids Res.* 50: e1. https://doi.org/10.1093/nar/gkab1099.

Varoquaux, N., Cole, B., Gao, C. et al. (2019). Transcriptomic analysis of field-droughted sorghum from seedling to maturity reveals biotic and metabolic responses. *Proc. Natl. Acad. Sci. U.S.A.* 116: 27124–27132.

3

Crop Improvement Under Climate Change

Shivendra Bajaj and Ratna Kumria

Federation of Seed Industry of India, New Delhi, India

Climate change is a reality of the modern world that needs to be recognized and its impact addressed in all aspects of life. It can be mitigated by sustainability in economic, social, and environmental contexts. It is briefly defined as "a change in climate which is attributed directly or indirectly to human activity that alters the composition of the global atmosphere and which is in addition to natural climate variability observed over comparable periods of time" (United Nations Framework Convention on Climate Change [UNFCCC] 1992). The UNFCCC was signed by about 150 countries in Rio de Janeiro in June 1992, indicating widespread recognition that climate change is potentially a major threat to the world's environment and economic development. Since then, climate change has been extensively researched across the globe, especially in vulnerable environments, and a mitigation strategy is being developed.

Agriculture is most impacted by climate change around the world, with far-reaching effects on survival and growth of communities. High temperatures and unpredictable and severe weather have reduced yields by impacting plant vigor; susceptibility to disease, weeds, and pests; and paucity of resources. Changes in weather patterns and harshness of weather have led to either crop failure or lower production, though some minor crops have now become dominant in farmers' fields due to their hardiness under adverse conditions and the farmer's need for a minimum assured income (Sulser et al. 2021).

A 2020 FAO study reported that 8.9% (690 million) of the global population went hungry. The food security challenge will further worsen by 2050, when 70% more food would be required to feed nine billion people (FAO 2020). The challenge becomes more crucial with climate change causing reduction in crop yield, nutritional quality of major cereals, and livestock productivity. Agriculture is also a major contributor to greenhouse gas (GHG) emissions; it is estimated to produce 19–29% of total GHG emissions. Over and above that, around 30% of the food produced is lost or wasted, further stressing the global food supply chain, food prices, and the environment. An unpredictable food supply makes the poor populations more vulnerable (Vermeulen et al. 2012). Therefore, both our farm and food systems need to be part of the solution to lower emissions and manage climate change.

Industrial scale agriculture with heavy inputs of fertilizers and pesticides will not be sustainable, especially for feeding a population of nine million with nutritious food. The climate change induced erratic rainfall and temperatures are leading to prolonged periods of drought, high temperatures, and flooding that are adversely impacting crop cycles, growth, and yields. Diseases and pests are also spreading in newer regions due to weather changes and associated changes in planting patterns. Pest dynamics and intensity of infestations are also reducing crop yields and quality. Farmers will have to adapt by changing cropping patterns and chosen crops while keeping in mind the multitude of variables so that global food supply chains are minimally impacted (UCSUSA 2019).

Health-related costs linked to mortality and noncommunicable diseases are projected to exceed \$1.3 trillion $year^{-1}$ by 2030. On the other hand, the diet-related social cost of GHG emissions is estimated to be more than \$1.7 trillion $year^{-1}$ by 2030. Therefore, adoption of healthy diets is projected to lead to a reduction of up to 97% in direct and indirect health costs and 41–74% in the social cost of GHG emissions by 2030 (FAO 2020).

Global Climate Change and Plant Stress Management, First Edition. Edited by Mohammad Wahid Ansari, Anil Kumar Singh, and Narendra Tuteja.

3.1 Crop Diversity to Mitigate Climate Change

Agriculture has adversely impacted biodiversity due to conversion of natural ecosystems to farmland, monocropping, and overuse of chemical fertilizers and pesticides. Industrial agriculture is based on monocropping and deals with the land under cultivation as a single ecosystem, negating the natural variations in the ecosystem. This simplification leads to greater risk for crops and amplifies changes in crop viability and encroaching pests. Of the 250,000 species amenable to agriculture, only 3% are used for food, feed, fuel, and fiber, and 12 of these provide 75% of the food. Mass planting of these 12 species has led to loss of habitat and reduced use of many more foods and other species (FAO 2018).

The devastation caused by unpredictable weather and damage caused by pollution and industrial agriculture have made our crops vulnerable to not only physical factors but also disease and pests. The intensity of pest attacks has increased, and newer pests have emerged for the industrially cultivated crops.

Moving forward, farmers will have to choose different crops that are better adapted to the changing environment or varieties resistant to pests and diseases. High yielding crops that require fewer resources and efficiently utilize inputs would be preferred. Tools like gene editing are being employed for de novo domestication of orphan and wild crops. These cultivars are better adapted to adverse growing conditions and pest attacks while being nutritionally rich and offer a great opportunity for germplasm conservation, diversity in food, nutrition, and sustainable production. GM technology has been used for imparting disease/pest resistance as well as abiotic stress tolerance to crop plants.

Conservation of land and germplasm is the key to tap into the diversity of plant species that can be used as food, fuel, and fiber. Conservation efforts at both local and regional levels will allow farmers to explore climate-resilient and pest-resistant plants. Both formal and informal germplasm conservation need investment for infrastructure and technology development.

The economics of genome sequencing and progress in the field of genomics have assured a tremendous increase in germplasm-associated data and knowledge that can be used in their conservation and re-establishment in eroded environments.

3.2 Technology to Mitigate Climate Change

The World Bank Group (WBG) has developed a Climate Change Action Plan 2016–2020 and 2021–2025 with a commitment to increased productivity, enhanced resilience, and reduced emissions. The project included multiple regions globally, and in India (Maharashtra), the project led to the adoption of climate-smart agriculture practices. About 56,602 ha of land have been brought under improved irrigation and drainage technologies using innovative solutions under this project (WorldBank 2021).

Scientists and plant breeders are supporting this transition in cropping patterns with efforts to improve the productivity of orphan and climate-resilient crops. Advanced technologies that allow in-depth study of the plant genes, metabolism, and stress responses are being employed to better understand newer species. Efficient trait deployment and breeding are helping in introducing improved crop species for cultivation. Technologies like genome sequencing, GWAS, marker-assisted breeding, genetic modification, and gene editing are being utilized to develop high-yielding, nutritionally enhanced, climate-resilient crops.

Most of the crop productivity improvement efforts are directed at seed improvement. Quality seed with enhanced traits that enable the crop to tide over adverse conditions will be the basic requirement for agriculture in the coming future. Improvements in accuracy and efficiency of plant breeding methods like field selection, cell and tissue culture, and genomics are contributing to the development of resource-efficient, quality, and high-yielding crops.

Technology like GM has been employed to develop insect-resistant and herbicide-tolerant crops, whereas gene editing is being used to develop nutritionally enhanced (soybean, tomatoes, and banana), resource-efficient (soybean and rice), and stress-tolerant (rice, soybean, and maize) crops (Alliance for Science n.d.; Le et al. 2022; Lin et al. 2022; Yan et al. 2022). Gene editing technology supported by genomics makes plant breeding more efficient and can offer quicker solutions to address climate change instead of traditional plant breeding that takes longer to bring the improved crop to the field. Adoption and acceptance of GM and gene-edited crops will help us manage climate change and mitigate its impact. The permissive regulations for gene editing across the globe are an example of acceptance of scientific data and forward thinking.

3.3 Farm Practices to Mitigate Climate Change

Conservation of ecosystem can be initiated by reducing the use of natural resources for crop production. Neglected bare soils have reduced water holding and increased vulnerability to erosion and water pollution, leading to lower organic content and a possible inorganic imbalance. Deteriorating soils in the long term are most impacted by drought and flooding.

Heavy reliance on chemical inputs degrades the soil and lowers the yields, which further amplifies the chemical inputs. Irrigation has also increased in response to rising temperature extremes and drought, depleting precious water, and impacting soil temperature and microbiome. Degraded soils and paucity of resources impact not only crop yields but also livestock and fisheries.

Sustainable agriculture, regenerative agriculture, or climate-smart agriculture are interchangeably used terms. They refer to farm practices that reduce erosion and improve water quality while increasing productivity, enhancing climate resilience, and lowering GHG emissions. Climate-smart agriculture is an opportunity for ecosystem conservation, lowering cultivation costs, and using earned carbon credits to improve farming income.

Carbon sequestration and no-till farming not only improve soil health and reduce atmospheric CO_2 but also reduce soil erosion. A corn field system in the United States had the return on expenses for no-till corn in 2022 field budget at $\$114\,\text{acre}^{-1}$ compared with $\$92\,\text{acre}^{-1}$ for conventionally tilled corn. The soil carbon was improved from less than 1% to over 3.5%. The addition of biochar and gypsum to the soil also improves carbon in the soil; this too is routinely used for sugarcane fields in Brazil and corn in the United States (Smith and Goodman 2021).

Another nutrient that is responsible for harmful emissions is nitrogen, and nutrient efficiency is being researched and improved. Nitrogen can be managed by growing legumes, cover crops, and nutrient-efficient varieties developed by using available technological tools. These sustainable practices of cover crops and no-till farming have reduced erosion rates from over $100\,\text{tons}\,\text{acre}^{-1}\,\text{year}^{-1}$ to less than 1 ton (Adotey et al. 2020).

Development of nutrient-efficient crops using biotechnology tools is being actively pursued by scientists across the globe to reduce farming costs and adopt climate-smart agricultural practices.

3.4 Conclusion

Climate change is an everyday challenge for farmers, and given the growing demand for food, feed, and fiber due to increasing population and economic growth, technology is the only way forward. Technology to improve crop varieties for stress tolerance, nutritional enhancement, and resource efficiency (both water and nutrients), improved cover crops for better carbon and nitrogen sequestration and soil health, *de novo* domestication of wild species, improvement in plant architecture for high density planting or amenability to mechanization, and resistance to pests, all these traits can help achieve sustainability.

Since climate change is a dynamic, unpredictable process, constant monitoring of the changes and the reaction of the crop to the challenges posed by it is essential for managing and mitigating the impact of climate change. Digitization of agriculture and technologies like AI are being employed in farms to make informed decisions about crops, farm inputs, irrigation, and pesticide usage. These practices will reduce farm-related pollution, reduce indiscriminate use of resources and inputs, improve the health of soil and water bodies, reduce emissions, and increase farm productivity.

References

Adotey, N., McClure, A., and Yin, X.(.F.). (2020). *Enhanced Efficiency Nitrogen Fertilizer as a Tool to Control Nitrogen Loss in Row Crop Production*. Real Life Solutions, UTIA https://extension.tennessee.edu/publications/Documents/PB1888.pdf (accessed 2 October 2022).

Alliance for Science (n.d.). https://allianceforscience.org/blog/2022/01/dhruv-patel-is-daring-plants-to-take-a-risk/.

FAO (2018). *Agrobiodiversity: A Training Manual for Farmer Groups in East Africa*, FAO, https://www.fao.org/3/I9307EN/i9307en.pdf (accessed 2 October 2022).

FAO (2020). The state of food security and nutrition in the world 2020. https://www.fao.org/3/ca9692en/online/ca9692en.html#chapter-executive_summary (accessed 2 October 2022).

Le, X.H., Lee, C.P., Monachello, D. et al. (2022). Metabolic evidence for distinct pyruvate pools inside plant mitochondria. *Nat. Plants* **8**: 694–705. https://doi.org/10.1038/s41477-022-01165-3.

Lin, M.T., Salihovic, H., Clark, F.K., and Hanson, M.R. (2022). Improving the efficiency of Rubisco by resurrecting its ancestors in the family *Solanaceae*. *Sci. Adv.* 8 (15): eabm6871. https://doi.org/10.1126/sciadv.abm6871.

Smith, V.H. and Goodman, B.K. (2021). *Climate Change and Agriculture: What Is the Role of Policy*. American Enterprise Institute https://www.aei.org/wp-content/uploads/2021/11/Climate-change-and-agriculture.pdf?x91208 (accessed 2 October 2022).

Sulser, T., Wiebe, K.D., Dunston, S. et al. (2021). *Climate Change and Hunger: Estimating Costs of Adaptation in the Agrifood System*, Food Policy Report. Washington, DC: International Food Policy Research Institute (IFPRI) https://doi.org/10.2499/9780896294165.

UCSUSA (2019). Climate-change-and-agriculture. https://www.ucsusa.org/resources/climate-change-and-agriculture.

UNFCCC (1992). https://unfccc.int/resource/docs/convkp/conveng.pdf (accessed 10 October 2022).

Vermeulen, S.J., Campbell, B.M., and Ingram, J.S.I. (2012). Climate change and food systems. *Annu. Rev. Environ. Resour.* 37: 195–222. https://doi.org/10.1146/annurev-environ-020411-130608.

WorldBank (2021). https://www.worldbank.org/en/topic/climate-smart-agriculture (accessed 2 October 2022).

Yan, D., Tajima, H., Cline, L.C. et al. (2022). Genetic modification of flavone biosynthesis in rice enhances biofilm formation of soil diazotrophic bacteria and biological nitrogen fixation. *Plant Biotechnol. J.* 20: 2135–2148. https://doi.org/10.1111/pbi.13894.

4

Reactive Nitrogen in Climate Change, Crop Stress, and Sustainable Agriculture

A Personal Journey

Nandula Raghuram

Centre for Sustainable Nitrogen and Nutrient Management, University School of Biotechnology, Guru Gobind Singh Indraprastha University, New Delhi, India

4.1 Introduction

Modern agriculture is not only affected by climate change but also contributes to it. Apart from methane emissions from agricultural soils, nitrous oxide and ammonia emissions from unused fertilizers, manures, and other nitrogenous wastes contribute to climate change, among other environmental and health impacts. While elemental nitrogen (N_2) is unreactive and harmless, many of its oxidized or reduced compounds are highly reactive and harmful if they accumulate in the environment. Nitrous oxide is a powerful greenhouse gas (GHG) with 265-fold higher global warming potential than carbon dioxide (IPCC 2021). Ammonia/ammonium ions, nitrates, nitrites, and NO_x (nitrogen dioxide and nitric oxide) are all harmful reactive nitrogen (N_r) compounds for human and environmental health. They also cause air and water pollution, affecting our health and altering biodiversity and ecosystem services. Their complex interactions with plants under changing agroclimatic conditions involving other nutrients, temperature, and water variability determine the crop's ability to deal with various stresses exacerbated by climate change. Therefore, improving crop nitrogen use efficiency (NUE) toward sustainable nitrogen (N) management in agriculture is extremely critical for climate change mitigation and adaptation, as a part of the overall strategy toward sustainable food systems.

4.2 Reactive Nitrogen in Climate Change, Agriculture, and Beyond

The successive assessment reports of the Intergovernmental Panel on Climate Change (IPCC) always listed carbon dioxide, methane, nitrous oxide, and short-lived climate forcers among the anthropogenically accumulated GHGs that cause climate change. In an attempt to simplify their impacts for the benefit of policy makers, they expressed the effects of all GHGs in terms of carbon dioxide (CO_2) equivalents, the most abundant and the most enduring of all GHGs, so much so that even methane is expressed as a CO_2 equivalent. But an unintended effect of this carbon-centric public communication of science has been that even many scientists, let alone governments, are ignorant of the role of other potent GHGs (Raghuram and Beig 2021). Nitrous oxide (N_2O) is the third most abundant GHG, with a global warming potential 265 times higher than CO_2 over a century. The latest IPCC report (AR6) projects that its concentration in the Earth's atmosphere could grow as fast as CO_2 or methane over the next few decades, unless dramatic interventions are undertaken. Its largest anthropogenic source is agriculture, most predominantly unused fertilizers/manures (Tian et al. 2020). In Indian agriculture, N_2O has grown faster than methane and has become the second-most important GHG (Abrol et al. 2017). This could be true in many other countries that rely on modern cropping and livestock. In other words, while we certainly need to save agriculture from climate change, it is equally important to save the climate from agriculture, and it is possible.

Why does this happen? Because modern agriculture has tipped the natural balance of nitrification and denitrification in the biogeochemical cycle of nitrogen. N_r compounds such as ammonia/ammonium are generated by lightning or bacterial nitrogen fixation, which are converted into nitrites and nitrates by bacterial nitrification in the soil. Some of them are taken up by plants and other microorganisms to make amino acids, nucleic acids, and other organic N-forms,

Global Climate Change and Plant Stress Management, First Edition. Edited by Mohammad Wahid Ansari, Anil Kumar Singh, and Narendra Tuteja.

on which all animals and humans depend to meet their nitrogen requirements through the food chain. Their excreta and the bacterial nitrification of organic N compounds from dead plants and animals again bring them back into circulation as inorganic N_r compounds. The ability of denitrifying bacteria to convert them back into elemental nitrogen (N_2) ensures that the nitrogen cycle is kept in balance. Even though nitrous oxide can be produced during both nitrifying and denitrifying bacterial conversions, the overall scale of environmental accumulation of N_r compounds was small. Traditional agriculture depended on legume-based crop rotations and cropping systems, which ensured that symbiotic bacteria fixed N_r compounds and their leftovers in the soil, supported non-legume crops like cereals, vegetables, fruits, and flowers. This sustained human civilization for over 11,000 years, until fertilizers tilted the balance heavily (Galloway et al. 2008).

Modern agriculture was facilitated by the advent of nitrogenous fertilizers manufactured by the Haber–Bosch process since 1908. The large-scale industrial nitrogen fixation surpassed biological nitrogen fixation, decoupled agriculture from legume-based cropping systems, and enabled monocultures. As soils were accumulating more N_r than ever, nitrous oxide production from both nitrification and denitrification was growing faster. The Haber–Bosch process was also used later for war munitions and other purposes, accelerating the anthropogenic conversion of atmospheric N_2 into N_r forms that continues till date. The industrial era also increased emissions of N_2O, NO_x, and others into the air, as by-products of the combustion of fossil fuels for energy and transportation. Further, nitrogenous effluents (nitrates, ammonium, and urea) were released from industrial and domestic wastes into water bodies. Together, they exceeded all of the Earth's terrestrial processes combined, by well over twofold (Galloway et al. 2008).

4.3 Nitrogen, Climate, and Planetary Boundaries of Sustainability

The fact that countries are still pledging net-zero GHG emissions within the next few decades indicates that they hope to control climate change. Indeed, while climate change itself has not yet crossed the planetary boundary of sustainability, various N_r forms have long crossed the planetary sustainability limits, and some of their impacts may never be fully reversed (Steffen et al. 2015). This is also happening to phosphorus and biodiversity loss, which itself is connected to nutrient pollution led by nitrogen and phosphorus. Therefore, nitrogen is not just the new carbon; it is much worse, as there are many effects of the accumulation of N_r compounds beyond climate change. Some of them even undo the benefits of carbon sequestration and climate change mitigation by crop plants. As Johan Rockstrom has been emphasizing lately, slowing down N_r accumulation and bringing it back within the boundaries of planetary tolerance and sustainability would increase the available margin to act on other anthropogenic threats to our planet's sustainability. In other words, our neglect of nitrogen management, despite its crossing the planetary boundaries, limits our ability to deal with other threats including climate change and biodiversity.

4.4 Emerging Global Response and India's Leadership in It

The last two decades of advocacy by the International Nitrogen Initiative (INI), a voluntary intercontinental community of scientists formed in 2003, led to the growing recognition of reactive nitrogen pollution and the importance of sustainable nitrogen management (www.initrogen.org). Its triennial conferences and their declarations, including the Nanjing declaration (2004), Delhi declaration (2010), Kampala declaration (2013), Melbourne declaration (2016), and most recently the Berlin declaration (2021), summarized the latest scientific evidence and called for action by all stakeholders. The European Nitrogen Assessment (2011) became the first ever comprehensive scientific effort to enable evidence-based policies for sustainable nitrogen management, leading several European countries to impose emission/effluent limits and taxes on N_r. Since 2019, UNEP resolutions (4/14 and 5/2) on sustainable nitrogen management have been mandating all member governments to work toward limiting N waste and pollution.

The early response of the Indian scientific community to the issues of N_r was prompted by the scientific literature of the 1990s, wherein even plant scientists were citing climate change as an important reason to study/improve the NUE of crops. Those were also the years when Indian scientists were learning to form interdisciplinary teams to collect evidence to counter IPCC reports on the Indian carbon emissions. The earliest IPCC reports extrapolated data from the developed countries and estimated methane emissions from paddy fields as ~30 Tg $year^{-1}$, whereas Indian scientists later measured and corrected it as 4 Tg $year^{-1}$. Such corrections in subsequent IPCC reports eased pressure on the Indian government and agriculture.

This was also the first major instance of domestic scientific capacities and inputs aiding India's positioning in the IPCC as well as international environmental diplomacy. India eventually chaired the IPCC and currently leads the global solar alliance.

But our biggest concern then was that the experience with carbon did not prepare India better for other national and global challenges such as nitrogen. As early as 2001, I discussed my concerns with Prof. Yash Pal Abrol, a retired scientist whose pioneering crop physiology research at the Indian Agricultural Research Institute led to N-fertilizer recommendations for rice and wheat during the Indian "green revolution." Soon, we realized the lack of an interdisciplinary network of scientific, industrial, policy, and other stakeholders working on different aspects of the biogeochemical cycle of nitrogen in India. We eventually launched the Indian Nitrogen Group (ING) and held various brainstorming events. They led to the first ever policy paper on the subject for the National Academy of Agricultural Sciences (NAAS 2005), a book on nitrogen in agriculture and the environment (Abrol et al. 2007), a report for the Indian National Science Academy and IGBP-SCOPE on the subject (Bijay-Singh et al. 2008), and a special issue of the Indian journal *Current Science* in its platinum jubilee year (Abrol et al. 2008). They established the nitrogen research and policy agenda in India very early, almost at par with the Europe and USA.

In the meantime, ING and INI discovered each other, and we established a South Asian regional directorate for INI in Delhi. We also joined the newly established UNEP Global Partnership on Nutrient Management (GPNM) as a scientific and civil society partner. With their help, ING hosted the Fifth International Nitrogen Conference in New Delhi in 2010 in which the Delhi Declaration was adopted (Raghuram et al. 2011). In addition, 20 ING Bulletins on sectoral assessment of reactive nitrogen were released, sowing the seeds for national and regional assessments and for India's international forays on nitrogen (Balasubramanian et al. 2010). The main emphasis has been on sustainable nitrogen management in fertilizer and fossil fuel use (Abrol et al. 2012). Simultaneously, we developed contacts with nitrogen scientists throughout the eight South Asian countries and with the intergovernmental South Asian Cooperative Environment Programme (SACEP). Our inputs to a SACEP study titled "Nutrient loading and Eutrophication of coastal waters of the South Asian seas" in 2013 led to the intergovernmental ministerial level recognition of reactive N pollution at the South Asian level. Later, we contributed to the UNEP-GPNM commissioned international report on nutrient pollution (Sutton et al. 2013) and its international projects, starting with the GEF-Global Nutrient Cycling project involving Chilika Lake and Manila Bay (www.nutrientchallenge.org).

Our publication of the Indian Nitrogen Assessment (Abrol et al. 2017) was a unique contribution in citizen science mode under the civil society umbrella, without being commissioned or funded by the Indian government or any other entity in India or abroad. It identified all the Indian sources and sinks of anthropogenic N_r in all sectors and the scope for reducing nitrogen pollution. It also mapped the relevant nitrogen policies and scenarios (Bhattacharya et al. 2017) for informed domestic actions and international negotiations. This led to envisaging "India as an emerging nitrogen champion" (Sutton et al. 2017; Raghuram et al. 2020). Consequently, Indian scientists became important partners in the $6 million GEF-UNEP-INI project titled "Targeted Research for Improving Understanding of the Global Nitrogen Cycle Towards the Establishment of an International Nitrogen Management System (INMS)." Its main effort is to understand the global N cycle and quantify N flows, threats, and benefits worldwide with seven regional case studies in all the INI regions, including the first ever South Asian Nitrogen Assessment being led by India (Raghuram et al. 2020, 2021).

The Government of India and UNEP soon recognized our pioneering efforts on nitrogen and showcased them during the World Environment Day celebrations hosted in Delhi in 2018. Further, our global efforts through INI, GPNM, and INMS prompted UNEP to identify reactive nitrogen as one of the five most important emerging threats to the planet Earth (Sutton et al. 2019). This further motivated the Indian government to lead the first ever UN resolution on Sustainable Nitrogen Management during the 4th UN Environment Assembly (UNEP 2019). In late 2021, the Indian government established a high-level interministerial national steering committee for the implementation of this UN resolution. These developments established India among the leaders of the global response to nitrogen pollution and its impact on climate, health, environment, and biodiversity (Raghuram et al. 2021). In late 2022, India was elected as a Co-Chair of the UNEP intergovernmental Nitrogen Working Group in 2022, along with Romania, to implement sustainable nitrogen management.

4.5 Regional and Global Partnerships for Effective Interventions

Addressing global problems requires committed global partnerships at the level of scientists and governments (Sutton et al. 2020a, b, c, d, 2021). The early initiative of the ING to identify its peers and partners in South Asia proved to be as important as building its global linkages through the INI. The draft resolution on sustainable nitrogen management was originally prepared and adopted during a workshop of the South Asian partners of the INMS project hosted by SACEP at Male' in 2017. Its subsequent approval by all the member countries of SACEP and the international support garnered by the INI facilitated its smooth passage when India led it in the 4th UN Environment Assembly.

A few years earlier, in 2015, the United Kingdom led intergovernmental discussions on International Research Partnerships on Agricultural Nitrogen, which led to the development of bilateral agreements for collaborative Virtual Nitrogen Centres between the United Kingdom and India, China and Brazil. They funded three-year projects to promote research on improving NUE in each of these countries between 2017 and 2019. Their success led to the UKRI directly funding the ongoing five-year project on "South Asian Nitrogen Hub" under its Global Challenges Research Fund, which included 40 institutions across the United Kingdom and all the eight countries of South Asia. Apart from scientific research into quantifying the status and trends of reactive nitrogen pollution in the region and improving NUE in agriculture, the project mapped the nitrogen policy landscape and produced national as well as regional policy briefs for informed actions. While India continues to play a prominent role in this project, Pakistan published its own national Nitrogen Assessment in 2021 (Aziz et al. 2021), making South Asia the only region in the world with two countries completing their national nitrogen assessments in quick succession.

The recent second UN resolution on sustainable nitrogen management led by Sri Lanka during the 5th UN Environment Assembly (UNEP 2022) also made South Asia the only region in the world to lead two successive resolutions on nitrogen. In 2023, the first ever international nitrogen assessment will be published, incorporating the first ever South Asian N assessment and case studies from all other continents. This will provide a rigorous global overview of the status of nitrogen pollution from various sources, identify the scope for policy interventions, and build scenarios for sustainability. It will also build on the growing understanding of these issues on a South Asian scale under the leadership of the Sustainable India Trust (SIT) (Bansal et al. 2022; Yang et al. 2022).

4.6 Building Crop NUE Paradigm Amidst Growing Focus on Stress

The growth of modern plant biology into the transgenic era in India (Raghuram 2002a, b, 2004) coincided with the research focus increasingly shifting toward abiotic and biotic stresses even among N-researchers. By then, despite half a century of global knowledge on the physiology of N-response and the transporters, enzymes, and genes involved in it, the NUE phenotype and the genetic targets for its improvement remained elusive. Early efforts toward transgenic overexpression of some of the nitrate transporters and nitrate-assimilatory genes in model plants and even some crop plants did not yield significant improvements in NUE (Raghuram et al. 2006, 2007; Lochab et al. 2007; Pathak et al. 2008, 2011; Chakraborty and Raghuram 2011). This ought to have been expected for quantitative traits like NUE, stress tolerance/resistance, and yield. Interestingly, low nutrient stress is as well known as high nutrient stress or the yield-inhibiting effect of overdosing nutrients. Yet, abiotic stress research rarely focused on its implications for nutrient use efficiency in general and NUE in particular (Jangam and Raghuram 2015).

Building a clear research paradigm for crop improvement toward NUE has been a genuine challenge for many reasons. The multitude of N-forms that exist in the soil and the different N-forms used in fertilizers made it difficult to build a common perspective in terms of plant functional biology and genetics. For example, ammonium nitrate is the most commonly used N-fertilizer in the developed world, whereas urea is far more predominant in the developing world. Even though nitrifying soil bacteria eventually convert all of them to nitrates under aerobic conditions, field soils often have dynamic mixtures of different N-forms, and different plants can use them to different extents. The many dozens of definitions of NUE only made the matters worse for plant biologists seeking universal solutions (Pathak et al. 2011). Moreover, crop geneticists were dealing with an increasingly narrowing gene pool of high-yielding varieties selected/bred under high input conditions. Crop improvement for NUE or any other input use efficiency for that matter requires germplasm screening under low input conditions, but such screening programs just do not exist in most countries even today (Mandal et al. 2018; Raghuram and Sharma 2019; Sinha et al. 2020; Madan et al. 2022).

Soon, we realized that we need to build our own paradigm. As early as 1998, we identified rice as our model crop, in view of its importance for India and the developing world, as well as its emergence as the first post-genomic model crop. We also characterized the organelle free cyanobacterium *Arthrospira (Spirulina) platensis* as a microbial model system for rice higher plants (Ali et al. 2008; Jha et al. 2007; Lochab et al. 2009, 2014). Soon we built our working hypothesis that the regulation of N-flux through the uptake, assimilation, and remobilization pathways was more important than the individual steps of these pathways themselves. We decided to use functional genomics as our main platform to identify N-responsive genes in rice and shortlist NUE targets from them. We also decided to characterize the NUE phenotype from scratch and use it to shortlist the N-responsive genes associated with it as candidates for NUE. For this purpose, we decided to use the agronomic definition of partial factor productivity (N harvested per unit N supplied) and sought to experimentally distinguish N-response from NUE using yield as the differentiator. We hypothesized that the targets for NUE improvement may not be found in the N-uptake, assimilation, and remobilization pathways but in their regulatory controls. Soon we found that the previously reported nitrate response elements were not a helpful start, as they were neither specific to nitrate nor universal to all nitrate-responsive genes in Arabidopsis and rice (Das et al. 2007; Pathak et al. 2009).

We built some early understanding of the N-responsive gene expression and enzyme activities and the involvement of G-proteins and protein kinases in maize (Raghuram and Sopory 1995a, b, 1999; Raghuram et al., 1999) and rice (Ali et al. 2007a, b, 2008). Our early efforts on the functional genomics of Arabidopsis mutants lacking G-protein α subunit, G-protein coupled receptor (GPCR), or both revealed their involvement in N- and P-responses, apart from stress (Chakraborty et al. 2015a, b, c, 2019). We later found similar results on the role of G-protein signaling in abiotic stress and nitrate response in Japonica rice, using transcriptomic analyses of the natural *d1* mutant (Jangam et al. 2016; Pathak et al. 2021). More recently, we compared the genome-wide nitrate response of this mutant with the wild type to provide firm genetic evidence of the role of G-protein signaling in N-response and NUE (Jangam et al. 2023). We also explored the common aspects of nutrient perception and signaling in plants from this perspective (Jaiswal and Raghuram 2019) and reviewed the role of protein phosphatases in N-response and NUE (Kumari and Raghuram 2020). We hypothesize that G-protein signaling may provide a common link between abiotic stress, nutrient response, and/or nutrient use efficiency, at least with respect to nitrogen.

4.7 From NUE Phenotype to Genotype in Rice

The lack of a clear ranking of rice germplasm in terms of NUE meant that we had to do it on our own to identify rice varieties with contrasting NUE for further work. Mere agronomic ranking was not enough, as we also needed some visually identifiable phenotypic trait(s) that strongly correlated with NUE to make biological sense of it. Building on the pioneering work of Prof. V. Sitaramam, who used seed germination as a phenotyping method for yield and stress in rice, we devised strategies to test its utility for NUE in rice. We initially compared 21 released Indica rice varieties for the time taken for 50% of the seeds to germinate in petri plates containing minimal media with or without N and ranked them on that basis, before checking their grain yields in the field at different N-levels (Sharma et al. 2018). We found that slow-germinating varieties of rice were long duration varieties that were more responsive to N, grew slowly with delayed flowering and seed-setting, but gave a higher yield and NUE. The fast-germinating varieties were short duration varieties that were less N-responsive, with early flowering, lesser yield, and lower NUE (Sharma et al. 2018).

We then used three of each of the contrasting rice genotypes from the above ranking by germination and conducted lifelong phenotyping for 25 traits using potted plants in a green house. In order to limit soil microbial effects, we used sterile, nutrient-free soil developed in-house (Sharma et al. 2019). This not only reconfirmed the role of germination but also yielded flowering time, root/shoot length, and biomass as universal NUE traits common to nitrate and urea fertilizers, whereas a couple of other traits were N-source specific (Sharma et al. 2021). By mining the literature for all the genes associated with each of these traits and performing Venn selections with the N-responsive genes that we identified earlier using microarrays (Pathak et al. 2020), we found 34 candidate genes for NUE (Sharma et al. 2021).

Using yield as a differentiator between N-response and NUE was of great help in shortlisting the NUE traits from N-responsive traits as well as genes in rice. We mined for all yield-associated genes in rice from the literature and identified over a thousand N-responsive and yield-related genes among them. We co-localized them onto NUE QTLs to hierarchically shortlist additional candidate genes for NUE (Kumari et al. 2021). We also expanded our N-responsive transcriptome data to identify genome-wide response to nitrate/urea in both Indica and Japonica rice (Pathak et al. 2020; Mandal et al. 2022), including comparative N-response response in two contrasting varieties of Indica rice for nitrate (Sharma et al. 2022) and urea (Sharma et al. 2023). We are currently consolidating all these different approaches to identify a priority list of candidate genes for NUE for further validation and crop improvement by marker development and/or genome editing.

4.8 Furthering the Research and Policy Agenda

In 2014, we co-founded a dedicated SIT to carry forward the work of the ING toward sustainable nitrogen management and expand it further to sustainable nutrient management (www.sustainableindiatrust.org). Soon, it became a partner in international projects such as INMS (2017) and SANH (2019) and also obtained UNEP accreditation as a NGO under S&T Major Stakeholder Groups in 2020. It also became a partner of INI and GPNM and facilitated the first ever UN resolution on sustainable nitrogen management (2019), apart from other scientific coordination, agenda development, and advocacy. It also collaborated in policy research on reactive nitrogen management at the global (Kanter et al. 2020) and South Asian levels (Bansal et al. 2022; Yang et al. 2022). Since 2020, SIT has also set up annual South Asian awards for excellence in research and/or policy on sustainable nitrogen management and runs Global Nitrogen awareness campaigns on WhatsApp and Twitter (@SusIndTrust). More recently, SIT undertook a UNEP project on nutrient recovery from sewage water plants, which includes recycling N, P, and other major nutrients recovered from the treated wastewater toward horticulture and agriculture to save fertilizers and improve overall full-chain NUE.

While some of these initiatives are best suited for NGOs, efforts to build research roadmaps continued at the professional level (Moring et al. 2021; Udvardi et al. 2021). More recently, to bring wider focus and visibility to the fast-growing N-research across the world, special thematic issues of prominent journals have been launched on various aspects of reactive nitrogen (Raghuram et al. 2022a, b; Winiwarter et al. 2022). This kind of synergy between professional research and voluntary organizations is extremely critical to attract public investments and deliver visible results for society and the environment. As the emphasis moves from yield-centric to efficiency-centric or productivity-centric crop improvement, low-input screening of crop germplasm for N and other inputs will expand the availability of donor genotypes for crop improvement toward NUE. As basic research on G-proteins increasingly moves toward applications, we will also witness increasing convergence between abiotic stress-tolerance/resistance, nutrient and water use efficiency as the integral focal point of crop improvement for climate resilient agriculture. These developments in crop genetics are set to break the agronomic barriers for improving NUE and other nutrient/water use efficiencies and usher the transition toward climate-smart agriculture and sustainable food systems.

References

Abrol, Y.P., Raghuram, N., and Sachdev, M.S. (ed.) (2007). *Agricultural Nitrogen Use and Its Environmental Implications*. I.K. International Pvt. Ltd., 552p.

Abrol, Y.P., Raghuram, N., and Hoysall, C. (ed.) (2008). Special issue on reactive nitrogen in agriculture, environment and health. *Curr. Sci.* 94 (11): 1343–1477. (Guest Editors)

Abrol, Y.P., Pandey, R., Raghuram, N., and Ahmad, A. (2012). Nitrogen cycle sustainability and sustainable technologies for nitrogen fertilizer and energy management. *Journal of the Indian Institute of Science* 92: 37–62.

Abrol, Y.P., Adhya, T.K., Aneja, V.P. et al. (ed.) (2017). *The Indian Nitrogen Assessment: Sources of Reactive Nitrogen, Environmental and Climate Effects, Management Options, and Policies*. UK: Elsevier. ISBN: 978-0-12-811836-8, 538p.

Ali, A., Sivakami, S., and Raghuram, N. (2007a). Effect of nitrate, nitrite, ammonium, glutamate, glutamine and 2-oxoglutarate on the RNA levels and enzyme activities of nitrate reductase and nitrite reductase in rice. *Physiol. Mol. Biol. Plants* 13 (1): 17.

Ali, A., Sivakami, S., and Raghuram, N. (2007b). Regulation of activity and transcript levels of NR in rice (*Oryza sativa*): roles of protein kinase and G-proteins. *Plant Sci.* 172 (2): 406–413.

Ali, A., Jha, P., Sandhu, K.S., and Raghuram, N. (2008). Spirulina nitrate-assimilating enzymes (NR, NiR, GS) have higher specific activities and are more stable than those of rice. *Physiol. Mol. Biol. Plants* 14 (3): 179–182.

Aziz, T., Wakeel, A., Watto, M. et al. (ed.) (2021). *Nitrogen Assessment. Pakistan as a Case-Study*, 214p. Academic Press.

Balasubramanian, V., Raghuram, N., Abrol, Y.P. et al. (2010). *Reactive Nitrogen: Good, Bad and Ugly. A Comprehensive Status Report* (ed. B. Singh). New Delhi: SCON-ING.

Bansal, S., Raghuram, N., Adhya, T.K. et al. (2022). Long-term trends of nitrous oxide emission from fuel combustion in South Asia. *Environ. Res. Lett.* 17 (4): 045028. https://doi.org/10.1088/1748-9326/ac5cf7.

Bhattacharya, S., Adhya, T.K., Pathak, H. et al. (2017). Issues and policies for reactive nitrogen management in the Indian region. In: *The Indian Nitrogen Assessment: Sources of Reactive Nitrogen, Environmental and Climate Effects, Management Options, and Policies* (ed. Y.P. Abrol, T.K. Adhya, V.P. Aneja, et al.), 491–513. UK. ISBN: 978-0-12-811836-8: Elsevier.

Chakraborty, N. and Raghuram, N. (2011). Nitrate sensing and signaling in genomewide plant N response. In: *Nitrogen Use Efficiency in Plants* (ed. V. Jain and P. Anandakumar), 45–62. New Delhi: New India Publishing Agency.

Chakraborty, N., Sharma, P., Kanyuka, K. et al. (2015a). Transcriptome analysis of *Arabidopsis* GCR1 mutant reveals its roles in stress, hormones, secondary metabolism and phosphate starvation. *PLoS One* 10 (2): e0117819.

Chakraborty, N., Sharma, P., Kanyuka, K. et al. (2015b). G-protein α-subunit (GPA1) regulates stress, nitrate and phosphate response, flavonoid biosynthesis, fruit/seed development and substantially shares GCR1 regulation in *A. thaliana*. *Plant Mol. Biol.* 89 (6): 559–576.

Chakraborty, N., Singh, N., Kaur, K., and Raghuram, N. (2015c). G-Protein signaling components GCR1 and GPA1 mediate responses to multiple abiotic stresses in Arabidopsis. *Front. Plant Sci.* 6: 1000.

Chakraborty, N., Kanyuka, K., Jaiswal, D.K. et al. (2019). GCR1 and GPA1 coupling regulates nitrate, cell wall, immunity and light responses in *Arabidopsis*. *Sci. Rep.* 9 (1): 1–17.

Das, S.K., Pathak, R.R., Choudhury, D., and Raghuram, N. (2007). Genomewide computational analysis of nitrate response elements in rice and *Arabidopsis*. *Mol. Gen. Genomics* 278 (5): 519–525.

Galloway, J., Raghuram, N., and Abrol, Y.P. (2008). A perspective on reactive nitrogen in a global, Asian and Indian context. *Current Sci.* 94 (11): 1375–1381.

IPCC (2021). *Climate Change 2021: The Physical Science Basis. Contribution of Working Group I to the Sixth Assessment Report of the Intergovernmental Panel on Climate Change* Masson-Delmotte, V., P. Zhai, A. Pirani, S.L. Connors, C. Péan, S. Berger, N. Caud, Y. Chen, L. Goldfarb, M.I. Gomis, M. Huang, K. Leitzell, E. Lonnoy, J.B.R. Matthews, T.K. Maycock, T. Waterfield, O. Yelekçi, R. Yu, and B. Zhou (eds.)]. Cambridge University Press, Cambridge, United Kingdom and New York, NY, USA. doi: https://doi.org/10.1017/9781009157896.

Jaiswal, D.K. and Raghuram, N. (2019). Nutrient perception and signaling in plants. In: *Sensory Biology of Plants*, 59–77. Singapore: Springer.

Jangam, A.P. and Raghuram, N. (2015). Nitrogen and stress. In: *Elucidation of Abiotic Stress Signaling in Plants*, 323–339. New York: Springer.

Jangam, A.P., Pathak, R.R., and Raghuram, N. (2016). Microarray analysis of rice d1 (RGA1) mutant reveals the potential role of G-protein alpha subunit in regulating multiple abiotic stresses such as drought, salinity, heat, and cold. *Front. Plant Sci.* 7: 11.

Jangam, A.P., Mandal, V.K., Kumar, D. et al. (2023). Nitrate-responsive transcriptome analysis of the rice RGA1 mutant reveals G-protein regulation of nitrogen use efficiency (NUE) and associated traits including water use efficiency, root/shoot growth, heading date, photosynthesis and yield. *Front. Plant Sci.* (Submitted).

Jha, P., Ali, A., and Raghuram, N. (2007). Nitrate-induction of nitrate reductase and its inhibition by nitrite and ammonium ions in *Spirulina platensis*. *Physiol. Mol. Biol. Plants* 13 (2): 163.

Kanter, D., Winiwarter, W., Bodirsky, B. et al. (2020). A framework for nitrogen futures in the shared socioeconomic pathways. *Glob. Environ. Chang.* 61: 102029.

Kumari, S. and Raghuram, N. (2020). Protein phosphatases in N response and NUE in crops. In: *Protein Phosphatases and Stress Management in Plants: Functional Genomic Perspective* (ed. G.K. Pandey), 233–244. Switzerland: Springer Nature https://doi.org/10.1007/978-3-030-48733-1_12.

Kumari, S., Sharma, N., and Raghuram, N. (2021). Meta-analysis of yield-related and N-responsive genes reveals chromosomal hotspots, key processes and candidate genes for nitrogen use efficiency (NUE) in rice. *Front. Plant Sci.* 11: 627955. https://doi.org/10.3389/fpls.2021.627955.

Lochab, S., Pathak, R.R., and Raghuram, N. (2007). Molecular approaches for enhancement of nitrogen use efficiency in plants. In: *Agricultural Nitrogen Use and Its Environmental Implications*, 327–350. New Delhi: I.K. International Pvt. Ltd.

Lochab, S., Oberoi, H.S., Gothwal, M. et al. (2009). Nitrate assimilatory enzymes of *Spirulina* (*Arthospira*) *platensis* are more thermotolerant than those of rice. *Physiol. Mol. Biol. Plants* 15 (3): 277.

Lochab, S., Kumar, P.A., and Raghuram, N. (2014). Molecular characterization of nitrate uptake and assimilatory pathway in *Arthrospira platensis* reveals nitrate induction and differential regulation. *Arch. Microbiol.* 196 (6): 385–394.

Madan, B., Malik, A., and Raghuram, N. (2022). Crop nitrogen use efficiency for sustainable food security and climate change mitigation. In: *Plant Nutrition and Food Security in the Era of Climate Change* (ed. V. Kumar, A. Srivastava, and S. Penna), 47–72. Elsevier, Academic Press. ISBN 978-0-12-822916-3.

Mandal, V., Sharma, N., and Raghuram, N. (2018). Molecular targets for improvement of crop nitrogen-use efficiency: current and emerging options. In: *Engineering Nitrogen Utilization in Crop Plants* (ed. A. Shrawat, A. Zayed, and D.A. Lightfoot), 77–93. Springer, ISBN: 978-3-319-92958-3.

Mandal, V.K., Jangam, A.P., Chakraborty, N., and Raghuram, N. (2022). Nitrate-responsive transcriptome analysis reveals additional genes/processes and associated traits viz. height, tillering, heading date, stomatal density and yield in *japonica* rice. *Planta* 255: 42.

Moring, A., Hooda, S., Raghuram, N. et al. (2021). Nitrogen challenges and opportunities for agricultural and environmental science in India. *Front. Sustain. Food Syst.* 5: 505347. https://doi.org/10.3389/fsufs.2021.505347.

NAAS (2005). *Policy Options for Efficient Nitrogen Use*, Policy Paper no. 33 (ed. Y.P. Abrol and B.N. Johri). New Delhi: National Academy of Agricultural Sciences.

Pathak, R.R., Ahmad, A., Lochab, S., and Raghuram, N. (2008). Molecular physiology of plant nitrogen use efficiency and biotechnological options for its enhancement. *Curr. Sci.* 1394–1403.

Pathak, R.R., Das, S.K., Choudhury, D., and Raghuram, N. (2009). Genomewide bioinformatic analysis negates any specific role for Dof, GATA and Ag/cTCA motifs in nitrate responsive gene expression in *Arabidopsis*. *Physiol. Mol. Biol. Plants* 15 (2): 145–150.

Pathak, R.R., Lochab, S., and Raghuram, N. (2011). Improving nitrogen use efficiency. In: *Comprehensive Biotechnology*, 2e, vol. 4 (ed. M. Moo-Young), 209–218. Oxford: Elsevier.

Pathak, R.R., Jangam, A.P., Malik, A. et al. (2020). Transcriptomic and network analyses reveal distinct nitrate responses in light and dark in rice leaves (*Oryza sativa* Indica var. Panvel1). *Sci. Rep.* 10: 12228. https://doi.org/10.1038/s41598-020-68917-z.

Pathak, R., Mandal, V., Jangam, A.P. et al. (2021). Heterotrimeric G-protein α subunit (RGA1) regulates tiller development, yield, cell wall, nitrogen response and biotic stress in rice. *Sci. Rep.* 11: 2323. https://doi.org/10.1038/s41598-021-81824-1.

Raghuram, N. (2002a). Indian plant biology enters the biotechnology era. *Trends Plant Sci.* 7: 92–94.

Raghuram, N. (2002b). India joins the GM club. *Trends Plant Sci.* 7 (7): 322–323.

Raghuram, N. (2004). Indian publishing. *Enduring the boom. Trends Plant Sci.* 9 (1): 9–12.

Raghuram, N. and Beig, G. (2021). Climate change is not just about carbon, there is nitrogen too! *Outlook* 12 August. https://www.outlookindia.com/website/story/india-news-climate-change-is-not-just-about-carbon-there-is-nitrogen-too/391230 (accessed 3 March 2023).

Raghuram, N. and Sharma, N. (2019). Improving crop nitrogen use efficiency. In: *Comprehensive Biotechnology*, vol. 4 (ed. M. Moo-Young), 211–220. Pergamon: Elsevier.

Raghuram, N. and Sopory, S.K. (1995a). Light regulation of nitrate reductase gene expression mechanism and signal response coupling. *Physiol. Mol. Biol. Plants* 1: 103–104.

Raghuram, N. and Sopory, S.K. (1995b). Evidence for some common signal transduction events for opposite regulation of nitrate reductase and phytochrome-I gene expression by light. *Plant Mol. Biol.* 29 (1): 25–35.

Raghuram, N. and Sopory, S.K. (1999). Roles of nitrate, nitrite and ammonium ion in phytochrome regulation of nitrate reductase gene expression in maize. *IUBMB Life* 47 (2): 239–249.

Raghuram, N., Chandok, M.R., and Sopory, S.K. (1999). Light regulation of nitrate reductase gene expression in maize involves a G-protein. *Mol. Cell Biol. Res. Commun.* 2 (2): 86–90.

Raghuram, N., Pathak, R.R., and Sharma, P. (2006). Signalling and the molecular aspects of N-use efficiency in higher plants. In: *Biotechnological Approaches to Improve Nitrogen Use Efficiency in Plants* (ed. R.P. Singh and P.K. Jaiwal), 19–40. Houston, TX: Studium Press.

Raghuram, N., Sachdev, M.S., and Abrol, Y.P. (2007). Towards an integrative understanding of reactive nitrogen. In: *Agricultural Nitrogen Use and Its Environmental Implications* (ed. Y.P. Abrol, N. Raghuram, and S. Sachdev), 1–6. New Delhi: I.K. International Pvt. Ltd.

Raghuram, N., Sangar, S., and Abrol, Y.P. (2011). Nitrogen management for sustainable development. *Curr. Sci.* 101 (9): 1126–1127.

Raghuram, N., Abrol, Y.P., Pathak, H.K. et al. (2020). South Asian Nitrogen Centre: capacity building for regional N assessment and management. In: *Just Enough Nitrogen. Perspectives on How to Get There for Regions with Too Much and Too Little Nitrogen* (ed. M. Sutton, K. Mason, A. Bleeker, et al.), 467–479. Cham. ISBN: 978-3-030-58064-3: Springer https://doi.org/10.1007/978-3-030-58065-0_32.

Raghuram, N., Sutton, M.A., Jeffery, R. et al. (2021). From South Asia to the world: embracing the challenge of global sustainable nitrogen management. *One Earth* 4 (1): 22–27. https://doi.org/10.1016/j.oneear.2020.12.017.

Raghuram, N., Aziz, T., Kant, S. et al. (2022a). Editorial: Nitrogen use efficiency and sustainable nitrogen management in crop plants. *Front. Plant Sci.* 13: 862091. https://doi.org/10.3389/fpls.2022.862091.

Raghuram, N., Schmidt, S., Aziz, T. et al. (ed.) (2022b). *Nitrogen Use Efficiency and Sustainable Nitrogen Management in Crop Plants*. Lausanne, Switzerland: Frontiers Media SA. ISBN 978-2-88974-284-4. https://doi.org/10.3389/978-2-88974-284-4.

Sharma, N., Sinha, V.B., Gupta, N., Rajpal, S., Kuchi, S., Sitaramam, V., Parsad R. and Raghuram, N. (2018). Phenotyping for nitrogen use efficiency (NUE): rice genotypes differ in N-responsive germination, oxygen consumption, seed urease activities, root growth, crop duration and yield at low N. *Front. Plant Sci.* 9:1452 doi: https://doi.org/10.3389/fpls.2018.01452

Sharma, N., Kuchi, S., Singh, V., and Raghuram, N. (2019). Method for preparation of nutrient-depleted soil for determination of plant nutrient requirements. *Commun. Soil Sci. Plant Anal.* 50 (15): 1878–1886.

Sharma, N., Sinha, V.B., Arun Prem Kumar, N. et al. (2021). Nitrogen use efficiency phenotype and associated genes: role of germination, flowering, root/shoot length and biomass. *Front. Plant Sci.* 11: 587464. https://doi.org/10.3389/fpls.2020.587464.

Sharma, N., Kumari, S., Jaiswal, D.K., and Raghuram, N. (2022). Comparative transcriptomic analyses of nitrate-response in rice genotypes with contrasting nitrogen use efficiency (NUE) reveals common and genotype-specific processes, molecular targets and NUE-candidates. *Front. Plant Sci.* 12: 881204. https://www.frontiersin.org/articles/10.3389/fpls.2022.881204.

Sharma, N., Jaiswal, D.K., Kumari, S. et al. (2023). Genomewide urea response in rice genotypes contrasting for nitrogen use efficiency. *Int. J. Mol. Sci.* 24(7): 6080 https://doi.org/10.3390/ijms24076080.

Singh, B., Tiwari, M.K., and Abrol, Y.P. (2008). *Reactive Nitrogen in Agriculture, Industry and Environment in India*, IGBP-WCRP. SCOPE Report Series, vol. 3. New Delhi: Indian National Science Academy, New Delhi.

Sinha, V.B., Jangam, A.P., and Raghuram, N. (2020). Biological determinants of crop N use efficiency and biotechnological avenues for improvement. In: *Just Enough Nitrogen. Perspectives on How to Get There for Regions with Too Much and Too Little Nitrogen* (ed. M. Sutton, K. Mason, A. Bleeker, et al.), 157–171. Cham. ISBN: 978-3-030-58064-3: Springer https://doi.org/10.1007/978-3-030-58065-0_11.

Steffen, W., Richardson, K., Rockström, J. et al. (2015). Planetary boundaries: guiding human development on a changing planet. *Science* 347 (6223): 1259855. https://doi.org/10.1126/science.1259855.

Sutton, M.A., Bleeker, A., Howard, C.M. et al. (2013). *Our Nutrient World: The Challenge to Produce More Food and Energy with Less Pollution*. Global Overview of Nutrient Management. Centre for Ecology & Hydrology, Edinburgh on behalf of the Global Partnership on Nutrient Management and the International Nitrogen Initiative. ISBN: 978-1-906698-40-9.

Sutton, M.A., Drewer, J., Moring, A. et al. (2017). The Indian nitrogen challenge in a global perspective. In: *The Indian Nitrogen Assessment* (ed. Y.P. Abrol, T.K. Adhya, V.P. Aneja, et al.), 9–28. UK: Elsevier ISBN: 978-0-12-811836-8.

Sutton, M., Raghuram, N., Adhya, T.K. et al. (2019). The nitrogen fix: from nitrogen cycle pollution to nitrogen circular economy. In: *Frontiers 2018/2019: Emerging Issues of Environmental Concern* (ed. P. Sarasas), 52–65. Nairobi: United Nations Environment Programme.

Sutton, M.A., Mason, K., Bleeker, A. et al. (ed.) (2020a). *Just Enough Nitrogen. Perspectives on How to Get There for Regions with Too Much and Too Little Nitrogen*. Cham: Springer, ISBN: 978-3-030-58064-3. 603p.

Sutton, M.A., Mason, K., Bleeker, A. et al. (2020b). Just enough nitrogen. Summary and synthesis of outcomes. In: *Just Enough Nitrogen. Perspectives on How to Get There for Regions with Too Much and Too Little Nitrogen* (ed. M. Sutton, K. Mason, A. Bleeker, et al.), 1–25. Cham. ISBN: 978-3-030-58064-3: Springer https://doi.org/10.1007/978-3-030-58065-0_1.

Sutton, M.A., Howard, C.M., Brownlie, W.J. et al. (2020c). Global challenges for nitrogen science-policy interactions: towards the International Nitrogen Management System (INMS) and improved coordination between multi-lateral environmental agreements. In: *Just Enough Nitrogen. Perspectives on How to Get There for Regions with Too Much and Too Little Nitrogen* (ed. M. Sutton, K. Mason, A. Bleeker, et al.), 517–560. Cham. ISBN: 978-3-030-58064-3: Springer https://doi.org/10.1007/978-3-030-58065-0_36.

Sutton, M.A., Ebanyat, P., Raghuram, N. et al. and Delegates of the 6th International Nitrogen Conference, Kampala (2020d). The Kampala statement-for-action on reactive nitrogen in Africa. In: *Just Enough Nitrogen. Perspectives on How to Get There for Regions with Too Much and Too Little Nitrogen* (ed. M. Sutton, K. Mason, A. Bleeker, et al.), 583–593. Cham, ISBN: 978-3-030-58064-3: Springer https://doi.org/10.1007/978-3-030-58065-0_38.

Sutton, M.A., Howard, C.M., Kanter, D.R. et al. (2021). The nitrogen decade: mobilizing global action on nitrogen to 2030 and beyond. *One Earth* 4 (1): 10–14. https://doi.org/10.1016/j.oneear.2020.12.016.

Tian, H., Xu, R., Canadell, J.G. et al. (2020). A comprehensive quantification of global nitrous oxide sources and sinks. *Nature* 586: 248–256. https://doi.org/10.1038/s41586-020-2780-0.

Udvardi, M., Below, F.E., Castellano, M. et al. (2021). A research road map for responsible use of agricultural nitrogen. *Front. Sustain. Food Syst.* 5: 660155. https://doi.org/10.3389/fsufs.2021.660155.

UNEP (2019). Sustainable Nitrogen Management. Resolution no. UNEP/EA.4/Res.14. Fourth United Nations Environment Assembly (15 March 2019), United Nations Environment Programme, Nairobi.

UNEP (2022). Sustainable Nitrogen Management. Resolution no. UNEP/EA.5/Res.2. Fifth United Nations Environment Assembly (2 March 2022), United Nations Environment Programme, Nairobi.

Winiwarter, W., Amon, B., Bodirsky, B.L. et al. (2022). Focus on reactive nitrogen and the UN sustainable development goals. Editorial: Reactive nitrogen in the environment: challenges, threats and solutions. *Environ. Res. Lett.* 17 (5): 050401.

Yang, A.L., Raghuram, N., Porter, S. et al. (2022). Policies to combat nitrogen pollution in South Asia: gaps and opportunities. *Environ. Res. Lett.* 17: 025007. https://doi.org/10.1088/1748-9326/ac48b2.

Part 2

Climate Change: Global Impact

5

Climate-Resilient Crops for CO_2 Rich-Warmer Environment

Opportunities and Challenges

Sayanta Kundu[1], Sudeshna Das[1], Satish K. Singh[2], Ratnesh K. Jha[1], and Rajeev Nayan Bahuguna[1,3]

[1] *Center for Advanced Studies on Climate Change, Dr. Rajendra Prasad Central Agricultural University, Samastipur, Bihar, India*
[2] *Department of Plant Breeding and Genetics, Dr. Rajendra Prasad Central Agricultural University, Samastipur, Bihar, India*
[3] *Agriculture Biotechnology, National Agri-Food Biotechnology Institute, Sector 81, SAS Nagar, Mohali, India*

Sayanta Kundu and Sudeshna Das contributed equally to this work

5.1 Introduction

The world population is increasing at a rapid pace. According to the Census Bureau of the United States, the estimated world population on 1 January 2022 was 7.8 billion, and it is expected to reach up to 9.7 billion by the year 2050 (Climate Reports United Nations 2019). Therefore, an additional 50–70% increase in food production would be required to sustain the future food and nutritional demand. On the other hand, the destruction of forests and agricultural lands due to urbanization and RAPID industrialization has affected agriculture as well as environmental conditions during the past century. Indeed, extensive release of greenhouse gases, particularly CO_2, in the environment has resulted in an unprecedented increase in global mean temperature, which is as the main driver of global climate change (Rangaswamy et al. 2021). Intergovernmental Panel on Climate Change (IPCC) has reported that the global mean temperature will increase by 3 °C by the end of 2100, resulting in more frequent incidence of severe heat waves (IPCC 2021). Increase in global mean temperature and heat spikes at regional levels are considered as a major threat to global crop production, including rice, wheat, and maize, thus threatening global food security (Kadam et al. 2014).

Cereal crops are sensitive to high temperatures; however, early vegetative, reproductive, and critical grain filling stages are more sensitive to heat stress (Kadam et al. 2014; Bahuguna et al. 2018a). Occurrence of heat stress during the vegetative stage ceases plants' growth, leading to delay in flowering and reduction in biomass accumulation, whereas heat stress during the reproductive stage advances flowering and reduces the life span (Shi et al. 2013; Sarsu et al. 2018). Additionally, heat stress also causes pollen sterility and reduces pollen viability, fertilization, and grain filling (Jagadish et al. 2010; Bahuguna et al. 2015; Jagadish 2020). Besides heat stress, occurrence of drought stress due to the increase in variability in rainfall patterns also affects crop production severely. Besides a direct impact on photosynthesis, moderate to severe drought during the vegetative stage has been reported to limit shoot elongation, tillering. Conversely, pollen and spikelet sterility (rice, wheat) and extended anthesis-silking interval (maize) are key traits affected by drought during the reproductive stage (Bahuguna et al. 2018b; Havrlentová et al. 2021).

In natural condition, crops seldom face individual stress condition, and often two or more stress may occur simultaneously, intermittently, or sequentially (Bahuguna et al. 2018a; Kumar et al. 2023). Eventually, plants' physiological and biochemical behaviors under stress combinations have been reported as a new stress type, which is characteristically different from the one under an isolated stress environment (Mittler 2006; Mittler and Blumwald 2010). Moreover, stresses in combinations have been documented to be more detrimental to growth, yield, and crop plants. Despite the occurrence of stress combinations in nature is inevitable, and some of the stress combinations are very obvious to co-occur such as heat and drought, our understanding of plant response and adaptation under combination stresses is clear (Yadav et al. 2022). Nevertheless, crop improvement strategies, including marker-assisted selection (MAS), and recent advances in genome

Global Climate Change and Plant Stress Management, First Edition. Edited by Mohammad Wahid Ansari, Anil Kumar Singh, and Narendra Tuteja.

editing techniques, including CRISPR-Cas9, could help in understanding genetic regulation and mechanistic basis of plant adaptation in complex stress environments under changing climate (Zafar et al. 2020; Das et al. 2021). Moreover, recent advancements in phenomics with the deployment of high throughput phenotyping platforms are remarkably helping in identification of novel traits for stress tolerance by exploring large number of natural and wild accessions of crops in very less time (Bahuguna et al. 2018a; Ostmeyer et al. 2022). The phenotyping data from a large and diverse germplasm can be used in identifying QTLs and loci responsible for stress tolerance through genome-wide association studies (GWAS) using molecular markers. Eventually, the identified loci/gene(s) can be introgressed into elite stress-sensitive cultivars for developing tolerant varieties through crop improvement programs (Challa and Neelapu 2018). Besides the advancement in developing stress tolerant varieties, identification of e[CO_2] responsive genotypes of C_3 crops including rice and wheat, and their inclusion into breeding programs could be a promising strategy to maintain to achieve the higher productivity under changed climatic condition (Bahuguna et al. 2022). In this chapter, we will summarize the recent trends of climate change, its impact on the major cereal crops, and approaches to achieve sustainable yield under an abruptly changing climate and associated stress factors.

5.2 Climate Change Trend and Abiotic Stress: Yield Losses Due to Major Climate Change Associated Stresses Heat, Drought and Their Combination

Temperature is a major driver of climate change and a major factor limiting growth and yield of crop plants when it goes beyond optimum levels. Temperatures beyond the optimum level affect crop growth stages, from the vegetative to the grain filling stage (Bahuguna and Jagadish 2015). For example, occurrence of heat stress during the vegetative stage reduces plants' photosynthetic rate and increases the rate of photorespiration. As a result, biomass accumulation in plants is reduced. In addition, water use efficiency (WUE) of plants is also reduced under higher ambient temperatures, as the rate of transpiration increases to cool down the plant tissue temperature. On the other hand, heat stress may directly affect pollen and spikelet sterility, reduce pollen viability, and extend ASI (in maize) during the reproductive stage. These lead to a reduction in fertilization and ultimately results in yield reduction. High temperature during nighttime particularly affects grain filling process and duration in cereals crops such as rice (Bahuguna et al. 2017). Thus, grain weight is reduced, and total biomass is reduced due to shorter grain filling period and higher C losses in respiration process (Shi et al. 2013; Bahuguna et al. 2017). Additionally, grain quality is also deteriorated due to the change in protein and starch composition (Xu et al. 2021).

Besides high temperatures, drought is a co-occurring and most obvious companion stress in field conditions, particularly in the rain-fed areas (Mittler and Blumwald 2010; Suzuki et al. 2014). It has been reported that rice and wheat production may reduce up to 32% and 26%, respectively, due to drought stress; however, this reduction may increase substantially if accompanied by a sudden heat wave at critical growth stages (Kadam et al. 2014). Occurrence of drought stress during different plant growth stages, viz. vegetative, reproductive, and grain filling stages causes substantial decrease in grain yield, and among these stages, the reproductive stage is the most vulnerable one (Zhang et al. 2018). The reduction in yield due to drought is majorly due to the decrease in photosynthetic rate, which is mostly due to the reduced activity of ribulose bisphosphate carboxylase/oxygenase (Rubisco). Conversely, the reduced uptake of CO_2 due to the restricted stomatal conductance is another reason for the reduced accumulation of photosynthates. In general, reduced stomatal conductance to reduce the rate of transpiration under water stress condition is a key trait in plants to improve WUE. However, this dominant feature is not effective under the combination of heat and drought, which is considered as the most obvious stress combinations in nature (Suzuki et al. 2014).

The level of vulnerability to a particular stress totally depends on its intensity, duration, and interaction with other environmental factors. For example, the cumulative interaction of heat stress with humidity, vapor pressure deficit (VPD), and soil moisture condition determine its severity (Bahuguna and Jagadish 2015). When adequate soil moisture is not available for plants to transpire for effective cooling under heat stress, plants face substantially higher yield loss (Thayil et al. 2020). It is also reported that a prolonged period of heat stress could cause physiological drought in plants due to the excessive release of water through transpiration (Seetharam et al. 2021). Conversely, the increasing atmospheric CO_2 can be advantageous under drought due to stomatal closure and lower stomatal conductivity reported under elevated CO_2 (Bahuguna et al. 2022). Moreover, stomatal opening has been suggested as major entry route for pathogenic microbes; thus, CO_2-led stomatal closure may help in reducing pathogen incidences in crops. Though elevated CO_2 plays a role in increasing water, light use efficiency, and decreased pest abundance (Skendžić et al. 2021), the interactive effect of CO_2 with other biotic and abiotic stresses warrants systematic studies to understand long-term effect of elevated CO_2 under current and more so in future environments.

5.3 Update on Crop Improvement Strategies Under Changing Climate

5.3.1 Advances in Breeding and Genomics

The green revolution era was majorly focused on improving grain yield. Eventually, crop breeding programs solely targeted yield-related traits for selection and development of high yielding varieties that are still grown as elite cultivars, having a narrow genetic base for stress tolerance (Sabagh et al. 2020). However, the rapid change in environment, frequent occurrence of severe stresses, and their combinations forced researchers shift crop improvement strategies toward developing climate-smart varieties for sustainable yield generation under suboptimal environments. Therefore, landraces and wild accessions with wider genetic background have been explored for the identification of targeted trait(s)-specific QTL(s) and dominant gene(s) responsible for multiple stress tolerance through GWAS (Würschum et al. 2022). A mapping population is developed by crossing contrasting parents such as stress-tolerant and sensitive genotypes. This helps in the segregation of QTLs/genes responsible for stress tolerance within the population. Subsequently, molecular markers are used to evaluate the association of genes/QTLs with targeted trait(s), and in this way, QTL(s)/gene(s) specific to targeted trait(s) such as stress tolerance are identified. These identified QTL(s)/gene(s) can eventually be introgressed into elite cultivars by either breeding or transgenic approach to develop tolerant varieties (Ahmed et al. 2022). Further, these mapping populations can be used in identifying traits associated with molecular markers, which can help in indirect selection of stress-specific trait of interest, process known as MAS. MAS is an effective and efficient technique for selecting trait of interest, and it is more reliable than conventional screening approaches based on morphological traits. Several molecular markers have been developed in the past decades, which are being used in marker-assisted screening and characterization of genotypes. Major molecular markers used are Restriction Fragment Length Polymorphism (RFLP), Amplified Fragment Length Polymorphism (AFLP), Random Amplified Polymorphic DNA (RAPD), Simple Sequence Repeats (SSR), and Single Nucleotide Polymorphism (SNP), and among these, SNPs are mostly preferred markers. SNPs are the highly abundant markers, present in the genomes, that help in getting the higher resolution of genome map (Mammadov et al. 2012). Besides, advanced genome editing tools including Clustered Regularly Interspaced Short Palindromic Repeats and CRISPR-associated Cas9 (CRISPR/Cas9) and Transcription Activator-like Effector Nuclease (TALEN) have been used for targeted modification of genomic region for developing stress-tolerant genotypes. Among these tools, CRISPR/Cas9 is the most widely used one, as it is the most accurate, efficient, fast, and cheaper genome editing tool (Khatodia et al. 2016). It is naturally present in bacteria for their defense against viral infection. CRISPR/Cas9 contains a short RNA sequence having guide RNA (gRNA), which specifically binds to the targeted genomic region and Cas9 is a nuclease for cleaving that specific region. In this way, CRISPR/Cas9 is used for adding, deleting, and altering targeted region for developing stress-tolerant genotypes. For example, CRISPR/Cas9-mediated targeted manipulation of Dehydration-responsive Element Binding 2 protein (TaDREB2) and *Triticum aestivum* Ethylene Responsive Factor 3 (TaERF3) genes helps in maintaining the expression of genes relevant under drought stress tolerance (Kim et al. 2018).

5.3.2 Advances in Phenomics and High Throughput Platforms

Scientists during the past decade have believed that genomics and gene technology could offer a solution to the challenges in plant breeding for the selection of stress-resilient genotypes. However, sequencing of the numerous crops, their modification through gene editing, and continuous selection of natural sources of genetic variation for crop improvement require analysis of hundreds of lines grown in diverse environmental conditions, which is time consuming, laborious, and often expensive. Furthermore, the success of any crop improvement program, including QTL analysis, fine mapping of genes/QTL, GWAS, and genomic selection (GS), largely depends on the meticulous, specific, and detailed recording of the phenotypic observations in crop plants (Singh et al. 2018). Thus, none of the crop genetic resources can be thoroughly utilized without the evolution and progress of the crop phenomics techniques (Cabrera-Bosquet et al. 2012). Nevertheless, the progress in the plant phenomics was not at par till the last decade, which was a major bottleneck in the success of crop improvement programs for yield and stress tolerance. The phenotype of a plant does not connect solely to its genotype. Rather, the phenotype of a plant interacts with its surrounding environment and exhibits modifications in accordance to the exposed environmental conditions. For example, day-length duration influences the number of leaves in *Arabidopsis thaliana* (Mishra et al. 2012), whereas light intensity and water and nitrogen status have a major impact on the number and size of the plant organs (Tardieu et al. 2017). Temperature, on an average summer day, may vary from 12–14 °C at dawn to 38–40 °C at dusk, which initiates rapid alterations in the plant growth and morphology (Chelle 2005). Conversely, CO_2 is the basic substrate for photosynthesis; hence, the unprecedented rise in atmospheric CO_2 concentration due to anthropogenic and other natural reasons would have a substantial impact on growth, morphology, physiology, and yield particularly in C_3 crops such as rice and wheat (Bahuguna et al. 2022). Orientation of leaf, sink size, and stomatal density are key features, which could be directly influenced by ambient CO_2 levels. Similarly, changes in root morphology as a result of soil water content are example of environment-influenced adaptation (Sebastian et al. 2016). Thus, phenomics strives to characterize the plant phenotypic plasticity at a temporal and spatial scale arising from any environmental influences or presence of genetic mutation.

Phenotyping for selection of an important agricultural trait (abiotic stress tolerance, yield potential, time of flowering, photosynthetic efficiency, or nutrient utilization efficiency) requires multiple trials over a large area under contrasting crop growing conditions with repeated and frequent destructive assays. These trials are usually unsuccessful due to the slow, laborious, and manual phenotyping. Additionally, if a propitious candidate gene in a mapping population is to be verified for allelic variation, the process should be free of any kind of error (Furbank and Tester 2011). However, phenotyping at the field level has a huge probability to be devoid of accuracy, prone to human errors and biasedness, and difficult to be conducted for a large number of genotypes. These drawbacks in field phenotyping have led the scientists and technologists to work toward combining novel technologies and basic science in a way so as to address the bottlenecks in field phenotyping and develop automated phenotyping platforms that could capture, store, and analyze maximum data in the shortest possible time. Eventually, the past decade has witnessed the development of non-destructive high-throughput platforms that combined the principles of non-invasive imaging, spectroscopy, image analysis, robotics, remote sensing, and high-performance computing to record, reserve, and interpret visible and non-visible changes in the plant phenotype at the temporal and spatial level with utmost precision and accuracy. These automated phenotyping facilities are capable of evaluating several traits ranging from developmental and morphological descriptors (flowering time, panicle development, height, tiller number) to physiological indicators (chlorophyll fluorescence, photosynthetic efficiency). An all-inclusive approach to crop phenotyping will enhance our knowledge about the environment-influenced crop phenotype and help breeders develop climate-resilient cultivars and crops through the identification of stress-tolerant genotypes (Singh et al. 2018).

5.3.3 Non-destructive Phenotyping to Exploit Untapped Potential of Natural Genetic Diversity

The absence of high-throughput non-destructive phenotyping has been the primary cause of low success rate in crop improvement programs. The main aim of utilizing non-destructive high-throughput phenotyping through novel technologies is that the data and observations for all desirable traits can be recorded from a single plant during the entire life cycle in a shorter period of time with higher precision and accuracy. This enhances the probability of better understanding of the underlying physiological and biochemical processes of the plant grown under diverse climatic conditions.

Infrared imaging is one of the most popular phenotyping techniques, which is used to detect the difference in leaf/surface temperature, air temperature, and canopy temperature under heat and drought stress. It has been put to use in several crop studies, such as those in melons, tomatoes, and lettuce (Qiu et al. 2009). Apart from heat and drought stress, infrared imaging has also been utilized in wheat and barley to detect the osmotic stress under salinity (Sirault et al. 2009). Besides measuring the stress severity, infrared imaging is utilized in detecting the flowering time in plants especially for those, which predominantly flower during the late evenings or at night (Sheehy et al. 2009; Aiqing et al. 2018). The imaging system captures the movement of the flower bristles, which occurs simultaneously with the extrusion of the anthers, and correlates it to the time of the flower opening (Desai et al. 2018).

Spectroscopic techniques involved in phenotyping can be used for detecting the photosynthetic rates at the leaf and canopy levels. The technique was utilized for measuring the photosynthetic electron transport feedback regulation in *Nicotiana sylvestris* (Kiirats et al. 2009). The technique of reflectance spectroscopy has been successfully used in pine and barley to phenotype its physiological process such as photosynthetic efficiency, its activity, and the involved biochemical pathway (Busch et al. 2008; Siebke and Ball 2009). Chlorophyll fluorescence is yet another significant and most common physiological process, which can indicate the stress severity through the estimation of the photosynthetic activity of the plant. This essential trait can be non-destructively measured by an easily available hand tool using the technique of spectral absorption and reflectance. Observing the 2D digital imaging of plant growth along with estimating the chlorophyll fluorescence of the plant could elucidate the changes taking place in the plant under drought and chilling stress (Jansen et al. 2009). Apart from capturing the changes during abiotic stress, the technique of fluorescence imaging can also be utilized to study the fate of herbicides in plants (Rungrat et al. 2016).

Yield and its attributes are the important parameters in a crop life cycle. Nevertheless, scientists are still not capable of accurately predicting the yield in stressed and non-stressed environments. Yield prediction is a complex process influenced by the dynamic environmental changes that are prone to variations on a large temporal and spatial scale and involves highly sophisticated and complicated systems, which utilizes a massive amount of data and multiple regression models or machine learning (Schwalbert et al. 2020). This aspect of yield prediction has led technology designers utilize various machine and deep learning tools, which attempt to identify traits such as number of panicles, spikelets through a convolutional neural network (CNN) classification (Xiong et al. 2015; Madec et al. 2019). However, the approach bears certain loopholes such as the fact that an increased number of unfilled grains in a panicle or a spikelet due to heat, drought, or any other stress is not accurately recorded by the current model. Thus, the model will require further improvisations before it can be engaged or used effectively to evaluate the seed loss under stressed environments.

5.4 Exploiting Climate-Smart Cultivation Practices

Climate-smart agriculture is an approach to amalgamate technologies, management practices, and strategies aiming to achieve food security through sustainable agriculture to deal with the changing climate. Climate-smart agriculture mainly focuses on three aspects: enhancing farm productivity and income, encouraging eco-friendly and sustainable farming practices, and establishing a climate-resilient agricultural system. Several interventions can be exploited to adapt the crops in an environment where the natural resources can be conserved or minimally used without hampering the ultimate yield of the crops. For instance, direct seeded rice is an important approach that eliminates puddling and flooding. This not only saves water but also cuts down the emission of greenhouse gases, which are a potential contributor to climate change. The second method that can be put to advantage in order to conserve water is the method of alternate wetting and drying, which requires setting up a perforated pipe in the field to determine the time of irrigation.

Water has always been an important factor for plant growth and development, and for drought-prone areas, agriculture has always been difficult. In such areas, altering the method of irrigation could be an apt technique to produce more crops with a limited amount of water. Micro-irrigation (drip-irrigation) is one such innovative technology that continuously deliver water to the plant rhizosphere in drop form. This leads to an effective utilization of the available water, which is capable of producing more yields per drop of water.

Nutrient is yet another significant factor that contributes to a healthy plant life cycle. Since different sites have different soil nutrient status, determination of the methodical and appropriate application of the nutrients to the plants is essential to enhance the productivity and profit of the growers. Thus, site-specific nutrient management (SSNM) system has been advocated for the effective supply of the required nutrients to the plant at a certain location. This system calculates the nutrients that should be supplied to the plant on the basis of the already available nutrients in the soil for plant uptake. More often than not, it has been observed that although a certain nutrient is present in abundance in the soil, the plant is unable to uptake the same due to its complexation with metal ions in the soil. This is usually found in the case of phosphorous. At times such as this, one could opt for a cost-effective and eco-friendly method such as application of phosphate-solubilizing microorganisms that, through the secretion of organic acids, enzymes, and siderophores, can solubilize the insoluble phosphates into plant absorbable form.

Apart from the climate-smart management practices, it is imperative to keep a track of the upcoming short-term weather changes. This information could be disseminated through mobile-based agro-advisory services. These agro-advisory services can keep the growers informed of the immediate weather change, if any, and thus the precautions for the crops can be planned accordingly. For instance, the sowing or harvest of a crop, the selection of crop for the upcoming season or the management practices to be undertaken can be strategically planned as per the forecasted short-term weather conditions. Thus, it can be deduced that the aim of crop resilience for the changing climate could be achieved only through the combination of technologies, management practices, and strategies (Figure 5.1). However, the accomplishment of the climate-resilient agricultural system with guaranteed food security is possible only when science works hand-in-hand with the policy makers and the growers to explore more options and learn from their experiences.

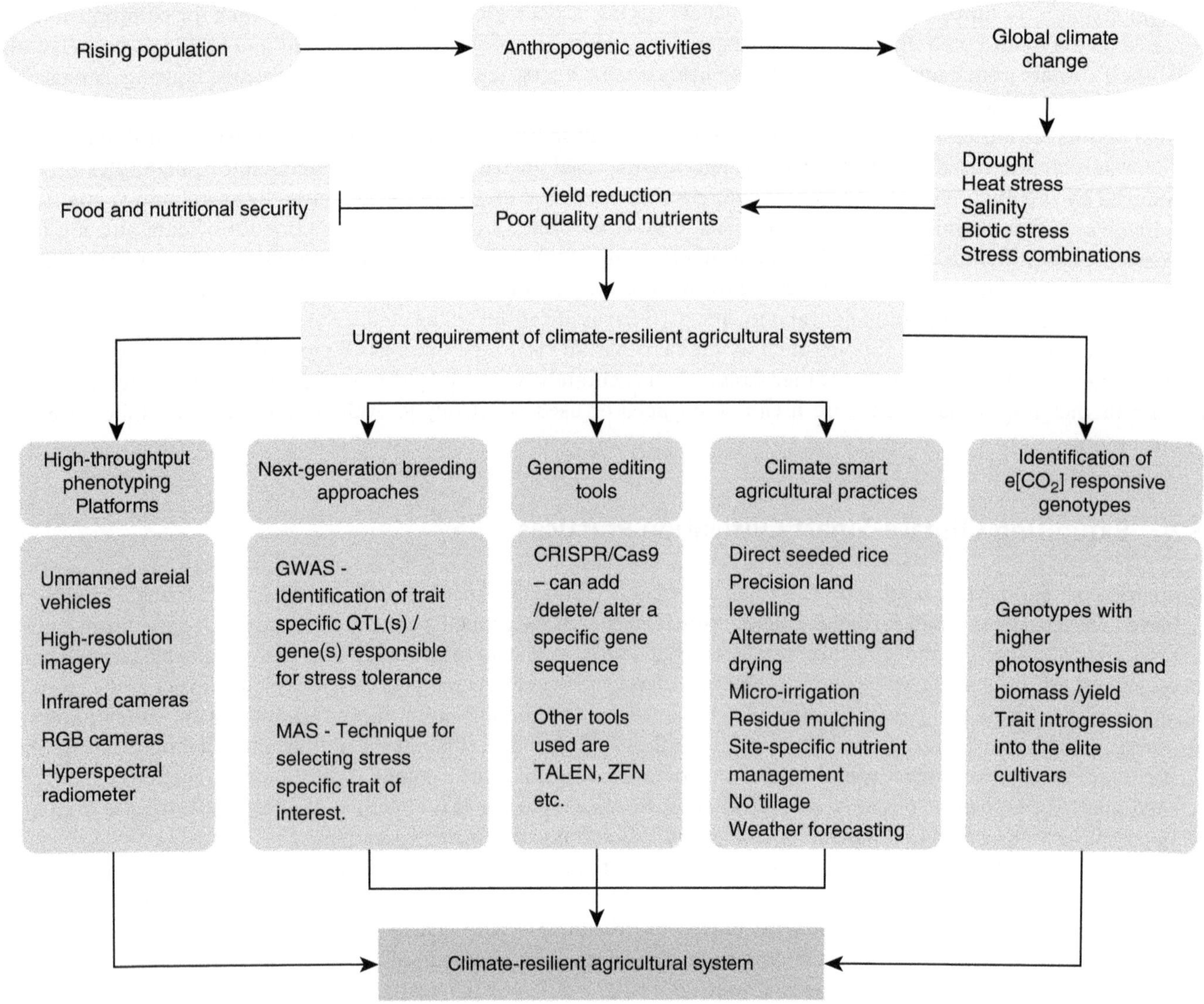

Figure 5.1 Schematic diagram showing the present global scenario of changing climate, its impact on global food and nutritional security, and approaches to develop climate-resilient agricultural system.

5.5 CO_2-Responsive C_3 Crops for Future Environment

In C_3 plants, the primary product of carbon assimilation is 3-phosphoglycerate (3PGA) that is produced through the action of Rubisco. Rubisco has a dual function that totally depends on the available CO_2 to O_2 ratio. Hence, the carboxylase activity of Rubisco increases under e[CO_2] condition, along with the decrease in photorespiration. This leads to enhanced level of photosynthate accumulation. Moreover, e[CO_2] also increases plant surface area for photosynthesis, tillering, and shoot biomass during the vegetative stage and improves yield. However, it was reported that spikelet sterility increases in rice under e[CO_2] condition. On the other hand, increased accumulation of carbohydrates in grains causes decrease in protein content that is also known as dilution effect (Jagadish et al. 2014). Recently, Bahuguna et al. (2022) demonstrated CO_2 responsiveness as a potential trait to increase the productivity of C_3 crops under e[CO_2], where genetic variation in CO_2 responsiveness has been found as a major factor to determine the productivity of C_3 crops under optimum and high night temperature environment. A major mechanism working under e[CO_2] was a positive carbon balance (high photosynthesis vs lower dark respiration), which led to compensate the carbon loss due to higher respiration under stress. Thus, identification of e[CO_2]-responsive genotypes and incorporation of the CO_2 responsiveness trait into breeding programs are potential routes to develop climate-resilient crops (Bahuguna et al. 2022).

5.6 Conclusion

The threat of climate change on crop production can be minimized by improving crop production strategies and adapting climate-smart agricultural practices. Moreover, the primary goal should be the development of climate-resilient crop varieties that can adapt and thrive well under changing climatic conditions. For this reason, improved techniques and tools, including high-throughput phenotyping platforms, next generation breeding approaches (GWAS and MAS), and genome editing tools, have been used for developing crops tolerant to different abiotic stresses and their combinations. On the other hand, gradual adaptation of climate-smart agricultural practices has been ensuring the improvement of economic, social, and environmental aspects of crop cultivation to confirm food security.

References

Ahmed, Z., Khalid, M., Ghafoor, A. et al. (2022). SNP-Based genome-wide association mapping of pollen viability under heat stress in tropical *Zea mays* L. inbred lines. *Front. Genet.* 13: 819849.

Aiqing, S., Somayanda, I., Sebastian, S.V. et al. (2018). Heat stress during flowering affects time of day of flowering, seed set, and grain quality in spring wheat. *Crop Sci.* 58: 380–392.

Bahuguna, R.N. and Jagadish, K.S.V. (2015). Temperature regulation of plant phenological development. *Environ. Exp. Bot.* 111: 83–90.

Bahuguna, R.N., Jha, J., Pal, M. et al. (2015). Physiological and biochemical characterization of NERICA-L-44: a novel source of heat tolerance at the vegetative and reproductive stages in rice. *Physiol. Plant.* 154: 543–559.

Bahuguna, R.N., Solis, C.A., Shi, W., and Jagadish, K.S.V. (2017). Post-flowering night respiration and altered sink activity account for high night temperature-induced grain yield and quality loss in rice (*Oryza sativa* L.). *Physiol. Plant.* 159: 59–73.

Bahuguna, R.N., Gupta, P., Bagri, J. et al. (2018a). Forward and reverse genetics approaches for combined stress tolerance in rice. *Ind. J. Plant Physiol.* 23: 630–646.

Bahuguna, R.N., Tamilselvan, A., Raveendran, M. et al. (2018b). Mild preflowering drought priming improves stress defences, assimilation and sink strength in rice under severe terminal drought. *Funct. Plant Biol.* 45: 827–839.

Bahuguna, R.N., Chaturvedi, A.K., Pal, M. et al. (2022). Carbon dioxide responsiveness mitigates rice yield loss under high night temperature. *Plant Physiol.* 188: 285–300.

Busch, F., Hüner, N.P.A., and Ensminger, I. (2008). Increased air temperature during simulated autumn conditions impairs photosynthetic electron transport between photosystem II and photosystem. *Plant Physiol.* 147: 402–414.

Cabrera-Bosquet, L., Crossa, J., von Zitzewitz, J. et al. (2012). High-throughput phenotyping and genomic selection: the frontiers of crop breeding converge. *J. Integr. Plant Biol.* 54: 312–320.

Challa, S. and Neelapu, N.R.R. (2018). Genome-wide association studies (GWAS) for abiotic stress tolerance in plants. In: *Biochemical, Physiological and Molecular Avenues for Combating Abiotic Stress in Plants* (ed. S.H. Wani), 135–150. Academic Press.

Chelle, M. (2005). Phylloclimate or the climate perceived by individual plant organs: What is it? How to model it? What for? *New Phytol.* 166: 781–790.

Climate Reports United Nations (2019). United Nations Climate Change Annual Report 2019. https://www.un.org/en/climatechange/reports (accessed 20 August 2020).

Das, E., Prasad, S., and Roy, I. (2021). *Saccharomyces cerevisiae* Fpr1 functions as a chaperone to inhibit protein aggregation. *Int. J. Biol. Macromol.* 191: 40–50.

Desai, J.S., Slabaugh, E., Liebelt, D.J. et al. (2018). Neural net classification combined with movement analysis to evaluate *Setaria viridis* as a model system for time of day of anther appearance. *Front. Plant Sci.* 9: 1585.

Furbank, R.T. and Tester, M. (2011). Phenomics – technologies to relieve the phenotyping bottleneck. *Trends Plant Sci.* 16: 635–644.

Havrlentová, M., Kraic, J., Gregusová, V., and Kovácsová, B. (2021). Drought stress in cereals – a review. *Agric. (Ponohospodárstvo)* 67: 47–60.

IPCC (2021). Summary for Policymakers of the Working Group I Contribution to the IPCC Sixth Assessment Report. NERC EDS Centre for Environmental Data Analysis.

Jagadish, S.V.K. (2020). Heat stress during flowering in cereals – effects and adaptation strategies. *New Phytol.* 226: 1567–1572.

Jagadish, S.V.K., Raveendran, M., Oane, R. et al. (2010). Physiological and proteomic approaches to address heat tolerance during anthesis in rice (*Oryza sativa* L.). *J. Exp. Bot.* 61: 143–156.

Jagadish, K.S.V., Kadam, N.N., Xiao, G. et al. (2014). Agronomic and physiological responses to high temperature, drought, and elevated CO_2 interactions in cereals. *Adv. Agron.* 127: 111–156.

Jansen, M., Gilmer, F., Biskup, B. et al. (2009). Simultaneous phenotyping of leaf growth and chlorophyll fluorescence via GROWSCREEN FLUORO allows detection of stress tolerance in *Arabidopsis thaliana* and other rosette plants. *Funct. Plant Biol.* 36: 902–914.

Kadam, N.N., Xiao, G., Melgar, R.J. et al. (2014). Agronomic and physiological responses to high temperature, drought, and elevated CO_2 interactions in cereals. *Adv. Agron.* 127: 111–156.

Khatodia, S., Bhatotia, K., Passricha, N. et al. (2016). The CRISPR/Cas genome-editing tool: application in improvement of crops. *Front. Plant Sci.* 7: 506.

Kiirats, O., Cruz, J.A., Edwards, G.E., and Kramer, D.M. (2009). Feedback limitation of photosynthesis at high CO_2 acts by modulating the activity of the chloroplast ATP synthase. *Funct. Plant Biol.* 36: 893–901.

Kim, D., Alptekin, B., and Budak, H. (2018). CRISPR/Cas9 genome editing in wheat. *Func. Integr. Genom.* 18: 31–41.

Kumar, A., Sengar, R.S., Pathak, R.K., and Singh, A.K. (2023). Integrated approaches to develop drought-tolerant rice: demand of era for global food security. *J. Plant Growth Regul.* 42: 96–120. https://doi.org/10.1007/s00344-021-10561-6.

Madec, S., Jin, X., Lu, H. et al. (2019). Ear density estimation from high resolution RGB imagery using deep learning technique. *Agric. For. Meteorol.* 264: 225–234.

Mammadov, J., Aggarwal, R., Buyyarapu, R., and Kumpatla, S. (2012). SNP markers and their impact on plant breeding. *Int. J. Plant Genom.* 2012: 1–11.

Mishra, Y., Johansson Jänkänpää, H., Kiss, A.Z. et al. (2012). Arabidopsis plants grown in the field and climate chambers significantly differ in leaf morphology and photosystem components. *BMC Plant Biol.* 12: 1–18.

Mittler, R. (2006). Abiotic stress, the field environment and stress combination. *Trends Plant Sci.* 11: 15–19.

Mittler, R. and Blumwald, E. (2010). Genetic engineering for modern agriculture: challenges and perspectives. *Ann. Rev. Plant Biol.* 61: 443–462.

Ostmeyer, T.J., Bahuguna, R.N., Kirkham, M.B. et al. (2022). Enhancing sorghum yield through efficient use of nitrogen – challenges and opportunities. *Front. Plant Sci.* 13: 478.

Qiu, G.Y., Omasa, K., and Sase, S. (2009). An infrared-based coefficient to screen plant environmental stress: concept, test and applications. *Funct. Plant Biol.* 36: 990–997.

Rangaswamy, T.C., Sridhara, S., Ramesh, N. et al. (2021). Article assessing the impact of higher levels of CO_2 and temperature and their interactions on tomato (*Solanum lycopersicum* L.). *Plants* 10: 1–14.

Rungrat, T., Awlia, M., Brown, T. et al. (2016). Using phenomic analysis of photosynthetic function for abiotic stress response gene discovery. *Arabidopsis Book* 14: e0185.

El Sabagh, A., Hossain, A., Iqbal, M.A. et al. (2020). Maize adaptability to heat stress under changing climate. In: (ed. A. Hossain). Plant Stress Physiology, London: IntechOpen.

Sarsu, F., Ghanim, A.M.A., Das, P., Bahuguna, R.N., Kusolwa, P.M., Ashraf, M., Singla-Pareek, S.L., Pareek, A., Forster, B.P., and Ingelbrecht, I. (2018). In: *Pre-Field Screening Protocols for Heat-Tolerant Mutants in Rice*, 39. Springer Nature. doi: https://doi.org/10.1007/978-3-319-77338-4

Schwalbert, R., Amado, T., Nieto, L. et al. (2020). Mid-season county-level corn yield forecast for US Corn Belt integrating satellite imagery and weather variables. *Crop Sci.* 60: 739–750.

Sebastian, J., Yee, M.C., Viana, W.G. et al. (2016). Grasses suppress shoot-borne roots to conserve water during drought. *Proc. Natl. Aca. Sci. U.S.A.* 113: 8861–8866.

Seetharam, K., Kuchanur, P.H., Koirala, K.B. et al. (2021). Genomic regions associated with heat stress tolerance in tropical maize (*Zea mays* L.). *Sci. Rep.* 11: 1–11.

Sheehy, J., Mabilangan, A., Dionora, M., and Pablico, P. (2009). Time of day of flowering in wild species of the genus *Oryza*. *Int. Rice Res. Notes* 32: 12–13.

Shi, W., Muthurajan, R., Rahman, H. et al. (2013). Source–sink dynamics and proteomic reprogramming under elevated night temperature and their impact on rice yield and grain quality. *New Phytol.* 197: 825–837.

Siebke, K. and Ball, M.C. (2009). Non-destructive measurement of chlorophyll b : a ratios and identification of photosynthetic pathways in grasses by reflectance spectroscopy. *Funct. Plant Biol.* 36: 857–866.

Singh, B., Mishra, S., Bohra, A. et al. (2018). Crop phenomics for abiotic stress tolerance in crop plants. In: *Biochemical, Physiological and Molecular Avenues for Combating Abiotic Stress in Plants* (ed. S.H. Wani), 277–296. Academic Press.

Sirault, X.R.R., James, R.A., and Furbank, R.T. (2009). A new screening method for osmotic component of salinity tolerance in cereals using infrared thermography. *Funct. Plant Biol.* 36: 970–977.

Skendžić, S., Zovko, M., Živković, I.P. et al. (2021). The impact of climate change on agricultural insect pests. *Insects* 12: 440.

Suzuki, N., Rivero, R.M., Shulaev, V. et al. (2014). Abiotic and biotic stress combinations. *New Phytol.* 203: 32–43.

Tardieu, F., Cabrera-Bosquet, L., Pridmore, T., and Bennett, M. (2017). Plant phenomics, from sensors to knowledge. *Curr. Biol.* 27: 770–783.

Thayil, V.M., Zaidi, P.H., Seetharam, K. et al. (2020). Genotype-by-environment interaction effects under heat stress in tropical maize. *Agronomy* 10: 1998.

Würschum, T., Weiß, T.M., Renner, J. et al. (2022). High-resolution association mapping with libraries of immortalized lines from ancestral landraces. *Theor. Appl. Genet.* 135: 243–256.

Xiong, D., Yu, T., Ling, X. et al. (2015). Sufficient leaf transpiration and nonstructural carbohydrates are beneficial for high-temperature tolerance in three rice (*Oryza sativa*) cultivars and two nitrogen treatments. *Funct. Plant Biol.* 42: 347–356.

Xu, Y., Chu, C., and Yao, S. (2021). The impact of high-temperature stress on rice: challenges and solutions. *Crop J.* 9: 963–976.

Yadav, C., Bahuguna, R.N., Dhankher, O.P. et al. (2022). Physiological and molecular signatures reveal differential response of rice genotypes to drought and drought combination with heat and salinity stress. *Physiol. Mol. Biol. Plants* 28: 899–910.

Zafar, S.A., Zaidi, S.S.E.A., Gaba, Y. et al. (2020). Engineering abiotic stress tolerance via CRISPR/Cas-mediated genome editing. *J. Exp. Bot.* 71: 470–479.

Zhang, J., Zhang, S., Cheng, M. et al. (2018). Effect of drought on agronomic traits of rice and wheat: a meta-analysis. *Int. J. Environ. Res. Public Health* 15: 839.

6

Potential Push of Climate Change on Crop Production, Crop Adaptation, and Possible Strategies to Mitigate This

Narendra Kumar[1] *and SM Paul Khurana*[2]

[1] *Department of Botany, Guru Ghasidas Vishwavidyalaya, a Central University, Bilaspur-495009(C. G.), India*
[2] *Amity Institute of Biotechnology, Amity University Haryana, Gurugram, Haryana, India*

6.1 Introduction

Agricultural production, human health, and natural ecosystems are getting badly affected because of devastating environmental changes (Arunanondchai et al. 2018). Since there is a rapid increase in the population, it necessitates a corresponding increase in food production where the stability of the global environment is of great concern.

The fertility of soil, air pollution, and water availability have potential impact on agricultural productivity (Noya et al. 2018). Owing to sudden changes in the environment, more negative impacts highly reduce plant productivity as direct and indirect effects of abiotic stresses. Because of excessive use of fossil fuels, the concentration of CO_2 has now increased from 280 to 400 μmol^{-1} in the atmosphere. It is projected that concentration of CO_2 will increase twofold, which will reach 800 μmol^{-1} by the end of this century. Thus, the emission of CO_2 is highly dangerous and is responsible for the increase in greenhouse effect. This is also true for warmer average global temperatures (Vaughan et al. 2018). The potential impacts of environmental variation and climate changes are mainly studied by their potential damage to agricultural crops and on daily life (FAO 2018). Agricultural yield in developing countries has suffered because of adverse environmental conditions. Excess CO_2 accumulation and high temperatures have forced scientists to develop new plans for coping with these challenges (Rosenzweig et al. 2014). To handle these limitations and get guaranteed food security, we need to produce new crop cultivars (Wheeler and Von Braun 2013). Abiotic stresses have huge impact on the growth of plant and yield. In natural climatic conditions, it has been observed that plants face salinity, cold, heat, drought, and water logging (Benevenuto et al. 2017; Ashraf et al. 2018).

The abiotic factors that induce more stresses also involve, viz., higher light intensity, flooding, UV-B, gas emissions, and chemical and physical factors (Suzuki et al. 2014). It has been assumed that in the twenty-first century the average temperature of the Earth may increase from 2 to 4.5 °C. As per IPCC-2014, the period that ranged between the nineteenth and twenty-first centuries experienced the most warming (Pachauri et al. 2014). It has been observed that extreme precipitation events cause destructions because of floods, but if there is scarcity or complete absence of rainfall for longer periods, it causes drought stresses (Khan et al. 2016). The global environment is changing continuously. Industrialization is a main factor that is responsible for the increase in temperature. Because of extreme weather events, global warming is expected to rise and disturb the ecosystem globally (Kanojia and Dijkwel 2018). All living organisms present on the Earth viz., animals, fishes, plants, and even humans get affected through conditions of extreme environment globally. These environmental fluctuations will increase the risk of food security. An investigation revealed that developed nations pose more vulnerability toward climatic changes (8–11%) in comparison with developing countries (Lesk et al. 2016; Altieri and Nicholls 2017). At this juncture it is well known that food insecurity and climate change depict two major issues of the twenty-first century. A data revealed that approximately 815 million people are affected by malnutrition that is responsible for creating hindrance in sustainable development programs for achieving the universal goal of eliminating hunger by 2030 (Richardson et al. 2018). The adverse climatic conditions are responsible for food insecurity and poor agricultural yields, which are reduced by elevation in temperatures (Ito et al. 2018). It has been observed that global production of major crops

Global Climate Change and Plant Stress Management, First Edition. Edited by Mohammad Wahid Ansari, Anil Kumar Singh, and Narendra Tuteja.

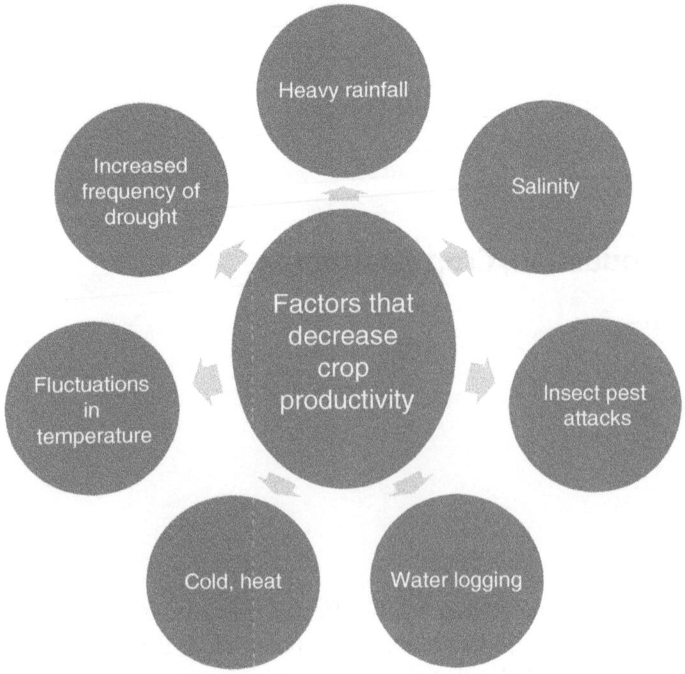

Figure 6.1 Potential factors that decrease crop productivity.

decreased as climatic severity increased from 2.6 to 4 °C. There occurs a reduction in crop productivity which is a main threat for food security which is a speedy increase in the world's population (Rogelj et al. 2016). World population may grow approximately to nine billion in 2050. Hence, the need for food can escalate by about 85% (FAOSTAT n.d.). Climate influences crop production that is highly unstable because of the environmental changes (Reckling et al. 2018). The factors that decrease crop productivity that leads to starvation are insect and pest attacks, salinity, heavy rainfall, increased frequency of drought, and fluctuations in temperature (Dhankher and Foyer 2018) (Figure 6.1). This is because of erratic rainfall due to which crop adaptability has suffered (Kang et al. 2009). So the main task to mitigate this problem is to think about the pressures posed on food security (Campbell et al. 2016). This chapter depicts potential impact of variations of weather on productivity of crops. It gives an overview of climate change, stresses posed because of climatic fluctuations, and their potential influence on agricultural crops. It also deals with the strategies for coping with extreme conditions of environmental changes along with recent approaches of genetic engineering in developing transgenic plants in mitigating abiotic stresses.

6.2 Influence of Climate Change on the Yield of Plants

Climatic variations have a potential to influence the physiology operating in plants by various means. The extremes of environmental conditions and variations in climatic conditions have enhanced chances of various stresses on plants (Thornton et al. 2014). Climatic fluctuations have profound influence on crop production by indirect, direct, and socioeconomic means (Figure 6.2). It has been observed that climate change (drought, flood, high temperature, storm, etc.) effects have increased dramatically (FAO 2017).

Boyer reported that the climatic fluctuations have reduced crop yield up to 70% since 1982 (Boyer 1982). A report of FAO 2007 states that cultivated areas of the entire world are affected by climatic fluctuations. Only 3.5% area in the world is safe from environmental limitations (Van Velthuizen 2007). Abiotic stress is the stress caused by environmental conditions such as heat, drought, salinity, cold, nutrient deficiency, and metal toxicity, which reduces growth and yield below the optimum levels (Gaur et al. 2018). However, the outcome of abiotic stresses on crop yield is difficult to calculate accurately. It may be supposed that abiotic stresses project a substantial influence on crop production. This depends upon the extent of damage in the total area that is under cultivation. It may be mentioned that major crop productivity decreases significantly in various parts of the world because of water shortage, other environmental impacts, and global warming (Bonan and Doney 2018; Tebaldi and Lobell 2018).

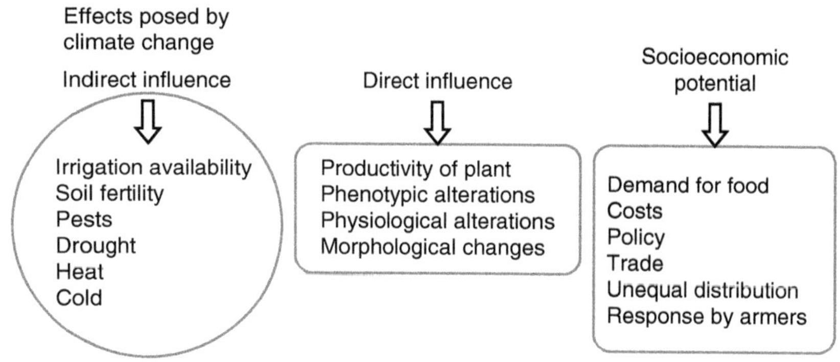

Figure 6.2 Potential impacts posed by alterations in climate.

It was observed that changing environmental conditions have affected crop yield. For northern Europe, cool temperature and short duration for crop development pose as major concern. But low rainfall and temperature extremes limit crop productivity in southern Europe. There were more negative effects in continental climate of Pannonian zone that covers Serbia, Hungary, Romania, and Bulgaria (Olesen et al. 2011). It has been predicted that abrupt climatic changes and increase in emissions of greenhouse gases will take place, which will increase crop yield in northwestern Europe but decrease crop yield in Mediterranean area (Olesen and Bindi 2002). It has been observed that wheat production is affected extremely because of temperature extremes and changes in climate in many parts of the world. This has reduced crop yield by 6% for each degree Celsius rise in temperature (Asseng et al. 2015). High temperatures and drought are stress factors that have high impact on the yield of cereals (Barnabás et al. 2008). It has been observed that RuBisCo has a key role in photosynthesis that gets disrupted when temperature increases above 35 °C. This checks the process of photosynthesis (Griffin et al. 2004). Gong et al. (1997) observed that heat stress has negative influence on antioxidant enzymes in *Zea mays*. It has also been observed that the combined effect of heat and drought stress has more damaging effects on maize, barley, and sorghum in comparison to individual stress (Wang and Huang 2004). The combined stresses of heat and drought given to *Leymus chinensis* revealed decrease in function of photosystem II (PSII) (Xu and Zhou 2006).

The reproductive phases of plant growth are adversely affected by deficit in water, climatic changes, and extremes of temperatures. Water stress adversely affects initiation of flowering and inflorescence development in cereals (Winkel et al. 1997). It was observed that when temperature increased above 30 °C at the time of floret development it caused sterility in cereals (Saini and Aspinall 1982). Another study demonstrated that there was 35–75% reduction in grain set because of deficit in water (Saini and Aspinall 1981; Sheoran and Saini 1996). Drought stress has profound effect on the process of fertilization and anthesis. Owing to the deficit in water harvest index, 60% reduction occurs in grain setting (Garrity and O'Toole 1994). The yield of cocoa reduced significantly due to drought spells in West Africa during the 1980s El N years (Ruf et al. 2015). A study stated that agricultural production may reduce to 25.7% by 2080 because of changes in climate. Maize is the most affected crop in Mexico (Hellin et al. 2014). A study based on ECHAM6 climate data was analyzed for North German Plains during two different time durations: 1981–2010 and 2041–2070. A data analysis revealed that yield of winter wheat may be sustained if water availability is guaranteed (Svoboda et al. 2015). Zhao et al. (2017) conducted an experiment to study climate change that influenced crop yields projecting yield reductions of 6%, 3.2%, 7.4%, and 3.1% in wheat, rice, maize, and soybean, respectively (Zhao et al. 2017). For handling climate change, new discoveries in genomics are pushing us for smart agriculture by enforcing climate-resilient crops (Scheben et al. 2016).

Drought stress influences wheat during all developmental stages, but grain formation and the reproductive stages are the most critical ones (Pradhan et al. 2012). Wheat yield was decreased from 1% to 30% during mild drought stress at post-anthesis, while this reduction increased up to 92% in case of prolonged mild drought stress at flowering and grain formation (Araus et al. 2002; De Oliveira et al. 2013). Drought stress greatly reduced the yield of important grain legumes. Mashbean (*Vigna mungo* L.) yield has been reduced by drought stress from 31% to 57% during flowering stage, while a 26% reduction was reported by drought stress during reproductive phase (Baroowa and Gogoi 2014). Maleki et al. (2013) reported that soybean yield was largely affected by drought stress and a 42% reduction was observed during the grain filling stage of soybean (Maleki et al. 2013). Schlenker and Roberts (2009) described that maize yield increased at an optimum temperature of 29 °C, but a further increase in temperature hampered the yield of maize. Every 1 °C rise in temperature was reported to negatively influence maize yield (Lobell et al. 2011). Similarly, it was observed that maize yield decreased by 8.3% through every 1 °C rise in temperature from its optimum temperature of growth (Lobell and Field 2007). Brown (2008) observed that wheat yield decreased by 10% if there is a rise in temperature for every 1 °C

(Brown 2008). Easterling et al. (2007) reported that if there was a 2 °C increase in temperature, it caused 7% reduction in yield, but if temperature is further increased to 4 °C, it reduced yield up to 34% in wheat. Similar is the reduction in yield of rice up to 2.6% for every 1 °C rise of temperature (Ray et al. 2015). In sorghum the yield gets reduced up to 7.8% because of a 1 °C increase in temperature (Kjellstrom et al. 2018). For sorghum, water shortage becomes a major issue in countries that have been reported as stock producers (Otto et al. 2017). Schlenker and Roberts (2009) described that threshold temperatures for soybean is 30 °C i.e. if there was a rise in temperature up to optimum level it will increase yield of soybean. Further rise in temperature will reduce the yields abruptly (Schlenker and Roberts 2009). Eastburn et al. (2010) reported that if there is rise in CO_2 and ozone concentration in the atmosphere, they will influence the type of disease. If there was continuous rise in temperature, there was an increase in the susceptibility to the disease (Eastburn et al. 2010).

An interest increased in studying on abiotic problems that is projected by Kitano on systems biology (Kitano 2002). Climate change has an impact on food security in a much complicated way. This checks yield of agriculture by changing agro-ecological environment. The influence of change of climate on food security has been traced in various ways (IPCC 2007).

There is a need for some advancements for establishing amplified rate of climate changes, for instance temperature extremes and drought in Mediterranean region or even massive rainfall spells in high temperate regions will increase the risk of floods (Schmidhuber and Tubiello 2007). But if we consider semiarid or arid areas, it will project a reduction in the growth of livestock. This will increase their death rate (IPCC 2000). There have been extensive rates of evapotranspiration. So it needs less moisture in soil in drier areas that have been depicted by climate models (IPCC 2007). In due course of time, the reason for changes in climatic conditions and the cultivated land become unsuitable for cultivation. If there is instability in temperature it will allow more insect pests on crops while it may boost at cold temperatures and emerging in the form of outbreaks in spring. Factors such as climate changes have negative impacts on the security and safety of global food production (Schmidhuber and Tubiello 2007).

6.3 Crop Adaptation in Mitigating Extreme Climatic Stresses

It has been well established that if there is an increase in the Earth's temperature, the climate undergoes several changes. This leads to a situation that is abiotically stressful. These changes result in the damage of crops posing threat to naturally present vegetation (Espeland and Kettenring 2018). In field conditions drought and heat become predominant stresses that have a major impact on plants (Pereira 2016). This has been observed that plants need optimum conditions for blooming and normal growth. The physiology of plants is highly affected due to fluctuations in temperature (Hatfield and Prueger 2015). Grain production is affected by heat stress, cold stress is responsible for sterility, and drought stress has negative impact on morphophysiology of plants (Barlow et al. 2015; Salehi-Lisar and Bakhshayeshan-Agdam 2016). This adversely affects the yield and development of plants. These give rise to significant responses that include biochemical, physiological, molecular, and modifications at morphological level (Zandalinas et al. 2018). It has been observed that global warming and climate changes have both positive and negative impacts on agricultural crops (Figure 6.3).

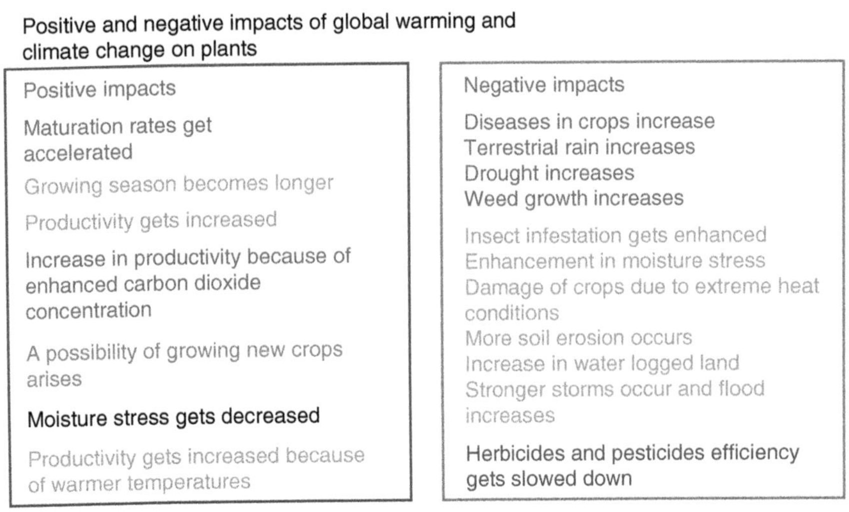

Figure 6.3 Impacts on crops due to changes in climate.

Studies on stress-resistant processes in plants have projected a difficult push for plant scientists for developing stress-resistant plants (Singh et al. 2018). Main cereal crops, for example wheat, rice, and maize, are crucial in meeting daily food demand. Among these wheat is the leading staple crop that is cultivated on a large scale (Tack et al. 2015). Wheat is cultivated on an area of 38.8%. This results in considerably high concentration of proteins that is 15% per gram in comparison with rice or maize that supplies 2–3% (FAO 2017). There was a reduction in the productivity of wheat with a 2 °C increase in temperature. Related research on environmental variability expected a 6% reduction in wheat yield (Abhinandan et al. 2018). Challinor and his colleagues described that due to the increase in temperature, the grain filling phase decrease is the major reason of crop productivity reduction in changing climatic conditions (Challinor et al. 2007). Therefore, sustaining crop yield and producing stress-tolerant crop plants are important tasks in current agricultural practices (Abhinandan et al. 2018).

6.4 Factors That Limit Crop Development

To achieve food security for increased population, it is essential to grow crops that are stress tolerant. So one needs to study crop responses in various conditions. For various climatic stresses, plants project variations in gene expression, metabolism, and physiology. Plants have the power to sense any type of variation in surroundings, even though only a few sensors have been studied (Zhu 2016). Because of various stresses, tissues and organs of plants are damaged. It has been observed that transcriptional responses vary for various stresses in specific tissue or cells of roots (Dinneny et al. 2008). The creation of stress-responsive proteins, elevated antioxidant ratio, and associated solutes are cellular signals. These may be formed because of drought, chemical effluence, and salinity. These parameters fall in the category of primary stresses. These are responsible for generating secondary stresses such as osmotic and oxidative stress (Carvalho and Amâncio 2018).

The initiation of reactive oxygen species (ROS) occurs because of CO_2 level in leaf due to drought. This has the potential to trigger multiple stresses in crops. Due to locked stomata, the CO_2 movement inside the leaf gets clogged. The occurrence of photosynthesis, development of plant, and respiration are disturbed through breakdown of membrane because of ROS production. It has been observed that building materials such as proteins, lipids, and carbohydrates get impaired by ROS under drought stress (Ahmad et al. 2018). It has been seen that development of plant is related with cellular metabolism. It has been observed that increased level of ROS is responsible for prompt cell apoptosis. So, a gentle balance in ROS development and decontamination can proceed in every oxygenated organism (Martinez et al. 2018).

Metabolic profiles have been worked out in many crop plants in conditions of water stress, for example in maize, rice, tomato, and soybean. In barley, various metabolomic studies were carried out to find the effect of water scarcity on free amino acids, abscisic acid (ABA), and oxidative phase. Barley cultivars were passed through a shortage of water in order to find the genetic variation at metabolomic level in grain formation phase (Wu et al. 2017). The inhibition in the production of protein is an initial metabolic signal that occurs in fighting abiotic factors (Vincent et al. 2007). The post-translational modifications also occurs because of abiotic stresses (Liu and Howell 2010). The coffee plant when subjected to constant drought stress showed an uptrend in the process of photosynthesis in comparison to plants with single event of drought stress if imposed on them. The plants showed advanced RuBisCo control and several enzymes related to metabolism (Menezes-Silva et al. 2017).

6.5 Influence of Climate Change on Plants' Morphobiochemical and Physiological Processes

Plants facing extreme fluctuations in environmental conditions find it limiting to adapt successfully. The changes in plant physiology would be useful in unique climatic conditions. However, if there is environmental variability, it will pose risk for plants (Becklin et al. 2006). It has been well observed that biological, morphological, and biochemical mechanisms of plants get affected much by abiotic stresses. However, due to expected conditions of weather in the future, the physiological reactions occurring in plants will propagate quickly having less variation in flowering and fruiting (DaMatta et al. 2010; Jan et al. 2016). The optimum temperature range for plant is 10–35 °C. But elevated temperature on specific point will permit generating excess energy. Larger increase in temperature will reduce photosynthesis rate and plant growth (Tkemaladze and Makhashvili 2016).

Drought stress affects turgor pressure, delaying the development of cell. Water shortage has a wider impact on the enzyme reactions occurring during photosynthesis. This reduces metabolic competency and destroys machinery of photosynthesis (Zargar et al. 2017). Due to fluctuations in environmental conditions CO_2 levels proliferate. This increases temperature level and reduces respiration. The rate of respiration gets elevated if temperature is raised from 15 to 40 °C. This influences morphological features of crops (Khan et al. 2017; Zargar et al. 2017). At the time of photosynthesis, *RuBisCo* enzyme is associated with carbon fixation. This is also involved in translation of CO_2 in a complex energy-rich compound (Nagarajan and Gill 2018). This is the RuBisCo activase that activates RuBisCo at optimum temperature. If there is an increase in temperature, *RuBisCo* will not work properly. The reason is because *RuBisCo* activase breaks down (Sage et al. 2008). ROS having H_2O_2 and OH are controlled through antioxidant defense mechanism. ROS is formed normally at minimum amount at optimum conditions, so environmental stress gets triggered (Nasim et al. 2017).

6.6 Responses of Plant Hormones in Abiotic Stresses

Under different abiotic stresses, the hormones are useful in controlling various signaling pathways and respond, for example salicylic acid (SA), ethylene, and ABA (Kurepin et al. 2017). The roles that these hormones play are presented in Figure 6.4. This is the ABA that controls climatic stresses in plants. This has role at various stages of plant development such as closing and opening of stomata, seed germination, dormancy, and drought stress. It has been observed that PYR/PYL/RCAR-PP2C-*SnRK2* may be depicted as a signaling cascade that is generated through ABA. This may regulate dormancy of seeds efficiently. If it is in drought, the growth of plant will be lower. This may increase the concentration of ABA in the cells. ABA accumulation during drought stress controls transpiration. This inhibits disclosure of stomata (Dong et al. 2018). The ABA has a role in triggering various physiological pathways in plants. This controls stomata to close down and is related with various stress-responsive genes (Kuromori et al. 2018). The ABA signaling pathways are now well studied. The signaling pathway framed consists of three units, PYR/PYL/RCAR proteins, protein kinase (SnRK2/OST1), and protein phosphatases (PP2C) (Takahashi et al. 2018).

Scientists have worked out potentials of ABA PYR/PYL/RCAR receptors (Ma et al. 2009; Park et al. 2009). The PP2C was observed first in abi1-1 and abi2-1 of *Arabidopsis* knockouts. These may be called as negative controllers of ABA (Leung and Giraudat 1998). Similarly protein kinase was obtained as *SnRK2*. This is an ABA activator (Yoshida et al. 2002). Salicylic acid has potential to control various physiological processes in climatic stresses. It has been seen that acetylsalicylic acid may regulate cell cycle (Carswell et al. 1989).

It was observed that SA may also control flowering initiation and bud development (Eberhard et al. 1989). The role of SA was well studied in the interaction of plant pathogens (Malamy et al. 1990). SA has a role in legume nodulation, stomata closing, fruit productivity, respiration, temperature resistance, and cell growth (Malamy et al. 1990) and controls production of stress-responsive hormones in plants (Verma et al. 2016; Arnao and Hernández-Ruiz 2017).

At the time of climatic fluctuations, the phytohormones project stress response through signal transduction mechanisms. In this the most peculiar is ethylene. It plays a role in the control of leaf growth, ripening, seed germination, and senescence in various biotic and abiotic climatic stresses. Various abiotic stresses, for example water logging, frost, nutrient deficiency, drought, salinity, high temperature, and heavy metal contact, are the causes that are responsible for modulating synthesis of ethylene (Dubois et al. 2018). The response factor for ethylene gets activated at the time of various environmental and physiological stresses (Klay et al. 2018).

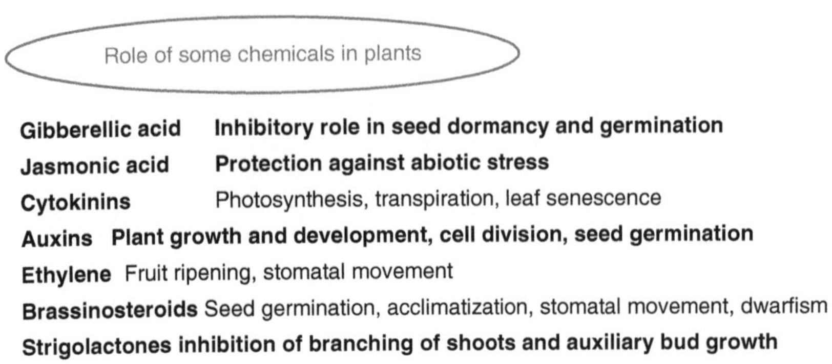

Figure 6.4 Role of some plant hormones in stress control.

6.7 Approaches to Combat Climate Changes

Variations in environment have a long-lasting influence on agriculture and food security globally. Food security and safety are threatened by severe weather conditions, and it is not a recent problem. But formerly, no consideration was adopted to tackle this problem. Therefore, to cope with these weather variations is now the top priority worldwide. For crops to adapt to changing environmental stresses the following approaches are required.

6.7.1 Cultural Methodologies

Farmers have adopted many strategies in order to handle climatic variations for plant adaptation. Various valuable approaches have been worked out by farmers, such as alteration in harvesting and planting time, collecting crops having short life cycles, irrigation methods, crop rotation, and even adopting alterations in cropping schemes. In conditions of stress, these are very useful in adaptability (Duku et al. 2018). The alterations in sowing time, using of cultivars that are drought resistant, and cultivating new crops are strategies for handling danger of climatic variability. This pushes for better adaptability in plants. This even ensures security and safety of food (Ali and Erenstein 2017). Next important thing may be adaptation of methods of crop management that have potential for enhancing crop development. Adopting a selection of sowing time, optimum irrigation, and planting density can be useful in handling weather stresses (Battisti et al. 2018). Fertilizers are useful in reducing effect of global warming. This even supports plants for better adaptability. This is helpful in maintaining fertility of the soil and increases productivity. So the value of fertilizers in the world for the purpose of nourishing is undeniable (Henderson et al. 2018).

6.7.2 Conventional Techniques

Plant breeding has a major role in handling various stresses. This is a dynamic technique for the betterment of crops. This can assure food safety and security in various harsh conditions of weather and be helpful in handling through the development of resistant cultivars. One popular method is to analyze genetic divergence. This is used for inbreeding, polymorphism, assortment, and recombination so as to obtain complete plant perfection. Genetic divergence study is considered a very good technique in developing new cultivars. This is based on genetic similarities and distances (Raza et al. 2018). It is a very useful source for genetic investigations on landraces. The wheat landrace that is kept in a data bank depicts broader genetic variance. This is a valuable basis for projecting stress resistance. This can give cultivars that may adjust in various environments (Lopes et al. 2015).

6.7.3 Strategies Concerned with Genetics and Genomics

6.7.3.1 Omics-Led Breeding and Marker-Assisted Selection (MAS)

It is evident from literature that omics approaches project useful resources in elucidating biological functions of any genetic information. This is just for development and upgrading of crops (Stinchcombe and Hoekstra 2008). Various molecular markers may be investigated in population genomics across environments in various individuals. This will be useful in finding out novel variation patterns. This will in turn help in finding genes that have roles in ecological traits (Keurentjes et al. 2008). The identification of phenotypes in various environmental variations/associations and the techniques of genetics and transcriptomic analysis are considered (Des Marais et al. 2013). Genomics is helpful in finding out molecular mechanisms involved that are concerned with abiotic stress resistance. These methods help in developing climate-smart crops. This helps in pushing yield and production in better form in various climatic changes (Roy et al. 2011). The data that is projected by various environmental extremes, quantitative trait loci (QTL), and DNA fingerprinting and mapping pushes for screening of elite germplasm under various abiotic stresses (Kole et al. 2015). QTL dealing with yield-related traits in different crops in stresses helps in developing new cultivars for better adaptability in various abiotic stresses (Collins et al. 2008). It can be well mentioned that molecular plant breeding projects an essential approach for pushing crop production and yield in various abiotic and biotic stresses. It has been well studied that marker-assisted selection (MAS) is helpful in speedy breeding progression. This is a very important part for achievement of better yielding crops and traits. The reason being so much discovery in the field of crop genomics has a lot of DNA markers. These are becoming a valuable tool for marker-assisted breeding (Wani et al. 2018).

Novel sequencing tools have solved the problems of researching genomic variants. They have helped in identifying huge amounts of DNA polymorphism. They are actually single nucleotide polymorphism (SNP) markers (D'Agostino and Tripodi 2017). The QTL mapping gave on average from 10–30 to <1 centimorgan (cM) (Kearsey and Farquhar 1998; Yu et al. 2011). Because of advances in linkage maps (Sehgal et al. 2016), high-throughput phenomics method also contributes in increasing accuracy of QTL mapping (Araus and Cairns 2014). The connection between genotypic and phenotypic data is important in enlightening various traits on genetics basis (Pikkuhookana and Sillanpää 2014). Through QTL mapping a wheat variety known as "Ripper" was developed that has potential to cope with conditions of drought without any effect on quality and grain yield (Haley et al. 2007). In another study QTL mapping was done in order to screen elite maize germplasm having high yield in drought stress (Badu-Apraku and Yallou 2009). An MAS conducted in *Triticum aestivum* L. (bread wheat) and *Triticum turgidum* L. (durum wheat) to know QTLs concern with drought stress (Merchuk-Ovnat et al. 2016). The cultivation of barley is done in a large area around the world although it is affected globally with drought stress. QTL studies projected that MASs are reliable genomic sections for regulation of malting feature (Kochevenko et al. 2018). Similarly, QTL study was conducted to find out epistatic mechanism and physiology. This was just for demonstrating the potential of targeted gene. This has potential impact on the productivity of grains (Dixit et al. 2017). Tahmasebi et al. (2016) investigated QTL mapping for recombinant inbred lines (RILs) population of wheat in various stresses of drought, heat, and flooding. This was also done by combining both heat and drought. The QTL mapping projected for 19.6% variation in the grain yield in various stress conditions. It is depicted that molecular markers might be used to find out a special allelic variation in wheat. This is just for enhancing the potential for screening of cultivars that are drought tolerant (Tahmasebi et al. 2016).

6.7.3.2 Genome-Wide Association Studies (GWAS) for Evaluating Stress Tolerance

Genome-wide association studies (GWAS) may be projected as a tool in finding out genetic variants in various crop cultivars. This is for recognizing allelic variants that may be linked with any specific trait (Manolio 2010). GWAS normally highlights linkage in SNPs and traits. This depends on genotyping tools, GWAS design, and statistical models for examination and interpretation of results (Bush and Moore 2012). In various crops GWAS have been made for exploiting genetic process that has role in genetic resistance under climate change (Mousavi-Derazmahalleh et al. 2018). In case of plants GWAS have wide applications that have been made in relation to abiotic and biotic stresses. GWAS have been used for finding heat, salt, and drought tolerance.

Verslues et al. (2014) investigated in *Arabidopsis thaliana* reverse genetic approaches for elucidating unique genes that accumulate prolines in drought stress. The linkage was studied in SNPs of both phenotypic and genotypic data. The concerned specific regions were studied that are concerned with regulation of proline accumulation. Similarly various proteins that control pro-accumulation, for example a MADS box protein, protein phosphatase 2A subunit A3, ribosomal protein RPL24A, Universal Stress Protein A domain proteins, thioredoxins, and even mitochondrial protease LON1, were studied well through techniques of reverse genetics. The findings project an idea of proline accumulation in drought stress (Verslues et al. 2014). It was observed that *Aegilops tauschii* contains various resistance genes that regulate abiotic stresses (Ashraf 2009). In another study Qin et al. (2016) observed 373 different varieties of *A. tauschii* to investigate 13 traits regulating drought stress. In GWAS, 7185 SNPs were investigated in order to find out phenotyping behavior (Qin et al. 2016).

6.7.3.3 Genome Selection (GS) Investigations for Crop Improvement

The genomic selection (GS) is an exciting tool for revolutionizing improvement of crops by using high-throughput phenotyping and marker densities for screening of elite germplasm. This is just for improvements in polygenic traits and development of economical breeding lines (Kumar et al. 2018). Now GS is projected for stimulating multi-environment designs useful in genomic estimation. Burgueño et al. (2012) first prepared statistical design through a linear mixed model to G × E model (Burgueño et al. 2012). Jarquín et al. (2014) showed a system of modeling connections that have markers and environment that integrate with each other (G × E) (Jarquín et al. 2014). Another model (GBLUP-type model) was given by Lopez-Cruz et al. (2015) where regression of phenotypes was considered for interaction of marker × environment (M × E). For genomic prediction, a modern multi-environment model was investigated by Cuevas et al. (2017) based on Bayesian model. The investigations were projected on four wheat and one maize cultivars. The CIMMYT data bank showed that G × E model may project high significance rate. This has better genomic predictions as compared to other models (Cuevas et al. 2017).

It has been observed that diversity array technology (DArT) is the most promising that has been used in GS. These are followed by SNP and genotyping by sequencing (GBS). The findings supported that GS might be used in cereal breeding

(Rutkoski et al. 2017). The GS designs were well studied in wheat to reveal the germplasms that have better ability to adapt to climate changes (Dong et al. 2018). Crain et al. (2018) investigated various GS methods for detecting phenotyping data from high throughput. At CIMMYT, stresses posed by drought and heat were examined in 1000 elite wheat cultivars. This was done by considering a high-throughput phenomics approach (Crain et al. 2018).

6.7.3.4 Genetic Engineering of Plants in Developing Stress Tolerance

It is now well recorded that biotechnology helps project genetic manipulation of the genome. This has concern for the betterment of human beings. So it will be a powerful technology for genetic modification. The data obtained from genetic investigations is quiet encouraging. This may be used significantly in various abiotic and biotic stresses, for example drought, cold, salinity, and heat. If identifying stress-responsive transcription factors (TFs), then it will be useful in developing stress-resistant crop cultivars. It has been observed that TFs may regulate phenotypes of genes in genetically engineered crops that have association in various stresses (Reynolds et al. 2015). It has been seen that various transgenic plants that were established through GE can handle biotic and abiotic stresses. The genetically produced plants may project resistance against climatic variations in comparison to normal plants (Shah et al. 2016; Nejat and Mantri 2017).

It has been seen that many TFs are plant specific (Riechmann and Meyerowitz 1998). The family of AP2/ERFBP TFs has roles in various plant growth pathways. This plays an important role in various biotic and abiotic stresses (Licausi et al. 2010). A study depicted that *DREB1* can also control cold stress in rice and *Arabidopsis* and *DREB2* as well as functions in salinity, drought, and stresses of high temperature (Liu et al. 1998; Lucas et al. 2011). It was well observed that *DREB1* TFs showed overexpression and developed transgenic *Arabidopsis* that was having better potential for withstanding drought, salinity, and freezing stresses (Gilmour et al. 1998; Jaglo-Ottosen et al. 1998). The *DREB1* genes were inserted in rice, rapeseed, tobacco, and tomato for developing resistance against cold stress (Hsieh et al. 2002; Kasuga et al. 2004). Various *DREB1* genes were obtained from rye, oil seed rape, wheat, and maize and was transformed for developing transgenic crops in fighting various abiotic stresses (Qin et al. 2004). Qin et al. (2007) studied gene *ZmDREB2A* isolated from maize and observed that this was overexpressed in *Arabidopsis*. This projected better resistance in fighting drought stress in transgenic *Arabidopsis*.

6.7.4 Strategies of Genome Editing

The plant genome manipulation can be carried out by genome editing (GE). This is a highly powerful strategy. This is manipulated through sequence-specific nucleases. Genome editing crop improvement projects its potential in solving problems of food insecurity. This may even develop a climate-smart agriculture system worldwide (Liu et al. 2013). In normal going on breeding approaches, the genes are searched to be associated by various important traits. This is through means of conventional breeding strategies and mutation, which are useful in developing/selecting high yielding germplasm (Abdelrahman et al. 2017). The methods of plant breeding have been highly influenced through GE tools. These are exploring new techniques of fast and accurate manipulations in crop genomes for protection in fighting various stresses and improving crop yield (Taranto et al. 2018). So it needs to bring novel changes in the gene pool in plant germplasm. These are required in various biotic and abiotic stresses for doing improvements in elite crop varieties. These have potential for producing high yielding crops (Kamburova et al. 2017). It is now well known that in GE technology the endonucleases used are highly site specific. These comprise zinc-finger nucleases (ZFNs), transcription activator-like effector nucleases (TALENs), and most popular CRISPR-Cas9 (Zhu et al. 2017). Now in comparison to TALENs and ZENs the CRISPR/Cas9 system is highly powerful GE strategy. The reason being it is rapid, economical, and accurate. This empowers various site-specific editing of the genome (Abdelrahman et al. 2018).

6.7.5 Involvement of CRISPR/Cas9

The CRISPR/Cas9 is a recent GE strategy. This depends on prokaryotic defense mechanism that is triggered through type II RNA organization. They protect prokaryotes from attacking viruses (Hussain et al. 2018; Lino et al. 2018). The editing in genome has been modernized through CRISPR-Cas9 assembly. This is by producing candidate gene mutants that may knock down single nucleotides in a genome (Larochelle 2018). In comparison to other tools of GE such as ZFNs/TALENs, the CRISPR-based techniques have been well exploited in plant genomes (Jaganathan et al. 2018). Actually this is the CRISPR/Cas9 tool that is responsible for changing to a comprehensible environment-friendly method establishing genome-edited non-transgenic plants in tackling environmental extremes. This being a support as guarantee for the food security (Khatodia et al. 2016; Gaur et al. 2018; Haque et al. 2018). These studies were done for disrupting the gene *TaERF3* and *TaDREB2* in producing abiotic stress resistance going through tool CRISPR GE (Kim et al. 2018). Likely, 21 KUP genes was confirmed in cassava that were hyperexpressed in conditions of abiotic stresses. The expression analysis of KUP genes depicted that they had drought resistance (Ou et al. 2018). For studying drought tolerance MAPKKK genes were studied through genome-wide analysis (Ye et al. 2017). The CRISPR/Cas9 was studied well in rice that produced triplet mutants. The genes *GW5*, *TGW6*, and *GW2* have a role in controlling the seed size. Through mutating this gene, seed size may be increased by 30% (Xu et al. 2016). CRISPR technology was used for mutation in *Brassica napus*. This was just by knocking down the *CLVTA3* gene that produced more seed. It was likely to be used in increasing seed size of wheat seed through knocking *TaGW2* gene. This has the power limiting the increase in the seed size (Wang et al. 2018). Recently methods have been developed for characterizing a target DNA present in a sample. The methods involve contacting the sample with a type V CRISPR/Cas effector protein and one or more guide RNAs. The contacting generates a cleavage product. This comprises a five overhang and ligating a double-stranded nucleic acid adapter to the cleavage product, to generate a ligation product. The ligation product includes the target DNA, which can be sequenced. The sample can be subjected to one or more amplification steps prior to the contacting step, with primers that provide for amplification of nucleic acids of, e.g. specific pathogens, categories of pathogens, two or more different pathogens, or two or more different categories of pathogens (Doudna et al. 2020).

6.8 Conclusions

The bell is ringing due to alterations in climate. Climate change is responsible for affecting production of agriculture and its products. This is the *industrialization* that releases poisonous gases that are responsible for global warming. That is in turn responsible for disturbing the world's environment. The alterations in climate result in devastating effects on plant's yield and growth. Abiotic factors project major stresses through which plants suffer greatly. In order to find out plant responses in various abiotic conditions, most pressing current need is to find out what is the genetic basis and underlying mechanisms. The physiological and molecular challenges present in plants need to be reinvestigated under abiotic stress conditions. Fluctuations in temperature and heavy rainfall play a crucial role in creating environmental stresses. Weather fluctuations collectively project both positive and negative outcomes even though it may be mentioned that negative effects are more provoking. It becomes difficult to overcome the imbalance in agriculture resulted by the changing conditions of climate. So how can we handle this situation and what may be the strategies to be used are questions that are still to be answered. So scientists should focus research on optimizing growth and development in plants under abiotic stresses. In order to achieve crop resistance in fighting biotic and abiotic stresses, there is a need to develop and try adoption of novel cultural methods. This needs implementation of different schemes of cropping. It also needs adoption of various conventional and non-conventional methods for saving agriculture in the future. Breeding techniques will definitely be most helpful in creating climate-resilient crops that will have better adaptability potential in stresses of drought and heat. The GS and GWAS having high-throughput phenotyping and genotyping strategies will play a highly useful role in confirming various genes for crop improvement in the changed climate. Genetic engineering techniques have been well developed and adopted in order to develop transgenic plants. This has enhanced resistance under various conditions of biotic and abiotic stresses. Further, it needs to focus on developing ecofriendly GE of crops using CRISPR/Cas9 or variant that will help better face the battle of changing climate.

Conflict of Interest Statement

We declare that there no conflicts of interest.

Acknowledgment

Authors are thankful to the Amity University Haryana authorities for the facilities and constant encouragement.

References

Abdelrahman, M., El-Sayed, M., Sato, S. et al. (2017). RNA-sequencing-based transcriptome and biochemical analyses of steroidal saponin pathway in a complete set of *Allium fistulosum, A. cepa* monosomic addition lines. *PLoS One* 12: e0181784.

Abdelrahman, M., Al-Sadi, A.M., Pour-Aboughadareh, A. et al. (2018). Genome editing using CRISPR/Cas9–targeted mutagenesis: an opportunity for yield improvements of crop plants grown under environmental stresses. *Plant Physiol. Biochem.* 131: 31–36.

Abhinandan K., Skori L., Stanic M., Hickerson N.M., Jamshed M., Samuel M.A (2018). Abiotic stress signaling in wheat—an inclusive overview of hormonal interactions during abiotic stress responses in wheat. *Front. Plant Sci.* 9:734. https://doi.org/10.3389/fpls.2018.00734.

Ahmad, Z., Anjum, S., Waraich, E.A. et al. (2018). Growth, physiology, and biochemical activities of plant responses with foliar potassium application under drought stress—a review. *J. Plant Nutr.* 41: 1734–1743.

Ali, A. and Erenstein, O. (2017). Assessing farmer use of climate change adaptation practices and impacts on food security and poverty in Pakistan. *Clim. Risk Manag.* 16: 183–194.

Altieri, M.A. and Nicholls, C.I. (2017). The adaptation and mitigation potential of traditional agriculture in a changing climate. *Clim. Chang.* 140: 33–45.

Araus, J.L. and Cairns, J.E. (2014). Field high-throughput phenotyping: the new crop breeding frontier. *Trends Plant Sci.* 19: 52–61.

Araus, J., Slafer, G., Reynolds, M., and Royo, C. (2002). Plant breeding and drought in C3 cereals: what should we breed for? *Ann. Bot.* 89: 925–940.

Arnao, M.B. and Hernández-Ruiz, J. (2017). Melatonin and its relationship to plant hormones. *Ann. Bot* 121: 195–207.

Arunanondchai, P., Fei, C., Fisher, A. et al. (2018). *The Routledge Handbook of Agricultural Economics*. Abingdon-on-Thames, UK: Routledge. How does climate change affect agriculture.

Ashraf, M. (2009). Biotechnological approach of improving plant salt tolerance using antioxidants as markers. *Biotechnol. Adv.* 27: 84–93.

Ashraf, M.A., Akbar, A., Askari, S.H. et al. (2018). Recent advances in abiotic stress tolerance of plants through chemical priming: an overview. In: *Advances in Seed Priming* (ed. A. Rakshit and H. Singh), 51–79. Berlin/Heidelberg, Germany: Springer.

Asseng, S., Ewert, F., Martre, P. et al. (2015). Rising temperatures reduce global wheat production. *Nat. Clim. Change* 5: 143.

Badu-Apraku, B. and Yallou, C. (2009). Registration of Striga-resistant and drought-tolerant tropical early maize populations TZE-W Pop DT STR C 4 and TZE-Y Pop DT STR C 4. *J. Plant Regist.* 3: 86–90.

Barlow, K., Christy, B., O'leary, G. et al. (2015). Simulating the impact of extreme heat and frost events on wheat crop production: a review. *Field Crops Res.* 171: 109–119.

Barnabás, B., Jäger, K., and Fehér, A. (2008). The effect of drought and heat stress on reproductive processes in cereals. *Plant Cell Environ.* 31: 11–38.

Baroowa, B. and Gogoi, N. (2014). Biochemical changes in black gram and green gram genotypes after imposition of drought stress. *J. Food Legum.* 27: 350–353.

Battisti, R., Sentelhas, P.C., Parker, P.S. et al. (2018). Assessment of crop-management strategies to improve soybean resilience to climate change in Southern Brazil. *Crop Pasture Sci.* 69: 154–162.

Becklin, K.M., Anderson, J.T., Gerhart, L.M. et al. (2006). Examining plant physiological responses to climate change through an evolutionary lens. *Plant Physiol.* 172: 635–649.

Benevenuto, R.F., Agapito-Tenfen, S.Z., Vilperte, V. et al. (2017). Molecular responses of genetically modified maize to abiotic stresses as determined through proteomic and metabolomic analyses. *PLoS One* 12: e0173069.

Bonan G.B., Doney S.C (2018). Climate, ecosystems, and planetary futures: the challenge to predict life in Earth system models. *Science* 359:eaam8328. https://doi.org/10.1126/science.aam8328.

Boyer, J.S. (1982). Plant productivity and environment. *Science* 218: 443–448.

Brown, L.R. (2008). *Mobilizing to Save Civilization (Substantially Revised)*. New York: WW Norton & Company.

Burgueño, J., de los Campos, G., Weigel, K., and Crossa, J. (2012). Genomic prediction of breeding values when modeling genotype× environment interaction using pedigree and dense molecular markers. *Crop Sci.* 52: 707–719.

Bush, W.S. and Moore, J.H. (2012). Genome-wide association studies. *PLoS Comput. Biol.* 8: e1002822.

Campbell, B.M., Vermeulen, S.J., Aggarwal, P.K. et al. (2016). Reducing risks to food security from climate change. *Glob. Food Sec.* 11: 34–43.

Carswell, G., Johnson, C., Shillito, R., and Harms, C. (1989). O-acetyl-salicylic acid promotes colony formation from protoplasts of an elite maize inbred. *Plant Cell Rep.* 8: 282–284.

Carvalho, L.C. and Amâncio, S. (2018). Cutting the Gordian Knot of abiotic stress in grapevine: from the test tube to climate change adaptation. *Physiol. Plant.* 2018: https://doi.org/10.1111/ppl.12857.

Challinor, A., Wheeler, T., Craufurd, P. et al. (2007). Adaptation of crops to climate change through genotypic responses to mean and extreme temperatures. *Agric. Ecosyst. Environ.* 119: 190–204.

Collins, N.C., Tardieu, F., and Tuberosa, R. (2008). Quantitative trait loci and crop performance under abiotic stress: where do we stand? *Plant Physiol.* 147: 469–486.

Crain, J., Mondal, S., Rutkoski, J. et al. (2018). Combining high-throughput phenotyping and genomic information to increase prediction and selection accuracy in wheat breeding. *Plant Genome.* 11: 1–14.

Cuevas, J., Crossa, J., Montesinos-López, O.A. et al. (2017). Bayesian genomic prediction with genotype× environment interaction kernel models. *G3: Genes. Genom. Genet.* 7: 41–53.

D'Agostino, N. and Tripodi, P. (2017). NGS-based genotyping, high-throughput phenotyping and genome-wide association studies laid the foundations for next-generation breeding in horticultural crops. *Diversity* 9: 38.

DaMatta, F.M., Grandis, A., Arenque, B.C., and Buckeridge, M.S. (2010). Impacts of climate changes on crop physiology and food quality. *Food Res. Int.* 43: 1814–1823.

De Oliveira, E.D., Bramley, H., Siddique, K.H. et al. (2013). Can elevated CO_2 combined with high temperature ameliorate the effect of terminal drought in wheat? *Funct. Plant Biol.* 40: 160–171.

Des Marais, D.L., Hernandez, K.M., and Juenger, T.E. (2013). Genotype-by-environment interaction and plasticity: exploring genomic responses of plants to the abiotic environment. *Annu. Rev. Ecol. Evol. Syst.* 44: 5–29.

Dhankher, O.P. and Foyer, C.H. (2018). Climate resilient crops for improving global food security and safety. *Plant Cell Environ.* 41: 877–884.

Dinneny, J.R., Long, T.A., Wang, J.Y. et al. (2008). Cell identity mediates the response of Arabidopsis roots to abiotic stress. *Science* 320: 942–945.

Dixit, S., Singh, A., Sandhu, N. et al. (2017). Combining drought and submergence tolerance in rice: marker-assisted breeding and QTL combination effects. *Mol. Breed.* 37: 143.

Dong, H., Bai, L., Chang, J., and Song, C.P. (2018). Chloroplast protein PLGG1 is involved in abscisic acid-regulated lateral root development and stomatal movement in Arabidopsis. *Biochem. Biophys. Res. Commun.* 495: 280–285.

Doudna, J.A., Harrington, L.B., Chen, J.S. et al. (2020). Methods for detecting and sequencing a target nucleic acid. Publication Number WO/2020/106630. https://patentscope.wipo.int/search/en/detail.jsf?docId (accessed 16 October 2020).

Dubois, M., Van den Broeck, L., and Inzé, D. (2018). The pivotal role of ethylene in plant growth. *Trends Plant Sci.* 23: 313–323.

Duku, C., Zwart, S.J., and Hein, L. (2018). Impacts of climate change on cropping patterns in a tropical, sub-humid watershed. *PLoS One* 13: e0192642.

Eastburn, D.M., Degennaro, M.M., Delucia, E.H. et al. (2010). Elevated atmospheric carbon dioxide and ozone alter soybean diseases at SoyFACE. *Glob. Chang. Biol.* 16: 320–330.

Easterling, W.E., Aggarwal, P.K., Batima, P. et al. (2007). Food, fibre and forest products. *Clim. Chang.* 273: 313.

Eberhard, S., Doubrava, N., Marfa, V. et al. (1989). Pectic cell wall fragments regulate tobacco thin-cell-layer explant morphogenesis. *Plant Cell.* 1: 747–755.

Espeland, E.K. and Kettenring, K.M. (2018). Strategic plant choices can alleviate climate change impacts: a review. *J. Environ. Manag.* 222: 316–324.

FAO (2017). FAOSTAT. Food Agriculture Organization. United Nations. http://www.fao.org/faostat/en/#home (accessed 2 August 2021).

FAO, UNICEF, WFP, WHO (2018). *The State of Food Security and Nutrition in the World 2017: Building Resilience for Peace and Food Security.* Rome, Italy: Food and Agriculture Organization of the United Nations (FAO).

FAOSTAT (n.d.). http://www.fao.org/faostat/en/#data (accessed 2 August 2021).

Garrity, D. and O'Toole, J. (1994). Screening rice for drought resistance at the reproductive phase. *Field Crops Res.* 39: 99–110.

Gaur, R.K., Verma, R.K., and Khurana, S.M.P. (2018). Genetic engineering of horticultural crops: present and future. In: *Genetic Engineering of Horticultural Crops* (ed. G.R. Rout and K.V. Peter), 23–46. Academic Press, Elsevier ISBN: 9780128104392.

Gilmour, S.J., Zarka, D.G., Stockinger, E.J. et al. (1998). Low temperature regulation of the Arabidopsis CBF family of AP2 transcriptional activators as an early step in cold-induced COR gene expression. *Plant J.* 16: 433–442.

Gong, M., Chen, S.-N., Song, Y.-Q., and Li, Z.-G. (1997). Effect of calcium and calmodulin on intrinsic heat tolerance in relation to antioxidant systems in maize seedlings. *Funct. Plant Biol.* 24: 371–379.

Griffin, J.J., Ranney, T.G., and Pharr, D.M. (2004). Heat and drought influence photosynthesis, water relations, and soluble carbohydrates of two ecotypes of redbud (*Cercis canadensis*). *J. Am. Soc. Hortic. Sci.* 129: 497–502.

Haley, S.D., Johnson, J.J., Peairs, F.B. et al. (2007). Registration of 'Ripper'wheat. *J. Plant Regist.* 1: 1–6.

Haque, E., Taniguchi, H., Hassan, M.M. et al. (2018). Application of CRISPR/Cas9 genome editing technology for the improvement of crops cultivated in tropical climates: recent progress, prospects, and challenges. *Front. Plant Sci.* 9: 617.

Hatfield, J.L. and Prueger, J.H. (2015). Temperature extremes: effect on plant growth and development. *Weather Clim. Extrem.* 10: 4–10.

Hellin, J., Bellon, M.R., and Hearne, S.J. (2014). Maize landraces and adaptation to climate change in Mexico. *J. Crop Improv.* 28: 484–501.

Henderson, B., Cacho, O., Thornton, P. et al. (2018). The economic potential of residue management and fertilizer use to address climate change impacts on mixed smallholder farmers in Burkina Faso. *Agric. Syst.* 167: 195–205.

Hsieh, T.H., Lee, J.T., Yang, P.T. et al. (2002). Heterology expression of the Arabidopsis C-repeat/dehydration response element binding Factor 1 gene confers elevated tolerance to chilling and oxidative stresses in transgenic tomato. *Plant Physiol.* 129: 1086–1094.

Hussain, B., Lucas, S.J., and Budak, H. (2018). CRISPR/Cas9 in plants: at play in the genome and at work for crop improvement. *Brief. Funct. Genom.* 17: 319–328.

Intergovernmental Panel on Climate Change (IPCC) (2000). *Special Report on Emissions Scenarios: A Special Report of Working Group III of the Intergovernmental Panel on Climate Change*. Cambridge, UK: Cambridge University Press.

Intergovernmental Panel on Climate Change (IPCC) (2007). *Climate Change 2007: Impacts, Adaptation, and Vulnerability: Working Group II Contribution to the Fourth Assessment Report of the Intergovernmental Panel on Climate Change*. Cambridge, UK: Cambridge University Press.

Ito, R., Vasconcelos, H.L., and Feeley, K.J. (2018). Global climate change increases risk of crop yield losses and food insecurity in the tropical Andes. *Glob. Chang. Biol.* 24: e592–e602.

Jaganathan, D., Ramasamy, K., Sellamuthu, G. et al. (2018). CRISPR for crop improvement: an update review. *Front. Plant Sci.* 9: 985.

Jaglo-Ottosen, K.R., Gilmour, S.J., Zarka, D.G. et al. (1998). Arabidopsis CBF1 overexpression induces COR genes and enhances freezing tolerance. *Science* 280: 104–106.

Jan, S.A., Shinwari, Z.K., and Rabbani, M.A. (2016). Morpho-biochemical evaluation of *Brassica rapa* sub-species for salt tolerance. *Genetika* 48: 323–338.

Jarquín, D., Crossa, J., Lacaze, X. et al. (2014). A reaction norm model for genomic selection using high-dimensional genomic and environmental data. *Appl. Genet.* 127: 595–607.

Kamburova, V.S., Nikitina, E.V., Shermatov, S.E. et al. (2017). Genome editing in plants: an overview of tools and applications. *Int. J. Agron.* 1–15. https://doi.org/10.1155/2017/7315351.

Kang, Y., Khan, S., and Ma, X. (2009). Climate change impacts on crop yield, crop water productivity and food security—a review. *Prog. Nat. Sci.* 19: 1665–1674.

Kanojia, A. and Dijkwel, P.P. (2018). Abiotic stress responses are governed by reactive oxygen species and age. *Annu. Plant Rev. 1*: 1–32.

Kasuga, M., Miura, S., Shinozaki, K., and Yamaguchi-Shinozaki, K. (2004). A combination of the Arabidopsis DREB1A gene and stress-inducible rd29A promoter improved drought-and low-temperature stress tolerance in tobacco by gene transfer. *Plant Cell Physiol.* 45: 346–350.

Kearsey, M. and Farquhar, A. (1998). QTL analysis in plants; where are we now? *Heredity* 80: 137.

Keurentjes, J.J., Koornneef, M., and Vreugdenhil, D. (2008). Quantitative genetics in the age of omics. *Curr. Opin. Plant Biol.* 11: 123–128.

Khan, A., Ijaz, M., Muhammad, J. et al. (2016). Climate change implications for wheat crop in Dera Ismail Khan district of Khyber Pakhtunkhwa. *Pak. J. Meteorol.* 13: 17–27.

Khan, A., Ali, M., Siddiqui, S.U. et al. (2017). Effect of various temperatures and duration on deterioration of rice seeds. *Science* 36: 79–83.

Khatodia, S., Bhatotia, K., Passricha, N. et al. (2016). The CRISPR/Cas genome-editing tool: application in improvement of crops. *Front. Plant Sci.* 7: 506.

Kim, D., Alptekin, B., and Budak, H. (2018). CRISPR/Cas9 genome editing in wheat. *Funct. Integr. Genom.* 18: 31–41.

Kitano, H. (2002). Systems biology: a brief overview. *Science* 295: 1662–1664.

Kjellstrom, E., Nikulin, G., Strandberg, G. et al. (2018). European climate change at global mean temperature increases of 1.5 and 2 degrees above pre-industrial conditions as simulated by the EURO-CORDEX regional climate models. *Earth. Syst. Dyn.* 9: 459–478.

Klay, I., Gouia, S., Liu, M. et al. (2018). Ethylene response factors (ERF) are differentially regulated by different abiotic stress types in tomato plants. *Plant Sci.* 274: 137–145.

Kochevenko, A., Jiang, Y., Seiler, C. et al. (2018). Identification of QTL hot spots for malting quality in two elite breeding lines with distinct tolerance to abiotic stress. *BMC Plant Biol.* 18: 106.

Kole, C., Muthamilarasan, M., Henry, R. et al. (2015). Application of genomics-assisted breeding for generation of climate resilient crops: progress and prospects. *Front. Plant Sci.* 6: 563.

Kumar, S., Muthusamy, S.K., Mishra, C.N. et al. (2018). Importance of genomic selection in crop improvement and future prospects. In: *Advanced Molecular Plant Breeding: Meeting the Challenge of Food Security* (ed. D.N. Bharadwaj), 275. Boca Raton, FL: CRC Press.

Kurepin L.V., Ivanov A.G., Zaman M., Pharis R.P., Hurry V., Hüner N.P (2017). *Photosynthesis: Structures, Mechanisms, and Applications.* Harvey JM. Hou, Mohammad Mahdi Najafpour, Gary F. Moore, Suleyman I. Allakhverdiev. Springer; Berlin/Heidelberg, Germany. Interaction of glycine betaine and plant hormones: protection of the photosynthetic apparatus during abiotic stress; pp. 185–202.

Kuromori, T., Seo, M., and Shinozaki, K. (2018). ABA transport and plant water stress responses. *Trends Plant Sci.* 23: 513–522.

Larochelle, S. (2018). Genomics: CRISPR–Cas goes RNA. *Nat. Methods.* 15: 312.

Lesk, C., Rowhani, P., and Ramankutty, N. (2016). Influence of extreme weather disasters on global crop production. *Nature* 529: 84.

Leung, J. and Giraudat, J. (1998). Abscisic acid signal transduction. *Annu. Rev. Plant Biol.* 49: 199–222.

Licausi, F., Giorgi, F.M., Zenoni, S. et al. (2010). Genomic and transcriptomic analysis of the AP2/ERF superfamily in *Vitis vinifera*. *BMC Genom.* 11: 719.

Lino, C.A., Harper, J.C., Carney, J.P., and Timlin, J.A. (2018). Delivering CRISPR: a review of the challenges and approaches. *Drug Deliv.* 25: 1234–1257.

Liu, J.-X. and Howell, S.H. (2010). Endoplasmic reticulum protein quality control and its relationship to environmental stress responses in plants. *Plant Cell.* 22: 2930–2942.

Liu, Q., Kasuga, M., Sakuma, Y. et al. (1998). Two transcription factors, DREB1 and DREB2, with an EREBP/AP2 DNA binding domain separate two cellular signal transduction pathways in drought-and low-temperature-responsive gene expression, respectively, in Arabidopsis. *Plant Cell.* 10: 1391–1406.

Liu, W., Yuan, J.S., and Stewart, C.N. (2013). Advanced genetic tools for plant biotechnology. *Nat. Rev. Genet.* 14: 781.

Lobell, D.B. and Field, C.B. (2007). Global scale climate–crop yield relationships and the impacts of recent warming. *Environ. Res. Lett.* 2: 014002.

Lobell, D.B., Bänziger, M., Magorokosho, C., and Vivek, B. (2011). Nonlinear heat effects on African maize as evidenced by historical yield trials. *Nat. Clim. Chang.* 1: 42.

Lopes, M.S., El-Basyoni, I., Baenziger, P.S. et al. (2015). Exploiting genetic diversity from landraces in wheat breeding for adaptation to climate change. *J. Exp. Bot.* 66: 3477–3486.

Lopez-Cruz, M., Crossa, J., Bonnett, D. et al. (2015). Increased prediction accuracy in wheat breeding trials using a marker × environment interaction genomic selection model. *G3: Genes Genom. Genet.* 5: 569–582.

Lucas, S., Durmaz, E., Akpınar, B.A., and Budak, H. (2011). The drought response displayed by a DRE-binding protein from *Triticum dicoccoides*. *Plant Physiol. Biochem.* 49: 346–351.

Ma, Y., Szostkiewicz, I., Korte, A. et al. (2009). Regulators of PP2C phosphatase activity function as abscisic acid sensors. *Science* 324: 1064–1068.

Malamy, J., Carr, J.P., Klessig, D.F., and Raskin, I. (1990). Salicylic acid: a likely endogenous signal in the resistance response of tobacco to viral infection. *Science* 250: 1002–1004.

Maleki, A., Naderi, A., Naseri, R. et al. (2013). Physiological performance of soybean cultivars under drought stress. *Bull. Environ. Pharmacol. Life Sci.* 2: 38–44.

Manolio, T.A. (2010). Genomewide association studies and assessment of the risk of disease. *N. Engl. J. Med.* 363: 166–176.

Martinez, V., Nieves-Cordones, M., Lopez-Delacalle, M. et al. (2018). Tolerance to stress combination in tomato plants: new insights in the protective role of melatonin. *Molecules* 23: 535.

Menezes-Silva, P.E., Sanglard, L.M., Ávila, R.T. et al. (2017). Photosynthetic and metabolic acclimation to repeated drought events play key roles in drought tolerance in coffee. *J. Exp. Bot.* 68: 4309–4322.

Merchuk-Ovnat, L., Barak, V., Fahima, T. et al. (2016). Ancestral QTL alleles from wild emmer wheat improve drought resistance and productivity in modern wheat cultivars. *Front. Plant Sci.* 7: 452.

Mousavi-Derazmahalleh, M., Bayer, P.E., Hane, J.K. et al. (2018). Adapting legume crops to climate change using genomic approaches. *Plant Cell Environ.* 42: 6–19.

Nagarajan, R. and Gill, K.S. (2018). Evolution of Rubisco activase gene in plants. *Plant Mol. Biol.* 96: 69–87.

Nasim, S., Shabbir, G., Ilyas, M. et al. (2017). Contemplation of wheat genotypes for enhanced antioxidant enzyme activity. *Pak. J. Bot.* 49: 647–653.

Nejat, N. and Mantri, N. (2017). Plant immune system: crosstalk between responses to biotic and abiotic stresses the missing link in understanding plant defence. *Curr. Issues Mol. Biol.* 23: 1–16.

Noya, I., González-García, S., Bacenetti, J. et al. (2018). Environmental impacts of the cultivation-phase associated with agricultural crops for feed production. *J. Clean. Prod.* 172: 3721–3733.

Olesen, J.E. and Bindi, M. (2002). Consequences of climate change for European agricultural productivity, land use and policy. *Eur. J. Agron.* 16: 239–262.

Olesen, J.E., Trnka, M., Kersebaum, K.C. et al. (2011). Impacts and adaptation of European crop production systems to climate change. *Eur. J. Agron.* 34: 96–112.

Otto, I.M., Reckien, D., Reyer, C.P. et al. (2017). Social vulnerability to climate change: a review of concepts and evidence. *Reg. Environ. Chang.* 17: 1651–1662.

Ou, W., Mao, X., Huang, C. et al. (2018). Genome-wide identification and expression analysis of the KUP family under abiotic stress in cassava (*Manihot esculenta* Crantz). *Front. Physiol.* 9: 17.

Pachauri, R.K., Allen, M.R., Barros, V.R. et al. (2014). *Climate Change 2014: Synthesis Report*. Geneva, Switzerland: IPCC. Contribution of Working Groups I, II and III to the Fifth Assessment Report of the Intergovernmental Panel on Climate Change.

Park, S.-Y., Fung, P., Nishimura, N. et al. (2009). Abscisic acid inhibits type 2C protein phosphatases via the PYR/PYL family of START proteins. *Science* 324: 1068–1071.

Pereira, A. (2016). Plant abiotic stress challenges from the changing environment. *Front. Plant Sci.* 7: 1123.

Pikkuhookana, P. and Sillanpää, M. (2014). Combined linkage disequilibrium and linkage mapping: Bayesian multilocus approach. *Heredity* 112: 351.

Pradhan, G.P., Prasad, P.V., Fritz, A.K. et al. (2012). Effects of drought and high temperature stress on synthetic hexaploid wheat. *Funct. Plant Biol.* 39: 190–198.

Qin, F., Sakuma, Y., Li, J. et al. (2004). Cloning and functional analysis of a novel DREB1/CBF transcription factor involved in cold-responsive gene expression in *Zea mays* L. *Plant Cell Physiol.* 45: 1042–1052.

Qin, F., Kakimoto, M., Sakuma, Y. et al. (2007). Regulation and functional analysis of ZmDREB2A in response to drought and heat stresses in *Zea mays* L. *Plant J.* 50: 54–69.

Qin, P., Lin, Y., Hu, Y. et al. (2016). Genome-wide association study of drought-related resistance traits in *Aegilops tauschii*. *Genet. Mol. Biol.* 39: 398–407.

Ray, D.K., Gerber, J.S., MacDonald, G.K., and West, P.C. (2015). Climate variation explains a third of global crop yield variability. *Nat. Commun.* 6: 5989.

Raza, A., Mehmood, S.S., Ashraf, F., and Khan, R.S.A. (2018). Genetic diversity analysis of *Brassica* species using PCR-based SSR markers. *Gesunde Pflanzen 71*: 1–7.

Reckling, M., Döring, T.F., Bergkvist, G. et al. (2018). Grain legume yield instability has increased over 60 years in long-term field experiments as measured by a scale-adjusted coefficient of variation. *Asp. Appl. Biol.* 138: 15–20.

Reynolds, M., Tattaris, M., Cossani, C.M. et al. (2015). Exploring genetic resources to increase adaptation of wheat to climate change. In: *Advances in Wheat Genetics: From Genome to Field* (ed. Y. Ogihara, S. Takumi, and H. Handa), 355–368. Berlin/Heidelberg, Germany: Springer.

Richardson, K.J., Lewis, K.H., Krishnamurthy, P.K. et al. (2018). Food security outcomes under a changing climate: impacts of mitigation and adaptation on vulnerability to food insecurity. *Clim. Chang.* 147: 327–341.

Riechmann, J.L. and Meyerowitz, E.M. (1998). The AP2/EREBP family of plant transcription factors. *Biol. Chem.* 379: 633–646.

Rogelj, J., Den Elzen, M., Höhne, N. et al. (2016). Paris Agreement climate proposals need a boost to keep warming well below 2 °C. *Nature* 534: 631.

Rosenzweig, C., Elliott, J., Deryng, D. et al. (2014). Assessing agricultural risks of climate change in the 21st century in a global gridded crop model intercomparison. *Proc. Natl. Acad. Sci. U.S.A.* 111: 3268–3273.

Roy, S.J., Tucker, E.J., and Tester, M. (2011). Genetic analysis of abiotic stress tolerance in crops. *Curr. Opin. Plant Biol.* 14: 232–239.

Ruf, F., Schroth, G., and Doffangui, K. (2015). Climate change, cocoa migrations and deforestation in West Africa: what does the past tell us about the future? *Sustain. Sci.* 10: 101–111.

Rutkoski, J.E., Crain, J., Poland, J., and Sorrells, M.E. (2017). Genomic selection for small grains improvement. In: *Genomic Selection for Crop Improvement* (ed. R.K. Varshney, M. Roorkiwal, and M.E. Sorrells), 99–130. Berlin/Heidelberg, Germany: Springer.

Sage, R.F., Way, D.A., and Kubien, D.S. (2008). Rubisco, Rubisco activase, and global climate change. *J. Exp. Bot.* 59: 1581–1595.

Saini, H. and Aspinall, D. (1981). Effect of water deficit on sporogenesis in wheat (*Triticum aestivum* L.). *Ann. Bot.* 48: 623–633.

Saini, H. and Aspinall, D. (1982). Abnormal sporogenesis in wheat (*Triticum aestivum* L.) induced by short periods of high temperature. *Ann. Bot.* 49: 835–846.

Salehi-Lisar, S.Y. and Bakhshayeshan-Agdam, H. (2016). Drought stress in plants: causes, consequences, and tolerance. In: *Drought Stress Tolerance in Plants*, vol. 1 (ed. M.A. Hossain and S.H. Wani), 1–16. Berlin/Heidelberg, Germany: Springer.

Scheben, A., Yuan, Y., and Edwards, D. (2016). Advances in genomics for adapting crops to climate change. *Curr. Plant Biol.* 6: 2–10.

Schlenker, W. and Roberts, M.J. (2009). Nonlinear temperature effects indicate severe damages to US crop yields under climate change. *Proc. Natl. Acad. Sci. U.S.A.* 106: 15594–15598.

Schmidhuber, J. and Tubiello, F.N. (2007). Global food security under climate change. *Proc. Natl. Acad. Sci. U.S.A.* 104: 19703–19708.

Sehgal, D., Singh, R., and Rajpal, V.R. (2016). Quantitative trait loci mapping in plants: concepts and approaches. In: *Molecular Breeding for Sustainable Crop Improvement* (ed. V.R. Rajpal, S.R. Rao, and S.N. Raina), 31–59. Berlin/Heidelberg, Germany: Springer.

Shah, S.H., Ali, S., Hussain, Z. et al. (2016). Genetic improvement of tomato (*Solanum lycopersicum*) with *AtDREB1A* dene for cold stress tolerance using optimized agrobacterium-mediated transformation system. *Int. J. Agric. Biol.* 18: 471–782.

Sheoran, I.S. and Saini, H.S. (1996). Drought-induced male sterility in rice: changes in carbohydrate levels and enzyme activities associated with the inhibition of starch accumulation in pollen. *Sex. Plant Reprod.* 9: 161–169.

Singh, P., Basu, S., and Kumar, G. (2018). Polyamines metabolism: a way ahead for abiotic stress tolerance in crop plants. In: *Biochemical, Physiological and Molecular Avenues for Combating Abiotic Stress Tolerance in Plants* (ed. S.H. Wani), 39–55. Amsterdam, the Netherlands: Elsevier.

Stinchcombe, J.R. and Hoekstra, H.E. (2008). Combining population genomics and quantitative genetics: finding the genes underlying ecologically important traits. *Heredity* 100: 158.

Suzuki, N., Rivero, R.M., Shulaev, V. et al. (2014). Abiotic and biotic stress combinations. *New Phytol.* 203: 32–43.

Svoboda, N., Strer, M., and Hufnagel, J. (2015). Rainfed winter wheat cultivation in the North German Plain will be water limited under climate change until 2070. *Environ. Sci. Eur.* 27: 29.

Tack, J., Barkley, A., and Nalley, L.L. (2015). Effect of warming temperatures on US wheat yields. *Proc. Natl. Acad. Sci. U.S.A.* 112: 6931. https://doi.org/10.1073/pnas.1415181112.

Tahmasebi, S., Heidari, B., Pakniyat, H., and McIntyre, C.L. (2016). Mapping QTLs associated with agronomic and physiological traits under terminal drought and heat stress conditions in wheat (*Triticum aestivum* L.). *Genome* 60: 26–45.

Takahashi, F., Kuromori, T., Sato, H., and Shinozaki, K. (2018). Regulatory gene networks in drought stress responses and resistance in plants. In: *Survival Strategies in Extreme Cold and Desiccation* (ed. M. Iwaya-Inoue, M. Sakurai, and M. Uemura), 189–214. Berlin/Heidelberg, Germany: Springer.

Taranto, F., Nicolia, A., Pavan, S. et al. (2018). Biotechnological and digital revolution for climate-smart plant breeding. *Agronomy* 8: 277.

Tebaldi, C. and Lobell, D. (2018). Estimated impacts of emission reductions on wheat and maize crops. *Clim. Chang.* 146: 533–545.

Thornton, P.K., Ericksen, P.J., Herrero, M., and Challinor, A.J. (2014). Climate variability and vulnerability to climate change: a review. *Glob. Chang. Biol.* 20: 3313–3328.

Tkemaladze, G.S. and Makhashvili, K. (2016). Climate changes and photosynthesis. *Ann. Agric. Sci.* 14: 119–126.

Van Velthuizen, H. (2007). *Mapping Biophysical Factors That Influence Agricultural Production and Rural Vulnerability*. Rome, Italy: Food & Agriculture Organization.

Vaughan, M.M., Block, A., Christensen, S.A. et al. (2018). The effects of climate change associated abiotic stresses on maize phytochemical defenses. *Phytochem. Rev.* 17: 37–49.

Verma, V., Ravindran, P., and Kumar, P.P. (2016). Plant hormone-mediated regulation of stress responses. *BMC Plant Biol.* 16: 86.

Verslues, P.E., Lasky, J.R., Juenger, T.E. et al. (2014). Genome-wide association mapping combined with reverse genetics identifies new effectors of low water potential-induced proline accumulation in Arabidopsis. *Plant Physiol.* 164: 144–159.

Vincent, D., Ergül, A., Bohlman, M.C. et al. (2007). Proteomic analysis reveals differences between *Vitis vinifera* L. cv. Chardonnay and cv. Cabernet Sauvignon and their responses to water deficit and salinity. *J. Exp. Bot.* 58: 1873–1892.

Wang, Z. and Huang, B. (2004). Physiological recovery of Kentucky bluegrass from simultaneous drought and heat stress. *Crop Sci.* 44: 1729–1736.

Wang, W., Pan, Q., He, F. et al. (2018). Transgenerational CRISPR-Cas9 activity facilitates multiplex gene editing in allopolyploid wheat. *CRISPR J.* 1: 65–74.

Wani, S.H., Choudhary, M., Kumar, P. et al. (2018). Marker-assisted breeding for abiotic stress tolerance in crop plants. In: *Biotechnologies of Crop Improvement*, vol. 3 (ed. S.S. Gosal and S.H. Wani), 1–23. Berlin/Heidelberg, Germany: Springer.

Wheeler, T. and Von Braun, J. (2013). Climate change impacts on global food security. *Science* 341: 508–513.

Winkel, T., Renno, J.-F., and Payne, W. (1997). Effect of the timing of water deficit on growth, phenology and yield of pearl millet (*Pennisetum glaucum* (L.) R. Br.) grown in Sahelian conditions. *J. Exp. Bot.* 48: 1001–1009.

Wu, X., Cai, K., Zhang, G., and Zeng, F. (2017). Metabolite profiling of barley grains subjected to water stress: to explain the genotypic difference in drought-induced impacts on malting quality. *Front. Plant Sci.* 8: 1547.

Xu, Z.Z. and Zhou, G.S. (2006). Combined effects of water stress and high temperature on photosynthesis, nitrogen metabolism and lipid peroxidation of a perennial grass *Leymus chinensis*. *Planta* 224: 1080–1090.

Xu, R., Yang, Y., Qin, R. et al. (2016). Rapid improvement of grain weight via highly efficient CRISPR/Cas9-mediated multiplex genome editing in rice. *J. Genet. Genom.* 43: 529–532.

Ye, J., Yang, H., Shi, H. et al. (2017). The MAPKKK gene family in cassava: genome-wide identification and expression analysis against drought stress. *Sci. Rep.* 7: 14939.

Yoshida, R., Hobo, T., Ichimura, K. et al. (2002). ABA-activated SnRK2 protein kinase is required for dehydration stress signaling in Arabidopsis. *Plant Cell Physiol.* 43: 1473–1483.

Yu, H., Xie, W., Wang, J. et al. (2011). Gains in QTL detection using an ultra-high density SNP map based on population sequencing relative to traditional RFLP/SSR markers. *PLoS One* 6: e17595.

Zandalinas, S.I., Mittler, R., Balfagón, D. et al. (2018). Plant adaptations to the combination of drought and high temperatures. *Physiol. Plant.* 162: 2–12.

Zargar, S.M., Gupta, N., Nazir, M. et al. (2017). Impact of drought on photosynthesis: molecular perspective. *Plant Gene.* 11: 154–159.

Zhao, C., Liu, B., Piao, S. et al. (2017). Temperature increase reduces global yields of major crops in four independent estimates. *Proc. Natl. Acad. Sci. U.S.A.* 114: 9326–9331.

Zhu, J.K. (2016). Abiotic stress signaling and responses in plants. *Cell* 167: 313–324.

Zhu, C., Bortesi, L., Baysal, C. et al. (2017). Characteristics of genome editing mutations in cereal crops. *Trends Plant Sci.* 22: 38–52.

7

Agrifood and Climate Change

Impact, Mitigation, and Adaptation Strategies

Sudarshna Kumari and Gurdeep Bains

Department of Plant Physiology, Govind Ballabh Pant University of Agriculture & Technology, Pantnagar, Uttarakhand, India

7.1 Introduction

One of the most important issues faced by the world today is climate change, which has significantly changed the environment of our planet. It is described as major shifts in the mean values of meteorological variables over a lengthy period of time, such as increasing temperature and unequal precipitation (Malhi et al. 2021). These changes in weather and climate negatively affect agricultural production and productivity. It leads to food and nutritional insecurity in the developing world, where a large section of the population is currently experiencing hunger, insecurity, and unwholesome food (Ahmed et al. 2019).

Agriculture holds the major land use across the globe. Approximately ≥1.2–1.5 billion hectares of land are under crops, with another 3.5 billion hectares being grazed. Another 4 billion hectares of forest are used by humans to differing degrees, whereas away from land, global fisheries are used very intensively, often beyond capacity (Field et al. 2012). To meet the food demand of an increasing population, agriculture also plays a major role in socioeconomic growth by providing a wide range of ecosystem services (Howden et al. 2007). It also provides food, which is the cornerstone of human existence. As a furnisher of industrial raw materials, it is an important contributor of the global economy. Most of the developing and underdeveloped countries' economy totally depends on agriculture. Agriculture plays a vital role in the Indian economy. Over 70% of rural households depend on agriculture. Agriculture is an important sector of the Indian economy, as it contributes about 17% to the total GDP and provides employment to around 58% of the population. Indian agriculture has registered impressive growth over the last few decades. The food grains production has increased from 51 million tons (MT) in 1950–1951 to 250 MT during 2011–2012, the highest ever since independence. The share of agriculture in GDP increased to 19.9% in 2020–2021 from 17.8% in 2019–2020. The last time the contribution of the agriculture sector to GDP was 20% in 2003–2004.

Past few decades indicate that agriculture is most vulnerable to climate change. The Intergovernmental Panel on Climatic Change (IPCC), which was established in the 1990s, has discovered evidence of accelerated global warming, climate variability, and change as a result of greenhouse gas (GHG) emissions. The concentration of GHGs such as methane (CH_4), carbon dioxide (CO_2), and nitrous oxide (N_2O) has increased by 150%, 40%, and 20%, respectively (IPCC 2014). Carbon dioxide emissions, which account for the maximum proportion of GHGs, rose to 36.14 billion metric tons in 2014 from 22.15 billion metric tons in 1990 (Malhi et al. 2021). Since 1975, the average global temperature has risen at a consistent pace of 0.15–0.20 °C every decade, and by 2021, it is predicted to have grown by 1.4–5.8 °C (Arora et al. 2005). The CO_2 concentration in the atmosphere had increased to 411.43 ppm in 2019 from 315.98 ppm in 1959. Many parts of the world, including India, are experiencing drought as a result of these changes. Temperature variations affected crop phenology, including leaf development, anthesis, harvest, fruit output, and pollinator asynchrony. High respiration rates, decreased pollen germination, a shorter grain filling period, lower biomass production, and low yields were all consequences of the temperature range. The sterility of cereal and fruit florets was caused by a 4 °C temperature increase brought on by high temperatures above 35 °C combined with high humidity and little wind speed (Singh et al. 2016).

Precipitation is usually seen as a benefit to crops and fields, but there is an "ideal" amount of rainfall in any given growing season for most crops. If the average rainfall is much lower or higher than the ideal, it can lead to significant problems, from drowned crops to lower yields (Ilbay-Yupa et al. 2021). Rainfall is also a good indicator of predicting common crop

Global Climate Change and Plant Stress Management, First Edition. Edited by Mohammad Wahid Ansari, Anil Kumar Singh, and Narendra Tuteja.

disease, as it can affect the spread of disease. Rain can spread pathogens, pests, and other diseases to plants, leading to massive diseased crops. This could affect its yield or cause the entire field to become unusable. Too much watering could drown the crops, especially if there has been rainfall. However, a combination of not enough watering and no rainfall could also lead to dying crops. Underwatering "starves" the plant of water, which can lead to crop death or low yield. Conversely, overwatering can lead to browned tips of crop leaves and also possible root rot and other issues. If crops are too wet, they could also start to mold or catch a fungus (Singh et al. 2016). The soil can also start to collect bacteria, mold, and fungus, which can then be absorbed by the plant. While this is not as common in crops as it is in indoor plants, poor drainage and irrigation systems can lead to these types of growths taking control over the crops. Along with mold or a fungus, disease can also spread among all crops.

To overpower the adversities of climate, climate impact assessment provides the scientific foundation for the development of mitigation and adaptation strategies to offset the negative impacts. These adaptation and mitigation strategies include cultural practices, water-smart practices (laser land leveling, rainwater harvesting, micro-irrigation, crop diversification, raised-bed planting, direct-seeded rice), nutrient-smart practices (precision nutrient application, leaf color charts, crop residue management) (Lipper et al. 2014; Zougmore et al. 2014), weather-smart activities (stress-tolerant varieties, ICT-based agrometeorological services), carbonsmart activities (zero tillage, legumes, crop residue management), and knowledge-smart activities (agricultural extensions to enhance capacity-building) (Malhi et al. 2021). Strategies for mitigation and adaptation are anticipated to boost farmers' incomes without endangering the sustainability of agricultural production. Keeping in views, this book chapter is written to assess the causes, impacts of climate change, and possible mitigation and adaptation strategies toward agriculture in industrial era.

7.2 Causes of Climate Change

The natural phenomenon and anthropogenic activities on the Earth initiate the concentration of greenhouse gases that led to global warming resulting in increased temperature and uneven precipitation. Fossil fuel combustion, greenhouse gas emissions, deforestation, agriculture, livestock, and nitrogen fertilizers are major causes of climate change.

7.2.1 Greenhouse Gases

Anthropogenic activities lead to the emission of four major GHGs such as carbon dioxide (CO_2), nitrous oxide (N_2O), and methane (CH_4) and the halocarbons lead to greenhouse effect. The greenhouse effect is a natural process that warms the Earth's surface and also led to ozone depletion in the atmosphere (Montzka et al. 2011). In the industrial era, the concentration of these gases increased. The increased CO_2 concentrations in the atmosphere directly affected microbial activities, and melting of glaciers resulted in increased water content in the soil. Therefore, the increasing atmospheric CO_2 (463–780 ppm) can cause methane and nitrous oxide emissions from wetlands and upland soil, respectively, which cancels the 16.6% climate change mitigation benefit projected by doing so (Mahato 2014).

7.2.2 Fossil Fuel Combustion

The burning of this fossil material returns this carbon back into atmosphere as carbon dioxide, at a rate that is hundreds to thousands of times faster than it took to bury and much faster than can be removed by the carbon cycle. Thus, the carbon dioxide released from the burning of fossil fuels accumulates in the atmosphere, some of which then dissolves in the ocean causing ocean acidification. The gases such as SO_2, NO_2, and CO_2 react with water vapor, oxygen, and other chemicals to form acid rain. Acid rain can contaminate freshwater sources, resulting in harmful algal blooms that reduce water oxygen levels and harm fish populations and other wildlife. Additionally, acid rain increases chemical weathering of rocks, including human-made structures (Debnath et al. 2021). The effects of acidic rains on the growth and development of plants are little understood, but nevertheless visible injuries and yield losses have been observed following treatments of simulated acid rain to crops in both the laboratory and field situations (Lal 2016). Simulated acid rain (SAR) exposure caused adverse effect on morphological, biochemical, and leaf epidermal parameters of wheat cv. HD-2329.

Plant growth (length, fresh and dry weight of shoot and root, tillers number, leaf area) and yield parameters (ear length, numbers of grains/ear, and weight of 100 grains) were suppressed significantly at SAR levels of 3.0, 4.0, and 5.0 pH. Highest suppressions were reduced at pH 3.0 level compared to control. Photosynthetic pigments such as Chl a, Chl b, total Chl, and carotenoids; seed carbohydrate; seed total protein content; and the number of stomata, stomatal aperture, and length of trichomes also decreased significantly with the increase of acidity (Kausar et al. 2010). The effects of SAR of different pH [distilled water-7.0 (control), 5.7, 4.5, and 3.0] in sunflower (*Helianthus annuus*) cv. "Morden" and its effects on plant root, shoot, and leaf at peak growth and maturity stages was evaluated. Their results revealed that biomass and lengths of the studied plant parts decreased with decreasing pH of acid rain solution. Comparison of biomass and length at peak growth and maturity stages recorded maximum difference in control and the difference narrowed with increasing acidity. Effects of SAR on sunflower increased more dramatically with the increase of SAR acidity and were correlated with exposure times and doses of SAR (Lal 2016).

7.2.3 Deforestation

Deforestation is also one of major contributor of climate change. Land use changes especially in the form of deforestation are the second largest anthropogenic source of atmospheric CO_2 emissions, after fossil fuel combustion. GHGs are emitted during combustion of forest biomass and decomposition of remaining plant material and soil carbon. Global models and national GHG inventories give similar results for deforestation emissions. As of 2019, deforestation is responsible for about 11% of global GHG emissions. Carbon emissions from tropical deforestation are accelerating (Feng et al. 2022). Growing forests are a carbon sink with additional potential to mitigate the effects of climate change. Some of the effects of climate change, such as more wildfires, may increase deforestation. Deforestation comes in many forms: agricultural clearcutting, wildfire, livestock ranching, and logging for timber, among others. The vast majority of agricultural activity resulting in deforestation is subsidized by government tax revenue. Forests cover 31% of the land area on the Earth and annually 75 700 km^2 (18.7 million acres) of the forest is lost. According to the World Resources Institute, there was a 12% increase in the loss of primary tropical forests from 2019 to 2020. Mass deforestation continues to threaten tropical forests, their biodiversity, and the ecosystem services they provide. The main area of concern of deforestation is tropical rain forests since they center on the majority of the planet's biodiversity.

7.2.4 Agricultural Expansion

Agricultural expansion is one of the worst offenders when it comes to deforestation in recent times. Since 1960, approximately 15% of the Amazon has been wiped out with the intention of replacing the land with agricultural practices consequently reducing the carbon sink (Grau et al. 2005). In 2019, the IPCC reported that 13–21% of anthropogenic GHGs came specifically from the agriculture, forestry, and other land uses sector. Emissions from agriculture such as N_2O, CH_4, and CO_2 make up to half of the GHGs produced by the overall food industry or 80% of agricultural emissions (Friel et al. 2009). The agricultural food system is responsible for a significant amount of GHG emissions. Agriculture consumes a lot of fossil fuels, uses a lot of land, and produces things like rice and cattle that directly contribute to GHG emissions (Steinfeld et al. 2006).

The consumption of high-value items like milk and meat is anticipated to change, leading to an even faster increase in emissions. Ruminant cattle for beef and dairy rank high in GHG emissions; monogastric, or pigs and poultry-related foods, are low. The consumption of the monogastric types may yield fewer emissions. Monogastric animals have higher feed-conversion efficiency, and they do not produce as much methane (Baumgard et al. 2012). The emission can be reduced either with technological mitigation or by reduced meat consumption or both. Enteric fermentation, fossil fuels, N_2O emissions, organic farming, liming, and fertilizer manufacturing are the primary sources of GHG emissions by the livestock industries (Lesschen et al. 2011). Emissions of GHGs are also caused by the usage of chemical fertilizers that contain nitrogen. Nitrogen fertilizer use can be reduced by 38% with better crop management. Additionally, improved crop management reduces GHG emissions by 20% while using 11% less energy input and producing 33% more yields (Malhi et al. 2021).

7.3 Impact of Climate Change on Agriculture

Agriculture is the most vulnerable sector affected by climate change, as variability in the pattern of rainfall and temperature significantly affects the yield of various agricultural crops. It will be one of the major deciding factors influencing the future food security of humankind on the Earth (Mendelsohn 2009). However, agriculture is not only sensitive to climate but also one of the major drivers of climate changes. The quality of fruits, vegetables, tea, coffee, aromatic and medicinal herbs, and basmati rice is significantly impacted by even little variations in temperature and rainfall. Temperature and humidity have a significant impact on insect and pathogen populations, and variations in these variables may alter the dynamics of these populations. Lower milk production from dairy cow and a drop in fish reproduction, migration, and harvests are two further effects on agriculture and allied industries. Global statistics predict that crop production would decline by 10–40% by 2100 (Arora 2019). The impact of climate change on agriculture is listed as follows.

7.3.1 Crop Productivity

Climate change may alter rainfall, soil water, runoff, yield variability, crop maturation, and the amount of land that is appropriate for growing certain crops (Kang et al. 2009; Ahmed et al. 2019). Climate change might limit crop production in many areas. Even the effect of climate change on the crop yield varies according to the irrigation, crop stage, crop location, and magnitude of weather parameters. Extreme weather conditions at the time of crop lead to reduction of crop yield due to various reasons such as poor pollination, pollen sterility, freezing injury, improper development, and physiological disease and disorder. The yield is observed to decrease when temperature rises; however, the influence of temperature increases is predicted to diminish if precipitation increases (Adams et al. 1998). In Iran, it has been reported that crop productivity depends on adaptation abilities and crop type, climate scenario, and CO_2 fertilization effect.

The significant decrease in precipitation or increase in temperature has been found to decrease the net revenue of farmers in Cameroon. According to statistical evidence, the temperature in Veracruz, Mexico, has an impact on coffee yield. Additionally, it was discovered that coffee production may not continue to be profitable for farmers in upcoming years because there are indications of a 34% decrease in present production (Gay et al. 2006). Climate change in general has more impact on tropical regions and tropical crops remain closer to their high-temperature optima and thereby experience high-temperature stress during elevated levels of temperature. If both the temperate and tropical regions warm by 2 °C, it is predicted that the total yield of wheat, rice, and maize will decline (Challinor et al. 2014).

Climate change can have a profoundly different effect on different crops and geographical areas. For instance, a 1 °C increase in the growing period temperature may result in a 3–10% decline in wheat productivity, a 5–35% decrease in winter wheat production, and a 2.4–45.6% loss in maize yield, depending on future warmer and drier conditions (Ahmed et al. 2019). Even if precipitation is unchanged, crop production may decrease by 15% on average due to the reduction in crop growth period and increased water stress as the result of higher temperature and evapotranspiration (You et al. 2009). Expected precipitation decreases can have a significant influence on crop output in arid and semiarid regions of the world where water is already scarce. For instance, due to the possibility of reduced growth periods brought on by higher temperatures and worsened by a loss in precipitation, winter wheat production in northwest Turkey may decrease by more than 20% as a result of future climate change (Özdoğan 2011) Table 7.1.

Higher reduction in wheat yield of 50% was found in Pakistan as shown in Table 7.2. In some other areas, climatic change might have positive influences on agricultural crop yield, i.e. in dry areas rainfall enhances under wet climatic warming and can lead to improved crop productions such as in Mexico the wheat yield would be increased by 25% in future (Table 7.2). Maize, rice, winter wheat, and potato crop yield can be enhanced with increasing air temperature and rainfall in the Plain of North China. In Ghana maize yield would be increased by 12% in the future (Table 7.2). The impact of climate change on sugarcane and cotton is shown in Table 7.3. Higher reduction in cotton yield of 17% and sugarcane yield of 40% was found in the United States (Table 7.3). However, some positive impacts of climate change on sugarcane yield were found in few countries such as Brazil and Australia (Table 7.3). The impacts of climate change on coarse grain, oilseed, and other minor crops such as pearl millet, sorghum, and sesame are shown in Table 7.1. Huge reduction in coarse grain and other crops were found in China by 2030. However, higher losses in oilseed crops were observed in India and Pakistan.

The rainfall along with wind speed also impact crop yields. It was found that climate change is likely to reduce the yields of wheat, corn, and rice in China by 18.26 ± 12.13, 45.10 ± 11.55, and $36.25 \pm 10.75\%$ until 2100 (Zhang et al. 2017). In majority of the world's areas, it has been predicted that there would be more droughts in the near future as a result of climate change, and that by 2100, the area impacted by droughts will have increased from 15.4% to 44.0%. The region considered to be most vulnerable is Africa. Major crops' yields in drought-affected areas are predicted to drop by nearly 90% by 2100 and by more than 50% by 2050 (Li et al. 2009). The loss of crop yields can increase food prices and can have an absurd effect on agriculture welfare globally, with a 0.3% annual loss of future GDP globally by 2100 (Stevanović et al. 2016). However, as per scientific evidences significant rainfall during active growth stage increases the crop yield of maize, rice, wheat, and apple (Singh et al. 2016).

Table 7.1 Impact of climate change on cereal crop production.

Crops	Country/continent	Yield reduction (%)
Wheat	Australia	−32
	Iran	−37
	Worldwide	−5.5
	Mexico	+25
	China	−17.5
	Asia	−7.7
	India	−5.2
	Pakistan	−50
	Turkey	−20
Rice	India	−7
	Indonesia	−11
	India	−8
	Asia	−6.3
	Italy	−12
	Japan	−11.3
	Nepal	−24
Maize	Portugal	−17
	Ghana	+12
	Africa	−20
	USA	−50
	Ethiopia	−4.7
	China	−46
	Africa	−32
	Pakistan	−27
	China	−30
	USA	−27

Source: Ahmed et al. (2019)/IntechOpen/Licensed under CC BY 3.0.

Table 7.2 Impact of climate change on other crop production.

Crops	Country/continent	Yield reduction (%)
Cotton	China	−5.5
	USA	−17
	Africa	−7
	USA	−9
	Pakistan	−8
	Burkina Faso	−13
	Australia	−17
Sugarcane	Brazil	+15
	Switzerland	−9
	Australia	+20
	Africa	+11
	USA	−40
	Brazil	−27
	India	−30

Source: Ahmed et al. (2019)/IntechOpen/Licensed under CC BY 3.0.

Table 7.3 Productivity shock due to climate change on rice, wheat, and coarse grains by 2030.

Countries	Coarse grains	Oilseeds	Other crops
China	−22 to 2	−12 to 12	−22 to 2
Philippines	−17 to −3	−10 to 4	−17 to −3
Thailand	−17 to −3	−10 to 4	−17 to −3
Rest Southeast Asia Asia	−17 to −3	−10 to 4	−17 to −3
Bangladesh	−17 to −3	−10 to 4	−17 to −3
India	−17 to −3	−15 to 4	−17 to −3
Pakistan	−17 to −3	−15 to 4	−17 to −3
Rest South Asia	−17 to −3	−15 to 4	−17 to −3

Source: Ahmed et al. (2019)/IntechOpen/Licensed under CC BY 3.0.

7.3.2 Disease Development

Arthropod insects including midges, ticks, fleas, mosquitoes, flies, and viruses are among the arthropods that are experiencing the most severe effects of climate change. It is projected that a crop will become more vulnerable to numerous pests, diseases, and weeds as a result of changes in an area's climate or weather pattern. In countries with high and middle latitudes, yields are anticipated to increase, while yields in lower latitudes are anticipated to decline (Shrestha et al.2019). However, estimates indicate that a one-degree increase in temperature will result in a 10–25% increase in losses from insect pest infestation. Given that insect population mostly depends on abiotic elements like humidity and temperature, climate change has the potential to increase pest population and its movement, which might negatively affect agricultural productivity and possibly viability. In Brazil, the infestation of coffee nematodes and leaf miners is expected to increase due to an increase in the number of generations in a month compared to the climatic conditions of 1961–1990 (Ghini et al. 2008). There is statistical evidence that increased rainfall and temperatures increased the costs of pesticides for crops such as corn, potatoes, and soybean, in contrast to a reduction in wheat in the United States. Through the use of Geographic Information System modeling, projection for potato tuber moth (*Phthorimaea operculella*) revealed a likely rise in the pest's capacity for damage in tropical and subtropical warmer climates, where it already predominates. Additionally, it is anticipated to spread into mountainous and temperate areas, where the potential for damage would be slightly higher. The life cycle of pathogens such as *Puccinia striiformis* f.sp. *trictici* is projected to be limited by increasing temperature, while an increase in the concentration of atmospheric CO_2 is estimated to provide advantageous conditions for *Fusarium pseudograminearum* (Luck et al. 2011). Particularly in temperate locations, climate change is anticipated to affect the development and metabolic rates of insects. As a result of climate change, more areas are suitable for pest infestation. Additionally, the threat of rice blast, late blight, and sheath blight is growing because of the increased CO_2 level and warming climate, which could represent a severe threat to global food security (Gautam et al. 2013).

7.3.3 Plant Responses to Climate Change

Physiological changes are more likely to be impacted by dramatic changes in these factors than by changes in mean climate because plant–water relations are very responsive to variations in temperature and precipitation. Plants' response to climate change varies according to the plant species and developmental stage of the plant. Depending on the plant's species and stage of growth, different plants react differently to climatic change. Different plants have species-specific thresholds, and their reactions, such as root elongation, change to the angle at which roots develop, and decreased yield varies between different plant species (Gray et al. 2006). Reduced transpiration in plants was observed with higher CO_2 level in the atmosphere, which increased air temperature by 0.42–0.02 K. The warming of terrestrial surfaces can be increased by 3.33–0.03 K due to the direct radiative effect and indirect physiological effect of elevated CO_2 (Cao et al. 2010). The quantity of crops that may be harvested is anticipated to increase as atmospheric CO_2 levels rise, and the type of crop will affect how plants develop. Although C3 crops are anticipated to produce more, C3 and C4 crops are anticipated to use less water in the absence of severe conditions. However, increased temperatures and altered precipitation are expected to cancel out these positive effects of increased CO_2.

7.3.4 Livestock

The potential impacts of climate change on livestock include changes in production and quality of feed crop and forage, water availability, animal growth and milk production (Henry et al. 2012), diseases (Thornton et al. 2009), reproduction (Nardone et al. 2010), and biodiversity. These impacts are primarily due to an increase in temperature and atmospheric CO_2 concentration, precipitation variation, and a combination of these factors (Polley et al. 2013). The impacts of climate change on livestock production factors are presented in Figure 7.1. Temperature affects most of the critical factors for livestock production, such as water availability, animal production, reproduction, and health. Forage quantity and quality are affected by a combination of increases in temperature, CO_2, and precipitation variation. Livestock diseases are mainly affected by an increase in temperature and precipitation variation.

7.3.5 Agriculture Economy

The Asian countries most vulnerable to climate change are Afghanistan, Bangladesh, Cambodia, India, Lao PDR, Myanmar, and Nepal. Afghanistan, Bangladesh, India, and Nepal are particularly vulnerable to declining crop yields due to glacial melting, floods, droughts, and erratic rainfall, among other factors. The agriculture industry may suffer since the temperature in India is expected to climb by 2.33–4.78 °C, CO_2 concentrations will double, and heat waves will last longer. Farmers in Pakistan's arid Rawalpindi region will suffer an annual loss of INR 4180/acre by 2100 due to a 1 °C rise in temperature, while their net income can improve by INR 377.4 and INR 649.21, respectively, with an increase in rainfall of 8% and 14% (Shakoor et al. 2011). The coastal regions of Indian agriculture such as Maharashtra, Gujarat, and Karnataka are most vulnerable, whereas small losses are also indicated for major food grain producing regions of Punjab, Haryana, and western Uttar Pradesh (Mendelsohn 2009). With a 1 °C increase in the world's mean surface temperature, it is predicted that the yield losses in three cereal grains (rice, maize, and wheat) will suffer by 10–25% (Gray et al. 2006). The average crop yield in sub-Saharan Africa is predicted to decline by 6–24% as a result of climate change. Additionally, it is anticipated that in 2050, overall fish demand on Solomon Islands would outpace fish output, having a significant impact on food security due to a decline in per capita consumption (Dey et al. 2016).

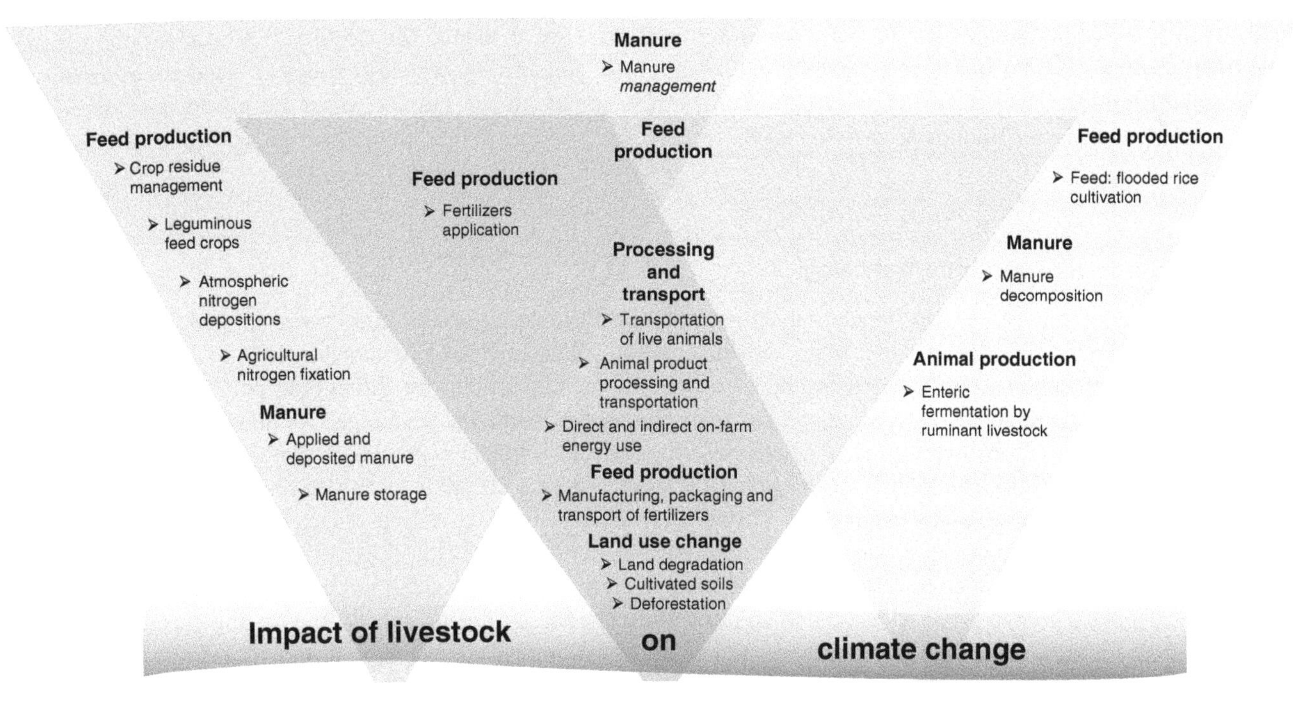

Figure 7.1 Impact of livestock on climate change. *Source:* Rojas-Downing et al. (2017)/with permission of Elsevier.

7.4 Mitigation and Adaptation to Climate Change

Crop development, growth, and production are all impacted by extreme weather conditions and shifting precipitation patterns. High temperatures throughout crucial growth phases could shorten the time grains fill before becoming sterile, which would diminish yields. Climate change can have a particularly negative impact on the agricultural sector. The main element that could help to lessen the negative impact of climate change is adaptation, which can help to reduce the risks in agriculture connected with climate change. The primary goal of adaptation strategy is to create climate change-resilient society (Lee et al. 2011). Adaptation strategies provide an opportunity to address the climate change challenges and to sustain crop production (Zhao et al. 2017).

In recent years, climate change adaptation has been explored by farmers in many ways by using scientific technologies including water-smart practices (laser land leveling, micro-irrigation, direct-seeded rice, crop diversification, rainwater harvesting, raised-bed planting), weather-smart activities (ICT-based agrometeorological services, stress-tolerant varieties), nutrient-smart practices (precision nutrient application, crop residue management, leaf color charts), carbon-smart activities (legumes, zero tillage, crop residue management), and knowledge-smart activities (agricultural extensions to enhance capacity building) (Malhi et al. 2021). By limiting unfavorable consequences, these technologies greatly reduce the effects of climate change on crops and improve their climatic suitability. Most farmers supported adaptations, but few favored GHG reduction, demonstrating the need to concentrate on treatments that combine adaptation and mitigation elements (Malhi et al. 2021). The main adaptation methods for mitigation of climate change are broadly classified into following categories.

7.4.1 Climate-Smart Cultural Practices

The adaptation of climate-smart cultural practices helps to reduce the vulnerabilities associated with crop production. Recently, farmers from Pakistan and Brazil have adapted the climate change variability by adjustment of planting time and optimization of plant populations (Ahmad et al. 2015). To utilize a crop's full potential, planting dates should be adjusted. High temperature during the grain-filling stage shortens the time required for grain filling and reduces yield loss. For crops, especially wheat, to yield more, the number of plants per unit area is crucial. Due to moisture stress and temperature fluctuations, a certain number of fertile tillers perish or remain unproductive. The optimal plant population makes up for the yield reduction. To maintain productivity in a changing environment, improved cultivars such as those that mature earlier and are more tolerant of heat and drought must be developed. Under conditions of moisture stress and high temperatures, the new cultivars would enhance productivity per unit area (Deressa et al. 2009). The rice that has been submerged produces methane gas. Flood water in rice prevents soil oxygen from penetrating, which provides an ideal environment for bacteria that produce methane gas. Therefore, methane emissions are decreased and water use efficiency is increased by innovative planting techniques as direct seeded rice and the system of rice intensification with alternate soaking and draying.

7.4.2 Climate-Smart Agriculture Technologies

Climate-smart agriculture (CSA) aims to adapt to climate change by adopting several interventions, including water-smart practices, nutrient-smart practices, weather-smart activities, carbon-smart activities, and knowledge-smart activities. Building evidence, boosting the efficiency of local institutions, encouraging climate-responsive agricultural policies, and connecting agricultural funding to climate are all ways that CSA promotes resistance to climate change (Malhi et al. 2021). In Punjab, Pakistan, CSA techniques were investigated, and increased cotton yield with higher returns and more efficient resource usage was noted (Imran et al. 2018). The Indo-Gangetic plain (IGP) is extremely susceptible to climate change, which has negative consequences on the region's rice and wheat crops. Farmers have demonstrated a readiness to use CSA technologies that can upgrade traditional farming methods to become more effective ones. Laser land leveling (LLL), weather-advisory services, and crop insurance are the most-preferred CSA technologies of the eastern IGP, while the farmers of the western IGP mainly prefer direct seeding, LLL, zero tillage, crop insurance, and irrigation scheduling (Taneja et al. 2019). These adaptability and mitigation potentials for these mitigation techniques are enormous. However, they are reliant on factors including a technology's applicability to the area, public image, economic viability, and technical intricacy. Additionally, these techniques are effective when several interventions are employed in tandem and in support of one another.

7.4.3 Stress-Tolerant Varieties

One method of coping with environmental challenges is through plant breeding to create new types. To determine whether a variety is suitable for the intended habitat, it will be necessary to choose the germplasm, shorten breeding cycles, and conduct trials in multiple locations (Atlin et al. 2017). The varieties are released in South Asian countries. It is crucial to create stress-tolerant cultivars as a mitigation technique because it is expected that abiotic stress will increase in frequency and intensity due to climate change. After being submerged for 18 days, certain high yield kinds of rice that have the SUB1A gene incorporated into them had better yields than the original varieties (Gregor and Hevner 2013).

7.4.4 Precision Management of Nutrients

By improving the effectiveness of fertilizers, precision nutrient management can boost a crop's resistance. Precision fertilizer management in crops, particularly maize, reduced the usage of fertilizers that would increase output and soil health, which in turn led to a drop in GHG emissions (Srinivasan 2006). The main driving force behind voluntary mitigation among farmers is their perceptions of the threat and severity of climate change. But the adaption is dependent on the accessibility of relevant information (Ahmed et al. 2019). Additionally, mitigation techniques will lessen the number of individuals who are subject to water stress, but those people who are still exposed to stress will need adaptation strategies.

7.4.5 Forestry and Agroforestry

Small farmers can adapt to climate change by using the ability of forestry and agroforestry sector to reduce atmospheric GHG concentrations. There are some straightforward methods for reducing GHG emissions, such as rice alternate drying, mid-season drainage, and better diets for cattle, enhanced N-use efficiency, and soil carbon. Changes in planting dates and types are examples of straightforward adaptation measures that could lessen the effects of climate change. Afforestation is also one of the most effective means of tackling climate change, particularly when it is designed to rely on green energy. This natural climate solution reduces the impact of desertification, supports ecosystems, and removes CO_2 from the atmosphere. Scientific evidences reported that trillion new trees could absorb one-third of CO_2 emissions made by human beings. In fact, an additional 25% of forested area could absorb 25% atmospheric carbon, making a significant impact on rising temperatures globally (Mishra et al. 2021). It is a countermeasure against deforestation, which has been contributing to climate change dramatically for the past few centuries. Recently, the UN Food and Agriculture Organization estimated that 100 billion square meters of forest were cut down each year over the past decade, while tropical deforestation has contributed 8% of global carbon emissions. IPCC 2021 report made it clear that deforestation is a direct cause of the increased presence of CO_2 in the air over the past decade, a rise incomparable to any other period in the past two million years. Forests are carbon sinks; they cycle damaging carbon out of the atmosphere and transform it into biomass through photosynthesis.

7.5 Conclusions and Future Prospects

Scientific evidences state that impact of climate change on agriculture will be one of the major deciding factor influencing the food and nutritional security of increasing population throughout the world. The key factors of climate such as GHGs, temperature, and precipitation significantly lead to disease infestation, drought, flood, soil infertility, and alter the physiology and metabolism of plant. To reversal deleterious impact of climate change on agriculture increases the demand of mitigation and adaptation strategies. These adaptation and mitigation strategies include cultural practices, water-smart practices (laser land leveling, rainwater harvesting, micro-irrigation, crop diversification, raised-bed planting, direct-seeded rice), nutrient-smart practices (precision nutrient application, leaf color charts, crop residue management), weather-smart activities (stress-tolerant varieties, ICT-based agrometeorological services), carbon-smart activities (zero tillage, legumes, crop residue management), and knowledge-smart activities (agricultural extensions to enhance capacity building). Strategies for mitigation and adaptation are anticipated to boost farmers' incomes without endangering the sustainability of agricultural production. Planning for mitigation and adaptation is rather difficult due to the very unpredictable nature of climate change's future effects. This calls for the development of climate-resilient technology using a regionally tailored transdisciplinary approach. Along with planned agronomic management and crop pest control, suitable cultivars that might adapt to climatic differences must be produced. In order to make the use of various climate-smart technologies by farmers easier, they must be made aware of them and given training.

References

Adams, R.M., Hurd, B.H., Lenhart, S., and Leary, N. (1998). Effects of global climate change on agriculture: an interpretative review. *Clim. Res.* 11 (1): 19–30.

Ahmad, A., Ashfaq, M., Rasul, G. et al. (2015). Impact of climate change on the rice–wheat cropping system of Pakistan. In: *Handbook of Climate Change and Agroecosystems: The Agricultural Model Intercomparison and Improvement Project Integrated Crop and Economic Assessments, Part 2* (ed. D. Hillel and C. Rosenzweig), 219–258. Imperial College Press.

Ahmed, I., Ullah, A., Rahman, M.H. et al. (2019). Climate change impacts and adaptation strategies for agronomic crops. In: *Climate Change and Agriculture*, 1–14. London: IntechOpen.

Arora, N.K. (2019). Impact of climate change on agriculture production and its sustainable solutions. *Environ. Sustain.* 2 (2): 95–96.

Arora, M., Goel, N.K., and Singh, P. (2005). Evaluation of temperature trends over India/Evaluation de tendances de temperature en Inde. *Hydrol. Sci. J.* 50: 81–93.

Atlin, G.N., Cairns, J.E., and Das, B. (2017). Rapid breeding and varietal replacement are critical to adaptation of cropping systems in the developing world to climate change. *Global Food Sec.* 12: 31–37.

Baumgard, L.H., Rhoads, R.P., Rhoads, M.L. et al. (ed.) (2012). Impact of climate change on livestock production. In: *Environmental Stress and Amelioration in Livestock Production*, 413–468. Berlin, Heidelberg: Springer.

Cao, L., Bala, G., Caldeira, K. et al. (2010). Importance of carbon dioxide physiological forcing to future climate change. *Proc. Nat. Acad. Sci.* 107 (21): 9513–9518.

Challinor, A.J., Watson, J., Lobell, D.B. et al. (2014). A meta-analysis of crop yield under climate change and adaptation. *Nat. Clim. Chang.* 4 (4): 287–291.

Debnath, B., Sikdar, A., Islam, S. et al. (2021). Physiological and molecular responses to acid rain stress in plants and the impact of melatonin, glutathione and silicon in the amendment of plant acid rain stress. *Molecules* 26 (4): 862.

Deressa, T.T., Hassan, R.M., Ringler, C. et al. (2009). Determinants of farmers' choice of adaptation methods to climate change in the Nile Basin of Ethiopia. *Glob. Environ. Chang.* 19 (2): 248–255.

Dey, M.M., Gosh, K., Valmonte-Santos, R. et al. (2016). Economic impact of climate change and climate change adaptation strategies for fisheries sector in Solomon Islands: implication for food security. *Mar. Policy* 67: 171–178.

Feng, Y., Zeng, Z., Searchinger, T.D. et al. (2022). Doubling of annual forest carbon loss over the tropics during the early twenty-first century. *Nat. Sustain.* 5 (5): 444–451.

Field, C.B., Barros, V., Stocker, T.F., and Dahe, Q. (ed.) (2012). *Managing the Risks of Extreme Events and Disasters to Advance Climate Change Adaptation: Special Report of the Intergovernmental Panel on Climate Change.* Cambridge University Press.

Friel, S., Dangour, A.D., Garnett, T. et al. (2009). Public health benefits of strategies to reduce greenhouse-gas emissions: food and agriculture. *The Lancet.* 374 (9706): 2016–2025.

Gautam, H.R., Bhardwaj, M.L., and Kumar, R. (2013). Climate change and its impact on plant diseases. *Current Sci.* 1685–1691.

Gay, C., Estrada, F., Conde, C. et al. (2006). Potential impacts of climate change on agriculture: a case of study of coffee production in Veracruz, Mexico. *Clim. Chang.* 79 (3): 259–288.

Ghini, R., Hamada, E., Pedro Júnior, M.J. et al. (2008). Risk analysis of climate change on coffee nematodes and leaf miner in Brazil. *Pesquisa agropecuária brasileira* 43: 187–194.

Grau, H.R., Gasparri, N.I., and Aide, T.M. (2005). Agriculture expansion and deforestation in seasonally dry forests of north-west Argentina. *Environ. Conserv.* 32 (2): 140–148.

Gregor, S. and Hevner, A.R. (2013). Positioning and presenting design science research for maximum impact. *MIS Quart.* 337–355.

Henry, B., Charmley, E., Eckard, R. et al. (2012). Livestock production in a changing climate: adaptation and mitigation research in Australia. *Crop Pasture Sci.* 63 (3): 191–202.

Howden, S.M., Soussana, J.F., Tubiello, F.N. et al. (2007). Adapting agriculture to climate change. *Proc. Natl. Acad. Sci. U.S.A.* 104 (50): 19691–19696.

Ilbay-Yupa, M., Ilbay, F., Zubieta, R. et al. (2021). Impacts of climate change on the precipitation and streamflow regimes in equatorial regions: Guayas river basin. *Water* 13 (21): 3138.

Imran, M.A., Ali, A., Ashfaq, M. et al. (2018). Impact of climate smart agriculture (CSA) practices on cotton production and livelihood of farmers in Punjab, Pakistan. *Sustainability* 10 (6): 2101.

IPCC (2014). Climate change 2014: synthesis report. In: *Contribution of Working Groups I, II and III to the Fifth Assessment Report of the Intergovernmental Panel on Climate Change* (ed. R.K. Pachauri and L.A. Meyer). Geneva, Switzerland: IPCC; 151p.

Kang, Y., Khan, S., and Ma, X. (2009). Climate change impacts on crop yield, crop water productivity and food security–a review. *Prog. Nat. Sci.* 19 (12): 1665–1674.

Kausar, S.H.A.M.E.E., Hussain, M.A., and Khan, A.A. (2010). Response of simulated acid rain on morphological, biochemical and leaf epidermal characteristics of wheat. *Trends Biosci.* 3: 34–36.

Lal, N. (2016). Effects of acid rain on plant growth and development. *J. Sci. Technol.* 11 (5): 85–108.

Lee, J., De Gryze, S., and Six, J. (2011). Effect of climate change on field crop production in California's Central Valley. *Clim. Chang.* 109 (1): 335–353.

Lesschen, J.P., van den Berg, M., Westhoek, H.J. et al. (2011). Greenhouse gas emission profiles of European livestock sectors. *Anim. Feed Sci. Technol.* 166: 16–28.

Li, Y., Ye, W., Wang, M., and Yan, X. (2009). Climate change and drought: a risk assessment of crop-yield impacts. *Clim. Res.* 39 (1): 31–46.

Lipper, L., Thornton, P., Campbell, B.M. et al. (2014). Climate-smart agriculture for food security. *Nat. Clim. Chang.* 4 (12): 1068–1072.

Luck, J., Spackman, M., Freeman, A. et al. (2011). Climate change and diseases of food crops. *Plant Pathol.* 60 (1): 113–121.

Mahato, A. (2014). Climate change and its impact on agriculture. *Int. J. Sci. Res. Publ.* 4 (4): 1–6.

Malhi, G.S., Kaur, M., and Kaushik, P. (2021). Impact of climate change on agriculture and its mitigation strategies: a review. *Sustainability* 13 (3): 1318.

Mendelsohn, R. (2009). The impact of climate change on agriculture in developing countries. *J. Nat. Resour. Policy Res.* 1 (1): 5–19.

Mishra, S., Ghosh, A., Rai, K. et al. (2021). Dimensions of climate change and its consequences on ecosystem functioning. In: *Global Climate Change* (ed. S. Singh, P. Singh, S. Rangabhashiyam, and K.K. Srivastava), 109–149. Elsevier.

Montzka, S.A., Dlugokencky, E.J., and Butler, J.H. (2011). Non-CO_2 greenhouse gases and climate change. *Nature* 476 (7358): 43–50.

Nardone, A., Ronchi, B., Lacetera, N. et al. (2010). Effects of climate changes on animal production and sustainability of livestock systems. *Livest. Sci.* 130 (1–3): 57–69.

Özdoğan, M. (2011). Modeling the impacts of climate change on wheat yields in Northwestern Turkey. *Agric. Ecosyst. Environ.* 141 (1-2): 1–12.

Polley, H.W., Briske, D.D., Morgan, J.A. et al. (2013). Climate change and North American rangelands: trends, projections, and implications. *Rangel. Ecol. Manag.* 66 (5): 493–511.

Rojas-Downing, M.M., Nejadhashemi, A.P., Harrigan, T., and Woznicki, S.A. (2017). Climate change and livestock: impacts, adaptation, and mitigation. *Clim. Risk Manag.* 16: 145–163.

Shakoor, U., Saboor, A., Ali, I., and Mohsin, A.Q. (2011). Impact of climate change on agriculture: empirical evidence from arid region. *Pak. J. Agric. Res.* 48 (4): 327–333.

Shrestha, S. (2019). Effects of climate change in agricultural insect pest. *Acta Sci. Agric.* 3 (12): 74–80.

Singh, N., Sharma, D.P., and Chand, H. (2016). Impact of climate change on apple production in India: a review. *Curr. World Environ.* 11 (1): 251.

Srinivasan, A. (ed.) (2006). Precision agriculture: an overview. In: *Handbook of Precision Agriculture*, 33–48. Boca Raton, FL: CRC Press.

Steinfeld, H., Gerber, P., Wassenaar, T.D. et al. (2006). *Livestock's Long Shadow: Environmental Issues and Options*. Rome: Food and Agriculture Organization of the United Nations.

Stevanović, M., Popp, A., Lotze-Campen, H. et al. (2016). The impact of high-end climate change on agricultural welfare. *Sci. Adv.* 2 (8): e1501452.

Taneja, G., Pal, B.D., Joshi, P.K. et al. (2019). Farmers' preferences for climate-smart agriculture-an assessment in the Indo-Gangetic Plain. In: *Clim. Smar. Agri. Sou. Asia*, 91–111. Singapore: Springer.

Thornton, P.K., van de Steeg, J., Notenbaert, A., and Herrero, M. (2009). The impacts of climate change on livestock and livestock systems in developing countries: a review of what we know and what we need to know. *Agric. Syst.* 101 (3): 113–127.

You, L., Rosegrant, M.W., Wood, S., and Sun, D. (2009). Impact of growing season temperature on wheat productivity in China. *Agric. For. Meteorol.* 149 (6–7): 1009–1014.

Zhang, P., Zhang, J., and Chen, M. (2017). Economic impacts of climate change on agriculture: the importance of additional climatic variables other than temperature and precipitation. *J. Environ. Econ. Manag.* 83: 8–31.

Zhao, C., Liu, B., Piao, S. et al. (2017). Temperature increase reduces global yields of major crops in four independent estimates. *Proc. Natl. Acad. Sci. U.S.A.* 114 (35): 9326–9331.

Zougmoré, R., Jalloh, A., and Tioro, A. (2014). Climate-smart soil water and nutrient management options in semiarid West Africa: a review of evidence and analysis of stone bunds and zaï techniques. *Agric. Food Secur.* 3 (1): 1–8.

8

Dynamic Photosynthetic Apparatus in Plants Combats Climate Change

Ramwant Gupta[1*] *and Ravinesh Rohit Prasad*[2]

[1] *Department of Biology, University of Guyana, Georgetown, South America*
[2] *Department of Geography, Fiji National University, Lautoka, Fiji Islands*

8.1 Introduction

In the twenty-first century, we face challenges such as global climate change, expanding global population, and recent stagnated yields of important crops, which has resulted in food insecurity (Ibrahimova et al. 2021; Li et al. 2021). According to the forecasts, feeding a world population of 9.1 billion people in 2050 will necessitate a 70% increase in overall food production between 2005 and 2050 (FAO 2009). In developing countries, production would have to nearly double. Crop yields would continue to increase, although at a slower pace than previously. This slowdown in growth has been going on for quite some time. Over the projection period, annual crop output growth would be nearly half (0.8%) of its historical rate (1.7%; 0.9 and 2.1% for developing nations) (FAO 2009). Furthermore, if management strategies such as precision irrigation and breeding for abiotic stress tolerance are not applied, extreme weather, viz., heat, drought, and flooding, among others, will reduce yields by up to 50% by 2050 (Bierbaum et al. 2007). Without expanding the quantity of land that may be used for agriculture and within the limitations of unpredictable weather, food production must be made more sustainable. The long-term viability of crop production, as well as the implications of the current scientific understanding of plant regulatory mechanisms, is one of the key objectives for photosynthesis research (Allakhverdiev 2020a, b; Ritchie et al. 2021). Photosynthesis is an energy transformation process that converts light energy into chemical energy and provides molecules of oxygen and a food source for sustaining life on the Earth. Photosynthesis is a complex process that could be divided into photochemical reactions driven by the passage of electrons by various protein complexes embedded in the thylakoid membrane and the Calvin cycle reaction of photosynthetic CO_2 fixation occurs in the stroma of the chloroplasts (Renger 2007; Rochaix 2011). The production of chemical energy from light is connected with electron flow from PSII to PSI in the photosynthetic apparatus (Rochaix 2011; Takahashi and Badger 2011; Yang et al. 2019). Photosynthetic apparatus comprises several pigments/protein complexes, viz., photosystem II (PSII), cytochrome b_6f (Cytb_6f), and photosystem I (PSI), situated in the thylakoid membrane of the chloroplast. Light energy caught by the light-reaping frameworks of PSII and PSI is moved to response chlorophyll, where charge division happens and electron transport is started (Krieger-Liszkay et al. 2019). Years of study have provided a thorough grasp of the difficulties in the photosynthetic process as well as recommendations for how to make it better. A wide range of potential for improving photosynthetic efficiency has been found (Orr et al. 2017; Brestic et al. 2021). This essential process of autotrophs leads to primary production on the planet, affected by extreme weather including high and low temperatures, drought, salinity, fluctuating light, flooding, and high CO_2 concentration (Brestic et al. 2021). Therefore, there is an urgent need to optimize photosynthesis by altering the photosynthetic apparatus and manipulating plant metabolism in response to global climate change. In this chapter, we summarized the attempts made to improve photosynthetic efficiency through molecular engineering that leads to plant growth and high yield in variable climatic conditions.

* Present Address: Department of Botany, Deen Dayal Upadhyaya Gorakhpur University, Gorakhpur, UP, India

Global Climate Change and Plant Stress Management, First Edition. Edited by Mohammad Wahid Ansari, Anil Kumar Singh, and Narendra Tuteja.

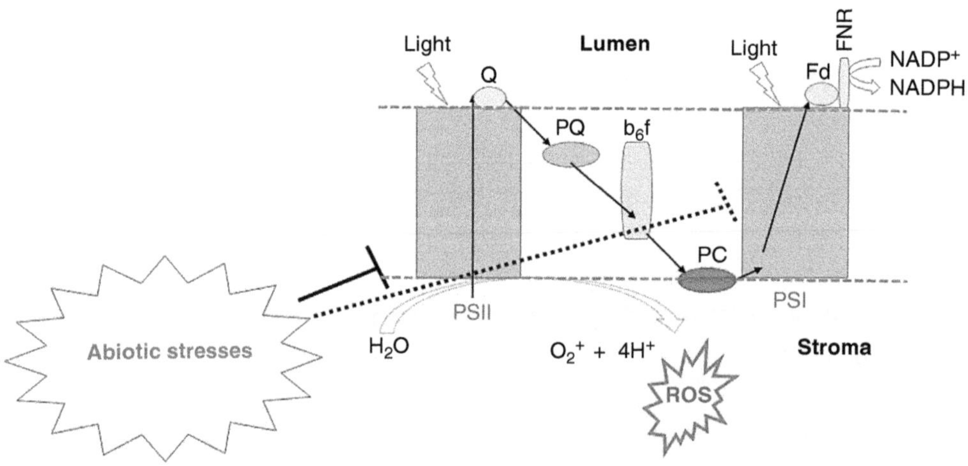

Figure 8.1 Effect of abiotic stresses viz., salinity, high and low temperature, drought, and high light intensity on photosynthetic apparatus in higher plants. The photosynthetic apparatus contains different proteins that are embedded in the thylakoid membrane and convert the light energy to chemical energy through a photosynthetic electron transport chain between PSII and PSI. *Source:* Based on Gupta et al. (2020). PSII is more vulnerable than PSI in extreme weather. Damaged PSII in response to climate change facilitates the production of reactive oxygen species (ROS) that adversely affect photosynthetic efficiency in plants. Here, PSII, photosystem II; Q_A, primary and secondary quinone electron acceptors of photosystem II; PQ, plastoquinone; b_6f, cytochrome b_6f complex; PC, plastocyanin; PSI, photosystem I; Fd, soluble ferredoxin; FNR, ferredoxin-$NADP^+$-reductase.

8.2 Climate Change and Photosynthetic Apparatus

Extreme weather, viz., chilling, heat, salinity, and high light intensity, hit the photosynthetic apparatus, PSII complex in particular (Sasi et al. 2018; Gupta 2020a; Huang et al. 2020) (Figure 8.1). Non-destructive Chl *a* fluorescence that derives OJIP transients access to valuable information on the structural and functional cohesion of the photosynthetic apparatus (Papageorgiou and Govindjee 2004). The efficiency of electron transport through PSII is determined after exposing the light to the dark-adapted leaves (Strasser et al. 2004; Stirbet and Govindjee 2011, 2012). In the OJIP transients, the OJ represents the reduction of primary electron acceptor quinone (Q_A) of PSII, JI means the reduction of secondary electron acceptors viz., quinone (Q_B), plastoquinone (PQ), cytochrome (Cyt b_6f), and plastocyanin (PC), and IP reveals the reduction of electron transporters of PSI ferredoxin (Fd), intermediate acceptors, and NADP (Kalaji et al. 2016). The PSII complex contains the water-splitting complex/oxygen-evolving complex (OEC), D_1, and D_2 proteins. The OEC is surrounded by extrinsic proteins, viz., PsbO, PsbP, PsbQ, and PsbR, which are easy to target for extreme weather conditions (Pérez-Bueno et al. 2011; Gupta 2020a). Disassociation of surrounding proteins of the OEC decreases the PSII activity and facilitates the production of the reactive oxygen species (ROS) resulting in photo-oxidation (Gururani et al. 2012; Gupta 2020a). OJIP transients revealed that PSI is more tolerant than PSII in response to all abiotic stresses (Gupta 2019, 2020a; Gupta et al. 2020, 2022). Generally, high light intensity inactivates the reaction center, damages the OEC, and declines the turnover of D1 protein, leading to photoinhibition of PSII in many plants (Kale et al. 2017; Gupta et al. 2020; Huang et al. 2020). During the heat, the donor site of the OEC is damaged, resulting in the appearance of a K peak at 0.3 ms on the OJIP transients (Brestic et al. 2012, 2016; Sasi et al. 2018; Gupta 2019; Gupta et al. 2021). Chilling stress inhibits the PSII repair mechanism while there is no evidence of photodamage to PSII (Allakhverdiev and Murata 2004; Murata et al. 2007; Gupta et al. 2021). Salinity devastates the manganese cluster of the OEC, leading to the cessation of oxygen production and reduction of PSII activity, reflecting a sharp decline in Fv/Fm values. Salinity disrupts the photosynthetic electron transport rate and overall photosynthetic activity of many plants (Mehta et al. 2010; Stefanov et al. 2011; Gupta 2020b).

8.3 Engineered Dynamic Photosynthetic Apparatus

More solar light is absorbed by plants than they can actually utilize resulting in the dissipation of excess light as heat, a process of non-photochemical quenching (NPQ) (Kromdijk et al. 2016). Tobacco was genetically modified to contain combinations of photoprotection genes, and those that accelerated the relaxation of NPQ increased biomass production rates by as much as 20% in glasshouse experiments and by about 15% in repeated field experiments (Kromdijk et al. 2016). The modifications included the addition of many genes that expressed important xanthophyll cycle and PsbS subunit of PSII genes. This combination allowed plants to use light for photosynthesis instead of photoprotection; they were able to more quickly respond to changing light conditions rather than continuing to emit heat after a drop in light levels. Conservation of NPQ throughout the plant suggests that this approach is also possible to promote the growth of other crops. Additionally, bacteriochlorophyll, which is present in a variety of anoxic photosynthetic species, can be incorporated into one of the photosystems to boost the intake of light energy and broaden the plant's light absorption spectrum to include wavelengths as far-red as 1100 nm (Blankenship et al. 2011). Changing the antenna size of the photosystem has also been proposed as a way to improve the conversion efficiency of solar light to reduced NPQ. This presumption is founded on the idea that an antenna can collect more light than is necessary for photochemistry. Based on this, plants with low numbers of photoharvest pigments (such as chlorophyll and carotenoids) per photosystem can significantly improve the efficiency of solar energy conversion. It has been suggested that replacing chlorophyll *a* with chlorophyll *d* from a cyanobacterium (*Acaryochloris marina*) and adding the machinery of a purple photosynthetic bacterium are two ways to improve the photosynthetic apparatus of plants (Brestic et al. 2021). Modifying the optical properties of the leaves reduces overabsorption and reduces the efficiency loss associated with wasted scattering of sunlight through the upper canopy. In addition, they have the potential to improve photosynthesis through higher light transmittance to the lower layer and improve canopy light distribution and canopy photosynthesis, resulting in higher productivity (Melis 2009; Ort and Melis 2011) (Figure 8.2).

A key challenge in designing Rubisco in plants is the 550 kDa giant hexadecameric complex consisting of eight large subunits (about 52 kDa each) and eight small subunits (about 13–15 kDa). Large chloroplast-encoded subunits with catalytic sites that can be easily altered in tobacco for functional genomic studies have generally been the focus of research (Sharwood 2017). Recent reports have highlighted the effect of small subunit manipulation on catalysis (Atkinson et al. 2017). Therefore, the nuclear-encoded small subunit (RbcS) gene family is growing as a target for genetic engineering, as transmutations have already been established in many species compared to a relatively small number of species amenable to chloroplast transformation (Bock 2015). In addition to initiatives to introduce higher plants with the cyanobacterial carbon concentrating mechanism (CCM), with the effective introduction of rubisco from *Synechococcus elongatus*, tobacco may now grow in environments with high CO_2 levels (Lin et al. 2014). It supports much higher catalytic rates than Rubisco with higher plants. Cyanobacterial Rubisco's CO_2 affinity is lower than its O_2 affinity (Occhialini et al. 2016). Therefore a functional CCM must therefore be jointly engineered in order for cyanobacteria Rubisco to assist plant growth at ambient CO_2 levels (Price and Howitt 2014). Other enzymes in the Calvin cycle have long been known to be useful targets for accelerating carbon fixation in plants. Recent efforts are based on previous success in overexpression of SBPase (sedoheptulose-1,7-bisphosphatase) to improve tobacco and rice growth (Rosenthal et al. 2011) (Figure 8.2).

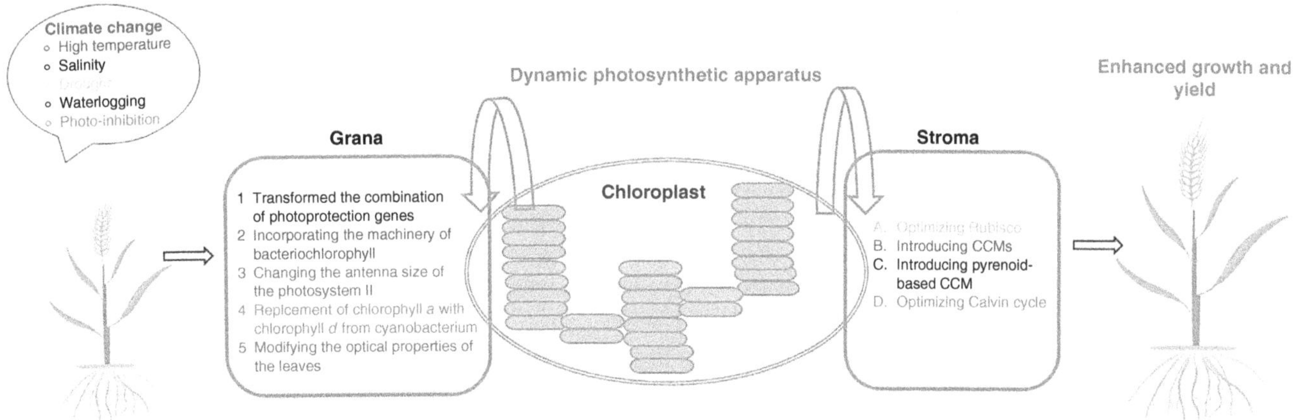

Figure 8.2 Several routes are being studied to improve photosynthesis in response to climate change. These include in grana (1) transforming the combination of photoprotection genes, (2) incorporating the machinery of bacteriochlorophyll, (3) changing the antenna size of the photosystem II, (4) replacement of chlorophyll *a* with chlorophyll *d* from a cyanobacterium, (5) modifying the optical properties of the leaves and in the stroma (A) optimizing Rubisco, (B) introducing CCMs, (C) introducing pyrenoid-based CCM, and (D) optimizing the Calvin cycle. *Source:* Model-based on Orr et al. (2017).

8.4 Conclusion and Prospects

The photosynthetic apparatus is more vulnerable to climate change, PSII in particular. Chlorophyll *a* fluorescence (OJIP transients) are used to understand the structural and functional relationships of photosynthetic apparatus. Improving photosynthetic efficiency appears to provide an exciting opportunity to tackle sustainable productivity challenges and increases the need to meet future food demand. Genetic engineering was used to develop simple traits such as insect resistance and weed resistance. Sophisticated molecular tinkering promises spectacular benefits when the basic signaling pathways are targeted, like photosynthesis. Therefore, the engineering of photosynthetic machinery can be a powerful technique to enhance the production and productivity of crops that can bring humanity closer to solving the future agricultural crisis. Achieving higher photosynthetic rates while using equal or less water and nitrogen resources could be a necessary turning point for transforming agriculture in the twenty-first century. The development of a more efficient photosynthesis and climate-resistant cultivation system is needed for both plant yields and nutritional value.

References

Allakhverdiev, S.I. (2020a). Editorial for the special issue on photosynthesis and hydrogen energy research for sustainability-2019. *Photosynth. Res.* 146: 1–3. https://doi.org/10.1007/s11120-020-00764.

Allakhverdiev, S.I. (2020b). Optimising photosynthesis for environmental fitness. *Funct. Plant Biol.* 47: iii–vii. https://doi.org/10.1071/FPv47n11_FO.

Allakhverdiev, S.I. and Murata, N. (2004). Environmental stress inhibits the synthesis *de novo* of proteins involved in the photodamage–repair cycle of Photosystem II in *Synechocystis* sp. PCC 6803. *Biochim. Biophys. Acta Bioenerg.* 1657: 23–32. https://doi.org/10.1016/j.bbabio.2004.03.003.

Atkinson, N., Leitão, N., Orr, D.J. et al. (2017). Rubisco small subunits from the unicellular green alga Chlamydomonas complement Rubisco-deficient mutants of *Arabidopsis*. *New Phytol.* 214 (2): 655–667.

Bierbaum, R.M., Holdren, J.P., MacCracken, M.C. et al. (2007). *Confronting Climate Change: Avoiding the Unmanageable and Managing the Unavoidable*. Washington, DC: The United Nations Foundation.

Blankenship, R.E., Tiede, D.M., Barber, J. et al. (2011). Comparing photosynthetic and photovoltaic efficiencies and recognizing the potential for improvement. *Science* 332 (6031): 805–809.

Bock, R. (2015). Engineering plastid genomes: methods, tools, and applications in basic research and biotechnology. *Annu. Rev. Plant Biol.* 66: 211–241.

Brestic, M., Zivcak, M., Kalaji, H.M. et al. (2012). Photosystem II thermostability *in situ*: environmentally induced acclimation and genotype-specific reactions in *Triticum aestivum* L. *Plant Physiol. Biochem.* 57: 93–105. https://doi.org/10.1016/j.plaphy.2012.05.012.

Brestic, M., Zivcak, M., Kunderlikova, K., and Allakhverdiev, S.I. (2016). High temperature specifically affects the photoprotective responses of chlorophyll b-deficient wheat mutant lines. *Photosynth. Res.* 130: 251–266. https://doi.org/10.1007/s11120-016-0249-7.

Brestic, M., Yang, X., Li, X., and Allakhverdiev, S.I. (2021). Crop photosynthesis for the twenty-first century. *Photosynth. Res.* 150 (1–3): 1–3. https://doi.org/10.1007/s11120-021-00869-5. PMID: 34674135.

FAO (2009). *Feeding the World in 2050. World Agricultural Summit on Food Security*. Rome: FAO.

Gupta, R. (2019). Tissue specific disruption of photosynthetic electron transport rate in pigeonpea (*Cajanuscajan* L.) under elevated temperature. *Plant Signal Behav.* 14 (6): 1601952. https://doi.org/10.1080/15592324.2019.1601952.

Gupta, R. (2020a). The oxygen-evolving complex: a super catalyst for life on earth, in response to abiotic stresses. *Plant Signal. Behav.* 15 (12): 1824721. https://doi.org/10.1080/15592324.2020.1824721.

Gupta, R. (2020b). Manganese repairs the oxygen-evolving complex (OEC) in maize (*Zea mays* L.) damage during seawater vulnerability. *J. Soil Sci. Plant Nutr.* 20: 1387–1396. https://doi.org/10.1007/s42729-020-00220-2.

Gupta, R., Sharma, R.D., and Singh, M. (2020). Energy dissipation and photosynthetic electron flow during the transition from juvenile red to mature green leaves in mango (*Mangifera indica* L.). *Plant Biosyst.* 155 (2): 1–12. https://doi.org/10.1080/11263504.2020.1810807.

Gupta, R., Sharma, R.D., Rao, Y.R. et al. (2021). Acclimation potential of Noni (*Morinda citrifolia* L.) plant to temperature stress is mediated through photosynthetic electron transport rate. *Plant Signal. Behav.* 16 (3): 1865687. https://doi.org/10.1080/15592324.2020.1865687.

Gupta, R., Ansari, A., and Vema, C.L. (2022). Physiological responses and adaptation mechanisms in plants to elevated temperature. In: *Understanding Abiotic Stress* (ed. V.D. Rajput, K.K. Verma, and T. Minkina), 133–152. Nova Science Publishers, Inc.

Gururani, M.A., Upadhyaya, C.P., Strasser, R.J. et al. (2012). Physiological and biochemical responses of transgenic potato plants with altered expression of PSII manganese stabilizing protein. *Plant Physiol. Biochem.* 58: 182–194.

Huang, S., Zuo, T., and Ni, W. (2020). Important roles of glycinebetaine in stabilizing the structure and function of the photosystem II complex under abiotic stresses. *Planta* 251: 36. https://doi.org/10.1007/s00425-019-03330-z.

Ibrahimova, U., Zivcak, M., Gasparovic, K. et al. (2021). Electron and proton transport in wheat exposed to salt stress: is the increase of the thylakoid membrane proton conductivity responsible for decreasing the photosynthetic activity in sensitive genotypes? *Photosynth. Res.* 150 (1–3): 195–211. https://doi.org/10.1007/s11120-021-00853-z.

Kalaji, H.M., Jajoo, A., Oukarroum, A. et al. (2016). Chlorophyll a fluorescence as a tool to monitor physiological status of plants under abiotic stress conditions. *Acta Physiol. Plant.* 38: 102.

Kale, R., Hebert, A.E., and Frankel, L.K. (2017). Amino acid oxidation of the D1 and D2 proteins by oxygen radicals during photoinhibition of photosystem II. *Proc. Natl. Acad. Sci. U.S.A.* 114: 2988–2993.

Krieger-Liszkay, A., Krupinska, K., and Shimakawa, G. (2019). The impact of photosynthesis on initiation of leaf senescence. *Physiol. Plant.* 166: 148–164. https://doi.org/10.1111/ppl.12921.

Kromdijk, J., Głowacka, K., Leonelli, L. et al. (2016). Improving photosynthesis and crop productivity by accelerating recovery from photoprotection. *Science* 354 (63): 857–861.

Li, D., Wang, M., Zhang, T. et al. (2021). Glycinebetaine mitigated the photoinhibition of photosystem II at high temperature in transgenic tomato plants. *Photosynth. Res.* 147: 301–315. https://doi.org/10.1007/s11120-020-00810-2.

Lin, M.T., Occhialini, A., Andralojc, P.J. et al. (2014). β-Carboxysomal proteins assemble into highly organized structures in Nicotiana chloroplasts. *Plant J.* 79 (1): 1–12.

Mehta, P., Jajoo, A., and Mathur, S. (2010). Chlorophyll a fluorescence study revealing effects of high salt stress on photosystem II in wheat leaves. *Plant Physiol. Biochem.* 48: 16–20.

Melis, A. (2009). Solar energy conversion efficiencies in photosynthesis: minimizing the chlorophyll antennae to maximize efficiency. *Plant Sci.* 177 (4): 272–280.

Murata, N., Takahashi, S., Nishiyama, Y., and Allakhverdiev, S.I. (2007). Photoinhibition of photosystem II under environmental stress. *Biochim. Biophys. Acta Bioenerg.* 1767: 414–421. https://doi.org/10.1016/j.bbabio.2006.11.019.

Occhialini, A., Lin, M.T., Andralojc, P.J. et al. (2016). Transgenic tobacco plants with improved cyanobacterial Rubisco expression but no extra assembly factors grow at near wild-type rates if provided with elevated CO_2. *Plant J.* 85 (1): 148–160.

Orr, D.J., Pereira, A.M., da Fonseca, P.P. et al. (2017). Engineering photosynthesis: progress and perspectives [version 1; peer review: 2 approved]. *F1000Research* 6: 1891. https://doi.org/10.12688/f1000research.12181.1.

Ort, D.R. and Melis, A. (2011). Optimizing antenna size to maximize photosynthetic efficiency. *Plant Physiol.* 155 (1): 79–85.

Papageorgiou, G.C. and Govindjee (ed.) (2004). *Chlorophyll a Fluorescence: A Signature of Photosynthesis*, Advances in Photosynthesis and Respiration, vol. 19, 820. Dordrecht: Springer.

Pérez-Bueno, M.L., Barón, M., and García-Luque, I. (2011). PsbO, PsbP, and PsbQ of photosystem II are encoded by gene families in *Nicotiana benthamiana*. Structure and functionality of their isoforms. *Photosynthetica* 49: 573–580.

Price, G.D. and Howitt, S.M. (2014). Plant science: towards turbocharged photosynthesis. *Nature* 513 (7519): 497–498.

Renger, G. (2007). *Primary Processes of Photosynthesis: Principles and Apparatus, 2 Parts.* Cambridge: Royal Society of Chemistry Publishers.

Ritchie, R.J., Sma-Air, S., Limsathapornkul, N. et al. (2021). Photosynthetic electron transport rate (ETR) in the littoral herb *Launaea sarmentosa* known as mole crab in Thailand. *Photosynth. Res.* 150 (1–3): 327–341. https://doi.org/10.1007/s11120-021-00826-2.

Rochaix, J.D. (2011). Reprint of: Regulation of photosynthetic electron transport. *Biochim. Biophys. Acta* 1807: 878–886. https://doi.org/10.1016/j.bbabio.2011.05.009.

Rosenthal, D.M., Locke, A.M., Khozaei, M. et al. (2011). Over-expressing the C3 photosynthesis cycle enzyme Sedoheptulose-1-7 Bisphosphatase improves photosynthetic carbon gain and yield under fully open air CO2 fumigation (FACE). *BMC Plant Biol.* 11: 123.

Sasi, S., Venkatesh, J., Daneshi, R.F., and Gururani, M.A. (2018). Photosystem II extrinsic proteins and their putative role in abiotic stress tolerance in higher plants. *Plants* 7: 100.

Sharwood, R.E. (2017). Engineering chloroplasts to improve Rubisco catalysis: prospects for translating improvements into food and fiber crops. *New Phytol.* 213 (2): 494–510.

Stefanov, D., Petkova, V., and Denev, I.D. (2011). Screening for heat tolerance in common bean (*Phaseolus vulgaris* L.) lines and cultivars using JIP-test. *Sci. Hortic.* 128 (1): 1–6.

Stirbet, A. and Govindjee (2011). On the relation between the Kautsky effect (chlorophyll a fluorescence induction) and photosystem II: basis and applications of the OJIP fluorescence transient. *J. Photochem. Photobiol. B* 104: 236–257.

Stirbet, A. and Govindjee (2012). Chlorophyll a fluorescence induction: a personal perspective of the thermal phase, the J-I-P rise. *Photosynth. Res.* 113: 15–61.

Strasser, R.J., Tsimilli-Michael, M., and Srivastava, A. (2004). Analysis of the chlorophyll a fluorescence transient. In: *Chlorophyll a Fluorescence: A Signature of Photosynthesis*, Advances in Photosynthesis and Respiration, vol. 19 (ed. G.C. Papageorgiou and Govindjee), 321–362. Dordrecht: Springer.

Takahashi, S. and Badger, M.R. (2011). Photoprotection in plants: a new light on photosystem II damage. *Trends. Plant Sci.* 16: 5359.

Yang, Y.-J., Liu, T., Zhang, S.-B., and Huang, W. (2019). Special issue in honor of Prof. Reto J. StrasserPhotoinhibition of oxygen-evolving complex and photosystem II at chilling stress in the tropical tree species *Dalbergia odorifera*. *Photosynthetica* https://doi.org/10.32615/ps.2019.138.

9

CRISPR/Cas Enables the Remodeling of Crops for Sustainable Climate-Smart Agriculture and Nutritional Security

Tanushri Kaul, Rachana Verma, Sonia Khan Sony, Jyotsna Bharti, Khaled Fathy Abdel Motelb, Arul Prakash Thangaraj, Rashmi Kaul, Mamta Nehra, and Murugesh Eswaran

Plant Biology and Biotechnology, Nutritional Improvement of Crops Group, International Centre for Genetic Engineering and Biotechnology (ICGEB), New Delhi, India

9.1 Introduction: CRISPR/Cas Facilitated Remodeling of Crops

In recent times, food and nutritional security have emerged as a prime challenge for underdeveloped and developing countries, especially in South Asia and Saharan Africa (Willett et al. 2019). According to the United Nations (UN) report, world population and food demand will rise to 10 billion and 13.6 billion tons, respectively, by 2050 (Ahmad et al. 2021). Globally, 2.37 billion people (one in three) were deprived of an adequate food supply in 2020. Moreover, approximately 57 million people on the Asian continent faced hunger in 2020 (FAO 2021). Food security is foremost in the United Nations 2030 Agenda for Sustainable Development Goals (SDG), which strives to culminate poverty and rescue the planet from environmental degradation. However, the COVID-19 pandemic that has outstretched rapidly since late 2019 has had intense inferences on food and nutritional security worldwide. This unfolding predicament has influenced food systems, threatening access to food via numerous dynamics. In 2020, about 7.8 billion of the world's population struggled with COVID-19 disease caused by a novel coronavirus, SARS-CoV-2 (Population Reference Bureau 2021). The circumstances are dynamic and fluid, indicating a high degree of uncertainty. According to the World Health Organization (WHO), the most destructive outcomes are yet to arrive (Ghebreyesus 2020; Khorsandi 2020). Besides, the ever-increasing world population poses severe challenges to world food security (Naidoo and Fisher 2020; Chiwona-Karltun et al. 2021; Wood et al. 2021), making it very difficult to meet the future nutritional food demands (Wallace et al. 2003; Bellard et al. 2012; Barrett 2021; Chiwona-Karltun et al. 2021). Moreover, there is a gap between the quantity of food supply and nutritional requirements. Thus, to bridge this gap, efforts need to be made to specify essential food sources or to develop new sources and make them available for consumption.

Plants are the ultimate source of nutrients for humans and livestock. However, crop yield is adversely influenced due to multiple stress factors (Fahad et al. 2017; Liliane and Charles 2020; Hossain et al. 2021). The insufficient crop yield with an inadequate nutritional value of crops due to numerous stresses hampered or already compromised the nutrition security for populations in developing and underdeveloped nations that have already been suffering from malnutrition (Kaul et al. 2020a). Furthermore, the blooming population has contributed enormously toward the deteriorating climate. The report considering the current trend of global warming given by the Intergovernmental Panel on Climate Change (IPCC) determined that continual anthropogenic activities will lead to an increase in the ambient temperature by a minimum of average 0.5–1 °C per decade (IPCC 2021). Drastically changing climate conditions elevate the risk of numerous natural calamities, for instance, high temperatures, changes in the precipitation patterns, floods, storms, cyclones, and drought, which lead to hampered agricultural land imparting adverse effect on the productivity of crop plant (Yohannes 2016; Aryal et al. 2020; Oo et al. 2020).

During the era of a drastically changing global climate, sustaining human life, and feeding the ever-increasing population is the prime apprehension guiding the adaptation of a novel approach that can aid in achieving the sustainable development of agriculture (Kaul et al. 2022). Over the last few decades, several attempts have been undertaken for crop improvement, most of which have relied on traditional or molecular breeding approaches (Adlak et al. 2019). Crop improvement

Global Climate Change and Plant Stress Management, First Edition. Edited by Mohammad Wahid Ansari, Anil Kumar Singh, and Narendra Tuteja.

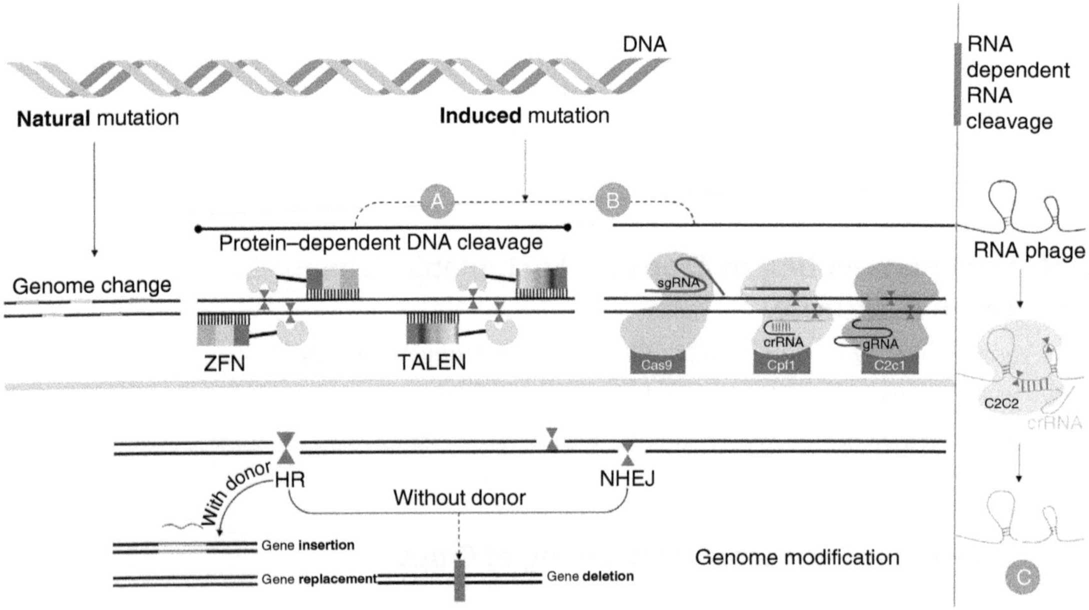

Figure 9.1 Novel approaches to genome editing conferring precise modification for the improvement of important traits.

has continuously aimed to achieve higher yields, enhanced nutritional quality, and resilience towards stresses. These strategies have saved us from the food crisis in the past, but they have reached a threshold where they have become incompetent to meet the present food demands. Moreover, the unavailability of the desired natural variance in a crop gene pool is another bottleneck that impedes trait improvements through conventional approaches (Allaby et al. 2019). However, due to the advancement of sophisticated genome sequencing platforms, the agriculturally important crops' genome sequences are now easily accessible. Therefore, it is possible to "edit" genes to improve key traits via utilizing and developing CRISPR/Cas-based genome editing techniques (Hua et al. 2019a; Kaul et al. 2019, 2020a; Gao et al. 2020a; Zaidi et al. 2020; Bhattacharya et al. 2021; Figure 9.1). The agricultural industry has grown significantly as a result of the CRISPR/Cas system, which has aided functional genomics research to develop high-yielding, biofortified, and oxidative stress-resistant crop lines, providing global food and nutritional security. Gene editing imparts essential alterations in crop plants, resulting in the sustainable improvement of future smart crops (Li and Siddique 2020). The application of precise genome editing techniques and the implementation of artificial intelligence (AI) attained via machine learning (ML) opened the door to a better understanding of the impact of genetics on living organisms by analyzing the underlying genetic blueprints (Almutiri et al. 2022). The idea of employing the computational power to predict the desired target genes and post-gene editing phenotypic influences can lead to optimized applications for improving crop plants and beyond (O'Brien et al. 2021). Policymakers, activists, and scientific community can believe in gene-editing technology as it has not only helped to produce sustainable, high-yielding varieties but also brought positive changes on human health (Karavolias et al. 2021). As genome-edited crops are considered as unsafe for general people, it is quintessential to be provided more information about the technology. Inevitably, this initiative would generate the roadmap for the acceptance of genome-edited crops.

This chapter emphasizes the use of the existing and evolving CRISPR tool kit as well as aids the documentation of some novel events for sustainable development of nutritionally secure climate-smart crop. Moreover, to access how the combination of next-generation deep sequencing (NGS) with genome editing can accelerate the modern molecular breeding approach. Furthermore, to recognize the stigma related to public acceptance of genome-edited crops in terms of human health. Therefore, highlighting the subsisting endeavors of the scientific communities and government ministries to provide a pitch positive turn to make a way forward for the release of the CRISPR generated new robust variety in the global market.

9.2 Impact of Climate Changes on Agriculture and Food Supply

A major sustainability concern of the twenty-first century is increasing agricultural output in order to meet the needs of a continually growing global population. According to the report published in 2020, approximately 8.9% of the total worldwide population – almost 690 million people – were famished, an upsurge of around 60 million in only five years.

By 2050, the world would require generating roughly 70% additional food to supply an estimated 10 billion people, rendering the food safety concern even more arduous. The challenge is exaggerated by agriculture's intense susceptibility to climate change. However, climate change projections include higher temperatures, higher CO_2 concentrations, and changes in echelons of precipitation in the atmosphere, all of which have a direct impact on yield (quality and quantity), moisture availability, photosynthesis, growth rate, and transpiration rate, agricultural inputs such as herbicides, insecticides, and agrochemicals, and all these factors result in decreased agricultural productivity. Moreover, crop production can be affected by several major and minor environmental factors, such as drainage frequency of soil and severity, that are responsible for nitrogen leaching, soil erosion, crop diversity loss, and land availability. As reported data have shown, these climate changes were estimated to have a direct impact on global food production. Since ancient times, agriculture has been more reliant on the monsoon. Any change in monsoon patterns has a significant impact on agriculture. An expansion in the average seasonal temperature could persist the growing season of countless crops such as rice, wheat, maize, and several others, reducing their final output. Warming will have a more significant effect on yields as both quantitative and qualitative traits were deteriorated in regions wherever temperatures are formerly near to biologically maximum for crops (IPCC 2021). Although there are many studies that emphasize the implications of a certain dimension of climate change in a special position, there are only a few that deliver a global evaluation. However, these findings have a propensity to focus on the explicit impact of variations in the average climate state on crop growth, somewhat than alterations in indirect or extreme consequences of climate change like a rise in the sea level or pests and diseases. CO_2 levels in the atmosphere are another factor for crop production. According to reports, climate change and rising CO_2 levels in the atmosphere may have an indirect impact on crops through pest and disease effects. These interfaces are complicated, and the full ramifications for crop output are unknown currently. Pests like weevil larvae (Staley and Johnson 2008) and aphids (Newman 2004) have been shown to respond favorably to high CO_2. According to Zhou et al. (1995) aphid overwintering mortality was reduced as temperatures rose, allowing for earlier and perhaps more widespread dispersal. However, these pre-monsoon changes were predominant in influencing the wheat crop in the Indo-Gangetic Plain >0.5 °C rise in time share 2010–2039; IPCC 2007. The shortfall in rice production throughout acute droughts over a period of one year, averages is about 40% of total production in the drought prone states of Chhattisgarh, Jharkhand, and Odisha, with an expected worth of $800 (Pandey et al. 2007). Rice, wheat, legume, and oilseed yields increase by 10–20% when CO_2 levels rise to 550 ppm. Wheat, soybean, mustard, peanut, and potato yields may be reduced by 3–7% with a 1 °C increase in temperature. At higher temperatures, there are a lot more losses. Most agricultural productivity would decline only slightly by 2020 but by 10–40% by 2100 because of increasing temperatures, changing rainfall patterns, and reduced irrigation water availability. Climate change will have the greatest impact on rain-fed or non-irrigated crops, which account for approximately 60% of cropland. In India, a 0.5 °C increase in winter temperature is expected to diminish rain-fed wheat yield by 0.45 tons/ha (Lal et al. 2007). Chickpea, sorghum, ragi, maize, and millet yields may improve, as well as coconut harvests on the west coast. Reduced frost damage has resulted in less damage to potatoes, mustard, and vegetables in northwestern India. Therefore, climate change has an immediate influence on food security globally. To address these environmental crises, such as increasing human population, biotic and abiotic stress in crop plants, climate change, there is a need of transition towards a resilient farming practice for feeding a growing population.

9.3 Nutritionally Secure Climate-Smart Crops

Agricultural food productivity is highly sensitive to the climatic aberrations, which manifest as global food insecurity and malnutrition (Swinburn et al. 2019; Karavolias et al. 2021). Significant changes in the ambient environment resulted in the crop yield losses annually. For instance, even an average increase in global temperature by 1 °C resulted in a reduction in yield of 3.2% in rice, 7.4% in maize, 6.0% in wheat, and 3.1% in soybean, globally (Zhao et al. 2017). Climate change and environmental aberrations not only limit the availability of arable land but also hamper the crop quality, i.e. reduced nutritional composition of the agricultural product under the projected pressure of pest and diseases (Godfray 2010; Wheeler and von Braun 2013; Springmann et al. 2016). According to Medek et al. (2017), a gradual increase in CO_2 level would cause approximately a 10% loss of zinc, 5% of iron, and 8% of protein contents in crops by 2050. Current global food habits render people susceptible to non-infectious chronic lifestyle ailments, for instance obesity, heart disease, diabetes, colon or rectal cancer, and others. Furthermore, with the recent outbreak of pandemic and an evident shift in the pattern of dietary preferences, there has been an alarming prevalence of chronic deprivation of essential nutritional components, which was projected as a huge chunk of 768 million globally undernourished population (FAO 2021). Currently, a lethal combination of malnutrition and climate change has enhanced the demand for high-yielding, nutritionally rich, and resilient plants, paving the way forward for developing future-ready agriculture. Therefore, CRISPR/Cas-based systems hold immense potential in generating adequate nutritionally balanced food in developing and under-developed nations, worldwide (Kaul et al. 2020b). Thus, pan-global improvements in crop yield, stress resilience, and nutritional value may be achieved via genome editing for food security and to mitigate the gap between the laboratory and field. In this context, innumerable studies have paved the way to incorporate genome editing into plant improvement, thus rendering them inseparable entities. Therefore, grains rich in amylose content offer a good source of resistant starch (RS) to cure lifestyle-related chronic disorders. In particular, RS resists digestion in stomach and small intestine and directly passes to large intestine; thus, it has benefitted human health by preventing incidences of insulin resistance and type 2 diabetes, thus curbing obesity (Sun et al. 2017). Thus, in rice, high and low amylose content varieties are generated via knockouts of the genes associated with starch and amylose biosynthesis, i.e. starch branching enzyme IIb (*OsBEIIb*) and granule-bound starch synthase I (*OsGBSSI*) genes, respectively (Sun et al. 2017). Similarly, in cassava, these two genes were targeted for reduced starchy content (Bull et al. 2018); in potatoes, KO of the GBSS gene has also led to altered starch content (Andersson et al. 2018); and altered sugar content in the strawberry variety was also generated via genome editing (Xing et al. 2020). Reportedly, macronutrient composition in rice grains (Abe et al. 2018) and oilseed rape (*Brassica napus*) was enhanced by KO of fatty acid desaturase 2 (FAD2) genes, *OsFAD2-1* and *OsFAD2*, to achieve significant increase in oleic acid content (Okuzaki et al. 2018). According to Li et al. (2018a), lycopene and β-carotene enriched tomato and banana have been developed by targeting lycopene epsilon-cyclase (*LCYε*) gene and in combination with stay green 1 (*SGR1*), beta-lycopene cyclase (*BLC*), and lycopene β-cyclase 2 (*LCY-B2*) genes that are involved in the lycopene biosynthesis simultaneously (Li et al. 2018a; Kaur et al. 2020). Kawaguchi et al. (2021) developed high sugar containing tomato by KO of invertase inhibitor (*SlINVINH1*). KO of homogentisate phytyltransferase (*HPT*) and homogentisate geranylgeranyl transferase (*HGGT*) genes exhibited micronutrient enhancement with high vitamin E (tocopherol) content in *Hordeum vulgare* (Zeng et al. 2020a). Crops biofortified with micronutrients and minerals such as iron, zinc, selenium, and iodine can curb the nutrient insufficiency. In addition, making anti-nutrient (heavy metals, phytate, and gluten)-devoid crops can make the unavailable nutrient available for absorption in the human body and protect humans from developing allergies, metabolic disorders, and chronic ailments. Phytic acid (PA) is the primary storage form of phosphorus in cereal grains and other plant seeds. However, PA reduces the nutrient absorption in the human body due to its complex formation, and thus reducing PA in cereal crops can boost the anti-nutrient property of the feed crops. Phytates are indigestible by humans due to the absence of the corresponding gene. Thus, with accumulation at high concentration, phytates can form complexes with nutrients and reduce the absorption of the minerals, proteins in rice (Khan et al. 2019; Liang et al. 2021), maize (Liang et al. 2014), and rapeseed (Sashidhar et al. 2020). Moreover, numerous endeavors world-wide have already reported the efficiency of the genome editing for the enhancement of the nutritional and functional quality of crop plants (Table 9.1). Based on the current progress of genome editing in plants, we can improve crops that are nutritionally abundant but yet have not been recognized on a commercial level due to numerous biotic/abiotic issues related to, i.e. disease, pest, palatability, flowering, and yield. Interestingly, a gene that is harmless to the ozone layer (*OsHOL1*) plays a major role in the production of methyl iodide; KO of this gene abolished methyl iodide emissions from rice plants (Carlessi et al. 2021), making it a climate-ready crop. The CRISPR-based technology may also efficaciously improve human health by altering numerous essential traits, especially improving the profile of nutritional metabolites, reducing anti-nutrients, and altering macro- and micro-nutrients.

9.4 Novel Game Changing Genome-Editing Approaches

Genome editing poses as a novel technological innovation and significant game-changer in crop improvement, which provides profundity into plant biology and transforms agriculture productivity. This technology offers innovative possibilities for increasing crop productivity via complementary techniques, for instance plant regeneration and genetic transformation, that many a times manifest as bottlenecks. Prompt deployment of genome editing strategies in plant genetics and plant breeding for future prospects is an untapped arena. It empowers plant scientists to interrogate the plant functional genomics by utilizing genetic transformation. Thus, the emergence of the CRISPR/Cas9 system and its improved versions like base editing and prime editing, in combination with two well-established genome editing procedures: zinc finger nucleases ZFNs and transcription activator-like effector nucleases TALENs, or site-specific double-stranded DNA breaks,

Table 9.1 CRISPR/Cas-based approaches for the improvement of majorly consumed crops.

Biofortification				
Organism	**Targeted gene**	**Editing technique**	**Associated trait**	**References**
Rice	Starch branching enzyme I (OsBEI), Starch branching enzyme IIb (OsBEIIb), Granule-bound starch synthase I (GBSSI)	Cas9: Knockout; base editing	High amylose content	Sun et al. (2017) and Xu et al. (2021)
	Phytoene synthase (PSY), Carotene desaturase (CrtI)	Cas9: Carotenoid biosynthesis cassettes inserted in genomic safe harbors (GSHs)	High β-carotene content	Dong et al. (2020)
	Glutamate decarboxylase 3 (OsGAD3)	Cas9: Deletion of C-terminal calmodulin binding domain of OsGAD3	High GABA content	Akama et al. (2020)
	Vacuolar Iron Transporter (OsVIT2)	Cas9: Knockout	High iron in rice grain	Che et al. (2021)
	Arsenite tolerant 1 (ASTOL1)	Cas9: Gene correction	High selenium content	Sun et al. (2021)
	Omega-6 fatty acid desaturase (OsFAD2-1)	Cas9: Knockout	High oleic acid proportion and reduced linolenic acid content	Abe et al. (2018)
	Phospholipase D gene (OsPLDα1)	Cas9: Knockout	Increased LPL content	Khan et al. (2019)
	Orange (OsOr)	Cas9: Knockout	Increased β-carotene synthesis (procitamin A)	Endo et al. (2019)
	Carotene hydroxylase (OsCYP97A4, OsDSM2), carotenoid cleavage dioxygenase 4a (OsCCD4a), carotenoid cleavage dioxygenase 4b (OsCCD4b), and carotenoid cleavage dioxygenase 7 (OsCCD7)	Cas9: Knockout	Increased carotenoid accumulation and quality	Yang et al. (2017)
	Isoamylase1 (OsISA1)	Cas9: Knockout	Affected starch synthesis and endosperm development	Chao et al. (2019)
	Phytoene desaturase (OsPDS)	Cas9: Knockout; Cas12a/Cpf1, FnCpf1 and CRISPR/LbCpf1; prime editors	Carotenoid and chlorophyll biosynthesis	Xu et al. (2020), Wang et al. (2017a), and Lu and Zhu (2017)
Wheat	Phytoene desaturase (TaPDS)	Cas9: Knockout	Carotenoid and chlorophyll biosynthesis	Howells et al. (2018)
Maize	Granule bound starch synthase (GBSSI aka Waxy)	Cas9: Knockout	High amylopectin content	Gao et al. (2020a)
	Phytoene synthase (PSY1)	Cas9: Knockout	Carotenoid biosynthesis	Zhu et al. (2016b)
Banana	Lycopene ε-cyclase (MaLCY-E)	Cas9: Knockout	High β-carotene content	Kaur et al. (2020)
Tomato	Glutamate decarboxylase 2 (SlGAD2), Glutamate decarboxylase 3 (SlGAD3)	Cas9: Introduction of stop codon before autoinhibitory domain	High GABA content	Nonaka et al. (2017)
	Stay-green 1 (SGR1), phytoene desaturase (slyPDS), lycopene β-cyclase (slLCY-B1 & 2), lycopene ε-cyclase (slLCY-E)	Cas9: Gene replacement of PDS and Knockout of SGR1, LCY-B and LCY-E	Increased lycopene content	Li et al. (2018)

(*Continued*)

Table 9.1 (Continued)

Biofortification				
Organism	**Targeted gene**	**Editing technique**	**Associated trait**	**References**
	Anthocyanin mutant 1 (ANT1), phytoene synthase (PSY1)	TALENs; CRISPR/Cas9; promoter insertion and gene ANT overexpression; HR-mediated repair of PSY	Increased anthocyanin levels	Cermák et al. (2015) and Filler et al. (2017)
	Pyruvate-dependent γ-aminobutyric acid transaminase (GABA-TP1, 2, 3), cationic amino acid transporter 9 (SlyCAT9), and succinate semialdehyde dehydrogenase (SSADH)	Cas9: Knockout	g-Aminobutyric acid synthesis and accumulation in fruits	Li et al. (2018d)
	Invertase inhibitor (SlINVINH1)	Cas9: Knockout	High sugar content	Kawaguchi et al. (2021)
	Mitogen-activated protein kinase 20 (MPK20)	Cas9: Knockout	Generation of pollen sterility with controlled sugar metabolism	Li et al. (2018a)
	Glutamate decarboxylase (GAD2, GAD3)	Cas9: Knockout	Increased GABA content	Nonaka et al. (2017)
	ALUMINUM-ACTIVATED MALATE TRANSPORTER 9 (ALMT9)	Cas9: Knockout	Enhanced maltate content	Yeo et al. (2018)
	GDP-l-galactose-phosphorylase 1 (SlGGP1)	Cas9: Knockout (editing in conserved upstream cis-acting open reading frames [uORFs])	Increased vitamin C level	Zhang et al. (2018b) and Deslous et al. (2021)
Rapeseed	Fatty acid desaturase (BnFAD2)	Cas9: Knockout	High oleic acid proportion	Okuzaki et al. (2018)
	Transparent testa 8 (BnTT8)	Cas9: Knockout	High oil production and GPC	Zhai et al. (2020)
Camelina	Fatty acid desaturase (CsFAD2)	Cas9: Knockout	High oleic acid proportion	Jiang et al. (2017)
Peanut	Fatty acid desaturase (AhFAD2, AhFAD2A and AhFAD2B)	TALEN: Knockout	High oleic acid content	Wen et al. (2018) and Yuan et al. (2019)
Potato	Starch-branching enzyme 1 (StSBE1), Starch-branching enzyme 2 (StSBE2), Granule-bound starch synthase (GBSS)	Cas9: Knockout	Increased quality of starch/High amylose content	Anderson et al. (2018), Kusano et al. (2018), and Tuncel et al. (2019)
Sweet potato	Granule-bound starch synthase I (IbGBSSI), Starch branching enzyme II (IbSBEII)	Cas9: Knockout	High amylose content	Wang et al. (2019a)
	Carotenoid biosynthesis related gene	Cas9: Knockout	High β-carotene content	Joseph (2021)
Cassava	PROTEIN TARGETING TO STARCH (PTST1), GRANULE-BOUND STARCH SYNTHASE (GBSS)	Cas9: Knockout	Altered starch content	Bull et al. (2018)
Soybean	Fatty acid desaturase (FAD2-1A, FAD2-1B, FAD3A)	TALEN: Knockout	Altered oil level	Al Amin et al. (2019) and Demorest et al. (2016)
Strawberry	Basic (region) leucine zipper proteins (FvebZIPs1.1)	CRISPR: Base editing (editing in conserved upstream open reading frames [uORFs])	Altered sugar content	Xing et al. (2020)

Barley	Homogentisate phytyltransferase (HvHPT), Homogentisate geranylgeranyl transferase (HvHGGT)	Cas9: Knockout	High vitamin E (α-tocopherol)	Zeng et al. (2020)
Sorghum	Phytoene desaturase (PDS)	Cas9: Knockout	Carotenoid and chlorophyll biosynthesis	Liu et al. (2019)
Kale (*Brassica oleracea*)	Carotenoid isomerase (BoaCRTISO)	Cas9: Knockout	Carotenoid biosynthesis	Sun et al. (2020)
Carrot	Flavanone-3-hydroxylase (DcF3H)	Cas9: Knockout	Development of anthocyanin	Chodacka et al. (2018)
Grape	Phytoene desaturase (VvPDS)	Cas9: Knockout	Carotenoid biosynthesis	Nakajima et al. (2017)
False flax (*Camelina sativa*)	CRUCIFERIN C (CsCRUC)	Cas9: Knockout	Enhancing seed proteome and fatty acid profile	Lyzenga et al. (2019)

Anti-nutrients

Organism	Targeted gene	Editing technique	Associated trait	References
Rapseed	Inositol tetrakisphosphate kinase (BnITPK)	Cas9: Knockout	Low phytic acid content	Sashidhar et al. (2020)
	Glucosinolate transporter (BnGTR1/2)	Cas9: Knockout	Low glucosinolates content	Li et al. (2021) and Tian et al. (2022)
Wheat	α-Gliadin genes, Gli-2 locus	Cas9: Knockout	Low gluten content	Sanchez-Leon et al. (2018)
	Inositol 1,3,4,5,6-pentakisphosphate 2-kinase 1(TaIPK1)	Cas9: Knockout	Low phytic acid content and improved accumulation of Fe and Zn	Ibrahim et al. (2021)
	Lipoxygenase (TaLOX2)	Microspore-mediated Cas9: Knockout	Reduced unpleasant flavor	Bhowmik et al. (2018) and Zhang et al. (2016)
Maize	Inositol tetrakisphosphate kinase (ZmIPK), Inositol tetrakisphosphate kinase 1 (ZmIPK1), Multidrug resistance-associated protein 4 (ZmMRP4)	ZFN, TALEN, Cas9	Reduced phytic acid	Liang et al. (2014)
Rice	Natural resistance associated macrophage protein 5 (OsNramp5)	Cas9: Knockout	Low Cd accumulations	Tang et al. (2017b)
	Low affinity cation transporter (OsLCT1)	Cas9: Knockout	Low Cd accumulations	
	Phospholipase D gene (OsPLDα1)	Cas9: Knockout	Low phytic acid content	Khan et al. (2019)
	Inositol 1,3,4-trisphosphate 5/6-kinase (OsITPK6)	Cas9: Knockout	Low phytic acid content	Jiang et al. (2019)
	Arsenite tolerant 1 (ASTOL1)	Cas9: Gene correction	Low arsenic content	Sun et al. (2021)
Grape	L-Idonate dehydrogenase gene (ldnDH)	Cas9: Knockout	Low tartaric acid	Ren et al. (2016)

(*Continued*)

Table 9.1 (Continued)

Anti-nutrients

Organism	Targeted gene	Editing technique	Associated trait	References
Sorghum	Kafirins gene family (K1C)	Cas9: Knockout (single gRNA gene family editing)	Reduced Kafirins	Li et al. (2018c)
Soybean	Lipoxygenases (GmLox1, GmLox2, and GmLox3)	Cas9: Knockout	Reduced cooked beans aroma, flavor and astringency with no adverse effect on other parameters like nutritive value, milky flavor, thickness, chalkiness or aftertaste	Wang et al., (2019b)
Mushroom	Polyphenol oxidase (PPO)	Cas9: Knockout	Anti-browning phenotype	Waltz (2016)
Onion	Fumonisin biosynthetic gene cluster (FUM1)	Cas9: Gene replacement of FUM1 to Hygromycin	Reduced fumonisin biosynthesis	Ferrara et al. (2019)
Lettuce	Brassinosteroid Insensitive 2 (LsBIN2)	Cas9: Knockout	Brassinosteroid synthesis	Woo et al. (2015)
Potato	Vacuolar invertase gene (VInv)	TALEN; Cas9: Knockout	Reduced sugar and acrylamide content	Clasen et al. (2016); ISAAA-CropBiotech-Update (2022)
	Sterol side chain reductase 2 (SSR2), 2-oxoglutarate-dependent dioxygenase (St16DOX)	TALEN; Cas9: Knockout	Prevent SGAs such as α-solanin and α-Chaconin accumulation	Friedman (2006), Nakayasu et al. (2018), and Yasumoto et al. (2019)
Ethiopian mustard (*Brassica carinata*)	Fatty acid elongase 1 (FAE1), Fatty acid desaturase (FAD2), and glucosinolate transporters (GTR1, GTR2)	Cas9: Knockout	Improve mustard oil quality with reduction of erucic acid content	Karembu (2021)

Abiotic stress

Organism	Targeted gene	Editing technique	Associated trait	References
Rice	Epidermal Patterning Factor like-9 (EPFL9)	Cas9 & Cpf1: Knockout	Drought tolerance	Yin et al. (2017)
	OsmiR535 (belongs to the miR156/miR529/miR535 superfamily)	Cas9: Knockout	Drought/salinity tolerance	Yue et al. (2020)
	Osmotic stress/ABA–activated protein kinase 2 (OsSAPK2)	Cas9: Knockout	Drought/salinity tolerance	Lou et al. (2017)
	Pyrabactin resistance like proteins (OsPYL9)	Cas9: Knockout	Drought tolerance	
	Enhanced response to ABA1 (OsERA1)	Cas9: Knockout	Drought tolerance	Ogata et al. (2020)
	Semi-rolled leaf 1,2 (OsSRL1, OsSRL2)	Cas9: Knockout	Drought tolerance	Liao et al. (2019)
	Drought and salt tolerance (DST)	Cas9: Knockout	Drought/salinity tolerance	Kumar et al. (2020)

	Heading date (HD2, HD4, HD5)	Cas9: Knockout	Early flowering	Li et al. (2017b)
	B-type response regulator (OsRR22)	Cas9: Knockout	Salinity tolerance	Zhang et al. (2019a)
	Annexin gene (OsAnn3)	Cas9: Knockout	Cold tolerance	Shen et al. (2017)
	R2R3-type MYB transcription factor (OsMYB30)	Cas9: Knockout	Cold tolerance	Zeng et al. (2020b)
	Cytochrome P450, Betain Aldehyde Hydrogenase 2 (OsBADH2)	Cas9: Knockout	Multi-stress tolerance	Usman et al. (2020b)
	AP2/ERF domain-containing RAV (related to ABI3/VP1) TF family (OsRAV2)	Cas9: Knockout	Salinity tolerance	Duan et al. (2016)
	Thermosensitive genic male sterile (TMS5)	Cas9: Knockout	Heat sensitive and yield improvement	Zhou et al. (2016) and Li et al. (2019b)
	NAC transcription factor coding gene (OsNAC041, OsNAC14)	Cas9: Knockout; Cas9-mutagenesis, overexpression of PGD1::OsNAC14 by transforming through Agrobacterium	Salinity tolerance, Drought tolerance	Bo et al. (2019) and Shim et al. (2018)
	Overly tolerant to salt 1 (OsOTS1)	Cas9: Knockout	Salinity tolerance	Morrell and Sadanandom (2019)
	Transcription factor consisting of conserved TIF[F/Y]XG domain (TIFYb and TIFYa)	Cas9: Knockout	Cold tolerance	Huang et al. (2017)
	Mitogen activated protein kinase (OsMPK5, OsMPK2)	Cas9: Knockout	Multi-stress tolerance	Xie and Yang (2013) and Shan et al. (2013)
Wheat	Dehydration responsive element binding protein 2,3 (TaDREB2,3) and ethylene responsive factor 3 (TaERF3)	Cas9: Knockout	Drought tolerance	Kim et al. (2018)
Maize	AGROS8	CRISPR-based promoter swap between native AGROS promoter to GOS2 promoter; Overexpression	Drought tolerance	Shi et al. (2017)
	MADS-box transcription factor, MYB (myeloblastosis) related protein (MYBR) and APETALA 2 (AP2)	Cas9: Knockout	Drought tolerance	Qi et al. (2016)
	High-affinity potassium transporter 1 (ZmHKT1)	Cas9: Knockout	Salinity tolerance	Zhang et al. (2018a)
Banana	Ma04g15900, Ma06g27710, Ma08g32850, Ma11g10500, Ma11g17210	Cas9: Knockout	Semi-dwarfed	Shao et al. (2020)
Tomato	Mitogen-activated protein kinase 3 (SlMAPK3)	Cas9: Knockout	Reduced drought tolerance	Wang et al. (2017a, b)
	Nonexpressor of pathogenesis-related gene 1 (SlNPR1)	Cas9: Knockout	Reduced drought tolerance	Li et al. (2019)
	C-repeat/dehydration responsive element binding factor 1 (SlCBF1)	Cas9: Knockout	Reduced chilling tolerance	Li et al. (2018b)
	High-affinity K+ 20 (SlHAK20)	Cas9: Knockout	Hypersensitivity to salt stress	Wang et al. (2020b)
	Salt Overly Sensitive 1 (SlSOS1)	Cas9: Knockout	Increased salt sensitivity	Wang et al. (2020a)
	Brassinazole-resistant 1 (SlBZR1)	Cas9: Knockout	Reduced heat tolerance	Yin et al. (2018)
	Hybrid proline-rich protein 1 (SlHyPRP1)	Cas9: Knockout	Salinity tolerance	Tran et al. (2020)

(Continued)

Table 9.1 (Continued)

Abiotic stress				
Organism	**Targeted gene**	**Editing technique**	**Associated trait**	**References**
Soybean	Double-stranded RNA-binding2 (Drb2a, Drb2b)	Cas9: Knockout	Drought and Salinity tolerance	Curtin et al. (2018)
Chilli	Zinc finger family protein (NAC72)	Base Editing	Drought tolerance	Joshi (2019)
Rapseed	REPRESSOR OF GA1-3 (BnaRGA)	Cas9: Knockout	Drought tolerance	Yang et al. (2017)
	BnaA9.WRKY47	Cas9: Knockout	Low Boron stress tolerance	Feng et al. (2018)

Biotic stress				
Organism	**Targeted gene**	**Editing technique**	**Associated trait**	**References**
Rice	Sugars will eventually be exported transporters (OsSWEET 14 promoter; OsSWEET 11 promoter; OsSWEET 13 promoter; OsSWEET 13)	TALEN, Cas9: Knockout	Bacterial blight (*Xanthomonas oryzae* pv. *Oryzae*) resistance	Jiang et al. (2013), Oliva et al. (2019), Xu et al. (2019), Zafar et al. (2020), Zeng et al. (2020b), and Zhou et al. (2015)
	Ethylene response factor (OsERF922)	Cas9: Knockout	Rice leaf blast resistance	Wang et al. (2016)
	Broad-spectrum resistance Kitaake-1 (Bsr-k1)	Cas9: Knockout	Broad spectrum	Kumar et al. (2018) and Zhou et al. (2018)
	Xanthomonas oryzae pv. *oryzae* resistance 13 (OsXa13)/NODULIN3 (N3) gene family located on chromosome 8 (Os8N3)/Sugars will eventually be exported transporters (SWEET11)	Cas9: Knockout	Bacterial leaf blight (Xoo) resistance	Kim et al. (2019), Li et al. (2019, 2020)
	Albino (OsALB1), Rosy (OsRSY1)	Cas9 Ribonucleo protein (RNP)	Rice leaf blast resistance	Foster et al. (2018)
	Pyricularia oryzae resistance-21 (OsPi21)	Cas9: Knockout	Rice leaf blast resistance	Li et al. (2019) and Nawaz et al. (2020)
	Mitogen activated kinase 5 (OsMPK5)	Cas9: Knockout	Resistance to Broad spectrum diseases	Xie and Yang (2013)
	Eukaryotic translation initiation factor 4 Gamma (eIF4G)	Cas9: Knockout	Rice tungro spherical virus (RTSV) resistance	Macovei et al. (2018)
	Distinct region of double strand RNA genome of Southern rice black-streaked dwarf virus (SRBSDV) and single-stranded RNA genome of Rice stripe mosaic virus (RSMV)	LshCas13a	Resistance to Southern rice black streaked dwarf virus (SRBDSV) and Rice stripe mosaic virus (RSMV)	Zhang et al. (2019a)
	Replication initiator protein (Rep/RepA)	Cas9	Resistance to Wheat Dwarf Virus (WDV)	Wang et al. (2017a)

	5-Enolpyruvylshikimate-3-phosphate synthase (EPSPS)	Cas9: Knock-in; Cas12a (Cpf1: FnCpf1 and LbCpf1)	Herbicide tolerance; weed management	Wang et al. (2017b) and Li et al. (2016a)
	Acetolactate synthase (ALS)	Cas9/Cpf1: Knock-in; Base editing; Prime editing	Herbicide tolerance; weed management	Kuang et al. (2020), Shimatani et al. (2018), Wang et al. (2021), Xu et al. (2020), Zhang et al. (2020)
	Acetyl-CoA carboxylase (ACCase)/acetyl-coenzyme A carboxylase 1 (ACC1)	Base editing; prime editing (prime-editing-library-mediated saturation mutagenesis [PLSM])	Herbicide tolerance; weed management	Liu et al. (2020a), and Xu et al. (2021)
	α-Tubulin homologue gene (TubA2)	Base editing (rice adenine base editor; rBE14)	Herbicide tolerance; weed management	Huang et al. (2022) and Liu et al. (2020b)
	Splicing factor 3b (SF3B1)	CRISPR/Cas-mediated directed evolution (CDE)	Herbicide tolerance; weed management	Butt et al. (2020)
	Bentazon-sensitive-lethal (OsBEL)	Cpf1: Knockout	Herbicide tolerance; weed management	Wang et al. (2017a)
Wheat	Mildew Locus O (TaMLO A1, B1, D1)	TALEN, Cas9:Knockout	Powdery mildew (PM) resistance	Wang et al. (2014) and Li et al. (2022)
	Enhanced disease resistance1 (TaEDR)	Cas9: Knockout	Powdery mildew resistance	Zhang et al. (2017)
	Acetolactate synthase (ALS)	Base editing	Herbicide tolerance; weed management	Zhang et al. (2019)
	Acetyl-CoA carboxylase (ACCase)	Base editing	Herbicide tolerance; weed management	Zhang et al. (2019)
	5-Enolpyruvylshikimate-3-phosphate synthase (EPSPS)	Cas9: Knock-in	Herbicide tolerance; weed management	Arndell et al. (2019)
Tomato	Downy mildew resistance 6-1 (SlDMR6-1)	Cas9: Knockout	Broad spectrum/Downy mildew resistance 6-1	de Toledo Thomazella et al. (2016)
	Mildew locus O (SlMLO1)	Cas9: Knockout	Resistance against *Oidium neolycopersici* (tomato powdery mildew)	Nekrasov et al. (2017)
	Powdery mildew resistance 4 (SlPMR4)		Resistance to powdery mildew	Santillan Martinez et al. (2020)
	Coat (CP) and replicase protein (REP) of Tomato yellow leaf curl virus (TYCV)	Cas9: Knockout	Tomato yellow leaf curl virus	Tashkandi et al. (2018)
	Jasmonate ZIM-domain 2 (SlJAZ2)	Cas9: Knockout	Resistance against bacterial speck disease	Ortigosa et al. (2019)
	Dicer like protein 2b (DCL2b)	Cas9: Knockout	Tomato mosaic virus	Wang et al. (2018c)
	SOLYC0g075770	Cas9: Knockout	Fusarium wilt disease susceptible	Prihantna et al. (2018)

(Continued)

Table 9.1 (Continued)

Biotic stress				
Organism	**Targeted gene**	**Editing technique**	**Associated trait**	**References**
	Mitogen-activated protein kinase 3 (SlMAPK3)	Cas9: Knockout	Gray mold disease susceptible	Zhang et al. (2018b)
	Carotenoid cleavage dioxygenase 8 (CCd8)	Cas9: Knockout	*Phelipanche aegyptiaca* Resistance	Bari et al. (2019)
	Acetolactate synthase (SlALS)	Cas9: Knockout; Base editing	Herbicide tolerance; weed management	Veillet et al. (2019) and Danilo et al. (2019)
	Eukaryotic translation initiation factor 4E 2 (elF4E2)	Cas9: Knockout	Pepper veinal motle virus (PYMV) resistance	Kuroiwa et al. (2022)
Soybean	Acetolactate synthase (ALS)	Cas9: Knockout	Herbicide tolerance; weed management	Li et al. (2015)
Maize	Acetolactate synthase (ALS)	Base editing; Prime editing	Herbicide tolerance; weed management	Li et al. (2020) and Jiang et al. (2020)
Rapseed	Calreticulin (BnCRT1a)	Cas9: Knockout	Resistance to *Verticillium longisporum* (Vl43)	Pröbsting et al. (2020)
	BnWRKY70	Cas9: Knockout	Resistance to *Sclerotinia sclerotiorum*	Sun et al. (2017)
	5-Enolpyruvylshikimate-3-phosphate synthase (EPSPS)	Cas9: Knock-in (with CRISPR associated RNA endoribonuclease Csy4 and a ribozyme/tRNA system)	Herbicide tolerance; weed management	Wang et al. (2021)
	Acetolactate synthase (ALS)	Base editing (CBE)	Herbicide tolerance; weed management	Wu et al. (2020)
Flaxseed	5-Enolpyruvylshikimate-3-phosphate synthase (EPSPS) and BFP	Cas9: Knock-in (single-stranded oligonucleotides; ssODNs)	Herbicide tolerance; weed management	Sauer et al. (2016)
Barley	Microtchidia (HvMORC1 and HvMORC6a) proteins	Cas9: Knockout	Broad spectrum resistance	Kumar et al. (2018) and Galli et al. (2022)
	Wheat dwarf virus (WDV): Movement protein (MP), Coat protein (CP), replication-associated protein (Rep/RepA), and long intergenic regions (LIR)	Cas9: Knockout	Resistance against dwarf virus and Geminiviridae	Kis et al. (2019)
	Hygromycin phosphotransferase (Hpt) and Phosphinotricin *N*-acetyltransferase (Pat)	Meganuclease I-SceI (HR)	Herbicide tolerance	Watanabe et al. (2016)
Cotton	Gh14-3-3d	Cas9: Knockout	Resistance to Cotton wilt (*Verticillium dahliae*)	Zhang et al. (2018b)
	Cotton leaf curl virus (CLCuV): Intergenic Regions (IR) & Replication-associated Protein (Rep) regions	Cas9: Knockout	Resistance to cotton leaf curl disease (CLCuD)	Iqbal et al. (2016) and Yin et al. (2019)

Potato	Coilin C-terminal	Cas9: Knockout	Resistance to Potato virus Y (PVY, Potyviridae)	Makhotenko et al. (2019)
	The third protein (P3), cylindrical inclusion protein (CI), nuclear inclusion "b" protein (NIb), and coat/capsid protein (CP)	LshCas13a: Knockout	Resistance to potato virus Y	Zhan et al. (2019b)
	Mildew resistance locus (MLO), Hydroxy-2-methyl-2-(*E*)-butenyl 4-diphosphate synthase (HDS), triphosphate tunnel metalloenzymes (TTM2)	Cas9:Knockout	Resistance to *Phytophthora palmivora*, *Phytophthora syringae*, *Hyaloperonospora* sp.	Kieu et al. (2021)
	Defense, no death 1 (StDND1), Cryptochrome-interacting basic-helix-loop-helix 1 (StCIB1)/homolog of BR enhanced expression 2 (BEE2) interacting with IBH 1 (HBI1)-like1 (StCHL1), Downy mildew resistance 6 (StDMR6-1/2)	Cas9: Knockout	Late blight resistance	Thomazella et al. (2016) and Kieu et al. (2021)
	Acetolactate synthase1 (ALS1)	Cas9: Knock-in; Base editing (CBE)	Herbicide tolerance; weed management	Butler et al. (2015, 2016) and Veillet et al. (2019)
Cassava	Nuclear cap-binding protein 1/2 (nCBP 1/2; elF4E family or isoforms)	Cas9: Knockout	Resistance to cassava brown streak disease (CBSD)	Gomez et al. (2019)
	Open reading frame of Begomovirus (AC2 and AC3)	Cas9: Knockout	African cassava mosaic virus (ACMV) resistance	Mehta et al. (2019)
	5-Enolpyruvylshikimate-3-phosphate synthase (EPSPS)	Cas9: Knock-in	Herbicide tolerance; weed management	Hummel et al. (2017)
Cacao	Non-expressor of pathogenesis-related 3 (TcNPR3)	Cas9: Knockout	Resistance to *Phytopthora*	Fister et al. (2018)
Cucumber	Eukaryotic translation initiation factor 4 e (eIF4e)	Cas9: Knockout	Broad-spectrum viral (potyviruses such as Cucumber vein yellowing virus (CVYV), Zucchini yellow mosaic virus (ZYMV) and Papaya ringspot virus (PRSV); no resistance against CMV or CGMMV) resistance	Chandrasekaran et al. (2016)
Lettuce	GDP-L-galactose phosphorylase (LsGGP2)	Cas9: Knockout (uORF)	Herbicide tolerance	Zhang et al. (2018c)
Banana	Endogenous banana streak virus (eBSV)	Cas9: Knockout	Resistance to banana streak virus (BSV)	Tripathi et al. (2019)
	Downy mildew resistance 6 (DMR6)	Cas9: Knockout	Resistance to Banana Xanthomonas Wilt (BXW)	Tripathi et al. (2020)
Grape	Mildew Locus O 3 & 7(VvMLO3, VvMLO7) and Downy mildew resistance 6 (DMR6)	Cas9: Knockout	Resistance against grape wine powdery mildew (*Erysiphe necator*) and downy mildew	Malnoy et al. (2016), Giacomelli et al. (2019), and Wan et al. (2020)

(*Continued*)

Table 9.1 (Continued)

Organism	Targeted gene	Editing technique	Associated trait	References
Biotic stress				
	Pathogenesis-related 4 (VvPR4b)	Cas9: Knockout	Editing decreases downy mildew (*Plasmopara viticola*) resistance	Li et al. (2020)
	VvWRKY52 & succinate dehydrogenase B (SDHB)	Cas9: Knockout (tpTEL vector based and RNP)	Resistance to graymold (*Botrytis cinerea*)	Wang et al. (2018b) and Leisen et al. (2020)
	Trans-acting small-interfering locus4 (TAS4b) and Myeloblastosis (MYBA)	Cas9: Knockout	No clear resistance gained against Pierce's disease (PD) caused by bacterium *Xylella fastidiosa* spread by xylem sap-sucking insects and to Red Blotch (RB) caused by Grapevine Red Blotch Virus (GRBV) transmitted by alfalfa leafhopper *Spissistilus festinus*.	Sunitha and Rock (2020)
	LATERAL ORGAN BOUNDARIES 1 (CsLOB1 promoter; CsLOB)	Cas9: Knockout	Resistance to citrus canker	Jia et al. (2016, 2017)
Watermelon	Acetolactate synthase (ALS)	Cas9: Base editing (cytidine deaminase, i.e. BE3)	Herbicide tolerance	Tian et al. (2018)
	Phytosulfokine (ClPSK1)	Cas9: Knockout	Resistance to *Fusarium oxysporum* f. sp. *Niveum* (FON)	Wu et al. (2020)
Citrus (*Citrus paradisi*, Sweet orange: *Citrus sinensis osbeck*)	LATERAL ORGAN BOUNDARIES 1 (CsLOB1), CsWRKY22	Cas9: Knockout; LbCas12a-D156R (ttLbCas12a): Knockout (Temperature insensitive Cpf1 variant)	Resistance to citrus canker *Xanthomonas citri* and *Xanthomonas citri* subsp. *Citric* (Xcc)	Peng et al. (2017), Jia et al. (2019, 2022), and Wang et al. (2019a)
	Downy mildew resistance 6 (DMR6)	Cas9:Knockout	Huanglongbing (yellow dragon disease) resistance	Zhang et al. (2018a)
Apple	1, 2, 4, DspE-interacting proteins of malus 1, 2, 4 (DIPM-1, 2, 4)	Cas9: Knockout (Ribonucleoproteins; RNPs)	Resistance to fire blight disease	Malnoy et al. (2016)
Papaya	Putative extracellular cystatin-like cysteine protease inhibitors 8 (alEPIC8)	Cas9: Knockout	Resistance to *Phytophra palmivora* fungus	Gumtow et al. (2017)

Yield

Organism	Targeted gene	Editing technique	Associated trait	References
Rice	Dense And Erect Panicle (DEP1), grain number 1a (Gn1A), ideal plant architecture1 (IPA1), grain size 3 (GS3)	Cas9 and Cpf1: Knockout	Land sparing, yield	Li et al. (2016b), Huang et al. (2018), Zhong et al. (2018), and Shen et al. (2018)

	pyrabactin resistance 1-like (PYL 1,4,6 and 9)	Cas9: Knockout	Land sparing, yield	Miao et al. (2018) and Usman et al. (2020b)
	Grain size 3 (OsGs3), grain width and weight 2 (OsGW2), grain number 1A (OsGn1A)	Cas9: Knockout	Land sparing, yield	Zhou et al. (2019)
	Semi-dwarf 1 (SD1)	Cas9: Knockout	Land sparing, yield	Hu et al. (2019)
	Ent-kaurene oxidase 2 (OsKO2)	Prime editing	Land sparing, yield	Tang et al. (2020)
	Grain width and weight 7 (OsGW7)	dCAS9: gene activation	Land sparing, yield	Xiong et al. (2021)
	Squamosa promoter binding protein-LIKE (SPL14, SLP17, SLP16, SLP18), rice DELLA protein (SLR1)	Base editing	Land sparing, yield	Hua et al. (2019b, 2020)
	fruit-weight 2.2-like (OsFWL4)	Cas9: Knockout	Land sparing, yield	Gao et al. (2020b)
	Auxin efflux carrier PIN FORMED-like gene (OsPIN5b)	Cas9: Knockout	Land sparing, yield	Zeng et al. (2020b)
	Grain width/weight (GW2, GW5, GW6) and Grain width 8 (qGW8)	Cas9: Knockout	Land sparing, yield	Xu et al. (2016) and Usman et al. (2021)
	LAZY1 (Lzy) or NEGATIVE GRAVITROPIC RESPONSE OF ROOTS (NGR), and DEEPER ROOTING (DRO)	Cas9: Knockout	Land sparing, yield	Miao et al. (2013)
	ABERRANT PANICLE ORGANIZATION 1 (APO1)	Prime editing: C42F modification.	Land sparing, yield	Hua et al. (2020b)
Wheat	Grain width/weight 2 (GW2), 9-LIPOXYGENASE (LPX-1)	Cas9: Knockout	Land sparing, yield	Wang et al. (2018d)
	Cytokinin oxidase/dehydrogenase (TaCKX2-1, TaCKX2-A1, TaCKX2-B1, TaCKX2-D1), grain length and weight on chromosome 7 (TaGLW7), grain width/weight 2 & 8 (TaGW2, TaGW8)	Cas9: Knockout	Increase the grain number	Zhang et al. (2019a)
	Snakin/GASA family member (TaGASR7A1, -B1, -D1 and TdGASR7), DENSE AND ERECT PANICLE1 (TaDEP1), **NAM;** no apical meristem **ATAF**; Arabidopsis transcription activation factor, **CUC**; cup-shaped cotyledon (TaNAC transcription factor family), Auxin efflux carrier PIN FORMED-like gene (TaPIN1) and Grain Width and Weight 2 (TaGW2)	Transiently expressing CRISPR/Cas9 DNA (TECCNA) or Transient expression of the in vitro transcripts (IVTs) of Cas9-coding sequence and guide RNA (TECCRNA)	Effect on grain length, number, size and weight	Zhang et al. (2016b)
Soybean	Glutamine Synthase (GS1) and chalcone-flavanone isomerase (CHI20)	Cas9: Knockout	Boost plant and soil health for higher yield	Michno et al. (2015)
	Photoperiod-responsive FT homologue (GmFT2a)	Cas9: Knockout	Yield benefits	Cai et al. (2018)
	SQUAMOSA PROMOTER BINDING PROTEIN-LIKE (GmSPL9a,b,c)	Cas9: Knockout	Yield Improvement	Bao et al. (2019)

(Continued)

Table 9.1 (Continued)

Yield				
Organism	**Targeted gene**	**Editing technique**	**Associated trait**	**References**
Maize	Defective kernel (Dek42)	Cas9: Knockout	Land sparing, yield	Zuo et al. (2019)
Tomato	Clavata 3 (SlCLV3), SELF-PRUNING 5g (SlSP5g),WUSCHE L (WUS)	Cas9: Knockout	Land sparing, yield	Li et al. (2018b), Rodríguez-Leal et al. (2017), Soyk (2017), and Ma et al. (2015)
	SELF-PRUNING (SP), MULTIFLORA, OVATE (O), FASCIATED, FRUIT WEIGHT 2.2 (FW2.2) and LYCOPENE BETA CYCLASE (CycB)	Cas9: Knockout	Yield; lycopene accumulation	Zsögön et al. (2018)
	Pectate Lyase (PL)	Cas9: Knockout	Long self-life	Uluisik et al. (2016)
	Alcobaca (ALC)	Cas9: Knock-in	Long self-life	Yu et al. (2017)
	BLADE-ON-PETIOLE (BOP1, BOP2, BOP3)	Cas9: Knockout	Land sparing, yield	Xu et al. (2016)
Ground Cherry	Argonaute (PprAGO7), SELF-PRUNING (PprSP), SELF-PRUNING 5G (PprSP5g), CLAVATA 1 (CLV1)	Cas9: Knockout	Highly nutritive crop modified for improved agronomic properties	Lemmon et al. (2018)
Kiwi fruit	CENTRORADIALIS like (CEN, CEN4)	Cas9: Knockout	Generation of a compact plant with rapid terminal flower and fruit development	Varkonyi-Gasic et al. (2018)
Cucumber	WPP (trp/pro/pro) domain Interacting Proteins (WIP1)	Cas9:Knockout	Generation of gynoecious plant; yield	Hu et al. (2019)
Rapseed	ENHANCER3 OF DA1 (BnaEOD3)	Cas9: Knockout	Yield	Khan et al. (2021)
	CLAVATA (CLV1, CLV2, CLV3)	Cas9: Knockout	Yield	Yang et al. (2017)
	More Axillary Growth 1 (MAX1)	Cas9: Knockout	Yield	Zheng et al. (2020)
	BREVIPEDICELLUS (BnBP)	Cas9: Knockout	Yield	Fan et al. (2021)
	ALCATRAZ (BnALC 1)	Cas9: Knockout	Shatter resistance	Braatz et al. (2017)
	INDEHISCENT (BnIND)	Cas9: Knockout	Shatter resistance	Zhai et al. (2020)
	JAGGED (BnJAG)	Cas9: Knockout	Shatter resistance	Zaman et al. (2019)
	SHATTERPROOF1 (BnSHP1) and SHATTERPROOF2 (BnSHP2)	Cas9: Knockout	Shatter resistance	Zaman et al. (2021)

render tremendous potential applications for crop improvement with desired traits Figure 9.1. These crop improvement techniques empowered fundamental alterations in crop native genes and their functionalities, as well as made genome editing prominently popular for widespread acceptance and significant discoveries. GE has immense prospects as a promising tool for plant functional genomics research and crop enhancement.

9.4.1 Knockout-Based Approach

Knockout approaches include modifying the DNA in such a way that the expression of the gene is completely stalled. This approach is applicable to all types of cells and organisms involving appropriate genetic techniques; CRISPR-based genome editing is currently the simplest and most direct method of accomplishing a precise gene knockout. Homologous recombination is the most effective way to create a gene knockout that allows a single gene to be eliminated without affecting all other genes in an organism. CRISPR/Cas9 technology has been employed in the past few years to generate knockout mutants of *OsSWEET13,* by targeting the promoter which resulted in progressed resistance against bacterial blight bug in Indica rice variety, IR24 (Li et al. 2020b). Similarly, in rice *OsAnn3* CRISPR knockouts, the role of the annexin gene *OsAnn3* during cold stress was demonstrated (Shen et al. 2017). Furthermore, CRISPR *TaMLO* knockdown was found to resist *Blumeria graminis* f. sp. Tritici powdery mildew disease Btg (Wang et al. 2014). In *Zea mays*, Liang et al. (2014) described the knockout of genes that were involved in the synthesis of PA (*ZmMRP4 ZmIPK1A,* and *ZmIPK*). Similarly, Feng et al. (2018) used a protoplast system in maize to demonstrate the efficacy of the CRISPR/Cas9 technique by targeting the albino marker gene *Zmzb7*. Consequently, Kapusi et al. (2017) reported knockout for endo-*N*-acetyl-β-D-glucosaminidase ENGase gene in barley, in addition to previous crops with CRISPR knockout. ENGase was knocked out by employing *Agrobacterium*-mediated and particle bombardment transformation with a set of five gRNAs. The mutational efficiency in barley lines of T0 and T1 was found to be 78%. In functional genetics, these knockout plants are beneficial for understanding the function of genes. Furthermore, CRISPR/Cas9-mediated gene knockout and several studies revealed the significance of Rj4, which is a predominant gene in soybeans, that prevents node formation by numerous strains of *Bradyrhizobium elkanii* (Tang et al. 2016). The gene SHORT-ROOT SHR was knocked out in the hairy roots of tomato by using CRISPR, indicating that gene function is conserved across tomato and *Arabidopsis*.

9.4.2 Knock-in-Based Approach

A gene knock-in is a genetic engineering technique that involves inserting a protein that codes for a cDNA sequence at a specific locus in an organism's chromosome. For plant research and breeding, proficient CRISPR-assisted gene knock-in are required. Several parameters can be studied to improve CRISPR/Cas-induced genome editing. The effective knock-in of genes through end-joining or homology-driven repair remains a major challenge in higher plants. Using knock-in and targeting a sequence between an endogenous target promoter and reporter gene, transgenic plants with an identical and preferred knock-in allele can be generated on a routine basis (Jefferson et al. 1987). In rice, using a robust gene-targeting method for both positive and negative selection was able to achieve transgenic knock-in plants with the fertile and promoter less GUS reporter gene that encodes β-glucuronidase coupled with the endogenous promoter of MET1a, and two rice MET1 genes encoding DNA methyltransferase (Yamauchi et al. 2009). In addition, these experiments resulted in primary transgenic T0 knock-in plants, showing a single copy of GUS, However the heterozygous lines linked to the target non-edited genes displayed phenotypic complexity of the heterozygous lines, and linked with gene targeting but with no events.

9.4.3 Activation or Repression-Based Approach

This category of CRISPR tools is characteristically based on nuclease-dead Cas9 (dCas9), which is guided by sgRNA. Even though dCas9 cannot create a double strand break in DNA, it will bind to the target DNA sequence and can be employed as a delivery vehicle for a number of gene transcription control strategies (Adli 2018). In plants, employing dCas9 to over-express and/or repress gene expression has become a widespread and robust approach for genetic and epigenetic modulation. CRISPR/dCas9 technology offers exceptional multiplexing capabilities by employment of several sgRNAs at the same time, in addition to single-gene regulation. dCas9-SunTag, dCas9-TV, and dCasEV2.1 are three techniques for multiplexed gene activation in plants (Papikian et al. 2019; Selma et al. 2019; Xiong et al. 2021). CRISPR-based activation (CRISPRa) and CRISPR-based interference (CRISPRi) can be achieved by combining transcriptional activators or repressors to the gRNA scaffold and/or dCas9, respectively (Li et al. 2017a; Selma et al. 2019). A protoplast-based promoter-luciferase assay was performed in *Arabidopsis* by utilizing a constitutive co-expression of dCas9-TV through pre-screened gRNAs targeting the promoters of more than six *Arabidopsis* genes, which resulted in 1.6- to 92-fold overexpression levels, whereas in comparison to the wild type, roughly 65% of CRISPRa-interfered *Arabidopsis* displayed a modest change in blooming time (Pan et al. 2021). Interestingly, a dCas9-based CRISPRa method, i.e. CRISPReAct3.0, was recently created by engaging numerous transcription activator-like activation or effector domains on the gRNA scaffold via an MS2-MCP interaction. In rice protoplasts, this approach was reported to accomplish increased transactivation levels, i.e. five- to sevenfold than dCas9-TV, operating mainly on two genes, *OsER1* and *OsGW7*, targeted by an identical gRNA (Knott and Doudna 2018). Additionally, CRISPReAct3.0 multiplexed gene activation targeting seven genes at the same time led to a spike in the activation (from less than 10 reaching to 20-fold) of all the targeted genes. However, the efficiency of gene activation in rice protoplasts differed significantly from that observed in transgenic seedlings. Furthermore, CRISPReAct3.0 was used to generate stable lines in *Arabidopsis* targeting two genes, TRICHOMELESS-1 and FLOWERING LOCUS-T, which resulted in stimulated expressions from 120- to 240-fold and three- to eightfold, correspondingly, in *Arabidopsis*. The distinctive phenotypes of transgenic plants were obtained up to the T_2 and T_3 generations (Lowder et al. 2015; Piatek et al. 2015; Li et al. 2017b; Ming et al. 2020). CRISPRi is a novel and prominent technique in gene suppression that targets specific genes (Lowder et al. 2015; Piatek et al. 2015). Very few occurrences of CRISPRi use in plants have been documented (Tang et al. 2017a; Ming et al. 2020). The transcriptional repressors including dCas9-SRDX and dCas9-3xSRDX-SUPERMAN Repressor Domain-X were employed in *Nicotiana benthamiana* and Arabidopsis, respectively, and resulted in decreased (~40% vs Control) transcript levels. Similarly, dCas12b- and dCas12a-mediated repressors were employed in *Arabidopsis* and rice (Tan et al. 2018; Yeo et al. 2018; Fontana et al. 2020; Son et al. 2020). Unfortunately, the limited efficacy of existing CRISPRi methods prevents it from being widely used in plants for controlled gene suppression. To maximize CRISPRi potency, techniques for enhancing CRISPRa efficiency might be used.

9.5 Genome Editing for Crop Enhancement: Ushering Towards Green Revolution 2.0

Genome editing for crop improvement has been revolutionized with the emergence of the CRISPR/Cas9 technology. Several new techniques to edit genomes have been established to assist scientists in precisely altering genome sequences. Due to its inherent simplicity and precision, genome editing has invoked substantial interest among agricultural scientists, as it offers novel prospects to advance enhanced crop varieties by expediting the removal of undesirable traits or the clear-cut addition of valued traits, thereby facilitating the expansion of crop varieties adapted to environmental changes. Employing this approach, scientists have explored the introduction of traits for instance high yield, insect and pest resistance, stress tolerance, as well as enhanced nutritional contents, which in turn decrease input costs. This technology is more precise than conventional crop breeding procedures or traditional genetic engineering approaches due to its on-target specificity. Gene editing can be used to improve crops in a variety of ways, including altering a few nucleotides out of the billions prevailing within the genomes of existing cells, changing the complete allele, or introducing a novel gene in a specific location of the genome. It is the most effective approach to generating pest-resistant crops that can grow in environmental cues and promoting their nutritional quotient. With genome editing, new crops can be generated more quickly and with a lower risk of off-target effects. This genome engineering technology paves the way for the Green Revolution's impact on agriculture, which was seen in the massive increase in crop yields, with world's grain output reaching millions of tons per year as one of the world's largest agricultural producers. During the Green Revolution, high-yielding wheat, rice, maize, and other crops were introduced, resulting in enhanced crop yields. Breeders are employing the newly discovered tool of gene editing to make crucial alterations to the crop plant's trait in order to make judicious use of resources while maintaining enough food for the people, ensuring nutrition, and limited land. The best scientific knowledge and modern biotechnology in improving the traits and enhancing the flavor using advanced gene-editing technologies is a prerequisite for Green Revolution 2.0 as a challenge to restructure the agricultural systems to meet existing challenges. It is time to focus on easing regulations around modern technologies that will help farmers improve their competitiveness and long-term viability.

9.5.1 Mitigation of Abiotic Stress

Abiotic stresses, such as drought, heavy metals, temperature, and salt, pose significant concerns for crop development and result in significant yield reductions globally. Advancements in genome editing technology open new avenues for crop improvement by utilizing precision genome engineering for targeted improvement of crop attributes. The identification of target genes, on the other hand, is crucial for the accomplishment of attaining the desired traits. These genes can be classified into two sorts: regulatory and structural. Regulatory genes achieve this implicitly by modulating the expression of additional genes involved in numerous cellular procedures. On the other hand, structural genes code for proteins that directly enhance stress tolerance. The availability of powerful genome editing techniques opens new possibilities for trait discovery. This would improve the series of new qualities, foremost to upgraded efficiency, particularly abiotic stress tolerance and effective agricultural trait advancement (Klap et al. 2017; Pompili et al. 2020). With respect to heat tolerance, recently CRISPR/Cas9-mediated genome editing was accomplished in tomatoes by targeting an S gene, *SlAGAMOUS-LIKE6* or SIAGL6, which led to increased fruit growth under heat stress (Klap et al. 2017). In addition, using CRISPR/Cas9 to target diverse genes in an individual organism has been efficiently optimized for a wide range of crops including rice (Miao et al. 2013), maize (Char et al. 2017), wheat (Wang et al. 2018a), and cotton (Gao et al. 2017). Thus, the stress-inducible CRISPR/Cas9 system may be counted as a useful approach for editing crops' genomes for abiotic stress resilience and other crucial traits.

9.5.2 Alleviation of Biotic Stress

Induced resistance to viral, fungal, and bacterial diseases falls under biotic stress tolerance, with umpteen studies focusing on viral disease tolerance. CRISPR/Cas systems are used in a series of applications, extending from biotic stress tolerance to improving the crop's yield. Frye et al. (2001) revealed that the enhanced disease resistance1 *(EDR1)* gene behaved as a negative regulator of powdery mildew resistance in *Arabidopsis* (Frye et al. 2001). Further, Zhang et al. (2016a) employed the CRISPR/Cas9 method in wheat (*Triticum aestivum*) to (KO) three homologs of *EDR1*, which at the same time resulted in Taedr1 plants with improved powdery mildew resistance. Consequently, both *eIF(iso)4E* and *eIF4E* genes were found as ideal candidates for employing the CRISPR/Cas9 method to create virus-resistant cucumber (Chandrasekaran et al. 2016) and Arabidopsis plants (Pyott et al. 2016), correspondingly. Also, CRISPR/Cas9 was used to KO the *OsERF922* gene, resulting in increased resistance to the blast fungus in rice (Macovei et al. 2018). Taking these findings together provides a robust indication that CRISPR/Cas9 was utilized to develop disease resistance in crucial crop seedlings.

9.5.3 Biofortification

Biofortification is reported to be a successful approach for increasing micronutrient levels in food crops. It is also a viable and sustainable technique for addressing micronutrient deficiencies in individuals who monotonously consume cereal and plant-based foods and have limited access to diverse food and adequate health facilities. Genetic engineering is regarded as a holistic method for the biofortification of grain crops in the recent era. The primary aim is to strengthen new grain varieties by incorporating targeted genes encoding attributes through overexpression and genome editing of negative regulators of grain yield via effective modulation of the biosynthesis of health-beneficial substances and minerals. The application of novel molecular and biotechnological approaches may override the limitations of traditional plant breeding, such as novel traits or variability, efficacy, and time. The extreme accuracy and simplicity of the CRISPR/Cas9 system have revolutionized genome editing for their effective utilization in the biofortification of a range of crops, including rice, wheat, barley, and millets, among several others. Rice (*Oryza sativa*), a staple crop, is the basis of nutrition for the world's population, accounting for up to 70% of daily calories. Solving hidden hunger by rice biofortification would be a long-term solution for individuals who eat mostly rice and have limited access to diverse foods (Datta and Bouis 2000). Significant increases in iron and zinc, as well as other essential minerals and provitamin-A (β-carotene), were accomplished in rice grains using the biofortification method through genetic engineering. Boonyaves et al. (2016) reported that NOD promoter-controlled *AtIRT1* expression works in concert with the NAS and FERRITIN genes to raise iron levels in rice grains. Overexpression of the *OsYSL2* gene in rice under the promoter of the sucrose transporter (*OsSUT1*) enhanced iron concentration in polished grain nearly fourfold (Masuda et al. 2012). Overexpression of the *OsSUT1* promoter-driven *OsYSL2* gene was discovered to be an effective method for iron biofortification. Several studies have found that overexpression of *OsNAS* genes increased iron and zinc contents in rice grains by several folds. Consequently, targeted knockout of the *HvHGGT* and *HvHPT* genes in barley fostered the investigation of their role in vitamin E biosynthesis pathways (Zeng et al. 2020a). Using the CRISPR/Cas9 strategy, new alleles of D-hordein were produced in barley, offering a new germplasm resource for understanding D-hordein protein function in barley and allowing the synthesis of novel cultivars with improved malt quality. Overexpression of zinc-regulated, iron-regulated transporter-like protein, ZIP genes has been shown to increase the zinc content of barley grains. In addition to barley, millets are another principle source of carbohydrates in the semiarid tropics and tropics (Shivran 2016). Despite the high quality of millet, India has exclusively chosen pearl millet as the crop optimal for iron biofortification. Millets are biofortified in two ways: (i) by increasing the concentration of nutrients in milled grains and (ii) by lowering antinutrients to boost mineral bioavailability. The presence of antinutrients such as polyphenols, tannins, and PA still limits biofortification in millets. Several biotechnological advances, including CRISPR, were employed to alleviate these antinutrients.

9.6 Harnessing the Potential of NGS and ML for Crop Design Target

Crop enhancement may be achieved by utilizing the species gene pool to identify genotypes and recombine genes to transfer improved traits or introduce new traits via mutations to crops for food, feed, fuel, and biomaterial production (Henry 2010). Whole-genome analysis using next-generation sequencing (NGS) can be used to determine the genetic basis of phenotypic variances in crop species (Henry 2005). NGS allows scientists to investigate genetic variations in plants and their wild relatives on a far greater scale than previously available and handle even the most complicated plant genomes (Ganal et al. 2009; Varshney et al. 2009; Edwards and Batley 2010). Therefore, whole-genome sequencing (WGS), metagenomics, transcriptome sequencing, and high-throughput genotyping are all possible with NGS that can be used for genome selection (GS). Until the advent of NGS, the identification of variations in plant genomes and the application of sequence-based selection to crop improvement were limited. The idea of employing a combination of NGS techniques and computational power (ML) to predict the pre- and post-gene editing phenotypic influences may be instrumental in developing optimized applications for improving crop plants and beyond. Despite being one of the most widely used genetic engineering tools, CRISPR/Cas9 genome editing is dependent on the precision of well-designed guide RNAs, which is a critical component of successful target gene editing (Cox et al. 2015). Various algorithms for evaluating CRISPR activity (on-target) and specificity (off-target), as well as web-based tools for *in silico* gRNA designing, have been developed in recent years (Henry et al. 2014; Zhu 2015). ML and AI offer revolutionary approaches to analyze CRISPR/Cas-edited crop lines with better attributes, such as grain size, palatability, higher nutrient value, root and flower architectures, stress resilience, etc. Several sgRNA designing tools (Table 9.2) have been developed to improve the efficiency and specificity of sgRNA design. These sgRNA designing tools, as well as off- and on-target prediction tools, have contributed to CRISPR genome technology's success.

Secondary structure, sequence composition, thermodynamics, and physicochemical properties have all been found to be essential for target gRNA activity, but for off-target predictions, the size, composition, and combination of discrepancies are the most critical factors. Numerous approaches have been established to represent the activity of CRISPR, which can be broadly divided into two categories, i.e. (i) **Machine-learning based**, including CRISPRater (Labuhn et al. 2017); CRISPRscan (Moreno-Mateos et al. 2015), sgRNA Designer (Doench et al. 2016), SSC (Xu et al. 2015), WU-CRISPR (Wong et al. 2015), Azimuth 2.0 & TUSCAN (Listgarten et al. 2018), and sgRNA Scorer (Chari et al. 2015) as well as (ii) **Deep learning-based**, including CNNstd (Lin and Wong 2018), DeepCas9 (Xue et al. 2019), DeepCRISPR (Chuai et al. 2018), CRISPRpred (Rahman and Rahman2017), and DeepCpf1 (Luo et al. 2019) predict sgRNA activity builds on automatic recognition of sequence characters using a Convolutional Nuclear Network (CNN). A group of researchers developed elevation from the Broad Institution and Microsoft, providing an exceptionally convenient platform for further development and application of CRISPR (Listgarten et al. 2018). Off-targets are estimated by the MIT server based on the distance and quantity of unpaired nucleotides (Hsu et al. 2013). Subsequently, a cutting frequency determination (CFD) score was calculated, which estimates off-target scores by replicating the frequencies of bases in the gRNA spacer sequence (Doench et al. 2016). Based on statistical and hypothetical methods, synergizing CRISPR merges the projection results of five different models, i.e. CFD, CCTop, MIT, CROPIT, and MIT, into an input function (Dobson et al. 2015; Singh et al. 2015). As a part of ML, a new algorithm called CRISPR target estimation (CRISTA) was established, which performed the critical job of identifying specific genomic locations that could be precisely excised using certain sgRNAs. The predictions of CRISTA have proven to be more accurate than earlier threshold predictions (Abadi et al. 2017; Listgarten et al. 2018). Therefore, machine learning and artificial intelligence assessed possible regression points that could converge or vary from on-target and off-target specificity charts. Moreover, each experimental design had specific predictive models and rules for CRISPR/Cas9 activity assessment. It remains unclear how efficient these techniques are at predicting gRNA activity in different species and cell types (Chuai et al. 2017). Different species-specific software has been developed due to huge variations between species, for instance, CRISPR-P: plants (Lei et al. 2014; Sanjana et al. 2014), flyCRISPR: fruit flies (Gratz et al. 2014), CRISPRscan: zebrafish (Moreno-Mateos et al. 2015), and EuPaGDT: pathogens (Peng and Tarleton 2015). Most of them are theoretical software, except CRISPRscan, which is generated based on ML. The development of web tools for *in silico* design of gRNAs, such as CRISPOR (Haeussler et al. 2016), CRISPRseek program (Zhu et al. 2014), ChopChop (Montague et al. 2014; Labun et al. 2016, 2019), and Cas-Designer (Park et al. 2015) as well as gRNA activity prediction algorithms,

Table 9.2 CRISPR/Cas-based designing tools.

Tool	Input	PAM	Website	References
Azimuth2.0	DNA sequence	NGG		Doench et al. (2016)
Benchling CRISPR gRNA design	Gene ID/Genome coordinates	User customizable	https://benchling.com/crispr	Doench et al. (2016)
Breaking-Cas	DNA sequence	User customizable	http://bioinfogp.cnb.csic.es/tools/breakingcas/	Oliveros et al. (2016)
Cas-designer	DNA sequence	NGG, NRG, NNAGAAW, NNNNGMTT	www.rgenome.net/cas-designer	Park et al. (2015)
Cas-OFFinder	crRNA sequence	20 PAMs (NGG, NRG, NNAGAAW, . . .)	http://www.rgenome.net/cas-offinder/	Bae et al. (2014)
CasEMBLR	DNA sequence	NGG	http://rna.tbi.univie.ac.at/cgi-bin/RNAWebSuite/RNAfold.cgi	Jakočiunas et al. (2015)
CasOT	DNA sequence	NGG, NAG, NNGG	http://eendb.zfgenetics.org/casot/	Xiao et al. (2014)
CASPER	DNA sequence	TTTN, NGG, NGCG	https://github.com/TrinhLab/CASPER	Mendoza and Trinh (2018)
CCTop	DNA sequence	NGG, NRG, NNGRRT, NNNNGATT, NNAGAAW, NAAAAC	https://crispr.cos.uni-heidelberg.de/	Stemmer et al. (2015)
CFD	DNA sequence	NGG, NAG, NCG, NGA	broadinstitute.org/rnai/public/software/index	Doench et al. (2016)
CGAT (CRISPR Genome Analysis Tool)	Gene ID/DNA sequence	NGG	http://cbc.gdcb.iastate.edu/cgat/	Brazelton et al. (2016)
CGBEs	DNA sequence	NGG	http://www.crisprbehive.design/	Koblan et al. (2021)
ChopChop	RefSeq, Genomic region, Gene ID	NGG, NGA, NAG, NRG, NNNNGANN, . . .), User customizable	https://chopchop.cbu.uib.no/	Montague et al. (2014)
ChopChop v2	RefSeq Gene ID Genomic region	User customizable	http://chopchop.cbu.uib.no/	Labun et al. (2016, 2019)
CINDEL	DNA sequence	TTTN, TTTA, TTTC, TTTG, TTTT, TTTV	http://big.hanyang.ac.kr/cindel	Kim et al. (2017)
CNN_std	DNA sequence	NAG, NGT, NTG, NGC, NGA, NGG, NAA, NCG	https://github.com/MichaelLinn/off_target_prediction	Lin and Wong (2018)
COD	DNA sequence	NGG, NRG, NNAGAAW, NNNNGMTT, NNGRRT	http://cas9.wicp.net sgRNAcas9	Park et al. (2015)
CrisFlash	DNA sequence	NGG	https://github.com/crisflash	Jacquin et al. (2019)
CRISPick	DNA sequence	NGG, CGGH, CGGT, TGGG	https://portals.broadinstitute.org/gpp/public/analysis-tools/sgrna-design	Doench et al. (2014)
CRISPOR	DNA Sequence/Genomic Region	NGG, NGA, NGCG, NGGNG, NNAGAA, NNGRRT, NNNRRT, NNNNACA,NNNNGMTT, TTTN	http://crispor.tefor.net	Haeussler et al. (2016)
CRISPR Configurator and Specificity Tool	DNA sequence/Genome coordinates	NGG, NAG	https://dharmacon.horizondiscovery.com/resources/tools-and-calculators/	Lei et al. (2014)

(Continued)

Table 9.2 (Continued)

Tool	Input	PAM	Website	References
CRISPR Finder	DNA sequence	NGG/User customizable	www.crispr.u-psud.fr/server	Kurtz (2003) and Doench et al. (2014)
CRISPR MultiTargeter	DNA sequence/Gene ID	NGG, User Customizable	http://www.multicrispr.net/	Prykhozhij et al. (2015)
CRISPR Primer Designer	DNA sequence	NGG	http://www.plantsignal.cn/	Yan et al. (2015)
CRISPR-ERA	DNA sequence	NGG	www.CRISPR-ERA.stanford.edu	Liu et al. (2015)
CRISPR-GE	DNA sequence/Gene ID	NGG, TTN, TTTN, User Customizable	http://skl.scau.edu.cn/	Xie et al. (2017)
CRISPR-P	DNA sequence/Gene locus/Genome coordinates	NGG, NAG	http://crispr.hzau.edu.cn/cgi-bin/CRISPR/CRISPR	Lei et al. (2014)
CRISPR-P 2.0	DNA sequence/Gene locus/Genome coordinates	14 PAMs (NGG, NNAGAAW, NNNNGMTT, TTTN, ...)	http://crispr.hzau.edu.cn/CRISPR2/	Liu et al. (2017)
Crispr-Plant	Gene locus/Genome coordinates	NGG	https://www.genome.arizona.edu/crispr/	Minkenberg et al. (2019)
CRISPRater	DNA sequence	NGG	https://crispr.cos.uni-heidelberg.de/	Labuhn et al. (2017)
CRISPRdirect	DNA sequence Genome coordinates	NNN, User customizable	http://crispr.dbcls.jp/	Naito et al. (2015)
CRISPRoff	DNA sequence	NGG, NAG, NGA	https://rth.dk/resources/crispr/	Alkan et al. (2018)
CRISPRpred	DNA sequence	NGG	https://github.com/khaled-buet/CRISPRpred	Rahman and Rahman (2017)
CRISPRscan	DNA sequence	NGG	www.crisprscan.org	Moreno-Mateos et al. (2015)
CRISPRseek	DNA sequence	NRG, NGG, User customizable	http://www.bioconductor.org/packages/release/bioc/html/CRISPRseek.html	Zhu et al. (2014)
CRISTA	DNA sequence	NGG	http://crista.tau.ac.il/pair_score.html	Abadi et al. (2017)
CRMAGE	DNA sequence	NGG	http://staff.biosustain.dtu.dk/laeb/crmage/	Ronda et al. (2016)
CROPIT	DNA sequence	NGG, NNG, GGG	http://cheetah.bioch.virginia.edu/AdliLab/CROP-IT/homepage.html	Singh et al. (2015)
CT-Finder	DNA sequence	NGG	http://bioinfolab.miamioh.edu/ct-finder	Zhu et al. (2016a)
DeepCas9	DNA sequence	NGG	https://github.com/lje00006/DeepCas9	Xue et al. (2019)
DeepCpf1	DNA sequence	TTTN	http://deepcrispr.info/	Luo et al. (2019)
DeepCRISPR	sgRNA sequence	NGG, NGT, NGA, NAG, NGC, NCG, NTG, NAA	http://www.deepcrispr.net/	Chuai et al. (2018)

E-CRISP	Gene ID/DNA sequence	NGG, User Customizable	http://www.e-crisp.org/E-CRISP/	Heigwer et al. (2014) and MacPherson and Scherf (2015)
Elevation	Gene ID Transcript ID Genomic region	NAG, NGA, NCG, NGC, NGG, NTG, NGT	https://crispr.ml/	Listgarten et al. (2018)
Elevation-search/dsNickFury	DNA sequence	NGG, NCG, NAG, NGA, NGG, NGC, NTG, NGT	https://github.com/michael-weinstein/dsNickFury3PlusOrchid	Listgarten et al. (2018)
EuPaGDT	DNA sequence	NGG, NAG, NGA	http://grna.ctegd.uga.edu/	Peng and Tarleton (2015)
FlashFry	DNA sequence	NGG	http://aaronmck.github.io/FlashFry/	McKenna and Shendure (2018)
Fly-CRISPR	DNA sequence	NGG	www.tools.flycrispr.molbio.wisc.edu/targetFinder	Gratz et al. (2014)
ge-CRISPR	DNA sequence	NGG	http://bioinfo.imtech.res.in/manojk/gecrispr/	Kaur et al. (2016)
GT-Scan	DNA sequence	User customizable	https://gt-scan.csiro.au/	O'Brien and Bailey (2014)
Off-Spotter	DNA sequence	NGG, NAG, NNGRRT, NNNNACA (R is A or G)	https://cm.jefferson.edu/Off-Spotter/	Pliatsik and Rigoutsos (2015)
Optimized CRISPR design	DNA sequence	NGG, NAG	https://crispr.mit.edu	Hsu et al. (2013)
Predict CRISPR	DNA sequence	NGG	https://github.com/penn-hui/OfftargetPredic	Peng et al. (2018)
Protospacer Workbench	Gene ID/DNA sequence	NGG	www.protospacer.com	MacPherson and Scherf (2015)
sgRNA designer	DNA sequence, Gene ID, Transcript ID	NGG	https://portals.broadinstitute.org/gpp/public/analysistools/sgrna-design	Doench et al. (2014)
sgRNA Scorer	DNA sequence	NGG, NAG, NNNNGMTT, NNAGAAW	https://crispr.med.harvard.edu/sgRNAScorerV2/	Chari et al. (2015)
sgRNAcas9	DNA sequence	NGG, NAG	www.biootools.com	Xie et al. (2014)
SSC	DNA sequence	NGG	www.crispr.dfci.harvard.edu/SSC/	Xu et al. (2015)
SSFinder	DNA sequence	NGG	https://code.google.com/archive/p/ssfinder/	Upadhyay and Sharma (2014)
SynergizingCRISPR	DNA sequence	NGG	https://github.com/Alexzsx/CRISPR	Zhang et al. (2019b)
Synthego design tool	DNA sequence	NGG	https://design.synthego.com/#/	Roginsky (2018)
TSAM	DNA sequence	NGG	http://www.aai-bioinfo.com/CRISPR	Peng et al. (2018)
TUSCAN	DNA sequence	NGG	https://github.com/BauerLab/TUSCAN	Wu et al. (2014)
uCRISPR	DNA sequence	NGG, NAG, NGA	https://github.com/Vfold-RNA/uCRISPR	Zhang (2019a)
WGE	DNA sequence	NGG	www.sanger.ac.uk/htgt/wge	Hodgkins et al. (2015)
WU-CRISPR	RNA sequence	NGG	http://crispr.wustl.edu/	Wong et al. (2015)
ZiFiT	DNA sequence	NGG	http://zifit.partners.org/ZiFiT	Sander et al. (2010)

have facilitated the use and advancement of CRISPR-Cas9 genome technologies. Researchers aimed to establish a new approach for measuring off-target action, termed CASPER, because organisms couldn't quickly limit the prior off-target scoring process (Mendoza and Trinh 2018). In addition, Cas-OFFinder (Baltes et al. 2014), CRISPR-P (Lei et al. 2014), SSFinder (Upadhyay and Sharma 2014), CasOT (Xiao et al. 2014), Elevation-search/dsNickFury (Listgarten et al. 2018), and FlashFry (McKenna and Shendure 2018), simplify the designing process of sgRNAs. Furthermore, an ML-based approach can accurately anticipate lethal sgRNA interactions and characterize target regions in certain gene combinations. However, there is still a lot of work to be done to test and optimize CRISPR-Cas9 gene editing in plant systems.

9.7 Does CRISPR/Cas Address the Snag of Genome Editing?

Gene-editing (GED) approaches have transformed classical and modern plant biology research. A series of breakthroughs, from mega-nucleases to prime editing approaches, have been included in gene-editing systems (Hua et al. (2020a); Nadakuduti and Enciso-Rodriguez 2021). The targeted improvement of crop varieties with superior agronomic features and enhanced nutritional value has been achieved using CRISPR-based genome-editing technologies. The versatility of the CRISPR-Cas technology has been demonstrated by the development of a range of crop plants resistant to biotic and abiotic factors with improved grain yield, thus facilitating the ideotyping of crop varieties by boosting photosynthetic parameters and plant architecture and thereby reducing the time to achieve food security (Arora and Narula 2017). Organelles, such as mitochondria and chloroplasts, contain their own genomes and hundreds of essential genes that may impact plant metabolism and photosynthesis, in addition to nuclear genes. CRISPR-based gene modifications within the genomes of specific crops would enhance photosynthetic machinery and, in turn, crop productivity. Conversely, in the near future, CRISPR-Cas technology could be employed to convert C_3 food crops (wheat, rice, and barley) into C_4-like or C_4 plants, which have greater NUE and water usage efficiency, as well as higher rates of photosynthesis. The generation of multiple hybrids by CRISPR would significantly reduce time and enhance yield. The conventional methods have limitations to deliver quick results; while adopting the CRISPR-Cas breeding method may accelerate the generation of climate-resilient crops, effective self-pollination in crucial food crops, and so on, in turn exacerbating grain yield production. Earlier artificial domestication of crops required many years to deliver a better crop for cultivation, but with the advent of CRISPR-mediated multiplex genome editing strategies, these challenges may be resolved for major food crops, including orphan crops, permitting them to play major roles in global food security and eradicate malnutrition. However, further research is a prerequisite to developing a deeper understanding of the CRISPR-Cas mechanism and its new innovative implications in plants. Researchers are currently confronted with many challenges in using CRISPR techniques in plant research, including GMO regulatory burdens. However, recent developments in India, wherein the government has embarked upon de-regulating CRISPR-edited SDN1 and SDN2 categories as well as GM-Mustard for field trials, are seen as significant steps Figure 9.2. Scientists are improving the traditional CRISPR-Cas workflow by employing varying modifications so that multiple copies of genes can be altered at once (Wilson et al. 2019; Jouanin et al. 2020; Smedley et al. 2021). Plant transformation is one of the most important tools for gene editing. Notwithstanding the advancements in techniques for delivering the CRISPR-Cas system into plant genomes, researchers must continue to develop more efficient delivery methods as transformation efficiency in plant species with complex genomes, for instance, wheat, cotton, and Brassica, species comes with its own set of challenges. As the number of chromosome copies expands, the possibility of achieving specific gene editing in each copy becomes low (Yang 2020). Furthermore, CRISPR/Cas-based gene editing necessitates tissue culture for plant regeneration from callus, which is time-consuming and labor-intensive. Besides, even well-established transformation and regeneration protocols for model crop plants may require several months to generate edited plants, and this could further expand by several months for recalcitrant crop species. As a result, genome editing techniques that rely on viral vector-mediated delivery systems are more efficacious than tissue culture based ones. Reducing off-target effects is another crucial aspect, whereby screening of accurate mutation and potential off-target sites poses significant challenges to gene-editing in crops. CRISPR systems have been reported to produce off-target effects, which could include harmful impacts on outcomes and may create safety issues. As a result, more precise and well-organized strategies are needed, such as biased and unbiased off-target detection methods, gRNA modification and engineering, improved Cas variants, efficient CRISPR system delivery methods, anti-CRISPR protein development, and effective base-editing and prime editing systems. A newly developed prime-editing (PE) double-strand break-free strategy that involves nCas9 (nickase) working in tandem with a viral reverse transcriptase domain has been revealed to be error-free and more acceptable in plants and animals, showcasing nearly zero off-targets and thereby rendering risk-free GE food crops.

Considering CRISPR-Cas9-edited plants as genetically modified organisms (GMOs) is the most significant impediment to genome editing. Numerous countries have invested in producing market-oriented crops employing genome editing; on the contrary, the edited plants are strictly regulated in some countries (Tiwari et al. 2021). The transformation of Cas9-encoding mRNA with sgRNAs or pre-assembled ribonucleoprotein (RNPs) complexes may be considered a by-pass to alleviate the obstacles related to plant genome editing (Kouranova et al. 2016). The USDA exempted strict GMO regulations in many CRISPR-edited crops, for instance, waxy corn, flavored tomato, and mushroom (Waltz 2016). However, to consider edited plants as non-GMOs, we needed modifications to the existing GMO legislation. Despite all these impediments, the CRISPR-Cas9 approach is considered the most promising tool due to its precision editing. Numerous heritable traits can be incorporated into plants via this approach, which may produce modified plants like those developed through conventional breeding. CRISPR-Cas9 strategy leads toward a progressive change via high-yielding crop plant production to meet food security globally.

9.8 Edited Plant Code: Security Risk Assessment

Globally, regulations relevant to plants and crops meant for food and feed were centered on securing its nation, community, and environment. Routinely, the GE crops are designated as either process-directed or product-directed (Callebaut 2015; Medvedieva and Blume 2018; Eckerstorfer et al. 2019). Process-directed regulations consider genome-edited crops as a novel approach as compared to traditional systems, necessitating the application of the statutory provision. The focus is on the technique that was employed to create the unique product. In contrast to those created by traditional breeding, product-oriented laws concentrate a higher emphasis on the generation of unique product features (McHughen 2016). The pattern that is used in Europe, India, New Zealand, and Australia governs the process of obtaining the product. As per this system, the outcome of new breeding procedures will depend on the process used. As a result, gene-edited crops are subjected to the same stringent regulations as applicable to GMOs. The other, less stringent regulatory paradigm focuses on the final product rather than the process of developing GE crops Figure 9.2. This framework, which is employed in the United States, Argentina, and Canada, is not uniformly applied, and works on a case-by-case basis, regardless of whether it is new or not.

The process-oriented legislation of GE crops in many countries is limiting their advancement. For instance, in the EU (European countries), global research and development enterprises are transferring their headquarters to nations having end-product-based regulations, such as the United States. The process-based model has been criticized by some EU member states,

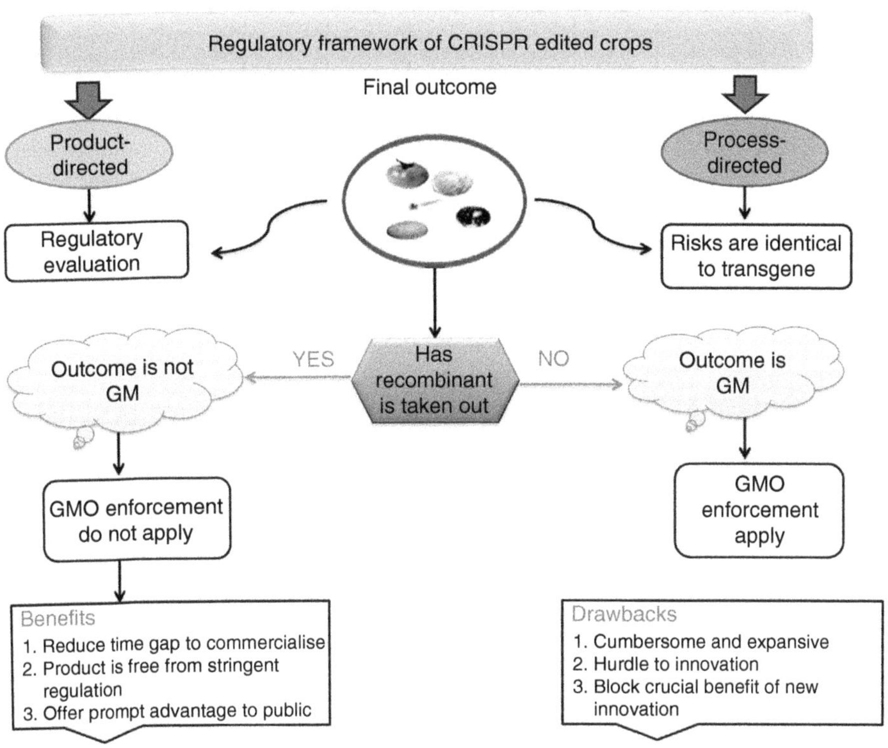

Figure 9.2 The regulatory roadmap for the CRISPR/Cas-based genome editing (GEd), including genome edited crops. Here, showing product/process based regulatory policy for GEd crops. GM: genetically modified; GMO: genetically modified organism.

notably Sweden and France. Canada and the United States have stated that GM regulation has centered on the outcome rather than the method. China established a commission in 2016 to set up guidelines for gene-edited products, which are still being disputed. In regards to R&D projects and sponsors for CRISPR projects, the world is anticipating a lenient attitude toward CRISPR-Cas-bred crops, with no rigorous regulations in place like those for conventionally bred crops.

The fundamental problem for global agriculture policy and the economy is meeting the food and nutritional demands of the world's rising population. Food and nutritional security issues necessitate creative solutions. Modern biotechnology offers significant and potential solutions that may astoundingly contribute to food and nutritional security, wealth, and sustainable development. Genetic engineering has provided tools to improve nutrition, increase yield, and enhance crop resilience. Novel genome editing tools provide ample means to overcome limitations inherent in conventional plant breeding, but their industrial applicability depends on the regulatory environment and decision-making. An alignment of goals between science and science-based policy can help in harvesting the potential fruits of modern biotechnology in order to contribute to food security, wealth, and sustainable development. The complexity of global issues in food security requires an interface between science and policy in order to ensure the synergetic effect of socio-technical integration consistent with socio-economic considerations and implemented in a socially responsible manner.

9.9 Conclusion: Food Security on the Verge of Climate change

The radically changing global climate creates an unpredictable environment for farmers. The unexpected occurrence of variable temperatures, drought, flood, disease, and insects poses a challenge to the sustainable development of agriculture for qualitative and quantitative aspects of food production. However, the application of genome editing techniques in the field of crop improvement has revolutionized the field of genetic engineering and administered novel ways to achieve nutritionally secure, climate-smart agriculture. Gene editing is a prominent and promising approach to developing smart crops that project resilience toward the anticipated future climatic changes. The traits like nutritional enhancement, yield improvement, and abiotic and biotic stress management can be targeted to achieve future-ready crops. However, to date, these traits do not represent the breadth that can be achieved via genome editing.

Over the last decade, we have witnessed the evolution of the CRISPR/Cas system with continued development and rigorous addition of path-breaking tools for genome editing in crop plants. Irrespective of the progress that the CRISPR/Cas system has achieved, the application of the same towards developing environment-friendly stress-resistant crops has not attained the real depth that this technique potentially can offers. Thereby, the utilization of machine learning algorithms to design models for reducing technical limitations and for screening large databases is essential to build a bridge between ML and genome editing. Globally, researchers are utilizing the new advanced fine-tuning ML models for designing large-scale experiments for quick outcomes, synchronized results, better comprehensibility, and improved reproducibility across boundaries. Invariably, this technique can aid in regaining the faith of the masses in this unrealistically powerful tool. However, to date, the global journey of genome-edited crops is not clear, considering the strict GMO regulatory aspects in numerous countries including India. However, current developments in India have brought in a breath of fresh air, whereby the government has embarked upon de-regulating CRISPR-edited SDN1 and SDN2 categories, in addition to GM-Mustard for field trials, which are seen as trailblazing steps within the subcontinent. As CRISPR-Cas-based system can bring Green Revolution 2.0, thus we require a positive policy roadmap aided by governmental influence to bridge the gap for genome-edited crops from laboratory to market. This comprehensive review offers a glimpse of the successful implementations of the CRISPR-Cas system in plant genome editing (Table 9.1) in order to generate potent and viable solutions for food security in the face of climate change, dwindling arable land, severe water shortages, and drastic soil degradation.

References

Abadi, S., Yan, W.X., Amar, D., and Mayrose, I. (2017). A machine learning approach for predicting CRISPR-Cas9 cleavage efficiencies and patterns underlying its mechanism of action. *PLoS Comput. Biol.* 13: 1–24. https://doi.org/10.1371/journal.pcbi.1005807.

Abe, K., Araki, E., Suzuki, Y. et al. (2018). Production of high oleic/low linoleic rice by genome editing. *Plant Physiol. Biochem.* 131: 58–62.

Adlak, T., Tiwari, S., Tripathi, M.K. et al. (2019). Biotechnology: an advanced tool for crop improvement. *Curr. J. Appl. Sci. Technol.* 33 (1): 1–11.

Adli, M. (2018). The CRISPR tool kit for genome editing and beyond. *Nat. Commun.* 9 (1): 1–13.

Ahmad, S., Tang, L., Shahzad, R. et al. (2021). CRISPR-based crop improvements: a way forward to achieve zero hunger. *J. Agric. Food Chem.* 69 (30): 8307–8323. https://doi.org/10.1021/acs.jafc.1c02653.

Akama, K., Akter, N., Endo, H. et al. (2020). An in vivo targeted deletion of the calmodulin-binding domain from rice glutamate decarboxylase 3 (OsGAD3) increases γ-aminobutyric acid content in grains. *Rice* 13: 20.

Al Amin, N., Ahmad, N., Wu, N. et al. (2019). CRISPR-Cas9 mediated targeted disruption of FAD2-2 microsomal omega-6 desaturase in soybean (*Glycine max* L.). *BMC Biotechnol.* 19: 1–10. https://doi.org/10.1186/s12896-019-0501.

Alkan, F., Wenzel, A., Anthon, C. et al. (2018). CRISPR-Cas9 off-targeting assessment with nucleic acid duplex energy parameters. *Genome Biol.* 19: 1–13. https://doi.org/10.1186/s13059-018-1534-x.

Allaby, R.G., Ware, R.L., and Kistler, L. (2019). A re-evaluation of the domestication bottleneck from archaeogenomic evidence. *Evol. Appl.* 12 (1): 29–37. https://doi.org/10.1111/eva.12680.

Almutiri, T., Saeed, F., and Alassaf, M. (2022). A survey of machine learning and deep learning applications in genome editing. In: *Advances on Smart and Soft Computing*, Advances in Intelligent Systems and Computing, vol. 1399 (ed. F. Saeed, T. Al-Hadhrami, E. Mohammed, and M. Al-Sarem). Singapore: Springer https://doi.org/10.1007/978-981-16-5559-3_13.

Andersson, M., Turesson, H., Olsson, N. et al. (2018). Genome editing in potato via CRISPR-Cas9 ribonucleoprotein delivery. *Physiol. Plant.* 164 (4): 378–384. https://doi.org/10.1111/ppl.12731.

Arndell, T., Sharma, N., and Langridge, P. (2019). gRNA validation for wheat genome editing with the CRISPR-Cas9 system. *BMC Biotechnol.* 19: 71. https://doi.org/10.1186/s12896-019-0565-z.

Arora, L. and Narula, A. (2017). Gene editing and crop improvement using CRISPR-Cas9 system. *Front. Plant Sci.* 8: 1932. https://doi.org/10.3389/fpls.2017.01932.

Aryal, J.P., Sapkota, T.B., Khurana, R. et al. (2020). Climate change and agriculture in South Asia: adaptation options in smallholder production systems. *Environ. Dev. Sustain.* 22: 5045–5075.

Bae, S., Park, J., and Kim, J.S. (2014). Cas-OFFinder: a fast and versatile algorithm that searches for potential off-target sites of Cas9 RNA-guided endonucleases. *Bioinformatics* 30: 1473–1475. https://doi.org/10.1093/bioinformatics/btu048.

Baltes, N.J., Gil-Humanes, J., Cermak, T. et al. (2014). DNA replicons for plant genome engineering. *Plant Cell* 26: 151–163. https://doi.org/10.1105/tpc.113.119792.

Bao, A., Chen, H., Chen, L. et al. (2019). CRISPR/Cas9-mediated targeted mutagenesis of GmSPL9 genes alters plant architecture in soybean. *BMC Plant Biol.* 19: 131. https://doi.org/10.1186/s12870-019-1746-6.

Bari, V.K., Nassar, J.A., and Kheredin, S.M. (2019). CRISPR/Cas9-mediated mutagenesis of CAROTENOID CLEAVAGE DIOXYGENASE 8 in tomato provides resistance against the parasitic weed *Phelipanche aegyptiaca*. *Sci. Rep.* 9: 11438. https://doi.org/10.1038/s41598-019-47893-z.

Barrett, C.B. (2021). Overcoming global food security challenges through science and Solidarity. *Am. J. Agric. Econ.* 103 (2): 422–447. https://doi.org/10.1111/ajae.12160.

Bellard, C., Bertelsmeier, C., Leadley, P. et al. (2012). Impacts of climate change on the future of biodiversity. *Ecol. Lett.* 15 (4): 365–377. https://doi.org/10.1111/j.1461-0248.2011.01736.x.

Bhattacharya, A., Parkhi, V., and Char, B. (2021). Genome editing for crop improvement: a perspective from India. *In Vitro Cell. Dev. Biol. Plant* 57: 565–573. https://doi.org/10.1007/s11627-021-10184-2.

Bhowmik, P., Ellison, E., Polley, B. et al. (2018). Targeted mutagenesis in wheat microspores using CRISPR/Cas9. *Sci. Rep.* 8 (1): 6502. https://doi.org/10.1038/s41598-018-24690-8.

Bo, W., Zhaohui, Z., Huanhuan, A.N.G. et al. (2019). Targeted mutagenesis of NAC transcription factor gene, OsNAC041 leading to salt sensitivity in rice. *Rice Sci.* 26 (2): 98–108.

Boonyaves, K., Gruissem, W., and Bhullar, N.K. (2016). NOD promoter-controlled AtIRT1 expression functions synergistically with NAS and FERRITIN genes to increase iron in rice grains. *Plant Mol. Biol.* 90: 207–215.

Braatz J, Harloff HJ, Mascher M, Stein N, Himmelbach A, Jung C. CRISPR-Cas9 targeted mutagenesis leads to simultaneous modification of different homoeologous gene copies in polyploid oilseed rape (Brassica napus). *Plant Physiol.* 2017 Jun;174(2):935–942. doi: https://doi.org/10.1104/pp.17.00426

Brazelton, V.A., Zarecor, S., Wright, D.A. et al. (2016). A quick guide to CRISPR sgRNA design tools. *GM Crops Food* 6: 266–276. https://doi.org/10.1080/21645698.2015.1137690.

Bull, S.E., Seung, D., Chanez, C. et al. (2018). Accelerated ex situ breeding of GBSS- and PTST1-edited cassava for modified starch. *Sci. Adv.* 4: 6086. https://doi.org/10.1126/sciadv.aat6086.

Butler, N.M., Atkins, P.A., Voytas, D.F., and Douches, D.S. (2015). Generation and inheritance of targeted mutations in potato (Solanum tuberosum L.) using the CRISPR/Cas system. *PloS One* 10: e0144591.

Butler, N.M., Baltes, N.J., Voytas, D.F. et al. (2016). Geminivirus-mediated genome editing in potato (Solanum tuberosum L.) using sequence-specific nucleases. Front. *Plant Sci* 7: 1045.

Butt, H., Rao, G.S., Sedeek, K. et al. (2020). Engineering herbicide resistance via prime editing in rice. *Plant Biotechnol. J.* 18: 2370–2372.

Cai, Y., Chen, L., Liu, X. et al. (2018). CRISPR/Cas9-mediated targeted mutagenesis of GmFT2a delays flowering time in soya bean. *Plant Biotechnol.* 16: 176–185. https://doi.org/10.1111/pbi.12758.

Callebaut, S. (2015). New developments in modern biotechnology: a survey and analysis of the regulatory status of plants produced through New Breeding techniques. https://lib.ugent.be/fulltxt/RUG01/002/213/647/RUG01-002213647_2015_0001 (accessed 22 January 2022).

Carlessi, M., Mariotti, L., Giaume, F. et al. (2021). Targeted knockout of the gene OsHOL1 removes methyl iodide emissions from rice plants. *Sci. Rep.* 11: 17010. https://doi.org/10.1038/s41598-021-95198-x.

Čermák, T., Baltes, N.J., Čegan, R. et al. (2015). High-frequency, precise modification of the tomato genome. *Genome Biol.* 16: 232. https://doi.org/10.1186/s13059-015-0796-9.

Chandrasekaran, J., Brumin, M., Wolf, D. et al. (2016). Development of broad virus resistance in non-transgenic cucumber using CRISPR/Cas9 technology. *Mol. Plant Pathol.* 17: 1140–1153. https://doi.org/10.1111/mpp.12375.

Chao, S., Cai, Y., Feng, B. et al. (2019). Editing of rice isoamylase gene ISA1 provides insights into its function in starch formation. *Rice Sci.* 26 (2): 77–87. https://doi.org/10.1016/j.rsci.2018.07.001.

Char, S.N., Neelakandan, A.K., Nahampun, H. et al. (2017). An Agrobacterium-delivered CRISPR/Cas9 system for high-frequency targeted mutagenesis in maize. *Plant Biotechnol.* 15 (2): 257–268.

Chari, R., Mali, P., Moosburner, M., and Church, G.M. (2015). Unraveling CRISPR-Cas9 genome engineering parameters via a library-on-library approach. *Nat. Methods* 12: 823–826. https://doi.org/10.1038/nmeth.3473.

Che, J., Yamaji, N., and Ma, J.F. (2021). Role of a vacuolar iron transporter OsVIT2 in the distribution of iron to rice grains. *New Phytol.* 230: 1049–1062.

Chiwona-Karltun, L., Amuakwa-Mensah, F., Wamala-Larsson, C. et al. (2021). COVID-19: from health crises to food security anxiety and policy implications. *Ambio* 50 (4): 794–811. https://doi.org/10.1007/s13280-020-01481-y.

Chuai, G., Wang, Q.L., and Liu, Q. (2017). In silico meets in vivo: towards computational CRISPR-based sgRNA design. *Trends Biotechnol.* 35: 12–21. https://doi.org/10.1016/j.tibtech.2016.06.008.

Chuai, G., Ma, H., Yan, J. et al. (2018). DeepCRISPR: optimized CRISPR guide RNA design by deep learning. *Genome Biol.* 19 (1): 80. https://doi.org/10.1186/s13059-018-1459-4.

Clasen, B.M., Stoddard, T.J., Luo, S. et al. (2016). Improving cold storage and processing traits in potato through targeted gene knockout. *Plant Biotechnol. J.* 14: 169–176. https://doi.org/10.1111/pbi.12370.

Cox, D.B.T., Platt, R.J., and Zhang, F. (2015). Therapeutic genome editing: prospects and challenges. *Nat. Med.* 21: 121–131. https://doi.org/10.1038/nm.3793.

Curtin, S.J., Xiong, Y., and Michno, J.M. (2018). CRISPR/Cas9 and TALENs generate heritable mutations for genes involved in small RNA processing of Glycine max and *Medicago truncatula*. *Plant Biotechnol.* 16 (6): 1125–1137. https://doi.org/10.1111/pbi.12857.

Danilo, B., Perrot, L., Mara, K. et al. (2019). Efficient and transgene-free gene targeting using Agrobacterium-mediated delivery of the CRISPR/Cas9 system in tomato. *Plant Cell Rep.* 38: 459–462. https://doi.org/10.1007/s00299-019-02373-6.

Datta, S.K. and Bouis, H.E. (2000). Application of biotechnology to improving the nutritional quality rice. *Food Nutr. Bull. 21*: 451–456.

Demorest, Z.L., Coffman, A., Baltes, N.J. et al. (2016). Direct stacking of sequence-specific nuclease-induced mutations to produce high oleic and low linolenic soybean oil. *BMC Plant Biol.* 16: 225. https://doi.org/10.1186/s12870-016-0906-1.

Deslous, P., Bournonville, C., Decros, G., Okabe, Y., Mauxion, J. P., Jorly, J., et al. (2021). "Overproduction of ascorbic acid impairs pollen fertility in tomato." *J. Exp. Bot.* 72 (8), 3091–3107. doi: https://doi.org/10.1093/jxb/erab040

Dobson, L., Reményi, I., and Tusnády, G.E. (2015). CCTOP: a consensus constrained TOPology prediction web server. *Nucleic Acids Res.* 43: 408–412. https://doi.org/10.1093/nar/gkv451.

Doench, J.G., Hartenian, E., Graham, D.B. et al. (2014). Rational design of highly active sgRNAs for CRISPR-Cas9–mediated gene inactivation. *Nat. Biotechnol.* 32: 1262–1267. https://doi.org/10.1038/nbt.3026.

Doench, J.G., Fusi, N., Sullender, M. et al. (2016). Optimized sgRNA design to maximize activity and minimize off-target effects of CRISPR-Cas9. *Nat. Biotechnol.* 34: 184–191. https://doi.org/10.1038/nbt.3437.

Dong, O.X., Yu, S., Jain, R. et al. (2020). Marker-free carotenoid enriched rice generated through targeted gene insertion using CRISPR-Cas9. *Nat. Commun.* 11: 1–10.

Duan, Y.B., Li, J., and Qin, R.Y. (2016). Identification of a regulatory element responsible for salt induction of rice OsRAV2 through *ex situ* and *in situ* promoter analysis. *Plant Mol. Biol.* 90: 49–62. https://doi.org/10.1007/s11103-015-0393-z.

Eckerstorfer, M.F., Engelhard, M., Heissenberger, A. et al. (2019). Plants developed by new genetic modification techniques – comparison of existing regulatory frameworks in the EU and Non-EU countries. *Front. Bioeng. Biotechnol.* 7: 26. https://doi.org/10.3389/fbioe.2019.0 0026.

Edwards, D. and Batley, J. (2010). Plant genome sequencing applications for crop improvement. *Plant Biotechnol. J.* 8: 2–9.

Endo, A., Saika, H., Takemura, M. et al. (2019). Novel approach to carotenoid accumulation in rice callus by mimicking the cauliflower Orange mutation via genome editing. *Rice* 12: 81. https://doi.org/10.1186/s12284-019-0345-3.

Fahad, S., Bajwa, A.A., Nazir, U. et al. (2017). Crop production under drought and heat stress: plant responses and management options. *Front. Plant Sci.* 8: 1147. https://doi.org/10.3389/fpls.2017.01147.

Feng, C., Su, H., Bai, H. et al. (2018). High-efficiency genome editing using a dmc1 promoter-controlled CRISPR/Cas9 system in maize. *Plant Biotechnol.* 16 (11): 1848–1857.

Ferrara, M., Haidukowski, M., Logrieco, A.F. et al. (2019). A CRISPR-Cas9 system for genome editing of *Fusarium proliferatum*. *Sci. Rep.* 9: 19836. https://doi.org/10.1038/s41598-019-56270-9.

Filler, H.S., Melamed, B.C., and Levy, A.A. (2017). Targeted recombination between homologous chromosomes for precise breeding in tomato. *Nat. Commun.* 8: 15605. https://doi.org/10.1038/ncomms15605.

Fister, A.S., Landherr, L., Maximova, S.N., and Guiltinan, M.J. (2018). Transient expression of CRISPR/Cas9 machinery targeting TcNPR3 enhances defense response in *Theobroma cacao*. *Front. Plant Sci.* 9: 268. https://doi.org/10.3389/fpls.2018.00268.

Fontana, J., Sparkman-Yager, D., Zalatan, J.G., and Carothers, J.M. (2020). Challenges and opportunities with CRISPR activation in bacteria for data-driven metabolic engineering. *Curr. Opin. Biotechnol.* 64: 190–198.

Food and Agriculture Organization FAO (2021). *International Fund for Agricultural Development IFAD, World Food Programme WFP. The State of Food Insecurity in the World 2021*. FAO.

Foster, A.J., Martin-Urdiroz, M., Yan, X. et al. (2018). CRISPR-Cas9 ribonucleo protein-mediated co-editing and counter selection in the rice blast fungus. *Sci. Rep.* 8: 14355.

Friedman, M. (2006). Potato glycoalkaloids and metabolites: roles in the plant and in the diet. *J. Agric. Food Chem.* 54 (23): 8655–8681.

Frye, C.A., Tang, D., and Innes, R.W. (2001). Negative regulation of defense responses in plants by a conserved MAPKK kinase. *Proc. Natl. Acad. Sci. U.S.A.* 98 (1): 373–378.

Ganal, M.V.V., Altmann, T., and Roder, M.S. (2009). SNP identification in crop plants. *Curr. Opin. Plant Biol.* 12: 211–217.

Gao, W., Long, L., Tian, X. et al. (2017). Genome editing in cotton with the CRISPR/Cas9 system. *Front. Plant Sci.* 8: 1364.

Gao, H., Gadlage, M.J., Lafitte, H.R. et al. (2020a). Superior field performance of waxy corn engineered using CRISPR–Cas9. *Nat. Biotechnol.* 38: 579–581. https://doi.org/10.1038/s41587-020-0444-0.

Gao, Q., Li, G., Sun, H. et al. (2020b). Targeted mutagenesis of the rice FW 2.2-like gene family using the CRISPR/Cas9 system reveals OsFWL4 as a regulator of tiller number and plant yield in rice. *Int. J. Mol. Sci.* 21: 809.

Ghebreyesus, T.A. (2020). WHO on Coronavirus pandemic: "The Worst Is Yet to Come". WWW.WABe.org

Godfray, H.C.J. (2010). Food security: the challenge of feeding 9 billion people. *Science* 327 (5967): 812–818.

Gomez, M.A., Lin, Z.D., Moll, T. et al. (2019). Simultaneous CRISPR/Cas9-mediated editing of cassava EIF4E isoforms NCBP-1 and NCBP-2 reduces cassava brown streak disease symptom severity and incidence. *Plant Biotechnol.* 17: 421–434. https://doi.org/10.1111/pbi.12987.

Gratz, S.J., Ukken, F.P., Rubinstein, C.D. et al. (2014). Highly specific and efficient CRISPR/Cas9-catalyzed homology-directed repair in drosophila. *Genetics* 196 (4): 961–971. https://doi.org/10.1534/genetics.113.160713.

Gumtow, R., Wu, D., Uchida, J., and Tian, M. (2017). A *Phytophthora palmivora* extracellular cystatin-like protease inhibitor targets papain to contribute to virulence on papaya. *Mol. Plant Microbe Interact.* 31 (3): 363–373. https://doi.org/10.1094/MPMI-06-17-0131-FI.

Haeussler, M., Schönig, K., Eckert, H. et al. (2016). Evaluation of off-target and on-target scoring algorithms and integration into the guide RNA selection tool CRISPOR. *Genome Biol.* 17: 1–12. https://doi.org/10.1186/s13059-016-1012-2.

Heigwer, F., Kerr, G., and Boutros, M. (2014). E-CRISP: fast CRISPR target site identification. *Nat. Methods* 11: 122–123. https://doi.org/10.1038/nmeth.2812.

Henry, R.J. (2005). *Plant Diversity and Evolution: Genotypic and Phenotypic Variation in Higher Plants*, 332. CABI Publishing.

Henry, R.J. (2010). *Plant Resources for Food, Fuel and Conservation*. London: Earthscan.

Henry, V.J., Bandrowski, A.E., Pepin, A.S. et al. (2014). OMICtools: an informative directory for multi-omic data analysis. *Database (Oxford)* 2014: 1–5. https://doi.org/10.1093/database/bau069.

Hodgkins, A., Farne, A., Perera, S. et al. (2015). WGE: a CRISPR database for genome engineering. *Bioinformatics* 31 (18): 3078–3080. https://doi.org/10.1093/bioinformatics/btv308.

Hossain, A., Skalicky, M., Brestic, M. et al. (2021). Consequences and mitigation strategies of abiotic stresses in wheat (*Triticum aestivum* L.) under the changing climate. *Agronomy* 11 (2): 241. https://doi.org/10.3390/agronomy11020241.

Howells, R.M., Craze, M., Bowden, S., and Wallington, E.J. (2018). Efficient generation of stable, heritable gene edits in wheat using CRISPR/Cas9. *BMC Plant Biol.* 18 (1): 215.

Hsu, P.D., Scott, D.A., Weinstein, J.A. et al. (2013). DNA targeting specificity of RNA-guided Cas9 nucleases. *Nat. Biotechnol.* 31: 827–832. https://doi.org/10.1038/nbt.2647.

Hu, X., Cui, Y., Dong, G. et al. (2019). Using CRISPR-Cas9 to generate semi-dwarf rice lines in elite landraces. *Sci. Rep.* 9: 19096.

Hua, K., Zhang, J., Botella, J.R. et al. (2019a). Perspectives on the application of genome-editing technologies in crop breeding. *Mol. Plant* 12: 1047–1059.

Hua, K., Tao, X., and Zhu, J.K. (2019b). Expanding the base editing scope in rice by using Cas9 variants. *Plant Biotechnol.* 17: 499–504.

Hua, K., Jiang, Y., Tao, X., and Zhu, J. (2020a). Precision genome engineering in rice using prime editing system. *Plant Biotechnol.* 18: 2167–2169.

Hua, K., Tao, X., Liang, W. et al. (2020b). Simplified adenine base editors improve adenine base editing efficiency in rice. *Plant Biotechnol.* 18: 770–778.

Huang, Q.N., Zeng, X., Li, J., and Zhao, D. (2017). Construction and analysis of tifya and tifyb mutants in rice (*Oryza sativa*) based on CRISPR/Cas9 technology. *J. Agric. Biotechnol.* 25: 1003–1012.

Huang, L., Zhang, R., Huang, G. et al. (2018). Developing superior alleles of yield genes in rice by artificial mutagenesis using the CRISPR/Cas9 system. *Crop J.* 6: 475–481. https://doi.org/10.1016/j.cj.2018.05.005.

Huang, X., Wang, Y., and Wang, N. (2022). Highly efficient generation of canker-resistant sweet orange enabled by an improved CRISPR/Cas9 system. *Front. Plant Sci.* 12: 769907. https://doi.org/10.3389/fpls.2021.769907.

Hummel, A.W., Chauhan, R.D., Cermak, T. et al. (2017). Allele exchange at the EPSPS locus confers glyphosate tolerance in cassava. *Plant Biotechnol. J.* 16: 1275–1282.

Ibrahim, S., Saleem, B., Rehman, N. et al. (2021). CRISPR/Cas9 mediated disruption of *Inositol Pentakisphosphate 2-Kinase 1* (*TaIPK1*) reduces phytic acid and improves iron and zinc accumulation in wheat grains. *J. Adv. Res.* 37: 33–41. https://doi.org/10.1016/j.jare.2021.07.006.

International Service for the Acquisition of Agri-biotech Applications (ISAAA) (2022). Less Browning and Lower Acrylamide Levels Achieved in Potatoes Thanks to CRISPR. Crop Biotech Update. www.isaaa.org/kc/cropbiotechupdate (accessed 27 January 2022).

IPCC (2021). Chapter 1: framing, context, methods. In: *Climate Change 2021: The Physical Science Basis*. Contribution of Working Group I to the Sixth Assessment Report of the Intergovernmental Panel on Climate Change (ed. V. Masson-Delmotte, P. Zhai, A. Pirani, et al.). Cambridge University Press.

Iqbal, Z., Sattar, M.N., and Shafiq, M. (2016). CRISPR/Cas9: a tool to circumscribe cotton leaf curl disease. *Front. Plant Sci.* 7: 475. https://doi.org/10.3389/fpls.2016.00475.

Jacquin, A.L.S., Odom, D.T., and Lukk, M. (2019). Crisflash: open-source software to generate CRISPR guide RNAs against genomes annotated with individual variation. *Bioinformatics* 35: 3146–3147. https://doi.org/10.1093/bioinformatics/btz019.

Jakočiūnas, T., Rajkumar, A.S., Zhang, J. et al. (2015). CasEMBLR: Cas9-facilitated multiloci genomic integration of *in vivo* assembled DNA parts in *Saccharomyces cerevisiae*. *ACS Synth. Biol.* 4: 1126–1134. https://doi.org/10.1021/acssynbio.5b0000.

Jefferson, R.A., Kavanagh, T.A., and Bevan, M.W. (1987). GUS fusions: beta-glucuronidase as a sensitive and versatile gene fusion marker in higher plants. *EMBO J.* 6 (13): 3901–3907.

Jia, H., Orbovic, V., Jones, J.B., and Wang, N. (2016). Modification of the PthA4 effector binding elements in type I CsLOB1 promoter using Cas9/SgRNA to produce transgenic Duncan grapefruit alleviating Xcc1pthA4:DCsLOB1.3 infection. *Plant Biotechnol.* 14: 1291–1301. https://doi.org/10.1111/pbi.12495.

Jia, H., Zhang, Y., Orbović, V. et al. (2017). Genome editing of the disease susceptibility gene CsLOB1 in citrus confers resistance to citrus canker. *Plant Biotechnol.* 15: 817–823. https://doi.org/10.1111/pbi.12677.

Jiang, W., Zhou, H., Bi, H. et al. (2013). Demonstration of CRISPR/Cas9/SgRNA-mediated targeted gene modification in *Arabidopsis*, tobacco, sorghum and rice. *Nucleic Acids Res.* 41: 188. https://doi.org/10.1093/nar/gkt780.

Jiang, W.Z., Henry, I.M., Lynagh, P.G. et al. (2017). Significant enhancement of fatty acid composition in seeds of the allohexaploid, *Camelina sativa*, using CRISPR/Cas9 gene editing. *Plant Biotechnol.* 15: 648–646.

Jiang, M., Liu, Y., Liu, Y. et al. (2019). Mutation of inositol 1,3,4-trisphosphate 5/6-kinase6 impairs plant growth and phytic acid synthesis in rice. *Plants* 8: 114. https://doi.org/10.3390/plants8050114.

Jiang, Y., Chai, Y., Lu, M. et al. (2020). Prime editing efficiently generates W542L and S621I double mutations in two ALS genes in maize. *Genome Biol.* 21: 257. https://doi.org/10.1186/s13059-020-02170-5.

Joseph, O.G. (2021). *Ghana Scientist Turns to Gene Editing to Improve Sweet Potato Crop*. Alliance for Science.

Joshi, R.K. (2019). Genome editing in chili pepper using a CRISPR/Cas9 cytidine base editing system. *Asian J. Plant Sci. Res.* 9: 16.

Jouanin, A., Gilissen, L.J.W.J., Schaart, J.G. et al. (2020). CRISPR/Cas9 gene editing of gluten in wheat to reduce gluten content and exposure—reviewing methods to screen for coeliac safety. *Front. Nutr.* 7: 51. https://doi.org/10.3389/fnut.2020.00051.

Kapusi, E., Corcuera-Gómez, M., Melnik, S., and Stoger, E. (2017). Heritable genomic fragment deletions and small indels in the putative ENGase gene induced by CRISPR/Cas9 in barley. *Front. Plant Sci.* 8: 540.

Karavolias, N.G., Horner, W., Abugu, M.N., and Evanega, S.N. (2021). Application of gene editing for climate change in agriculture. *Front. Sustain. Food Syst.* 5: 685801. https://doi.org/10.3389/fsufs.2021.685801.

Karembu, M. (2021). "Genome editing in Africa's agriculture 2021: an early take-of," in International Service for the Acquisition of Agri-biotech Applications (ISAAA AfriCenter) (Nairobi). https://africenter.isaaa.org/ (accessed 31 August 2022).

Kaul, T., Raman, N.M., Eswaran, M. et al. (2019). Data mining by pluralistic approach on CRISPR gene editing in plants. *Front. Plant Sci.* 10: 801.

Kaul, T., Sony, S.K., Raman, N.M. et al. (2020a). How crisp is CRISPR? CRISPR/Cas mediated crop improvement with special focus on nutritional traits. In: *Advancement in Crop Improvement Techniques*, vol. 20 (ed. N. Tuteja, R. Tuteja, N. Passricha, and S. Saifi), 159–197. New Delhi: Woodhead Printing.

Kaul, T., Sony, S.K., Verma, R. et al. (2020b). Revisiting CRISPR-Cas mediated crop improvement: special focus on nutrition. *J. Biosci.* 45: 137.

Kaul, T., Sony, S.K., Bharti, J. et al. (2022). CRISPR genome editing brings global food security into the first lane: enhancing nutrition and stress resilience in crops. In: *Next-Generation Plant Breed Approaches for Stress Resilience in Cereal Crops* (ed. M.G. Mallikarjuna, S.C. Nayaka, and T. Kaul), 288–344. Springer.

Kaur, K., Gupta, A.K., Rajput, A., and Kumar, M. (2016). Ge-CRISPR: an integrated pipeline for the prediction and analysis of sgRNAs genome editing efficiency for CRISPR/Cas system. *Sci. Rep.* 6: 1–12. https://doi.org/10.1038/srep30870.

Kaur, N., Alok, A., Shivani, K.P. et al. (2020). CRISPR/Cas9 directed editing of lycopene epsilon-cyclase modulates metabolic flux for Î²-carotene biosynthesis in banana fruit. *Metab. Eng.* 59: 76–86. https://doi.org/10.1016/j.ymben.2020.01.008.

Kawaguchi, K., Takei-Hoshi, R., and Yoshikawa, I. (2021). Functional disruption of cell wall invertase inhibitor by genome editing increases sugar content of tomato fruit without decrease fruit weight. *Sci. Rep.* 11: 21534. https://doi.org/10.1038/s41598-021-00966-4.

Khan, M.S.S., Basnet, R., Islam, S.A., and Shu, Q. (2019). Mutational analysis of OsPLDα1 reveals its involvement in phytic acid biosynthesiin rice grains. *J. Agric. Food Chem.* 67: 11436–11443.

Khan, M., Hu, L., Zhu, M. et al. (2021). Targeted mutagenesis of EOD3 gene in Brassica napus L. regulates seed production. *J. Cell. Physiol.* 236 (3): 1996–2007.

Khorsandi, P. (2020). WFP chief warns of 'hunger pandemic' as Global Food Crises Report launched. *World Food Programme Insight, 22 April 2020*. https://www.wfp.org/ (accessed 24 June 2022).

Kieu, N.P., Lenman, M., Wang, E.S. et al. (2021). Mutations introduced in susceptibility genes through CRISPR/Cas9 genome editing confer increased late blight resistance in potatoes. *Sci. Rep.* 11: 4487. https://doi.org/10.1038/s41598-021-83972-w.

Kim, D., Alptekin, B., and Budak, H. (2017). CRISPR/Cas9 genome editing in wheat. *Funct. Integr. Genomics* 18 (1): 31–41. https://doi.org/10.1007/s10142-017-0572-x.

Kim D, Alptekin B, Budak H. CRISPR/Cas9 genome editing in wheat. *Funct. Integr. Genomics* 2018 18(1):31–41. doi: https://doi.org/10.1007/s10142-017-0572-x.

Kim, Y.A., Moon, H., and Park, C.J. (2019). CRISPR/Cas9-targeted mutagenesis of Os8N3 in rice to confer resistance to *xanthomonas oryzae* pv. oryzae. *Rice* 12: 67. https://doi.org/10.1186/s12284-019-0325-7.

Kis, A., Hamar, É., Tholt, G. et al. (2019). Creating highly efficient resistance against wheat dwarf virus in barley by employing CRISPR/Cas9 system. *Plant Biotechnol.* 17: 1004–1006.

Klap, C., Yeshayahou, E., Bolger, A.M. et al. (2017). Tomato facultative parthenocarpy results from Sl AGAMOUS-LIKE 6 loss of function. *Plant Biotechnol.* 15 (5): 634–647.

Klimek-Chodacka, M., Oleszkiewicz, T., Lowder, L.G. et al. (2018). Efficient CRISPR/Cas9-based genome editing in carrot cells. *Plant Cell Rep.* 37 (4): 575–586. https://doi.org/10.1007/s00299-018-2252-2.

Knott, G.J. and Doudna, J.A. (2018). CRISPR-Cas guides the future of genetic engineering. *Science* 361: 866–869.

Koblan, L.W., Arbab, M., Shen, M.W. et al. (2021). Efficient C•G-to-G•C base editors developed using CRISPRi screens, target-library analysis, and machine learning. *Nat. Biotechnol.* 39 (11): 1414–1425. https://doi.org/10.1038/s41587-021-00938-z.

Kouranova, E., Forbes, K., Zhao, G. et al. (2016). CRISPRs for optimal targeting: delivery of CRISPR components as DNA, RNA, and protein into cultured cells and single-cell embryos. *Hum. Gene Ther.* 27 (6): 464–475.

Kuang, Y., Li, S., Ren, B. et al. (2020). Base-editing mediated artificial evolution of OsALS1 in planta to develop novel herbicide-tolerant rice germplasms. *Mol. Plant* 13: 565–572.

Kumar, N., Galli, M., Ordon, J. et al. (2018). Further analysis of barley MORC1 using a highly efficient RNA-guided Cas9 Gene-editing system. *Plant Biotechnol.* 16: 1892–1903. https://doi.org/10.1111/pbi.12924.

Kumar, V.V.S., Verma, R.K., Yadav, S.K. et al. (2020). CRISPR-Cas9 mediated genome editing of drought and salt tolerance (OsDST) gene in indica mega rice cultivar MTU1010. *Physiol. Mol. Biol. Plants* 26: 1099–1110.

Kuroiwa, K., Thenault, C., Nogué, F. et al. (2022). CRISPR-based knock-out of eIF4E2 in a cherry tomato background successfully recapitulates resistance to pepper veinal mottle virus. *Plant Sci.* 316: 111160. https://doi.org/10.1016/j.plantsci.2021.111160.

Kurtz, S. (2003). The Vmatch large scale sequence analysis software. *Comput. Progr.* 4–12.

Kusano, H., Ohnuma, M., Mutsuro-Aoki, H. et al. (2018). Establishment of a modified CRISPR/Cas9 system with increased mutagenesis frequency using the translational enhancer dMac3 and multiple guide RNAs in potato. *Sci. Rep.* 8: 13753. https://doi.org/10.1038/s41598-018-32049-2.

Labuhn, M., Adams, F.F., Ng, M. et al. (2017). Refined sgRNA efficacy prediction improves large- and small-scale CRISPR-Cas9 applications. *Nucleic Acids Res.* 46: 1375–1385. https://doi.org/10.1093/nar/gkx1268.

Labun, K., Montague, T.G., Gagnon, J.A. et al. (2016). CHOPCHOP v2: a web tool for the next-generation of CRISPR genome engineering. *Nucleic Acids Res.* 44: 272–276. https://doi.org/10.1093/nar/gkw398.

Labun, K., Montague, T.G., Krause, M. et al. (2019). CHOPCHOP v3: expanding the CRISPR web toolbox beyond genome editing. *Nucleic Acids Res.* 47: 171–174. https://doi.org/10.1093/nar/gkz365.

Lal, R., Follett, R.F., Stewart, B.A., and Kimble, J.M. (2007). Soil carbon sequestration to mitigate climate change and advance food security. *Soil Sci.* 172 (12): 943–956.

Lei, Y., Lu, L., Liu, H.Y. et al. (2014). CRISPR-P: a web tool for synthetic single-guide RNA design of CRISPR-system in plants. *Mol. Plant* 7: 1494–1496. https://doi.org/10.1093/mp/ssu044.

Lemmon, Z.H., Reem, N.T., Dalrymple, J. et al. (2018). Rapid improvement of domestication traits in an orphan crop by genome editing. *Nat. Plants* 4: 766–770. https://doi.org/10.1038/s41477-018-0259-x.

Li, X. and Siddique, K.H.M. (2020). Future Smart Food: Harnessing the potential of neglected and underutilized species for Zero Hunger. *Matern. Child Nutr.* 16 (S3): 13008. https://doi.org/10.1111/mcn.13008.

Li, Z., Liu, Z.B., Xing, A. et al. (2015). Cas9-guide RNA directed genome editing in soybean. *Plant Physiol.* 169 (2): 960–970.

Li, J., Meng, X., Zong, Y. et al. (2016a). Gene replacements and insertions in rice by intron targeting using CRISPR-Cas9. *Nat. Plants* 2: 16139.

Li, M., Li, X., Zhou, Z. et al. (2016b). Reassessment of the four yield-related genes Gn1a, DEP1, GS3, and IPA1 in rice using a CRISPR/Cas9 system. *Front. Plant Sci.* 7: 377.

Li, Z., Zhang, D., Xiong, X. et al. (2017a). A potent Cas9-derived gene activator for plant and mammalian cells. *Nat. Plants* 3: 930–936.

Li, X., Zhou, W., Ren, Y. et al. (2017b). High-efficiency breeding of early-maturing rice cultivars via CRISPR/Cas9-mediated genome editing. *J. Genet. Genomics* 44: 175–178. https://doi.org/10.1016/j.jgg.2017.02.001.

Li, X., Wang, Y., Chen, S. et al. (2018a). Lycopene is enriched in tomato fruit by CRISPR/Cas9-mediated multiplex genome editing. *Front. Plant Sci.* 9: 559.

Li, R., Zhang, L., Wang, L. et al. (2018b). Reduction of tomato-plant chilling tolerance by CRISPR-Cas9-mediated SlCBF1 mutagenesis. *J. Agric. Food Chem.* 66: 9042–9051. https://doi.org/10.1021/acs.jafc.8b02177.

Li, A., Jia, S., Yobi, A. et al. (2018c). Editing of an alphakafirin gene family increases, digestibility and protein quality in sorghum. *Plant Physiol.* 177: 1425–1438. https://doi.org/10.1104/pp.18.00200.

Li, R., Li, R., Li, X. et al. (2018d). Multiplexed CRISPR/Cas9-mediated metabolic engineering of γ-aminobutyric acid levels *in Solanum lycopersicum*. *Plant Biotechnol.* 16 (2): 415–427. https://doi.org/10.1111/pbi.12781.

Li, R., Liu, C., Zhao, R. et al. (2019a). CRISPR/Cas9-mediated SlNPR1 mutagenesis reduces tomato plant drought tolerance. *BMC Plant Biol.* 19: 38. https://doi.org/10.1186/s12870-018-1627-4.

Li, S., Shen, L., Hu, P. et al. (2019b). Developing disease-resistant thermosensitive male sterile rice by multiplex gene editing. *J. Integr. Plant Biol.* 61: 1201–1205.

Li, C., Li, W., Zhou, Z. et al. (2020). A new rice breeding method: CRISPR/Cas9 system editing of the Xa13 promoter to cultivate transgene-free bacterial blight-resistant rice. *Plant Biotechnol.* 18: 313–315.

Li X, Sandgrind S, Moss O, Guan R, Ivarson E, Wang ES, Kanagarajan S, Zhu LH. Efficient protoplast regeneration protocol and CRISPR/Cas9-mediated editing of glucosinolate transporter (GTR) genes in rapeseed (Brassica napus L.). Front *Plant Sci.* 2021 7;12:680859. doi: https://doi.org/10.3389/fpls.2021.680859.

Li, S., Lin, D., Zhang, Y. et al. (2022). Genome-edited powdery mildew resistance in wheat without growth penalties. *Nature* 602: 455–460. https://doi.org/10.1038/s41586-022-04395-9.

Liang, Z., Zhang, K., Chen, K., and Gao, C. (2014). Targeted mutagenesis in *Zea mays* using TALENs and the CRISPR/Cas system. *J. Genet. Genomics* 41 (2): 63–68.

Liang, J., Shen, Q., Wang, L. et al. (2021). Rice contains a biosynthetic gene cluster associated with production of the casbane-type diterpenoid phytoalexin ent-10-oxodepressin. *New Phytol.* 231 (1): 85–93.

Liao, S., Qin, X., Luo, L. et al. (2019). CRISPR/Cas9-induced mutagenesis of semi-rolled leaf1,2 confers curled leaf phenotype and drought tolerance by influencing protein expression patterns and ROS scavenging in rice (*Oryza sativa* L.). *Agronomy* 9 (11): 728. https://doi.org/10.3390/agronomy9110728.

Liliane, T.N. and Charles, M.S. (2020). Factors affecting yield of crops. In: *Agronomy – Climate Change & Food Security* (ed. A. Amanullah). London: IntechOpen https://doi.org/10.5772/intechopen.90672.

Lin, J. and Wong, K.C. (2018). Off-target predictions in CRISPR-Cas9 gene editing using deep learning. *Bioinformatics* 34: 656–663. https://doi.org/10.1093/bioinformatics/bty554.

Listgarten, J., Weinstein, M., Kleinstiver, B.P. et al. (2018). Prediction of off-target activities for the end-to-end design of CRISPR guide RNAs. *Nat. Biomed. Eng.* 2: 38–47. https://doi.org/10.1038/s41551-017-0178-6.

Liu, H., Wei, Z., Dominguez, A. et al. (2015). CRISPR-ERA: a comprehensive design tool for CRISPR-mediated gene editing, repression and activation. *Bioinformatics* 31: 36763678. https://doi.org/10.1093/bioinformatics/btv423.

Liu, H., Ding, Y., Zhou, Y. et al. (2017). CRISPR-P 2.0: an improved CRISPR-Cas9 tool for genome editing in plants. *Mol. Plant* 10: 530–532. https://doi.org/10.1016/j.molp.2017.01.003.

Liu, G., Li, J., and Godwin, I.D. (2019). Genome editing by CRISPR/Cas9 in sorghum through biolistic bombardment. *Methods Mol. Biol.* 1931: 169–183. https://doi.org/10.1007/978-1-4939-9039-9_12.

Liu, X., Qin, R., Li, J. et al. (2020a). A CRISPR-Cas9-mediated domain-specific base-editing screen enables functional assessment of ACCase variants in rice. *Plant Biotechnol.* 18: 1845–1847.

Liu, Y., Kuang, F., Yan, S. et al. (2020b). Developing a novel artificial rice germplasm for dinitroaniline herbicide resistance by base editing of OsTubA2. *Plant Biotechnol. J.* https://doi.org/10.1111/pbi.13430.

Lou, D., Wang, H., Liang, G., and Yu, D. (2017). OsSAPK2 confers abscisic acid sensitivity and tolerance to drought stress in rice. *Front. Plant Sci.* 8: 1–15.

Lowder, L.G., Zhang, D., Baltes, N.J. et al. (2015). A CRISPR/Cas9 toolbox for multiplexed plant genome editing and transcriptional regulation. *Plant Physiol.* 169: 971–985.

Lu, Y. and Zhu, J.-K. (2017). Precise editing of a target base in the rice genome using a modified CRISPR/Cas9 system. *Mol. Plant* 10: 523–525.

Luo, J., Chen, W., Xue, L., and Tang, B. (2019). Prediction of activity and specificity of CRISPR-Cpf1 using convolutional deep learning neural networks. *BMC Bioinformatics* 20: 1–10. https://doi.org/10.1186/s12859-019-2939-6.

Lyzenga, W.J., Harrington, M., Bekkaoui, D. et al. (2019). CRISPR/Cas9 editing of three CRUCIFERIN C homoeologues alters the seed protein profile in Camelina sativa. *BMC Plant Biol* 19: 292.

Ma, X., Zhang, Q., Zhu, Q. et al. (2015). A robust CRISPR-Cas9 system for convenient, high-efficiency multiplex genome editing in monocot and dicot plants. *Mol. Plant* 8: 1274–1284.

Macovei, A., Sevilla, N.R., Cantos, C. et al. (2018). Novel alleles of rice EIF4G generated by CRISPR/Cas9-targeted mutagenesis confer resistance to Rice tungro spherical virus. *Plant Biotechnol.* 16: 1918–1927. https://doi.org/10.1111/pbi.12927.

MacPherson, C.R. and Scherf, A. (2015). Flexible guide-RNA design for CRISPR applications using Protospacer Workbench. *Nat. Biotechnol.* 33: 805–806. https://doi.org/10.1038/nbt.3291.

Makhotenko, A.V., Khromov, A.V., Snigir, E.A. et al. (2019). Functional analysis of coilin in virus resistance and stress tolerance of potato *solanum tuberosum* using CRISPR-Cas9 editing. *Dokl. Biochem. Biophys.* 484: 88–91. https://doi.org/10.1134/S1607672919010241.

Malnoy, M., Viola, R., Jung, M.-H. et al. (2016). DNA-free genetically edited grapevine and apple protoplast using CRISPR/Cas9 ribonucleoproteins. *Front. Plant Sci.* 7: 1904. https://doi.org/10.3389/fpls.2016.01904.

Masuda, H., Ishimaru, Y., Aung, M.S. et al. (2012). Iron biofortification in rice by the introduction of multiple genes involved in iron nutrition. *Sci. Rep.* 2 (1): 1–7.

McHughen, A. (2016). A critical assessment of regulatory triggers for products of biotechnology: product vs. process. *GM Crops Food* 7: 3–4. https://doi.org/10.1080/21645698.2016.1228516.

McKenna, A. and Shendure, J. (2018). FlashFry: a fast and flexible tool for large-scale CRISPR target design. *BMC Biol.* 16: 4–9. https://doi.org/10.1186/s12915-018-0545-0.

Medek, D.E., Schwartz, J., and Myers, S.S. (2017). Estimated effects of future atmospheric CO_2 concentrations on protein intake and the risk of protein deficiency by country and region. *Environ. Health Perspect.* 125 (8): 087002. https://doi.org/10.1289/EHP41.

Medvedieva, M.O. and Blume, Y.B. (2018). Legal regulation of plant genome editing with the CRISPR/Cas9 technology as an example. *Cytol. Genet.* 52: 204–212. https://doi.org/10.3103/S0095452718030106.

Mehta, D., Stürchler, A., Anjanappa, R.B. et al. (2019). Linking CRISPR-Cas9 interference in cassava to the evolution of editingresistant geminiviruses. *Genome Biol.* 20: 80.

Mendoza, B.J. and Trinh, C.T. (2018). Enhanced guide-RNA design and targeting analysis for precise CRISPR genome editing of single and consortia of industrially relevant and non-model organisms. *Bioinformatics* 34: 16–23. https://doi.org/10.1093/bioinformatics/btx564.

Miao, J., Guo, D., Zhang, J. et al. (2013). Targeted mutagenesis in rice using CRISPR-Cas system. *Cell Res.* 23 (10): 1233–1236.

Miao, C., Xiao, L., Hua, K. et al. (2018). Mutations in a subfamily of abscisic acid receptor genes promote rice growth and productivity. *Proc. Natl. Acad. Sci. U.S.A.* 115: 6058–6063. https://doi.org/10.1073/pnas.1804774115.

Michno, J.M., Wang, X., Liu, J. et al. (2015). CRISPR/Cas mutagenesis of soybean and *Medicago truncatula* using a new web-tool and a modified Cas9 enzyme. *GM Crops Food* 6: 243–252. https://doi.org/10.1080/21645698.2015.1106063.

Ming, M., Ren, Q., Pan, C. et al. (2020). CRISPR–Cas12b enables efficient plant genome engineering. *Nat. Plants* 6: 202–208.

Minkenberg, B., Zhang, J., Xie, K., and Yang, Y. (2019). CRISPR-PLANT v2: an online resource for highly specific guide RNA spacers based on improved off-target analysis. *Plant Biotechnol.* 17: 5–8. https://doi.org/10.1111/pbi.13025.

Montague, T.G., Cruz, J.M., Gagnon, J.A. et al. (2014). CHOPCHOP: a CRISPR/Cas9 and TALEN web tool for genome editing. *Nucleic Acids Res.* 42: 401–407. https://doi.org/10.1093/nar/gku410.

Moreno-Mateos, M.A., Vejnar, C.E., Beaudoin, J.-D. et al. (2015). CRISPRscan: designing highly efficient sgRNAs for CRISPR-Cas9 targeting in vivo. *Nat. Methods* 12: 982–988. https://doi.org/10.1038/nmeth.3543.

Morrell, R. and Sadanandom, A. (2019). Dealing with stress: a review of plant SUMO proteases. *Front. Plant Sci.* 10: 1122.

Nadakuduti, S.S. and Enciso-Rodriguez, F. (2021). Advances in genome editing with CRISPR systems and transformation technologies for plant DNA manipulation. *Front. Plant Sci.* 11: 637159. https://doi.org/10.3389/fpls.2020.637159.

Naidoo, R. and Fisher, B. (2020). Reset sustainable development goals for a pandemic world. *Nature* 583: 198–201. https://doi.org/10.1038/d41586-020-01999-x.

Naito, Y., Hino, K., Bono, H., and Ui-Tei, K. (2015). CRISPRdirect: software for designing CRISPR/Cas guide RNA with reduced off-target sites. *Bioinformatics* 31: 1120–1123. https://doi.org/10.1093/bioinformatics/btu743.

Nakajima, I., Ban, Y., Azuma, A. et al. (2017). CRISPR/Cas9-mediated targeted mutagenesis in grape. *PLoS One* 12 (5): 0177966. https://doi.org/10.1371/journal.pone.0177966.

Nakayasu, M., Akiyama, R., Lee, H.J. et al. (2018). Generation of α-solanine-free hairy roots of potato by CRISPR/Cas9 mediated genome editing of the St16DOX gene. *Plant Physiol. Biochem.* 131: 70–77. https://doi.org/10.1016/j.plaphy.2018.04.026.

Nawaz, G., Usman, B., Peng, H. et al. (2020). Knockout of Pi21 by CRISPR/Cas9 and iTRAQ-based proteomic analysis of mutants revealed new insights into *M. oryzae* resistance in elite rice line. *Genes (Basel)* 11: 735.

Nekrasov, V., Wang, C., Win, J. et al. (2017). Rapid generation of a transgene-free powdery mildew resistant tomato by genome deletion. *Sci. Rep.* 7: 482. https://doi.org/10.1038/s41598-017-00578-x.

Newman, J.A. (2004). Climate change and cereal aphids: the relative effects of increasing CO_2 and temperature on aphid population dynamics. *Glob. Chang. Biol.* 10 (1): 5–15.

Nonaka, S., Arai, C., Takayama, M. et al. (2017). Efficient increase of ɣ-aminobutyric acid (GABA) content in tomato fruits by targeted mutagenesis. *Sci. Rep.* 7: 7057.

O'Brien, A. and Bailey, T.L. (2014). GT-scan: identifying unique genomic targets. *Bioinformatics* 30: 2673–2675. https://doi.org/10.1093/bioinformatics/btu354.

O'Brien, A.R., Burgio, G., and Bauer, D.C. (2021). Domain-specific introduction to machine learning terminology, pitfalls and opportunities in CRISPR-based gene editing. *Brief. Bioinform.* 22 (1): 308–314. https://doi.org/10.1093/bib/bbz145.

Ogata, T., Ishizaki, T., Fujita, M., and Fujita, Y. (2020). CRISPR/Cas9-targeted mutagenesis of OsERA1 confers enhanced responses to abscisic acid and drought stress and increased primary root growth under nonstressed conditions in rice. *PLoS One* 15 (12): 0243376. https://doi.org/10.1371/journal.pone.0243376.

Okuzaki, A., Ogawa, T., Koizuka, C. et al. (2018). CRISPR/Cas9-mediated genome editing of the fatty acid desaturase 2 gene in *Brassica napus*. *Plant Physiol. Biochem.* 131: 63–69.

Oliva, R., Ji, C., Atienza-Grande, G. et al. (2019). Broad-spectrum resistance to bacterial blight in rice using genome editing. *Nat. Biotechnol.* 37: 1344–1350. https://doi.org/10.1038/s41587-019-0267-z.

Oliveros, J.C., Franch, M., Tabas-Madrid, D. et al. (2016). Breaking-Cas—interactive design of guide RNAs for CRISPR-Cas experiments for ENSEMBL genomes. *Nucleic Acids Res.* 44: 267–271. https://doi.org/10.1093/nar/gkw407.

Oo, A.T., Van Huylenbroeck, G., and Speelman, S. (2020). Measuring the economic impact of climate change on crop production in the dry zone of Myanmar: a Ricardian approach. *Climate* 8: 9.

Ortigosa, A., Gimenez-Ibanez, S., Leonhardt, N., and Solano, R. (2019). Design of a bacterial speck resistant tomato by CRISPR/Cas9-mediated editing of SlJAZ2. *Plant Biotechnol.* 17: 665–673. https://doi.org/10.1111/pbi.13006.

Pan, C., Wu, X., Markel, K. et al. (2021). CRISPR-Act3.0 for highly efficient multiplexed gene activation in plants. *Nat. Plants.* 7 (7): 942–953. https://doi.org/10.1038/s41477-021-00953-7.

Pandey, V., Patel, H.R., and Patel, V.J. (2007). Impact assessment of climate change on wheat yield in Gujarat using CERES-wheat model. *J. Agrometeorol.* 9 (2): 149–157.

Papikian, A., Liu, W., Gallego-Bartolomé, J., and Jacobsen, S.E. (2019). Site specific manipulation of *Arabidopsis loci* using CRISPR-Cas9 SunTag systems. *Nat. Commun.* 10: 729.

Park, J., Bae, S., and Kim, J.S. (2015). Cas-designer: a web-based tool for choice of CRISPR-Cas9 target sites. *Bioinformatics* 31: 4014–4016. https://doi.org/10.1093/bioinformatics/btv537.

Peng, D. and Tarleton, R. (2015). EuPaGDT: a web tool tailored to design CRISPR guide RNAs for eukaryotic pathogens. *Microb. Genom.* 1: 1–7. https://doi.org/10.1099/mgen.0.000033.

Peng, A., Chen, S., Lei, T. et al. (2017). Engineering canker-resistant plants through CRISPR/Cas9-targeted editing of the susceptibility gene CsLOB1 promoter in citrus. *Plant Biotechnol.* 15 (12): 1509–1519. https://doi.org/10.1111/pbi.12733.

Peng, H., Zheng, Y., Blumenstein, M. et al. (2018). CRISPR/Cas9 cleavage efficiency regression through boosting algorithms and Markov sequence profiling. *Bioinformatics* 34: 3069–3077. https://doi.org/10.1093/bioinformatics/bty298.

Piatek, A., Ali, Z., Baazim, H. et al. (2015). RNA-guided transcriptional regulation in planta via synthetic dCas9-based transcription factors. *Plant Biotechnol.* 13: 578–589.

Pliatsik, V. and Rigoutsos, I. (2015). "Off-Spotter": very fast and exhaustive enumeration of genomic lookalikes for designing CRISPR/Cas guide RNAs. *Biol. Direct* 10: 1–10. https://doi.org/10.1186/s13062-015-0035-z.

Pompili, V., Dalla Costa, L., Piazza, S. et al. (2020). Reduced fire blight susceptibility in apple cultivars using a high-efficiency CRISPR/Cas9-FLP/FRT-based gene editing system. *Plant Biotechnol.* 18 (3): 845–858.

Population Reference Bureau (2021). World Population Data Sheet. https://www.prb.org/ (accessed 27 June 2022).

Prihatna, C., Barbetti, M.J., and Barker, S.J. (2018). A novel tomato Fusarium wilt tolerance gene. *Front. Microbiol.* 9: 1226. https://doi.org/10.3389/fmicb.2018.01226.

Probsting, M., Schenke, D., Hossain, R. et al. (2020). Loss of function of CRT1a (calreticulin) reduces plant susceptibility to Verticillium longisporum in both Arabidopsis thaliana and oilseed rape (Brassica napus). *Plant Biotechnol. J.* 18: 2328–2344.

Prykhozhij, S.V., Rajan, V., Gaston, D., and Berman, J.N. (2015). CRISPR multitargeter: a web tool to find common and unique CRISPR single guide RNA targets in a set of similar sequences. *PLoS One* 10: 1–18. https://doi.org/10.1371/journal.pone.0119372.

Pyott, D.E., Sheehan, E., and Molnar, A. (2016). Engineering of CRISPR/Cas9-mediated potyvirus resistance in transgene-free *Arabidopsis* plants. *Mol. Plant Pathol.* 17 (8): 1276–1288.

Qi, W., Zhu, T., Tian, Z. et al. (2016). High-efficiency CRISPR/Cas9 multiplex gene editing using the glycine tRNA-processing system-based strategy in maize. *BMC Biotechnol.* 16 (1): 58. Published 11 August 2016. https://doi.org/10.1186/s12896-016-0289-2.

Rahman, M.K. and Rahman, M.S. (2017). CRISPRpred: a flexible and efficient tool for sgRNAs on-target activity prediction in CRISPR/Cas9 systems. *PLoS One* 12: 1–14. https://doi.org/10.1371/journal.pone.0181943.

Ren, C., Liu, X., Zhang, Z. et al. (2016). CRISPR/Cas9-mediated efficient targeted mutagenesis in Chardonnay (*Vitis vinifera* L.). *Sci. Rep.* 6: 1–9.

Rodríguez-Leal, D., Lemmon, Z.H., Man, J. et al. (2017). Engineering quantitative trait variation for crop improvement by genome editing. *Cell* 171: 470–480.

Roginsky, J. (2018). Analyzing CRISPR editing results. *Genet. Eng. Biotechnol. News* 38: S24–S26. https://doi.org/10.1089/gen.38.11.13.

Ronda, C., Pedersen, L.E., Sommer, M.O.A., and Nielsen, A.T. (2016). CRMAGE: CRISPR optimized MAGE recombineering. *Sci. Rep.* 6: 1–11. https://doi.org/10.1038/srep19452.

Sanchez-Leon, S., Gil-Humanes, J., Ozuna, C.V. et al. (2018). Low-gluten, nontransgenic wheat engineered with CRISPR/Cas9. *Plant Biotechnol.* 16: 902–910.

Sander, J.D., Maeder, M.L., Reyon, D. et al. (2010). ZiFiT (Zinc Finger Targeter): an updated zinc finger engineering tool. *Nucleic Acids Res.* 38: 462–468.

Sanjana, N.E., Shalem, O., and Zhang, F. (2014). Improved vectors and genome-wide libraries for CRISPR screening. *Nat. Methods* 11: 783–784.

Santillan Martínez, M.I., Bracuto, V., Koseoglou, E. et al. (2020). CRISPR/Cas9-targeted mutagenesis of the tomato susceptibility gene PMR4 for resistance against powdery mildew. *BMC Plant Biol.* 20: 284. https://doi.org/10.1186/s12870-020-02497-y.

Sashidhar, N., Harloff, H.J., Potgieter, L., and Jung, C. (2020). Gene editing of three BnITPK genes in tetraploid oilseed rape leads to significant reduction of phytic acid in seeds. *Plant Biotechnol.* 18: 2241–2250.

Sauer, N.J., Narvaez-Vasquez, J., Mozoruk, J. et al. (2016). Oligonucleotide-mediated genome editing provides precision and function to engineered nucleases and antibiotics in plants. *Plant Physiol.* 170: 1917–1928.

Selma, S., Bernabé-Orts, J.M., Vazquez-Vilar, M. et al. (2019). Strong gene activation in plants with genome-wide specificity using a new orthogonal CRISPR/Cas9-based programmable transcriptional activator. *Plant Biotechnol.* 17: 1703–1705.

Shan, Q., Wang, Y., Li, J. et al. (2013). Targeted genome modification of crop plants using a CRISPR-Cas system. *Nat. Biotechnol.* 31 (8): 686–688.

Shao, X., Wu, S., Dou, T. et al. (2020). Using CRISPR/Cas9 genome editing system to create MaGA20ox2 gene-modified semi-dwarf banana. *Plant Biotechnol.* 18: 17–19. https://doi.org/10.1111/pbi.13216.

Shen, C., Que, Z., Xia, Y. et al. (2017). Knock out of the annexin gene *OsAnn3* via CRISPR/Cas9-mediated genome editing decreased cold tolerance in rice. *J. Plant Biol.* 60: 539–547.

Shen, L., Wang, C., Fu, Y. et al. (2018). QTL editing confers opposing yield performance in different rice varieties. *J. Integr. Plant Biol.* 60: 89–93.

Shi, J., Gao, H., Wang, H. et al. (2017). ARGOS8 Variants generated by CRISPR-Cas9 improve maize grain yield under field drought stress conditions. *Plant Biotechnol.* 15: 207–216. https://doi.org/10.1111/pbi.12603.

Shim, J.S., Oh, N., Chung, P.J. et al. (2018). Overexpression of OsNAC14 improves drought tolerance in rice. *Front. Plant Sci.* 9: 310. https://doi.org/10.3389/fpls.2018.00310.

Shimatani, Z., Fujikura, U., Ishii, H. et al. (2018). Inheritance of co-edited genes by CRISPR-based targeted nucleotide substitutions in rice. *Plant Physiol. Biochem.* 131: 78–83.

Shivran, A.C. (2016). Biofortification for nutrient-rich millets. In: *Biofortification of Food Crops* (ed. U. Singh, C.S. Praharaj, S.S. Singh, and N.P. Singh), 409–420. https://doi.org/10.1007/978-81-322-2716-8_30.

Singh, R., Kuscu, C., Quinlan, A. et al. (2015). Cas9-chromatin binding information enables more accurate CRISPR off-target prediction. *Nucleic Acids Res.* 43: 1–8.

Smedley, M.A., Hayta, S., Clarke, M., and Harwood, W.A. (2021). CRISPR-Cas9 based genome editing in wheat. *Curr. Protoc.* 1 (3): 65.

Son, J., Jang, S.H., Cha, J.W., and Jeong, K.J. (2020). Development of CRISPR interference (CRISPRi) platform for metabolic engineering of *Leuconostoc citreum* and its application for engineering riboflavin biosynthesis. *Int. J. Mol. Sci.* 21: 5614.

Soyk, S. (2017). Variation in the flowering gene SELF PRUNING 5G promotes day-neutrality and early yield in tomato. *Nat. Genet.* 49: 162–168.

Springmann, M., Godfray, H., Charles, J. et al. (2016). Analysis and valuation of the health and climate change co benefits of dietary change. *Proc. Natl. Acad. Sci. U.S.A.* 113: 4146. https://doi.org/10.1073/pnas.1523119113.

Staley, J.T. and Johnson, S.N. (2008). Climate change impacts on root herbivores. In: *Root Feeders: An Ecosystem Perspective* (ed. S.N. Johnson and P.J. Murray), 192–213. Wallingford, UK: CABI.

Stemmer, M., Thumberger, T., Del Sol, K.M. et al. (2015). CCTop: an intuitive, flexible and reliable CRISPR/Cas9 target prediction tool. *PLoS One* 10: 1–11.

Sun, Y., Jiao, G., Liu, Z. et al. (2017). Generation of high-amylose rice through CRISPR/Cas9-mediated targeted mutagenesis of starch branching enzymes. *Front. Plant Sci.* 8: 1–15.

Sun, B., Jiang, M., and Zheng, H. (2020). Color-related chlorophyll and carotenoid concentrations of Chinese kale can be altered through CRISPR/Cas9 targeted editing of the carotenoid isomerase gene BoaCRTISO. *Hortic. Res.* 7: 161. https://doi.org/10.1038/s41438-020-00379-w.

Sun, S.K., Xu, X.J., Tang, Z. et al. (2021). A molecular switch in sulfur metabolism to reduce arsenic and enrich selenium in rice grain. *Nat. Commun.* 12: 1392.

Sunitha, S. and Rock, C.D. (2020). CRISPR/Cas9-mediated targeted mutagenesis of TAS4 and MYBA7 loci in grapevine rootstock 101–14. *Transgenic Res.* 29: 355–367. https://doi.org/10.1007/s11248-020-00196-w.

Swinburn BA, Kraak VI, Allender S, Atkins VJ, Baker PI, Bogard JR, Brinsden H, Calvillo A, et al. (2019) The global syndemic of obesity, undernutrition, and climate change: the Lancet Commission report. *Lancet.* 393(10173):791–846. doi: https://doi.org/10.1016/S0140

Tan, S.Z., Reisch, C.R., and Prather, K.L.J. (2018). A robust CRISPR interference gene repression system in Pseudomonas. *J. Bacteriol.* 200: 1–12.

Tang, F., Yang, S., Liu, J., and Zhu, H. (2016). Rj4, a gene controlling nodulation specificity in soybeans, encodes a thaumatin-like protein but not the one previously reported. *Plant Physiol.* 170 (1): 26–32. https://doi.org/10.1104/pp.15.01661.

Tang, X., Lowder, L.G., Zhang, T. et al. (2017a). A CRISPR-Cpf1 system for efficient genome editing and transcriptional repression in plants. *Nat. Plants* 3: 1–5.

Tang, L., Mao, B., Li, Y. et al. (2017b). Knockout of OsNramp5 using the CRISPR/Cas9 system produces low Cd-accumulating indica rice without compromising yield. *Sci. Rep.* 7: 14438.

Tang, X., Sretenovic, S., Ren, Q. et al. (2020). Plant prime editors enable precise gene editing in rice cells. *Mol. Plant* 13: 667–670.

Tashkandi, M., Ali, Z., Aljedaani, F. et al. (2018). Engineering resistance against tomato yellow leaf curl virus via the CRISPR/Cas9 system in tomato. *Plant Signal. Behav.* 13: 1525996. https://doi.org/10.1080/15592324.2018.1525996.

Tian, S., Jiang, L., Cui, X. et al. (2018). Engineering herbicide-resistant watermelon variety through CRISPR/Cas9-mediated base-editing. *Plant Cell Rep.* 37 (9): 1353–1356. https://doi.org/10.1007/s00299-018-2299-0.

Tian, Q., Li, B., Feng, Y. et al. (2022). Application of CRISPR/Cas9 in rapeseed for gene function research and genetic improvement. *Agronomy* 12 (4): 824. https://doi.org/10.3390/agronomy12040824.

Tiwari, M., Kumar Trivedi, P., and Pandey, A. (2021). Emerging tools and paradigm shift of gene editing in cereals, fruits, and horticultural crops for enhancing nutritional value and food security. *Food Energy Secur.* 10 (1): 258.

de Toledo Thomazella, D.P., Brail, Q., Dahlbeck, D., and Staskawicz, B. (2016). CRISPR-Cas9 mediated mutagenesis of a DMR6 ortholog in tomato confers broad-spectrum disease resistance. *BioRxiv [Preprint].* 1–23. https://doi.org/10.1101/064824.

Tran, M.T., Doan, D.T.H., Kim, J. et al. (2020). CRISPR/Cas9-based precise excision of SlHyPRP1 domain(s) to obtain salt stress-tolerant tomato. *Plant Cell Rep.* 40 (6): 999–1011. https://doi.org/10.1007/s00299-020-02622-z.

Tripathi, J.N., Ntui, V.O., Ron, M. et al. (2019). CRISPR/Cas9 editing of endogenous banana streak virus in the B genome of *Musa* Spp. overcomes a major challenge in banana breeding. *Commun. Biol.* 2: 1–11. https://doi.org/10.1038/s42003-019-0288-7.

Tripathi, L., Ntui, V.O., and Tripathi, J.N. (2020). CRISPR/Cas9-based genome editing of banana for disease resistance Curr. *Opin. Plant Biol.* 56: 118–126.

Tuncel, A., Corbin, K.R., Ahn-Jarvis, J. et al. (2019). Cas9-mediated mutagenesis of potato starch-branching enzymes generates a range of tuber starch phenotypes. *Plant Biotechnol.* 17: 2259–2271.

Uluisik S, Chapman NH, Smith R, Poole M, Adams G, Gillis RB, Besong TM, Sheldon J, Stiegelmeyer S, Perez L, Samsulrizal N, Wang D, Fisk ID, Yang N, Baxter C, Rickett D, Fray R, Blanco-Ulate B, Powell AL, Harding SE, Craigon J, Rose JK, Fich EA, Sun L, Domozych DS, Fraser PD, Tucker GA, Grierson D, Seymour GB (2016) Genetic improvement of tomato by targeted control of fruit softening. *Nat. Biotechnol.* 34(9):950–2. doi: https://doi.org/10.1038/nbt.3602.

Upadhyay, S.K. and Sharma, S. (2014). SSFinder: high throughput CRISPR-Cas target sites prediction tool. *Biomed. Res. Int.* 2014: 742482. https://doi.org/10.1155/2014/742482.

Usman, B., Nawaz, G., Zhao, N. et al. (2020a). Precise editing of the *OsPYL9* gene by RNA-guided Cas9 nuclease confers enhanced drought tolerance and grain yield in rice (*Oryza sativa* L.) by regulating circadian rhythm and abiotic stress responsive proteins. *Int. J. Mol. Sci.* 21: 7854.

Usman, B., Nawaz, G., Zhao, N. et al. (2020b). Generation of high yielding and fragrant rice (*Oryza sativa* L.) lines by CRISPR/Cas9 targeted mutagenesis of three homoeologs of cytochrome P450 gene family and OsBADH2 and transcriptome and proteome profiling of revealed changes triggered by mutations. *Plants* 9 (6): 788. https://doi.org/10.3390/plants9060788.

Usman, B., Nawaz, G., Zhao, N. et al. (2021). Programmed editing of rice (*Oryza sativa*) osspl16 gene using crispr/cas9 improves grain yield by modulating the expression of pyruvate enzymes and cell cycle proteins. *Int. J. Mol. Sci.* 22: 249.

Varkonyi-Gasic E, Wang T, Voogd C, Jeon S, Drummond RSM, Gleave AP, Allan AC (2018) Mutagenesis of kiwifruit Centrodialis-like genes transforms a climbing woody perennial with long juvenility and axillary flowering into a compact plant with rapid terminal flowering. *Plant Biotechnol.* doi:https://doi.org/10.1111/pbi.13021.

Varshney, R.K., Nayak, S.N., and May, G.D. (2009). Next generation sequencing technologies and their implications for crop genetics and breeding. *Trends Biotechnol.* 27: 522–530.

Veillet, F., Perrot, L., Chauvin, L. et al. (2019). Transgene-free genome editing in tomato and potato plants using agrobacterium-mediated delivery of a CRISPR/Cas9 cytidine base editor. *Int. J. Mol. Sci.* 20: 402. https://doi.org/10.3390/ijms20020402.

Wallace, J.S., Acreman, M.C., and Sullivan, C.A. (2003). The sharing of water between society and ecosystems: from conflict to catchment–based co-management. *Philos. Trans. R. Soc. Lond. Ser. B Biol. Sci.* 358 (1440): 2011–2026. https://doi.org/10.1098/rstb.2003.1383.

Waltz, E. (2016). Gene-edited CRISPR mushroom escapes US regulation. *Nature* 532 (7599): 293. https://doi.org/10.1038/nature.2016.19754.

Wan, D.Y., Guo, Y., Cheng, Y. et al. (2020). CRISPR/Cas9-mediated mutagenesis of VvMLO3 results in enhanced resistance to powdery mildew in grapevine (*Vitis Vinifera*). *Hortic. Res.* 7: 1–14. https://doi.org/10.1038/s41438-020-0339-8.

Wang, Y., Cheng, X., Shan, Q. et al. (2014). Simultaneous editing of three homoeoalleles in hexaploid bread wheat confers heritable resistance to powdery mildew. *Nat. Biotechnol.* 32 (9): 947–951. https://doi.org/10.1038/nbt.2969.

Wang, F., Wang, C., Liu, P. et al. (2016). Enhanced rice blast resistance by CRISPR/Cas9-targeted mutagenesis of the ERF transcription factor gene OsERF922. *PLoS One* 11: e0154027. https://doi.org/10.1371/journal.pone.0154027.

Wang, M., Mao, Y., Lu, Y. et al. (2017a). Multiplex gene editing in rice using the CRISPR-Cpf1 system. *Mol. Plant* 10: 1011–1013.

Wang, L., Chen, L., Li, R. et al. (2017b). Reduced drought tolerance by CRISPR/Cas9-mediated SlMAPK3 mutagenesis in tomato plants. *J. Agric. Food Chem.* 65: 8674–8682. https://doi.org/10.1021/acs.jafc.7bv.

Wang, W., Pan, Q., He, F. et al. (2018a). Transgenerational CRISPR-Cas9 activity facilitates multiplex gene editing in allopolyploid wheat. *CRISPR* 1: 65–74. https://doi.org/10.1089/crispr.2017.0010.

Wang, Q., Cobine, P.A., and Coleman, J.J. (2018b). Efficient genome editing in *Fusarium oxysporum* based on CRISPR/Cas9 ribonucleoprotein complexes. *Fungal Genet. Biol.* 117: 21–29.

Wang, L., Wu, J., Wang, Q. et al. (2018c). Programmed self-elimination of the CRISPR/Cas9 construct greatly accelerates the isolation of edited and transgene-free rice plants. *Mol. Plant* 11 (9): 1210–1213.

Wang, X., Tu, M., Wang, D. et al. (2018d). CRISPR/Cas9-mediated efficient targeted mutagenesis in grape in the first generation. *Plant Biotechnol.* 16 (4): 844–855. https://doi.org/10.1111/pbi.12832.

Wang, H., Wu, Y., Zhang, Y. et al. (2019a). CRISPR/Cas9-based mutagenesis of starch biosynthetic genes in sweet potato (*Ipomoea batatas*) for the improvement of starch quality. *Int. J. Mol. Sci.* 20: 4702.

Wang, J., Kuang, H., Zhang, Z. et al. (2019b). Generation of seed lipoxygenase-free soybean using CRISPR-Cas9. *Crop J.* 8 (3): 432–439. https://doi.org/10.1016/j.cj.2019.08.008.

Wang, Z., Hong, Y., Li, Y. et al. (2020a). Natural variations in SlSOS1 contribute to the loss of salt tolerance during tomato domestication. *Plant Biotechnol. J.* 19 (1): 20–22. https://doi.org/10.1111/pbi.13443.

Wang, Z., Hong, Y., Zhu, G. et al. (2020b). Loss of salt tolerance during tomato domestication conferred by variation in a Na (+) /K(+) transporter. *EMBO* 39: 103256. https://doi.org/10.15252/embj.201913256.

Wang, F., Xu, Y., Li, W. et al. (2021). Creating a novel herbicide-tolerance OsALS allele using CRISPR/Cas9-mediated gene editing. *Crop J.* 9: 305–312. https://doi.org/10.1016/j.cj.2020.06.001.

Watanabe, K., Ulrike, B., Götz, H. et al. (2016). Stable gene replacement in barley by targeted double-strand break induction. *J. Exp. Bot.* 67 (5): 1433–1445. https://doi.org/10.1093/jxb/erv537.

Wen, S., Liu, H., Li, X. et al. (2018). TALEN-mediated targeted mutagenesis of fatty acid desaturase 2 (FAD2) in peanut (*Arachis hypogaea* L.) promotes the accumulation of oleic acid. *Plant Mol. Biol.* 97: 177–185. https://doi.org/10.1007/s11103-018-0731-z.

Wheeler, T. and von Braun, J. (2013). Climate change impacts on global food security. *Science* 341 (6145): 508–513. https://doi.org/10.1126/science.1239402.

Willett, W., Rockström, J., Loken, B. et al. (2019). Food in the Anthropocene: the EAT–Lancet Commission on healthy diets from sustainable food systems. *Lancet* 393 (10170): 447–492. https://doi.org/10.1016/S0140-6736(18)31788-4.

Wilson, F.M., Harrison, K., Armitage, A.D. et al. (2019). CRISPR/Cas9-mediated mutagenesis of phytoene desaturase in diploid and octoploid strawberry. *Plant Methods* 15 (1): 1–13.

Wong, N., Liu, W., and Wang, X. (2015). WU-CRISPR: characteristics of functional guide RNAs for the CRISPR/Cas9 system. *Genome Biol.* 16: 1–8.

Woo, J.W., Kim, J., Kwon, S.I. et al. (2015). DNA-free genome editing in plants with preassembled CRISPR-Cas9 ribonucleo proteins. *Nat. Biotechnol.* 33 (11): 1162–1164. https://doi.org/10.1038/nbt.3389.

Wood, J., Wong, C., and Paturi, S. (2021). Addressing food security in constrained urban environments. In: *Tropical Constrained Environments and Sustainable Adaptations* (ed. S. Azzali and K. Thirumaran), 169–186. Singapore: Springer https://doi.org/10.1007/978-981-33-4631-4_10.

Wu, X., Scott, D.A., Kriz, A.J. et al. (2014). Genome-wide binding of the CRISPR endonuclease Cas9 in mammalian cells. *Nat. Biotechnol.* 32: 670–676.

Wu, J., Chen, C., Xian, G., Liu, D., Lin, L., Yin, S., et al. (2020). Engineering herbicide-resistant oilseed rape by CRISPR/Cas9-mediated cytosine base-editing. *Plant. Biotechnol. J.* 18: 1857–1859. doi: https://doi.org/10.1111/pbi.13368

Xiao, A., Cheng, Z., Kong, L. et al. (2014). CasOT: a genome-wide Cas9/gRNA off-target searching tool. *Bioinformatics* 30 (8): 1180–1182.

Xie, K. and Yang, Y. (2013). RNA-guided genome editing in plants using a CRISPR–Cas system. *Mol. Plant* 6: 1975–1983.

Xie, S., Shen, B., Zhang, C. et al. (2014). SgRNAcas9: a software package for designing CRISPR sgRNA and evaluating potential off-target cleavage sites. *PLoS One* 9: 1–9.

Xie, X., Ma, X., Zhu, Q. et al. (2017). CRISPR-GE: a convenient software toolkit for CRISPR-based genome editing. *Mol. Plant* 10: 1246–1249.

Xing, S., Chen, K., Zhu, H. et al. (2020). Fine-tuning sugar content in strawberry. *Genome Biol.* 21: 230. https://doi.org/10.1186/s13059-020-02146-5.

Xiong, X., Liang, J., Li, Z. et al. (2021). Multiplex and optimization of dCas9-TV-mediated gene activation in plants. *J. Integr. Plant Biol.* 63: 634–645.

Xu, H., Xiao, T., Chen, C.H. et al. (2015). Sequence determinants of improved CRISPR sgRNA design. *Genome Res.* 25: 1147–1157.

Xu, R., Yang, Y., Qin, R. et al. (2016). Rapid improvement of grain weight via highly efficient CRISPR/Cas9-mediated multiplex genome editing in rice. *J. Genet. Genomics* 43: 529–532.

Xu, Z., Xu, X., Gong, Q. et al. (2019). Engineering broad spectrum bacterial blight resistance by simultaneously disrupting variable TALE-binding elements of multiple susceptibility genes in rice. *Mol. Plant* 12: 1434–1446. https://doi.org/10.1016/j.molp.2019.08.006.

Xu, R., Li, J., Liu, X. et al. (2020). Development of plant prime-editing systems for precise genome editing. *Plant Commun.* 1: 100043.

Xu, Y., Lin, Q., Li, X. et al. (2021). Fine-tuning the amylose content of rice by precise base editing of the Wx gene. *Plant Biotechnol.* 19: 11–13.

Xue, L., Tang, B., Chen, W., and Luo, J. (2019). Prediction of CRISPR sgRNA activity using a deep convolutional neural network. *J. Chem. Inf. Model* 59: 615–624.

Yamauchi, T., Yasuyo, J., Sachiko, F. et al. (2009). Homologous recombination-mediated knock-in targeting of the MET1a gene for a maintenance DNA methyl transferase reproducibly reveals dosage-dependent spatiotemporal gene expression in rice. *Plant J.* 60 (2): 386–396. https://doi.org/10.1111/j.1365-313x.2009.03947.x.

Yan, L., Wei, S., Wu, Y. et al. (2015). High-efficiency genome editing in *Arabidopsis* using YAO promoter-driven CRISPR/Cas9 system. *Mol. Plant* 8: 1820–1823.

Yang, B. (2020). Grand challenges in genome editing in plants. *Front. Genome Ed.* 2: 2.

Yang, X., Chen, L., He, J., and Yu, W. (2017). Knocking out of carotenoid catabolic genes in rice fails to boost carotenoid accumulation, but reveals a mutation in strigolactone biosynthesis. *Plant Cell Rep.* 36 (10): 1533–1545. https://doi.org/10.1007/s00299-017-2172-6.

Yasumoto, S., Umemoto, N., Lee, H.J. et al. (2019). Efficient genome engineering using Platinum TALEN in potato. *Plant Biotechnol.* 36: 167–173.

Yeo, N.C., Chavez, A., Lance-Byrne, A. et al. (2018). An enhanced CRISPR repressor for targeted mammalian gene regulation. *Nat. Methods* 15: 611–616.

Yin, X., Biswal, A.K., Dionora, J. et al. (2017). CRISPR-Cas9 and CRISPR-Cpf1 mediated targeting of a stomatal developmental gene EPFL9 in rice. *Plant Cell Rep.* 36: 745–757. https://doi.org/10.1007/s00299-017-2118-z.

Yin, Y., Qin, K., Song, X. et al. (2018). BZR1 transcription factor regulates heat stress tolerance through FERONIA receptor like kinase-mediated reactive oxygen species signaling in tomato. *Plant Cell Physiol.* 59: 2239–2254. https://doi.org/10.1093/pcp/pcy146.

Yohannes, H.A. (2016). Review on relationship between climate change and agriculture. *J. Earth Sci. Clim. Chang.* 7: 335.

Yu Q, Wang B, Li N, Tang Y, Yang S, Yang T, Xu J, Guo C, Yan P, Wang Q, Asmutola P (2017) CRISPR/Cas9-induced targeted mutagenesis and gene replacement to generate long-shelf life tomato lines. *Sci. Rep.* 7(1):11874. doi:https://doi.org/10.1038/s41598-017-12262-1.

Yuan, M., Zhu, J., Gong, L. et al. (2019). Mutagenesis of FAD2 genes in peanut with CRISPR/Cas9 based gene editing. *BMC Biotechnol.* 19: 24. https://doi.org/10.1186/s12896-019-0516-8.

Yue, E., Cao, H., and Liu, B. (2020). OsmiR535, a potential genetic editing target for drought and salinity stress tolerance in *Oryza sativa*. *Plants* 9 (10): 1337. https://doi.org/10.3390/plants9101337.

Zafar, K., Khan, M.Z., Amin, I. et al. (2020). Precise CRISPR-Cas9 mediated genome editing in super basmati rice for resistance against bacterial blight by targeting the major susceptibility gene. *Front. Plant Sci.* 11: 575.

Zaidi, S.SeA., Mahas, A., Vanderschuren, H. *et al*. Engineering crops of the future: CRISPR approaches to develop climate-resilient and disease-resistant plants. *Genome Biol.* **21**, 289 (2020). https://doi.org/10.1186/s13059-020-02204-y

Zaman QU, Li C, Cheng H, et al. Genome editing opens a new era of genetic improvement in polyploid crops. *Crop J.* 2019;7:141–150. DOI: https://doi.org/10.1016/j.cj.2018.07.004.

Zaman QU, Wen C, Yuqin S, Mengyu H, Desheng M, Jacqueline B, Baohong Z, Chao L, Qiong H. Characterization of SHATTERPROOF homoeologs and CRISPR-Cas9-mediated genome editing enhances pod-shattering resistance in Brassica napus L. *CRISPR J.* 2021 4(3):360–370. doi: https://doi.org/10.1089/crispr.2020.0129.

Zeng, Z., Han, N., Liu, C. et al. (2020a). Functional dissection of HGGT and HPT in barley vitamin E biosynthesis via CRISPR/Cas9-enabled genome editing. *Ann. Bot.* 126 (5): 929–942. https://doi.org/10.1093/aob/mcaa115.

Zeng, Y., Wen, J., Zhao, W. et al. (2020b). Rational improvement of rice yield and cold tolerance by editing the three genes OsPIN5b, GS3, and OsMYB30 with the CRISPR–Cas9 system. *Front. Plant Sci.* 10: 1663.

Zhai, Y., Yu, K., Cai, S. et al. (2020). Targeted mutagenesis of BnTT8 homologs controls yellow seed coat development for effective oil production in *Brassica napus* L. *Plant Biotechnol.* 18: 1153–1168.

Zhang, B., Yang, X., and Yang, C. (2016a). Exploiting the CRISPR/Cas9 system for targeted genome mutagenesis in petunia. *Sci. Rep.* 6: 20315. https://doi.org/10.1038/srep20315.

Zhang, Y., Liang, Z., Zong, Y. et al. (2016b). Efficient and transgene-free genome editing in wheat through transient expression of CRISPR/Cas9 DNA or RNA. *Nat. Commun.* 7 (1): 1–8.

Zhang, Y., Bai, Y., Wu, G. et al. (2017). Simultaneous modification of three homoeologs of TaEDR1 by genome editing enhances powdery mildew resistance in wheat. *Plant J. Cell Mol. Biol.* 91: 714–724. https://doi.org/10.1111/tpj.13599.

Zhang, M., Cao, Y., Wang, Z. et al. (2018a). A retrotransposon in an HKT1 family sodium transporter causes variation of leaf Na+ exclusion and salt tolerance in maize. *New Phytol.* 217 (3): 1161–1176.

Zhang, Z., Ge, X., Luo, X. et al. (2018b). Simultaneous editing of two copies of Gh14-3-3d confers enhanced transgene-clean plant defense against *Verticillium dahliae* in allotetraploid upland cotton. *Front. Plant Sci.* 9: 842. https://doi.org/10.3389/fpls.2018.00842.

Zhang, H.W., Si, X.W., Ji, X. et al. (2018c). Genome editing of upstream open reading frames enables translational control in plants. *Nat Biotechnol* 36: 894–898.

Zhang, A., Liu, Y., Wang, F. et al. (2019a). Enhanced rice salinity tolerance via CRISPR/Cas9-targeted mutagenesis of the OsRR22 gene. *Mol. Breed.* 39: 47. https://doi.org/10.1007/s11032-019-0954-y.

Zhang, S., Li, X., Lin, Q., and Wong, K.C. (2019b). Synergizing CRISPR/Cas9 off-target predictions for ensemble insights and practical applications. *Bioinformatics* 35 (7): 1108–1115.

Zhang, R., Chen, S., Meng, X. et al. (2020). Generating broad-spectrum tolerance to ALS-inhibiting herbicides in rice by base editing. *Sci. China Life Sci.* 64 (10): 1624–1633.

Zhao, C., Liu, B., Piao, S. et al. (2017). Temperature increase reduces global yields of major crops in four independent estimates. *Proc. Natl. Acad. Sci. U.S.A.* 114 (35): 9326–9331. https://doi.org/10.1073/pnas.1701762114.

Zheng, M., Zhang, L., Tang, M. et al. (2020). Knockout of two BnaMAX1 homologs by CRISPR/Cas9-targeted mutagenesis improves plant architecture and increases yield in rapeseed (*Brassica napus* L.). *Plant Biotechnol.* 18 (3): 644–654.

Zhong, Z., Zhang, Y., You, Q. et al. (2018). Plant genome editing using FnCpf1 and LbCpf1 nucleases at redefined and altered PAM sites. *Mol. Plant* 11: 999–1002.

Zhou, X., Richard, H., Ian, P.W. et al. (1995). Effects of temperature on aphid phenology. *Global Change Biol.* 1 (4): 303–313. https://doi.org/10.1111/j.1365-2486. 1995.tb00029.x.

Zhou, J., Peng, Z., Long, J. et al. (2015). Gene targeting by the TAL effector PthXo2 reveals cryptic resistance gene for bacterial blight of rice. *Plant J.* 82: 632–643. https://doi.org/10.1111/tpj.12838.

Zhou, H., He, M., Li, J. et al. (2016). Development of commercial thermo-sensitive genic male sterile rice accelerates hybrid rice breeding using the CRISPR-Cas9-mediated TMS5 editing system. *Sci. Rep.* 6: 37395.

Zhou, X., Liao, H., Chern, M. et al. (2018). Loss of function of a rice TPR-Domain RNA-binding protein confers broadspectrum disease resistance. *Proc. Natl. Acad. Sci. U.S.A.* 115: 3174–3179. https://doi.org/10.1073/pnas.1705927115.

Zhou, J., Xin, X., He, Y. et al. (2019). Multiplex QTL editing of grain-related genes improves yield in elite rice varieties. *Plant Cell Rep.* 38: 475–485. https://doi.org/10.1007/s00299-018-2340-3.

Zhu, L.J. (2015). Overview of guide RNA design tools for CRISPR-Cas9 genome editing technology. *Front Biol. (Beijing)* 10: 289–296.

Zhu LJ, Holmes BR, Aronin N, Brodsky MH. CRISPRseek: a bioconductor package to identify target-specific guide RNAs for CRISPR-Cas9 genome-editing systems. *PLoS One.* 2014 9(9):e108424. doi: https://doi.org/10.1371/

Zhu, H., Misel, L., Graham, M. et al. (2016a). CT-Finder: a web service for CRISPR optimal target prediction and visualization. *Sci. Rep.* 6: 1–8. https://doi.org/10.1038/srep25516.

Zhu, J., Song, N., Sun, S. et al. (2016b). Efficiency and inheritance of targeted mutagenesis in maize using CRISPR-Cas9. *J. Genet. Genomics* 43 (1): 25–36. https://doi.org/10.1016/j.jgg.2015.10.006.

Zsögön, A., Tomáš, Č., Rezende, N.E. et al. (2018). *De novo* domestication of wild tomato using genome editing. *Nat. Biotechnol.* 36: 1211–1216.

Zuo, Y., Feng, F., Qi, W., and Song, R. (2019). Dek42 encodes an RNA-binding protein that affects alternative Pre-MRNA splicing and maize kernel development. *J. Integr. Plant Biol.* 61: 728–748. https://doi.org/10.1111/jipb.12798.

Part 3

Socioeconomic Aspects of Climate Change

10

Perspective of Evolution of the C_4 Plants to Develop Climate Designer C_4 Rice as a Strategy for Abiotic Stress Management

Shuvobrata Majumder, Karabi Datta, and Swapan K. Datta

Department of Botany, University of Calcutta, Kolkata, India

10.1 Introduction

Addressing the effects of climate change in the agriculture sector is the need of the hour. Adequate amount of food production for the increasing population is a global challenge (Figure 10.1). Agriculture-dependent countries and specifically regions dependent on small-scale farming are directly affected by climate change, causing serious socioeconomic imbalance. Current plant breeding systems are dedicated to improving crop yields to secure food for us under rapidly changing climatic conditions. A continuous search for improved and upgraded strategies to deal with adverse climate conditions like temperatures, carbon dioxide concentration, aridity, salinity, and drought is much needed.

If we look at the evolution of the plant species for the last 60–70 million years, from grass species and chenopods first appearing on the earth to modern crop plants, we can understand plant adaptation based on changing environmental conditions (Figure 10.2). Initially, the atmospheric carbon dioxide concentration was extremely high, around 3600 ppm, and primitive plants assimilated ambient CO_2 to reduce into organic matter by the process of C_3 photosynthesis (Calvin–Benson cycle). Several important crops like rice (*Oryza sativa*), wheat (*Triticum aestivum*), cotton (*Gossypium* sp.), jute (*Corchorus* sp.), soybean (*Glycine max*), potato (*Solanum tuberosum*), tomato (*Solanum lycopersicum*), alfalfa (*Medicago sativa*), sunflower (*Helianthus annuus*), and sugar beets (*Beta vulgaris*) are examples of C_3 plants. Rubisco (ribulose-1,5-bisphosphate carboxylase oxygenase), the main carboxylating enzyme, evolved in C_3 plants (Hayes 1994). More knowledge and understanding of this adaptation pattern are required for the relevant formulation of new crop stress management strategies.

About 35 million years ago, atmospheric conditions started changing such that the level of CO_2 became low (by more than 10 times) from 3600 to 350 ppm and oxygen surged; photorespiration was thriving (Berner and Kothavala 2001). Photorespiration can inhibit photosynthesis by over 30% when conditions of warmth prevail (Jordan and Ogren 1984). Therefore, metabolic changes took place in plants to scale up photosynthesis by gradually shifting from the C_3 to the C_4 photosynthesis cycle about 34–24 million years ago. It was essentially an adaptation to compensate for the high rate of photorespiration and carbon deficiency. C_4 photosynthesis is believed to have evolved more than 62 times in the past 60 million years (Sage et al. 2011). Crops like maize (*Zea mays*), sorghum (*Sorghum bicolor*), switchgrass (*Panicum virgatum*), sugarcane (*Saccharum officinarum*), and millet (*Paniceae* sp.) are examples of C_4 crop plants.

The C_4 cycle is a combination of complex physiological and biochemical processes that maintain CO_2 concentration around Rubisco in the bundle sheath (Dengler and Nelson 1999) and suppress the negative effect of photorespiration on plant photosynthesis. Bundle sheath cells (BSCs) and mesophyll cells (MCs) are the two types of photosynthetic cells that harbor photosynthetic enzymes in the C_4 leaf, while the photosynthetic enzymes in the C_3 leaf are encompassed only in the mesophyll cells (Table 10.1). The C_4 leaf has ample venation, and each vein is surrounded by a circular arrangement of BSCs that is again surrounded by MCs. This looks like a wreath-like arrangement of cells under the microscope and is known as the Kranz anatomy, found in many variations in C_4 monocots and dicots (Dengler and Nelson 1999; Kadereit et al. 2003).

Global Climate Change and Plant Stress Management, First Edition. Edited by Mohammad Wahid Ansari, Anil Kumar Singh, and Narendra Tuteja.

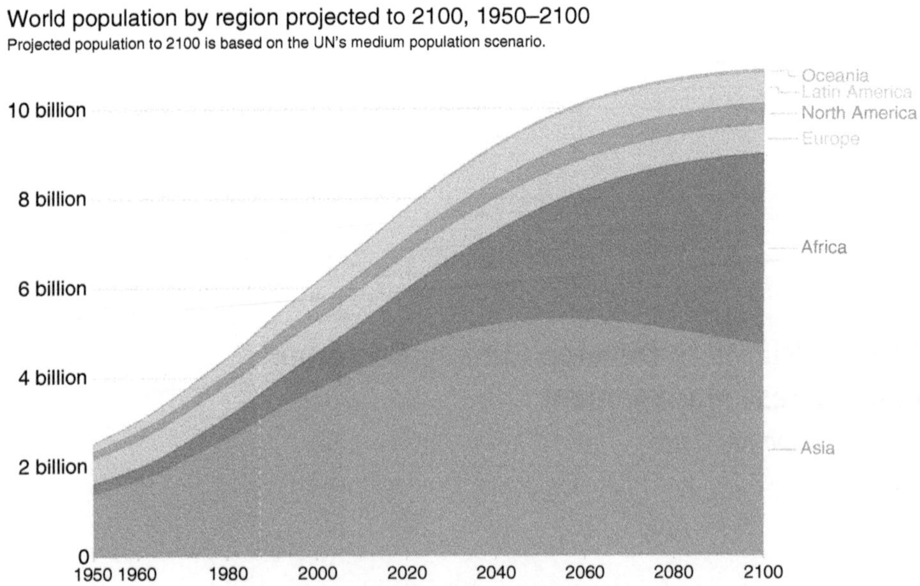

Figure 10.1 Current and projected world population. *Source:* www.ourworldindata.org/last accessed 1 December 2022/Licensed under CC BY 4.0.

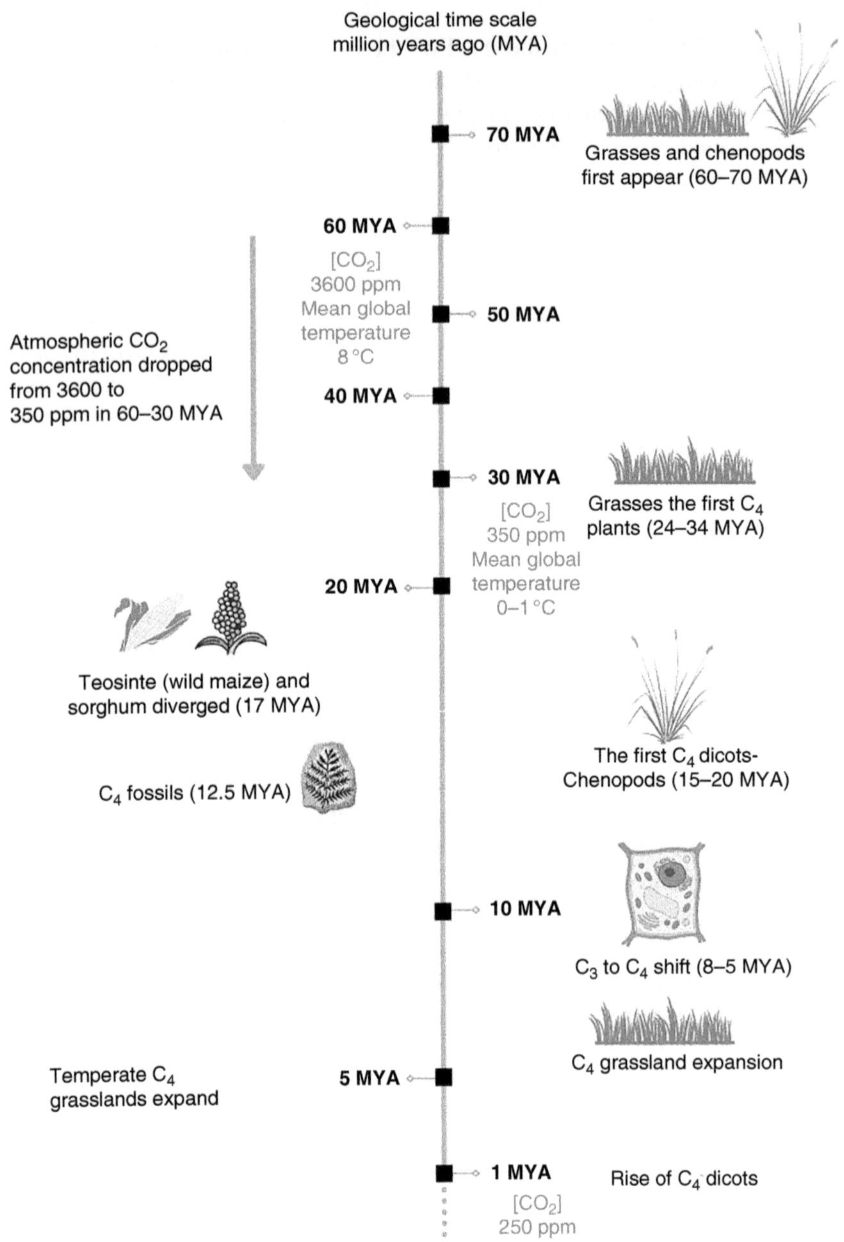

Figure 10.2 Landmarks in evolution, from C_3 to C_4 plants, in the geological time scale in million years. This infographic has been created. *Source:* Adapted from the information by Sage (2004).

Table 10.1 Major photosynthetic enzymes in C_4 plant.

Enzyme	Location in leaves	Main function
Rubisco (C_3)	Bundle sheath cells, chloroplasts	Net CO_2 fixation
PEPC (C_4)	Mesophyll cells, cytosol	Initial CO_2 (HCO_3^-) fixation
PPDK (C_4)	Mesophyll cells, chloroplasts	Regeneration of PEP
NADP-MDH (C_4)	Mesophyll cells, chloroplasts	OAA to malate conversion
NADP-ME (C_4)	Bundle sheath cells, chloroplasts	Decarboxylation of malate
NAD-ME (C_4)	Bundle sheath cells, mitochondria	Decarboxylation of malate
AspAT (C_4)	Mesophyll cells, cytosol Bundle sheath cells, mitochondria	OAA to asparate conversion in Mesophyll cells and asparate to OAA conversion in bundle sheath cells
PEP-CK (C_4)	Bundle sheath cells, cytosol	Decarboxylation of OAA
CA (C_4)	Mesophyll cells, cytosol	CO_2 to HCO_3^- conversion

Source: Adapted from Matsuoka et al. (2001).

In the C_4 pathway, PEPC starts carbon dioxide fixation in the mesophyll cytosol and produces oxaloacetate (OAA). A further reduction by NADP-malate dehydrogenase (NADP-MDH) converts OAA to malate. OAA could also be transaminated by aspartate aminotransferase (AspAT) to aspartate. The C_4 acid thus produced travels to the BSCs to be decarboxylated, producing CO_2 that concentrates around Rubisco. This conversion of OAA to a particular C_4 acid depends on the type of enzyme – NAD-malic enzyme (NAD-ME), NADP-malic enzyme (NADP-ME), or phosphoenolpyruvate carboxykinase (PEP-CK) – used in a particular evolutionary lineage, and hence the three biochemical subtypes of C_4 photosynthesis.

In the following sections, we briefly discuss how the C_4 system evolved and what are the advantages of C_4 plants over C_3 plants. This information could help us understand one of the recent areas of modern science that is engineering of a C_4 photosynthetic pathway into C_3 rice to enhance yield potential under abiotic stress conditions.

10.2 How Did Plants Evolve to the C_4 System?

It would be apt to consider the C_4 system as one of the highest convergents of the complex events in the history of evolution, owing to the 62 recognizable lineages of C_4 photosynthesis (Sage et al. 2011). This extraordinary system holds the secret to evolutionary adaptation and is worth studying. Evolving from a C_3 to a C_4 system was neither easy nor rapid. This mammoth morphosis manifested slowly, subsequently involving a certain order of changes. These have been discussed in brief here.

10.2.1 Gene Amplification and Modification

Abundant multiple copies of genes facilitated modification of the copies and at the same time allowed to conserve the original function of the transcribed protein (Lynch and Conery 2000). This could be achieved through sexual recombination and intense selection pressure on the duplicated genes, stimulating to evolve C_4 like traits. *Perennials* and other short-cycle plants, with new mutations are screened for stable adaptations to the changing environment, and once it has been established reproductive barriers commence the isolation process of the populations (Lynch and Conery 2000; Monson 2003).

10.2.2 Anatomical Preconditioning

Concentrating CO_2 in the bundle sheath is an important step in C_4 photosynthesis. This requires an effective CO_2 concentration mechanism. The foremost step in this direction would be to create the least possible distance between the MCs and BSCs for the metabolites to move rapidly (Raghavendra 1980; Ehleringer et al. 1997). This could be achieved by reducing the interveinal distance or by enhancing the bundle sheath layer size. This feat is more easily achievable in grass species with parallel venetian and difficult for dicots with reticulate venation, explaining why C_4 is more popular in the grass family (Ehleringer et al. 1997).

10.2.3 Increase in Bundle Sheath Organelles

Increasing the number of chloroplasts and mitochondria in the bundle sheath would lead to better metabolism of glycine and C_4 acids. With the new C_4 anatomy of reduced interveinal space and enhanced bundle sheath size in the leaf, chloroplast numbers increase to support light absorbance for better photosynthetic activity facilitating glycine movement (Brown and Hattersley 1989; Rawsthorne 1992).

10.2.4 Glycine Shuttles and Photorespiratory CO_2 Pumps

Efficient transportation and metabolism of glycine are required in the C_4 system. A glycine shuttle that transports glycine from the mesophyll vicinity to the Rubisco in the bundle sheath would be immensely beneficial in this scenario. This new and improved feature would help bring down the CO_2 compensation point and reduce the diffusion distance, thereby increasing intercellular transport.

10.2.5 Enhancement of PEPC and PPDK Activity in the Mesophyll Tissue

A rise in PEPC activity in the C_4 system dealt efficiently with the escaping CO_2 from the bundle sheath. The C_4 acids produced could be efficiently directed back to the bundle sheath for refixation by the increased PEPC activity. A 40 times increase in PEPC activity has been documented in *Flaveria* species C_3 to C_4 intermediates (Monson and Rawsthorne 2000).

A vital enzyme in the C_4 carbon assimilation pathway is PPDK. The synthesis of phosphoenolpyruvate, catalyzed by PPDK (acting as a primary CO_2 acceptor), is a rate-limiting step in the C_4 pathway (Hatch and Slack 1968). Although PPDK activity is scarce and not green leaf-specific in C_3 plants, it is impressively high in C_4 and CAM leaves (Edwards and Andreo 1992).

10.2.6 Integration of C_3 and C_4 Cycles

In C_4 evolution, the activity of all components of photosynthesis is increased, and a competition for CO_2 and ATP arises in the Rubisco and the C_3 cycles. This demands a major reorganizing of the enzymatic expression patterns for the merger of the C_3 and C_4 cycles. Differential screening approach revealed genes in the C_4 *Sorghum bicolor*, that are expressed in the MCs or BSCs (Wyrich et al. 1998). 25 cDNAs with mesophyll-specific expression and 8 cDNAs with bundle sheath-specific expression have been identified. Expression specific to the mesophyll included the enzymes vital to the C_4 cycle – PEPC, PPDK, CA, Photosystem II proteins, NADP oxidoreductase, and ferredoxin. Enzymes specific to the bundle sheath include Rubisco, NADP-ME, Rubisco activase, and many Calvin cycle enzymes. Integration of the cycles becomes undisputed when leaf PEPC activity equals Rubisco activity, as seen in *Flaveria* (Monsoon and Rawsthorne 2000).

10.3 What Are the Advantages of C_4 Plants over C_3 Plants?

Global temperature and CO_2 are gradually increasing (Figure 10.3), and so are abiotic stresses that are related to and induced by temperature changes. C_4 photosynthesis is more efficient than C_3 photosynthesis when environmental factors like heat, drought, salinity, and low humidity become high. In C_3 plants, high temperatures pose a direct stimulant for photorespiration and dark respiration. The availability of CO_2 declines in warm climates as it reduces the solubility of CO_2 over O_2. Under such conditions, C_4 photosynthesis proves to be more efficient.

Aridity and salinity in the soil facilitate stomatal closure. Stomatal closure reduces intercellular CO_2 level, which promotes photorespiration in an increased CO_2 deficiency condition. Stomatal closure was also reported in genetically engineered rice with stress inducible *rd29* promoter driven *DREB* genes for the development of drought-tolerant rice (Datta et al. 2012). Hot, arid regions are often accompanied by low relative humidity, which further reduces stomatal conductance, especially if the conditions are those of drought stress. In such adverse conditions, shifting to a C_4 system could make

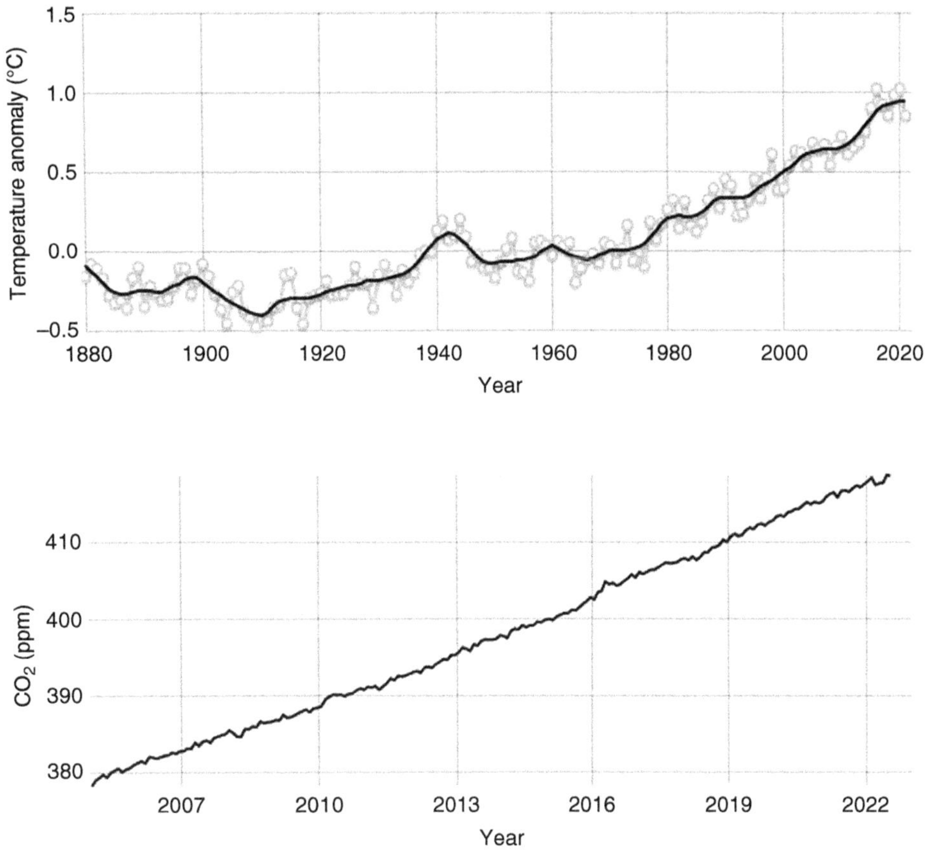

Figure 10.3 Global temperature and CO_2 increase in recent times. *Source:* www.climate.nasa.gov/last accessed 1 December 2022/ Public Domain.

a marked difference and scale up photosynthesis. Water use efficiency in the C_4 photosynthetic system is greater than the C_3 system. This enhances stomatal ability and functions efficiently even in low CO_2 and light (Schulze and Hall 1982; Huxman and Monson 2003).

In hot, humid environments such as low-latitude wetlands and flooded soil, the photorespiratory activity of the plants growing in these regions becomes high. C_3–C_4 intermediates are found amply in such regions. Aquatic C_4 evolution probably was favorable in similar conditions.

Conditions like low CO_2 levels, limited water supply, and low nitrogen use efficiency adversely affect C_3 photosynthesis, such that plant growth rate and recovery become slow. Plants are affected negatively in terms of flowering and reproduction, which leads to thinning of vegetation. C_4 photosynthesis is significantly efficient over C_3 photosynthesis under such conditions.

Water use efficiency improves in the C_4 photosynthetic system than the C_3 system, which assists in two vital whole-plant developments: increased stomatal sensitivity to photons and CO_2 (Schulze and Hall 1982; Huxman and Monson 2003) and flexibility in the xylem function to make way for structural alterations in the xylem. Leaf-specific conductivity in C_4 plants is low. This allows for an increased canopy, both in dimension and number (Kocacinar and Sage 2003).

These environment-resilient characteristics of C_4 plants attract plant researchers to develop strategies that protect from abiotic stress and produce sustainable plants.

10.4 Molecular Engineering of C_4 Enzymes in Rice

Rice is a C_3 plant. It is considered as an important staple food for more than half of the world's population (Majumder et al. 2019). Maintaining its yield (Figure 10.4a) and improving its production (Figure 10.4b) in the world is important for ensuring food security. Enzymes from C_4 plants have been introduced into rice to develop C_4 rice plants. A few such researches successfully improved photosynthetic ability and enhanced yield; on the other hand, some were not found to be significant. In the following section, we discuss some selected results from such experiments on the development of C_4 rice plants.

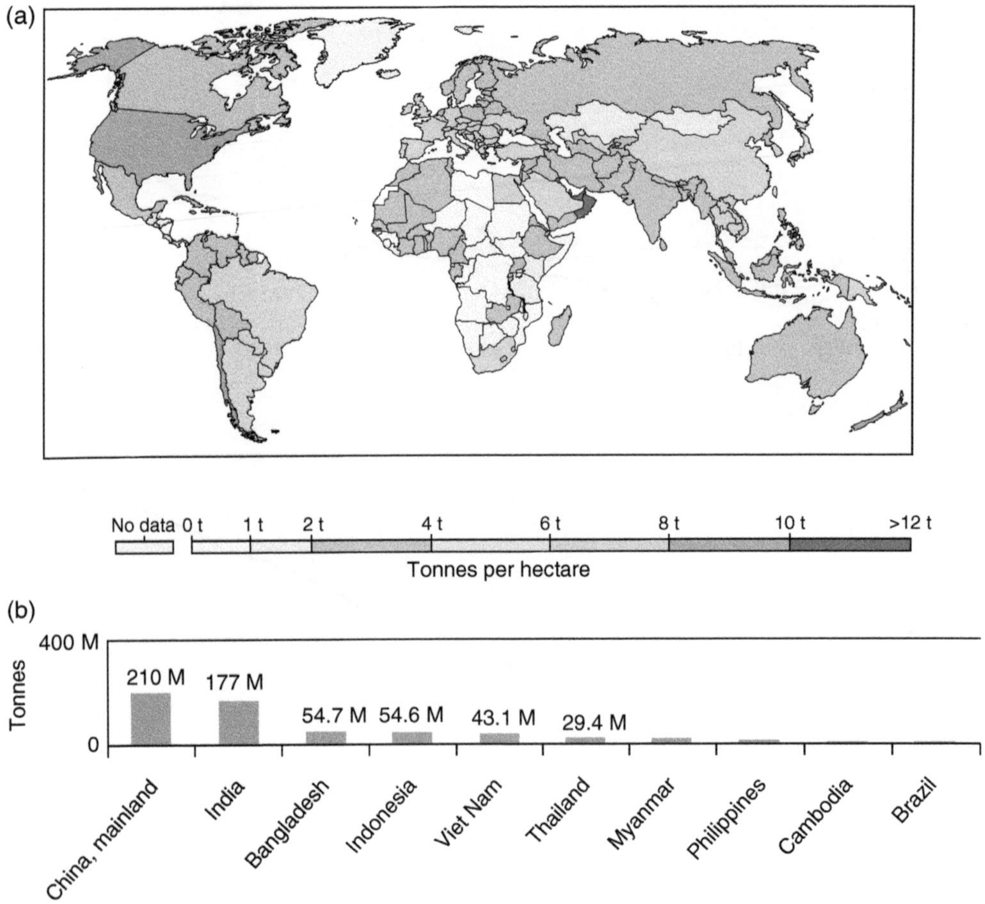

Figure 10.4 Global yield of rice (a) and the top 10 global producers of rice (b) according to the average of 2019–2020 data. *Source:* Data from FAO, www.fao.org/faostat.

10.4.1 Green Tissue-Specific Promoters

Green tissue is the primary location for plant photosynthesis. Generally, promoters of light-inducible genes are good sources of tissue-specific promoters. Several such genes and their promoters from rice and maize have been characterized like *ribulose-1,5-bisphosphate carboxylase/oxygenase (rbcS), leaf panicle 2 (LP2), PPDK, PEPC, rice light-harvesting complex II chlorophyll-a/b binding* gene, and pD_{540} (Molla et al. 2013; Karmakar et al. 2016; Sen et al. 2017; Swain et al. 2021). Promoters of these genes have been used for C_4 enzyme-related gene expression in rice.

10.4.2 Expressing C_4 Enzyme, PEPC in Rice

PEPC is known to play an important role in the initial CO_2 fixation in C_4 plants like maize. Studies on the overproduction of maize PEPC showed its positive effects on photosynthesis and yield (Ku et al. 1999; Hatch and Slack 2002). The gene of maize PEPC was expressed in totality in rice, and this enzyme accounted for 12% of the total leaf protein in transgenic rice with enhanced photosynthetic rate and reduced O_2 inhibition of photosynthesis (Ku et al. 1999). A similar finding encountered in indica rice expressing maize PEPC with enhanced photosynthesis and increased crop yield was reported by Bandyopadhyay et al. (2007). The CA enzyme provides substrate to PEPC by catalyzing the reversible hydration of CO_2 into bicarbonate in mesophylls. Expression of CA combined with PEPC helps increase CO_2 concentration within the cells and improves photosynthesis in transgenic rice (Ku et al. 2012).

10.4.3 Expressing C_4 Enzyme, PPDK in Rice

Intact maize *PPDK* gene with its promoter region was introduced in a rice variety, Kitaake, to study its functionality by Fukayama et al. (2001). It was found that in rice, this gene is also expressed in leaves, and it functionally regulates light/dark.

Intact *PPDK* gene with its own promoter was also introduced in IR64 (Zhang et al. 2010). Resulting transgenic rice plants showed higher CO_2 assimilation rates and photosynthetic activities compared to untransformed rice plants. A higher number of tillers and enhanced yield were reported for *PPDK* transformed IR64 rice in the greenhouse.

Sen et al. (2017) expressed three C_4 enzymes, PPDK, CA, and PEPC, which were expressed in indica rice Khitish under the direction of promoters exclusive to green tissues. Transgenic rice showed increased photosynthetic efficiency with a 12% increase in yield in greenhouse conditions. The transformed rice plants showed phenotypic changes like wider leaves, stronger stalks, and higher root biomass than the non-transformed rice controls.

PPDK and *ME* from the C_4 plant foxtail millet (*Setaria italica*) were transformed into the indica rice variety Naveen under the control of maize PPDK promoter (Swain et al. 2021). Transgenic rice showed increased photosynthesis with plant height, panicle length, panicle weight, and improvement in yield. Overexpression of PPDK also facilitates ME-mediated decarboxylation by lowering pyruvate concentration in transgenic rice (Taniguchi et al. 2008).

10.4.4 Expressing C_4 Enzyme, ME and NADP-ME in Rice

Overexpression of C_4 enzymes in C_3 plants was not always found to be suitable for the plants. Overexpressing maize C_4-specific ME in rice made it susceptible to photoinhibition (Lin et al. 2020). Due to the elevated NADPH/$NADP^+$ ratio in the stroma of the maize-ME overexpressed rice plants, they become susceptible to photoinhibition (Tsuchida et al. 2001). Similar result of photoinhibition and stunted growth in rice by maize NADP-ME overexpression was reported by Takeuchi et al. (2000). Interestingly, when maize ME was overexpressed in rice with PEPC, no photoinhibition was observed.

In a recent study, it was demonstrated that the expression of millet's ME (*SiME*) in transgenic rice through maize PPDK green tissue-specific promoter increased chlorophyll accumulation in transgenic rice without detrimental effects on plant growth (Swain et al. 2021).

10.4.5 Expressing Multiple C_4 Enzymes in Rice

A combination of four C_4 enzymes, namely *PEPC, PPDK*, NADP-*ME*, and NADP-*MDH*, have been introduced in a single MC of rice. The transgenic rice showed positive but limited effects of these combined enzymes on photosynthesis (Taniguchi et al. 2008; Miyao et al. 2011). As a combined effect of these four enzymes, CO_2 assimilation rate increased in the transgenic samples compared to control rice.

10.5 Application of CRISPR for Enhanced Photosynthesis

As an alternative to expressing C_4 enzymes in C_3 plants, some works have been reported to have improved photosynthesis efficiency and crop yield through CRISPR/Cas-mediated genome engineering of already available genes in C_3 plants. Recently, in three indica rice varieties, Huanghuazhan, Meixiangzhan, and Wushansimiao, the CRISPR/Cas9 gene-editing technique was used to eliminate the rice hexokinase (*OsHXK1*) gene (Zheng et al. 2021). The CRISPR-ed rice varieties showed increased light saturation points, light tolerance, stomatal conductance, and rice yields. This finding could be beneficial in developing high-photosynthetic efficiency and high-yielding rice varieties in the near future.

CRISPR/Cas gene-editing in Nipponbare rice for *STOMAGEN* and its paralog, *EPFL10* gene, created rice with reduced stomatal density while maintaining photosynthesis and conserving more water (Karavolias et al. 2021). Reduction in stomatal density can be one of the climate-adaptive strategies.

10.6 Single-Cell C_4 Species

Some of the simplest organizations of the complex C_4 photosynthesis system is present on this planet, where there is no requirement for the Kranz anatomy or compartment system for successful photosynthesis. The halophyte *Suaeda aralocaspica* is one such example where complete C_4 photosynthesis occurs within individual cells. Draft genome of *S. aralocaspica* revealed that its 452 Mb genome has 29,604 protein-coding genes with a 146,654 bp circular chloroplast genome (Wang et al. 2019). Cao et al. (2021) commenced a genome-wide search for *PEPC* gene in *S. aralocaspica and* reported that its genome has a bacterial type PEPC, which catalyzes the first step in the C_4 pathway.

Certain aquatic angiosperms, like *Hydrilla verticillata,* performed single-cell C_4 photosynthesis in the absence of the Kranz anatomy (Magnin et al. 1997). Primitive types of C_4 photosynthesis found in a terrestrial succulent *Borszczowia aralocaspica* also lacked the Kranz anatomy (Freitag and Stichler 2000). Such findings give us some idea and hope that important C_3 crops can be converted into C_4 crops provided the right combination of genes have been expressed. Complete omics data from such single-cell C_4 species will improve our knowledge of the C_4 system and allow us to apply it to C_3 crops.

10.7 Conclusion

The increasing world population, global warming, and climate change are undeniable truths that compel us to think about our future. Production of adequate amounts of food under adverse environmental conditions is a challenge at the moment. Plants adapted according to environmental conditions as they evolved from C_3 to C_4 system to minimize photorespiration effect on photosynthesis under adverse abiotic stress conditions. At this point in the time line of the C_4 evolution, the C_4 photosynthetic pathway is fully functional and can thrive well in abiotic stress conditions, including drought. Many commercially important crops like rice have a C_3 photosynthetic system and are vulnerable to abiotic stress. With respect to human life and the rate of global population change, the process of evolution is extremely slow, and we cannot simply wait to see how C_3 crops are naturally adapting under global warming. It could be too late to address the global food security issue to feed 10 billion people by 2050. This is a prime time when we have cutting-edge genome sequencing technology, advanced omics platforms, CRISPR/Cas tools, and established gene transformation methods for rice improvement. We can use the C_4 photosynthesis genes as a strategy for C_4 rice development, but we should not limit ourselves to exploring new strategies and policies for climate-resilient rice development as well. Interestingly, diversified plants (C_3, C_4, and intermediates) exist in the same ecosystem today with preferential natural conditions. In response to environmental change, the present and the future innovative science and technology would help us in making designer crops, including rice, grow on our natural earth or in artificial conditions considering the evolutionary process learned and the requirement of future foods.

Acknowledgments

This work was supported by the Department of Biotechnology (DBT), Government of India, and the award (Prof. Swapan K. Datta) of "Distinguished Biotechnology Research Professor (DBT)" at the University of Calcutta, Kolkata, India.

References

Bandyopadhyay, A., Datta, K., Zhang, J. et al. (2007). Enhanced photosynthesis rate in genetically engineered indica rice expressing *pepc* gene cloned from maize. *Plant Sci.* 172 (6): 1204–1209.

Berner, R.A. and Kothavala, Z. (2001). GEOCARB III: a revised model of atmospheric CO_2 over Phanerozoic time. *Am. J. Sci.* 301: 182–204.

Brown, R.H. and Hattersley, P.W. (1989). Leaf anatomy of C_3–C_4 species as related to evolution of C_4 photosynthesis. *Plant Physiol.* 91: 1543–1550.

Cao, J., Cheng, G., Wang, L. et al. (2021). Genome-wide identification and analysis of the *phosphoenolpyruvate carboxylase* gene family in *Suaeda aralocaspica*, an annual halophyte with single-cellular C_4 anatomy. *Front. Plant Sci.* 12: 665279.

Datta, K., Baisakh, N., Ganguly, M. et al. (2012). Overexpression of *Arabidopsis* and rice stress genes inducible transcription factor confers drought and salinity tolerance to rice. *Plant Biotechnol. J.* 10: 579–586.

Dengler, N.G. and Nelson, T. (1999). Leaf structure and development in C_4 plants. In: *C4 Plant Biology* (ed. R.F. Sage and R.K. Monson), 133–172. San Diego, CA: Academic Press.

Edwards, G.E. and Andreo, C.S. (1992). NADP-malic enzyme from plants. *Phytochemistry* 31: 1845–1857.

Ehleringer, J.R., Cerling, T.E., and Helliker, B.R. (1997). C_4 photosynthesis, atmospheric CO_2 and climate. *Oecologia* 112: 285–299.

Freitag, H. and Stichler, W. (2000). A remarkable new leaf type with unusual photosynthetic tissue in a central Asiatic genus of Chenopodiaceae. *Plant Biol.* 2: 154–160.

Fukayama, H., Tsuchida, H., Agarie, S. et al. (2001). Significant accumulation of C_4 specific pyruvate, orthophosphate dikinase in a C_3 plant rice. *Plant Physiol.* 127 (3): 1136–1146.

Hatch, M.D. and Slack, C.R. (1968). A new enzyme for the interconversion of pyruvate and phosphopyruvate and its role in the C_4 dicarboxylic acid pathway of photosynthesis. *Biochem. J.* 106 (1): 141–146.

Hatch, M.D. and Slack, C.R. (2002). Pyruvate, P1 dikinase from leaves. *Methods Enzymol.* 42: 212–219.

Hayes, J.M. (1994). Global methanothophy at the Archean-Proterozoic transition. In: *Early Life on Earth* (ed. S. Bengston), 220–236. New York: Columbia University Press.

Huxman, T.E. and Monson, R.K. (2003). Stomatal responses of C_3, C_3–C_4 and C_4 *Flaveria* species to light and intercellular CO_2 concentration: implications for the evolution of stomatal behaviour. *Plant Cell Environ.* 26: 313–322.

Jordan, D.B. and Ogren, W.L. (1984). The CO_2/O_2 specificity of ribulose 1,5-bisphosphate carboxylase/oxygenase. *Planta* 161: 308–313.

Kadereit, G., Borsch, T., Weising, K. et al. (2003). Phylogeny of Amaranthaceae and Chenopodiaceae and the evolution of C_4 photosynthesis. *Int. J. Plant Sci.* 164 (6): 959–986.

Karavolias, N.G., Pate, D., Seong, K. et al. (2021). CRISPR/Cas9 knockout of EPFL10 reduces stomatal density while maintaining photosynthesis and enhancing water conservation in rice. *bioRxiv*. https://doi.org/10.1101/2021.12.21.473329

Karmakar, S., Molla, K.A., Chanda, P.K. et al. (2016). Green tissue-specific co-expression of *chitinase* and *oxalate oxidase 4* genes in rice for enhanced resistance against sheath blight. *Planta* 243 (1): 115–130.

Kocacinar, F. and Sage, R.F. (2003). Photosynthetic pathway alters xylem structure and hydraulic function in annual plants. *Plant Cell Environ.* 26: 2015–2066.

Ku, M., Agarie, S., Nomura, M. et al. (1999). High-level expression of maize phosphoenolpyruvate carboxylase in transgenic rice plants. *Nat. Biotechnol.* 17: 76–80.

Ku, M.S.B., Cho, D., Ranade, U. et al. (2012). Carbonic anhydrase and the molecular evolution of C_4 photosynthesis. *Plant Cell Environ.* 35 (1): 22–37.

Lin, H., Arrivault, S., Coe, R.A. et al. (2020). A partial C_4 photosynthetic biochemical pathway in rice. *Front. Plant Sci.* 11: 564463.

Lynch, M. and Conery, J.S. (2000). The evolutionary fate and consequences of duplicate genes. *Science* 290: 1151–1155.

Magnin, N.C., Cooley, B.A., Reiskind, J.B. et al. (1997). Regulation and localization of key enzymes during the induction of Kranz-less, C_4 photosynthesis in *Hydrilla verticillate*. *Plant Physiol.* 115 (4): 1681–1689.

Majumder, S., Datta, K., and Datta, S.K. (2019). Rice biofortification: high iron, zinc, and vitamin-A to fight against "hidden hunger". *Agronomy* 9: 803.

Matsuoka, M., Furbank, R.T., Fukayama, H. et al. (2001). Molecular engineering of C_4 photosynthesis. *Annu. Rev. Plant Physiol. Plant Mol. Biol.* 52: 297–314.

Miyao, M., Masumoto, C., Miyazawa, S.I. et al. (2011). Lessons from engineering a single-cell C_4 photosynthetic pathway into rice. *J. Exp. Bot.* 62: 3021–3029.

Molla, K.A., Karmakar, S., Chanda, P.K. et al. (2013). Rice *oxalate oxidase* gene driven by green tissue-specific promoter increases tolerance to sheath blight pathogen (*Rhizoctonia solani*) in transgenic rice. *Mol. Plant Pathol.* 14 (9): 910–922.

Monson, R.K. (2003). Gene duplication, neofunctionalization, and the evolution of C_4 photosynthesis. *Int. J. Plant Sci.* 164: S43–S54.

Monson, R.K. and Rawsthorne, S. (2000). CO_2 assimilation in C_3–C_4 intermediate plants. In: *Photosynthesis: Physiology and Metabolism* (ed. R.C. Leegood, T.D. Sharkey, and S. von Caemmerer), 533–550. Dordrecht: Kluwer Academic.

Raghavendra, A.S. (1980). Characteristics of plant species intermediate between C_3 and C_4 pathways of photosynthesis: their focus of mechanism and evolution of C_4 syndrome. *Photosynthetica* 14: 271–273.

Rawsthorne, S. (1992). C_3–C_4 intermediate photosynthesis: linking physiology to gene expression. *Plant J.* 2: 267–274.

Sage, R.F. (2004). The evolution of C_4 photosynthesis. *New phytol.* 161: 341–370.

Sage, R.F., Christin, P.A., and Edwards, E.J. (2011). The C_4 plant lineages of planet Earth. *J. Exp. Bot.* 62: 3155–3169.

Schulze, E.-D. and Hall, A.E. (1982). Stomatal responses, water loss and CO_2 assimilation rates of plants in contrasting environments. In: *Physiological Plant Ecology II: Water Relations and Carbon Assimilation* (ed. O.L. Lange, P.S. Nobel, C.B. Osmond, and H. Ziegler), 181–230. Berlin: Springer-Verlag.

Sen, P., Ghosh, S., Sarkar, S.N. et al. (2017). Pyramiding of three C_4 specific genes towards yield enhancement in rice. *Plant Cell Tissue Org. Cult.* 128: 145–160.

Swain, A., Behera, D., Karmakar, S. et al. (2021). Morphophysiological alterations in transgenic rice lines expressing *PPDK* and *ME* genes from the C_4 model *Setaria italica*. *J. Plant Physiol.* 264: 153482.

Takeuchi, Y., Akagi, H., Kamasawa, N. et al. (2000). Aberrant chloroplasts in transgenic rice plants expressing a high level of maize NADP-dependent malic enzyme. *Planta* 211: 265–274.

Taniguchi, Y., Ohkawa, H., Masumoto, C. et al. (2008). Overproduction of C_4 photosynthetic enzymes in transgenic rice plants: an approach to introduce the C_4-like photosynthetic pathway into rice. *J. Exp. Bot.* 59 (7): 1799–1809.

Tsuchida, H., Tamai, T., Fukayama, H. et al. (2001). High level expression of C_4-specific NADP-malic enzyme in leaves and impairment of photoautotrophic growth in a C_3 plant, rice. *Plant Cell Physiol.* 42 (2): 138–145.

Wang, L., Ma, G., Wang, H. et al. (2019). A draft genome assembly of halophyte *Suaeda aralocaspica*, a plant that performs C_4 photosynthesis within individual cells. *GigaScience* 8 (9): giz116.

Wyrich, R., Dressen, U., Brockmann, S. et al. (1998). The molecular basis of C_4 photosynthesis in sorghum: isolation, characterization and RFLP mapping of mesophyll- and bundle-sheath-specific cDNAs obtained by differential screening. *Plant Mol. Bio.* 37: 319–335.

Zhang, J., Bandyopadhyay, A., Krisnan, S. et al. (2010). Characterization of a C_4 maize pyruvate orthophosphate dikinase expressed in C_3 transgenic rice plants. *Afr. J. Biotechnol.* 9 (2): 234–242.

Zheng, S., Ye, C., Lu, J. et al. (2021). Improving the rice photosynthetic efficiency and yield by editing *OsHXK1* via CRISPR/Cas9 system. *Int. J. Mol. Sci.* 22 (17): 9554.

11

Role of Legume Genetic Resources in Climate Resilience

Ruchi Bansal[1], Swati Priya[2], and H. K. Dikshit[3]

[1] *Division of Plant Physiology, ICAR-Indian Agricultural Research Institute, New Delhi, India*
[2] *Department of Botany, Kurukshetra University, Kurukshetra, Haryana, India*
[3] *Department of Genetics, ICAR-Indian Agricultural Research Institute, New Delhi, India*

11.1 Introduction

PGRs provide raw material for screening and improvement by breeding, to ensure food availability for increasing worldwide human population that tends to reach 9.6 billion by 2050 from 7.2 billion in the present. The major causes for rapid population rise are the ecosystem disturbances and climatic change. To enhance and sustain agricultural productivity with rapid cropping pattern changes, in situ and ex situ conservation plays a crucial role.

Climate change is a serious threat to global food security, limiting agricultural productivity worldwide (Mickelbart et al. 2015). Plants show a vonses to deal with the different environmental factors (Hassan et al. 2017). Since the past few decades, breeders have developed different stress-tolerant varieties using conventional and molecular breeding approaches; more vigorous efforts are required to achieve the target (Mittler and Blumwald 2010). Pedigree analysis of recently developed varieties reveals that they have a very narrow genetic base. To overcome the problem, we need to incorporate genetic variability into our breeding program.

PGRs are the foundation of a breeding program. Characterization and evaluation of crop germplasm to identify stress-tolerant donors and their use in the breeding cycle is the best way to deal with the bottleneck caused by poor gene pool of the cultivated popular varieties (Kudapa et al. 2018). Existence of cultivated species and crop improvement programs depends on the existing variability in the gene pool. Therefore, conservation becomes vital for generating raw material for diverse germplasm in present as well as future use. Modern practices in agriculture have led to decreased agricultural biodiversity; now most of humankind lives only on 12 crops, where wheat, rice, maize, and potatoes account for 60% of diet (Swamy et al. 2021). Therefore, the primary food sources are now genetically more vulnerable than ever. As we face these climate change problems, the need to adapt crops becomes important (Jha et al. 2014; Varshney et al. 2014). PGRs, including landraces, have long been identified as potential sources of traits for seed nutrition, stress tolerance, local adaptation, and yield stability. Based on ways of gene flow from the wild relative into the cultivated species, wild relatives of crop plants can be divided as primary, secondary, and tertiary gene pools of legumes. Thus, representing potential genetic diversity that can be explored in the background of usually grown types to deal with abiotic stress effects. Limited genetic recombination, reproductive isolation, hybrid sterility and embryo breakdown are major drawbacks of exploiting wild species beyond their primary gene pool. To overcome it, desired genes from cross-incompatible wild species can be utilized, new approaches must be considered as to make legumes one of the alternatives to cereals.

Legumes are globally the third most widely grown crops followed by cereals and oilseeds, having inherited resistance to various abiotic stresses. Legumes are a climate-smart crop that needs minimum irrigation and indicates tolerance to several abiotic stresses like cold, drought, heat, waterlogging, and salinity by adapting morphological changes such as reduced leaf surface, rolling leaves, having higher water use efficiency, increased root length, and various other inbuilt antioxidant enzyme mechanisms in the crop. Legumes are the wonderful and cheap substitute for being inclusive in the cropping patterns under climate change regime for its high nutritive value like protein, folic acid, and rich iron source. At present, 33% of thc nitrogen (N) requirement in human dietary is fulfilled by legumes (Song et al. 2016; Zhang et al. 2017). Therefore,

Global Climate Change and Plant Stress Management, First Edition. Edited by Mohammad Wahid Ansari, Anil Kumar Singh, and Narendra Tuteja.

legumes have an important role in maintaining global food security (Pandey et al. 2016). Moreover, shifting cultivated land use toward production of legumes can lower down the carbon footprint of protein production required for human consumption, enhance N use efficiency, and increase soil fertility (Foyer et al. 2016; Song et al. 2016). However, in legumes, the sustainability of crop production is at risk from increased environmental stress, including prolonged drought, heat waves, increased soil salinity, and poor soil nutrient availability (Maqbool et al. 2017; Zhang et al. 2017). Therefore, it is necessary to study resources on legume genetic variability, to develop legume cultivars that possess genetic stability to environmental stresses. Although temperature rise due to global warming may favor legume cultivation in colder regions of Northern hemisphere, such as Northern Europe, Canada, and Russia, it may decrease legume production in zones of warm climates, such as in tropical and subtropical regions, including Ethiopia, India, and Pakistan (Foyer et al. 2016). Such decline in regional legume productivity may lead to food shortages, especially in populated developing countries. Thus, abiotic stress-resistant legume cultivars that can replace abiotic stress-sensitive cultivars must be discovered early. The present chapter emphasizes the role of some important legume species with varied gene pool, showing multifaceted response to various abiotic stress. The area-wise usage of varied genetic resources will open an era of genetic variability within species to know, identify, and unravel the abiotic stress-tolerant gene pools in legumes.

11.2 Legumes Under Abiotic Stress

11.2.1 Legumes Under Drought Stress

Drought is imposed by low precipitation or inappropriate irrigation. Approximately 60% of crop productivity suffers due to it (Naeem et al. 2013). Drought stress in legumes negatively impacts the total biomass, number of pods, number of seeds, seed test weight, quality of seed, and per-plant yield (Toker et al. 2007). Urdbean and mungbean are known to be drought-susceptive, whereas chickpea and lentil are known as drought-tolerant genera (Toker et al. 2007). Singh et al. (1999) arranged kharif-season food legumes in increasing order of drought tolerance: soybean < urdbean < mungbean < groundnut < pea. Sinclair and Serraj (1995) observed that chickpeas, which export amides (principally glutamine and asparagine) in the xylem, are generally more resistant to drought stress than pigeon peas, which export ureides (allantoin and allantoic acid). Under drought stress, the nitrogen fixation rate in legumes decreases rapidly due to (i) decline in shoot N demand, (ii) the accumulation of ureides in both shoots and nodules (Vadez et al. 2000), and (iii) lower xylem translocation rate because of decrease in transpiration rate, and (iv) reduction in metabolic enzyme activity (Valentine et al. 2010). Various studies have indicated that drought stress negatively regulates nodule formation, growth, and development in plants (Vadez et al. 2000; Valentine et al. 2010). In legumes, the decrease in nitrogen content under drought stress was linked with a reduced rate of photosynthesis (Valentine et al. 2010). In many legumes, drought stress has resulted in total sugars and sucrose stimulation (Valentine et al. 2010). This observation was similar to a study on pea mutants that suggested sucrose synthase (SS) is essential for the development and function of nodules (Gordon et al. 1999). Several reports suggest that legumes when increases enzymatic antioxidant defense can show greater drought resistance than chickpeas (Kaur et al. 2009).

11.2.2 Legumes Under Waterlogging

Waterlogging occurs when water present in between soil particles is more than its draining capacity under gravity. It is one of the major causes of abiotic stress that causes losses to leguminous crops (Ashraf 2010; El-Enany et al. 2013). It affects seedling germination, vigor, root growth, and shoot growth by up to 80% and also causes severe seedling diseases (Toker et al. 2007). When mMungbean plants were exposed to waterlogging stress, the activity of various antioxidant enzyme like APX, GR, SOD, and CAT was reduced (Ahmed et al. 2002). Studies have also shown that oxidative damage is not directly responsible for affecting photosynthetic machinery of plants under waterlogging stress. But, in pigeon pea genotypes, increases in enzymatic antioxidants, SOD, CAT, peroxidase (POD), and APX, activity were observed when they were subjected to waterlogging at different levels (Kumutha et al. 2009). A pot experiment was carried out by El-Enany et al. (2013) on cowpea with triplicates of onefold field capacity waterlogging (WL) and 75% and 50% water deficit (WD) plants. Data revealed that both stresses significantly decreased the fresh weight as well as dry weight of roots and shoots, nitrogenase activity, and nodule number per plant. Antioxidative metabolites like ascorbic acids, phenolic compounds, proline, H_2O_2, and MDA were significantly increased under WD and WL conditions (El-Enany et al. 2013). Thus, waterlogging is found to reduce the endogenous nutrient level in different plant parts (Ashraf 2010). Deficiency of oxygen in the submerged root leads to remarkable decline in the uptake of K^+/Na^+ ion and restricts the transport of K^+ to the shoots (Ashraf 2010). Among winter-season food legumes, faba bean is more tolerant to waterlogging than pea, lentil, and chickpea (Siddique et al. 2000). Various management practices are used in minimizing the waterlogging effects involved in paddock selection, sowing time, seeding rate, and proper drainage plan (Toker et al. 2007). Thus, to deal with it, the role of genetic variation in tolerating waterlogging in legumes deserves attention (Toker et al. 2007).

11.2.3 Legumes Under Salinity Stress

Salinity stress has adverse effect on germination, seedling vigor, and crop yield, particularly in arid and semi-arid areas (Munns and Tester 2008; Aggarwal et al. 2012). Currently, 20% of cultivated land is globally affected due to salinity stress, especially in desert and arid regions, comprising 25% of the total available land area (Rasool et al. 2013). Higher salinity affects crops in many ways: ion toxicity, drought stress, nutritional disorders, oxidative stress, metabolic process alteration, reduced cell division and expansion, membrane disorganization, and genotoxicity (Munns 2002; Djanaguiraman and Prasad 2013). These effects together reduce crop survival, growth, and development (Hameed et al. 2014). Leguminous crops are relatively salt susceptible in comparison to cereal crops, thus farmers avoid growing it in salinity affected areas (Saxena et al. 1993; Toker et al. 2007; Egamberdieva and Lugtenberg 2014). The susceptibility of legumes may be due to salinity effect on bacterial activity and nitrogen fixation (Toker et al. 2007; Egamberdieva and Lugtenberg 2014). Salt stress led to reduced shoot growth of chickpea and mungbean plants (Saha et al. 2010; Hussain et al. 2011; Rasool et al. 2013). Chickpea has shown a higher tolerance to salinity due to effective antioxidant enzymatic defense (Kaur et al. 2009). Salinity stress significantly reduces the activities of phosphate enzymes (acid and alkaline) and nitrogenase, as reported in faba beans (Hussain et al. 2011). The salinity stress effect on growth and metabolic activities of mungbeans was studied by Saha et al. (2010). They suggested that salinity stress suppresses germination and seedling growth in mungbean. It also affects the photosynthetic machinery under lower chlorophyll content and induces malondialdehyde, proline, and hydrogen peroxide accumulation in the roots and shoots of the mungbean.

Moreover, salinity stress increased the activities of superoxide dismutase (SOD), catalase (CAT), and catechol peroxidase (CPX) in the plant roots and shoots of mungbean. Rasool et al. (2013) reported that in chickpea genotypes (SKUA-06 and SKUA-07) tolerance to salinity stress is related to the efficiency of the antioxidant enzymes ascorbate peroxidase (APX), glutathione reductase (GR), SOD, and CAT accumulation against reactive oxygen species, which maintains the integrity of cellular components and redox homeostasis. Various methods have been used in salinity tolerance screening, including seedling survival, seedling length, dry biomass production, shoot length, number of pods, seed yield, and K^+/Na^+ ratio (Toker et al. 2007).

11.2.4 Legumes Under Extreme Temperature

Temperature is one of the factors causing abiotic stress that affects productivity and legume quality (Christophe et al. 2011). Heat stress is often explained as the temperature rise at which irreversible damage is caused to plant development and function after an exposure for certain duration (Bhattacharya and Vijaylaxmi 2010). Damage to plants occurs in many ways, either by higher day or night temperatures or by high air or soil temperatures. Crop species and their cultivars also differ in their response to high temperatures sensitivity. That may affect respiration, photosynthesis, membrane stability, and relative water content and also regulate availability of primary and secondary metabolites and phytohormones. Moreover, throughout plant ontogeny, production of ROS, increased expression of a variety of heat-shock proteins, other stress-related proteins provide major responses to plant against high temperature stress (Bhattacharya and Vijaylaxmi 2010). Winter annual species are more sensitive to heat stress than warm-season annuals. Heat affects assimilation of nitrates in legumes by lowering the ureides synthesis, nitrate reductase activity as well as glutamate synthase up to minimal levels (Christophe et al. 2011). Heat stress may also target nitrogenase activity in nodules, leading to decreased efficiency of N_2 fixation or accelerating nodule senescence, resulting in reduced nodule longevity (Christophe et al. 2011). While understanding the mechanism of heat tolerance, cell membrane stability can be chosen as an important index (Sullivan 1972), as found in various literatures available from different legumes (Srinivasan et al. 1996). Other factors like heat shock proteins and lipid deposition in the pollen can also help in denotation of heat-tolerant genotypes (Mittler and Blumwald 2010). Moreover, osmoregulatory contents can also provide defense against heat stress (Christrophe et al. 2011).

Chilling stress usually affects crops that are natively inhabited or grow in tropical or subtropical regions of the world. Legumes vary greatly in their sensitivity to chilling stress. Plants that are sensitive to cold stress are those that could not survive or get injured when the temperature rises 15–20 °C above the freezing point of the tissues (Bhattacharya and Vijaylaxmi 2010). Cold-resistant plants can survive at temperatures near 0 °C (Bhattacharya and Vijaylaxmi 2010). In Australia, and other colder countries where chickpeas are sown in autumn/winter, yield gets adversely affected by floral abortion because of low spring temperatures. In contrary, Berger et al. (2005) reported that some annual wild *Cicer* species (including *C. reticulatum* and *C. echinospermum*) were almost insensitive to temperature change experienced at the time of flowering when strategically located at a cool site; interspecific hybrids even resulted in a significant resistance to chilling temperatures as reflected in the reproductive period from anthesis to pod appearance.

Oktem et al. (2008) investigated the effects of cold as well as drought stress on growing parameters as well as antioxidative defenses in the roots and shoots of lentil saplings exposed to cold (4 °C) and drought stress for five days. Study showed that the shoot length and its fresh weight reduced remarkably under both drought and cold stress, but increases in these growth parameters were noted in roots under similar conditions. The proline levels in roots and shoots increased even more when subjected to cold stress. The oxidative damage led to increase of hydrogen peroxide (H_2O_2) and higher malondialdehyde (MDA) in shoots under cold stress. An increase in lipid levels in root tissues was observed under both stress conditions. SOD activity altered shoot and root tissues differently under cold and drought stress. The CAT activity increased in roots when given drought stress, whereas APX activity was higher in roots under cold stress (Oktem et al. 2008; Bhattacharya and Vijaylaxmi 2010).

11.3 Genetic Resources for Legume Improvement

Information on the level of genetic variation in legumes is essential within and among populations of crops for any breeding program. Natural genetic variability is a key to respond to the climate change effects on crop yields (Rubio Teso et al. 2022). Rigorous selection and genetic erosion happened during crop domestication. In past few decades, development of high yielding varieties has further narrowed down the genetic base of the crop. Use of new trait donors is required to incorporate climate resilience against various abiotic stresses (Jha et al. 2014). Hence, the broadening of genetic base in different legumes is very much needed (Raina et al. 2019).

Germplasm accessions of legumes are preserved globally in different genebanks. Globally, ex situ lentils collection is recorded at 58,405 stored at several national and international genebanks along with duplicates of sizeable number (FAO 2010). Presently, a total of 14,577 lens accessions are conserved at the International Centre for Agricultural Research in the Dry Areas (ICARDA) genebank, which includes 602 wild accessions, 11,203 landraces, and 2,755 breeding lines. Larger lentil germplasm collections are available at the National Bureau of Plant Genetic Resources (NBPGR), New Delhi, India (7712 accessions), European Cooperative Programme for Plant Genetic Resources (ECPGR) (4598 accessions), Australian Grains Genebank (AGG), Horsham, Australia (5254 accessions), USDA-ARS (3187 accessions), and Vavilov Institute, Russia (2556 accessions). For various morphological and phenological traits, ICARDA has characterized 11,165 lentils accessions to date.

Similarly, about 43,000 mungbean germplasm are globally preserved ex situ and can play a crucial role in improving yield as well as the stability of the crop against biotic and abiotic stresses (Gayacharan et al. 2020). Russian genebanks hold over 863 accessions of mungbean. In India, the ICAR-NBPGR national genebank has reported larger collection of 11,000 mungbean accessions, comprising 7453 indigenous accessions and 3588 exotic accessions (Gayacharan et al. 2020). Of these, at present, 3927 accessions consisting of 3392 indigenous collections from 28 states of India and 535 exotic collections from 12 countries across the world are conserved in long-term storage conditions (−20 °C).

Moreover, the chickpea collection consists of 98,766 germplasm accessions (including 1444 wild-type and 97,322 cultivated accessions) that are preserved in >118 genebanks in 62 countries. For chickpea improvement, adequate genetic variability has been reported in the genus *Cicer*, consisting of 44 species (cultivated *Cicer arietinum*, 10 annuals, and 36 perennial wild *Cicer* species) belonging to the family *Leguminosae* (Van der Maesen 1972; Muehlbauer 1993). Similarly, for pigeon pea, 43,394 germplasm accessions (3147 wild-type and 40,247 cultivated accessions) are stored in >104 genebanks of 63 countries.

In the pigeon pea crop, sufficient genetic variability is noted in the genus *Cajanus*, belonging to the *Leguminosae* family. 562 accessions from 68 wild species in the genus *Cajanus* and its related groups are stored in the RS Paroda genebank as a useful resource for crop improvement.

In the following sections, we will get an overview of our current understanding of the role of genetic variability in abiotic stress responses and adaptation in legumes. We have also tried to summarize the reports where crop wild relatives serve as a source of stress tolerance/quality. These genetic resources are of great importance to legume breeding for sustainable development.

11.3.1 Lentil

Lentil (*Lens culinaris*) is one of the most ancient domesticated cool-season legumes grown widely across the globe and belongs to the Mediterranean area (Muehlbauer 1993). The distribution of all PGRs, specifically CWRs, is found to overlap in Turkey and then spread geographically (Davies et al. 2007; Singh et al. 2014). *Lens culinaris* subsp. *orientalis* expands in the Fertile Crescent (Lebanon, Israel, Jordan, Syria, and Cyprus), then spread over northern part of globe (Armenia, Azerbaijan, Czech Republic, and Russia) and then east of Iran, Turkmenistan, Tajikistan, and Uzbekistan with open to partially shady habitats on basalt to calcareous soils at 500–1700 m. *Lens culinaris* subsp. *odemensis* has limited distribution in Israel, Syria, Palestine in open grassland habitats, in pine groves on shallow calcareous soils in Turkey, or on basaltic gravel, at 700–1400 m. *Lens culinaris* subsp. *tomentosus* is found in Syria and Turkey. *Leptotyphlops nigricans* is reported in Syria and west to Italy, Greece, Spain, France, Croatia, Montenegro, and northern Ukraine and Crimea, and the eastern part of Bahrain, in fully open or partially shaded stony habitats of granitic soils and limestone, on abandoned terraces, settlements, and plantations, up to 1.200 m. *Lens lamottei* has a limited distribution in France and Spain. In semi-arid areas mostly, it is sown under rainfed conditions, which makes it suffer from water stress regularly at the reproductive stage (Silim et al. 1993; Erskine et al. 1994). It is also subjected to cold stress, to a lesser degree, at the vegetative phase (Erskine et al. 1994). *Lens ervoides* grows in Jordan, Palestine, Syria, and westward to Croatia, Montenegro, Italy, and north to Azerbaijan, Armenia, Ukraine, and Russia, in shady/partly shady habitats with shrubs and trees. With increased demand for lentils in market, efforts taken to improvise cultivars for high-yielding ability along with resistance to abiotic stresses. Several existing varieties that have been introduced recently under breeding programs through hybridization and pure line breeds mostly belong to *L. culinaris* subsp. *culinaris*, leading to narrowing of genetic bases, stagnating potential yield, and sensitivity to abiotic (cold, heat, drought, and salinity) stress.

The genus *Lens* comprises seven taxa in four species, namely *L. culinaris* Medikus [with subspecies *culinaris*, *tomentosus* (Ladizinsky) Fergusan, Maxted, Slageren, and Robertson, *orientalis* (Boiss.) Ponert, and *odemensis* (Ladizinsky) Fergusan, Maxted, Slageren & Robertson], *L. nigricans* (M.Bieb.) Godron, *L. lamottei* Czefr, and *L. ervoides* (Brign.) Grande (Dikshit et al. 2015). As genetic variability has been found to be higher in wild relatives, cross-breeding considering wild accessions of *L. culinaris* ssp. *Tomentosus*, *L. culinaris* ssp. *Orientalis*, *L. ervoides, L. nigricans*, *L. lamottei*, and *L. odemensis* favors the inheritance of desirable traits as well as increases genetic variability, making cultivated lentil more usable. Significant variation has been reported in flowering response to temperature and photoperiod (Erskine et al. 1989; Hamdi and Erskine 1996), winter-hardiness, yield traits (Erskine et al. 1989; Kumar 2015), iron deficiency, chlorosis and boron imbalances (Srivastava et al. 2000), and drought and heat tolerance (Hamdi and Erskine 1996). Attempts have been taken at various research levels to screen wild genetic resources of lentil for resistance to biotic/abiotic stresses (Table 11.1).

Table 11.1 Wild gene pool of lentil reported for tolerance against abiotic stresses.

Lens species/subspecies	Tolerant against	References
L. culinaris ssp. *Orientalis*	Drought stress, cold stress	Erskine et al. (1994), Hamdi and Erskine (1996), and Srivastava et al. (2000)
L. culinaris ssp. *Odemensis*	Drought stress	Hamdi and Erskine (1996) and Gorim and Vandenberg (2017)
L. culinaris ssp. *Tomentosus*	Drought stress	Pratap et al. (2014) and Gorim and Vandenberg (2017)
L. nigricans	Drought stress, cold stress	Coyne et al. (2020)
L. ervoides	Drought stress, cold stress, waterlogging stress	Hamdi and Erskine (1996), Singh et al. (2016), and Gorim and Vandenberg (2017)
L. lamottei	Drought stress, salinity stress	Van Der Maesen (1987) and Thormann et al. (2016)

Early maturity, with crop average duration of 100 days (over years and locations), supports escaping terminal drought and heat stress in the Central Zone. With the expansion of cropping in the Near East Fertile Crescent, the potential within it has long been recognized, and people have started seeking ways to utilize its high nutritive potential and reduced water use, making it well adapted to cereal-based cropping patterns in arid regions. Under the CGIAR global mandate for breeding lentils, efforts by ICARDA focused on introducing improvised varieties having resistance against one or more specific abiotic stress types, improved seed quality, and a short crop maturity cycle for fitting lentils into several cropping systems, like paddy-based systems in South Asia. Hence, rather than focusing on only drought and heat resistance, efforts should be towards developing multiple stress tolerance also reported in the arid regions and biotic stress-tolerant and machine-harvestable varieties for the West Asia and North Africa region. Despite being this large collection, there are gaps at genotype and species levels of major germplasm; thus, a continuum is must to cover these gaps in wild gene pool from the neglected areas of diversity in the genebank looking for abiotic stress resilient PGRs.

11.3.2 Mungbean

Mungbean (*Vigna radiata* (L.) Wilczek), is a small-seeded crop, indigenous to the Indo-Burma region of Asia, lesser known in the Americas (Jain and Mehra 1978; Lawn and Ahn 1985; Poehlman 1991). The wild species of mungbean has been named *V. radiata* var. *sublobata* and has its origin in India (Arora 1988). The wild *V. radiata* var. *sublobata* appears in sub-Himalayan tract, Tarai region, and in eastern and western peninsular tracts of India (Arora 1988). Mungbean diversity is well distributed across the continent from Himalayas in the north. Central Asia appears as secondary center of origin and is now introduced to some extent in the United States and in Australia (Lawn and Ahn 1985; Poehlman 1991). The crop is known to perform well under minimal soil moisture conditions. On contrary, it is widely grown as subsistence crops with lesser inputs and lower average yields, around 400 kg/ha (Lawn and Ahn 1985; Lambrides and Godwin 2007). The mungbean yield potential under experimental conditions seems relatively low, exceeding rarely to 2500 kg/ha to southern peninsular and north-eastern region. The Indo-Gangetic Plains are the secondary center of diversity for mungbeans (Poehlman 1991). In the early days, mungbean seed was transported by emigrants and traders from Asia to the Middle East, East Africa, Latin America, parts of South America, and Australia (Poehlman 1991). The Asiatic Vigna species are most widely distributed, and their production is continuously increasing (Kim et al. 2015). Many tolerant cultivars were developed against abiotic (drought, salinity, and heat) stress. These soon become responsive irresistant to the changing environmental stress due to similar genetic base and almost stagnant yield from the last few decades. Most of the mungbean accessions exhibit sensitivity to photoperiod, pod shattering, and some degree of lodging, as well a continuous pod production and sometimes weather damage during the harvest period (Lawn and Ahn 1985; Lambrides and Godwin 2007). Its wild relatives can be utilized to widen up narrowed genetic variability by incorporating desirable traits into the mungbean genotypes. Of these wild relatives, *Vigna aridicola* showed high drought resistance, while *Vigna vexillata* var. *ovate, Vigna trilobata*, and *Vigna monantha* are moderately drought tolerant. High salinity tolerance is found in *Vigna marina*, *Vigna luteola*, *Vigna nakashimae*, *V. vexillata* var. *macrosperma*, and *Vigna riukiuensis* (Table 11.2). Salinity tolerance was also found in *V. monantha*, *V. trilobata* and *V. vexillata* var. *ovate*. However, lesser amount of work is done and is thus needed to be spread for further exploitation of the wild species of mungbean. Breeding programs have emphasized quality of seed

Table 11.2 Wild gene pool of mungbean reported for tolerance against abiotic stresses.

Vigna species/subspecies	Tolerant against	References
V. aridicola	Drought stress	Gayacharan et al. (2020) and Asghar et al. (2021)
V. trilobata	Drought stress, salinity stress	Asghar et al. (2021)
V. vexillata	Drought stress, salinity stress	Takahashi et al. (2018), Nair et al. (2020), and Asghar et al. (2021)
V. monantha	Drought stress, salinity stress	Asghar et al. (2021)
V. luteola	Salinity stress	Gayacharan et al. (2020) and Asghar et al. (2021)
V. marina	Salinity stress	Gayacharan et al. (2020) and Asghar et al. (2021)
V. nakashimae	Salinity stress	Asghar et al. (2021)
V. riukiuensis	Salinity stress	Iseki et al. (2016)

characteristics (Lambrides and Godwin 2007) and, to some extent, characteristics that relate to an easier harvest through mechanization such as an erect and determinate growth habit in order to face climate extremes.

11.3.3 Pigeon Pea

Pigeon pea (*Cajanus cajan*) is native to India probably, but later distributed to Africa before 2000 BCE, as considered to be secondary center of diversity (Khoury et al. 2015). It then spread from West Africa to West Indies, where it is a widely consumed legume (Giller 2001). In India, its production is between 70% and 85%, but the crop has a promising hold on yield productivity in Africa, as it is very much tolerant to drought by legume standards (Giller 2001; Odeny 2007) and has decreased the pressure from the weed Striga in subsequent cereal crops (Odeny 2007). It is also becoming more common in the United States and Australia (Flower and Ludlow 1987). This versatile crop has multiple uses: whole seeds are used in dhal, pods and green seeds are used as vegetables, crushed seeds and leaves are used to feed livestock, stems are used as fuel wood or for making huts and baskets (Khoury et al. 2015).

Wild *Cajanus* species like, *Cajanus acutifolius, C. scarabaeoides, C. platycarpus, C. sericeus, C. albicans*, and *C. reticulatus* provide resistance to abiotic stress (Table 11.3) (Sharma et al. 2009). Screening of wild *Cajanus* species has found a new potential parental wild gene pool showing resistance against salinity, drought, and photoperiod insensitivity (Upadhyaya et al. 2013). Wild species are thus identified as the potential donors for developing genome-wide introgression (GWI) lines for the genetic improvement of pigeon pea as well as chickpea and groundnut. Early-maturing cultivars have also been developed but seem to be more susceptible to drought (Subbarao et al. 2000).

11.3.4 Chickpea

Chickpea belongs to the genus *Cicer*; among the annual species, *Cicer arietinum* species is the commonly cultivated and domesticated species worldwide (Toker et al. 2021). Vavilov suggested that the Fertile Crescent and the Mediterranean region as primary origin centers of chickpea, whereas South Asia and Ethiopia were the secondary centers of origin (Vavilov 1926; Van der Maesen 1987). Based on whole-genome resequencing, a comprehensive study suggests that the migration route of chickpea occurred through the Mediterranean region/Fertile Crescent to southern Asia (India), and then toward East Africa and Central Asia in parallel. The migration to America even occurred from Central Asia or Eastern Africa and not only from Mediterranean basin alone (Varshney et al. 2019). The global collection of chickpea germplasm consists of 99,877 accessions, constituting 1476 wild types, conserved in 120 genebanks, which is distributed in around 64 nations (Raina et al. 2019). The three main chickpea germplasm banks are the International Crops Research Institute for the Semi-Arid Tropic (ICRISAT) (18,963 accessions) in India, NBPGR in India (15,986 accessions), and ICARDA in Lebanon (13,065 accessions) (Waqas et al. 2019).

Table 11.3 Wild gene pool of pigeon pea reported for tolerance against abiotic stresses.

Cajanus species/subspecies	Tolerant against	References
C. scarabaeoides	Drought stress, heat stress, and salt stress	Upadhyaya (2006), Choudhary et al. (2011), and Khoury et al. (2015)
C. acutifolius	Drought stress, heat stress	Mallikarjuna et al. (2011) and Khoury et al. (2015)
C. platycarpus	Heat stress, salinity stress, cold stress	Subbarao et al. (2000), Choudhary et al. (2011), and Khoury et al. (2015)
C. reticulatus	Drought stress and heat stress	Khoury et al. (2015)
C. sericeus	Heat stress, drought stress, and salinity stress	Choudhary et al. (2011) and Khoury et al. (2015)
C. albicans	Drought stress and salinity stress	Choudhary et al. (2011)
C. cajanifolius (Haines)	Heat stress, salt stress	Khoury et al. (2015)
C. lineatus	Drought stress and salinity stress	Khoury et al. (2015)

Being grown in tropical settings during winter helps chickpea to avoid severe drought and temperature stress, but, sometimes its productivity gets reduced due to cold winters, especially at the time of anthesis (Bhattarai and Fettig 2005). The desi chickpea type is found to be more drought resilient than the kabuli type, partly due to improved carbon and nitrogen availability in seeds under stress (Davies et al. 2007). Therefore, inclusion of cold temperature tolerance genes into cultivated higher yield producing cultivars is necessary to obtain high productivity during winters.

Wild annual *Cicer* species are categorized into three gene pools (primary, secondary, and tertiary) based on the studies of crossing ability and hybrids potential for fertility using interspecific crosses with used chickpeas cultivars, that reflects their potential at par with the cultivated species (Toker et al. 2021). The primary gene pool of chickpea constitutes the cultivated *C. arietinum* and its immediate progenitors, *C. reticulatum*, which is used readily in crossing with *C. arietinum*; their offspring, due to superior chromosomal pairing, are fully fertile (Toker et al. 2021). Therefore, it is a potential source that can be served for broadening the genetic base and enhancing components of yield and other related traits for tolerance to abiotic stress in the cultivated genetic pool (Shan et al. 2007; Bajaj et al. 2015; Singh et al. 2021). The secondary gene pool comprises *C. echinospermum*, which also has the potential to be crossed along cultivated species but has a low fertility rate in the resulting hybrids in consequent offspring. Utilizing new techniques, interspecific hybrids have been produced between *C. arietinum* × *C. judaicum* (Verma et al. 1990, 1995), *C. arietinum* × *C. pinnatifidum* (Verma et al. 1990; Mallikarjuna 1999; Mallikarjuna and Jadhav 2008), *C. arietinum* × *C. cuneatum* (Singh and Singh 1989), and *C. arietinum* × *C. bijugum* (Verma et al. 1990; Mallikarjuna et al. 2007), to introduce studied essential alien genes from these cross-incompatible wild *Cicer* species into cultivated chickpea. Subsequently, such species are potential valuable alleles/genes donors for chickpea improvement. The tertiary gene pool includes the remaining *Cicer* species, consisting of annual species such as *C. judaicum, C. bijugum, C. cuneatum*, and *C. pinnatifidum*. Therefore, through interspecific hybridization, the genetic improvement program was sped up for understanding the molecular diversity of cultivated and wild *Cicer* species (Bajaj et al. 2015).

Higher variation has been studied genetically among wild *Cicer* species. The levels of genetic diversity were recorded higher in South Asian populations, while the levels were reported lowest in American populations, similar to the process of evolution and migration of chickpea (Singh et al. 2021). Moreover, the Fertile Crescent population has higher genetic diversity compared to those from Central Asia (Shan et al. 2007). However, the cultivated chickpea has deeply narrow genetic diversity because of various bottlenecks occurring during their domestication process, making it difficult for breeders to produce new elite cultivars of annual species of *Cicer* belonging to differential gene pools (Toker et al. 2021).

In chickpea, hundreds of high-yielding cultivars tolerant to major abiotic stresses have been developed through methods of conventional breeding (Chaturvedi et al. 2014). The genetic diversity of wild *Cicer* species, specifically *C. microphyllum*, has been explored and collected by NBPGR from several ecological habitats, particularly from Himachal Pradesh (Mayar Valley in Lahaul and Lossar, Kukumseri, Trilokinath, and Throat and Tabo in Spiti region). Various other wild *Cicer* species, like *C. microphyllum, C. nuristanicum, C. reticulatum, C. macrocanthum, C. judaicum, C. yamashitae*, and *C. pinnatifidum*, were collected from Afghanistan, Turkey, Pakistan, Syria, and adjoining areas by ICARDA and ICRISAT. Conservation and usage of PGRs for cultivated and wild species and their germplasm utilization are essential for the future of chickpea production.

The genetic resources for tolerance to cold stress were classified in a trial comprising 3276 lines. Cold tolerance was also present at significant levels in *C. reticulatum, C. bijugum, C. pinnatifidum*, and *C. echinospermum* (Singh et al. 1990) (Table 11.4). In *C. reticulatum* accessions, the responsible genes for vernalization are present but were absent in cultigens (Berger et al. 2005). The breeding improvements for drought tolerance are very steady due to many limiting factors, like change in available moisture across the years, root traits, vast genetic basis of the character (Turner et al. 2001). Canci and Toker (2009) On evaluating eight wild annual species, it was found that one accession of *C. pinnatifidum* and four accessions of *C. reticulatum* were resistant to heat and drought stress (Table 11.4). Chickpea accessions of CWR types have the longest range for genetic diversity as they facilitate a broad germplasm resource for screening several traits of agroeconomic importance (Singh et al. 2014). Among others, *C. reticulatum* represents a potential source of cultivated gene pools for widening the genetic base due to its useful genetic variations. *C. echinospermum* lies next in having such a source of potential for increased genetic variation. The improvement in transpiration efficiency is one of the major components of yield traits for increasing the resistance to drought and heat stress (Kashiwagi et al. 2006b). Various post-reproductive traits such as plant biomass, root length density, and root system depth are a few essential traits that contribute significantly to the avoidance of drought and heat stress mechanism, along with higher yield. Kashiwagi et al. (2006) evaluated and characterized the root traits

Table 11.4 Wild gene pool of chickpea reported for tolerance against abiotic stresses.

Cicer species/ subspecies	Tolerant against	References
C. reticulatum	Drought stress, heat stress, cold stress, salinity stress	Singh et al. (1995), Canci and Toker (2009, Toker (2005), Berger et al. (2012), and Arriagada et al. (2022)
C. pinnatifidum	Drought stress, heat stress	Bhattarai and Fettig (2005) and Jha et al. (2014)
C. bijugum	Cold stress	Toker (2005), Berger et al. (2012), and Arriagada et al. (2022)
C. echinospermum	Cold stress	Singh et al. (1995), Toker (2005), Berger et al. (2012), and Arriagada et al. (2022)
C. microphyllum	Cold stress	Singh et al. (2016) and Arriagada et al. (2022)
C. judaicum	Cold stress, drought stress	Saxena and Singh (1987), Robertson et al. (1997), Berger et al. (2012), and Arriagada et al. (2022)

of 216 chickpea lines and reported higher amount of variability in root length density (RLD) (Serraj et al. 2004). Therefore, it is needed to explore the germplasms further for identifying more such landraces with improved root traits to proceed with the selection and breeding of chickpeas against abiotic stress.

11.4 Conclusion

Crop-improvement programs are to an extent, determined by the level of genetic variation that exists in crop germplasm. Viewing this, a large accession numbers have been screened regularly for several traits particularly to involve tolerance to abiotic stress in legumes (lentil, mungbean, pigeon pea, and chick pea). For any crop species, variation within species are the natural ailment is abiotic stress resistance. During the crop selection and domestication process, these PGRs are very important for studying genetic variation that has gone unnoticed.

Present-day plant breeding has emphasized wide adaptation, resulting in modern agriculture leading to a small number of cultivars for major crops. For instance, 74% of Indonesian rice cultivars are derived from the same parental stock; similarly in the United States 75% of potatoes, 50% of wheat, and 50% of soybeans are from 4, 9, and 6 cultivars, respectively (World Conservation Monitoring Centre 1992). There was a significant loss of genetic variability when analyzed after data collection missions in Albania in 1941 and 1993 and in southern Italy in 1950 and the late 1980s. The estimated genetic erosion was 72.4% and 72.8%, respectively (Hammer et al. 1996). For example, before colonialism, the number of rice cultivars reduced on Indian farms from around 400,000 to 30,000 by mid-nineteenth century, and thousands more, known or unknown were lost after Green Revolution. Greece also lost wheat landraces by up to 95% after encouraging farmers to replace old ones with modern cultivars (Lopez 1994). The plant breeding evolution may cause genetic erosion and gradual shifting toward agriculture model based on uniformity. Consequences of diversity loss are becoming more important as we need the crops to adapt the climate change (Upadhyaya et al. 2013). For instance, chickpea, perennial *Cicer* species (wild) viz. *C. microphyllum, C. anatolicum, C. montbretti, C. songaricum*, and *C. oxydon*, were evaluated using a scale of 1 (highly tolerant) to 5 (highly susceptible) for drought tolerance (Toker et al. 2007). The findings showed a wide range of variation in shoot biomass, days to 50% flowering, maturity, and seed yield under drought stress, indicating their potential to withhold stress duration (Krishnamurthy et al. 2010). Similarly, harnessing genetic variability in terms of germplasm variability is important for developing a heat tolerant genotype. Understanding the importance of tolerance to heat, in-depth studies are to be carried out for discovering the patterns of inheritance in legumes for heat tolerance. Upadhyaya et al. (2013) even suggested that heat tolerance in legumes is in control of multigene. As a substitute to direct selection of the heat-tolerant genotypes, early flowering and maturity in tolerant genotypes can be made as a choice. For example, opting out phenotyping for heat stress in Mediterranean spring sown and south Indian sown chickpea (Toker et al. 2007; Berger et al. 2012). Cross incompatibility, reproductive isolation, embryo abortion, limited level of genetic recombination, and hybrid sterility are also some constraints in using these PGRs as resources in crop breeding programs at a larger extent.

In the last five years, developments have been made in studying the advanced genetic variation of many legume crops, including lentil, mungbean, chickpea, and pigeon pea (Jain and Mehra 1978; Song et al. 2016; Garg et al. 2016;

Varshney et al. 2019). This has also brought advancements in transcriptomics, discovery of SNPs, and genes for several important traits for legume productivity under both controlled and stressed conditions (Garg et al. 2016). Despite such advances in developing stress-tolerant legume cultivars, it still lags with respect to cereals. The noted increase in production of legumes over the past decade has been contributed more to the increase in area of plantations rather than to the development of noble or genetically improved cultivars with higher yield potential (Foyer et al. 2016; Rubio Teso et al. 2022).

Greater efforts are required for our understanding toward the role of PGRs against abiotic stress effects and responses in legumes. For these modern approaches, like transcriptome analysis of key legume crops exposed to various environmental stressors, individually, or in combination, are needed to gain an insight into the molecular mechanisms associated with abiotic stress mitigation. Also, a better understanding of the crossing ability, barriers to post-fertilization, embryo rescue technique, and use of phytohormones can enhance the possibilities of utilizing these variable genetic resources. However, a lot is still required in utilizing PGRs for genetic transformation across genome boundaries. There is always a need for continuous exploration and identification of resistant germplasm for abiotic stresses because of the continuous appearance of new physiological races derived from previously released cultivars. Recent experience with interspecific crosses between common bean and *P. coccineus* may shed light on ways to employ such species for the crop-related improvement. Selection of drought-resistant germplasm has been reported to produce lines with higher yield under stress or in favorable conditions (Beebe et al. 2008). Such improved yield performance across changing environments was suggested to be due to improved partitioning of the seed. Therefore, we believe that the PGRs collection, conservation, characterization, and evaluation must be keys to recover superior progeny in pulse production, and this experience could be extended to crosses with other species of similar physiology. The knowledge gained from this widespread study in the future should enable the more rapid development of new stress-resistant legume cultivars and ensure greater food security in the future.

References

Aggarwal, A., Kadian, N., Karishma, K. et al. (2012). Arbuscular mycorrhizal symbiosis and alleviation of salinity stress. *J. Appl. Nat. Sci.* 4 (1): 144–155. https://doi.org/10.31018/jans.v4i1.239.

Ahmed, S., Nawata, E., Hosokawa, M. et al. (2002). Alterations in photosynthesis and some antioxidant enzymatic activities of mungbean subjected to waterlogging. *Plant Sci.* 163 (1): 117–123. https://doi.org/10.1016/S0168-9452(02)00080-8.

Arora, R.K. (1988). The Indian gene centre. Priorities and prospects for collection. In: *Plant Genetic Resources: Indian Perspectives* (ed. R.S. Paroda, R.K. Arora, and K.P.S. Chandel), 545. NBPGR.

Arriagada, O., Cacciuttolo, F., Cabeza, R.A. et al. (2022). A comprehensive review on chickpea (*Cicer arietinum* L.) breeding for abiotic stress tolerance and climate change resilience. *Int. J. Mol. Sci.* 23 (12): 6794. https://doi.org/10.3390/ijms23126794.

Asghar, M.J., Hameed, A., Shahid, M. et al. (2021). Mungbean wild genetic resource: a potential source of genetic improvement for biotic and abiotic stress tolerance. In: *Wild Germplasm for Genetic Improvement in Crop Plants* (ed. M.T. Azhar and S.H. Wani), 299–319. Academic Press.

Ashraf, M. (2010). Inducing drought tolerance in plants: recent advances biotechnology. *Advances* 28 (1): 169–183. https://doi.org/10.1016/j.biotechadv.2009.11.005.

Bajaj, D., Das, S., Badoni, S. et al. (2015). Genome-wide high-throughput SNP discovery and genotyping for understanding natural (functional) allelic diversity and domestication patterns in wild chickpea. *Sci. Rep.* 5: 12468. https://doi.org/10.1038/srep12468.

Beebe, S.E., Rao, I.M., Cajiao, C., and Grajales, M. (2008). Selection for drought resistance in common bean also improves yield in phosphorus limited and favourable environments. *Crop Sci.* 48 (2): 582–592. https://doi.org/10.2135/cropsci2007.07.0404.

Berger, J.D., Buck, R., Henzell, J.M., and Turner, N.C. (2005). Evolution in the genus *Cicer*-vernalisation response and low temperature pod set in chickpea (*C. arietinum* L.) and its annual wild relatives. *Aust. J. Agric. Res.* 56 (11): 1191e1200. https://doi.org/10.1071/AR05089.

Berger, J.D., Kumar, S., Nayyar, H. et al. (2012). Temperature-stratified screening of chickpea (*Cicer arietinum* L.) genetic resource collections reveals very limited reproductive chilling tolerance compared to its annual wild relatives. *Field Crop Res.* 126: 119–129. https://doi.org/10.1016/j.fcr.2011.09.020.

Bhattacharya, A. and Vijaylaxmi. (2010). Physiological responses of grain legumes to stress environments. In: *Chickpea Breeding and Management* (ed. S.S. Yadav), 35–86. CAB International.

Bhattarai, T. and Fettig, S. (2005). Isolation and characterization of a dehydrin gene from *Cicer pinnatifidum*, a drought-resistant wild relative of chickpea. *Physiol. Plant.* 123 (4): 452–458. https://doi.org/10.1111/j.1399-3054.2005.00478.x.

Canci, H. and Toker, C. (2009). Evaluation of yield criteria for drought and heat resistance in chickpea (*Cicer arietinum* L.). *J. Agron. Crop Sci.* 19: 47–54.

Chaturvedi, S., Singh, N., Gaur, P. et al. (2014). New challenges in breeding chickpea under changing climate. In: *National Conference on Pulses: Challenges and Opportunities Under Changing Climatic Scenario* (ed. G.P. Dixit, J. Singh, and N.P. Singh), 102–113. Indian Society of Pulses Research and Development.

Choudhary, A.K., Sultana, R., Pratap, A. et al. (2011). Breeding for abiotic stresses in pigeon pea. *J. Food Legumes* 24 (3): 165–174.

Christophe, S., Jean-Christophe, A., Annabelle, L. et al. (2011). Plant N fluxes and modulation by nitrogen, heat and water stresses: a review. Based on comparison of legumes and non-legume plants. In: *Abiotic Stress in Plants – Mechanisms and Adaptations* (ed. A.K. Shanker and B. Venkateswarlu), 79–119. IntechOpen.

Coyne, C.J., Kumar, S., von Wettberg, E.J.B. et al. (2020). Potential and limits of exploitation of crop wild relatives for pea, lentil, and chickpea improvement. *Legume Science* 2 (2): e36. https://doi.org/10.1002/leg3.36.

Davies, P.A., Lulsdorf, M.M., and Ahmad, M. (2007). Wild relatives and biotechnical approaches. In: *Lentil: An Ancient Crop for Modern Times, Wild Relatives and Biotechnical Approaches* (ed. S.S. Yadav, D. McNeil, and P.C. Stevenson), 225–240. Springer.

Dikshit, H.K., Singh, A., Singh, D. et al. (2015). Genetic diversity in lens species revealed by EST and genomic simple sequence repeat analysis. *PLoS One* 10 (9): e0138101. https://doi.org/10.1371/journal.pone.0138101.

Djanaguiraman, M. and Prasad, P.V.V. (2013). Effects of salinity on ion transport, water relations and oxidative damage. In: *Ecophysiology and Responses of Plants Under Salt Stress* (ed. P. Ahmad, M.M. Azooz, and M.N.V. Prasad), 89–114. Springer Science+Business Media.

Egamberdieva, D. and Lugtenberg, B. (2014). Use of plant growth-promoting rhizobacteria to alleviate salinity stress in plants. In: *Use of Microbes for the Alleviation of Soil Stresses* (ed. M. Miransari), 73–96. Springer Science+Business Media.

El-Enany, A.E., Al-Anazi, A.D., Dief, N. et al. (2013). Role of antioxidant enzymes in amelioration of water deficit and waterlogging stresses on *Vigna sinensis* plants. *J. Biol. Earth Sci.* 3 (1): B144–B153.

Erskine, W., Adham, Y., and Holly, L. (1989). Geographic distribution of variation in quantitative characters in a world lentil collection. *Euphytica* 43 (1–2): 97–103. https://doi.org/10.1007/BF00037901.

Erskine, W., Hussain, A., Tahir, M. et al. (1994). Field evaluation of a model of photothermal flowering responses in a world lentil collection. *Theor. Appl. Genet.* 88 (3–4): 423–428. https://doi.org/10.1007/BF00223655.

Flower, D.J. and Ludlow, M.M. (1987). Variation among accessions of pigeonpea (*Cajanus cajan*) in osmotic adjustment and dehydration tolerance of leaves. *Field Crop Res.* 17 (3–4): 229–243. https://doi.org/10.1016/0378-4290(87)90037-2.

Food and Agriculture Organization (Food and Agriculture Organization of the United Nations) (2010). *The Second Report on the State of the World's Plant Genetic Resources for Food and Agriculture*. Food and Agriculture Organization.

Foyer, C.H., Lam, H.M., Nguyen, H.T. et al. (2016). Neglecting legumes has compromised human health and sustainable food production. *Nat. Plants* 2 (8): 16112. https://doi.org/10.1038/nplants.2016.112.

Garg, R., Shankar, R., Thakkar, B. et al. (2016). Transcriptome analyses reveal genotype- and developmental stage-specific molecular responses to drought and salinity stresses in chickpea. *Sci. Rep.* 6: 19228. https://doi.org/10.1038/srep19228.

Gayacharan, S.A., Gupta, K., Gupta, V. et al. (2020). Mungbean genetic resources and utilization. In: *The Mungbean Genome*, Compendium of Plant Genomes (ed. R.M. Nair, R. Schafleitner, and S.-H. Lee). https://doi.org/10.1007/978-3-030-20008-4.

Giller, K.E. (2001). *Nitrogen Fixation in Tropical Cropping Systems*, 423. CAB International.

Gordon, A.J., Minchin, F.R., James, C.L., and Komina, O. (1999). Sucrose synthase in legume nodules is essential for nitrogen fixation. *Plant Physiol.* 120 (3): 867–878. https://doi.org/10.1104/pp.120.3.867.

Gorim, L.Y. and Vandenberg, A. (2017). Evaluation of wild lentil species as genetic resources to improve drought tolerance in cultivated lentil. *Front. Plant Sci.* 8: 1129. https://doi.org/10.3389/fpls.2017.01129.

Hamdi, A. and Erskine, W. (1996). Reaction of wild species of the genus Lens to drought. *Euphytica* 91 (2): 173–179. https://doi.org/10.1007/BF00021067.

Hameed, A., Dilfuza, E., Abd-Allah, E.F. et al. (2014). Salinity stress and arbuscular mycorrhizal symbiosis in plants. In: *Use of Microbes for the Alleviation of Soil Stresses* (ed. M. Miransari), 139–159. New York: Springer Science+Business Media.

Hammer, K., Knpffer, H., Xhuveli, L., and Perrino, P. (1996). Estimating genetic erosion in landraces – two case studies. *Genet. Resour. Crop. Evol.* 43 (4): 329–336. https://doi.org/10.1007/BF00132952.

Hassan, M.U., Aamer, M., Chattha, M.U. et al. (2017). The role of potassium in plants under drought stress: mini review. *J. Basic Appl. Sci.* 13: 268–271. https://doi.org/10.6000/1927-5129.2017.13.44.

Hussain, N., Sarwar, G., Schmeisky, H. et al. (2011). Salinity and drought management in legume crops. In: *Climate Change and Management of Cool Season Grain Legume Crops* (ed. S.S. Yadav, D.L. McNeil, R. Redden, and S.A. Patil), 171–192. Springer.

Iseki, K., Takahashi, Y., Muto, C. et al. (2016). Diversity and evolution of salt tolerance in the genus Vigna. *PLoS One* 11 (10): e0164711. https://doi.org/10.1371/journal.pone.0164711.

Jain, H.K. and Mehra, K.L. (1978). Evolution, adaptation, relationships and uses of the species of Vigna cultivated in India. In: *Advances in Legume Science. Proceedings of the International Legume Conference* (ed. R.J. Summerfield and A.H. Bunting), 459–468. Kew: Royal Botanical Gardens.

Jha, U.C., Bohra, A., and Singh, N.P. (2014). Heat stress in crop plants: its nature, impacts and integrated breeding strategies to improve heat tolerance. *Plant Breed.* 133 (6): 679–701. https://doi.org/10.1111/pbr.12217.

Kashiwagi, J., Krishnamurthy, L., Crouch, J.H., and Serraj, R. (2006). Variability of root length density and its contribution to seed yield in chickpea (*Cicer arietinum* L.) under terminal drought stress. *Field Crop Res.* 95 (2–3): 171–181. https://doi.org/10.1016/j.fcr.2005.02.012.

Kashiwagi, J., Krishnamurthy, L., Singh, S. et al. (2006b). Relationships between transpiration efficiency and carbon isotope discrimination in chickpea (*C. arietinum* L.). *J. SAT Agric. Res.* 2 (1): 1–3.

Kaur, S., Gupta, A.K., Kaur, N. et al. (2009). Antioxidative enzymes and sucrose synthase contribute to cold stress tolerance in chickpea. *J. Agron. Crop Sci.* 195 (5): 393–397. https://doi.org/10.1111/j.1439-037X.2009.00383.x.

Khoury, C.K., Castañeda-Alvarez, N.P., Achicanoy, H.A. et al. (2015). Crop wild relatives of pigeon pea [*Cajanus cajan* (L.) Millsp.]: distributions, ex situ conservation status, and potential genetic resources for abiotic stress tolerance. *Biol. Conserv.* 184: 259–270. https://doi.org/10.1016/j.biocon.2015.01.032.

Kim, S.K., Nair, R.M., Lee, J., and Lee, S.H. (2015). Genomic resources in mungbean for future breeding programs. *Front. Plant Sci.* 6: 626. https://doi.org/10.3389/fpls.2015.00626.

Krishnamurthy, L., Kashiwagi, J., Gaur, P.M. et al. (2010). Sources of tolerance to terminal drought in the chickpea (*Cicer arietinum* L.) minicore germplasm. *Field Crop Res.* 119 (2–3): 322–330. https://doi.org/10.1016/j.fcr.2010.08.002.

Kudapa, H., Garg, V., Chitikineni, A., and Varshney, R.K. (2018). The RNA-Seq-based high resolution gene expression atlas of chickpea (*Cicer arietinum* L.) reveals dynamic spatio-temporal changes associated with growth and development. *Plant Cell Environ.* 41 (9): 2209–2225. https://doi.org/10.1111/pce.13210.

Kumar, J. (2015). Identification of early flowering genotype from wild taxa of lineil. *J Food Legumres* 28 (3): 264–265.

Kumutha, D., Ezhilmathi, K., Sairam, R.K. et al. (2009). Waterlogging induced oxidative stress and antioxidant activity in pigeonpea genotypes. *Biol. Plant.* 53 (1): 75–84. https://doi.org/10.1007/s10535-009-0011-5.

Lambrides, C.J. and Godwin, I. (2007). Mungbean. In: *Genome Mapping and Molecular Breeding in Plants Pulses, Sugar and Tuber Crops*, vol. 3 (ed. C. Kole), 69–90. Springer.

Lawn, R.J. and Ahn, C.S. (1985). Mungbean (*Vigna radiata* (L.) Wilczek/*Vigna mungo* (L.) Hepper). In: *Grain Legume Crops* (ed. R.J. Summerfield and E.H. Roberts), 584–623. London: Collins Sons & Co. Ltd.

Lopez, P.B. (1994). A new plant disease: Uniformity. *Ceres* 26: 41–47.

Mallikarjuna, N. (1999). Ovule and embryo culture to obtain hybrids from interspecific incompatible pollinations in chickpea. *Euphytica* 110 (1): 1–6. https://doi.org/10.1023/A:1003621908663.

Mallikarjuna, N. and Jadhav, D.R. (2008). Techniques to produce hybrid between *Cicer arietinum* L. × *C. pinnatifidum*Jaub. *Indian J. Genet. Plant Breed.* 68 (4): 398–405.

Mallikarjuna, N., Sharma, H.C., and Upadhyaya, H.D. (2007). Exploitation of wild relatives of pigeon pea and chickpea for resistance to *Helicoverpa armigera*. *J. SAT Agric. Res.* 3 (1): 4.

Mallikarjuna, N., Saxena, K.B., and Jadhav, D.R. (2011). *Cajanus*. In: *Wild Crop Relatives: Genomic and Breeding Resources, Legume Crops and Forages* (ed. C. Kole), 21–33. Springer-Verlag.

Maqbool, M.A., Aslam, M., and Ali, H. (2017). Breeding for improved drought tolerance in Chickpea (*Cicer arietinum* L.). *Plant Breed.* 136 (3): 300–318. https://doi.org/10.1111/pbr.12477.

Mickelbart, M.V., Hasegawa, P.M., and Bailey-Serres, J. (2015). Genetic mechanisms of abiotic stress tolerance that translate to crop yield stability. *Nat. Rev. Genet.* 16 (4): 237–251. https://doi.org/10.1038/nrg3901. Epub 10 March 2015. PubMed: 25752530.

Mittler, R. and Blumwald, E. (2010). Genetic engineering for modern agriculture: challenges and perspectives. *Annu. Rev. Plant Biol.* 61: 443–462. https://doi.org/10.1146/annurev-arplant-042809-112116.

Muehlbauer, F.J. (1993). Use of wild species as a source of resistance in cool-season food legume crops. In: *Breeding for Stress Tolerance in Cool-Season Food Legumes, ICARDA* (ed. K.B.S.M.C. Singh), 359–372. Baffins Lane, Chichester: A Wiley-Sayce Co-Publication, John Wiley and Sons.

Munns, R. (2002). Comparative physiology of salt and water stress. *Plant Cell Environ.* 25 (2): 239–250. https://doi.org/10.1046/j.0016-8025.2001.00808.x.

Munns, R. and Tester, M. (2008). Mechanisms of salinity tolerance. *Annu. Rev. Plant Biol.* 59: 651–681. https://doi.org/10.1146/annurev.arplant.59.032607.092911.

Naeem, M., Khan, M.N., and Khan, M. (2013). Adverse effects of abiotic stresses on medicinal and aromatic plants and their alleviation by calcium. In: *Plant Acclimation to Environmental Stress* (ed. N. Tuteja and S. Singh Gill), 101–146. Springer.

Nair, R.M., Schafleitner, R., and Lee, S.-H. (ed.) (2020). *The Mungbean Genome*, Compendium of Plant Genomes. Cham: Springer https://doi.org/10.1007/978-3-030-20008-4.

Odeny, D.A. (2007). The potential of pigeonpea (*Cajanus cajan* (L.) Millsp.) in Africa. *Nat. Res. Forum* 31 (4): 297–305. https://doi.org/10.1111/j.1477-8947.2007.00157.x.

Öktem, H.A., Eyidoðan, F., Demirba, D. et al. (2008). Antioxidant responses of lentil to cold and drought stress. *J. Plant Biochem. Biotechnol.* 17 (1): 15–21. https://doi.org/10.1007/BF03263254.

Pandey, G., Sharma, N., Sahu, P.P., and Prasad, M. (2016). Chromatin-based epigenetic regulation of plant abiotic stress response. *Curr. Genomics* 17 (6): 490–498. https://doi.org/10.2174/1389202917666160520103914.

Poehlman, J.M. (1991). *The Mungbean*. New Delhi: Oxford and IBH Publishers.

Pratap, A., Kumar, J., and Kumar, S. (2014). Evaluation of wild species of lentil for agro-morphological traits. *Legum. Res.* 37 (1): 11–18. https://doi.org/10.5958/j.0976-0571.37.1.002.

Raina, A., Khan, S., Wani, M.R. et al. (2019). Chickpea (*Cicer arietinum* L.) cytogenetics, genetic diversity and breeding. In: *Advances in Plant Breeding Strategies: Legumes* (ed. J. Al-Khayri, S. Jain, and D. Johnson), 53–112. Springer.

Rasool, S., Ahmad, A., Siddiqi, T.O., and Ahmad, P. (2013). Changes in growth, lipid peroxidation and some key antioxidant enzymes in chickpea genotypes under salt stress. *Acta Physiol. Plant.* 35 (4): 1039–1050. https://doi.org/10.1007/s11738-012-1142-4.

Robertson, L.D., Ocampo, B., and Singh, K.B. (1997). Morphological variation in wild annual *Cicer* species in comparison to the cultigen. *Euphytica* 95 (3): 309–319. https://doi.org/10.1023/A:1003004516921.

Rubio Teso, M.L., Lara-Romero, C., Rubiales, D. et al. (2022). Searching for abiotic tolerant and biotic stress resistant wild lentils for introgression breeding through predictive characterization. *Front. Plant Sci.* 13: 817849. https://doi.org/10.3389/fpls.2022.817849.

Saha, P., Chatterjee, P., and Biswas, A.K. (2010). NaCl pretreatment alleviates salt stress by enhancement of antioxidant defense system and osmolyte accumulation in mungbean (*Vigna radiata* L. Wilczek). *Indian J. Exp. Biol.* 48 (6): 593–600.

Saxena, M.C. and Singh, K.B. (1987). *Experimental Agriculture*, 563–564. CAB International, 409.£ 35.00, 25(4).

Saxena, N.P., Krishnamurthy, L., and Johansen, C. (1993). Registration of a drought-resistant chickpea germplasm. *Crop Sci.* 33 (6): 1424. https://doi.org/10.2135/cropsci1993.0011183X003300060088x.

Serraj, R., Krishnamurthy, L., Kashiwagi, J. et al. (2004). Variation in root traits of chickpea (*Cicer arietinum* L.) grown under terminal drought. *Field Crop Res.* 88 (2–3): 115–127. https://doi.org/10.1016/j.fcr.2003.12.001.

Shan, F., Clarke, H.J., Yan, G. et al. (2007). Identification of duplicates and fingerprinting of primary and secondary wild annual *Cicer* gene pools using AFLP markers. *Genet. Resour. Crop. Evol.* 54 (3): 519–527. https://doi.org/10.1007/s10722-006-0008-2.

Sharma, S., Khan, I.A., Ali, I. et al. (2009). Evaluation of the antimicrobial, antioxidant, and anti-inflammatory activities of hydroxychavicol for its potential use as an oral care agent. *Antimicrob. Agents Chemother.* 53 (1): 216–222. https://doi.org/10.1128/AAC.00045-08.

Siddique, K.H.M., Brinsmead, R.B., Knight, R. et al. (2000). Adaptation of chickpea (*Cicer arietinum* L.) and faba bean (*Viciafaba* L.) to Australia. In: *Linking Research and Marketing Opportunities for Pulses in the 21st Century* (ed. R. Knight), 289.e303. Springer.

Silim, S.N., Saxena, M.C., and Erskine, W. (1993). Adaptation of lentil to the Mediterranean environment. I. Factors affecting yield under drought conditions. *Exp. Agric.* 27: 155–175.

Sinclair, T.R. and Serraj, R. (1995). Legume nitrogen fixation and drought. *Nature* 378 (6555): 344–344. https://doi.org/10.1038/378344a0.

Singh, R.P. and Singh, B.D. (1989). Recovery of rare interspecific hybrids of gram *Cicer arietinum* × *C. cuneatum* L. through tissue culture. *Curr. Sci.* 58 (15): 874–876.

Singh, K.B., Malhotra, R.S., and Saxena, M.C. (1990). Sources for tolerance to cold in *Cicer* species. *Crop Sci.* 30 (5): 1136. https://doi.org/10.2135/cropsci1990.0011183X003000050036x.

Singh, K.B., Malhotra, R.S., and Saxena, M.C. (1995). Additional sources of tolerance to cold in cultivated and wild *Cicer* species. *Crop Sci.* 35 (5): 1491–1497. https://doi.org/10.2135/cropsci1995.0011183X003500050037x.

Singh, B.B., Mai-Kodomi, Y., and Terao, T. (1999). Relative drought tolerance of major rainfed crops of the semi-arid tropics. *Indian J. Genet.* 59 (4): 437–477.

Singh, M., Gautam, N.K., Rana, M.K. et al. (2014). Pigeon pea genetic resources and its utilization in India, current status and future prospects. *J. Plant Sci. Res.* 1 (2): 107.

Singh, D., Singh, C.K., Tomar, R.S.S. et al. (2016). Exploring genetic diversity for heat tolerance among lentil (*Lens culinaris* Medik.) genotypes of variant habitats by simple sequence repeat markers. *Plant Breed.* 135 (2): 215–223. https://doi.org/10.1111/pbr.12341.

Singh, M., Malhotra, N., and Singh, K. (2021). Broadening the genetic base of cultivated chickpea following introgression of wild *Cicer* species-progress, constraints and prospects. *Genet. Resour. Crop. Evol.* 68 (6): 2181–2205. https://doi.org/10.1007/s10722-021-01173-w.

Song, M., Hu, F.B., Wu, K. et al. (2016). Trajectory of body shape in early and middle life and all cause and cause specific mortality: results from two prospective US cohort studies. *BMJ* 353: i2195. https://doi.org/10.1136/bmj.i2195.

Srinivasan, A., Takeda, H., and Senboku, T. (1996). Heat tolerance in food legumes as evaluated by cell membrane thermo stability and chlorophyll fluorescence techniques. *Euphytica* 88 (1): 35–45. https://doi.org/10.1007/BF00029263.

Srivastava, S.P., Bhandari, T.M.S., Yadav, C.R. et al. (2000). Boron deficiency in lentil, yield loss and geographic distribution in a germplasm collection. *Plant Soil* 219 (1/2): 147–151. https://doi.org/10.1023/A:1004708700075.

Subbarao, G.V., Chauhan, Y.S., and Johansen, C. (2000). Patterns of osmotic adjustment in pigeonpea – its importance as a mechanism of drought resistance. *Eur. J. Agron.* 12 (3–4): 239–249. https://doi.org/10.1016/S1161-0301(00)00050-2.

Sullivan, C.Y. (1972). Mechanisms of heat and drought resistance in grain sorghum and methods of measurements. In: *Sorghum in the Seventies* (ed. N.G.P. Rao and R.L. House), 247–264. New Delhi: Oxford & IBH Publishing Co.

Swamy, B.P.M., Marathi, B., Ribeiro-Barros, A.I.F., and Ricachenevsky, F.K. (2021). Editorial: development of healthy and nutritious cereals: recent insights on molecular advances in breeding. *Front. Genet.* 12: 635006. https://doi.org/10.3389/fgene.2021.635006.

Takahashi, Y., Muto, C., Iseki, K. et al. (2018). A new taxonomic treatment for some wild relatives of mungbean (*Vigna radiata* (L.) Wilcz.) based on their molecular phylogenetic relationships and morphological variations. *Genet. Resour. Crop. Evol.* 65 (4): 1109–1121. https://doi.org/10.1007/s10722-017-0599-9.

Thormann, I., Parra-Quijano, M., Rubio Teso, M.L. et al. (2016). Predictive characterization methods for accessing and using CWR diversity. In: *Enhancing Crop Genepool Use: Capturing Wild Relative and Landrace Diversity for Crop Improvement* (ed. N. Maxted, M.E. Dulloo, and B. Ford-Loyd), 64–77. CAB International.

Toker, C. (2005). Preliminary screening and selection for cold tolerance in annual wild *Cicer* species. *Genet. Resour. Crop. Evol.* 52 (1): 1–5. https://doi.org/10.1007/s10722-005-1743-5.

Toker, C., Lluch, C., Tejera, N.A. et al. (2007). Abiotic stress. In: *Chickpea Breeding and Management* (ed. S.S. Yadav, R. Redden, W. Chen, and B. Sharma), 474–496. CAB International.

Toker, C., Berger, J., Eker, T. et al. (2021). *Cicer turcicum*: a new *Cicer* species and its potential to improve chickpea. *Front. Plant Sci.* 12: 662891. https://doi.org/10.3389/fpls.2021.662891, PubMed: 33936152.

Turner, N.C., Wright, G.C., and Siddique, K.H.M. (2001). Adaptation of grain legumes (pulses) to water-limited environments. *Adv. Agron.* 71: 193–231. https://doi.org/10.1016/S0065-2113(01)71015-2.

Upadhyaya, H.D. (2006). Improving pigeon pea with the wild. *SAT. Trend. ICRISAT's Mon. Newsl.* 62: 1–1.

Upadhyaya, H.D., Reddy, K.N., Pundir, R.P.S. et al. (2013). Diversity and geographical gaps in *Cajanus scarabaeoides* (L.) *Thou.* germplasm conserved at the ICRISAT GeneBank. *Plant Genet. Res.* 11 (1): 3–14. https://doi.org/10.1017/S1479262111000736.

Vadez, V., Sinclair, T.R., and Serraj, R. (2000). Asparagine and ureide accumulation in nodules and shoots as feedback inhibitors of N2 fixation in soybean. *Physiol. Plant.* 110 (2): 215–223. https://doi.org/10.1034/j.1399-3054.2000.110211.x.

Valentine, A.J., Benedito, V.A., and Kang, Y. (2010). Legume nitrogen fixation and soil abiotic stress: from physiology to genomics and beyond. *Annu. Plant Rev.* 42: 207–248.

Van der Maesen, L.J.G. (1972). *Cicer L., A Monograph of the Genus, with Special Reference to the Chickpea (Cicer arietinum L.), Its Ecology and Distribution*. Wageningen, The Netherlands: Mendelingen Land Bou Who Geschool Wageningen, 1.e341.

Van der Maesen, L.J.G. (1987). Origin, history and taxonomy of chickpea. In: *The Chickpea* (ed. M.C. Saxena and K.B. Singh), 11–34. CAB International. https://edepot.wur.nl/304694 (accessed 17 December 2021).

Varshney, R.K., Thudi, M., Nayak, S.N. et al. (2014). Genetic dissection of drought tolerance in chickpea (*Cicer arietinum* L.). *Theor. Appl. Genet.* 127 (2): 445–462. https://doi.org/10.1007/s00122-013-2230-6, PubMed: 24326458.

Varshney, R.K., Thudi, M., Roorkiwal, M. et al. (2019). Resequencing of 429 chickpea accessions from 45 countries provides insights into genome diversity, domestication and agronomic traits. *Nat. Genet.* 51 (5): 857–864. https://doi.org/10.1038/s41588-019-0401-3.

Vavilov, N.I. (1926). Centres of origin of cultivated plants. *Bull. Appl. Bot. Genet. Plant Breed.* 16: 1–248.

Verma, M.M., Sandhu, J.S., Rrar, H.S., and Brar, J.S. (1990). Crossability studies in different species of *Cicer* (L.). *Crop. Improv.* 17: 179–181.

Verma, M.M., Ravi, S.J.S., and Sandhu, J.S. (1995). Characterization of the interspecific cross *Cicer arietinum* L. × *C. judaicum* (Boiss). *Plant Breed.* 114 (6): 549–551. https://doi.org/10.1111/j.1439-0523.1995.tb00856.x.

Waqas, M., Azhar, M.T., Rana, I.A. et al. (2019). Drought stress in chickpea: physiological, breeding, and omics perspectives. In: *Recent Approaches in Omics for Plant Resilience to Climate Change* (ed. W.S. Hussain), 189–227. Springer.

Wood, R. (1992). *Global Biodiversity: Status of the Earth's Living Resources* (ed. B. Groombridge), 585. London: A Report Compiled by the World Conservation Monitoring Centre, Chapman & Hall.

Zhang, B., Tian, H., Lu, C. et al. (2017). Global manure nitrogen production and application in cropland during 1860–2014: a 5 arcmin gridded global dataset for Earth system modeling. *Earth Syst. Sci. Data* 9 (2): 667–678. https://doi.org/10.5194/essd-9-667-2017.

12

Oxygenic Photosynthesis – a Major Driver of Climate Change and Stress Tolerance

Baishnab C. Tripathy

Department of Biotechnology, Sharda University, Greater Noida, India

12.1 Introduction

The planet Earth had its origin 4.5 billion years ago, and it did not have water. Due to collision of water-rich asteroids with early Earth, water reached our planet. The early Earth atmosphere had a lot of CO_2 and methane and very small amounts of oxygen. Therefore, early lifeforms that evolved 3.5 billions of years ago in our planet were anaerobic. The earliest lifeforms were anaerobic archaea and methanogenic bacteria as there was little oxygen in the atmosphere of the early Earth. With the evolution of bacteriochlorophyll containing anoxygenic photosynthetic bacteria, H_2S and other compounds prevalent in the reducing atmosphere of early Earth were used to reduce highly prevalent CO_2 to carbohydrates at the expense of solar energy that liberated S as the by-product.

12.2 Evolution of Chlorophyll

Evolution of chlorophyll a most likely from bacteriochlorophyll g took place by isomerization reaction in the tetrapyrrole moiety. Chlorophyll a that absorbed the red photon at 663 nm of the solar spectrum had sufficient energy to break the covalent bonds of H_2O to evolve O_2. Chlorophyll d and other longer wavelength absorbing forms of chlorophyll probably evolved before the evolution of chlorophyll a. Transition of anoxygenic photosynthesis of bacteriochlorophyll containing photosynthetic bacteria to cyanobacteria-like organisms was accompanied by recruitment of composite Mn-containing oxygen evolving complex proteins that could break H_2O to O_2 utilizing the solar energy absorbed by chlorophylls. In the process of evolution, the longer wavelength absorbing forms of bacteriochlorophyll metamorphosed to shorter wavelength forms of Chl f, Chl d, Chl a, and Chl b; their infrared/red absorption maxima shifted to shorter wavelengths.

One of the important mysteries of nature remains unsolved. Why plants evolved green. Ideally the chlorophylls should be absorbing all the wavelengths of light to optimally utilize the solar radiation in photosynthesis to manufacture its own food and consequently should have been black. However, all the forms of chlorophylls do not absorb green light and therefore do not effectively utilize the green part of the solar spectrum for photosynthesis. The bacteriochlorophylls and chlorophylls evolved billions of years ago most likely in an archaea-dominated habitat. Many archaean organisms (e.g. *Halobacterium salinarium*, formerly called *H. halobium*) have the retinal protein bacteriorhodopsin as the major chromophore. It usually absorbs maximally in the green (560 nm) and uses the energy of green light to drive the proton pump and create an electrochemical proton gradient across the membrane that is utilized for ATP production by ATP synthase (Figure 12.1). Therefore, Chl containing cyanobacteria-like organisms when evolved in anaerobic world below the archaean bloom in the ocean could not perceive the green light that was preferentially absorbed by archaea. Consequently, the cyanobacteria-like organisms that originated beneath the archaea evolved the light-absorbing pigment chlorophyll that could absorb most of the visible light other than green. Although, different isoforms of Chl evolved in the oxygenic world, none of them could absorb green light as the structural changes that occurred among different forms of Chl took place in the side chains; the basic tetrapyrrole macrocycle having Mg at the center was conserved throughout the evolutionary process of the photosynthetic pigment.

Global Climate Change and Plant Stress Management, First Edition. Edited by Mohammad Wahid Ansari, Anil Kumar Singh, and Narendra Tuteja.

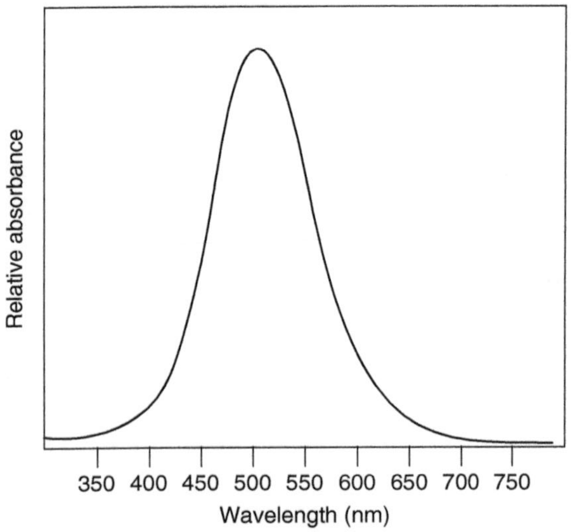

Figure 12.1 Relative probability of absorption of photons of different wavelengths by rhodopsin.

The photosynthetic cyanobacteria-like organisms evolved nearly three billion years ago and easily oxidized ocean water to liberate oxygen via oxygenic photosynthesis utilizing shorter wavelength forms of Chl. This changed the nature of the atmosphere from anaerobic to aerobic i.e. from a reducing to the oxidizing atmosphere that we have today.

12.3 The Great Oxygenation Event

When the O_2 concentration created by oxygenic photosynthetic microorganisms reached 0.2%, high in the atmosphere, the formation ozone layer began. Some oxygen (O_2) molecules absorbed energy from the sun's ultraviolet (UV) rays and split to form single oxygen atoms. These atoms combined with molecular oxygen (O_2) to form ozone (O_3) molecules, which are very effective at absorbing UV rays. It took millions of years to form effective ozone layer. The ozone layer required to prevent lethal UV (200–300 nm) penetration took millions of years to be fully formed and was probably completed when the aquatic photosynthetic organisms converted photolysed ocean water to O_2 and its concentration reached around 2%. Because of the UV shield, the penetration of solar UV light to the surface of the Earth slowly diminished, and consequently, the UV-induced damage to the DNA of autotrophic and heterotrophic organisms was prevented. This ensured survival of species and reduced the extinction events leading to increased biodiversity.

Nearly 1.5 billion years ago, the first eukaryotic cells evolved that initiated intracellular complexity. About 700 million years ago, several single-celled eukaryotes aggregated to form pre-multicellular to multicellular organisms. The Volvocales algal forms represent evolution from the unicellular organism *Chlamydomonas* to aggregated colony of unicellular organisms. *Pandorina, Gonium, Eudorina, Pleodorina*, and primitive multicellular forms of *Volvox*-like organisms have cells that are connected by cytoplasmic bridges. Multicellular algae could grow and photosynthesize adding to the diversity of photosynthetic apparatus and water oxidation to oxygen.

The marine aquatic plants migrated to land first becoming terrestrial plants about 500 million years ago. The presence of UV-absorbing ozone shield enabled organisms to develop and live on the land. With their wider habitats, organisms proliferated greatly on the land mass. This further increased the O_2 concentration in the atmosphere.

12.4 Role of Forest in the Regulation of O_2 and CO_2 Concentrations in the Atmosphere

Massive amounts of forest that dominated carboniferous era evolved 360 million years ago in the cooling phase or ice house of planet Earth and performed massive photosynthesis that resulted in a slow decrease in the atmospheric CO_2 concentration and further rise in the $[O_2]$ in the atmosphere. The increased photosynthetically derived O_2 in the atmosphere promoted migration of sea animals to land that resulted in the dominance of amphibians in the carboniferous era. The carboniferous rainforest collapsed due to an extinction event that took place around 300 million years ago in the late carboniferous era. This resulted in the forest biomass being converted to fossil fuel coal and petroleum. It also altered the huge forests covering the equatorial parts of Euramerica: Europe and America. In the forests of Mesozoic era (65–245 million years ago), biological diversity increased greatly. The Mesozoic era had most damaging extinction event recorded after Cambrian period. Perhaps 500,000 species survived this extinction event, and they have gradually given rise to millions of complex living species.

12.5 Evolution of C_4 Plants

One group of plant species that evolved after this extinction event is the evolution of C_4 plants. Nearly 40–50 million years ago, the Earth was hot and the CO_2 concentration fell to a minimum, and this prompted O_2 to compete for the active site of ribulose bisphosphate carboxylase oxygenase (Rubisco) to dilute the carbon pool of plants to liberate CO_2 via photorespiration that was often detrimental for biomass accumulation. To obviate photorespiration, C_4 plants evolved to concentrate CO_2 at the vicinity of Rubisco via carbon pump to minimize oxygenation reaction of Rubisco. Rubisco that constitutes nearly 50% of C_3 leaf protein evolved in photosynthetic bacteria in the anaerobic world billions of years ago, in the absence of O_2 and therefore, performed only carboxylase function. Over billions of years of evolution, although Rubisco has evolved to a complex 16-mer protein made up of 8 large and 8 small subunits, chloroplasts of eukaryotic plants either could not or did not change the active site of Rubisco in favor of CO_2 to promote only carboxylation. Rather, it came up with CO_2 concentrating pump/process in the vicinity of Rubisco to outcompete O_2. To accomplish higher photosynthesis in low CO_2 and high temperature world, the C_4 form of phosphoenolpyruvate carboxylase (PEPC) evolved from its cytoplasmic C_3 isoform after undergoing a lot of changes in the base sequence of its promoter to achieve its increased expression and only a few sequence changes in its coding region for better enzyme function. As C_4 form of PEPC evolved 40–50 million years ago in the aerobic world, unlike Rubisco, its active site is protected from oxygen attack and therefore, the substrate phosphoenolpyruvate (PEP) is not oxidized by O_2. The PEP carboxylation product, the C_4 oxaloacetic acid, is often converted to malate that migrates to the bundle sheath (BS) cell to be decarboxylated by the malic enzyme to pyruvate and CO_2. This decarboxylation event in the Rubisco-containing BS cells increases the CO_2 concentration in the vicinity of Rubisco to favor carboxylation over oxygenation. Therefore, C_4 plants have a higher rate of photosynthesis, very low photorespiration, and increased plant productivity as compared to C_3 plants. As C_4 plants evolved in a warmer climate that prevailed 40 million years ago, they are usually genetically tolerant to high temperatures and could optimally photosynthesize at around 35 °C at higher light intensities. However, most C_3 plants have diminished photosynthesis at this temperature (Figure 12.2).

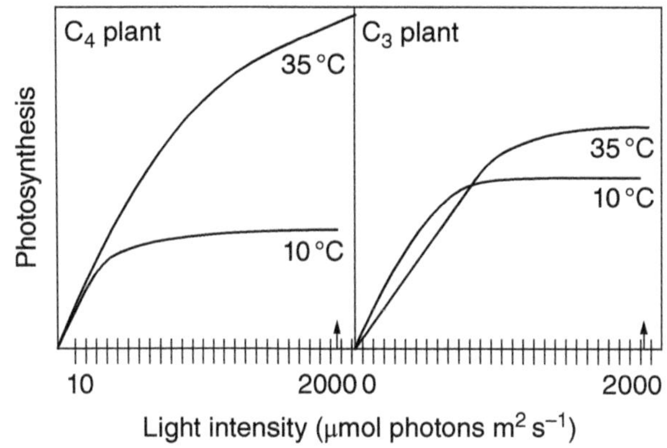

Figure 12.2 Average photosynthetic rates of C_4 and C_3 plants grown in 10 and 35 °C.

The green plants reduced the greenhouse gas from the atmosphere by fixing CO_2 for millions of years and mostly became fossils after carboniferous era due to the several extinction events to form coal and petroleum. Postindustrial age, accelerated burning of fossil fuels resulted in the increase of CO_2 in the atmosphere. The current high CO_2 world may not favor further evolution of C_4 crop plants. As C_4 plants have minimal O_2 inhibition of photosynthesis, they have a low CO_2 compensation point and their photosynthesis is usually half-saturated at CO_2 compensation point of C_3 plants. Therefore, C_4 photosynthesis is already saturated at the prevailing [CO_2], and it is unlikely that its photosynthetic carbon assimilation rate shall increase much in higher CO_2 world.

However, in future, an increased [CO_2] in the atmosphere shall effectively compete with O_2 for the active site of Rubisco to promote carboxylation over oxygenation of ribulose bisphosphate (Ru-P2). Therefore, the rate of photorespiration shall decrease and consequently the rate of photosynthesis and biomass accumulation by C_3 plants shall increase, albeit a partial downregulation of Rubisco content that may take place due to their continuous growth in high CO_2 environment. The C_3 photosynthesis is usually saturated around 1000 μmol mol^{-1} of CO_2, and it is unlikely that [CO_2] in the atmosphere is going to reach this level after 100 years when fossil fuel shall be substantially depleted from the Earth and solar energy shall be widely used as an alternate fuel reducing the carbon footprint. Therefore, C_3 photosynthesis may not be saturated within 100 years, unless the volcanic activity abruptly increases.

12.6 The Impact of High Temperature

As we are presently in the interglacial period, the world temperature has increased slowly for past 11,000 years due to geological reasons over which we do not have any control. This increase in temperature is being further accentuated by greenhouse gas emission from anthropogenic sources. A new glaciation, i.e. a cooling period has been expected to begin now; however, due to human-induced climate change, the next glaciation is being delayed for thousands of years. Therefore, it is expected that the present Holocene interglacial period may last at least another 10,000 years, i.e. global warming will continue for a very long time.

A moderate increase in temperature stress can induce substantial pollen sterility and reduced grain filling. Therefore, the reproductive development is going to be highly impacted in C_3 plants in higher CO_2 and temperature world that shall lead to reduced pollination and grain filling impacting grain yield. Consequently, the food output shall decrease from the heat-sensitive C_3 crop plants. Besides, the increased rate of photosynthesis in C_3 plants in high atmospheric [CO_2] shall increase the leaf and grain carbohydrate content of C_3 plants. However, the rate of N assimilation by nonleguminous plants cannot keep pace with increased photosynthesis due to limitation of soil N and its assimilation. Therefore, the protein content of grains per gram seed weight may decline resulting in malnutrition problem in the developing world.

The C_4 species evolved in warmer climatic conditions nearly 40–50 million years ago. Although C_4 species constitute only a very small percentage (2–3%) of flowering plants, they account for nearly 20% of the primary production of the Earth (Ehleringer et al. 1997; Sage and Monson 1999). Usually C_4 plants dominate the tropics and other warm climatic areas. They can also perform better in higher light intensities than C_3 plants. Under high temperature and high light environment, C_4 plants possess higher photosynthetic rates and biomass accumulation (Figure 12.2). In stress environment, i.e. high temperature and high light conditions, C_4 plants usually have higher water use efficiency (WUE), carbon assimilation rates, and nitrogen assimilation capacity than C_3 plants. Among most prevalent C_4 crop species of the world are maize (*Zea mays*), sugarcane (*Saccharum officinarum*), sorghum (*Sorghum bicolor*), and other millets (e.g. *Pennisetum* sp.). Similarly, several fodder species possess C_4 pathway mostly belonging to genera *Miscanthus, Paspalum, Bouteloua, Amaranthus, Chloris*, and *Cynodon*.

The future climatic condition will bring us higher temperature stress, water stress, and salt stress due to rise in sea levels that will flood coastal regions of the world. The C_4 plants can optimally function at higher temperature (38 °C) and high light, can tolerate drought and salt, and possess higher N use efficiency and radiation use efficiency than C_3 plants. Therefore, in the present and future scenarios, the crop and fodder improvement strategies should focus more on C_4 plants.

In C_3 plants the rate of photorespiration at higher temperature is higher due to an increase in the ratio of O_2/CO_2 dissolved in the cytoplasm and the specificity of Rubisco for O_2 favoring oxygenation over carboxylation. This is one of the reasons for sensitivity of C_3 plant net photosynthesis to high temperature. Besides, the photosynthetic membranes that carryout the electron transport to produce NADPH and ATP and the stromal enzyme system that is instrumental for carbon reduction are affected at temperatures around 32 °C.

Conversely, the C_4 photosynthesis usually can function optimally around 38 °C of leaf temperature, whereas the C_3 photosynthesis is reduced around 30 °C. The net photosynthesis (An) of maize is inhibited at leaf temperatures above 38 °C. The maximum quantum yield of photosystem II (F_v/F_m) in the dark-adapted leaves of maize is not seriously affected by leaf temperatures up to 45 °C. Furthermore, the primary carboxylating enzyme PEPC evolved in C_4 plants from its non-photosynthetic C_3 isoforms could nicely tolerate higher temperature (40 °C). The pyruvate phosphate dikinase is also not affected by the leaf temperature of 45 °C. The acclimation of C_4 photosynthesis to high temperature is accompanied by the expression of a new Rubisco activase polypeptide that protects Rubisco of C_4 plants localized in the BS cells. To summarize, the structure and function of chloroplasts of C_4 plants are naturally protected from the impact of high temperature. Similarly, the onset of pollen sterility for C_4 plants occurs at a higher temperature than C_3 plants (Crafts-Brandner and Salvucci 2002).

Therefore, C_4 plants have relatively higher vegetative growth rate and fertility (%) than C_3 plants at higher temperature. The C_4 Rubisco localized in the BS cells is already adapted to high [CO_2], and it is unlikely to be affected in high carbon environment. Therefore, attempts need to be made to target C_4 plants for crop improvement. It is important to breed maize, sugarcane, and millets for higher yield, resistance to pests, and high temperature to create crop for our future. Therefore, crop improvement strategy should target both C_3 and C_4 crop plants. Therefore, transfer of C_4 traits into C_3 plants is expected to improve photosynthetic rate, yield, and other agronomic features such as nutrient use efficiency (NUE) and WUE in C_3 plants (von Caemmerer et al. 2012, 2014; Cui 2021; Ermakova et al. 2020a, b; Karki et al. 2013; Niklaus and Kelly 2019; Sage and Zhu 2011; Schuler et al. 2016).

12.7 C_4 Plants Are Tolerant to Salt Stress

The evolution of salt tolerance has taken place in grasses at least 22 times (Sage 2001, 2004). However, only a few monocots are tolerant to salinity although the salt tolerance evolved many times in nature (Bromham and Bennet 2014). Salt tolerance evolves more frequently in C_4 grass lineages (Bromham and Bennet 2014). However, it has evolved more frequently among C_4 species suggesting that this metabolic pathway aids the salt tolerance. It is because it generates C_4 carboxylic acids, reduces transpiration, increases water-use efficiency, and limits the uptake of toxic ions. Therefore, it is not surprising that many halophytes are C_4. The salt tolerance has evolved more frequently in lineages than C_3 plants. This may be due to the ability of certain C_4 plants to use Na^+ as an osmoticum for plant growth in saline environment (Kronzucker et al. 2013).

The Na^+ is used for carbon concentration in plants having C_4 photosynthesis to reduce photorespiration. The pyruvate transport process is coupled to Na^+ in chloroplasts (Furumoto et al. 2011). However, unlike mangroves several C_4 crop plants, i.e. maize and sugarcane, can tolerate salt stress but do not possess growth benefits in the presence of Na^+ (Subbarao et al. 2003).

Salt-treated maize plants had increased activities of cytoplasmic phospho*enol*pyruvate carboxylase (PEPC), chloroplastic pyruvate orthophosphate dikinase (PPDK), and malate dehydrogenase (NADP-MDH) and mitochondrial malate dehydrogenase (NAD-MDH) present in mesophyll cells. The increase in these enzymatic activities is an adaptive response of C_4 plants i.e. *Miscanthus sinensis* to salinity (Sun et al. 2020). Intrinsically, the C_4 plant maize has a robust advanced antioxidant defense response. Therefore, the membrane lipid peroxidation is lower than wheat in saline environment. The efficiency of PS II-dependent electron transport chain was less affected in maize as compared to wheat.

Similarly, C_4 genes, i.e. PEPC, NADP-dependent malate dehydrogenase, and malic enzyme when overexpressed in C_3 plants generate excess proline that confers tolerance to transgenics (Omoto et al. 2010).

The salt tolerance of most C_4 monocot plants suggests that crop improvement strategies in saline environment may focus on C_4 plants and by mimicking C_4 photosynthesis in C_3 crop plants by overexpressing specific or combination of genes in C_3 plants.

12.8 Converting C_3 Plants into C_4 – A Himalayan Challenge

Transformation of C_3 plants into C_4 have focused on plant breeding and transgenic approach. Genetic engineering approaches in rice and other C_3 plants have mostly focused establishing a C_4 cycle in a single cell (Miyao et al. 2011, Kanoi ..., for a review see Singh et al. 2021; Taniguchi et al. 2008). Plants overexpressing C_4 enzymes in rice had severe defects such as increased photoinhibition and stunted growth due to incongruous gene regulation (Miyao et al. 2011; Taniguchi et al. 2008). However, in several instances increased photosynthetic rates are reported in *Arabidopsis*, tobacco, wheat, and rice when certain C_4 photosynthetic enzymes such as CA, PEPC, and PPDK are overexpressed individually. Overexpression of NADP-ME from maize in rice or *Arabidopsis* resulted in downregulation of photosynthesis and plant growth. Although, several reports demonstrate an increased photosynthetic rate in combined overexpression of C_4 genes improved photosynthesis, there are several reports to the contrary. A good carbon concentrating mechanism (CCM) in C_3 plants are not yet achieved (Ermakova et al. 2020a; Hu et al. 2012; Kandoi et al. 2016, 2018, 2022; Lin et al. 2020; Ren et al. 2014; Sen et al. 2017; Wang et al. 2012; Yadav et al. 2020; Zhang et al. 2014, 2021a, b). Actually, a drastic increase in the expression and activities of carboxylating and decarboxylating enzymes is needed for an operational C_4 pathway in C_3 plants.

12.9 Carbonic Anhydrase

CA is an enzyme for establishing CCM in C_4 plants. It catalyzes the reversible hydration of CO_2. It is a ubiquitous enzyme found in plant and animal cells. Plant CAs have multiple isoforms of α, β, ϒ, δ, and ζ – CA family, which share little sequence homology among themselves (Ludwig 2011; Moroney et al. 2004; Yu et al. 2007). The β isoforms are the most intensely studied CAs in plants as they are the most prevalent among land plants (DiMario et al. 2016). In C_4 plant species, the cytosolic β isoform of CA is recruited to operate in C_4 photosynthesis (Ludwig 2016). C_3 plants also contain the orthologs of β CA occurring in C_4 plants, but their exact functions are relatively unknown (Ludwig 2016). The role of CA in maintaining the steady-state C_4 photosynthetic rates has remained a subject of research. Different groups have reported inconsistent results that vary among C_4 plant species. Maize double mutants of *βca1* and *βca2* (*ca1ca2*) possessing 3% CA activity of the wild type (WT) did not show any reduction in photosynthetic rate or growth at ambient or elevated CO_2 concentrations. However, growth and photosynthetic activity of *ca1ca2* mutants was significantly reduced at suboptimal CO_2 levels. RNAi transformed *Setaria viridis* plants with decreased leaf CA activity had no effect on the CO_2 assimilation rates at ambient pCO_2. At low pCO_2 transformed plants had lower CO_2 assimilation rates than the null plants (Osborn et al. 2017; Studer et al. 2014). Alongside reduced CA activity, mesophyll conductance was a significant factor determining CO_2 assimilation rates at low pCO_2 (Osborn et al. 2017). Enhancing mesophyll conductance could improve CO_2 assimilation rates at low intercellular CO_2 concentrations prevailing in drought conditions. In transgenic *Flaveria bidentis* plants with reduced cytosolic CA activity, steady-state photosynthesis was unaltered until CA activity was below 20% of WT. CA activity less than 10% of WT drastically reduced CO_2 fixation rates and growth of the transformants at ambient CO_2 levels (Von Caemmerer et al. 2004).

CAs have also been associated with stomatal movement and development in response to varying CO_2 concentrations (Engineer et al. 2014; Hu et al. 2010, 2015). It has been shown that mutations in *Arabidopsis βCA1* and *βCA4* genes led to higher stomatal index and density than the WTs at elevated CO_2 levels. This was reversed when complemented by the expression of either *AtβCA1* or *AtβCA4* or the unrelated human *αCAII* under the guard cell specific promoter *pGC1*. The complementation test suggests that the catalytic activity of *CA* is important for stomatal closure in response to high CO_2 (Hu et al. 2010). Also, it was found that CAs regulate stomatal development via an extracellular signaling pathway (Engineer et al. 2014). Therefore, enhanced activity of CAs in guard cells should result in decreased stomatal density and consequently reduced transpiration resulting in improved WUE, which is desired in the current atmosphere of high CO_2 and water scarcity. Transgenic *Arabidopsis* plants overexpressing either *AtβCA1* or *βCA4* in guard cells exhibited improved instantaneous WUE with no growth penalty (Hu et al. 2010).

The overexpression of *FbβCA3* of the dicot C_4 *F. bidentis* possessing a very low K_m for CO_2 in the cytosol of the C_3 dicot plant *Arabidopsis thaliana* resulted in increased production of HCO_3^- that served as the substrate for the endogenous C_3 form of PEPC (Kandoi et al. 2022). Higher PEPC activity increased the supply of the carboxylic acid malate that plays an anaplerotic role in the tricarboxylic acid (TCA) cycle. These C_4 carboxylic acids served as additional carbon skeleton for augmented synthesis of amino acids aspartate, asparagine, glutamate, and glutamine. Consequently, protein content increased in the transgenics. Further, *FbβCA3* expression in *A. thaliana* resulted in higher chlorophyll content, increased electron transport rate, and carbon assimilation in the transgenics that had a larger biomass. Increased C assimilation resulted in higher sugar and starch content in the transgenics. The transgenics had reduced stomatal conductance, lower transpiration rate, and enhanced water-use efficiency. These demonstrate that expression of C_4 *CA* in the cytosol of C_3 crop plants could improve their photosynthetic efficiency, biomass, and water-use efficiency.

12.10 Phosphoenolpyruvate Carboxylase

Instead of Rubisco, PEPC is the primary CO_2 acceptor enzyme in C_4 plants. The C_4 PEPC is probably evolved from C_3 PEPC by few amino acid substitutions and post-translational modifications (Aldous et al. 2014; Bläsing et al. 2000; Chollet et al. 1996; Nimmo et al. 2001; Paulus et al. 2013). Due to the lack of oxygenase activity, the PEPC fixes CO_2 in C_4 species very efficiently. Therefore, the C_4 PEPC has been extensively used to genetically transform different C_3 plants, i.e. *Arabidopsis*, rice, and wheat, to impart improved photosynthetic parameters such as higher CO_2 assimilation rates, nitrogen content, and dry matter accumulation (Kandoi et al. 2016; Ku et al. 1999; Peng et al. 2018). The increased synthesis of oxaloacetic acid (OAA) played an anaplerotic role in TCA cycle of *PEPC* overexpressors. It was used to synthesize amino acids such as glutamate, aspartate, proline, glutamine, and asparagine that resulted in increased protein content and biomass of the transgenics (Kandoi et al. 2016; Peng et al. 2018). Further, the upregulated C_4 PEPC independently or in association with other C_4 photosynthetic enzymes such as PPDK, transferred to C_3 plants, can refix the CO_2 released during respiration and/or photorespiration into C_4 organic acids and can therefore improve C_3 photosynthesis (Gu et al. 2013; Kaachra et al. 2018). Refixation of respired CO_2 is a common phenomenon in stems of C_3 plants and trees (Alessio et al. 2005; Berveiller and Damesin 2008; Bloemen et al. 2013; Hibberd and Quick 2002). C_3 plants, rice, and wheat refix the respired and photorespired CO_2 to augment photosynthetic efficiency (Busch et al. 2013). High activities of PEPC, NAD-ME, NADP-ME, phosphoenolpyruvate carboxykinase (PEPCK), and PPDK have been detected in vascular tissues surrounding stems of C_3 species (Berveiller and Damesin 2008; Hibberd and Quick 2002; Ivanov et al. 2006; Wittmann et al. 2006). These enzymes are thought to be involved in the refixation of CO_2 released during respiration in tree stems (Berveiller and Damesin 2008) or the assimilation of malate produced in the roots of C_3 plants and transported to stem via xylem (Hibberd and Quick 2002).

Similar to C_4 plants, the transgenics carrying C_4 *PEPC* shows tolerance to certain abiotic stresses, high temperature and salinity, and high light albeit to a reduced extent (Bandyopadhyay et al. 2007; Ding et al. 2015; Gu et al. 2013; Kandoi et al. 2016; Lian et al. 2014; Liu et al. 2016; Qi et al. 2017; Qian et al. 2015; Qin et al. 2016; Shen et al. 2015). Gene expression of reactive oxygen species (ROS) defense enzymes, superoxide dismutase (SOD), peroxidase, and catalase along with compatible solutes, proline, and sucrose increased in transgenics that protected them from abiotic stresses (Ding et al. 2015; Kandoi et al. 2016; Qi et al. 2017).

The stomatal closure under water stress decreases CO_2 conductance to chloroplasts and increases the rate of ROS generation via Mehler reaction due to overreduction of chloroplastic electron transport chain. Enhanced NO and Ca^{2+} in PEPC overexpressing plants in drought conditions increases the PEPC reaction to produce organic acids and certain compatible solutes for increased drought tolerance (Liu et al. 2016, 2017). Further, the C_4 organic acids can be utilized by TCA cycle to increase intracellular CO_2 concentrations (Shen et al. 2015). Higher CO_2 levels inside cells raise the electron sink capacity of the electron transport chain (ETC) in chloroplasts and reduces ROS production by preventing O_2 reduction, i.e. Mehler reaction.

C_4 PEPC transformed plants have higher carbon assimilation rates than the control plants leading to higher sugar contents. Sucrose is converted to glucose and fructose by invertase and sucrose synthase. Both these enzymes are involved in signaling pathways important for plant development (Braun et al. 2014). Sucrose, glucose, and fructose help in maintaining osmotic potential under different stress conditions. It is suggested that glucose acts as a substrate for mitochondrial hexokinase, which is crucial for maintaining the rates of mitochondrial ETC and consequently ROS production therein (Xiang et al. 2011).

12.11 Malate Dehydrogenase

The malate dehydrogenases occur quite ubiquitously catalyzing the interconversion of OAA and malate. There are several isoforms of MDH with different specificities for the cofactors NADP, NADPH, NAD, or NADH (Gietl 1992a; Martinoia and Rentsch 1994; Musrati et al. 1998). NAD-MDH-dependent isoforms are present in the cytosol, mitochondria, and peroxisomes (Gietl 1990, 1992b; Gietl et al. 1990).

Chloroplast contains both the NAD- and NADP-dependent isoforms (Beeler et al. 2014; Berkemeyer et al. 1998). Both NADP- and NAD-dependent isoforms are homodimers and possess similar catalytic properties. NADP-MDHs contain eight conserved cysteine residues responsible for its redox regulation (Carr et al. 1999). Chloroplastic MDHs play vital roles in both C_3 and C_4 type plants. Plants performing C_4 photosynthesis have recruited the NADP-MDH isoform for the conversion of OAA to malate in mesophyll cells of NADP-ME type plants (McGonigle and Nelson 1995).

In C_4 plants the activation state of NADP-MDH is regulated by NADPH/NADP that demonstrates its regulation by redox status of the chloroplast (Ashton and Hatch 1983). In day time, the NADP-MDH located in chloroplast stroma is activated via reduction of S-S bond by reduced thioredoxin generated from reduced ferredoxin produced via photosynthetic ETC (Carr et al. 1999). The ATP and NADPH produced during day via light reactions are utilized for CO_2 fixation. Excess NADPH is utilized to reduce OAA to malate by light-activated NADP-MDH to regenerate NADP that is reused as the terminal electron acceptor of photosystem 1 (PSI). This prevents the generation of O_2^- via Mehler reaction. During night in the absence of ferredoxin-thioredoxin redox cycle the NADP-MDH remains inactive. This shuts down the operation of malate valve in night and preserves NADPH pool for assimilatory processes in plastids (Scheibe et al. 2005). Malate is a vital player in plant metabolism and balances the redox status of the cell by controlling the levels of reducing power in different organelles through malate shuttles (Scheibe et al. 2005). Overexpression of the plastidic maize C_4 NADP-MDH conferred salt tolerance to *Arabidopsis* plants by maintaining the appropriate cellular redox environment by efficient operation of malate valve (Kandoi et al. 2018). Transgenic plants also had higher amounts of the osmoprotectant, proline, than the WTs (Kandoi et al. 2018).

Malate oxidation by mNAD-MDH in mitochondria results in the formation of OAA, which is substrate for energy generation in TCA cycle. In salt-stressed maize plants, enhanced activities of the marker enzyme of the TCA cycle, succinate dehydrogenase (SDH), and MDH suggest that more energy is required by cells to withstand stress conditions (Eprintsev and Fedorina 2007). Increase in ATP production was also observed in tomato plants harboring an apple cytosolic *MDH* gene (*MdMDH*). Overexpression of this gene in tomato and apple conferred tolerance to cold and salinity stress by modulating the redox status and salicylic acid (SA) content of the cells (Wang et al. 2016; Yao et al. 2011). In transgenic apple plants, the contents of ascorbic acid (AsA), reduced glutathione (GSH), and SA were increased, which are key players in plant defense against both biotic and abiotic stresses (Jayakannan et al. 2015; Wang et al. 2016). Increasing the amounts of NAD has been shown to induce SA biosynthesis. The increased SA content might be due to higher NAD levels caused by the activity of cytosolic *MdMDH* (Wang et al. 2016).

12.12 Decarboxylating Enzymes

The rate of decarboxylation must be higher than the carboxylation reaction to have higher rates of carboxylation than oxygenation. Further, the rate of decarboxylation must compensate for CO_2 leakage. Therefore, the decarboxylating enzyme has to be highly efficient. Plants performing C_4 photosynthesis have three known decarboxylating enzymes, i.e. NADP-ME,

NAD-ME, and PEPCK (Rao and Dixon 2016). Maize, sorghum, and sugarcane are classified as NADP-ME type C_4 species, whereas species such as pearl millet, switchgrass, and amaranth belong to NAD-ME type. C_4 plants like *Urochloa panicoides*, *Panicum maximum*, and *Leymus chinensis* are grouped into PEPCK subtype.

12.12.1 NAD/NADP-Malic Enzyme

The NADP-ME subtype C_4 plant maize harbors three NADP-ME isoforms, i.e. chloroplastic C_4 NADP-ME, non-photosynthetic C_3 like plastidic NADP-ME, and cytosolic NADP-ME (Detarsio et al. 2008). The NAD-ME subtype C_4 species switchgrass contains four NAD-ME genes. Three of the four NAD-MEs clustered with the known photosynthetic NAD-ME2 isoforms, while the fourth clustered with the non-photosynthetic NAD-ME1 proteins of different plant species (Rao et al. 2016). The NAD and NADP-ME involved in the CCM of C_4 plants catalyze the oxidative decarboxylation of malate to pyruvate in mitochondria and chloroplasts, respectively. The CO_2 generated in the reaction is used by the Rubisco enzyme for the C_3 cycle. Therefore, attempts to transfer the C_4 cycle NADP-ME in C_3 plants were focused to improve the photosynthetic rate by enriching CO_2 near Rubisco. Overexpression of C_4 NADP-ME in C_3 plants led to various aberrations such as abnormal chloroplasts, bleaching of leaf color, growth hindrance, and senescence in the dark (Chi et al. 2004; Fahnenstich et al. 2007b; Takeuchi et al. 2000; Tsuchida et al. 2001). However, transgenic tobacco plants expressing maize C_4 NADP-ME in chloroplasts had reduced stomatal conductance and gained more fresh mass per unit of water consumed than the WTs with no growth penalty (Laporte et al. 2002). The conflicting outcomes could be a result of the use of different promoters, which led to different levels of NADP-ME expression and consequently altered malate metabolism. Laporte et al. (2002) used a modified mannopine synthase promoter, which ensured expression of the gene in different tissues including guard cells. Stomata open when there is high concentration of K^+, malate, and Cl^- in guard cells. Utilization of malate by NADP-ME decreases its concentration in guard cells leading to shrinkage of stomatal aperture. Therefore, C_4 NADP-ME (with a higher V_m and lower K_m value) can be used as an effective drought avoidance strategy if expressed in guard cells where malate levels remain high (Laporte et al. 2002).

Genome of C_3 plant *Arabidopsis* codes for 4 NADP-ME(1–4) and 2 NAD-ME isoforms. Out of these NADP-ME(1–3) are cytosolic and NADP-ME4 is located in plastids. Both the NAD-ME proteins are localized in mitochondria (Badia et al. 2017; Gerrard Wheeler et al. 2008). The non-photosynthetic isoforms of NAD/NADP-ME in C_3 and C_4 plants are engaged in performing housekeeping and non-photosynthetic functions. The non-photosynthetic plastidic NADP-ME and the cytosolic NADP-ME are involved in general plant metabolism such as lipid, protein, and lignin biosynthesis; nitrogen assimilation in plastids; and pH homeostasis by managing the levels of malate, pyruvate, and NADPH essential for these cellular activities (Brown Naomi et al. 2009; Eastmond et al. 1997; Gerrard Wheeler et al. 2005, 2008; Lai et al. 2002a, b; Maier et al. 2011; Schaaf et al. 1995; Smith et al. 1992). Both the cytosolic and chloroplastic NADP-ME play an important role in stress responses against water stress, low temperature, salt stress, and osmotic stress in different plants (Alvarez et al. 2013; Cheng and Long 2007; Detarsio et al. 2008; Doubnerová Hýsková et al. 2014; Doubnerová and Ryšlavá 2011; Emami et al. 2016; Fu et al. 2009, 2011; Fushimi et al. 1994; Guo et al. 2009; Liu et al. 2007; Nguyen et al. 2009; Pinto et al. 1999; Sicher et al. 2015; Walter Michael et al. 1994). Further, in tobacco and *Arabidopsis*, the C_4 acid decarboxylases function in supplying CO_2 to cells near the veins by decarboxylating malate present in the transpiration stream (Brown Naomi et al. 2009; Hibberd and Quick 2002). The non-photosynthetic NAD-ME isoforms are involved in regulating carbon flux through TCA cycle and glycolysis (Brown Naomi et al. 2009; Jenner et al. 2001; Tronconi et al. 2008). Transgenic *Arabidopsis* plants expressing rice C_3 NADP-ME and NAD-ME1 isoforms have shown improved growth characteristics under salt, drought, and other oxidative stresses than the WTs (Chen et al. 2015; Cheng and Long 2007; Liu et al. 2007; Zhou et al. 2011). However, *Arabidopsis* plants constitutively overexpressing NADP-ME2, the most abundant cytosolic isoform in *Arabidopsis*, showed impaired growth, imbalanced redox state, and increased sensitivity to osmotic stress as compared to WTs (Badia et al. 2015). NADP-ME2 is a pyruvate carboxylating ME in *Arabidopsis* and is thought to be involved in reassimilation of CO_2 in C_4 acids released during photorespiration (Badia et al. 2017).

NADP-ME produces NADPH and pyruvate by decarboxylating malate and using NADP as its cofactor. NADPH thus produced could be utilized to detoxify the cellular environment via the ascorbate-glutathione and the thioredoxin system, in the production of osmoprotectants such as proline and mannitol and lipid, flavonoid, and lignin biosynthesis. Flavonoids protect the photosynthetic apparatus by absorbing UV-B light. These may also scavenge hydroxyl radicals and H_2O_2 (Casati et al. 1999; Pinto et al. 1999), whereas lignin deposition in cell walls acts as physical barriers for both biotic and abiotic stresses (Dicko et al. 2005). Malic enzyme isoforms such as SgME1 and GmME1, which synthesize malate from carboxylation of pyruvate, confer resistance to Al stress by increasing malate and citrate concentration in the cytoplasm and root exudates (Sun et al. 2014; Zhou et al. 2011).

12.12.2 Phosphoenolpyruvate Carboxykinase

In BS cells, the cytoplasmic PEPCK decarboxylates OAA to generate PEP and CO_2 that diffuses to chloroplast to concentrate carbon. Attempts were made to improve photosynthetic rates in C_3 plants by increasing the CO_2 concentration near Rubisco by using C_4 plant-specific PEPCK (Suzuki et al. 2000). Overexpression of PEPCK from *U. panicoides* hooked to *rbcs* transit peptide that directs it to chloroplasts of rice resulted in carbon flow akin to C_4 photosynthesis, where high amounts of carbon was fixed in four-carbon acids such as malate, OAA, and aspartate (Suzuki et al. 2000). Transgenic rice plants overexpressing *ZmPEPCK* targeted to mesophyll chloroplasts had improved growth (20–50%) and yield (10–50%) than control plants under field conditions (Gupta and Huang 2014). Overexpression of PEPCK in the chloroplasts along with PEPC in the cytoplasm of rice led to increased incorporation of $^{14}CO_2$ into C_4 compounds and sucrose. However, these transgenics had reduced Chl content and swollen thylakoids (Suzuki et al. 2006).

In C_3 plants PEPCK plays important roles in a variety of physiological processes like seedling establishment, plant defense, nitrogen metabolism, and pH regulation (Aubry et al. 2011; Penfield et al. 2004). In *Arabidopsis*, *PEPCK1* is involved in regulating stomatal conductance via malate metabolism in guard cells and has implications in water retention in dark. The *pepck1* mutant plants were found susceptible to drought as compared to WTs due to increased stomatal conductance and increased transpiration, which could be a result of impaired malate metabolism (Penfield et al. 2011). Overexpression of *SlPEPCK* in *Solanum lycopersicum* also enhanced seedling growth and sugar content in ripened fruits as compared to WTs, pointing toward its role in gluconeogenesis during fruit ripening (Huang et al. 2015). The role of PEPCK in gluconeogenesis in seeds of *Arabidopsis* and castor beans and cotyledons of *Cucurbita pepo* has already been confirmed (Benedict and Beevers 1961; Leegood and ap Rees 1978; Rylott et al. 2003). PEPCK converts the C_4 acid OAA, obtained predominantly from the breakdown product of lipid reserves via the glyoxylate cycle, to PEP (Eastmond et al. 2000, 2015). Although the possibility of the presence of an alternate route other than the glyoxalate cycle for the respiration of fatty acids in plants is also suggested (Eastmond et al. 2000). The formation of PEP is the first step of this gluconeogenic pathway to produce sucrose that is essential for germination and seedling establishment (Eastmond et al. 2000; Rylott et al. 2003). More recently, phloem parenchyma-localized PEPCK is shown to be important reutilization of plant nitrogen resources present in the form of amino acids (asparagine and glutamine) and amides in the xylem transpiration stream (Bailey and Leegood 2016). The metabolism of asparagine/aspartate generates OAA that is converted by PEPCK to PEP (Bailey and Leegood 2016). PEP so formed could be used in developing flowers and seeds and for the secondary metabolite synthesis via Shikimate pathway (Bailey and Leegood 2016; Chen et al. 2004; Delgado-Alvarado et al. 2007; Hibberd and Quick 2002). Increased PEPCK activity in rice leaves reduced guttation (Bailey and Leegood 2016). Decrease in the amount of guttation fluid could also increase the WUE by reducing water loss through hydathodes.

12.13 Pyruvate Orthophosphate Dikinase

PPDK is present in all C_3 and C_4 plants in cytoplasm and chloroplast (Hyskova and Ryslava 2016). It is relatively less abundant in C_3 plants. In C_4 plants, chloroplastic form is the predominant one. The chief role of PPDK in C_4 photosynthesis is the regeneration of PEP, the CO_2 acceptor molecule, along with AMP and PPi, from pyruvate by using ATP in chloroplasts of mesophyll cells, thus potentially serving as a rate limiting enzyme in CO_2 assimilation (Chastain et al. 2011). C_4 PPDK activity is regulated by the phosphorylated/dephosphorylated state of an active site threonine residue by PPDK REGULATORY PROTEIN (Chastain et al. 2011; Chen et al. 2014).

Two genes named *PPDK1* and *PPDK2* are present in maize. *PPDK2* is the predominant enzyme in endosperm, whereas *PPDK1* is prevalent in mesophyll cells (Lappe et al. 2018). Mutation in *PPDK1* is seedling lethal, suggesting the obligate C_4 photosynthesis in maize (Lappe et al. 2018). PPDK in endosperm affects glycolytic flux and is associated with kernel hardness (Lappe et al. 2018). In maize, it had been found previously that PPDK is accumulated in significantly higher amounts at later stages of endosperm development and might have a role in balancing the ratio of starch and protein in endosperm (Méchin et al. 2007). Overexpression of maize C_4 PPDK gene in rice and *Arabidopsis* has been shown to improve photosynthetic rate, yield, and resistance to drought stress in transformed plants as compared to WTs (Gu et al. 2013; Ku et al. 2000;

Wang et al. 2012). Under desiccation stress when stomata are closed, CO_2 released by PPDK helps in maintaining photosynthetic rates and minimizing water loss (Gu et al. 2013; Hyskova and Ryslava 2016; Wang et al. 2012).

In C_4 grass *Miscanthus × giganteus*, the expression of enzymes involved in photosynthesis, especially the C_4 specific PPDK, is noticeably increased and is thought to provide tolerance against chilling stress (Spence et al. 2014). However, no statistically significant difference in biomass yields or in photosynthetic capacity was observed in cool conditions between WT and transformants of switchgrass (*Panicum virgatum*) harboring *M × giganteus PPDK* (Halter et al. 2015). Recently, it has been shown that the chilling tolerance of *M × giganteus* is due to its ability to accumulate higher water soluble carbohydrates such as raffinose and sucrose during early growth period in low temperatures. But later in the growing season when temperature rises, the photosynthetic capacity increases due to enhanced PPDK activity (Fonteyne et al. 2017). In another study, the C-terminal amino acid sequence of the maize PPDK was altered by point mutations to mimic the PPDK sequence of a cold tolerant C_3/C_4 intermediate species, *Flaveria brownie* (Ohta et al. 2006). These alterations resulted in significant improvement in cold tolerance of transgenic maize plants that had 23% higher photosynthetic rate at 8 °C than the control plants (Ohta et al. 2006). Variable observations in the two studies indicate that further work is needed to elucidate the role of PPDK in cold tolerance of some C_4 plant species.

12.14 Regulation of C_4 Photosynthetic Gene Expression

Although all the key enzymes of C_4 photosynthesis are already present in C_3 plants, their expression levels and tissue specificity vary significantly when compared to the enzymes present in C_4 plants. These differences in the C_4 enzymes present in C_3 and C_4 plants are due to transcriptional regulation by *cis*-elements (promoter, enhancers, and UTRs), *trans*-factors, post-transcriptional and post-translational controls, and epigenetic changes such as DNA methylation and histone modifications (Hibberd and Covshoff 2010; Kajala et al. 2012; Miyao 2003; Offermann et al. 2008; Perduns et al. 2015; Reyna-Llorens et al. 2018; Wang et al. 2014). Recently, it has been shown that duplication of genes involved in C_4 photosynthesis and their neofunctionalization could be a key factor in the evolution of C_4 photosynthetic traits and differential expression pattern of these genes (Emms et al. 2016).

Mesophyll specific expression of C_4 genes: In maize, mesophyll cell specific expression of PEPC is associated with changes in the promoter sequence, demethylation of a PvuII site at 3.1 kb 5′ of the transcription start site, and trimethylation of histone H3K4 (Danker et al. 2008; Kausch et al. 2001; Langdale et al. 1991; Taniguchi et al. 2000). In C_4 plant *Gynandropsis gynandra*, 5′ or 3′ UTR of *PPDK* and *CA4* gene is sufficient for generating mesophyll cell specificity and these elements are also present in C_3 plant *Arabidopsis* (Kajala et al. 2012). In C_4 *Flaveria trinervia*, the presence of a mesophyll enhancing module 1 (MEM1) in the promoter region is responsible for the mesophyll specific expression of PEPC (Akyildiz et al. 2007; Gowik et al. 2004). An MEM has also been identified in the promoter region of *CA3* gene in *F. bidentis,* whose function is yet to be confirmed (Tanz et al. 2009). Very recently, another MEM named as MEM2 in the UTRs of *CA2*, *CA4*, and *PPDK* has been identified in *G. gynandra* and *Arabidopsis* (Williams et al. 2016). The presence of MEM2 and loss of some *cis*-elements in the promoter and introns, respectively, is suggested to increase the expression of *CA4* in mesophyll cells of *G. gynandra* as compared to *Arabidopsis* (Williams et al. 2016). Similarly, the presence of *cis*-acting DNA sequences in the promoter region of PPDK in maize and *F. trinervia* is responsible for its high level of expression in mesophyll cells (Matsuoka and Numazawa 1991; Rosche et al. 1998). In maize, multiple presence of a putative *cis*-element (RGCGR; R = A/G) in the promoter region of C_4 genes such as *PPDK*, *ZmPDRP1* (*PPDK REGULATORY PROTEIN 1*), *PEPC*, and *CA* is thought to be associated with their mesophyll specific expression patterns (Wang et al. 2014). This *cis*-element is also conserved in other C_4 plant species, sorghum and *Setaria italica*. However, it is absent in rice (Wang et al. 2014).

Bundle sheath specific expression of C_4 genes: The exact mechanism of BS specific expression of C_4 genes is not well defined. Strong BS specific expression of C_4 *NADP-ME1* in *F. bidentis* requires both 5′ and 3′ UTRs (Marshall et al. 1997). The *cis*-elements required for BS specific accumulation of NAD-ME and NADP-ME in C_3 plants such as rice and *Arabidopsis* are already present in their gene structure. Most likely the presence of *trans*-factor(s) in C_4 BS cells are responsible for their BS restricted expression. However, these *trans*-acting factors are not yet identified (Brown et al. 2011). The BS specificity of *NADME1*, *NADME2*, and *mMDH* in C_4 *G. gynandra* is controlled by the presence of two *cis*-elements, BSM1a and BSM1b (bundle Sheath Motif 1a and 1b), present in the first exon of their coding region (Reyna-Llorens et al. 2018). These *cis*-elements are named as duons due to their dual function of determining amino acid sequence and regulating the spatial expression pattern of these genes in C_4 plants. Despite the presence of BSM1a and BSM1b in the orthologous *NAD-ME* genes in *A. thaliana*, the closest C_3 relative of *G. gynandra*, BS cell preferential expression is not observed in *Arabidopsis*

(Reyna-Llorens et al. 2018). Interaction of these motifs with *trans*-acting factor/s present in *G. gynandra* but not in *Arabidopsis* might be the reason of the transcription repression of *NAD-ME* genes in mesophyll cells of *G. gynandra* (Reyna-Llorens et al. 2018).

12.15 Use of C_3 Orthologs of C_4 Enzymes

The C_4 genes are evolved from C_3 precursors, and all the C_4 gene orthologs are present in C_3 plants. However, their functions in C_3 plants are different from C_4. The genetic manipulation of C_3 orthologs of C_4 enzymes could improve the flux through C_4 enzymes (Gowik et al. 2004; Hibberd and Covshoff 2010; Kajala et al. 2012; Reyna-Llorens and Hibberd 2017; Reyna-Llorens et al. 2018; Williams et al. 2016). The kinetic properties of some C_4 enzymes such as PEPC, PPDK, and NADP-ME are optimized for C_4 photosynthesis by changes in few amino acid residues in the C_3 ancestral isoforms (Alvarez et al. 2013; Bläsing et al. 2000; DiMario et al. 2020; Paulus et al. 2013; Shi et al. 2020). However, the promoters of C_4 genes are highly modified for increased expression and appropriate localization in the leaves (Gowik et al. 2004; Reyna-Llorens and Hibberd 2017; Williams et al. 2016). Therefore, it will be appropriate to modify the C_3 promoters to C_4 equivalent. It is also important to perform DNA base editing to modify the catalytic site of C_3 ortholog genes to have higher activity. The genome editing technology using CRISPR/Cas9 is successfully used to change certain bases and long DNA motifs including promoters in a few plant species including rice (Dong and Ronald 2021; Lu et al. 2020). The gene editing of certain C_4 orthologs in C_3 plants shall minimize the number of genes required for transformation to generate a C_4 pathway in rice. It shall also overcome the problems of gene silencing associated with insertion of several genes in a one construct.

The Kranz anatomy that contains thick walled BS cells helps the CCM in BS by preventing the escape or diffusion of CO_2 in its high CO_2 environment to other neighboring cells. For the operation of C_4 pathway, there are a few prerequisites, i.e. transport of C_4 acids from mesophyll to BS cells, increased vein density, higher chloroplast number in BS cells, suppression of Rubisco and other enzymes of carbon reduction cycle in the mesophyll cells, and suppression of glycine decarboxylase expression in BS. These need to be accomplished to introduce an effective CCM around Rubisco in BS cells of C_3 plants via gene editing technology (Danila et al. 2021; Ermakova et al. 2020b; Eva et al. 2019; Giuliani et al. 2019; Karki et al. 2013; Lin et al. 2016; Maheshwari et al. 2021; Schuler et al. 2016; ShirinZamani-Nour et al. 2021). These available literature suggest that modifying the expression of genes of important metabolic processes, i.e. Calvin–Benson cycle and photorespiration, might be disadvantageous for plants.

Although certain important progresses are made in the C_4 rice project, an engineered C_4 rice remains elusive. Our knowledge of C_4 metabolic regulatory network is still not clear and needs further basic studies. The C_4 plants are better suited in tropics and in cold climate and low light regime in temperate zones, and C_3 plants often outperform C_4 plants.

12.16 Conclusions and Future Directions

Extreme hot and dry conditions are becoming more common due to the rise in global temperatures. The major food crops in world, rice, wheat, potato, soyabean, and mustard, belong to the C_3 type. As C_3 plants are less tolerant to stress conditions (Stepien and Klobus 2005; Uzilday et al. 2012), they are vulnerable to unpredicted changes in climatic conditions. The C_4 plants have evolved nearly 40 million years ago in low CO_2 and high temperature environment. Therefore, the C_4 plants perform far better than the C_3 in high temperature and low CO_2. Although a huge C_4 rice consortium at IRRI, Philippines, along with many global partners is attempting to convert rice, a C_3 crop, into a C_4 crop, it has not yet been achieved (von Caemmerer et al. 2012). Also, efforts to introduce Kranz anatomy into C_3 plants have not succeeded so far. However, genes like *SCARECROW*, *SHORTROOT1*, *CYP90D2*, and *Golden2-like (GLK)* could be candidates in increasing the vein density and establishing Kranz-like anatomy in C_3 plants, a key feature of C_4 plants (Rizal et al. 2015; Slewinski et al. 2012, 2014; Wang et al. 2017). Transformation of rice with maize *GLK* transcription factors *ZmG2* and *ZmGLK1* led to the emergence of traits peculiar of a proto Kranz anatomy such as increased organelle volume in terms of chloroplasts and mitochondria, increased plasmodesmatal connections, and higher accumulation of photosynthetic enzymes in sheath cells around leaf veins (Wang et al. 2017). In single cell C_4 approach, C_3 plants transformed with C_4 photosynthetic genes may not be able to transform C_3 plant to C_4 in the very near future. It may have better agronomic traits such as higher yield or NUE or WUE and abiotic stress tolerance than the untransformed C_3 plants. Therefore, C_3 crops equipped with such features shall be helpful to address the concern of short-term goals of world food and energy security in the prevailing fluctuating environmental conditions.

Comparative molecular, in silico, and evolutionary studies in different C_3 and C_4 plants have led to the identification of factors such as *cis*-elements, *trans*-acting factors, epigenetic modifications, and amino acid substitutions, which are responsible for the distinctness in the expression level, tissue specificity, and catalytic properties of different C_4 photosynthetic enzymes (Bräutigam et al. 2011, 2014; Brown et al. 2011; Rangan et al. 2016; Wang et al. 2009, 2014). With the advent of genome editing technologies like CRISPR/Cas and TALENs, precise nucleotide engineering and epigenetic modifications including DNA methylation and acetylation are achievable at a much higher rate (Enríquez 2016). Therefore, it is possible to tailor the C_3 photosynthetic genes to make them more C_4-like and/or more competent that would highly increase their tissue specific expression and enzymatic activities. Such genetic manipulations would preclude the need of generating de novo C_3 transgenic crops from C_4 genes and thereby prevent the consequent unintended effects (Bortesi and Fischer 2015). Moreover, these genetically edited plants would help in understanding the regulatory mechanisms of C_4 enzymes in a C_3 background and could be used as a base material for further incorporation of genes required to transform partially or completely a C_3 plant into C_4 type. Such plants shall be tolerant to high temperature, high light, salt stress, and shall possess high water-use efficiency to conserve water for sustainable climate resilient agriculture.

Acknowledgment

Author thanks Pragya Yadav, Shalu Chaudhury, Jitender Singh, and Deepika Kandoi for their help in the preparation of the manuscript.

References

Akyildiz, M., Gowik, U., Engelmann, S. et al. (2007). Evolution and function of a cis-regulatory module for mesophyll-specific gene expression in the C_4 dicot *Flaveria trinervia*. *Plant Cell* 19: 3391–3402.

Aldous, S.H., Weise, S.E., Sharkey, T.D. et al. (2014). Evolution of the phosphoenolpyruvate carboxylase protein kinase family in C_3 and C_4 *Flaveria* spp. *Plant Physiol.* 165: 1076–1091.

Alessio, G.A., Pietrini, F., Brilli, F., and Loreto, F. (2005). Characteristics of CO_2 exchange between peach stems and the atmosphere. *Funct. Plant Biol.* 32: 787–795.

Alvarez, C.E., Saigo, M., Margarit, E. et al. (2013). Kinetics and functional diversity among the five members of the NADP-malic enzyme family from *Zea mays*, a C_4 species. *Photosynth. Res.* 115: 65–80.

Ashton, A.R. and Hatch, M.D. (1983). Regulation of C_4 photosynthesis: regulation of activation and inactivation of NADP-malate dehydrogenase by NADP and NADPH. *Arch. Biochem. Biophys.* 227: 416–424.

Aubry, S., Brown, N.J., and Hibberd, J.M. (2011). The role of proteins in C_3 plants prior to their recruitment into the C_4 pathway. *J. Exp. Bot.* 62: 3049–3059.

Badia, M.B., Arias, C.L., Tronconi, M.A. et al. (2015). Enhanced cytosolic NADP-ME2 activity in *A. thaliana* affects plant development, stress tolerance and specific diurnal and nocturnal cellular processes. *Plant Sci.* 240: 193–203.

Badia, M.B., Mans, R., Lis, A.V. et al. (2017). Specific *Arabidopsis thaliana* malic enzyme isoforms can provide anaplerotic pyruvate carboxylation function in *Saccharomyces cerevisiae*. *FEBS J.* 284: 654–665.

Bailey, K.J. and Leegood, R.C. (2016). Nitrogen recycling from the xylem in rice leaves: dependence upon metabolism and associated changes in xylem hydraulics. *J. Exp. Bot.* 67: 2901–2911.

Bandyopadhyay, A., Datta, K., Zhang, J. et al. (2007). Enhanced photosynthesis rate in genetically engineered indica rice expressing pepc gene cloned from maize. *Plant Sci.* 172: 1204–1209.

Beeler, S., Liu, H.C., Stadler, M. et al. (2014). Plastidial NAD-dependent malate dehydrogenase is critical for embryo development and heterotrophic metabolism in *Arabidopsis*. *Plant Physiol.* 164: 1175.

Benedict, C.R. and Beevers, H. (1961). Formation of sucrose from malate in germinating castor beans I conversion of malate to phosphoenol-pyruvate. *Plant Physiol.* 36: 540–544.

Berkemeyer, M., Scheibe, R., and Ocheretina, O. (1998). A novel, non-redox-regulated NAD-dependent malate dehydrogenase from chloroplasts of *Arabidopsis thaliana* L. *J. Biol. Chem.* 273: 27927–27933.

Berveiller, D. and Damesin, C. (2008). Carbon assimilation by tree stems: potential involvement of phosphoenolpyruvate carboxylase trees. *Struct. Funct.* 22: 149–157.

Bläsing, O.E., Westhoff, P., and Svensson, P. (2000). Evolution of C_4 phosphoenolpyruvate carboxylase in *Flaveria*, a conserved serine residue in the carboxyl-terminal part of the enzyme is a major determinant for C_4-specific characteristics. *J. Biol. Chem.* 275: 27917–27923.

Bloemen, J., Mcguire, M.A., Aubrey, D.P. et al. (2013). Transport of root-respired CO_2 via the transpiration stream affects aboveground carbon assimilation and CO_2 efux in trees. *New Phytol.* 197: 555–556.

Bortesi, L., & Fischer, R. (2015). The CRISPR/Cas9 system for plant genome editing and beyond. Biotechnology advances, 33(1), 41–52.

Braun, D.M., Wang, L., and Ruan, Y.L. (2014). Understanding and manipulating sucrose phloem loading, unloading, metabolism, and signalling to enhance crop yield and food security. *J. Exp. Bot.* 65: 1713–1735.

Bräutigam, A., Kajala, K., Wullenweber, J. et al. (2011). An mRNA blueprint for C_4 photosynthesis derived from comparative transcriptomics of closely related C_3 and C_4 species. *Plant Physiol.* 155: 142–156.

Bräutigam, A., Schliesky, S., Külahoglu, C., Osborne, C. P., & Weber, A. P. (2014). Towards an integrative model of C_4 photosynthetic subtypes: insights from comparative transcriptome analysis of NAD-ME, NADP-ME, and PEP-CK C_4 species. Journal of experimental botany, 65(13), 3579–3593.

Bromham, L. and Bennet, T.H. (2014). Salt tolerance evolves more frequently in C_4 grass lineages. *J. Evol. Biol.* 27: 653–659.

Brown Naomi, J., Palmer, B.G., Stanley, S. et al. (2009). C_4 acid decarboxylases required for C_4 photosynthesis are active in the mid-vein of the C_3 species *Arabidopsis thaliana,* and are important in sugar and amino acid metabolism. *Plant J.* 61: 122–133.

Brown, N.J., Newell, C.A., Stanley, S. et al. (2011). Independent and parallel recruitment of pre-existing mechanisms underlying C_4 photosynthesis. *Science* 331: 1436–1439.

Busch, F.A., Sage, T.L., Cousins, A.B., and Sage, R.F. (2013). C_3 plants enhance rates of photosynthesis by reassimilating photorespired and respired CO_2. *Plant Cell Environ.* 36: 200–212.

von Caemmerer, S., Quinn, V., Hancock, N.C. et al. (2004). Carbonic anhydrase and C_4 photosynthesis: a transgenic analysis. *Plant Cell Environ.* 27: 697–703.

von Caemmerer, S., Quick, W.P., and Furbank, R.T. (2012). The development of C_4 rice: current progress and future challenges. *Science* 336: 1671–1672.

von Caemmerer, S., Edwards, G.E., Koteyeva, N., and Cousins, A.B. (2014). Single cell C_4 photosynthesis in aquatic and terrestrial plants: a gas exchange perspective. *Aquat. Bot.* 118: 71–80.

Carr, P.D., Verger, D., Ashton, A.R., and Ollis, D.L. (1999). Chloroplast NADP-malate dehydrogenase: structural basis of light-dependent regulation of activity by thiol oxidation and reduction. *Structure* 7: 461–475.

Casati, P., Drincovich, M.F., Edwards, G.E., and Andreo, C.S. (1999). Malate metabolism by NADP-malic enzyme in plant defense. *Photosynth. Res.* 61: 99–105.

Chastain, C.J., Failing, C.J., Manandhar, L. et al. (2011). Functional evolution of C_4 pyruvate, orthophosphate dikinase. *J. Exp. Bot.* 62: 3083–3091.

Chen, Z.-H., Walker, R.P., Técsi, L.I. et al. (2004). Phosphoenolpyruvate carboxykinase in cucumber plants is increased both by ammonium and by acidification, and is present in the phloem. *Planta* 219: 48–58.

Chen, Y.-B., Lu, T.-C., Wang, H.-X. et al. (2014). Posttranslational modification of maize chloroplast pyruvate orthophosphate dikinase reveals the precise regulatory mechanism of its enzymatic activity. *Plant Physiol.* 165: 534–549.

Chen, L., Tsugama, D., Takano, T., and Liu, S. (2015). Rice (*Oryza sativa* L.) OsNADP-ME4 gene responds to adversity stresses. *Cell Biol. Biophys.* 4: 1–7.

Cheng, Y. and Long, M. (2007). A cytosolic NADP-malic enzyme gene from rice (*Oryza sativa* L.) confers salt tolerance in transgenic *Arabidopsis*. *Biotechnol. Lett.* 29: 1129–1134.

Chi, W., Zhou, J.S., Zhang, F., and Wu, N.H. (2004). Photosynthetic features of transgenic rice expressing sorghum C_4 type NADP-ME. *Acta Bot. Sin.* 46: 873–882.

Chollet, R., Vidal, J., and O'Leary, M.H. (1996). Phosphoenolpyruvate carboxylase: a ubiquitous, highly regulated enzyme in plants. *Annu. Rev. Plant Physiol. Plant Mol. Biol.* 47: 273–298.

Crafts-Brandner, S.J. and Salvucci, M.E. (2002). Sensitivity of photosynthesis in a C4 plant, maize, to heat stress. *Plant Physiol.* 129: 1773–1780.

Cui, H. (2021). Challenges and approaches to crop improvement through C3-to-C4 engineering. *Front. Plant Sci.* 12.

Danila, F.R., Thakur, V., Chatterjee, J. et al. (2021). Bundle sheath suberisation is required for C4 photosynthesis in a Setaria viridis mutant. *Commun. Biol.* 4: 254.

Danker, T., Dreesen, B., Offermann, S. et al. (2008). Developmental information but not promoter activity controls the methylation state of histone H3 lysine 4 on two photosynthetic genes in maize. *Plant J.* 53: 465–474.

Delgado-Alvarado, A., Walker Robert, P., and Leegood Richard, C. (2007). Phosphoenolpyruvate carboxykinase in developing pea seeds is associated with tissues involved in solute transport and is nitrogen-responsive. *Plant Cell Environ.* 30: 225–235.

Detarsio, E., Maurino, V.G., Alvarez, C.E. et al. (2008). Maize cytosolic NADP-malic enzyme (ZmCytNADP-ME): a phylogenetically distant isoform specifically expressed in embryo and emerging roots. *Plant Mol. Biol.* 68: 355.

Dicko, M.H., Gruppen, H., Barro, C. et al. (2005). Impact of phenolic compounds and related enzymes in sorghum varieties for resistance and susceptibility to biotic and abiotic stresses. *J. Chem. Ecol.* 31: 2671–2688.

DiMario, R.J., Quebedeaux, J.C., Longstreth, D. et al. (2016). The cytoplasmic carbonic anhydrases βCA2 and βCA4 are required for optimal plant growth at low CO_2. *Plant Physiol.* 171: 01990.

DiMario, R.J., Kophs, A.N., Pathare, V.S. et al. (2020). Kinetic variation in grass phospho enol pyruvate carboxylases provides opportunity to enhance C4 photosynthetic efficiency. *Plant J.* 105: 1677–1688.

Ding, Z.S., Sun, X.F., Huang, S.H. et al. (2015). Response of photosynthesis to short-term drought stress in rice seedlings overexpressing C_4 phosphoenolpyruvate carboxylase from maize and millet. *Photosynthetica* 53: 481–488.

Dong, O.X. and Ronald, P.C. (2021). Targeted DNA insertion in plants. *Proc. Natl. Acad. Sci.* 118 (22): e2004834117.

Doubnerová Hýsková, V., Miedzińska, L., Dobrá, J. et al. (2014). Phosphoenolpyruvate carboxylase, NADP-malic enzyme, and pyruvate, phosphate dikinase are involved in the acclimation of *Nicotiana tabacum* L. to drought stress. *J. Plant Physiol.* 171: 19–25.

Doubnerová, V. and Ryšlavá, H. (2011). What can enzymes of C_4 photosynthesis do for C_3 plants under stress? *Plant Sci.* 180: 575–583.

Eastmond, P.J., Dennis, D.T., and Rawsthorne, S. (1997). Evidence that a malate/inorganic phosphate exchange translocator imports carbon across the leucoplast envelope for fatty acid synthesis in developing castor seed endosperm. *Plant Physiol.* 114: 851–856.

Eastmond, P.J., Germain, V., Lange, P.R. et al. (2000). Postgerminative growth and lipid catabolism in oilseeds lacking the glyoxylate cycle. *Proc. Natl. Acad. Sci. U.S.A.* 97: 5669–5674.

Eastmond, P.J., Astley, H.M., Parsley, K. et al. (2015). *Arabidopsis* uses two gluconeogenic gateways for organic acids to fuel seedling establishment. *Nat. Commun.* 6: 6659.

Ehleringer, J. R., Cerling, T. E., & Helliker, B. R. (1997). C_4 photosynthesis, atmospheric CO_2, and climate. Oecologia, 112, 285–299.

Emami, P., Razi, H., and Dastfal, M. (2016). Effect of salt stress on NADP-malic enzyme activity, proline and ionic contents of durum wheat genotypes. *Biol. Forum Int. J.* 8: 112–119.

Emms, D.M., Covshoff, S., Hibberd, J.M., and Kelly, S. (2016). Independent and parallel evolution of new genes by gene duplication in two origins of C_4 photosynthesis provides new insight into the mechanism of phloem loading in C_4 species. *Mol. Biol. Evol.* 33: 1796–1806.

Engineer, C.B., Ghassemian, M., Anderson, J.C. et al. (2014). Carbonic anhydrases, EPF2 and a novel protease mediate CO_2 control of stomatal development. *Nature* 513: 246.

Enríquez, P. (2016). CRISPR-mediated epigenome editing Yale. *J. Biol. Med.* 89: 471–486.

Eprintsev, A.T. and Fedorina, O.S. (2007). Functioning of malate dehydrogenase system in mesophyll and bundle sheath cells of maize leaves under salt stress conditions. *Russ. J. Plant. Physiol.* 54: 728–735.

Ermakova, M., Arrivault, S., Giuliani, R. et al. (2020a). Installation of C4 photosynthetic pathway enzymes in rice using a single construct. *Plant Biotechnol. J.*.

Ermakova, M., Danila, F.R., Furbank, R.T., and von Caemmerer, S. (2020b). On the road to C4 rice: advances and perspectives. *Plant J.*.

Éva, C., Oszvald, M., and Tamás, L. (2019). Current and possible approaches for improving photosynthetic efficiency. *Plant Sci.* 280: 433–440.

Fahnenstich, H., Saigo, M., Niessen, M. et al. (2007b). Alteration of organic acid metabolism in *Arabidopsis* overexpressing the maize C_4 NADP-malic enzyme causes accelerated senescence during extended darkness. *Plant Physiol.* 145: 640–652.

Fonteyne, S., Staelens, A., Muylle, H. et al. (2017). Physiological basis of chilling tolerance and early-season growth in *Miscanthus*. *Ann. Bot.* 121: 281–295.

Fu, Z.-Y., Zhang, Z.-B., Hu, X.-J. et al. (2009). Cloning, identification, expression analysis and phylogenetic relevance of two NADP-dependent malic enzyme genes from hexaploid wheat. *C. R. Biol.* 332: 591–602.

Fu, Z.Y., Zhang, Z.B., Liu, Z.H. et al. (2011). The effects of abiotic stresses on the NADP-dependent malic enzyme in the leaves of the hexaploid wheat. *Biol. Plant.* 55: 196–200.

Furumoto, T., Yamaguchi, T., Ohshima-Ichie, Y. et al. (2011). A plastidial sodium-dependent pyruvate transporter. *Nature* 476: 472–475.

Fushimi, T., Umeda, M., Shimazaki, T. et al. (1994). Nucleotide sequence of a rice cDNA similar to a maize NADP-dependent malic enzyme. *Plant Mol. Biol.* 24: 965–967.

Gerrard Wheeler, M.C., Tronconi, M.A., Drincovich, M.F. et al. (2005). A comprehensive analysis of the NADP-malic enzyme gene family of *Arabidopsis*. *Plant Physiol.* 139: 39–51.

Gerrard Wheeler, M.C., Arias, C.L., Tronconi, M.A. et al. (2008). *Arabidopsis thaliana* NADP-malic enzyme isoforms: high degree of identity but clearly distinct properties. *Plant Mol. Biol.* 67: 231–242.

Gietl, C. (1990). Glyoxysomal malate dehydrogenase from watermelon is synthesized with an amino-terminal transit peptide. *Proc. Natl. Acad. Sci. U.S.A.* 87: 5773.

Gietl, C. (1992a). Malate dehydrogenase isoenzymes: cellular locations and role in the flow of metabolites between the cytoplasm and cell organelles. *Biochim. Biophys. Acta Bioenerg.* 1100: 217–234.

Gietl, C. (1992b). Partitioning of malate dehydrogenase isoenzymes into glyoxysomes, mitochondria, and chloroplasts. *Plant Physiol.* 100: 557–559.

Gietl, C., Lehnerer, M., and Olsen, O. (1990). Mitochondrial malate dehydrogenase from watermelon: sequence of cDNA clones and primary structure of the higher-plant precursor protein. *Plant Mol. Biol.* 14: 1019–1030.

Giuliani, R., Karki, S., Covshof, S. et al. (2019). Knockdown of glycine decarboxylase complex alters photorespiratory carbon isotope fractionation in Oryza sativa leaves. *J. Exp. Bot.* 70: 2773–2786.

Gowik, U., Burscheidt, J., Akyildiz, M. et al. (2004). cis-Regulatory elements for mesophyll-specific gene expression in the C_4 plant *Flaveria trinervia*, the promoter of the C_4 phosphoenolpyruvate carboxylase gene. *Plant Cell* 16: 1077–1090.

Gu, J.-F., Qiu, M., and Yang, J.-C. (2013). Enhanced tolerance to drought in transgenic rice plants overexpressing C_4 photosynthesis enzymes. *Crop J.* 1: 105–114.

Guo, P., Baum, M., Grando, S. et al. (2009). Differentially expressed genes between drought-tolerant and drought-sensitive barley genotypes in response to drought stress during the reproductive stage. *J. Exp. Bot.* 60: 3531–3544.

Gupta, B. and Huang, B. (2014). Mechanism of salinity tolerance in plants: physiological, biochemical, and molecular characterization. *Int. J. Genomics* 2014: 701596.

Halter, M., Mitchell, J., Mann, D.G.J. et al. (2015). Photosynthetic parameters of switchgrass (*Panicum virgatum*) under low radiation: influence of stable overexpression of *Miscanthus×giganteus* PPDK on responses to light and CO_2 under warm and cool growing conditions new negatives. *Plant Sci* 1–2: 23–32.

Hibberd, J.M. and Covshoff, S. (2010). The regulation of gene expression required for C_4. *Photosynthesis* 61: 181–207.

Hibberd, J.M. and Quick, W.P. (2002). Characteristics of C_4 photosynthesis in stems and petioles of C_3 flowering plants. *Nature* 415: 451.

Hu, H., Boisson-Dernier, A., Israelsson-Nordström, M. et al. (2010). Carbonic anhydrases are upstream regulators of CO_2-controlled stomatal movements in guard cells. *Nat. Cell Biol.* 12: 87–93.

Hu, L., Li, Y., Xu, W. et al. (2012). Improvement of the photosynthetic characteristics of transgenic wheat plants by transformation with the maize C 4 phosphoenolpyruvate carboxylase gene. *Plant Breed* 131: 385–391.

Hu, H., Rappel, W.-J., Occhipinti, R. et al. (2015). Distinct cellular locations of carbonic anhydrases mediate carbon dioxide control of stomatal movements. *Plant Physiol.* 169: 1168–1178.

Huang, Y.-X., Goto, Y., Nonaka, S. et al. (2015). Overexpression of the phosphoenolpyruvate carboxykinase gene (SlPEPCK) promotes soluble sugar accumulation in fruit and post-germination growth of tomato (*Solanum lycopersicum* L.). *Plant Biotechnol.* 32: 281–289.

Hyskova, V. and Ryslava, H. (2016). Unusual properties and functions of plant pyruvate, orthophosphate dikinase. *Biochem. Anal. Biochem.* 5: e161.

Ivanov, A.G., Krol, M., Sveshnikov, D. et al. (2006). Characterization of the photosynthetic apparatus in cortical bark chlorenchyma of Scots pine. *Planta* 223: 1165–1177.

Jayakannan, M., Bose, J., Babourina, O. et al. (2015). The NPR1-dependent salicylic acid signalling pathway is pivotal for enhanced salt and oxidative stress tolerance in *Arabidopsis*. *J. Exp. Bot.* 66: 1865–1875.

Jenner, H.L., Winning, B.M., Millar, A.H. et al. (2001). NAD malic enzyme and the control of carbohydrate metabolism in potato tubers. *Plant Physiol.* 126: 1139–1149.

Kaachra, A., Vats, S.K., and Kumar, S. (2018). Heterologous expression of key C and N metabolic enzymes improves re-assimilation of photorespired CO_2 and NH_3, and growth. *Plant Physiol.* 177: 1396–1409.

Kajala, K., Brown, N.J., Williams, B.P. et al. (2012). Multiple *Arabidopsis* genes primed for recruitment into C_4 photosynthesis. *Plant J.* 69: 47–56.

Kandoi, D., Mohanty, S., Govindjee, and Tripathy, B.C. (2016). Towards efficient photosynthesis: overexpression of *Zea mays* phosphoenolpyruvate carboxylase in *Arabidopsis thaliana*. *Photosynth. Res.* 130: 47–72.

Kandoi, D., Mohanty, S., and Tripathy, B.C. (2018). Overexpression of plastidic maize NADP-malate dehydrogenase (ZmNADP-MDH) in *Arabidopsis thaliana* confers tolerance to salt stress. *Protoplasma* 255: 547–563.

Kandoi, D., Ruhil, K., Govindjee, G., and Tripathy, B.C. (2022). Overexpression of cytoplasmic C4 *Flaveria bidentis* carbonic anhydrase in C3 Arabidopsis thaliana increases amino acids, photosynthetic potential, and biomass. *Plant Biotechnol. J.* 20: 1518–1532.

Karki, S., Rizal, G., and Quick, W.P. (2013). Improvement of photosynthesis in rice (Oryza sativa L.) by inserting the C4 pathway. *Rice* 6: 1–8.

Kausch, A.P., Page Owen, T., Zachwieja, S.J. et al. (2001). Mesophyll-specific, light and metabolic regulation of the C_4 PPCZm1 promoter in transgenic maize. *Plant Mol. Biol.* 45: 1–15.

Kronzucker, H.J., Coskun, D., Schulze, L.M. et al. (2013). Sodium as nutrient and toxicant. *Plant Soil* 369: 1–23.

Ku, M.S.B., Agarie, S., Nomura, M. et al. (1999). High-level expression of maize phosphoenolpyruvate carboxylase in transgenic rice plants. *Nat. Biotechnol.* 17: 76.

Ku, M. S., Ranade, U., Hsu, T. P., Cho, D., Li, X., Jiao, D. M., ... & Matsuoka, M. (2000). Photosynthetic performance of transgenic rice plants overexpressing maize C4 photosynthesis enzymes. In Studies in plant science (Vol. 7, pp. 193-204). Elsevier.

Lai, L.B., Tausta, S.L., and Nelson, T.M. (2002a). Differential regulation of transcripts encoding cytosolic NADP-malic enzyme in C_3 and C_4 *Flaveria* species. *Plant Physiol.* 128: 140–149.

Lai, L.B., Wang, L., and Nelson, T.M. (2002b). Distinct but conserved functions for two chloroplastic NADP-malic enzyme isoforms in C_3 and C_4 *Flaveria* species. *Plant Physiol.* 128: 125–139.

Langdale, J.A., Taylor, W.C., and Nelson, T. (1991). Cell-specific accumulation of maize phosphoenolpyruvate carboxylase is correlated with demethylation at a specific site >3 kb upstream of the gene. *Mol. Gen. Genet. MGG* 225: 49–55.

Laporte, M.M., Shen, B., and Tarczynski, M.C. (2002). Engineering for drought avoidance: expression of maize NADP-malic enzyme in tobacco results in altered stomatal function. *J. Exp. Bot.* 53: 699–705.

Lappe, R.R., Baier, J.W., Boehlein, S.K. et al. (2018). Functions of maize genes encoding pyruvate phosphate dikinase in developing endosperm. *Proc. Natl. Acad. Sci. U.S.A.* 115: E24–E33.

Leegood, R.C. and Ap Rees, T. (1978). Dark fixation of CO_2 during gluconeogenesis by the cotyledons of *Cucurbita pepo* L. *Planta* 140: 275–282.

Lian, L., Wang, X., Zhu, Y. et al. (2014). Physiological and photosynthetic characteristics of indica Hang2 expressing the sugarcane PEPC gene. *Mol. Biol. Rep.* 41: 2189–2197.

Lin, H., Karki, S., Coe, R.A. et al. (2016). Targeted knockdown of GDCH in rice leads to a photorespiratory-defcient phenotype useful as a building block for C4 rice. *Plant Cell Physiol.* 57: 919–932.

Lin, H.C., Arrivault, S., Coe, R.A. et al. (2020). A partial C4 photosynthetic biochemical pathway in rice. *Front. Plant Sci.*.

Liu, S., Cheng, Y., Zhang, X. et al. (2007). Expression of an NADP-malic enzyme gene in rice (*Oryza sativa* L.) is induced by environmental stresses; over-expression of the gene in *Arabidopsis* confers salt and osmotic stress tolerance. *Plant Mol. Biol.* 64: 49.

Liu, X., Li, X., Zhang, C. et al. (2016). Phosphoenolpyruvate carboxylase regulation in C_4-PEPC-expressing transgenic rice during early responses to drought stress. *Physiol. Plant* 159: 178–200.

Liu, X., Li, X., Dai, C. et al. (2017). Improved short-term drought response of transgenic rice over-expressing maize C_4 phosphoenolpyruvate carboxylase via calcium signal cascade. *J. Plant Physiol.* 218: 206–221.

Lu, Y., Tian, Y., Shen, R. et al. (2020). Targeted, efcient sequence insertion and replacement in rice. *Nat. Biotechnol.* 38: 1402–1407.

Ludwig, M. (2011). The molecular evolution of β-carbonic anhydrase in *Flaveria*. *J. Exp. Bot.* 62: 3071–3081.

Ludwig, M. (2016). Evolution of carbonic anhydrase in C_4 plants. *Curr. Opin. Plant Biol.* 31: 16–22.

Maier, A., Zell, M.B., and Maurino, V.G. (2011). Malate decarboxylases: evolution and roles of NAD(P)-ME isoforms in species performing C_4 and C_3 photosynthesis. *J. Exp. Bot.* 62: 3061–3069.

Maheshwari, C., Coe, R.A., Karki, S. et al. (2021). Targeted knockdown of ribulose-1, 5-bisphosphate carboxylase-oxygenase in rice mesophyll cells. *J. Plant Physiol.* 260: 153395.

Marshall, J.S., Stubbs, J.D., Chitty, J.A. et al. (1997). Expression of the C_4 Me1 gene from *Flaveria bidentis* requires an interaction between 5[prime] and 3[prime] sequences. *Plant Cell* 9: 1515–1525.

Martinoia, E. and Rentsch, D. (1994). Malate compartmentation-responses to a complex metabolism. *Annu. Rev. Plant Physiol. Plant Mol. Biol.* 45: 447–467.

Matsuoka, M. and Numazawa, T.J.M. (1991). CIS-acting elements in the pyruvate, orthophosphate dikinase gene from maize. *Mol. Gen. Genet. MGG* 228: 143–152.

McGonigle, B. and Nelson, T. (1995). C_4 Isoform of NADP-malate dehydrogenase (cDNA cloning and expression in leaves of C_4, C_3, and C_3-C_4 intermediate species of *Flaveria*). *Plant Physiol.* 108: 1119.

Méchin, V., Thévenot, C., Le Guilloux, M. et al. (2007). Developmental analysis of maize endosperm proteome suggests a pivotal role for pyruvate orthophosphate dikinase. *Plant Physiol.* 143: 1203–1219.

Miyao, M. (2003). Molecular evolution and genetic engineering of C_4 photosynthetic enzymes. *J. Exp. Bot.* 54: 179–189.

Moroney, J.V., Bartlett, S.G., and Samuelsson, G. (2004). Carbonic anhydrases in plants and algae. *Plant Cell Environ.* 24: 141–153.

Miyao, M., Masumoto, C., Miyazawa, S.I., and Fukayama, H. (2011). Lessons from engineering a single-cell C 4 photosynthetic pathway into rice. *J. Exp. Bot.* 62: 3021–3029.

Musrati, R.A., Kollarova, M., Mernik, N., and Mikulasova, D. (1998). Malate dehydrogenase: distribution, function and properties. *Gen. Physiol. Biophys.* 17: 193–210.

Nguyen, H.T., Leipner, J., Stamp, P., and Guerra-Peraza, O. (2009). Low temperature stress in maize (*Zea mays* L.) induces genes involved in photosynthesis and signal transduction as studied by suppression subtractive hybridization. *Plant Physiol. Biochem.* 47: 116–122.

Niklaus, M., & Kelly, S. (2019). The molecular evolution of C_4 photosynthesis: opportunities for understanding and improving the world's most productive plants. Journal of Experimental Botany, 70(3), 795–804.

Nimmo, H.G., Fontaine, V., Hartwell, J. et al. (2001). PEP carboxylase kinase is a novel protein kinase controlled at the level of expression. *New Phytol.* 151: 91–97.

Offermann, S., Dreesen, B., Horst, I. et al. (2008). Developmental and environmental signals induce distinct histone acetylation profiles on distal and proximal promoter elements of the C_4-Pepc gene in maize. *Genetics* 179: 1891–1901.

Ohta, S., Ishida, Y., and Usami, S. (2006). High-level expression of cold-tolerant pyruvate, orthophosphate dikinase from a genomic clone with site-directed mutations in transgenic maize. *Mol. Breed.* 18: 29.

Omoto, E., Taniguchi, M., and Miyake, H. (2010). Effects of salinity stress on the structure of bundle sheath and mesophyll chloroplasts in NAD-malic enzyme and PCK type C_4 plants. *Plant Prod. Sci.* 13: 169–176.

Osborn, H.L., Alonso-Cantabrana, H., Sharwood, R.E. et al. (2017). Effects of reduced carbonic anhydrase activity on CO_2 assimilation rates in *Setaria viridis*: a transgenic analysis. *J. Exp. Bot.* 68: 299–310.

Paulus, J.K., Schlieper, D., and Groth, G. (2013). Greater efficiency of photosynthetic carbon fixation due to single amino-acid substitution. *Nat. Commun.* 4: 1–7.

Penfield, S., Rylott, E.L., Gilday, A.D. et al. (2004). Reserve mobilization in the *Arabidopsis* endosperm fuels hypocotyl elongation in the dark, is independent of abscisic acid, and requires phosphoenolpyruvate CARBOXYKINASE1. *Plant Cell* 16: 2705.

Penfield, S., Clements, S., Bailey Karen, J. et al. (2011). Expression and manipulation of phosphoenolpyruvate carboxykinase 1 identifies a role for malate metabolism in stomatal closure. *Plant J.* 69: 679–688.

Peng, C., Xu, W., Hu, L. et al. (2018). Effects of the maize C_4 phosphoenolpyruvate carboxylase (ZmPEPC) gene on nitrogen assimilation in transgenic wheat. *Plant Growth Regul.* 84: 191–205.

Perduns, R., Horst-Niessen, I., and Peterhansel, C. (2015). Photosynthetic genes and genes associated with the C_4 trait in maize are characterized by a unique class of highly regulated histone acetylation peaks on upstream promoters. *Plant Physiol.* 168: 1378–1388.

Pinto, M.E., Casati, P., Hsu, T.-P. et al. (1999). Effects of UV-B radiation on growth, photosynthesis, UV-B-absorbing compounds and NADP-malic enzyme in bean (*Phaseolus vulgaris* L.) grown under different nitrogen conditions. *J. Photochem. Photobiol. B Biol.* 48: 200–209.

Qi, X., Xu, W., Zhang, J. et al. (2017). Physiological characteristics and metabolomics of transgenic wheat containing the maize C_4 phosphoenolpyruvate carboxylase (PEPC) gene under high temperature stress. *Protoplasma* 254: 1017–1030.

Qian, B., Li, X., Liu, X. et al. (2015). Enhanced drought tolerance in transgenic rice over-expressing of maize C_4 phosphoenolpyruvate carboxylase gene via NO and Ca^{2+}. *J. Plant Physiol.* 175: 9–20.

Qin, N., Xu, W., Hu, L. et al. (2016). Drought tolerance and proteomics studies of transgenic wheat containing the maize C_4 phosphoenolpyruvate carboxylase (PEPC) gene. *Protoplasma* 253: 1503–1512.

Rangan, P., Furtado, A., and Henry, R.J. (2016). New evidence for grain specific C_4 photosynthesis in wheat. *Sci. Rep.* 6: 31721.

Rao, X. and Dixon, R.A. (2016). The differences between NAD-ME and NADP-ME subtypes of C_4 photosynthesis: more than decarboxylating enzymes. *Front. Plant Sci.* 7: 1525–1533.

Rao, X., Lu, N., Li, G. et al. (2016). Comparative cell-specific transcriptomics reveals differentiation of C_4 photosynthesis pathways in switchgrass and other C_4 lineages. *J. Exp. Bot.* 67: 1649–1662.

Ren, C.G., Li, X., Liu, X.L. et al. (2014). Hydrogen peroxide regulated photosynthesis in C4-pepc transgenic rice. *Plant Physiol. Biochem.* 74: 218–229.

Reyna-Llorens, I. and Hibberd, J.M. (2017). Recruitment of pre-existing networks during the evolution of C_4 photosynthesis. *Philos. Trans. R. Soc. B. Biol. Sci.* 372: 20160386.

Reyna-Llorens, I., Burgess, S.J., Reeves, G. et al. (2018). Ancient duons may underpin spatial patterning of gene expression in C_4 leaves. *Proc. Natl. Acad. Sci. U.S.A.* 115: 1931–1936.

Rizal, G., Thakur, V., Dionora, J., Karki, S., Wanchana, S., Acebron, K., ... & Quick, W. P. (2015). Two forward genetic screens for vein density mutants in sorghum converge on a cytochrome P450 gene in the brassinosteroid pathway. The Plant Journal, 84(2), 257–266.

Rosche, E., Chitty, J., Westhoff, P., and Taylor, W.C. (1998). Analysis of promoter activity for the gene encoding pyruvate orthophosphate dikinase in stably transformed C_4 *Flaveria* species. *Plant Physiol.* 117: 821–829.

Rylott, E.L., Gilday, A.D., and Graham, I.A. (2003). The gluconeogenic enzyme Phospho*enol*pyruvate carboxykinase in *Arabidopsis* is essential for seedling establishment. *Plant Physiol.* 131: 1834–1842.

Sage, R. (2001). Environmental and evolutionary preconditions for the origin and diversification of the C_4 photosynthetic syndrome. *Plant Biol.* 3: 202–213.

Sage, R. (2004). The evolution of C_4 photosynthesis. *New Phytol.* 161: 341–370.

Sage, R. F. (1999). Why C_4 photosynthesis. C_4 plant biology, 3.

Sage, R.F. and Zhu, X.-G. (2011). Exploiting the engine of C4 photosynthesis. *J. Exp. Bot.* 62: 2989–3000.

Schaaf, J., Walter, M.H., and Hess, D. (1995). Primary metabolism in plant defense (regulation of a bean malic enzyme gene promoter in transgenic tobacco by developmental and environmental cues). *Plant Physiol.* 108: 949.

Scheibe, R., Backhausen, J.E., Holtgrefe, S., and Emmerlich, V. (2005). Strategies to maintain redox homeostasis during photosynthesis under changing conditions. *J. Exp. Bot.* 56: 1481–1489.

Schuler, M.L., Mantegazza, O., and Weber, A.P.M. (2016). Engineering C_4 photosynthesis into C_3 chassis in the synthetic biology age. *Plant ssss1J.* 87: 51–65.

Sen, P., Ghosh, S., Sarkar, S.N. et al. (2017). Pyramiding of three C4 specific genes towards yield enhancement in rice. *Plant Cell Tissue Organ Cult.* 128: 145–160.

Shen, W.J., Chen, G.X., Xu, J.G. et al. (2015). Overexpression of maize phosphoenolpyruvate carboxylase improves drought tolerance in rice by stabilization the function and structure of thylakoid membrane. *Photosynthetica* 53: 436–446.

Shi, W., Yue, L., Guo, J. et al. (2020). Identification and evolution of C4 photosynthetic pathway genes in plants. *BMC Plant Biol.* 20: 132.

Sicher, R., Bunce, J., Barnaby, J., and Bailey, B. (2015). Water-deficiency effects on single leaf gas exchange and on C_4 pathway enzymes of maize genotypes with differing abiotic stress tolerance. *Photosynthetica* 53: 3–10.

Singh, J., James, D., Achary, V.D.M. et al. (2021). Coordinated overexpression of *OsSUT1, OsSWEET11* and *OsSWEET14* in rice impairs carbohydrate metabolism that has implications in plant growth, yield and susceptibility to *Xanthomonas oryzae pv oryzae (Xoo). BioRxivdoi* https://doi.org/10.1101/2021.01.07.425507.

Slewinski, T.L., Anderson, A.A., Zhang, C., and Turgeon, R. (2012). Scarecrow plays a role in establishing kranz anatomy in maize leaves. *Plant Cell Physiol.* 53: 2030–2037.

Slewinski, T.L., Anderson, A.A., Price, S. et al. (2014). Short-root1 plays a role in the development of vascular tissue and kranz anatomy in maize leaves. *Mol. Plant* 7: 1388–1392.

Smith, R.G., Gauthier, D.A., Dennis, D.T., and Turpin, D.H. (1992). Malate- and pyruvate-dependent fatty acid synthesis in leucoplasts from developing castor endosperm. *Plant Physiol.* 98: 1233–1238.

Spence, A.K., Boddu, J., Wang, D. et al. (2014). Transcriptional responses indicate maintenance of photosynthetic proteins as key to the exceptional chilling tolerance of C_4 photosynthesis in *Miscanthus* × *giganteus*. *J. Exp. Bot.* 65: 3737–3747.

Stepien, P. and Klobus, G. (2005). Antioxidant defense in the leaves of C_3 and C_4 plants under salinity stress. *Physiol. Plant* 125: 31–40.

Studer, A.J., Gandin, A., Kolbe, A.R. et al. (2014). A limited role for carbonic anhydrase in C_4 photosynthesis as revealed by a ca1ca2 double mutant in maize. *Plant Physiol* 165: 608.

Subbarao, G., Ito, O., Berry, W., and Wheeler, R. (2003). Sodium, a functional plant nutrient. *Crit. Rev. Plant Sci.* 22: 391–416.

Sun, L., Liang, C., Chen, Z. et al. (2014). *Superior aluminium* (Al) tolerance of *Stylosanthes* is achieved mainly by malate synthesis through an Al-enhanced malic enzyme, SgME1. *New Phytol.* 202: 209–219.

Sun, Q., Yamada, Y., Han, Y., and Takano, T. (2020). Influence of salt stress on C_4 photosynthesis in *Miscanthus sinensis* Anderss. *Plant Biol.* 23: 44–56.

Suzuki, S., Murai, N., Burnell, J.N., and Arai, M. (2000). Changes in photosynthetic carbon flow in transgenic rice plants that express C_4-type phosphoenolpyruvate carboxykinase from *Urochloa panicoides*. *Plant Physiol.* 124: 163.

Suzuki, S., Murai, N., Kasaoka, K. et al. (2006). Carbon metabolism in transgenic rice plants that express phosphoenolpyruvate carboxylase and/or phosphoenolpyruvate carboxykinase. *Plant Sci.* 170: 1010–1019.

Takeuchi, Y., Akagi, H., Kamasawa, N. et al. (2000). Aberrant chloroplasts in transgenic rice plants expressing a high level of maize NADP-dependent malic enzyme. *Planta* 211: 265–274.

Taniguchi, M., Izawa, K., Ku, M.S.B. et al. (2000). Binding of cell type-specific nuclear proteins to the 5′-flanking region of maize C_4 phosphoenolpyruvate carboxylase gene confers its differential transcription in mesophyll cells. *Plant Mol. Biol.* 44: 543–557.

Taniguchi, Y., Ohkawa, H., Masumoto, C. et al. (2008). Overproduction of C_4 photosynthetic enzymes in transgenic rice plants: an approach to introduce the C_4-like photosynthetic pathway into rice. *J. Exp. Bot.* 59: 1799–1809.

Tanz, S.K., Tetu, S.G., Vella, N.G.F., and Ludwig, M. (2009). Loss of the transit peptide and an increase in gene expression of an ancestral chloroplastic carbonic anhydrase were instrumental in the evolution of the cytosolic C_4 carbonic anhydrase in *Flaveria*. *Plant Physiol.* 150: 1515–1529.

Tronconi, M.A., Fahnenstich, H., Gerrard Weehler, M.C. et al. (2008). *Arabidopsis* NAD-malic enzyme functions as a homodimer and heterodimer and has a major impact on nocturnal metabolism. *Plant Physiol.* 146: 1540.

Tsuchida, H., Tamai, T., Fukayama, H. et al. (2001). High level expression of C_4-specific NADP-malic enzyme in leaves and impairment of photoautotrophic growth in a C_3 plant. *Rice Plant. Cell Physiol.* 42: 138–145.

Uzilday, B., Turkan, I., Sekmen, A.H. et al. (2012). Comparison of ROS formation and antioxidant enzymes in *Cleome gynandra* (C_4) and *Cleome spinosa* (C_3) under drought stress. *Plant Sci.* 182: 59–70.

Walter Michael, H., Grima-Pettenati, J., and Feuillet, C. (1994). Characterization of a bean (*Phaseolus vulgaris* L.) malic-enzyme gene. *Eur. J. Biochem.* 224: 999–1009.

Wang, J. Z., Cui, L. J., Wang, Y., & Li, J. L. (2009). Growth, lipid peroxidation and photosynthesis in two tall fescue cultivars differing in heat tolerance. Biologia Plantarum, 53, 237–242.

Wang, Y.-M., Xu, W.-G., Hu, L. et al. (2012). Expression of maize gene encoding C_4-pyruvate orthophosphate dikinase (PPDK) and C_4-phosphoenolpyruvate carboxylase (PEPC) in transgenic *Arabidopsis*. *Plant Mol. Biol. Report* 30: 1367–1374.

Wang, Y., Bräutigam, A., Weber, A.P.M., and Zhu, X.-G. (2014). Three distinct biochemical subtypes of C_4 photosynthesis? A modelling analysis. *J. Exp. Bot.* 65: 3567–3578.

Wang, Q.J., Sun, H., Dong, Q.L. et al. (2016). The enhancement of tolerance to salt and cold stresses by modifying the redox state and salicylic acid content via the cytosolic malate dehydrogenase gene in transgenic apple plants. *Plant Biotechnol. J.* 14: 1986–1997.

Wang, S., Tholen, D., and Zhu, X.G. (2017). C4 photosynthesis in C3 rice: a theoretical analysis of biochemical and anatomical factors. *Plant Cell Environ.* 40 (1): 80–94.

Williams, B.P., Burgess, S.J., Reyna-Llorens, I. et al. (2016). An untranslated cis-element regulates the accumulation of multiple C_4 enzymes in *Gynandropsis gynandra* mesophyll cells. *Plant Cell* 28: 454–465.

Wittmann, C., Pfanz, H., Loreto, F. et al. (2006). Stem CO_2 release under illumination: corticular photosynthesis, photorespiration or inhibition of mitochondrial respiration? *Plant Cell Environ.* 29: 1149–1158.

Xiang, L., Le Roy, K., Vanhaecke, M. et al. (2011). Exploring the neutral invertase–oxidative stress defence connection in *Arabidopsis thaliana*. *J. Exp. Bot.* 62: 3849–3862.

Yadav, S., Rathore, M.S., and Mishra, A. (2020). The pyruvate-phosphate dikinase (C4-SmPPDK) gene from suaeda monoica enhances photosynthesis, carbon assimilation, and abiotic stress tolerance in a C3 plant under elevated CO_2 conditions. *Front. Plant Sci.* 11.

Yao, Y.-X., Dong, Q.-L., Zhai, H. et al. (2011). The functions of an apple cytosolic malate dehydrogenase gene in growth and tolerance to cold and salt stresses. *Plant Physiol. Biochem.* 49: 257–264.

Yu, S., Zhang, X., Guan, Q. et al. (2007). Expression of a carbonic anhydrase gene is induced by environmental stresses in Rice (Oryza sativa L.). *Biotechnol. Lett.* 29: 89–94.

Zamani-Nour, S., Lin, H. C., Walker, B. J., Mettler-Altmann, T., Khoshravesh, R., Karki, S., ... & Weber, A. P. (2021). Overexpression of the chloroplastic 2-oxoglutarate/malate transporter disturbs carbon and nitrogen homeostasis in rice. Journal of Experimental Botany, 72(1), 137–152.

Zhang, Q., Li, Y., Xu, W. et al. (2021a). Joint expression of Zmpepc, Zmppdk, and Zmnadp-me is more efcient than expression of one or two of those genes in improving the photosynthesis of Arabidopsis. *Plant Physiol. Biochem.* 158: 410–419. https://doi.org/10.1016/j.plaphy.2020.11.030.

Zhang Q, Qi X, Xu W et al (2021b) Response of transgenic Arabidopsis expressing maize C4 photosynthetic enzyme genes to high light. *Plant Signal Behav.* https://doi.org/10.1080/15592324.2021.1885894

Zhang, H.F., Xu, W.G., Wang, H.W. et al. (2014). Pyramiding expression of maize genes encoding phosphoenolpyruvate carboxylase (PEPC) and pyruvate orthophosphate dikinase (PPDK) synergistically improve the photosynthetic characteristics of transgenic wheat. *Protoplasma* 251: 1163–1173.

Zhou, H., Liu, S.K., and Yang, C.P. (2011). Over-expression of a NAD-malic enzyme gene from rice in *Arabidopsis thaliana* confers tolerances to several abiotic stresses. *Adv. Mater. Res.* 393: 863–866.

13

Expand the Survival Limits of Crop Plants Under Cold Climate Region

Bhuvnesh Sareen[1] *and Rohit Joshi*[1,2]

[1] *Division of Biotechnology, CSIR-Institute of Himalayan Bioresource Technology, Palampur, Himachal Pradesh, India*
[2] *Academy of Scientific and Innovative Research (AcSIR), CSIR-HRDC Campus, Ghaziabad, Uttar Pradesh, India*

13.1 Introduction

Cold regions with low temperatures are characterized by stressful conditions that change dramatically not only following the yearly cycles but also on a daily basis (Alonso-Amelot 2008; Gusain et al., 2023). The highly changing and unpredictable climates in these regions result in the coalescence of four seasons (spring, summer, autumn, and winter), and it appears that only winter and summer are the dominating ones. In these climate regions, cold stress is a major constraint to global food production, agriculture productivity, preharvest as well as postharvest damages, plant growth, yield, and geographical distribution (Sanghera et al. 2011; Zhang et al. 2017). It is also affecting the photosynthetic machinery, including light reactions, photosystems, Calvin cycle, and photorespiration, which ultimately affects the energy utilization in plants (Ruelland and Zachowski 2010; Friesen and Sage 2016). Growth phases (vegetative and reproductive) of the plant life cycle are also adversely affected in cold climate region. Impaired pollen or ovule fertility, flower abortion, lack of seed set, grain yield, etc., are the major traits affected by cold stress (Thakur et al. 2010). In response to this, plants adapt themselves to protect against damages imposed by the environment in which they exist. In response to wide fluctuations in temperature, plants are expected to evolve with mechanisms of tolerance and maintenance of cellular activities. Thus, various physiological and ecological mechanisms are adopted by plants to manage environmental stresses.

Exposure of plants to low temperatures and outside their optimum conditions can be considered stressful. Thus, the abilities to persist at such temperatures regimes differ among the plants (Levitt 1980). Those from tropical and sub-tropical regions, including agriculturally important ones, are mostly chilling-sensitive as compared to plants from temperate regions with variable degree of tolerance (Fowler et al. 2007). Moreover, they deal with the specific environmental demands in order to increase freezing tolerance by (i) evolutionary (phylogenetic) adaptation (Körner 2016), (ii) non-reversible ontogenic modifications (Sierra-Almeida and Cavieres 2012), or (iii) reversible adjustments termed as cold acclimation (Thomashow 1999). Whereas cold acclimation are often associated with biochemical as well as physiological changes, accumulation of small molecules, changes in the lipid accumulation, and finally changes in gene expression (Levitt 1980; Sakai and Larcher 1987; Thomashow 1998; Yamaguchi and Shinozaki 2006). Interestingly, the profound changes in the expression result ultimately affecting their transcriptome, proteome, and metabolomics of the plant (Shriram et al. 2016). The results decipher the complex network of metabolism and plants under stress conditions (Pérez-Clemente et al. 2013). Further, considerable progress has been also made for the isolation and characterization of key genes those are specifically activated under low temperature conditions. Thus, the action of such important genes paved the way for the identification of freezing tolerance mechanism with cold acclimation of plants. Although the mechanism, importance, and the molecular basis of cold acclimation extensively studied for imparting tolerance, plants from tropical and sub-tropical origins are lacking in these. Further, the conventional breeding approach and variations induced under *in vitro* conditions also been studied to expand the tolerance in other crop plants. However, major drawbacks related to complexity of traits, genetic variation, limitation in efficient selection methods are responsible for the limited accomplishment of cold tolerance during conventional breeding

Global Climate Change and Plant Stress Management, First Edition. Edited by Mohammad Wahid Ansari, Anil Kumar Singh, and Narendra Tuteja.

(Sanghera et al. 2011). Later according to various reports published, biotechnological tools and engineering of crop plants are alternative strategy for developing crop plants as freezing tolerant.

13.2 Physiology of Cold Stress Tolerant Plants

The temperature below 20 °C as chilling and below 0 °C (sub-zero temperatures) as freezing stress are defined as low temperatures. Under such harsh environmental conditions, structure of proteins, cellular membranes, biological activity, and germination response get altered. More extremely, the formation of ice crystals at sub-zero temperatures is further associated with dehydrations stress, cellular as well as electrolyte leakage, and membrane lipid phase changes (Zhang et al. 2011; Demidchik et al. 2014). The severity of cold stress increases during the reproductive stages, which leads to huge yield loses and imparts sterility in crop plants (Sanghera et al. 2001). Moreover, germination rate is also adversely affected under freezing temperatures. Thus, the plants can tolerate or avoid exposure to such conditions by displaying morpho-physiological changes, cellular changes, preferences for different growth forms, and flexible reproductive strategies. However, in order to survive under adverse or fluctuated stress condition, maintenance of cellular activity, stability and integrity of cellular membrane, and functional and biological activities are important. The plants exhibits (i) altered chlorophyll accumulation, (ii) reduced sensitivity to photosynthesis, (iii) osmotic adjustments (Körner 2003), (iv) large accumulation of carbohydrate and lipid reserves (Körner and Renhardt 1987), (v) accumulation of low molecular weight compatible solutes, for example proline, sucrose etc. (Rudolph and Crowe 1985), (vi) accumulation of cryoprotective polypeptides (Hincha et al. 1988, 1990), (vii) supercooling to avoid ice crystallization (Squeo et al. 1991), (viii) efficient ROS scavenging and antioxidant defence system.

Research with rice plant in low temperature established the differences in the accumulation of chlorophyll content between cold-tolerant (*japonica*) and sensitive lines (*indica*). Former accumulates more chlorophyll than the latter (Glaszmann et al. 1990). While Rasolofo et al. (1986) identified tolerant line through leaf discoloration score, Sanghera et al. (2001) identified cold-tolerant genotypes based on dark green color. At low temperature, changes in the number of chloroplast, arrangements of grana, and lamellar structures are responsible for more accumulation of chlorophyll. Moreover, the reduced sensitivity in photosynthesis at low temperature related to enzymes was also reported in maize inbred (Singh 2000). Prolonged exposure in *Jatropha curcas* for chilling at 40 hours and freezing for 2 hours at −1 to −3 °C significantly reduces leaf photosynthesis (Ploschuk et al. 2014). Alternation of lipid composition and fatty acid unsaturation are also responsible for imparting tolerance to cold stress. Increase in the number of membrane lipid unsaturation at low temperature affects the physiological functioning of membrane (Sanghera et al. 2011). In addition, the lipid asymmetry maintains membrane physical structure toward cold stress (Gomès et al. 2000). Accumulation of solutes with low molecular weight is involved in acclimation at colder regions. Accumulation of sugars resulted in enhancing the stability of membranes and protection against damages (Anchordoguy et al. 1987). It has been observed that, cold-adaptive plants reserve sugars such as D-Glucose, amylose, starch, and maltose in their underground tissues (Khan et al. 2015). Moreover, plants such as *Crassula lacteal, Bryophyllum pinnatum, Yucca aloifolia*, and *Aloe vera* reported with increased contents of sugar after cold stress (Janská et al. 2010; Londo et al. 2018). Importantly, not only sugars but cryoprotective proteins and polypeptides also increase the stability of membranes against freeze induced injury. One such an important example is LEA proteins and polypeptides (*COR15a*) as well as hydrophilic polypeptides those act against cellular damages and imparts stabilization of membranes. Mechanisms of polypeptides involve the formation of amphipathic α helices having hexagonal II phase while affecting the internal curvature of monolayers (Epand et al. 1995). Other molecules, such as nitric oxide, GABA, polyamines, and proline, were also reported to increase in tea plants during cold stress (Wang et al. 2020). At low temperature, reactive oxygen species (ROS) also contributes toward protein denaturation, decreased photosynthetic efficiency, failure of electron transport chain, and finally cell rupture (Guy 1990). In such conditions, increased malondialdehyde (MDA) content, H_2O_2 content and Ca^{2+} in the cytoplasm alter lipid degradation and affect the enzymes activity of catalase, superoxide dismutase, ascorbate peroxidase, and peroxidase subsequently the signaling cascade (Longo et al. 2017; Sun et al. 2019). Thus, activation of an efficient antioxidant defense mechanism increases tolerance against ROS stress condition (Ritonga and Chen 2020). Study on bell pepper and *Arabidopsis* reported that the treatment of methyl jasmonate and ethylene, respectively, imparts more tolerance toward cold while altering the antioxidant enzyme activity (Sun et al. 2019). Chilling in cucumber reported the increased activity of SOD, APX, glutathione reductase, and GPX, while in pepper, increased peroxidase activity was recorded (Lee and Lee 2000; Guo et al. 2012). Recently, Rezaiae et al. (2020) reported the enhanced tolerance to cold stress in *Ocimum basilicum* while increasing the activity of SOD and GPX.

13.3 Stress Perception and Signaling

Cold signaling pathway has been extensively studied in many plants and contributes significantly to our understanding of signal perception and transduction pathway (Ma et al. 2015; Behera et al. 2015; Huo et al. 2016; Li et al. 2017a; Liu et al. 2017; Zhang et al. 2017; Shi et al. 2018; Maurya et al. 2020). For this, the cell membrane, the primary interface, senses stress and programs into the multifaceted signal transduction pathways, which downstream respond towards adaptive mechanisms (Figure 13.1). Due to stress signal of cold temperature, membrane fluidity changes followed by an influx of Ca^{2+} ions led to initiate the cascade of genes for accumulation of solutes, cryoprotective molecules, antioxidative enzymes, cold-regulated (COR), and antifreeze proteins (Uemura and Steponkus 2003; Chen et al. 2015). The crucial sensors for cold are the influx channels, histidine kinases two-component system, and receptors of G-proteins (Xiong et al. 2002; Pareek et al. 2017). The early evidences for this studied in *Medicago sativa* and *Alfalfa,* displaying low temperature induced influx of Ca^{2+} ions with the positive correlation between Ca^{2+} influx and the rate of temperature drop (Monroy et al. 1993; Monroy and Dhindsa 1995). Further, the calcium ion intracellular gradient is established, leading to the activation of calcium-dependent protein kinases (*CDPKs*), calmodulin, and calcium-sensitive protein phosphatases.

Importantly, CBF-COR signaling pathway has been characterized and is considered the most accepted model for cold stress (Figure 13.1). In this study, CBF/DREB1, (C-REPEAT BINDING FACTOR/DEHYDRATION RESPONSIVE ELEMENT BINDING PROTEIN1) an important transcription factor of the family AP2/ERF, showed a predominant role in cold acclimatization (Thomashow 1999; Lee et al. 2021). CBFs are the key proteins that act on CRT/DRE (C-repeat/dehydration responsive element), responsible for the cold as well as for other stresses (Stockinger et al. 1997). In *Arabidopsis,* up-accumulation of the *CBF/DREB1* gene resulted in increased tolerance and involved in the introduction of *COR* (cold-responsive) genes (Jaglo-Ottosen et al. 1998; Liu et al. 1998). Similarly, CBF genes are also isolated and characterized in other crops (Shi et al. 2018). Interestingly, COLD-REGULATED (*COR15A, COR47*), LOW TEMPERATURE INDUCED (*LTI*), Desiccation-responsive protein 29A (*RD29a*), and Knotted1 Induced1 (*KIN1*) are the COR class of genes, which

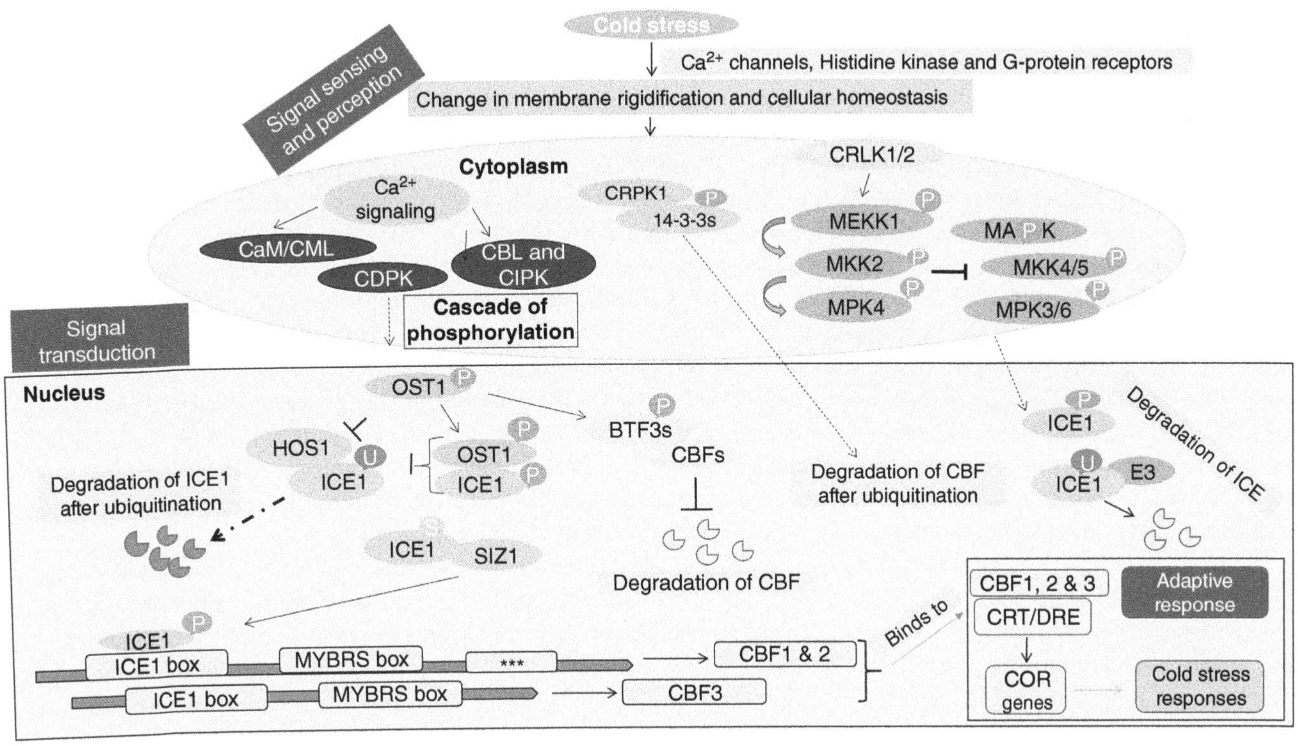

Figure 13.1 Schematic representation of cold responsive genes involved in stress perception and signaling mechanism of cold response in plants. The cold stress under low temperature is sensed by the membrane receptors on the cell membrane modulating the membrane rigidity and cellular homeostasis for the signal transduction. The signals from the membranes induce Ca^{2+}, histidine kinase, and G-protein receptor mediated signal transduction via Calcineurin-B Like proteins (CBL), *CPKs, CIPKs, CDPKs* to the nucleus through a pathway mediated by ICE-CBF/DREB transcription factors. The final product of the COR genes results in physiological responses for cold acclimation.

ultimately on activation through CBFs display multiple mechanisms of protection in plants against freezing stress injury (Yamaguchi and Shinozaki 1994).

The upstream regulation of CBFs occurs either positively or negatively through various transcription factors. Among them, Inducer of CBF Expression (*ICE*), Calmodulin-Binding Transcription Activator 3 (*CAMTA3*), and Ca^{2+}/Calmodulin-Regulated Receptor-Like Kinase (*CRLK1*) are the master transcriptional activator of *CBF* genes reported in a large number of plants including *Arabidopsis* (Chinnusamy et al. 2003). Importantly, members of MYC-type bHLH transcription factor such as *ICE1* and *ICE 2* bind to *CBF3* and *CBF1* promoters, respectively, on sumoylation by SUMO E3 ligase (*SIZ1*) and phosphorylation by OPEN STOMATA 1 (*OST1*) (Ding et al. 2015). Similarly, *CAMTA3* transcription factor activates *CBF2* through directly binding to the *CBF2* promoter (Doherty et al. 2009). Whereas Ca^{2+}-binding *CRLK1* also improves tolerance by regulating *CBF1* through inhibition of MAPK kinase activity. Recently, a cytosolic redox protein, Thioredoxin h2 (*Trx-h2*), identified in *Arabidopsis,* showed interaction with *CBF1* (Lee et al. 2021). The protein function by binding with a glycine residue at the second position of cytoplasmic endomembranes myristoylation (Traverso et al. 2013). However, the protein on activation through cold stress is demyristoylated and moves toward nucleus, where it act to reduce the CBF residues to regulate the expression of COR genes (Figure 13.1). In addition to these, E3 ubiquitin-protein ligase HOS1 (*HOS1*), Myb transcription factor 15 (*MYB15*), MAPK kinase (*MPK3/6*), 14-3-3s, PSEUDO RESPONSE REGULATORs (*PRRs*), and Phytochrome interacting factor 3 (*PIF3*) are the negative regulators of *CBF*-COR pathway. Whereas *HOS1* interacts with *ICE1* protein and ubiquitinates to the proteasomal degradation (Dong et al. 2006a). Similarly, another member of protein kinase family that is COLD-RESPONSIVE PROTEIN KINASE 1 (*CRPK1*) negatively regulates the CBF proteins (Liu et al. 2017) through phosphorylation and nuclear translocation of 14-3-3 proteins.

In addition to COR genes regulation through the CBF pathway, there are many other regulatory pathways and regulators to control the COR gene expression during stress such as translational and post-translational regulator, epigenetic regulators, hormonal regulators, and circadian pathway regulators.

13.4 Plant Survival Mechanism

Plants vary in their sensitivity and response to stress conditions with established survival mechanisms. Plants from tropical and sub-tropical regions are subtle to chilling, while temperate plants can tolerate cold temperature. Crops such as rice, corn, soybean, and potato are chilling sensitive, while oats are tolerant to chilling and barley, wheat, and rye are well sustained at freezing temperature (Chinnusamy et al. 2007; Zhang et al. 2011). The actual mechanism for survival in plants involves the encoding of stress perception and signaling cascade as well as gene transcription networks in response to environmental signals (Sanghera et al. 2011; Fu and Dong 2013; Joshi et al. 2016). An active role of large number of genes and transcription factors was studied and found upregulated in response to cold temperature in various plants (Tolosa and Zhang 2020). These include *AtMYB15, OsMYB4, OsMYB2, MlNAC5* genes for improving tolerance in plants such as *Oryza, Arabidopsis* (Vannini et al. 2004; Agarwal et al. 2006; Yang et al. 2012, 2015). In addition to these, *CBF1, CBF2, CBF3, ICE1, ICE2, CAMTA3, COR15a*, and *COR15b* are important transcription factors related to cold (Wang et al. 2018).

For imparting cold tolerance, various mechanisms play an effective role (Devi et al. 2017). These include (i) physical barriers such as cuticle, cell wall, (ii) signaling mechanism through ROS, (iii) signaling through hormones, (iv) changes in redox status and inorganic ion fluxes, (v) physiological changes for preventing formation of ice crystals, (vi) epigenetic and molecular regulations, (vii) involvement of R-genes and systematic acquired resistance to improve crop resilience (Kissoudis et al. 2014), and (viii) accumulation of metabolites and cryoprotectants at low temperature.

Modification of morphological structures and adopting compact growth forms are the morphological adaption strategies adopted by plants under cold conditions. Whereas, in secondary metabolites, increased content of phenolics and other phenylpropanoid compounds and pigments like anthocyanins were reported to increase during chilling of potato, apple, and maize (Graham and Patterson 1982; Christie et al. 1994). Increased levels of chlorogenic acid were reported in the adaptation of apple tree to cold temperature (Pérez-Ilzarbe et al. 1997). Resistance to cold temperature was also reported during the increased phenylpropanoids and their accumulation into the cell wall (Griffith and Yaish 2004). One important application of suberin and lignin incorporation into the cell wall is for freezing resistance (Griffith and Yaish 2004). Deposition of compounds and formation of the secondary wall as well as lignification have also been reported as a survival mechanism under cold stress (Moura et al. 2010).

At the molecular level, overexpression of *CBF* and other downstream genes supports the survival of plants. Importantly among them are cold-regulated *COR* genes, cold-induced *KIN* genes, response to dehydration *RD* genes, low temperature induced *LT1* genes, early response to dehydration *ERD* genes (Thomashow 1998). These genes belong to the family of LEA or dehydrins proteins; however, the functional activity of most proteins remain putative (Wise and Tunnacliffe 2004).

Overexpression of *RDM4* gene in *Arabidopsis* (Chan et al. 2016) and expression of *OsGH3-2*, *OsMADS57* in rice (Du et al. 2012; Chen et al. 2018), *GhDREB1* in *Gossypium hirsutum* (Shan et al. 2007), *CPK* in *Camellia sinensis* (Ding et al. 2019), *TaTPS11* in *Triticum aestivum* (Liu et al. 2019), *MdMYB108L* and *MdHY5* in *Malus domestica* (Wang et al. 2019), *DlICE1* in *Dimocarpus longan* (Yang et al. 2019b), *VvCBF* in *Vitis vinifera* (Rubio and Pérez 2019), *StCH4* in *Arabidopsis* (Puhakainen et al. 2004), *AtGRXS17* in *Solanum lycopersicum* (Hu et al. 2015) increase survival rate and improved tolerance to cold and chilling stress under low temperature conditions. Also, Cold Induced Small Protein 1 (*CISP1*) is the effector gene that comes into effect for the cold acclimation process in Poaceae family (Ying and Kidou 2017). Polypeptide encoded by *COR15a* protects and stabilizes the membrane from freezing injury (Steponkus et al. 1998; Thalhammer and Hincha 2014). It is also speculative that COR/LEA/dehydrin proteins, in addition to their role, in stabilization of membrane, also protect other proteins from freeze-thaw inactivation and could act as hydration buffers (Bravo et al. 2003; Wise and Tunnacliffe 2004).

In association with CBF-COR pathway, hormones significantly contribute toward defense while activating the genes and transcription factors for tolerance (Wang et al. 2017; Yang and Huang 2018). These include ABA, GA, brassinosteroids (BR), jasmonates (JA), and auxins, which regulates the expression of *CBF* genes (Wang et al. 2017). Importantly among them, ABA is primarily been engineered for improving cold stress tolerance in plants (Wani and Kumar 2015). ABA-dependent and CBF-COR pathways share the genetic machinery for imparting cold tolerance. Induction of ABA subsequently resulted in upregulated expression of CBF genes while binding with CRT/DRE element (Ritonga and Chen 2020). ABA also regulates COR gene expression, which strengthens the acclimation process. *AtHAP5A* of *Arabidopsis* binds to CCAAT motif of *AtXTH21* for regulating the freezing stress tolerance (Shi et al. 2014; Ritonga and Chen 2020). On the other hand, GA and JA also play an active role in stimulating CBF-COR pathway genes and transcription factors for cold tolerance (Wang et al. 2017). The studies stated that jasmonates activate the transcription factors, the latter regulating the response of target genes while interacting with CIS-acting element of the promoter (Kim et al. 2021). In species such as *Arabidopsis, Citrus sinensis, Lycopersicon esculentum, Nicotiana tabacum, T. aestivum* and *Triticum turgidum*, an ethylene responsive (ERF) genes were induced in response to ethylene, ABA, and Jasmonate hormone at low temperature (Müller and Munné-Bosch 2015). In these plants, the mechanism of tolerance to cold stress involves the binding of ERF to GCC box and DRE elements. Importantly, brassinazole-resistance 1 and CESTA are the two Brassinosteroids hormones regulating the expression of CBF genes (Barrero-Gil and Salinas 2017; Wang et al. 2017). The GA enhances the tolerance of rice plants in stressful conditions.

In addition to significant studies at the mRNA level for expanding the limits of stress tolerance, Wang et al. (2021) also reported the response in tea plant at chromatin accessibility and post-transcriptional level. For this, ATAC-seq, RNA-seq, and Ribo-seq technologies were used in combination. The study revealed that under chilling stress, translation of uORFs was enhanced, and chromatin landscapes of tea were changed (Wang et al. 2021). Histone modification (*H3K4me3* and *H3K27me3*) and chromatin accessibility in active gens of potato were also reported (Zeng et al. 2019).

13.5 Engineering Cold Stress Tolerance

Biotechnological interventions of plants for improved tolerance to cold stress expand their limits for survival under low temperature conditions. Despite the trait improvement for cold stress is complex approach, an array of genes were identified and characterized for imparting freezing tolerance. Importantly, these genes are involved in regulating the molecular mechanism for acclimation and tolerance in cold conditions (Chinnusamy et al. 2010). In view of this, a large number of studies were accomplished to develop the transgenic lines of various crop plants with high degree of lipid unsaturation, increased membrane tolerance, changes in lipid composition, detoxification enzymes, LEA proteins, and stabilization of membranes (Table 13.1). Modifications in the unsaturation level of fatty acids were studied in tobacco plant while increasing the expression of *GPAT* gene (glycerol-3-phosphate acyltransferase) related to phosphatidyl glycerol fatty acid desaturation from *Arabidopsis thaliana* (Kodama et al. 1994). Expression of *GPAT* gene not only improved tolerance in tobacco but also showed importance in other crop plants (Sui et al. 2007). On the other hand, several genes were identified those intricate in the synthesis of osmoprotectant, sugars, and sucrose that contribute toward the stabilization of membranes and osmotic adjustments. To overcome the cellular damages at low temperature, studies related to transcription factors that control the expression of many genes are important ones. Since the freezing tolerance belongs to multigenic complex trait, therefore, transformation with one gene is replaced with transcriptional regulatory factors for the production of crop

Table 13.1 Identified genes and their functions in plants for improving cold, chilling and freezing stress tolerance.

Genes	Species	Transgenic plant/mutant lines/overexpressed lines/induced in response to cold stress	Function	References
GPAT	*Cucurbita maxima*, *Arabidopsis thaliana*	*Nicotiana tabacum*	Increase in the number of unsaturated fatty acids and a corresponding decrease in the chilling sensitivity	Murata et al. (1992)
NtGRP	*N. tabacum*	—	Cold shock protein	Kingsley and Palis (1994)
CBF1	*A. thaliana*	—	Improved tolerance to chilling, stimulates transcription in response to low temperature	Stockinger et al. (1997)
AtCBF3	*A. thaliana*	Overexpression in *A. thaliana*	Biochemical changes associated with cold acclimation	Gilmour et al. (2000)
WCS19	*Triticum aestivum*	*Arabidopsis*	Increased freezing tolerance	Gong et al. (2002)
AtGRP2b	*A. thaliana*	*A. thaliana*	Induced in response to cold stress	Karlson and Imai (2003)
OsbHLH1	*Oryza sativa*	*O. sativa*	Cold stress response	Wang et al. (2003)
WCOR410	*Wheat dehydrin gene*	*Strawberry*	Enhanced freezing tolerance of leaves	Houde et al. (2004)
WCSP1	*T. aestivum*	—	Cold acclimation	Nakaminami et al. (2006)
OsDREB1	*O. sativa*	*O. sativa*	Increased cold tolerance	Ito et al. (2006)
AtGRP2	*A. thaliana*	*A. thaliana*	Cold shock protein, cold response and flowering, involved in RNA processing and/or translation ability under cold conditions	Fusaro et al. (2007)
HvCBF4	Barley	*O. sativa*	Increased cold tolerance	Oh et al. (2007)
Ring Zinc finger protein *RDCPt*	*Capsicum annuum*	*N. tabacum*	Improved cold tolerance	Kim et al. (2007a)
OsMYB3R-2	*O. sativa*	*Arabidopsis*	Increased tolerance to freezing when exposed to −8 °C for 10 h	Dai et al. (2007)
OsCOIN	*O. sativa*	*O. sativa*	Increased tolerance to chilling	Liu et al. (2007)
OsCSP1	*O. sativa*	—	Cold stress adaptation	Chaikam and Karlson (2008)
OsCSP2	*O. sativa*	—	Cold stress adaptation and development	Chaikam and Karlson (2008)
TPS and TPP	*O. sativa*	*Rice and tobacco*	Increased accumulation of trehalose and imparts tolerance to cold stress	Ge et al. (2008)
OsISAP8	*O. sativa*	*O. sativa*	Cold stress tolerance	Kanneganti and Gupta (2008)
AtCSP3	*A. thaliana*	Overexpressed in *A. thaliana*	Involved in the acquisition of freezing tolerance in plants	Kim et al. (2009)
CSDP1	*A. thaliana*	*A. thaliana*	Freezing tolerance and seed germination	Park et al. (2009)
AtCSP3	*A. thaliana*	Overexpression in *A. thaliana*	Freezing tolerance	Kim et al. (2009)
GhDREB	Cotton	Barley	Enhanced cold tolerance	Gao et al. (2009)
AtCSP4	*A. thaliana*	Overexpression in *A. thaliana*	Silique development	Yang and Karlson (2011)

Table 13.1 (Continued)

Genes	Species	Transgenic plant/mutant lines/overexpressed lines/ induced in response to cold stress	Function	References
TaCBF3	*T. aestivum*	Barley and Wheat	Increased frost and cold tolerance	Morran et al. (2011)
ZmCBF3	Maize	Rice	Increased cold tolerance	Xu et al. (2011)
OsPIP2 and *OsPIP5*	*O. sativa*	—	Cold stress adaptation	Ahamed et al. (2012)
TaDREB3	*T. aestivum*	Barley	Enhanced frost tolerance	Kovalchuk et al. (2013)
TaCBF14 and *TaCBF15*	*T. aestivum*	Barley	Enhanced frost tolerance	Soltész et al. (2013)
COR15a and *COR15b*	*Arabidopsis*	*A. thaliana*	Promote freezing tolerance	Thalhammer and Hincha (2014)
HvCBF2a	Barley	Overexpression in barley	Enhanced frost tolerance	Jeknić et al. (2014)
DaCBF7	*Deschampsia antartica*	*O. sativa*	Enhanced cold tolerance	Byun et al. (2015)
OsABA8o1	*O. sativa*	*A. thaliana* and *O. sativa*	Improved seedling vigour under cold stress	Mega et al. (2015)
BZR1	*A. thaliana*	*A. thaliana*	Enhanced freezing tolerance	Li et al. (2017b)
ERF105	*A. thaliana*	*A. thaliana*	Freezing tolerance and cold acclimation	Bolt et al. (2017)
DaCBF4	*D. antartica*	*O. sativa*	Resulted in a cold-specific phenotype, increased cold tolerance	Byun et al. (2018)
GNOM	*A. thaliana*	*A. thaliana*	Cold resistance	Ashraf and Rahman (2019)
PICKLE	*A. thaliana*	*A. thaliana*	Chilling and freezing tolerance	Yang et al. (2019a)
STCH4/REIL2	*Arabidopsis*	*Arabidopsis*	Confers chilling and freezing tolerance by promoting rRNA processing and CBF protein translation	Yu et al. (2020)
OsCAF1B	*O. sativa*	*O. sativa*	Confers tolerance of low temperature to rice seedlings	Fang et al. (2021)

varieties with improved characteristics (Thomashow 2001). Importantly among them are the proteins related to cold shock, such as *CspA* and *CspB*, that accumulate or are upregulated at cold stress that act as translational enhancers and RNA chaperons by affecting the RNA structure and subsequently translation (Nakaminami et al. 2006; Kim et al. 2007b). Other important class of genes for low temperature adaptation includes *CBF* genes, which further activates the cold-regulated *COR* genes. The expression of some of the low-temperature inducible genes is observed to be regulated through CBF/DREB1 transcription factors (Jaglo et al. 2001). Among them, overexpression of *CBF1* transcription factor observed in transgenic *Arabidopsis* for enhancing stress tolerance without affecting the growth of the plant. However, the expression of *CBF1* in tomato enhances tolerance to multiple stresses on one side, while affecting the growth on the other (Hsieh et al. 2002). In wheat and barley, CBF is found in the frost resistance locus (*FR2*) (Mickelbart et al. 2015). Moreover, enhanced frost tolerance in barley was observed while overexpressing the *TaDREB3*, *TaCBF14*, and *TaCBF15* genes, whereas in wheat, overexpression of *HvCBF2a* resulted in the same. Importantly, the enhanced frost tolerance in barley and wheat was attained by increased expression of downstream genes *COR14b* and *DHN5* (Kovalchuk et al. 2013; Soltész et al. 2013; Jeknić et al. 2014). On the other hand, increased expression of *TaCBF3* from wheat resulted in frost and cold tolerance in both wheat and barley, while the overexpression of *GhDREB* from cotton increased cold tolerance in barley (Gao et al. 2009; Morran et al. 2011). Also, cold tolerance in rice was reported to increase with the overexpression of *OsDREB1*, *HvCBF4*, and *ZmCBF4* genes (Ito et al. 2006; Oh et al. 2007; Xu et al. 2011). In addition to this, zincfinger-containing glycine-rich RNA binding protein and DEAD-box RNA helicase genes are upregulated in response to cold stress (Yu and Owttrim 2000; Kim et al. 2005). Important examples include *LOS4* (low expression of osmotically responsive gene 4) and *AtNUP160* (*Arabidopsis* nucleophorin) suppressor of auxin resistance 1 (*SAR1*), regulating RNA control for increasing freezing tolerance (Gong et al. 2005; Dong et al. 2006b). Interestingly, Yu et al. (2018) reported the first transgenic pipeline for water lilies using pollen tube pathway. For this, the choline oxidase (*CodA*) gene was transformed into water lilies, and the expression of exogenous *CodA* significantly improved to combat cold stress.

Also, CDPKs from the protein kinase family importantly involved in enhancing cold tolerance. The overexpression of *OsCDPK7* in rice has been reported at low temperatures to enhance chilling tolerance in seedlings (Saijo et al. 2000). In addition to CDPKs, MAPK pathways are also involved in freezing tolerance in plants (Smallwood and Bowles 2002).

13.6 Future Directions

Low temperature is a major constraints toward the productivity, yield, and metabolism of crop plants. However, the extent to which it affects a particular climate zone depends on the duration and the plant type. Therefore, plant response and its adaptive mechanisms to understand the development of tolerance toward low temperature need to be strengthened for agriculture crop plants. Engineering of crop plants, either by introduction or regulation of the expression of vital genes, seems to be a potential possibility to impart cold tolerance in crop plants. This approach also overcomes the limitations of conventional or molecular breeding while targeting beneficial genes for crop improvement. Advancement in omics approaches further helps in understanding the complex quantitative traits related to temperature extremes. This may help in understanding which class of genes and proteins enhances freezing tolerance most effectively and determining their specific role in different climatic regions. Although the regulation mediated through CBF/DREB is extensively studied, studies on other transcription factors will probably accelerate progress in this area. In addition, advances in fundamental research may provide information regarding specialized molecules that provide the way understanding of cold and frost tolerance mechanisms in plants.

Acknowledgment

BS and RJ gratefully acknowledge the Director, CSIR-Institute of Himalayan Bioresource Technology, Palampur, for providing the facilities to carry out this work. CSIR support in the form of project MLP0201, MLP0170, MLP0172, SSP0125, and SSP-0126 for this study is highly acknowledged. This manuscript represents CSIR-IHBT communication no. 4940.

References

Agarwal, M., Hao, Y., Kapoor, A. et al. (2006). A R2R3 type MYB transcription factor is involved in the cold regulation of CBF genes and in acquired freezing tolerance. *J. Biol. Chem.* 281 (49): 37636–37645.

Ahamed, A., Murai-Hatano, M., Ishikawa-Sakurai, J. et al. (2012). Cold stress-induced acclimation in rice is mediated by root-specific aquaporins. *Plant Cell Physiol.* 53 (8): 1445–1456.

Alonso-Amelot, M.E. (2008). High altitude plants, chemistry of acclimation and adaptation. In: *Studies in Natural Products Chemistry*, vol. 34, 883–982. Elsevier.

Anchordoguy, T.J., Rudolph, A.S., Carpenter, J.F., and Crowe, J.H. (1987). Modes of interaction of cryoprotectants with membrane phospholipids during freezing. *Cryobiology* 24 (4): 324–331.

Ashraf, M.A. and Rahman, A. (2019). Cold stress response in *Arabidopsis thaliana* is mediated by GNOM ARF-GEF. *Plant J.* 97 (3): 500–516.

Barrero-Gil, J. and Salinas, J. (2017). CBFs at the crossroads of plant hormone signaling in cold stress response. *Mol. Plant* 10: 542–544.

Behera, S., Wang, N., Zhang, C. et al. (2015). Analyses of Ca^{2+} dynamics using a ubiquitin-10 promoter-driven Yellow Cameleon 3.6 indicator reveal reliable transgene expression and differences in cytoplasmic Ca^{2+} responses in *Arabidopsis* and rice (*Oryza sativa*) roots. *New Phytol.* 206 (2): 751–760.

Bolt, S., Zuther, E., Zintl, S. et al. (2017). ERF105 is a transcription factor gene of *Arabidopsis thaliana* required for freezing tolerance and cold acclimation. *Plant Cell Environ.* 40 (1): 108–120.

Bravo, L.A., Gallardo, J., Navarrete, A. et al. (2003). Cryoprotective activity of a cold-induced dehydrin purified from barley. *Physiol. Plant.* 118 (2): 262–269.

Byun, M.Y., Lee, J., Cui, L.H. et al. (2015). Constitutive expression of DaCBF7, an Antarctic vascular plant *Deschampsia antarctica* CBF homolog, resulted in improved cold tolerance in transgenic rice plants. *Plant Sci.* 236: 61–74.

Byun, M.Y., Cui, L.H., Lee, J. et al. (2018). Identification of rice genes associated with enhanced cold tolerance by comparative transcriptome analysis with two transgenic rice plants overexpressing *DaCBF4* or *DaCBF7*, isolated from antarctic flowering plant *Deschampsia antarctica*. *Front. Plant Sci.* 9: 601.

Chaikam, V. and Karlson, D. (2008). Functional characterization of two cold shock domain proteins from *Oryza sativa*. *Plant Cell Environ.* 31 (7): 995–1006.

Chan, Z., Wang, Y., Cao, M. et al. (2016). RDM 4 modulates cold stress resistance in *Arabidopsis* partially through the CBF-mediated pathway. *New Phytol.* 209 (4): 1527–1539.

Chen, H., Chen, X., Chen, D. et al. (2015). A comparison of the low temperature transcriptomes of two tomato genotypes that differ in freezing tolerance: *Solanum lycopersicum* and *Solanum habrochaites*. *BMC Plant Biol.* 15 (1): 1–16.

Chen, L., Zhao, Y., Xu, S. et al. (2018). Os MADS 57 together with Os TB 1 coordinates transcription of its target Os WRKY 94 and D14 to switch its organogenesis to defense for cold adaptation in rice. *New Phytol.* 218 (1): 219–231.

Chinnusamy, V., Ohta, M., Kanrar, S. et al. (2003). ICE1: a regulator of cold-induced transcriptome and freezing tolerance in *Arabidopsis*. *Genes Dev.* 17 (8): 1043–1054. https://doi.org/10.1101/gad.1077503.

Chinnusamy, V., Zhu, J., and Zhu, J.K. (2007). Cold stress regulation of gene expression in plants. *Trends Plant Sci.* 12 (10): 444–451.

Chinnusamy, V., Jhu, J.K., and Sunkar, R. (2010). Gene regulation during cold stress acclimation in plants. *Methods Mol. Biol.* 639: 39–55.

Christie, P.J., Alfenito, M.R., and Walbot, V. (1994). Impact of low-temperature stress on general phenylpropanoid and anthocyanin pathways: enhancement of transcript abundance and anthocyanin pigmentation in maize seedlings. *Planta* 194 (4): 541–549.

Dai, X., Xu, Y., Ma, Q. et al. (2007). Overexpression of an R1R2R3 MYB gene, OsMYB3R-2, increases tolerance to freezing, drought, and salt stress in transgenic *Arabidopsis*. *Plant Physiol.* 143 (4): 1739–1751.

Demidchik, V., Straltsova, D., Medvedev, S.S. et al. (2014). Stress-induced electrolyte leakage: the role of K^+-permeable channels and involvement in programmed cell death and metabolic adjustment. *J. Exp. Bot.* 65 (5): 1259–1270.

Devi, E.L., Kumar, S., Singh, T.B. et al. (2017). Adaptation strategies and defence mechanisms of plants during environmental stress. In: *Medicinal Plants and Environmental Challenges*, 359–413. Cham: Springer.

Ding, Y., Li, H., Zhang, X. et al. (2015). OST1 kinase modulates freezing tolerance by enhancing ICE1 stability in Arabidopsis. *Dev. Cell* 32 (3): 278–289.

Ding, C., Lei, L., Yao, L. et al. (2019). The involvements of calcium-dependent protein kinases and catechins in tea plant [*Camellia sinensis* (L.) *O. Kuntze*] cold responses. *Plant Physiol. Biochem.* 143: 190–202.

Doherty, C.J., Van Buskirk, H.A., Myers, S.J., and Thomashow, M.F. (2009). Roles for *Arabidopsis CAMTA* transcription factors in cold-regulated gene expression and freezing tolerance. *Plant Cell* 21 (3): 972–984.

Dong, C.H., Agarwal, M., Zhang, Y. et al. (2006a). The negative regulator of plant cold responses, HOS1, is a RING E3 ligase that mediates the ubiquitination and degradation of ICE1. *Proc. Natl. Acad. Sci.* 103 (21): 8281–8286.

Dong, C.H., Hu, X., Tang, W. et al. (2006b). A putative *Arabidopsis* nucleoporin, AtNUP160, is critical for RNA export and required for plant tolerance to cold stress. *Mol. Cell. Biol.* 26 (24): 9533–9543.

Du, H., Wu, N., Fu, J. et al. (2012). A GH3 family member, OsGH3-2, modulates auxin and methylation and chromatin patterning abscisic acid levels and differentially affects drought and cold tolerance in rice. *J. Exp. Bot.* 63: 6467–6480.

Epand, R.M., Shai, Y., Segrest, J.P., and Anantharamiah, G.M. (1995). Mechanisms for the modulation of membrane bilayer properties by amphipathic helical peptides. *Biopolym. Orig. Res. Biomol.* 37 (5): 319–338.

Fang, J.C., Tsai, Y.C., Chou, W.L. et al. (2021). A CCR4-associated factor 1, OsCAF1B, confers tolerance of low-temperature stress to rice seedlings. *Plant Mol. Biol.* 105 (1): 177–192.

Fowler, S., Cook, D., and Thomashow, M.F. (2007). The CBF cold-response pathway. In: *Plant Abiotic Stress* (ed. M.A. Jenks and P.M. Hasegawa), 71–99. Chichester: Wiley-Blackwell.

Friesen, P.C. and Sage, R.F. (2016). Photosynthetic responses to chilling in a chilling-tolerant and chilling-sensitive Miscanthus hybrid. *Plant Cell Environ.* 39 (7): 1420–1431.

Fu, Z.Q. and Dong, X. (2013). Systemic acquired resistance: turning local infection into global defense. *Annu. Rev. Plant Biol.* 64: 839–863.

Fusaro, A.F., Bocca, S.N., Ramos, R.L.B. et al. (2007). *AtGRP2*, a cold-induced nucleo-cytoplasmic RNA-binding protein, has a role in flower and seed development. *Planta* 225 (6): 1339–1351.

Gao, S.Q., Chen, M., Xia, L.Q. et al. (2009). A cotton (*Gossypium hirsutum*) DRE-binding transcription factor gene, *GhDREB*, confers enhanced tolerance to drought, high salt, and freezing stresses in transgenic wheat. *Plant Cell Rep.* 28 (2): 301–311.

Ge, L.F., Chao, D.Y., Shi, M. et al. (2008). Overexpression of the trehalose-6-phosphate phosphatase gene *OsTPP1* confers stress tolerance in rice and results in the activation of stress responsive genes. *Planta* 228 (1): 191–201.

Gilmour, S.J., Sebolt, A.M., Salazar, M.P. et al. (2000). Overexpression of the Arabidopsis CBF3 transcriptional activator mimics multiple biochemical changes associated with cold acclimation. *Plant Physiol.* 124 (4): 1854–1865.

Glaszmann, J.C., Kaw, R.N., and Khush, G.S. (1990). Genetic divergence among cold tolerant rices (*Oryza sativa* L.). *Euphytica* 45 (2): 95–104.

Gomès, E., Jakobsen, M.K., Axelsen, K.B. et al. (2000). Chilling tolerance in *Arabidopsis* involves ALA1, a member of a new family of putative aminophospholipid translocases. *Plant Cell* 12 (12): 2441–2453.

Gong, Z., Lee, H., Xiong, L. et al. (2002). RNA helicase-like protein as an early regulator of transcription factors for plant chilling and freezing tolerance. *Proc. Natl. Acad. Sci.* 99 (17): 11507–11512.

Gong, Z., Dong, C.H., Lee, H. et al. (2005). A DEAD box RNA helicase is essential for mRNA export and important for development and stress responses in *Arabidopsis*. *Plant Cell* 17 (1): 256–267.

Graham, D. and Patterson, B.D. (1982). Responses of plants to low, nonfreezing temperatures: proteins, metabolism, and acclimation. *Annu. Rev. Plant Physiol.* 33 (1): 347–372.

Griffith, M. and Yaish, M.W. (2004). Antifreeze proteins in overwintering plants: a tale of two activities. *Trends Plant Sci.* 9 (8): 399–405.

Guo, W.L., Chen, R.G., Gong, Z.H. et al. (2012). Exogenous abscisic acid increases antioxidant enzymes and related gene expression in pepper (*Capsicum annuum*) leaves subjected to chilling stress. *Genet. Mol. Res.* 11 (4): 4063–4080.

Gusain, S., Joshi, S. and Joshi, R.(2023). Sensing, signalling, and regulatory mechanism of cold-stress tolerance in plants. Plant Physiol. Biochem. 197: 107646.

Guy, C.L. (1990). Cold acclimation and freezing stress tolerance: role of protein metabolism. *Annu. Rev. Plant Biol.* 41 (1): 187–223.

Hincha, D.K., Schmitt, J.M., and Heber, U. (1988). Cryoprotective proteins in plants? *Cryobiology* 25 (6): 557–558.

Hincha, D.K., Heber, U., and Schmitt, J.M. (1990). Proteins from frost-hardy leaves protect thylakoids against mechanical freeze-thaw damage *in vitro*. *Planta* 180 (3): 416–419.

Houde, M., Dallaire, S., N'Dong, D., and Sarhan, F. (2004). Overexpression of the acidic dehydrin WCOR410 improves freezing tolerance in transgenic strawberry leaves. *Plant Biotechnol. J.* 2 (5): 381–387.

Hsieh, T.H., Lee, J.T., Yang, P.T. et al. (2002). Heterology expression of the *Arabidopsis* C-repeat/dehydration response element binding factor 1 gene confers elevated tolerance to chilling and oxidative stresses in transgenic tomato. *Plant Physiol.* 129 (3): 1086–1094.

Hu, Y., Wu, Q., Sprague, S.A. et al. (2015). Tomato expressing *Arabidopsis* glutaredoxin gene *AtGRXS17* confers tolerance to chilling stress via modulating cold responsive components. *Hortic. Res.* 2 (1): 1–11.

Huo, C., Zhang, B., Wang, H. et al. (2016). Comparative study of early cold-regulated proteins by two-dimensional difference gel electrophoresis reveals a key role for phospholipase Dα1 in mediating cold acclimation signaling pathway in rice. *Mol. Cell. Proteomic.* 15 (4): 1397–1411.

Ito, Y., Katsura, K., Maruyama, K. et al. (2006). Functional analysis of rice DREB1/CBF-type transcription factors involved in cold-responsive gene expression in transgenic rice. *Plant Cell Physiol.* 47 (1): 141–153.

Jaglo, K.R., Kleff, S., Amundsen, K.L. et al. (2001). Components of the *Arabidopsis* C-repeat/dehydration-responsive element binding factor cold-response pathway are conserved in *Brassica napus* and other plant species. *Plant Physiol.* 127 (3): 910–917.

Jaglo-Ottosen, K.R., Gilmour, S.J., Zarka, D.G. et al. (1998). *Arabidopsis* CBF1 overexpression induces COR genes and enhances freezing tolerance. *Science* 280 (5360): 104–106.

Janská, A., Maršík, P., Zelenková, S., and Ovesná, J. (2010). Cold stress and acclimation–what is important for metabolic adjustment? *Plant Biol.* 12 (3): 395–405.

Jeknić, Z., Pillman, K.A., Dhillon, T. et al. (2014). Hv-CBF2A overexpression in barley accelerates COR gene transcript accumulation and acquisition of freezing tolerance during cold acclimation. *Plant Mol. Biol.* 84 (1-2): 67–82.

Joshi, R., Wani, S.H., Singh, B. et al. (2016). Transcription factors and plants response to drought stress: current understanding and future directions. *Front. Plant Sci.* 7: 1029.

Kanneganti, V. and Gupta, A.K. (2008). Overexpression of OsiSAP8, a member of stress associated protein (SAP) gene family of rice confers tolerance to salt, drought and cold stress in transgenic tobacco and rice. *Plant Mol. Biol.* 66 (5): 445–462.

Karlson, D. and Imai, R. (2003). Conservation of the cold shock domain protein family in plants. *Plant Physiol.* 131 (1): 12–15.

Khan, H., Shah, S.H., Uddin, N. et al. (2015). Biochemical and physiological changes of different plants species in response to heat and cold stress. *ARPN J. Agric. Biol. Sci.* 10: 213–216.

Kim, Y.O., Kim, J.S., and Kang, H. (2005). Cold-inducible zinc finger-containing glycine-rich RNA-binding protein contributes to the enhancement of freezing tolerance in *Arabidopsis thaliana*. *Plant J.* 42 (6): 890–900.

Kim, J.S., Park, S.J., Kwak, K.J. et al. (2007a). Cold shock domain proteins and glycine-rich RNA-binding proteins from *Arabidopsis thaliana* can promote the cold adaptation process in *Escherichia coli*. *Nucleic Acids Res.* 35 (2): 506–516.

Kim, J.Y., Park, S.J., Jang, B. et al. (2007b). Functional characterization of a glycine-rich RNA-binding protein 2 in *Arabidopsis thaliana* under abiotic stress conditions. *Plant J.* 50 (3): 439–451.

Kim, M.H., Sasaki, K., and Imai, R. (2009). Cold shock domain protein 3 regulates freezing tolerance in *Arabidopsis thaliana*. *J. Biol. Chem.* 284 (35): 23454–23460.

Kim, H.J., Seomun, S., Yoon, Y., and Jang, G. (2021). Jasmonic acid in plant abiotic stress tolerance and interaction with abscisic acid. *Agronomy* 11 (9): 1886.

Kingsley, P.D. and Palis, J. (1994). GRP2 proteins contain both CCHC zinc fingers and a cold shock domain. *Plant Cell* 6 (11): 1522.

Kissoudis, C., van de Wiel, C., Visser, R.G., and van der Linden, G. (2014). Enhancing crop resilience to combined abiotic and biotic stress through the dissection of physiological and molecular crosstalk. *Front. Plant Sci.* 5: 207.

Kodama, H., Hamada, T., Horiguchi, G. et al. (1994). Genetic enhancement of cold tolerance by expression of a gene for chloroplast [omega]-3 fatty acid desaturase in transgenic tobacco. *Plant Physiol.* 105 (2): 601–605.

Körner, C. (2003). *Alpine Plant Life*, 2e. Springer.

Körner, C. (2016). Plant adaptation to cold climates. *F1000Research* 5: F1000.

Körner, C.H. and Renhardt, U. (1987). Dry matter partitioning and root length/leaf area ratios in herbaceous perennial plants with diverse altitudinal distribution. *Oecologia* 74 (3): 411–418.

Kovalchuk, N., Jia, W., Eini, O. et al. (2013). Optimization of *TaDREB3* gene expression in transgenic barley using cold-inducible promoters. *Plant Biotechnol. J.* 11 (6): 659–670.

Lee, D.H. and Lee, C.B. (2000). Chilling stress-induced changes of antioxidant enzymes in the leaves of cucumber: in gel enzyme activity assays. *Plant Sci.* 159 (1): 75–85.

Lee, E.S., Park, J.H., Wi, S.D. et al. (2021). Redox-dependent structural switch and CBF activation confer freezing tolerance in plants. *Nat. Plants* 7: 1–9.

Levitt, J. (1980). *Response of Plants to Environmental Stress; Chilling Freezing and High Temperature Stress*. New York: Academic Press.

Li, H., Ding, Y., Shi, Y. et al. (2017a). MPK3-and MPK6-mediated ICE1 phosphorylation negatively regulates ICE1 stability and freezing tolerance in *Arabidopsis*. *Dev. Cell* 43 (5): 630–642.

Li, H., Ye, K., Shi, Y. et al. (2017b). BZR1 positively regulates freezing tolerance via CBF-dependent and CBF-independent pathways in *Arabidopsis*. *Mol. Plant* 10 (4): 545–559.

Liu, Q., Kasuga, M., Sakuma, Y. et al. (1998). Two transcription factors, DREB1 and DREB2, with an EREBP/AP2 DNA binding domain separate two cellular signal transduction pathways in drought-and low-temperature-responsive gene expression, respectively, in *Arabidopsis*. *Plant Cell* 10 (8): 1391–1406.

Liu, K., Wang, L., Xu, Y. et al. (2007). Overexpression of OsCOIN, a putative cold inducible zinc finger protein, increased tolerance to chilling, salt and drought, and enhanced proline level in rice. *Planta* 226 (4): 1007–1016.

Liu, Z., Jia, Y., Ding, Y. et al. (2017). Plasma membrane CRPK1-mediated phosphorylation of 14-3-3 proteins induces their nuclear import to fine-tune CBF signaling during cold response. *Mol. Cell* 66 (1): 117–128.

Liu, X., Fu, L., Qin, P. et al. (2019). Overexpression of the wheat trehalose 6-phosphate synthase 11 gene enhances cold tolerance in *Arabidopsis thaliana*. *Gene* 710: 210–217.

Londo, J.P., Kovaleski, A.P., and Lillis, J.A. (2018). Divergence in the transcriptional landscape between low temperature and freeze shock in cultivated grapevine (*Vitis vinifera*). *Hortic. Res.* 5 (1): 1–14.

Longo, V., Kamran, R.V., Michaletti, A. et al. (2017). Proteomic and physiological response of spring barley leaves to cold stress. *Cell* 6 (7).

Ma, Y., Dai, X., Xu, Y. et al. (2015). COLD1 confers chilling tolerance in rice. *Cell* 160 (6): 1209–1221.

Maurya, N.K., Goswami, A.K., Singh, S.K. et al. (2020). Studies on expression of *CBF1* and *CBF2* genes and anti-oxidant enzyme activities in papaya genotypes exposed to low temperature stress. *Sci. Hortic.* 261: 108914.

Mega, R., Meguro-Maoka, A., Endo, A. et al. (2015). Sustained low abscisic acid levels increase seedling vigor under cold stress in rice (*Oryza sativa* L.). *Sci. Rep.* 5 (1): 1–13.

Mickelbart, M.V., Hasegawa, P.M., and Bailey-Serres, J. (2015). Genetic mechanisms of abiotic stress tolerance that translate to crop yield stability. *Nat. Rev. Genet.* 16 (4): 237–251.

Monroy, A.F. and Dhindsa, R.S. (1995). Low temperature signal transduction—induction of genes of alfalfa by calcium at 25 °C. *Plant Cell* 7: 321–331.

Monroy, A.F., Sarhan, F., and Dhindsa, R.S. (1993). Coldinduced changes in freezing tolerance, protein phosphorylation and gene expression: evidence for a role of calcium. *Plant Physiol.* 102: 1227–1235.

Morran, S., Eini, O., Pyvovarenko, T. et al. (2011). Improvement of stress tolerance of wheat and barley by modulation of expression of DREB/CBF factors. *Plant Biotechnol. J.* 9 (2): 230–249.

Moura, J.C.M.S., Bonine, C.A.V., de Oliveira Fernandes Viana, J. et al. (2010). Abiotic and biotic stresses and changes in the lignin content and composition in plants. *J. Integr. Plant Biol.* 52 (4): 360–376.

Müller, M. and Munné-Bosch, S. (2015). Ethylene response factors: a key regulatory hub in hormone and stress signaling. *Plant Physiol.* 169 (1): 32–41.

Murata, N., Ishizaki-Nishizawa, O., Higashi, S. et al. (1992). Genetically engineered alteration in the chilling sensitivity of plants. *Nature* 356 (6371): 710–713.

Nakaminami, K., Karlson, D.T., and Imai, R. (2006). Functional conservation of cold shock domains in bacteria and higher plants. *Proc. Natl. Acad. Sci.* 103 (26): 10122–10127.

Oh, S.J., Kwon, C.W., Choi, D.W. et al. (2007). Expression of barley HvCBF4 enhances tolerance to abiotic stress in transgenic rice. *Plant Biotechnol. J.* 5 (5): 646–656.

Pareek, A., Khurana, A., Sharma, A.K., and Kumar, R. (2017). An overview of signaling regulons during cold stress tolerance in plants. *Curr. Genomics* 18 (6): 498–511.

Park, S.J., Kwak, K.J., Oh, T.R. et al. (2009). Cold shock domain proteins affect seed germination and growth of *Arabidopsis thaliana* under abiotic stress conditions. *Plant Cell Physiol.* 50 (4): 869–878.

Pérez-Clemente, R.M., Vives, V., Zandalinas, S.I. et al. (2013). Biotechnological approaches to study plant responses to stress. *BioMed Res. Intl.* 2013: 654120.

Pérez-Ilzarbe, J., Hernández, T., Estrella, I., and Vendrell, M. (1997). Cold storage of apples (cv. Granny Smith) and changes in phenolic compounds. *Z. Lebensmittelunters. Forsch. A* 204 (1): 52–55.

Ploschuk, E.L., Bado, L.A., Salinas, M. et al. (2014). Photosynthesis and fluorescence responses of *Jatropha curcas* to chilling and freezing stress during early vegetative stages. *Environ. Expt. Bot.* 102: 18–26.

Puhakainen, T., Li, C., Boije-Malm, M. et al. (2004). Short-day potentiation of low temperature-induced gene expression of a C-repeat-binding factor-controlled gene during cold acclimation in silver birch. *Plant Physiol.* 136 (4): 4299–4307.

Rasolofo, P.R., Vergara, B.S., and Visperas, R.M. (1986). Screening rice cultivars at seedling stage and anthesis for low temperature tolerance in Madagascar. *Intl. Rice Res. Newsl. Philipp.* 11: 12–13.

Rezaie, R., Mandoulakani, B.A., and Fattahi, M. (2020). Cold stress changes antioxidant defense system, phenylpropanoid contents and expression of genes involved in their biosynthesis in *Ocimum basilicum* L. *Sci. Rep.* 10 (1): 1–10.

Ritonga, F.N. and Chen, S. (2020). Physiological and molecular mechanism involved in cold stress tolerance in plants. *Plants* 9 (5): 560.

Rubio, S. and Pérez, F.J. (2019). ABA and its signaling pathway are involved in the cold acclimation and deacclimation of grapevine buds. *Sci. Hortic.* 256: 108565.

Rudolph, A.S. and Crowe, J.H. (1985). Membrane stabilization during freezing: the role of two natural cryoprotectants, trehalose and proline. *Cryobiology* 22 (4): 367–377.

Ruelland, E. and Zachowski, A. (2010). How plants sense temperature. *Environ. Exp. Bot.* 69 (3): 225–232.

Saijo, Y., Hata, S., Kyozuka, J. et al. (2000). Over-expression of a single Ca^{2+}-dependent protein kinase confers both cold and salt/drought tolerance on rice plants. *Plant J.* 23 (3): 319–327.

Sakai, A. and Larcher, W. (1987). *Frost Survival of Plants: Responses and Adaptation to Freezing Stress*. Berlin: Springer-Verlag.

Sanghera, G.S., Zarger, M.A., Anwar, A. et al. (2001). Studies on spikelet fertility and incidence of leaf blast on certain IRCTN rice genotypes under temperate conditions. *National Symposium on Plant Protection Strategies for Sustainable Agri-Horticulture, SKAUST-Jammu,* Jammu (12–13 October 2001), p. 125.

Sanghera, G.S., Wani, S.H., Hussain, W., and Singh, N.B. (2011). Engineering cold stress tolerance in crop plants. *Curr. Genomics* 12 (1): 30.

Shan, D.P., Huang, J.G., Yang, Y.T. et al. (2007). Cotton GhDREB1 increases plant tolerance to low temperature and is negatively regulated by gibberellic acid. *New Phytol.* 176 (1): 70–81.

Shi, H., Ye, T., Zhong, B. et al. (2014). At HAP 5A modulates freezing stress resistance in Arabidopsis through binding to CCAAT motif of AtXTH21. *New Phytol.* 203 (2): 554–567.

Shi, Y., Ding, Y., and Yang, S. (2018). Molecular regulation of CBF signaling in cold acclimation. *Trends Plant Sci.* 23 (7): 623–637.

Shriram, V., Kumar, V., Devarumath, R.M. et al. (2016). MicroRNAs as potential targets for abiotic stress tolerance in plants. *Front. Plant Sci.* 7: 817.

Sierra-Almeida, A. and Cavieres, L.A. (2012). Summer freezing resistance of high-elevation plant species changes with ontogeny. *Environ. Exp. Bot.* 80: 10–15.

Singh, B.D. (2000). *Plant Breeding*, 443–460. Ludhiana: Kalyani Publishers.

Smallwood, M. and Bowles, D.J. (2002). Plants in a cold climate. *Philos. Trans. R. Soc. Lond. Ser. B Biol. Sci.* 357 (1423): 831–847.

Soltész, A., Smedley, M., Vashegyi, I. et al. (2013). Transgenic barley lines prove the involvement of *TaCBF14* and *TaCBF15* in the cold acclimation process and in frost tolerance. *J. Exp. Bot.* 64 (7): 1849–1862.

Squeo, F.A., Rada, F., Azocar, A., and Goldstein, G. (1991). Freezing tolerance and avoidance in high tropical Andean plants: is it equally represented in species with different plant height? *Oecologia* 86: 378–382.

Steponkus, P.L., Uemura, M., Joseph, R.A. et al. (1998). Mode of action of the COR15a gene on the freezing tolerance of *Arabidopsis thaliana*. *Proc. Natl. Acad. Sci.* 95 (24): 14570–14575.

Stockinger, E.J., Gilmour, S.J., and Thomashow, M.F. (1997). *Arabidopsis thaliana CBF1* encodes an AP2 domain-containing transcriptional activator that binds to the C-repeat/DRE, a *cis*-acting DNA regulatory element that stimulates transcription in response to low temperature and water deficit. *Proc. Natl. Acad. Sci.* 94 (3): 1035–1040.

Sui, N., Li, M., Zhao, S.J. et al. (2007). Overexpression of glycerol-3-phosphate acyltransferase gene improves chilling tolerance in tomato. *Planta* 226 (5): 1097–1108.

Sun, X., Zhu, Z., Zhang, L. et al. (2019). Overexpression of ethylene response factors VaERF080 and VaERF087 from *Vitis amurensis* enhances cold tolerance in *Arabidopsis*. *Sci. Hortic.* 243: 320–326.

Thakur, P., Kumar, S., Malik, J.A. et al. (2010). Cold stress effects on reproductive development in grain crops: an overview. *Environ. Expt. Bot.* 67 (3): 429–443.

Thalhammer, A. and Hincha, D.K. (2014). A mechanistic model of *COR15* protein function in plant freezing tolerance: integration of structural and functional characteristics. *Plant Signal. Behav.* 9 (12): e977722.

Thomashow, M.F. (1998). Role of cold-responsive genes in plant freezing tolerance. *Plant Physiol.* 118 (1): 1–8.

Thomashow, M.F. (1999). Plant cold acclimation: freezing tolerance genes and regulatory mechanisms. *Annu. Rev. Plant Biol.* 50 (1): 571–599.

Thomashow, M.F. (2001). So what's new in the field of plant cold acclimation? Lots! *Plant Physiol.* 125 (1): 89–93.

Tolosa, L.N. and Zhang, Z. (2020). The role of major transcription factors in *Solanaceous* food crops under different stress conditions: current and future perspectives. *Plants* 9 (1): 56.

Traverso, J.A., Micalella, C., Martinez, A. et al. (2013). Roles of N-terminal fatty acid acylations in membrane compartment partitioning: *Arabidopsis* h-type thioredoxins as a case study. *Plant Cell* 25: 1056–1077.

Uemura, M. and Steponkus, P.L. (2003). Modification of the intracellular sugar content alters the incident of freeze-induced membrane lesions of protoplasts isolated from *Arabidopsis thaliana* leaves. *Plant Cell Environ.* 26: 1083–1096.

Vannini, C., Locatelli, F., Bracale, M. et al. (2004). Overexpression of the rice *Osmyb4* gene increases chilling and freezing tolerance of *Arabidopsis thaliana* plants. *Plant J.* 37 (1): 115–127.

Wang, Y.J., Zhang, Z.G., He, X.J. et al. (2003). A rice transcription factor *OsbHLH1* is involved in cold stress response. *Theor. Appl. Genet.* 107 (8): 1402–1409.

Wang, D.Z., Jin, Y.N., Ding, X.H. et al. (2017). Gene regulation and signal transduction in the ICE–CBF–COR signaling pathway during cold stress in plants. *Biochem. Mos.* 82 (10): 1103–1117.

Wang, L., Ma, R., Yin, Y., and Jiao, Z. (2018). Role of carbon ion beams irradiation in mitigating cold stress in *Arabidopsis thaliana*. *Ecotoxicol. Environ. Saf.* 162: 341–347.

Wang, Y., Mao, Z., Jiang, H. et al. (2019). Feedback loop involving *MdMYB108L* and *MdHY5* controls apple cold tolerance. *Biochem. Biophys. Res. Commun.* 512: 381–386.

Wang, Y., Xiong, F., Nong, S. et al. (2020). Effects of nitric oxide on the GABA, polyamines, and proline in tea (*Camellia sinensis*) roots under cold stress. *Sci. Rep.* 10 (1): 1–10.

Wang, P., Jin, S., Chen, X. et al. (2021). Chromatin accessibility and translational landscapes of tea plants under chilling stress. *Hortic. Res.* 8 (1): 1–15.

Wani, S.H. and Kumar, V. (2015). Plant stress tolerance: engineering ABA: a potent phytohormone. *Transcriptomics* 3 (2): 1000113.

Wise, M.J. and Tunnacliffe, A. (2004). POPP the question: what do LEA proteins do? *Trends Plant Sci.* 9 (1): 13–17.

Xiong, L., Lee, H., Ishitani, M. et al. (2002). Repression of stress-responsive genes by FIERY2, a novel transcriptional regulator in *Arabidopsis*. *Proc. Natl. Acad. Sci.* 99 (16): 10899–10904.

Xu, M., Li, L., Fan, Y. et al. (2011). ZmCBF3 overexpression improves tolerance to abiotic stress in transgenic rice (*Oryza sativa*) without yield penalty. *Plant Cell Rep.* 30 (10): 1949–1957.

Yamaguchi-Shinozaki, K. and Shinozaki, K. (1994). A novel cis-acting element in an *Arabidopsis* gene is involved in responsiveness to drought, low-temperature, or high-salt stress. *Plant Cell* 6 (2): 251–264.

Yamaguchi-Shinozaki, K. and Shinozaki, K. (2006). Transcriptional regulatory networks in cellular responses and tolerance to dehydration and cold stresses. *Annu. Rev. Plant Biol.* 57: 781–803.

Yang, T. and Huang, X.S. (2018). Deep sequencing-based characterization of transcriptome of *Pyrus ussuriensis* in response to cold stress. *Gene* 661: 109–118.

Yang, Y. and Karlson, D.T. (2011). Overexpression of *AtCSP4* affects late stages of embryo development in *Arabidopsis*. *J. Exp. Bot.* 62 (6): 2079–2091.

Yang, A., Dai, X., and Zhang, W.H. (2012). A R2R3-type MYB gene, OsMYB2, is involved in salt, cold, and dehydration tolerance in rice. *J. Exp. Bot. 63* (7): 2541–2556.

Yang, X., Wang, X., Ji, L. et al. (2015). Overexpression of a *Miscanthus lutarioriparius* NAC gene MlNAC5 confers enhanced drought and cold tolerance in *Arabidopsis*. *Plant Cell Rep.* 34 (6): 943–958.

Yang, R., Hong, Y., Ren, Z. et al. (2019a). A role for PICKLE in the regulation of cold and salt stress tolerance in *Arabidopsis*. *Front. Plant Sci.* 10: 900.

Yang, X., Wang, R., Hu, Q. et al. (2019b). *DlICE1*, a stress-responsive gene from *Dimocarpus longan*, enhances cold tolerance in transgenic *Arabidopsis*. *Plant Physiol. Biochem.* 142: 490–499.

Ying, M. and Kidou, S.I. (2017). Discovery of novel cold-induced CISP genes encoding small RNA-binding proteins related to cold adaptation in barley. *Plant Sci.* 260: 129–138.

Yu, E. and Owttrim, G.W. (2000). Characterization of the cold stress-induced cyanobacterial DEAD-box protein CrhC as an RNA helicase. *Nucleic Acids Res.* 28 (20): 3926–3934.

Yu, C., Qiao, G., Qiu, W. et al. (2018). Molecular breeding of water lily: engineering cold stress tolerance into tropical water lily. *Hortic. Res.* 5 (1): 1–11.

Yu, H., Kong, X., Huang, H. et al. (2020). STCH4/REIL2 confers cold stress tolerance in *Arabidopsis* by promoting rRNA processing and CBF protein translation. *Cell Rep.* 30 (1): 229–242.

Zeng, Z., Zhang, W., Marand, A.P. et al. (2019). Cold stress induces enhanced chromatin accessibility and bivalent histone modifications H3K4me3 and H3K27me3 of active genes in potato. *Genome Biol.* 20 (1): 1–17.

Zhang, F., Jiang, Y., Bai, L. et al. (2011). The ICE-CBF-COR pathway in cold acclimation and AFPs in plants. *Middle East J. Sci. Res.* 8 (2): 493–498.

Zhang, X., da Silva, J.A.T., Niu, M. et al. (2017). Physiological and transcriptomic analyses reveal a response mechanism to cold stress in *Santalum album* L. leaves. *Sci. Rep.* 7 (1): 1–18.

14

Arbuscular Mycorrhizal Fungi (AMF) and Climate-Smart Agriculture

Prospects and Challenges

Sharma Deepika[1,2], Vikrant Goswami[2], and David Kothamasi[2,3]

[1] Department of Botany, Zakir Husain Delhi College, University of Delhi, New Delhi, India
[2] Laboratory of Soil Biology and Microbial Ecology, Department of Environmental studies, University of Delhi, New Delhi, India
[3] Strathclyde Centre for Environmental Law and Governance, University of Strathclyde, Glasgow, United Kingdom

14.1 Introduction

Food security is a serious concern across the world. This is especially true in developing countries (Nelson et al. 2009; Noya et al. 2018). Climate change, rising global population, and modern agricultural practices are the greatest challenges undermining food security. Modern agricultural practices are high-input practices that involve usage of high-yield crops, intensive application of chemical fertilizers, and irrigation dependent on groundwater extraction (Sarkar 2015). This affects global environmental stability. Land degradation, soil erosion, loss of biodiversity, and dwindling sources of freshwater for irrigation impact agricultural productivity. They form hurdles to sustainable food production and food security (Tilman et al. 2011; Noya et al. 2018).

Global human population is projected to reach 9.7 billion by the year 2050 (UN 2015). It is estimated that approximately 36 million people (12% of the current global population) lack access to food and suffer from hunger (Magdoff and Tokar 2009). Global food production must increase by more than 50% from 2012 levels to the meet the food requirements of 2050. It is estimated that by 2050, nearly two-thirds of the global population will be urban. In comparison, only a little over half of the current population is urban (WHO 2013). Such demographic changes over the next few decades will further exacerbate food insecurity. The transition to urbanization and higher average incomes will have serious implications on the carbon footprint (Godfray and Garnett 2014). Increase in average wealth will lead to a "nutritional shift" from staple tubers and grains to energy-intensive foods like meat and dairy (Drewnowski et al. 1997; Popkin 1998). Rising inequality in food entitlements affects a large population in developing countries and is a major obstacle to achieving food security.

Global warming affects food security through its impact on food availability, accessibility, utilization, and food system stability. Anthropogenic warming is expected to negatively affect food production through increase in incidences of extreme weather events, gradual changes in temperature and rainfall, and geographical redistribution of pests and diseases (Gornall et al. 2010). Major grains and legume crops were earlier predicted to benefit from moderate increase in temperatures due to CO_2 fertilization effect (enhanced photosynthetic rates and productivity due to increased CO_2 concentration) in the temperate zones but decline in the tropics (Long et al. 2005). However, benefits of CO_2 fertilization on crop productivity are offset by interactions between climatic factors (CO_2, O_3, temperature, and soil moisture), nutrient scarcity, and pollution (Gregory et al. 2005). The anticipated detrimental effects of global warming will be further aggravated by socioeconomic development, changing demographics, and nutritional transition of the growing population.

Green revolution witnessed a dramatic increase in per capita production of major cereal crops, alleviating poverty and malnutrition throughout the world. The per capita production of cereals almost tripled during green revolution despite just 30% increase in cultivable land area (Wik et al. 2008). This was achieved chiefly with the use of high yielding disease-resistant varieties and agrochemicals combined with better irrigation practices. The world achieved food security and belied Malthusian catastrophe that had predicted food deficits due to overpopulation. However, water-intensive crops and extensive use of agrochemicals under green revolution have caused serious environmental damage. Agricultural yields witnessed a sustained increase for ~30 years after the green revolution (Jain 2018). Since then, yields of wheat and rice have

Global Climate Change and Plant Stress Management, First Edition. Edited by Mohammad Wahid Ansari, Anil Kumar Singh, and Narendra Tuteja.

dropped. This is a major fallout of monoculture cropping practices and decrease in genetic potential of crops (Handral et al. 2017; John and Babu 2021). Lands becoming degraded and infertile due to chemical-intensive farming practices may cause a further drop in agricultural yield (Godfray and Garnett 2014). Monoculture cropping practices and heavy reliance on crops pollinated by insects or other animals pose a serious threat to food security in the event of crop failures due to a catastrophic pest or disease outbreak. Green revolution has also diminished the diversity of crops globally. For instance, India alone has lost ~100,000 indigenous rice varieties due to the green revolution. This unintended damage is one of the reasons leading to increase in the ecological footprint of agriculture (Prasad 2016).

Agriculture is both a contributor and victim of the consequences of global warming. It accounts for 38% of the planet's land surface (5 billion ha) and contributes 24–35% of global greenhouse gas emissions (Foley et al. 2011; Mcdougall et al. 2020). Agriculture generates around 50% of all anthropogenic CH_4 and close to three quarters of total human-caused N_2O emissions globally (Lynch et al. 2021). An integrated approach is required to address the interlinked effects of agriculture and global warming on food security. In view of the current scenario, it is necessary to adopt climate-smart sustainable agricultural practices to mitigate and strengthen resilience to the adverse impacts of global warming for achieving food security.

14.2 What Is Climate-Smart Agriculture?

Climate-smart agriculture (CSA) integrates the connected challenges of food security and climate change. CSA, like sustainable agriculture, is based on the principles of increased productivity and sustainability. Sustainable agriculture combines long-term human stewardship of natural resources with short-term economic gains from food production. CSA differs by integrating climate change mitigation and adaptation in planning and development of sustainable agricultural systems. The three major objectives of CSA are sustainable increase in agricultural productivity and farmer incomes, adaptation and building resilience to climate change; and reducing agricultural greenhouse gas (GHG) emissions (Figure 14.1) (FAO 2013).

CSA like sustainable agriculture can improve human health, which has deteriorated due to the consumption of synthetic materials present in the chemical fertilizers and pesticides via the food chain. Crops grown sustainably are cleaner and more nutritious food sources compared to those grown in chemical treated agricultural fields (Crinnion 2010). This practice can significantly reduce the release of agrochemicals contaminating the soil, subsoil, and groundwater. CSA can further make the environment beneficial for humans by improving soil health, reducing soil degradation and erosion, and water conservation. The healthier soils in turn promote the growth of beneficial soil microbes such as nitrogen-fixing bacteria and phosphate solubilizers, which can be exploited for their plant growth promoting potential as biofertilizers.

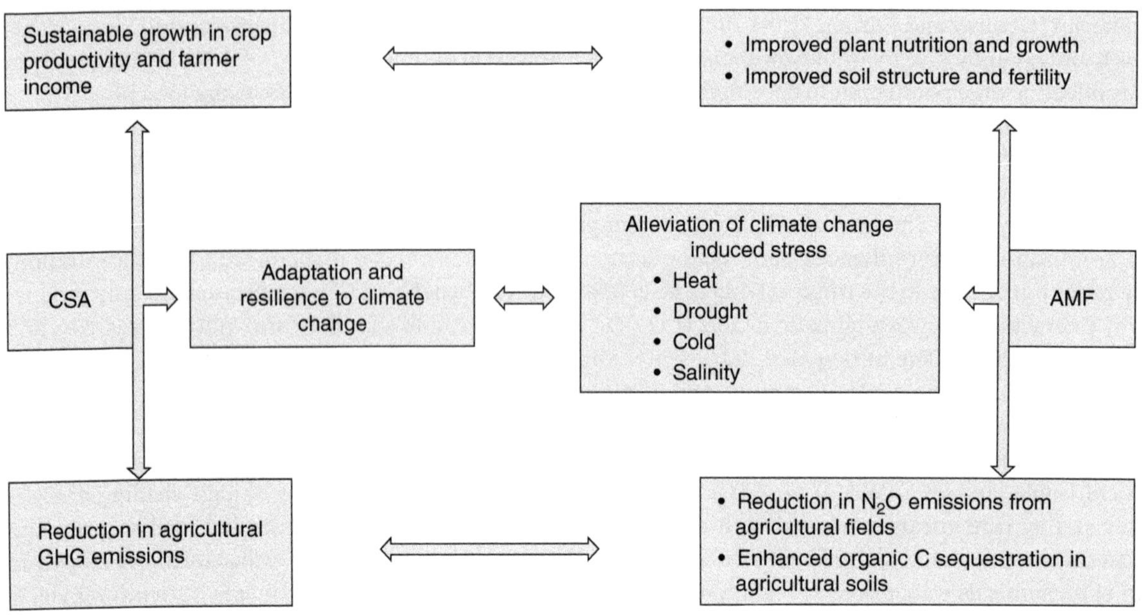

Figure 14.1 Arbuscular mycorrhizal fungi (AMF) are an important constituent of climate smart agriculture (CSA). They reduce agricultural greenhouse gas emissions and contribute to sustainable crop productivity and famer income by improving plant growth and alleviating climate induced stresses on the crop.

Microbial biofertilizers are an integral component of CSA. They promote plant growth and make them resilient to biotic and abiotic stresses (Martínez-Garcia et al. 2017; Fasusi et al. 2021). The need to increase sustainable food production shifts focus to the role of soil biodiversity in general (Bender et al. 2016) and arbuscular mycorrhizal fungi (AMF) in particular (Rillig et al. 2016; Thirkell et al. 2017). Among the several microbes employed as biofertilizers, AMF are especially important due to their ubiquitous presence in agricultural soils and ability to form symbioses with most crop plants.

14.3 AMF as a Tool to Practice Climate-Smart Agriculture

AMF are obligate root symbionts of the phylum Glomeromycota (Schüßler et al. 2001). They form mutualistic associations with the roots of ~80% of terrestrial plants including several agriculturally important crop species (Smith and Read 2008). Members of family Brassicaceae and Chenopodiaceae do not form AMF associations (Lambers and Teste 2013). AMF hyphae in plant roots form vesicles and arbuscules inside the root cortical region. Arbuscules are a characteristic feature of AMF symbioses. They are the site of exchange for inorganic nutrients and water absorbed by the fungus from the soil and plant photosynthates. Plants may allocate up to 20% of their photosynthates for maintaining the AMF symbiosis (Smith and Read 2008). A major benefit of enhanced uptake of immobile nutrients like phosphate and zinc by AMF is the balancing of carbon costs incurred by the plant to sustain the fungus (Parniske 2008). Improved tolerance to biotic and abiotic stresses is an additional benefit conferred by AMF to plant hosts.

AMF form an interface between plant roots and surrounding living and non-living components of the soil (Lenoir et al. 2016; Rasmussen et al. 2018) and determine diversity and productivity of ecosystems. AMF play an important role in nutrient cycling and reduce leaching losses in agroecosystems (Bender and Van der Heijden 2015; Cavagnaro et al. 2015; Martínez-Garcia et al. 2017). AMF hyphae in the soil produce Glomalin-related soil proteins (GRSPs), which facilitate soil carbon sequestration mitigating climate change.

The positive effects of AMF on plant growth and ecosystems and their significant role in soil carbon sequestration make them ideal biological inputs for CSA. Biofertilization with AMF has the potential to improve the nutritional status of plants and their ability to adapt to the changing climate (Figure 14.1). The low host and niche specificity of AMF is an added advantage for their use in sustainable agricultural systems under different environmental conditions.

14.3.1 AMF in Increasing Productivity of Agricultural Systems

AMF, a key component of natural soils, provide a multitude of ecological services (e.g. as mentioned earlier). The multiple benefits of AMF inoculation on soil health and crop productivity make them an ideal choice to be used as a bioinoculants in CSA. Biofertilization with AMF is estimated to reduce the consumption of inorganic fertilizers in agriculture by up to 50% depending on the plant species and environmental stresses (Begum et al. 2019a).

14.3.1.1 Plant Nutrition and Growth

Plant growth and productivity are dependent on nutrient availability in the rhizosphere. Phytoavailability of nutrients is affected by physical and chemical properties of soil, climate, and plant species. AMF form mutualistic associations with most major agricultural and horticultural crops. Inoculation with AMF improves the uptake of various macro- and micronutrients in exchange for plant photosynthates leading to improved plant nutrition and growth. The AMF-mediated plant growth promotion is due to improved absorption and translocation of mineral nutrients from soil through the subterranean network of AMF mycelium known as the common mycorrhizal network (CMN). AMF hyphae can explore more soil volumes as compared to plant roots alone due to their large surface areas as they are 100× thinner than plant root hairs, leading to better absorption of nutrients and water. The benefits of AMF fertilization have been observed in several agriculturally and horticulturally important crops even under stressed environments (Table 14.1).

A major contribution of AMF symbiosis is improved uptake of P. The availability of P is limited in various soil systems due to its low solubility and mobility (Zhu et al. 2018; Wang and Lambers 2020). There are two pathways (Figure 14.2) for P acquisition from the soil: (i) a direct pathway by root epidermal cells and root hairs (Gordon-Weeks et al. 2003); and (ii) mycorrhizal pathway by AMF symbiosis (Smith et al. 2003; Vance et al. 2003; Chiu and Paszkowski 2019; Ferrol et al. 2019). In the direct pathway, P is absorbed from the rhizosphere through P translocators located in the plant cell of root apex and in root hairs (Smith et al. 2003; Bayle et al. 2011; Chiu and Paszkowski 2019). In the mycorrhizal pathway, AMF hyphae absorb P from areas inaccessible to host plant roots in the soil and solubilize it by releasing organic acids (Sharif et al. 2011).

Table 14.1 Responses of AMF inoculation in crop plants exposed to different abiotic stresses.

Abiotic stress	Host plant	AMF species used	Inference	References
Drought	*Lycopersicon esculentum*	*Rhizophagus clarus (previously Glomus clarum)*	Improved leaf area, dry mass, stomatal conductance, photosynthetic activity, and root hydraulic conductivity	Dell'Amico et al. (2002)
	Zea mays, Solanum lycopersicum	*R. irregularis*	Enhanced apoplastic water flow	Bárzana et al. (2012)
	Hordeum vulgare	*R. intraradices (previously G. intraradices)*	Increased root volume, P content, and activity of phosphatase enzyme	Bayani et al. (2015)
	Phaseolus vulgaris	Mixed inoculum consisting of *Entrophospora etunicata* (previously *G. etunicatum), R. intraradices* and *Funneliformis monosporus* (previously *G. monosporum)*	Improved vegetative growth characteristics, yield, total proteins and total sugars, N, P, K, Mg, Ca, Fe, Zn, Mn and Cu concentrations, and reduced proline concentration	Salim and Abou El-Yazied (2015)
	Solanum lycopersicum	Mixed inoculum consisting of *F. mosseae* (previously *G. mosseae), E. etunicata, E. claroidea (previously G. claroideum), Rhizoglomus microaggregatum (previously G. microaggregatum), F. geosporus (previously G. geosporum)* and *R. irregularis*	Enhanced fruit yield, water use efficiency, biomass and leaf water potential	Bakr et al. (2016)
	Solanum lycopersicum	*F. mosseae, R. irregularis*	Increased plant height and biomass, intrinsic water use efficiency (iWUE) index, stomatal density, capacity to absorb CO_2 and proline concentrations, and reduced hydrogen peroxide, leaf and root ABA contents	Chitarra et al. (2016)
	Solanum lycopersicum, Lactuca sativa	*R. irregularis*	Improved rate of growth and more efficient photosystem II, increased production of strigolactone for symbiosis promotion	Ruiz-Lozano et al. (2016)
	Cucumis melo	Mixed inoculum consisting of *R. intraradices* and *Glomus* spp.	Improved physiological parameters and photosynthesis, increased chlorophyll content	Cakmakci et al. (2017)
	Cicer arietinum	Mixed inoculum consisting of *Claroideoglomus etunicatum, R. irregularis* and *F. mosseae*	Increased nitrogen fixation, leghemoglobin content and activity of nitrate reductase enzyme	Hashem et al. (2019a)
	Triticum aestivum	Mixed inoculum consisting of *R. intraradices, F. mosseae* and *F. geosporum*	Alleviation of damage to PSI and PSII	Mathur et al. (2019)
	Zea mays	*G. versiforme*	Reduction in oxidative damage, improve growth, photosynthesis, chlorophyll content, mineral uptake and assimilation	Begum et al. (2019b)
	Glycine max	*R. clarus*	Maintain water status of plant, improve yield and increase in concentration of photosynthetic pigment	Oliveira et al. (2022)
Cold	*Cucumis sativus*	*F. mosseae*	Improved dry weight and fresh weight of seedlings, enhanced secondary metabolites content and enzymatic activity, reduced hydrogen peroxide content	Chen et al. (2013)
	Solanum lycopersicum	*F. mosseae*	Reduced cell membrane injuries, more efficient photosystem II, improved growth of seedlings	Caradonia et al. (2019)
	Hordeum vulgare	Mixed inoculum consisting of *G. versiforme* and *R. irregularis*	Increased survival rate of plant, alleviation of low-temperature stress	Hajiboland et al. (2019)

Table 14.1 (Continued)

Abiotic stress	Host plant	AMF species used	Inference	References
	Cucumis sativus	*R. irregularis*	Improved efficiency of photosynthesis and enhanced carbon sink strength, sugar content of leaves and non-photochemical quenching	Ma et al. (2019)
	Solanum melongena	Mixed inoculum consisting of *F. mosseae, C. etunicatum, R. irregularis,* and *Diversispora versiformis*	Improved photochemical reactions, activation of antioxidant defense system, accumulation of protecting molecules and reducing membrane damage	Pasbani et al. (2020)
	Citrullus lanatus	*R. intraradices*	Restoration of photosynthetic efficiency, decreased accumulation of H_2O_2 and malondialdehyde (MDA) and amelioration of electrolyte leakage	Shirani Bidabadi and Mehralian (2020)
	Hordeum vulgare	Mixed inoculum consisting of *D. versiformis* and *R. irregularis*	Activation of antioxidant defense, phenolics accumulation, reduction in hydrogen peroxide content and accumulation of proline	Joudmand et al. (2021)
Heat	*Zea mays*	*E. etunicata*	Increased net photosynthetic rate, stomatal conductance and transpiration rate in leaves; improved fluorescence, maximum quantum efficiency of PSII photochemistry, potential photochemical efficiency, chlorophyll content, water use efficiency, water holding capacity and relative water content	Zhu et al. (2011)
	Cucumis sativus	*F. mosseae*	Increased root and shoot fresh weight, enhanced antioxidant activity and phenol content	Haghighi et al. (2015)
	Triticum aestivum	Mixed inoculum consisting of *R. irregularis, F mosseae, F. geosporum* and *C. claroideum*	Increased nutrient content in plant and enhanced yield	Cabral et al. (2016)
	Cucumis sativus	*R. irregularis*	Increased photosynthetic efficiency of plant seedlings, protecting photosynthetic apparatus of plant	Ma et al. (2019)
	Zea mays	*Funneliformis* sp.	Regulated photosystem (PS) II heterogeneity	Mathur and Jajoo (2020)
	Solanum lycopersicum, Capsicum annum and *Cucumis sativus*	*R. irregularis*	Increased vigour, productivity and fruit quality	Reva et al. (2021)
Salinity	*Gossypium arboreum*	Mixed inoculum consisting of *F. mosseae* (isolate GM1) and *F. mosseae* (isolate GM2)	Increased biomass and phosphorus concentrations	Tian et al. (2004)
	Triticum aestivum	Mixed inoculum consisting of *E. etunicata, F. mosseae* and *R. intraradices*	Increased plant growth, nutrient uptake and grain yield, and reduced concentrations of Na^+ and Cl^-	Daei et al. (2009)
	Lycopersicon esculentum	Mixed inoculum consisting of *R. intraradices, F. mosseae, R. aggregatus (previously G. aggregatum), R. clarus, F. monosporus, Septoglomus deserticola (previously G. deserticola), Paraglomus brasilianum (previously G. brasilianum), E. etunicata* and *Gigaspora margarita*	Increased plant growth, fresh stem weight, enhanced chlorophyll a and b content	Basak et al. (2011)

(*Continued*)

Table 14.1 (Continued)

Abiotic stress	Host plant	AMF species used	Inference	References
	Solanum melogena	*S. deserticola*	Enhanced growth and aggressiveness colonization in roots	Mohammad and Mittra (2013)
	Oryza sativa	*C. etunicatum*	Increased shoot and root biomass, stomatal conductance, soluble sugars, free α-amino acids, and Na^+ and K^+ uptake	Porcel et al. (2015)
	Cucumis sativus	Mixed inoculum consisting of *E. etunicata, R. intraradices* and *F. mosseae*	Enhancement of pigment synthesis, antioxidant enzyme activity, lower lipid peroxidation and electrolyte leakage, accumulation of K, Ca, Zn, Fe, Mg, Mn, Cu and reduction of Na^+ ions	Hashem et al. (2018)
	Glycine max	Mixed inoculum consisting of *F. mosseae, R. intraradices* and *C. etunicatum*	Protected plant membrane damage, reduced the production of H_2O_2 and lipid peroxidation, improved plant growth and symbiotic performance by stimulating the endogenous level of auxins that contribute to improved root systems and nutrient acquisition	Hashem et al. (2019b)
	Lycopersicon esculentum	*R. fasciculatus (previously G. fasciculatum)*	Improved P, phosphatase enzyme activity and reduced pH	Sivakumar et al. (2019)
	Lactuca sativa	*C. claroideum*	Reduction of oxidative damage and enhanced growth, development and yield	Santander et al. (2019)
	Pisum sativum	Single species (*Rh. intraradices*), Formulated species (*F. mosseae* and *R. intraradices*), and multispecies (*Rh. fasciculatum* and *Gi.* sp.)	Enhanced nutrient uptake, increased biomass, synthesis of chlorophyll, compatible osmolytes accumulation, higher growth attributes and yield	Parihar et al. (2020)
Drought + Heat (combined stress)	*Solanum lycopersicum*	*S. deserticola* and *S. constrictum*	Increased leaf water potential, stomatal conductance and enhanced biomass production	Duc et al. (2018)

The extraradical mycelium has H^+/Pi or Na^+/Pi symporters, which mediate P absorption (Chiu and Paszkowski 2019). The extraradical AMF mycelium translocates it to the intraradical mycelium from where it is exchanged with the plant for carbohydrates (Smith et al. 2003).

Nitrogen is an essential macronutrient required for plant growth as well as yield (Daei et al. 2009). AMF associations have been documented to promote N acquisition from both organic and inorganic sources in the soils. AMF also promote biological nitrogen fixation in the rhizosphere through their stimulatory effects on the nitrate reductase activity (Miransari 2011; Devi et al. 2021) and promoting growth, nodulation, and nitrogen fixation in the legume-Rhizobium symbiosis (Devi et al. 2021). It is estimated that ~50% of the total nitrogen requirement of plants is supplied by AMF (McFarland et al. 2010). In addition, AMF also boosts the phytoavailability of immobile micronutrients such as zinc and copper.

Nutrient leaching is a significant concern in agriculture with significant economic as well as environmental consequences. AMF can reduce nutrient leaching from soil systems (Bender et al. 2015; Cavagnaro et al. 2015; Martínez-Garcia et al. 2017; Goswami et al. 2023), particularly P and N. It increases nutrient uptake, improves soil structure, and helps the other microbial communities improve nutrient immobilization (Cavagnaro et al. 2015). N lost through leaching and gaseous emissions from agricultural fields, leads to nutrient loss, contamination of groundwater sources, and soil acidification. N is lost from agricultural lands in its highly mobile form, NO_3^-, which makes it vulnerable to leaching. AMF absorbing N in the form of NH_4^+ reduces the risk of nitrification and hence reduces the mobility of N. It can also intercept N that has

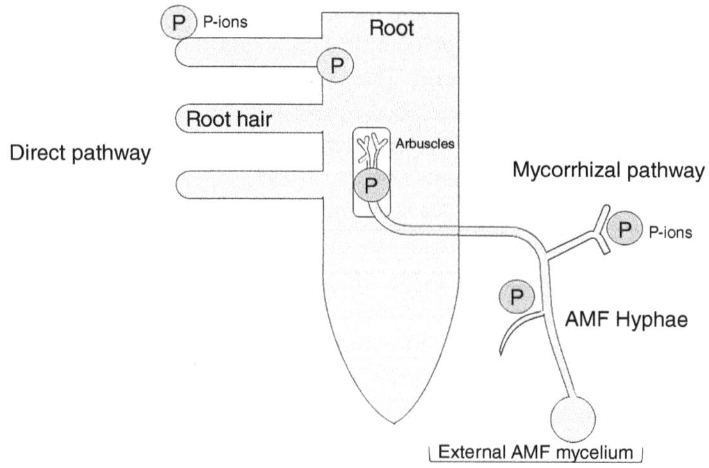

Figure 14.2 Direct and mycorrhizal pathways of P-uptake in plants. Root hairs and root epidermal cells absorb P from the rhizosphere in the direct pathway. In the mycorrhizal pathway, arbuscular mycorrhizal fungal (AMF) hyphae absorb P from areas inaccessible to the roots. *Source:* Figure modified from Smith and Smith (2011).

migrated down to the subsoil and transport it to the host plant (Kautz et al. 2013). P being immobile in soils is believed to be resistant to leaching. However, excessive fertilization may lead to significant increase in P leaching in the soil ecosystem (Djodjic et al. 2004; Schoumans 2015). AMF have been documented to reduce the need of extensive fertilization due to enhanced P uptake (Smith and Smith 2011). AMF improve water retention capacity, reduce leaching losses, and intercept P that has migrated down to the subsoil, and transport it to the host plant (Sosa-Hernández et al. 2019). It is estimated that AMF reduce the leaching of N and P by 40 and 50%, respectively (Martínez-Garcia et al. 2017).

Crop inoculation with AMF can promote sustainable nutrient cycling due to their potential in reducing nutrient losses through nutrient leaching in soils leading to improved fertilizer use efficiency and reduced threat of environmental degradation.

The natural reservoirs of inorganic fertilizers (especially P) used in modern agriculture are gradually declining globally. The last decade witnessed an increase in the price of inorganic fertilizers (Cordell et al. 2009). Organic fertilization with AMF can be used as a sustainable alternative to expensive inorganic fertilizer in CSA.

14.3.1.2 Improved Soil Structure and Fertility

Soil erosion, reduced soil fertility, changes in organic carbon transformation, soil compaction, and nutrient leaching are climate-induced threats to food security and environmental sustainability (Hamidov et al. 2018). Soil aggregates and pore spaces, which are a dimensional arrangement of organo-mineral complexes, determine soil fertility (Rillig and Mummey 2006; Carneiro et al. 2015). By influencing the physical, chemical, and biological properties of soil, AMF perform a crucial role in maintaining the structure and fertility of soils.

AMF have an important role in the formation and maintenance of soil aggregates. The extraradical AMF mycelium secretes glomalin, a glycoprotein which through its cementing action binds soil particles together. This improves stability of soil aggregates. Glomalin is a polysaccharide. It sequestrates carbon, facilitates organic matter formation, adheres to the microaggregates, and facilitates the formation of larger macroaggregates, which are essential factors in soil formation (Miller and Jastrow 1990; Bossuyt et al. 2001; Hoorman et al. 2011). Glomalin and GRSPs improve soil quality and act as carbon sink (Rillig et al. 2001). Soil aggregate stability improves crop productivity through better retention of water and minerals, reduced surface runoff, lower soil erosion, and less nutrient leaching. AMF hyphal networks contribute resistance to soil erosion by wind and water by strengthening the root system and growth of the host plant (Gutjahr and Paszkowski 2013). Soils of dry lands particularly in arid regions often have low fertility and higher vulnerability to erosion by wind and rain (Chen et al. 2018).

AMF are dominant microbial components of the three major soil biogeochemical cycles – P, N, and C. They improve soil chemical properties through their role in the uptake and translocation of limiting and immobile mineral nutrients. Inorganic and organic P are the two available forms of P in soils. Most P in soils is accumulated from fertilizers called "legacy P." It is estimated that legacy P is sufficient for sustaining crop demand worldwide for ~100 years (Khan and Zaidi 2007). Plants absorb inorganic P in the form of orthophosphates. Inorganic orthophosphate is adsorbed to soil Fe, Ca, and Al cations, rendering it unavailable to plants. This affects the supply of P in the soil, and <1% of the legacy P is available to plants. Low availability and mobility of P in the rhizosphere creates a phosphate depletion zone around plant roots. AMF hyphae in the soil through chemical reactions and biological interaction with phosphate solubilizing bacteria (PSB) convert P into phytoavailable forms. The extraradical mycelium of AMF absorbs the soluble P for plants. The PSBs produce phosphatases and organic acids that mineralize the insoluble P into soluble form of P that can be easily assimilated by plants. AMF through their metabolic activities also produce phosphatases and acids, which solubilize the insoluble forms of P and make it accessible to plants (Kalayu 2019).

AMF mycelium absorbs N in the form of NH_4^+, NO_3^-, and amino acids through transporters in AMF hyphae (Jansa et al. 2019). They also play an important role in the decomposition and mineralization of organic matter (Lambers et al. 2008).

AMF are an integral component of the global C cycle. AMF colonized roots create a sink demand for C, which leads to C translocation in the soil. The extraradical AMF hyphae, which represent 20–80% of total microbial biomass in soil, contain 15% of soil organic carbon (SOC) (Fall et al. 2022).

AMF interact with various soil microbes in the rhizosphere. This interaction plays an important role in the formation and stability of soil aggregates and fertility. AMF regulate the composition and activity of other microbes in the rhizosphere through their exudates and quantitative and qualitative modification of rhizodeposition products. Positive interactions of AMF with soil microbes result in better nutrient uptake, tolerance to abiotic stresses, root pathogens, and soil fertility.

14.3.2 AMF-Induced Resilience in Crops to Climate Change

Global climate is predicted to undergo significant changes in temperature and precipitation patterns throughout the world (IPCC 2018). During the twentieth century, 76% of the global land area witnessed a significant increase in temperature. This is expected to increase by a further 2.4–4 °C by 2100 (Gonzalez et al. 2010). Between 1950 and 1980, prolonged heat waves affected <1% of the global land area. However, this has increased from 10% to 22.21% in 2010 (Hansen et al. 2012). This rise is expected to continue into the twenty-first century (Wu et al. 2020). Climate change threatens food security through its effects on crop production. Plants encounter a variety of biotic, abiotic, and sometimes even a combination of different stresses during their life cycle that negatively affect their growth and productivity. Anthropogenic warming has further exacerbated the vulnerability of crops to abiotic stresses.

Drought, heat, and salinity are the major abiotic stressors affecting crop production. Extreme temperatures, drought, deficiency (plant unable to access sufficient essential nutrients for its growth), and toxicity (decrease in plant growth due to excess essential nutrients) of plant nutrients cause 51–82% loss in global crop yield annually (Oshunsanya et al. 2019). Drought alone has caused 21% and 40% reduction in yields of wheat and maize, respectively, between 1980 and 2015 (Daryanto et al. 2016). Among the various combined stresses, the probability of plants encountering drought and heat stress simultaneously is significantly higher in the current scenario (Prasad et al. 2011; Jedmowski et al. 2013). In arid and semi-arid regions of the world, salinity and heat stress combination are highly probable (Pandey et al. 2017). Similarly, plants may encounter more than one biotic stress simultaneously or sequentially due to co-infection with fungal, bacterial, or viral pathogens. A plant may even be affected by a biotic and abiotic stressor simultaneously.

Inoculation with AMF can be used as a tool to alleviate the deleterious effects of climate change on crop production. They have been shown to protect plants against several biotic and abiotic stresses. Plants inoculated with AMF have higher nutrient uptake and productivity even in stressed environments (Table 14.1). Due to their potential to improve crop productivity and alleviate major climate-induced stresses, AMF make crops resilient to climate change and are good candidates as bioprotectors in CSA (Figure 14.1).

14.3.2.1 AMF and Salinity Stress

Physical, chemical, biological, and biochemical degradation of soils are the predicted hydrological consequences of climate change (Varallyay 2010). Around 120 countries, which represent 20% of the total global irrigated land area, are affected by soil salinization (Pardeep and Sharma 2020). Climate-induced salinization is a major threat to agriculture in arid, semi-arid and coastal areas of the world (Corwin 2021). Changes in hydrological balance leading to accumulation of soluble salts, sea salt intrusion, and wind-borne deposition of salt are likely effects of climate change (IPCC 1995). Climate-induced alteration of precipitation patterns and increased rates of evapotranspiration will further contribute to soil salinization due to reduced soil leaching efficiency especially in the arid and semiarid regions of the world (Hassani et al. 2021). Higher demand for irrigation water in a warming world along with poor drainage and irrigation practices may contribute to secondary salinization (Yeo 1998). In coastal areas, rising sea levels and unsustainable use of aquifers increase the risk of soil salinization due to sea water ingress (Dasgupta et al. 2015; Karmakar et al. 2016). For instance, sea water intrusion due to excessive extraction of groundwater in coastal soils of Bangladesh has triggered a 32% increase in saline soils between 1990 and 2015 (Rahman et al. 2018).

AMF colonization has been shown to confer tolerance to salinity stress in plants through various mechanisms (Table 14.1). AMF-induced amelioration of salinity stress includes alteration of the root architecture leading to improved exploration of non-saline areas by roots for absorption of water and minerals (Evelin et al. 2019). High salt concentrations alter the ratios of Na^+:Ca^{2+}, Na^+:K^+, and Ca^{2+}:Mg^{2+} respectively, in soil solution (Abdel-Fattah and Asrar 2012). This causes an imbalance in plant ionic composition, which eventually affects its physiology (Hasegawa et al. 2000; Munns et al. 2006). Biofertilization with AMF has been shown to improve nutrient uptake and maintain ionic homeostasis in salt-stressed plants. Moreover, saline soils lower the turgor pressure of cells leading to cellular dehydration (Evelin et al. 2019). Plants respond to cellular dehydration through osmoregulation by accumulating water-soluble and nontoxic solutes called osmolytes at high concentrations (Chen and Murata 2011). Osmolytes are osmoprotectants that protect plant cells by osmotic adjustments, maintenance of membrane integrity, reactive oxygen species (ROS) quenching, and stabilizing cellular proteins and enzymes (Ashraf and Foolad 2007). AMF-inoculated plants have been observed to induce higher accumulation of osmolytes such as trehalose (Garg and Pandey 2016), organic acids (Sheng et al. 2011), polyamines (Evelin et al. 2013), sugars (Garg and Bharti 2018), and proline (Evelin et al. 2019) to protect plants cells from cellular dehydration due to salt stress. However, studies on AMF-induced proline accumulation in stressed plants show conflicting results (Evelin et al. 2019). Salt stress induces oxidative trauma in plants. This disrupts the ROS equilibrium and affects cellular functions (Foyer and Noctor 2005; Gill and Tuteja 2010). Plant response to shield plant cells from cellular oxidative damage due to ROS is through activation of antioxidative pathways (Porcel et al. 2012). Mycorrhization has been shown to augment the ROS scavenging system of plants by stimulating activities and production of antioxidant enzymes and molecules (Evelin et al. 2019). Saline rhizospheres make water physiologically unavailable to plants by creating physiological aridity in soils (Kothamasi et al. 2006; Deepika and Kothamasi 2021). AMF improve water status of stressed plants through improved water uptake (Schnepf et al. 2011), adjustment of osmotic potential by solute accumulation (Graham and Syvertsen 1984), and regulation of aquaporin genes in roots (Evelin et al. 2019). AMF-induced amelioration of salinity stress also includes protection of photosynthetic apparatus, improved photosynthetic efficiency, and regulation of plant growth hormones (Evelin et al. 2019).

14.3.2.2 AMF and Drought Stress

Drought is amongst the major abiotic stresses diminishing crop productivity globally. Increasing temperatures and erratic precipitation patterns caused by climate change are predicted to amplify the frequency and intensity of droughts in different parts of the world, especially in arid and semiarid regions (Kling et al. 2005). Frequent droughts during the growing season along with loss of water due to greater soil and surface water evaporation and transpiration caused by higher temperatures are a significant stress for plants (Wang et al. 2011). This is predicted to nullify the benefit of enhanced photosynthesis achieved by plants because of CO_2 fertilization effect due to stomatal closure by plants to conserve water in response to high temperatures (Ruiz-Vera et al. 2013). A recent study on the vulnerability of major agricultural crops (wheat, rice, maize, and soy) to drought revealed that approximately three-fourths of the global harvested area (454 million hectares) experienced drought-induced production losses between 1983 and 2009 (Kim et al. 2019).

Water deficit negatively affects the various physiological parameters of plants including photosynthesis, structure of enzymes, and uptake and transport of nutrients, which triggers a state of hormonal and nutritional imbalance within the plant (Selmar and Kleinwächter 2013; Ruiz-lozano et al. 2016; Xie et al. 2018). Drought may also lead to plant growth inhibition and development due to turgor loss because of osmotic stress caused by water deficiency (Selmar and Kleinwächter 2013). Drought-induced production of ROS may cause oxidative damage to carbohydrates, protein, and lipid metabolism, resulting in cell death and membrane damage in plant tissues (Gill and Tuteja 2010; Atkinson and Urwin 2012).

AMF mitigate the effects of water deficit through a combination of mechanisms operating at the physical, nutritional, physiological, and cellular levels (Auge 2001; Ruiz-Lozano 2003). AMF mitigate drought stress through various mechanisms including improvement in the plant water status and water flow within the plant by increased water absorption via the external AMF hyphae and regulation of aquaporin expression in the host and the AMF symbiont (Auge et al. 2007; Barzana et al. 2014). AMF reduce the oxidative damage caused by ROS under water deficit conditions by altering the osmotic adjustment ability and leaf water potential of host plants (Abbaspour et al. 2012; Ruiz-Lozano 2003). In addition to improved water and nutrient uptake, AMF-mediated amelioration of drought also includes improved stomatal regulation, increased antioxidant activity (Goicoechea et al. 2005) and modifications in the content on plant hormones such as Jasmonic acid, Strigolactones, and Abscissic acid (ABA) (Bahadur et al. 2019).

14.3.2.3 AMF and Heat Stress

Climate change induced rise in global temperatures is predicted to have a negative effect on agriculture due to the deleterious effect of high temperatures on plant growth and development. Global temperatures are predicted to increase by a further 1.5 °C between 2030 and 2052 if warming continues at the current rate (IPCC 2018). A 3–4 °C rise in temperature is predicted to cause 15–35% and 25–35% decrease in crop yields in Africa, Asia and Middle East, respectively (Ortiz et al. 2008). Rising temperature poses a significant threat to food security as temperature affects all stages of plant development from seed germination to reproduction. Reproductive organs of plants and the process of seed set are highly vulnerable to high temperatures (Pareek et al. 2020), which may lead to catastrophic losses in crop production globally. For instance, an estimated decrease of 4–6% has been predicted on global wheat production for every 1 °C rise in global temperature (Asseng et al. 2015). This will cause a sharp climate change induced decline in calorie availability by 2050 and an estimated 20% increase in malnutrition in children (Bita and Gerats 2013).

Plant susceptibility to heat stress varies with the stage of plant development and the species and genotype of the plant (Sakata and Higashitani 2008; Bita and Gerats 2013). High temperatures affect the molecular, cellular, and physiological parameters of plants. In general, heat stress reduces plant growth through its effect on net shoot assimilation rates and total dry weight of plants (Wahid 2007). Scorching of leaves and stems, inhibition of shoot and root growth, leaf abscission and senescence, and fruit damage are the various physiological injuries observed in plants because of heat stress (Vollenweider and Gunthardt-Goerg 2005). At the cellular level, elevated temperature causes changes in the organization of organelles and the cytoskeleton and membrane functions (Weis and Berry 1988). A decline in normal protein synthesis, increased production of phytohormones such as ABA, antioxidants, and other protective molecules (Maestri et al. 2002), and higher rates of transcription and translation of heat shock proteins (HSPs) (Bray et al. 2000) are some of the major metabolic responses of heat stress in plants. The major cause of productivity decline in plants exposed to heat stress is because of changes in respiration and photosynthesis as higher temperatures have been shown to cause structural changes in chloroplast protein complexes and enzyme activity (Ahmad et al. 2010). The activities of ribulose 1,5-bisphosphate carboxylase (Rubisco) and Rubisco activase are especially highly vulnerable to temperature stress and get inhibited even at a low level of temperature stress (Maestri et al. 2002; Morales et al. 2003). Modification in the activities of enzymes involved in carbon metabolism, accumulation of primary metabolites such as proline, glycine betaine, soluble sugars (Wahid 2007), and osmolytes such as sugar alcohols (polyols), tertiary and quaternary ammonium compounds (Sairam and Tyagi 2004) are some of the major molecular responses to heat stress in plants. A change in the levels of phytohormones, such as ABA, ethylene, salicylic acid, and gibberellic acid, has been observed in plants exposed to elevated temperatures (Bita and Gerats 2013).

AMF ameliorate heat stress through improved plant growth and biomass production, thereby improving the fitness of plants. AMF inoculation has been shown to improve water status, activity, and hydraulic conductivity of roots, stomatal conductance and gas exchange capacity in plants (Zhu et al. 2017). AMF enhance the uptake of water and stomatal conductance of plants by regulating the expression of plant aquaporin genes (Aroca et al. 2007; Liu et al. 2014). AMF improves the photosynthetic efficiency of heat-stressed plants through ameliorating chloroplast damage in the mesophyll and by increasing the production of carotenoids, the accessory light-harvesting pigments that have an important role in the photoprotection of the photosynthetic apparatus (Evelin et al. 2009; Zhu et al. 2011). At the cellular level, AMF protect plants against oxidative damage from temperature stress by increasing the production of antioxidative enzymes such as super oxide dismutase, ascorbate peroxidase, peroxidase, catalase, glutathione reductase, and other antioxidants (low molecular weight) such as ascorbic acid, glutathione, and polyphenols (Evelin et al. 2009; Zhu et al. 2010; Maya and Matsubara 2013). Improved nutrient uptake (Hu et al. 2015a; Liu et al. 2015), accumulation of osmolytes (proline, polyamines, betaines, sugars, and acylatedsterols) (Theocharis et al. 2012) and cryoprotectants (sugars) (Zhu et al. 2017), accumulation of secondary metabolites, and activation of enzymes related to secondary metabolism of plants (Chen et al. 2013) are some of the other possible mechanisms of AMF-mediated alleviation of temperature stress in plants.

14.3.2.4 AMF and Cold Stress

Climate change induced variability in temperature extremes is primarily linked with an increase in global temperatures. However, persistent extreme cold weather driven by anthropogenic warming holds serious concerns for the future. Recent research indicated that arctic warming leading to stratospheric polar vortex disruption is responsible for incidences of extreme cold events in the northern midlatitudes (Cohen et al. 2021). Global agricultural landscapes have been witnessing incessant cold extremes of varying frequency, intensity and duration (Kodra et al. 2011; Augspurger 2013). Extreme cold temperatures stress plants and affect their growth and productivity (Jha et al. 2017; Xin and Browse 2000). Cold stress includes both chilling (0–15 °C) and freezing (<0 °C). Chilling leads to deleterious effects on the morphology, physiology, biochemistry and triggers molecular changes in plants. Extreme low temperatures (freezing) lead to ice crystal formation in plants which disrupt their metabolism and cause mechanical injury (Yadav 2010). Cold stress affects major agricultural crops such as, wheat (Hassan et al. 2021) and rice (Zhang et al. 2014) across the globe. Many species of tropical and subtropical origin are injured or killed by cold stress and exhibit symptoms of chlorosis, necrosis or growth retardation (Sanghera et al. 2011). Rice, the world's most staple food crop is more vulnerable to cold stress, than wheat and barley (Zhang et al. 2014). Temperatures <15–20 °C have been reported to cause substantial yield loss of rice (Yoshida et al. 1996; Nakagahra et al. 1997) in Australia, Japan, China and Korea (Li 1993; Farrell et al. 2001; Lee 2001; Singh et al. 2005; Shimono et al. 2007). A majority of the wheat grown in China (Xiao et al. 2018), Europe (Trnka et al. 2014), Australia (Zheng et al. 2015; Crimp et al. 2016) and the United States (Holman et al. 2011) sustain cold stress. Indeed, every year during the early booting stage, 85% of the global wheat crop gets affected by spring frost (Yue et al. 2016). Severe frost damage occurs when canopy temperature falls below 0 °C during the spring season (Frederiks et al. 2015; Zheng et al. 2015). Chilling (low temperatures) affect plants from seed germination to maturity. Both vegetative and reproductive phases of the plant's life cycle are impacted by cold stress. However, the reproductive phase is particularly more vulnerable to cold stress. This can have serious economic consequences as cold stress affects fruit set through its negative impacts on several reproductive parameters including induction of flower abscission, pollen sterility, distortion of pollen tubes and ovule abortion (Thakur et al. 2010). The global distribution of plants is also affected by cold stress as plants have variable tolerance to chilling. Increased incidences of erratic extreme cold events driven by climate change during the growing season can compromise the sustainability of the agricultural produce globally. This can seriously hamper our goal to achieve food security in the changing climatic scenario and rising population.

Exposure to cold stress leads to poor seed germination, stunted seedlings (Yadav 2010), wilting (Bagnall et al. 1983), cracking of stems (Adhikari et al. 2022), discoloration, reduced leaf expansion, curling and growth (Sowiński et al. 2005; Rymen et al. 2007) chlorosis (Yoshida et al. 1996), tissue damage (Goering et al. 2021), necrosis and membrane damage (Atayee and Noori 2020) in susceptible plants. Disruption in the integrity of intracellular organelles and DNA strands, reduced photosynthetic efficiency, destabilization of protein complexes, reduced enzymatic activity, electrolyte leakage across membrane, slower metabolism, lower absorption of nutrients, dehydration, increased oxidative stress and accumulation of ROS are some of the other major cellular and metabolic impacts of low temperature stress in plants (Thakur and Nayyar 2013; Atayee and Noori 2020; Adhikari et al. 2022). Freezing on the other hand induces lesions, shrinkage of protoplast, cell deformation and ice formation in the intra- and intercellular spaces (Uemura et al. 1995).

Benefits of AMF inoculation have been observed in several agricultural and horticultural crops species exposed to low temperature stress (Table 14.1). Biopriming with AMF benefits plants exposed to cold stress through regulation of root hydraulic conductance, improved respiration, nutrient uptake, photosynthetic efficiency, activation of ROS scavenging system and regulation at the transcriptome level (Devi et al. 2019). AMF regulate the osmotic potential of the cold-stressed plants through accumulation of cryo/osmoprotectants such as, soluble sugars and proline (Kishor et al. 2005; Theocharis et al. 2012). Plants in symbiotic association with AMF benefit through improved uptake of water under water-deficit conditions compared to non-AMF plants (Khalvati et al. 2005). In *Phaseolus vulgare*, cold stress did not have any effect on hydraulic conductance in AMF inoculated plants. In comparison, the hydraulic conductance decreased after cold exposure in non-AMF plants (Aroca et al. 2007). Increased photosynthetic pigment (chlorophyll) content and improved leaf gaseous exchange contributing to improved photosynthetic efficiency have been reported on AMF inoculation in cold-stressed plants (Devi et al. 2019). AMF improves plant respiratory rate by increasing the concentrations of respiratory substrates such as, starch, soluble sugars and pyruvate thereby improving the efficiency of TCA cycle and consequently ATP production in plants exposed to low temperatures (Devi et al. 2019). Low temperature induced inability of water and nutrient uptake in plants is mitigated by AMF through induction of specific Pi-transporter genes (Ma et al. 2015, 2018) and improved potassium uptake (Liu et al. 2016) in cold stressed plants. AMF ameliorate oxidative stress caused due to low temperatures by triggering the plant antioxidative response involving improved superoxide dismutase, ascorbate peroxidase and catalase activity (Abdel Latef and Chaoxing 2011). H_2O_2 produced during oxidative stress is neutralized by ascorbate and glutathione recycling in AMF inoculated plants in response to cold stress (Liu et al. 2016). In addition, upregulation of AMF induced differentially expressed genes (DEGs) involved in glycolysis, metabolism (pyruvate, phenylalanine, glutathione and ascorbate), biosynthetic pathways (flavanoids, diterpenoids, zeatin) and hormonal signalling have been identified in response to cold stress in AMF colonized plants (Devi et al. 2019). For instance, 180 DEGs significantly upregulated by low temperature stress have been identified in AMF inoculated *Cucumis sativus* plants (Ma et al. 2018).

14.3.3 AMF-Mediated Mitigation of Climate Change

Agricultural practices contribute approximately 12% of the total global anthropogenic GHG emissions (Linquist et al. 2012). CO_2, N_2O, and CH_4 are the key global pollutants driving anthropogenic climate change (Myhre et al. 2013). The contribution of the world's food production system to the annual GHG emissions is approximately 21–37% (Lynch et al. 2021).

The Paris Conference of Parties' (COP-21) goal of limiting temperature rise to below 2 °C and preferably at 1.5 °C above preindustrial levels (UNFCCC 2015) will require significant reduction of GHGs from agriculture. This may further burden the food production system which is already under pressure to feed the growing population in the changing climate scenario. CSA provides a way forward to address the dual challenges of food security and climate change for sustainable food production (Figure 14.1). AMF inoculation of agricultural crops has the potential to increase plant growth and productivity and protect them from the adverse impacts of climate change. AMF have also been shown to play a significant role in reducing the emissions of two GHGs: N_2O and CO_2 from agriculture (Figure 14.1).

N_2O, a gaseous product of the nitrogen cycle, is a potent GHG with 298× more warming potential than CO_2 (Skiba and Rees 2014) and has a perturbation lifetime of 118–131 years (IPCC 2001; Fleming et al. 2011). Agricultural landscapes are the primary anthropogenic sources of N_2O in the atmosphere. Use of synthetic nitrogen-based fertilizers and manures in agriculture is the major contributor of agricultural N_2O emissions. N_2O concentration in the atmosphere has increased from 270 ppb in 1750 to 331 ppb in 2018 (20% increase from preindustrial levels). As per an estimate, for every 1000 kg of nitrogen fertilizers used in agriculture, 10–50 kg of nitrogen are lost to the atmosphere as N_2O (Hu et al. 2015b). The global N_2O concentrations are predicted to rise continuously in the coming decades (Reay et al. 2012) as fertilizer application and agricultural land areas are both expected to rise by 35–60% before 2030 (IPCC 2007). It is estimated that by 2030, agricultural practices will contribute up to 59% of the total N_2O emissions (Hu et al. 2015b). Microbially mediated processes causing significant N_2O emissions from agricultural soils include ammonia oxidation, dissimilatory nitrate reduction to ammonium (DNRA) and other denitrification pathways (Baggs 2011; Zhu et al. 2013). N_2O formation in soil ecosystems is majorly from nitrification related pathways (ammonia oxidation and nitrifier denitrification) and heterotrophic denitrification in water-limited and high-water content soils, respectively (Zhu et al. 2013; Mathieu et al. 2006). The growing food demand due to overpopulation will accelerate N_2O emissions from agricultural landscapes further and will jeopardize the efforts to meet the goals of Paris agreement. AMF have been suggested to play a key role in controlling N_2O emissions from agricultural soils by negatively affecting DNRA and denitrification rates by decreasing NO_3 leaching in the subsoil (Martínez-Garcia et al. 2017). Beneficial effects of AMF inoculation on N_2O emissions have been observed experimentally in a recent study using agricultural soils (Storer et al. 2018). The study showed a reduction in N_2O emissions in the presence of AMF hyphae and identified AMF competing with slow growing nitrifiers for ammonium to be the mechanism for the observed AMF effect (Storer et al. 2018).

CO_2 is the other important GHG responsible for climate change. Industry, transportation, and domestic use are key contributors to atmospheric CO_2 levels. They account for annual emissions of 10 Gt carbon (Amundson and Biardeau 2018). Agricultural CO_2 emissions are relatively minor, difficult to quantify, and uncertain (Vermeulen et al. 2012). Application of urea and lime, emissions from agricultural operations, fertilizer manufacture, and transportation are the major sources of agricultural CO_2 emissions (Lynch et al. 2021). Conversion of forested land to pastures or agricultural land (land use change) is an important source of CO_2 emissions from agriculture or food production systems (Lynch et al. 2021). Land use change alone contributes to 14% of the annual anthropogenic emissions of CO_2 (Le Quere et al. 2018) out of which, 10% is directly related to agriculture (Lynch et al. 2021). Once emitted, CO_2 persists in the atmosphere for a long time. A substantial portion of every ton of anthropogenic CO_2 emitted can remain in the atmosphere for millennia (Archer and Brovkin 2008) and cause long-term environmental damage. In view of the scale, long-term persistence and accumulating behavior of CO_2, urgent mitigation efforts are required to counterbalance global carbon emissions before the carbon sinks get saturated. Capture and long-term storage of atmospheric CO_2 as organic carbon in the soil is a viable "negative emission strategy" of CO_2 mitigation through the management of terrestrial ecosystems. Soil carbon sequestration can be achieved in agriculture by increasing biomass production, application of biofertilizers and low-cost plant growth regulators, reduced pesticide use, nitrogen fixation by legumes, crop rotation, and agricultural conservation practices (no tillage, application of manure, and biochar) (Paustian et al. 2019). Manure management and optimized animal feeding practices can further reduce emissions. Carbon reservoir in the soil is 1500 GtC at 1 m depth and 2400 GtC at 2 m depth (Batjes 1996). Carbon reservoir in soils is much higher compared to the carbon present in vegetation and atmosphere combined. Organic carbon content of soils is determined as a balance between the rate at which carbon is added to soils in the form of plant residues including roots, organic amendments, such as manure and compost, and the rate at which C is lost from soils. Organic carbon is lost from soils as CO_2 mainly due to heterotrophic soil respiration, as CH_4 from anaerobic systems (flooded) and due to leaching of dissolved organic carbon (Paustian et al. 2019). AMF can be used as a biological tool to improve carbon sequestration capacity of agricultural soils. Improvement in the photosynthetic rate and photosynthate production and its transfer from aboveground parts of plant to roots or microbes is the major mechanism of AMF-mediated enhancement of organic carbon sequestration in soils (Wang et al. 2016).

14.3.4 Agricultural Practices and AMF Symbiosis – Crop Rotations, Tillage, and Agrochemicals

AMF are most abundant symbionts in both natural and agricultural landscapes and contribute immensely to sustainable agriculture (van der Heijden et al. 2008; Bender et al. 2016). However, conventional agriculture practices such as, low crop diversity (monocropping), intensive tillage, and extensive use of agrochemicals compromise the benefits of AMF inoculation due to their negative effects on diversity of AMF in agricultural soils (Helgason et al. 1998; Verbruggen and Toby Kiers 2010; Rillig et al. 2016; Hontoria et al. 2019). Conventional farming practices despite having higher productivity (Tilman et al. 2001) are environmentally unsustainable (Foley et al. 2005). The damaging effects of conventional farming practices on the environment include soil degradation (Bowman et al. 1999; Campbell et al. 2000), GHG emissions (Dusenbury et al. 2008; Guo et al. 2010), reduction in soil organic matter (Janzen 2001) and negative effects on natural environment caused by extensive use of agrochemicals (Carpenter et al. 1998; Tilman et al. 2001). The vulnerability of AMF to agricultural practices is a major limitation for their use in agriculture as a bioinoculant.

Tillage is a common practice in agriculture across the world. It is majorly practiced for seedbed preparation, soil aeration, weed suppression, turning over cover crops and forages, mixing crop residue, and incorporating organic residues and fertilizers (Rasmussen 1999). Tillage alters the physical, chemical, and biological characteristics of the soil, which affect plant growth and yield (Wasaya et al. 2011; Rashidi and Keshavarzpour 2007). However, excessive and unsuitable tillage practice disrupts soil structure, water cycle, and organic carbon, increases surface run-off and soil erosion and causes nutrients and moisture loss from soil (Liu et al. 2010; Dumanski and Peiretti 2013). Intensive tillage leads to less diverse AMF communities (Chagnon et al. 2013) and decreased AMF colonization due to disruption of AMF hyphal networks in the soil (Johnson et al. 2001; Garcia et al. 2007).

Monocropping/monoculture in space and/or time leads to decrease in AMF diversity due to low diversity of plant hosts available for AMF to colonize (Burrows and Pfleger 2002; Oehl et al. 2003; Johnson et al. 2004; Strom et al. 2020). In contrast, polycropping/polyculture/intercropping/crop rotation promotes AMF diversity due to higher plant host diversity (Guzman et al. 2021). For instance, higher AMF sporulation was reported from a maize-common bean crop rotation compared to maize grown as a monoculture (Oruru and Njeru 2016). The higher AMF diversity in polycultures/crop rotation contributes to better nutrient uptake and improved plant growth (Higo et al. 2010).

The excessive use of agrochemicals causes N and P accumulation in agroecosystems. High nutrient availability in the soil leads to abundant but less mutualistic associations (Johnson et al. 1997; Johnson 2010). Heavy fertilization select aggressive AMF competitors for plant carbohydrates as plant's dependence on AMF decreases when nutrients are available in abundance This leads to a higher net cost for plant hosts (Johnson 2010; Thirkell et al. 2017). Moreover, higher nutrient availability in the rhizosphere cause a decrease in AMF diversity and lowers the benefits provided by the symbiont to the host (Williams et al. 2017; Smith et al. 2018; Higo et al. 2020). Soil acidification caused due to fertilizer application (Geisseler and Scow 2014) also affects soil AMF communities (Guzman et al. 2021). Continuous usage of fungicides has been shown to impede propagule production and colonization by AMF (Hage-Ahmed et al. 2018).

14.3.5 AMF Symbiosis and Climate Change

The multiple benefits associated with the use of AMF as bioinoculants make them ideal candidates for CSA (Figure 14.1). However, in an era impacted by climate change, it is necessary to understand the effects, if any, of climate change on AMF symbiosis. Factors related to climate change have been implicated to alter the root exudate composition, C/N ratio, or nutrient content through increased carbon allocation to the root system, which may eventually affect the abundance, composition, and activity of microbial communities associated with plants. Elevated CO_2, global warming, and drought have been shown to affect microbial communities beneficial to plants in a variety of ways. However, the effects are variable depending on plant species, ecosystem type, soil type, the microbial genotype, and the climate change factor being studied (Compant et al. 2010).

Elevated CO_2 is predicted to affect AMF community richness and diversity (Cotton 2018). Two conflicting hypotheses have been put forward to explain the effect of increasing atmospheric CO_2 levels on AMF communities. Rising CO_2 levels are suggested to increase the carbon supply to AMF which will eventually lead to decreased AMF diversity in soils due to disproportionate benefits to only a few AMF taxa (Cotton et al. 2015). The other hypothesis predicts elevated CO_2 to cause an increase in the AMF richness as a consequence of carbon becoming less limiting to microbes and below-ground microbial resources becoming more limiting to plants (Antoninka et al. 2011). Most studies do not report any effect of elevated CO_2 on AMF richness (Wolf et al. 2003; Klironomos et al. 2005; Kimball et al. 2007; Antoninka et al. 2011; Cotton et al. 2015; Mueller and Bohannan 2015). However, compositional shifts in AMF communities were observed within plant roots (Treseder et al. 2003; Wolf et al. 2003; Antoninka et al. 2011; Drigo et al. 2013; Cotton et al. 2015; Mueller and Bohannan 2015). Elevated CO_2 levels favored members of the order Glomerales (Klironomos et al. 2005). The relative abundance of the family Gigasporaceae, on the contrary decreased in an elevated CO_2 environment (Klironomos et al. 2005).

Nitrogen deposition due to climate change has been linked to negative effects on richness, taxonomic and phylogenetic diversity of AMF in most studies conducted so far (Egerton-Warburton and Allen 2000; Wolf et al. 2003; Egerton-Warburton et al. 2007; Antoninka et al. 2011; Liu et al. 2012; Tian et al. 2013; Camenzind et al. 2014; Chen et al. 2014, 2017; Williams et al. 2017). However, trends similar to the effect of elevated CO_2 were observed with respect to the relative abundances of Glomeraceae and Gigasporaceae (Treseder et al. 2003). *Rhizophagus irregularis*, which responded positively and had higher relative abundance under elevated CO_2 levels, is nitrophilous (Jumpponen et al. 2005) and favored by high nitrogen levels (Cotton et al. 2015).

Global warming (high temperature) effects on AMF richness varied greatly ranging from unaffected in a majority of studies (Heinemeyer and Fitter 2004; Sun et al. 2013; Yang et al. 2013; Kim et al. 2015; Gao et al. 2016) to increased (Yang et al. 2013; Kim et al. 2014) and decreased AMF richness under high temperature conditions (Shi et al. 2017). High temperature effects on AMF composition followed a similar pattern being unaffected in some studies (Heinemeyer and Fitter 2004; Kim et al. 2014, 2015; Gao et al. 2016) to altered AMF composition observed in other studies (Yang et al. 2013; Kim et al. 2014; Shi et al. 2017).

Studies on the effects of changing moisture regimes on AMF communities are few and present variable results (Deepika and Kothamasi 2015; Deveautour et al. 2015; Furze et al. 2017). For instance, drought did not affect AMF composition in a pot culture with *Sorghum* and a soybean monoculture but affected AMF composition in an agroforest and grassland (Deepika and Kothamasi 2015; Deveautour et al. 2015; Furze et al. 2017). Similar mixed results with no real pattern were found with the effect of increased rainfall and flooding on AMF (Deepika and Kothamasi 2015; Deveautour et al. 2015; Li et al. 2015; Gao et al. 2016; Chen et al. 2017). Effect of elevated tropospheric ozone on AMF communities is also poorly studied and does not reveal any effect (Cotton et al. 2015).

It is evident that the effect of climate change on AMF symbiosis and the richness and diversity of AMF are still in infancy with no clear patterns emerging. However, these studies highlight the vulnerability of AMF communities to climate change. The inconclusive results observed with most of the climate change factors may be due to different methodologies or experimental conditions in different studies. Moreover, the obligate symbiotic nature of AMF makes it challenging to subject AMF alone to different climate stressors (Barrett et al. 2014).

14.3.6 Conclusions and Future Perspectives

Climate change is a major hurdle for the humanity's quest of achieving food security due to its catastrophic effects on agriculture. Climate-induced changes in temperature and rainfall patterns are affecting the yields of major agricultural crops globally. Food insecurity driven by rising global population is leading to a "resource-exhaustive agriculture," which is further contributing to the deterioration of the environment. CSA is a green, climate resilient practice which has the potential to transform agricultural systems, making them more productive and sustainable with less carbon footprint. Use of AMF as bioinoculants in agriculture are a viable strategy owing to their multiple roles as biofertilizers and bioprotectants and the myriad of benefits associated with them. AMF technology is already in use as a component of sustainable agricultural systems. However, the AMF role in making plants resilient to climate change aligns them with the objectives of CSA.

AMF play a key role in the global carbon cycle as they receive 4–20% of the plant photosynthates. An estimated 5 billion tons carbon flow through them globally. Research on the effects of climate change on AMF symbiosis is still in its infancy. However, studies have highlighted the vulnerability of AMF to climate change but the effect of different climatic stressors on the AMF symbiosis is not well understood. Therefore, it is difficult to predict how AMF will respond and perform in future environmental conditions. Owing to their strong influence on the global carbon cycle, they may help mitigate or exacerbate climate change in the future.

AMF are ubiquitous and are not host specific. However, the benefits associated with AMF inoculation are highly variable from species to species and under different environmental conditions. There are even reports of AMF not being successful in agriculture. This may pose difficulties for the farmers in selecting specific AMF species/strain/combination to be used in agricultural farms with a specific ambient conditions. Moreover, indigenous AMF have been shown to perform better compared to inoculation with exotic AMF species. Therefore, future research should focus on isolating native AMF isolates adapted to a particular abiotic/biotic stress for crop inoculation in CSA. Selection of the ideal crop-AMF symbiont combination for a particular soil/climate is even more challenging in view of the continuously evolving climatic scenario. Large-scale cultivation of high-quality AMF inocula tolerant to different abiotic/biotic stresses for agricultural use is an additional challenge.

Acknowledgment

DK received a Marie Skłodowska-Curie Actions Horizon 2020 Individual Fellowship under the project MICROB-COM, grant ID 894188.

References

Abbaspour, H., Saeidi-Sar, S., Afshari, H., and Abdel-Wahhab, M.A. (2012). Tolerance of mycorrhiza infected pistachio (*Pistacia vera* L.) seedling to drought stress under glasshouse conditions. *J. Plant Physiol.* 169 (7): 704–709.

Abdel Latef, A.A.H. and Chaoxing, H. (2011). Arbuscular mycorrhizal influence on growth, photosynthetic pigments, osmotic adjustment and oxidative stress in tomato plants subjected to low temperature stress. *Acta Physiol. Plant.* 33 (4): 1217–1225.

Abdel-Fattah, G.M. and Asrar, A.W.A. (2012). Arbuscular mycorrhizal fungal application to improve growth and tolerance of wheat (*Triticum aestivum* L.) plants grown in saline soil. *Acta Physiol. Plant.* 34 (1): 267–277.

Adhikari, L., Baral, R., Paudel, D.R. et al. (2022). Cold stress in plants: strategies to improve cold tolerance in forage species. *Plant Stress* 4: 100081.

Ahmad, A., Diwan, H., and Abrol, Y.P. (2010). Global climate change, stress and plant productivity. In: *Abiotic Stress Adaptation in Plants: Physiological, Molecular and Genome Foundation* (ed. A. Pareek, S.K. Sopory, H.J. Bohnert, and Govindjee), 503–521. Springer Science+Business Media B.V.

Amundson, R. and Biardeau, L. (2018). Opinion: soil carbon sequestration is an elusive climate mitigation tool. *Proc. Natl. Acad. Sci. U. S. A.* 115: 11652–11656.

Antoninka, A., Reich, P.B., and Johnson, N.C. (2011). Seven years of carbon dioxide enrichment, nitrogen fertilization and plant diversity influence arbuscular mycorrhizal fungi in a grassland ecosystem. *New Phytol.* 192 (1): 200–214.

Archer, D. and Brovkin, V. (2008). The millennial atmospheric lifetime of anthropogenic CO_2. *Clim. Chang.* 90 (3): 283–297.

Aroca, R., Porcel, R., and Ruiz-Lozano, J.M. (2007). How does arbuscular mycorrhizal symbiosis regulate root hydraulic properties and plasma membrane aquaporins in *Phaseolus vulgaris* under drought, cold or salinity stresses? *New Phytol.* 173 (4): 808–816.

Ashraf, M.F.M.R. and Foolad, M.R. (2007). Roles of glycine betaine and proline in improving plant abiotic stress resistance. *Environ. Exp. Bot.* 59 (2): 206–216.

Asseng, S., Ewert, F., Martre, P. et al. (2015). Rising temperatures reduce global wheat production. *Nat. Clim. Chang.* 5 (2): 143–147.

Atayee, A.R. and Noori, M.S. (2020). Alleviation of cold stress in vegetable crops. *J. Sci. Agric.* 4: 38–44.

Atkinson, N.J. and Urwin, P.E. (2012). The interaction of plant biotic and abiotic stresses: from genes to the field. *J. Exp. Bot.* 63 (10): 3523–3543.

Augé, R.M. (2001). Water relations, drought and vesicular-arbuscular mycorrhizal symbiosis. *Mycorrhiza* 11 (1): 3–42.

Augé, R.M., Toler, H.D., Moore, J.L. et al. (2007). Comparing contributions of soil versus root colonization to variations in stomatal behavior and soil drying in mycorrhizal *Sorghum bicolor* and *Cucurbita pepo*. *J. Plant Physiol.* 164 (10): 1289–1299.

Augspurger, C.K. (2013). Reconstructing patterns of temperature, phenology, and frost damage over 124 years: spring damage risk is increasing. *Ecology* 94: 41–50.

Baggs, E.M. (2011). Soil microbial sources of nitrous oxide: recent advances in knowledge, emerging challenges and future direction. *Curr. Opin. Environ. Sustain.* 3 (5): 321–327.

Bagnall, D., Wolfe, J.O.E., and King, R.W. (1983). Chill-induced wilting and hydraulic recovery in mung bean plants. *Plant Cell Environ.* 6 (6): 457–464.

Bahadur, A., Batool, A., Nasir, F. et al. (2019). Mechanistic insights into arbuscular mycorrhizal fungi-mediated drought stress tolerance in plants. *Int. J. Mol. Sci.* 20 (17): 4199.

Bakr, J., Daood, H.G., Pék, Z. et al. (2016). Yield and quality of mycorrhized processing tomato under water scarcity. *Appl. Ecol. Environ. Res.* 15 (1): 401–413.

Barrett, G., Campbell, C.D., and Hodge, A. (2014). The direct response of the external mycelium of arbuscular mycorrhizal fungi to temperature and the implications for nutrient transfer. *Soil Biol. Biochem.* 78: 109–117.

Bárzana, G., Aroca, R., Paz, J.A. et al. (2012). Arbuscular mycorrhizal symbiosis increases relative apoplastic water flow in roots of the host plant under both well-watered and drought stress conditions. *Ann. Bot.* 109 (5): 1009–1017.

Bárzana, G., Aroca, R., Bienert, G.P. et al. (2014). New insights into the regulation of aquaporins by the arbuscular mycorrhizal symbiosis in maize plants under drought stress and possible implications for plant performance. *Mol. Plant-Microbe Interact.* 27 (4): 349–363.

Basak, H., Demir, K., Kasim, R., and Okay, F.Y. (2011). The effect of endo-mycorrhiza (VAM) treatment on growth of tomato seedling grown under saline conditions. *Afr. J. Agric. Res.* 6: 3326–3332.

Batjes, N.H. (1996). Total carbon and nitrogen in the soils of the world. *Eur. J. Soil Sci.* 47 (2): 151–163.

Bayani, R., Saateyi, A., and Faghani, E. (2015). Influence of arbuscular mycorrhiza in phosphorus acquisition efficiency and drought-tolerance mechanisms in barley *Hordeum vulgare* L. *Int. J. Biosci.* 7: 86–94.

Bayle, V., Arrighi, J.F., Creff, A. et al. (2011). *Arabidopsis thaliana* high-affinity phosphate transporters exhibit multiple levels of posttranslational regulation. *Plant Cell* 23 (4): 1523–1535.

Begum, N., Qin, C., Ahanger, M.A. et al. (2019a). Role of arbuscular mycorrhizal fungi in plant growth regulation: implications in abiotic stress tolerance. *Front. Plant Sci.* 10: 1068.

Begum, N., Ahanger, M.A., Su, Y. et al. (2019b). Improved drought tolerance by AMF inoculation in maize (*Zea mays*) involves physiological and biochemical implications. *Plants* 8 (12): 579.

Bender, S.F. and van der Heijden, M.G. (2015). Soil biota enhance agricultural sustainability by improving crop yield, nutrient uptake and reducing nitrogen leaching losses. *J. Appl. Ecol.* 52 (1): 228–239.

Bender, S.F., Conen, F., and Van der Heijden, M.G. (2015). Mycorrhizal effects on nutrient cycling, nutrient leaching and N_2O production in experimental grassland. *Soil Biol. Biochem.* 80: 283–292.

Bender, S.F., Wagg, C., and van der Heijden, M.G. (2016). An underground revolution: biodiversity and soil ecological engineering for agricultural sustainability. *Trends Ecol. Evol.* 31 (6): 440–452.

Bita, C. and Gerats, T. (2013). Plant tolerance to high temperature in a changing environment: scientific fundamentals and production of heat stress-tolerant crops. *Front. Plant Sci.* 4: 273.

Bossuyt, H., Denef, K., Six, J. et al. (2001). Influence of microbial populations and residue quality on aggregate stability. *Appl. Soil Ecol.* 16 (3): 195–208.

Bowman, R.A., Vigil, M.F., Nielsen, D.C., and Anderson, R.L. (1999). Soil organic matter changes in intensively cropped dryland systems. *Soil Sci. Soc. Am. J.* 63 (1): 186–191.

Bray, E.A., Bailey-Serres, J., and Weretilnyk, E. (2000). *Biochemistry and Molecular Biology of Plants*, 1158–1203. Rockville: American Society of Plant Physiologists.

Burrows, R.L. and Pfleger, F.L. (2002). Host responses to AMF from plots differing in plant diversity. *Plant Soil* 240 (1): 169–180.

Cabral, C., Ravnskov, S., Tringovska, I., and Wollenweber, B. (2016). Arbuscular mycorrhizal fungi modify nutrient allocation and composition in wheat (*Triticum aestivum* L.) subjected to heat-stress. *Plant Soil* 408 (1): 385–399.

Cakmakci, O., Cakmakci, T., Durak, E.D. et al. (2017). Effects of arbuscular mycorrhizal fungi in melon (*Cucumis melo* L.) seedling under deficit irrigation. *Fresenius Environ. Bull.* 26 (12): 7513–7520.

Camenzind, T., Hempel, S., Homeier, J. et al. (2014). Nitrogen and phosphorus additions impact arbuscular mycorrhizal abundance and molecular diversity in a tropical montane forest. *Glob. Chang. Biol.* 20 (12): 3646–3659.

Campbell, C.A., Zentner, R.P., Liang, B.C. et al. (2000). Organic C accumulation in soil over 30 years in semiarid southwestern Saskatchewan–effect of crop rotations and fertilizers. *Can. J. Soil Sci.* 80 (1): 179–192.

Caradonia, F., Francia, E., Morcia, C. et al. (2019). Arbuscular mycorrhizal fungi and plant growth promoting rhizobacteria avoid processing tomato leaf damage during chilling stress. *Agronomy* 9 (6): 299.

Carneiro, M.A.C., Ferreira, D.A., Souza, E.D.D. et al. (2015). Arbuscular mycorrhizal fungi in soil aggregates from fields of "murundus" converted to agriculture. *Pesq. Agrop. Brasileira* 50: 313–321.

Carpenter, S.R., Caraco, N.F., Correll, D.L. et al. (1998). Nonpoint pollution of surface waters with phosphorus and nitrogen. *Ecol. Appl.* 8 (3): 559–568.

Cavagnaro, T.R., Bender, S.F., Asghari, H.R., and van der Heijden, M.G. (2015). The role of arbuscular mycorrhizas in reducing soil nutrient loss. *Trends Plant Sci.* 20 (5): 283–290.

Chagnon, P.L., Bradley, R.L., Maherali, H., and Klironomos, J.N. (2013). A trait-based framework to understand life history of mycorrhizal fungi. *Trends Plant Sci.* 18 (9): 484–491.

Chen, T.H. and Murata, N. (2011). Glycinebetaine protects plants against abiotic stress: mechanisms and biotechnological applications. *Plant Cell Environ.* 34 (1): 1–20.

Chen, S., Jin, W., Liu, A. et al. (2013). Arbuscular mycorrhizal fungi (AMF) increase growth and secondary metabolism in cucumber subjected to low temperature stress. *Sci. Hortic.* 160: 222–229.

Chen, Y., Zhang, X., Ye, J. et al. (2014). Six-year fertilization modifies the biodiversity of arbuscular mycorrhizal fungi in a temperate steppe in Inner Mongolia. *Soil Biol. Biochem.* 69: 371–381.

Chen, Y.-L., Xu, Z.-W., Xu, T.-L. et al. (2017). Nitrogen deposition and precipitation induced phylogenetic clustering of arbuscular mycorrhizal fungal communities. *Soil Biol. Biochem.* 115: 233–242.

Chen, M., Arato, M., Borghi, L. et al. (2018). Beneficial services of arbuscular mycorrhizal fungi–from ecology to application. *Front. Plant Sci.* 1270.

Chitarra, W., Pagliarani, C., Maserti, B. et al. (2016). Insights on the impact of arbuscular mycorrhizal symbiosis on tomato tolerance to water stress. *Plant Physiol.* 171 (2): 1009–1023.

Chiu, C.H. and Paszkowski, U. (2019). Mechanisms and impact of symbiotic phosphate acquisition. *Cold Spring Harb. Perspect. Biol.* 11 (6): a034603.

Cohen, J., Agel, L., Barlow, M. et al. (2021). Linking Arctic variability and change with extreme winter weather in the United States. *Science* 373 (6559): 1116–1121.

Compant, S., van der Heijden, M.G.A., and Sessitsch, A. (2010). Climate change effects on beneficial plant-microorganism interactions. *FEMS Microbiol. Ecol.* 73: 197–214.

Cordell, D., Drangert, J.O., and White, S. (2009). The story of phosphorus: global food security and food for thought. *Glob. Environ. Chang.* 19 (2): 292–305.

Corwin, D.L. (2021). Climate change impacts on soil salinity in agricultural areas. *Eur. J. Soil Sci.* 72 (2): 842–862. https://doi.org/10.1111/ejss.13010.

Cotton, T.A. (2018). Arbuscular mycorrhizal fungal communities and global change: an uncertain future. *FEMS Microbiol. Ecol.* 94 (11): 179.

Cotton, T.A., Fitter, A.H., Miller, R.M. et al. (2015). Fungi in the future: interannual variation and effects of atmospheric change on arbuscular mycorrhizal fungal communities. *New Phytol.* 205 (4): 1598–1607.

Crimp, S.J., Zheng, B., Khimashia, N. et al. (2016). Recent changes in southern Australian frost occurrence: implications for wheat production risk. *Crop Pasture Sci.* 67 (8): 801–811.

Crinnion, W.J. (2010). Organic foods contain higher levels of certain nutrients, lower levels of pesticides, and may provide health benefits for the consumer. *Altern. Med. Rev.* 15 (1): 4–12.

Daei, G., Ardekani, M., Rejali, F. et al. (2009). Alleviation of salinity stress on wheat yield, yield components, and nutrient uptake using arbuscular mycorrhizal fungi under field conditions. *J. Plant Physiol.* 166: 617–625.

Daryanto, S., Wang, L., and Jacinthe, P.A. (2016). Global synthesis of drought effects on maize and wheat production. *PLoS One* 11 (5): e0156362.

Dasgupta, S., Hossain, M., Huq, M., and Wheeler, D. (2015). Climate change and soil salinity: the case of coastal Bangladesh. *Ambio* 44 (8): 815–826.

Deepika, S. and Kothamasi, D. (2015). Soil moisture – a regulator of arbuscular mycorrhizal fungal community assembly and symbiotic phosphorus uptake. *Mycorrhiza* 25: 67–75.

Deepika, S. and Kothamasi, D. (2021). Plant hosts may influence arbuscular mycorrhizal fungal community composition in mangrove estuaries. *Mycorrhiza* 31 (6): 699–711.

Dell'Amico, J., Torrecillas, A., Rodriguez, P. et al. (2002). Responses of tomato plants associated with the arbuscular mycorrhizal fungus Glomus clarum during drought and recovery. *J. Agric. Sci.* 138 (4): 387–393.

Deveautour, C., Donn, S., Power, S.A. et al. (2015). Experimentally altered rainfall regimes and host root traits affect grassland arbuscular mycorrhizal fungal communities. *Mol. Ecol.* 27 (8): 2152–2163.

Devi, T.S., Gupta, S., and Kapoor, R. (2019). Arbuscular mycorrhizal fungi in alleviation of cold stress in plants. In: *Advancing Frontiers in Mycology & Mycotechnology* (ed. T. Satyanarayana, S. Deshmukh, and M. Deshpande), 435–455. Singapore: Springer.

Devi, S.H., Bhupenchandra, I., Sinyorita, S., Chongtham, S.K., Devi, E.L. (2021). Mycorrhizal fungi and sustainable agriculture. Ohyama, T., Inubushi, K. (Eds.): Nitrogen in Agriculture - Physiological, Agricultural and Ecological Aspects, Intech Open, 1–19. 10.5772/intechopen.99262.

Djodjic, F., Börling, K., and Bergström, L. (2004). Phosphorus leaching in relation to soil type and soil phosphorus content. *J. Environ. Qual.* 33 (2): 678–684.

Drewnowski, A., Renderson, S.A., Driscoll, A., and Rolls, B.J. (1997). The dietary variety score: assessing diet quality in healthy young and older adults. *J. Am. Diet. Assoc.* 97 (3): 266–271.

Drigo, B., Kowalchuk, G.A., Knapp, B.A. et al. (2013). Impacts of 3 years of elevated atmospheric CO_2 on rhizosphere carbon flow and microbial community dynamics. *Glob. Chang. Biol.* 19 (2): 621–636.

Duc, N.H., Csintalan, Z., and Posta, K. (2018). Arbuscular mycfraorrhizal fungi mitigate negative effects of combined drought and heat stress on tomato plants. *Plant Physiol. Biochem.* 132: 297–307.

Dumanski, J. and Peiretti, R. (2013). Modern concepts of soil conservation. *Intl. Soil Water Conserv. Res.* 1 (1): 19–23.

Dusenbury, M.P., Engel, R.E., Miller, P.R. et al. (2008). Nitrous oxide emissions from a northern great plains soil as influenced by nitrogen management and cropping systems. *J. Environ. Qual.* 37 (2): 542–550.

Egerton-Warburton, L.M. and Allen, E.B. (2000). Shifts in arbuscular mycorrhizal communities along an anthropogenic nitrogen deposition gradient. *Ecol. Appl.* 10 (2): 484–496.

Egerton-Warburton, L.M., Johnson, N.C., and Allen, E.B. (2007). Mycorrhizal community dynamics following nitrogen fertilization: a cross-site test in five grasslands. *Ecol. Monogr.* 77 (4): 527–544.

Evelin, H., Kapoor, R., and Giri, B. (2009). Arbuscular mycorrhizal fungi in alleviation of salt stress: a review. *Ann. Bot.* 104 (7): 1263–1280.

Evelin, H., Giri, B., and Kapoor, R. (2013). Ultrastructural evidence for AMF mediated salt stress mitigation in *Trigonella foenum-graecum*. *Mycorrhiza* 23: 71–86.

Evelin, H., Devi, T.S., Gupta, S., and Kapoor, R. (2019). Mitigation of salinity stress in plants by arbuscular mycorrhizal symbiosis: current understanding and new challenges. *Front. Plant Sci.* 10: 470.

Fall, A.F., Nakabonge, G., Ssekandi, J., Founoune-Mboup, H., Apori, S.O., Ndiaye, A., Badji, A. and Ngom, K.(2022). Roles of arbuscular mycorrhizal fungi on soil fertility: contribution in the improvement of physical, chemical, and biological properties of the soil. *Front. Fungal Biol.***3**:723892.

Farrell, T.C., Williams, R.L., and Fukai, S. (2001). The cost of low temperature to the NSW rice industry. *Proceedings of the 10th Australian Agronomy Conference*, Hobart, Tasmania (28 January to 1 February 2001).

Fasusi, O.A., Cruz, C., and Babalola, O.O. (2021). Agricultural sustainability: microbial biofertilizers in rhizosphere management. *Agriculture* 11 (2): 163.

Ferrol, N., Azcón-Aguilar, C., and Pérez-Tienda, J. (2019). Arbuscular mycorrhizas as key players in sustainable plant phosphorus acquisition: an overview on the mechanisms involved. *Plant Sci.* 280: 441–447.

Fleming, E.L., Jackman, C.H., Stolarski, R.S., and Douglas, A.R. (2011). A model study of the impact of source gas changes on the stratosphere for 1850–2100. *Atmos. Chem. Phys.* 11 (16): 8515–8541.

Foley, J.A., DeFries, R., Asner, G.P. et al. (2005). Global consequences of land use. *Science* 309 (5734): 570–574.

Foley, J.A., Ramankutty, N., Brauman, K.A. et al. (2011). Solutions for a cultivated planet. *Nature* 478 (7369): 337–342.

Food and Agriculture Organization (2013). *Climate-Smart Agriculture Sourcebook*. Rome, Italy: *Food and Agriculture Organization of the United Nations (FAO)*.

Foyer, C.H. and Noctor, G. (2005). Redox homeostasis and antioxidant signaling: a metabolic interface between stress perception and physiological responses. *Plant Cell* 17: 1866–1875.

Frederiks, T.M., Christopher, J.T., Sutherland, M.W., and Borrell, A.K. (2015). Post-head-emergence frost in wheat and barley: defining the problem, assessing the damage, and identifying resistance. *J. Exp. Bot.* 66: 3487–3498.

Furze, J.R., Martin, A.R., Nasielski, J. et al. (2017). Resistance and resilience of root fungal communities to water limitation in a temperate agroecosystem. *Ecol. Evol.* 7: 3443–3454.

Gao, C., Kim, Y.-C., Zheng, Y. et al. (2016). Increased precipitation, rather than warming, exerts a strong influence on arbuscular mycorrhizal fungal community in a semiarid steppe ecosystem. *Botany* 94: 459–469.

Garcia, J.P., Wortmann, C.S., Mamo, M. et al. (2007). *One-Time Tillage of No-Till: Effects on Nutrients, Mycorrhizae, and Phosphorus Uptake*, 21. North Platte: *West Central Research and Extension Center*.

Garg, N. and Bharti, A. (2018). Salicylic acid improves arbuscular mycorrhizal symbiosis, and chickpea growth and yield by modulating carbohydrate metabolism under salt stress. *Mycorrhiza* 28: 727–746.

Garg, N. and Pandey, R. (2016). High effectiveness of exotic arbuscular mycorrhizal fungi is reflected in improved rhizobial symbiosis and trehalose turnover in *Cajanus cajan* genotypes grown under salinity stress. *Fungal Ecol.* 21: 57–67.

Geisseler, D. and Scow, K.M. (2014). Long-term effects of mineral fertilizers on soil microorganisms: a review. *Soil Biol. Biochem.* 75: 54–63.

Gill, S.S. and Tuteja, N. (2010). Reactive oxygen species and antioxidant machinery in abiotic stress tolerance in crop plants. *Plant Physiol. Biochem.* 48: 909–930.

Godfray, H.C.J. and Garnett, T. (2014). Food security and sustainable intensification. *Philos. Trans. R. Soc., B* 369 (1639): 20120273.

Goering, R., Larsen, S., Tan, J. et al. (2021). QTL mapping of seedling tolerance to exposure to low temperature in the maize IBM RIL population. *PLoS One* 16 (7): e0254437.

Goicoechea, N., Merino, S., and Sánchez-Díaz, M. (2005). Arbuscular mycorrhizal fungi can contribute to maintain antioxidant and carbon metabolism in nodules of *Anthyllis cytisoides* L. subjected to drought. *J. Plant Physiol.* 162 (1): 27–35.

Gonzalez, P., Neilson, R.P., Lenihan, J.M., and Drapek, R.J. (2010). Global patterns in the vulnerability of ecosystems to vegetation shifts due to climate change. *Glob. Ecol. Biogeogr.* 19 (6): 755–768.

Gordon-Weeks, R., Tong, Y., Davies, T.E., and Leggewie, G. (2003). Restricted spatial expression of a high-affinity phosphate transporter in potato roots. *J. Cell Sci.* 116 (15): 3135–3144.

Gornall, J., Betts, R., Burke, E. et al. (2010). Implications of climate change for agricultural productivity in the early twenty-first century. *Philos. Trans. R. Soc., B* 365 (1554): 2973–2989.

Goswami, V., Deepika, S., Chandra, R., Babu, C.R., Kothamasi, D.(2023). Arbuscular mycorrhizas accelerate the degradation of colour containing organic pollutants present in distillery spent wash leachates. *J. Hazard. Mater.* **452**:131291

Graham, J.H. and Syvertsen, J.P. (1984). Influence of vesicular–arbuscular mycorrhiza on the hydraulic conductivity of roots of two citrus rootstocks. *New Phytol.* 97 (2): 277–284.

Gregory, P.J., Ingram, J.S., and Brklacich, M. (2005). Climate change and food security. *Philos. Trans. R. Soc., B* 360 (1463): 2139–2148.

Guo, J.H., Liu, X.J., Zhang, Y. et al. (2010). Significant acidification in major Chinese croplands. *Science* 327 (5968): 1008–1010.

Gutjahr, C. and Paszkowski, U. (2013). Multiple control levels of root system remodeling in arbuscular mycorrhizal symbiosis. *Front. Plant Sci.* 4: 204.

Guzman, A., Montes, M., Hutchins, L. et al. (2021). Crop diversity enriches arbuscular mycorrhizal fungal communities in an intensive agricultural landscape. *New Phytol.* 231 (1): 447–459.

Hage-Ahmed, K., Rosner, K., and Steinkellner, S. (2018). Arbuscular mycorrhizal fungi and their response to pesticides. *Pest Manag. Sci.* 75: 583–590.

Haghighi, M., Mozafariyan, M., and Abdolahipour, B. (2015). Effect of cucumber mycorrhiza inoculation under low and high root temperature grown on hydroponic conditions. *J. Crop Sci. Biotechnol.* 18 (2): 89–96.

Hajiboland, R., Joudmand, A., Aliasgharzad, N. et al. (2019). Arbuscular mycorrhizal fungi alleviate low-temperature stress and increase freezing resistance as a substitute for acclimation treatment in barley. *Crop. Pasture Sci.* 70: 218–233.

Hamidov, A., Helming, K., Bellocchi, G. et al. (2018). Impacts of climate change adaptation options on soil functions: a review of European case-studies. *Land Degrad. Dev.* 29 (8): 2378–2389.

Handral, A.R., Singh, A., Singh, D.R. et al. (2017). Scenario of changing dynamics in production and productivity of major cereals in India. *Indian J. Agric. Sci.* 87 (10): 1371–1376.

Hansen, J., Sato, M., and Ruedy, R. (2012). Perception of climate change. *Proc. Natl. Acad. Sci.* 109 (37): E2415–E2423.

Hasegawa, P.M., Bressan, R.A., Zhu, J.K., and Bohnert, H.J. (2000). Plant cellular and molecular responses to high salinity. *Annu. Rev. Plant Biol.* 51 (1): 463–499.

Hashem, A., Alqarawi, A.A., Radhakrishnan, R. et al. (2018). Arbuscular mycorrhizal fungi regulate the oxidative system, hormones and ionic equilibrium to trigger salt stress tolerance in *Cucumis sativus* L. *Saudi J. Biol. Sci.* 25: 1102–1114.

Hashem, A., Kumar, A., Al-Dbass, A.M. et al. (2019a). Arbuscular mycorrhizal fungi and biochar improves drought tolerance in chickpea. *J. Biol. Sci.* 26 (3): 614–624.

Hashem, A., Abd_Allah, E.F., Alqarawi, A.A. et al. (2019b). Comparing symbiotic performance and physiological responses of two soybean cultivars to arbuscular mycorrhizal fungi under salt stress. *Saudi J. Biol. Sci.* 26 (1): 38–48.

Hassan, M.A., Xiang, C., Farooq, M. et al. (2021). Cold stress in wheat: plant acclimation responses and management strategies. *Front. Plant Sci.* 12: 1234.

Hassani, A., Azapagic, A., and Shokri, N. (2021). Global predictions of primary soil salinization under changing climate in the 21st century. *Nat. Commun.* 12 (1): 1–17.

Heinemeyer, A. and Fitter, A.H. (2004). Impact of temperature on the arbuscular mycorrhizal causes minimal changes in arbuscular mycorrhizal colonization but shifts community composition—evidence from rDNA data. *Biol. Fertil. Soils* 41: 217–224.

Helgason, T., Daniell, T.J., Husband, R. et al. (1998). Ploughing up the wood-wide web? *Nature* 394 (6692): 431–431.

Higo, M., Isobe, K., Kang, D.J. et al. (2010). Inoculation with arbuscular mycorrhizal fungi or crop rotation with mycorrhizal plants improves the growth of maize in limed acid sulfate soil. *Plant Prod. Sci.* 13 (1): 74–79.

Higo, M., Azuma, M., Kamiyoshihara, Y. et al. (2020). Impact of phosphorus fertilization on tomato growth and arbuscular mycorrhizal fungal communities. *Microorganisms* 8: 178.

Holman, J.D., Schlegel, A.J., Thompson, C.R., and Lingenfelser, J.E. (2011). Influence of precipitation, temperature, and 56 years on winter wheat yields in Western Kansas. *Crop Manag.* 4: 1–15.

Hontoria, C., García-González, I., Quemada, M. et al. (2019). The cover crop determines the AMF community composition in soil and in roots of maize after a ten-year continuous crop rotation. *Sci. Total Environ.* 660: 913–922.

Hoorman, J.J., Sá, J.C.M., and Reeder, R. (2011). The biology of soil compaction. *Soil Tillage Res.* 68: 49–57.

Hu, Y., Wu, S., Sun, Y. et al. (2015a). Arbuscular mycorrhizal symbiosis can mitigate the negative effects of night warming on physiological traits of *Medicago truncatula* L. *Mycorrhiza* 25 (2): 131–142.

Hu, H.W., Chen, D., and He, J.Z. (2015b). Microbial regulation of terrestrial nitrous oxide formation: understanding the biological pathways for prediction of emission rates. *FEMS Microbiol. Rev.* 39 (5): 729–749.

Intergovernmental Panel on Climate Change. Climate Change 1995: *IPCC Second Assessment Report*. IPCC, Brighton, March 1995.

Intergovernmental Panel on Climate Change (2001). *Climate Change 2001: The Scientific Basis. Contribution of Working Group Ito the Third Assessment Report of the Intergovernmental Panel on Climate Change*, 881. New York: Cambridge University Press.

Intergovernmental Panel on Climate Change (2007). *Climate Change 2007: The Scientific Basis. Contribution of Working Group Ito the Fourth Assessment Report of the Intergovernmental Panel on Climate Change*. New York: Cambridge University Press.

Intergovernmental Panel on Climate Change, 2018: Global warming of 1.5°C. An IPCC Special Report on the impacts of global warming of 1.5°C above pre-industrial levels and related global greenhouse gas emission pathways, in the context of strengthening the global response to the threat of climate change, sustainable development, and efforts to eradicate poverty [Masson-Delmotte, V., Zhai, P., Pörtner, H.O., Roberts, D., Skea, J., Shukla, P.R., Pirani, A., Moufouma-Okia, W., Péan, C., Pidcock, R., Connors, S., Matthews, J.B.R., Chen, Y., Zhou, X., Gomis, M.I., Lonnoy, E., Maycock, T., Tignor, M., Waterfield, T. (eds.)]. IPCC, Incheon, South Korea, 8 Oct 2018.

Jain, A. (2018). Analysis of growth and instability in area, production, yield and price of rice in India. *Soc. Change Dev.* 15 (2): 46–66.

Jansa, J., Forczek, S.T., Rozmoš, M. et al. (2019). Arbuscular mycorrhiza and soil organic nitrogen: network of players and interactions. *Chem. Biol. Technol. Agric.* 6: 1–10.

Janzen, H.H. (2001). Soil science on the Canadian prairies-peering into the future from a century ago. *Can. J. Soil Sci.* 81 (4): 489–503.

Jedmowski, C., Ashoub, A., and Brüggemann, W. (2013). Reactions of Egyptian landraces of *Hordeum vulgare* and *Sorghum bicolor* to drought stress, evaluated by the OJIP fluorescence transient analysis. *Acta Physiol. Plant.* 35 (2): 345–354.

Jha, U.C., Bohra, A., and Jha, R. (2017). Breeding approaches and genomics technologies to increase crop yield under low-temperature stress. *Plant Cell Rep.* 36 (1): 1–35.

John, D.A. and Babu, G.R. (2021). Lessons from the aftermaths of green revolution on food system and health. *Front. Sustain. Food Syst.* 5: 644559.

Johnson, D., Leake, J.R., and Read, D.J. (2001). Novel in-growth core system enables functional studies of grassland mycorrhizal mycelial networks. *New Phytol.* 152: 555–562.

Johnson, D., Vandenkoornhuyse, P.J., Leake, J.R. et al. (2004). Plant communities affect arbuscular mycorrhizal fungal diversity and community composition in grassland microcosms. *New Phytol.* 161 (2): 503–515.

Johnson, N.C. (2010). Resource stoichiometry elucidates the structure and function of arbuscular mycorrhizas across scales. *New Phytol.* 185 (3): 631–647.

Johnson, N.C., Graham, J.H., and Smith, F.A. (1997). Functioning of mycorrhizal associations along the mutualism–parasitism continuum. *New Phytol.* 135 (4): 575–585.

Joudmand, A., Hajiboland, R., and Aliasgharzad, N. (2021). Mitigation of chilling and freezing stresses through colonization with arbuscular mycorrhizal fungi in spring barley. *J. Plant Process Funct.* 9 (40): 13–24.

Jumpponen, A., Trowbridge, J., Mandyam, K. and Johnson, L. (2005). Nitrogen enrichment causes minimal changes in arbuscular mycorrhizal colonization but shifts community composition - evidence from rDNA data. Biology and Fertility of Soils 41:217–224.

Kalayu, G. (2019). Phosphate solubilizing microorganisms: promising approach as biofertilizers. *Intl. J. Agron.* 2019: 4917256. https://doi.org/10.1155/2019/4917256.

Karmakar, R., Das, I., Dutta, D., and Rakshit, A. (2016). Potential effects of climate change on soil properties: a review. *Sci. Intl.* 4: 51–73.

Kautz, T., Amelung, W., Ewert, F. et al. (2013). Nutrient acquisition from arable subsoils in temperate climates: a review. *Soil Biol. Biochem.* 57: 1003–1022.

Khalvati, M.A., Hu, Y., Mozafar, A., and Schmidhalter, U. (2005). Quantification of water uptake by arbuscular mycorrhizal hyphae and its significance for leaf growth, water relations, and gas exchange of barley subjected to drought stress. *Plant Biol.* 7 (06): 706–712.

Khan, M.S. and Zaidi, A. (2007). Synergistic effects of the inoculation with plant growth-promoting rhizobacteria and an arbuscular mycorrhizal fungus on the performance of wheat. *Turk. J. Agric. For.* 31 (6): 355–362.

Kim, Y., Gao, C., Zheng, Y. et al. (2014). Different responses of arbuscular mycorrhizal fungal community to day-time and night-time warming in a semiarid steppe. *Chin. Sci. Bull.* 59: 5080–5089.

Kim, Y.-C., Gao, C., Zheng, Y. et al. (2015). Arbuscular mycorrhizal fungal community response to warming and nitrogen addition in a semiarid steppe ecosystem. *Mycorrhiza* 25: 267–276.

Kim, W., Iizumi, T., and Nishimori, M. (2019). Global patterns of crop production losses associated with droughts from 1983 to 2009. *J. Appl. Meteorol. Climatol.* 58 (6): 1233–1244.

Kimball, B.A., Idso, S.B., Johnson, S., and Rillig, M.C. (2007). Seventeen years of carbon dioxide enrichment of sour orange trees: final results. *Glob. Chang. Biol.* 13 (10): 2171–2183.

Kishor, P.K., Sangam, S., Amrutha, R.N. et al. (2005). Regulation of proline biosynthesis, degradation, uptake and transport in higher plants: its implications in plant growth and abiotic stress tolerance. *Current Sci.* 88: 424–438.

Kling, G., Hayhoe, K., Johnson, L.B. et al. (2005). *Confronting Climate Change in the Great Lakes Region: Impacts on Our Communities and Ecosystems*, Technical Report. Cambridge, MA; Washington, DC: Union of Concerned Scientists; Ecological Society of America.

Klironomos, J.N., Allen, M.F., Rillig, M.C. et al. (2005). Abrupt rise in atmospheric CO_2 overestimates community response in a model plant–soil system. *Nature* 433 (7026): 621–624.

Kodra, E., Steinhaeuser, K., and Ganguly, A.R. (2011). Persisting cold extremes under 21st-century warming scenarios. *Geophys. Res. Lett.* 38: 47–103.

Kothamasi, D., Kothamasi, S., Bhattacharyya, A. et al. (2006). Arbuscular mycorrhizae and phosphate solubilising bacteria of the rhizosphere of the mangrove ecosystem of Great Nicobar island, India. *Biol. Fertil. Soils* 42 (4): 358–361.

Kumar, P. and Sharma, P. K. (2020) Soil salinity and food security in India. *Frontiers in Sustainable Food Systems* 4:533781 https://doi.org/10.3389/fsufs.2020.533781

Lambers, H. and Teste, F.P. (2013). Interactions between arbuscular mycorrhizal and non-mycorrhizal plants: do non-mycorrhizal species at both extremes of nutrient availability play the same game. *Plant Cell Environ.* 36 (11): 1911–1915.

Lambers, H., Chapin, F.S., and Pons, T.L. (2008). *Plant Physiological Ecology*, vol. 2. New York: Springer.

Le Quéré, C., Andrew, R.M., Friedlingstein, P. et al. (2018). Global carbon budget 2017. *Earth Syst. Sci. Data* 10 (1): 405–448.

Lee, M.H. (2001). Low temperature tolerance in rice: the Korean experience. *Increased Lowland Rice Production in the Mekong Region*, Vientiane, Laos (30 October to 2 November 2000), 109–117.

Lenoir, I., Fontaine, J., and Sahraoui, A.L.H. (2016). Arbuscular mycorrhizal fungal responses to abiotic stresses: a review. *Phytochemistry* 123: 4–15.

Li TG, Guo W.M. (1993). Identification and study on tolerance in main stresses of China cultivated rice germplasm resource. In: Ying CS (ed) Rice germplasm resources in China. China Agricultural Science and Technology Press, Beijing, 71–75.

Li, X., Zhu, T., Peng, F. et al. (2015). Inner Mongolian steppe arbuscular mycorrhizal fungal communities respond more strongly to water availability than to nitrogen fertilization. *Environ. Microbiol.* 17: 3051–3068.

Linquist, B., Van Groenigen, K.J., Adviento-Borbe, M.A. et al. (2012). An agronomic assessment of greenhouse gas emissions from major cereal crops. *Glob. Chang. Biol.* 18 (1): 194–209.

Liu, X.B., Zhang, X.Y., Wang, Y.X. et al. (2010). Soil degradation: a problem threatening the sustainable development of agriculture in Northeast China. *Plant Soil Environ.* 56 (2): 87–97.

Liu, Y., Shi, G., Mao, L. et al. (2012). Direct and indirect influences of 8 yr of nitrogen and (AM) symbiosis: growth responses of the host plant and its AM fungal partner. *J. Exp. Bot.* 55: 525–534.

Liu, S., Liu, Y., Yang, X. et al. (2014). The *Brassica oleracea* genome reveals the asymmetrical evolution of polyploid genomes. *Nat. Commun.* 5 (1): 1–11.

Liu, T., Sheng, M., Wang, C.Y. et al. (2015). Impact of arbuscular mycorrhizal fungi on the growth, water status, and photosynthesis of hybrid poplar under drought stress and recovery. *Photosynthetica* 53 (2): 250–258.

Liu, A., Chen, S., Wang, M. et al. (2016). Arbuscular mycorrhizal fungus alleviates chilling stress by boosting redox poise and antioxidant potential of tomato seedlings. *J. Plant Growth Regul.* 35 (1): 109–120.

Long, S.P., Ainsworth, E.A., Leakey, A.D., and Morgan, P.B. (2005). Global food insecurity. Treatment of major food crops with elevated carbon dioxide or ozone under large-scale fully open-air conditions suggests recent models may have overestimated future yields. *Philos. Trans. R. Soc., B* 360 (1463): 2011–2020.

Lynch, J., Cain, M., Frame, D., and Pierrehumbert, R. (2021). Agriculture's contribution to climate change and role in mitigation is distinct from predominantly fossil CO_2-emitting sectors. *Front. Sustain. Food Syst.* 4: 518039. https://doi.org/10.3389/fsufs.2020.518039.

Ma, J., Janoušková, M., Li, Y. et al. (2015). Impact of arbuscular mycorrhizal fungi (AMF) on cucumber growth and phosphorus uptake under cold stress. *Funct Plant Biol.* 42: 1158–1167.

Ma, J., Sun, C., Bai, L. et al. (2018). Transcriptome analysis of cucumber roots reveals key cold-resistance genes induced by AM fungi. *Plant Mol. Biol. Report.* 36: 135–148.

Ma, J., Janouskova, M., Ye, L. et al. (2019). Role of arbuscular mycorrhiza in alleviating the effect of cold on the photosynthesis of cucumber seedlings. *Photosynthetica* 57 (1): 86–95.

Maestri, E., Klueva, N., Perrotta, C. et al. (2002). Molecular genetics of heat tolerance and heat shock proteins in cereals. *Plant Mol. Biol.* 48 (5): 667–681.

Magdoff, F. and Tokar, B. (2009). Agriculture and food in crisis. *Mon. Rev.* 61 (3): 1–16.

Martínez-García, L.B., De Deyn, G.B., Pugnaire, F.I. et al. (2017). Symbiotic soil fungi enhance ecosystem resilience to climate change. *Glob. Chang. Biol.* 23 (12): 5228–5236.

Mathieu, O., Hénault, C., Lévêque, J. et al. (2006). Quantifying the contribution of nitrification and denitrification to the nitrous oxide flux using 15N tracers. *Environ. Pollut.* 144 (3): 933–940.

Mathur, S. and Jajoo, A. (2020). Arbuscular mycorrhizal fungi protects maize plants from high temperature stress by regulating photosystem II heterogeneity. *Ind. Crop. Prod.* 143: 111934.

Mathur, S., Tomar, R.S., and Jajoo, A. (2019). Arbuscular Mycorrhizal fungi (AMF) protects photosynthetic apparatus of wheat under drought stress. *Photosynth. Res.* 139 (1): 227–238.

Maya, M.A. and Matsubara, Y.I. (2013). Influence of arbuscular mycorrhiza on the growth and antioxidative activity in cyclamen under heat stress. *Mycorrhiza* 23 (5): 381–390.

Mcdougall, R., Rader, R., and Kristiansen, P. (2020). Urban agriculture could provide 15% of food supply to Sydney, Australia, under expanded land use scenarios. *Land Use Policy* 94: 104554.

McFarland, J.W., Ruess, R.W., Kielland, K. et al. (2010). Cross-ecosystem comparisons of in situ plant uptake of amino acid-N and NH_4^+. *Ecosystems* 13 (2): 177–193.

Miller, R.M. and Jastrow, J.D. (1990). Hierarchy of root and mycorrhizal fungal interactions with soil aggregation. *Soil Biol. Biochem.* 22 (5): 579–584.

Miransari, M. (2011). Arbuscular mycorrhizal fungi and nitrogen uptake. *Arch. Microbiol.* 193 (2): 77–81.

Mohammad, A. and Mittra, B. (2013). Effects of inoculation with stress-adapted arbuscular mycorrhizal fungus *Glomus deserticola* on growth of *Solanum melogena* L. and *Sorghum sudanese* Staph. seedlings under salinity and heavy metal stress conditions. *Arch. Agron. Soil Sci.* 59: 173–183.

Morales, D., Rodríguez, P., Dell'Amico, J. et al. (2003). High-temperature preconditioning and thermal shock imposition affects water relations, gas exchange and root hydraulic conductivity in tomato. *Biol. Plant.* 47 (2): 203–208.

Mueller, R.C. and Bohannan, B.J. (2015). Shifts in the phylogenetic structure of arbuscular mycorrhizal fungi in response to experimental nitrogen and carbon dioxide additions. *Oecologia* 179 (1): 175–185.

Munns, R., James, R.A., and Läuchli, A. (2006). Approaches to increasing the salt tolerance of wheat and other cereals. *J. Exp. Bot.* 57 (5): 1025–1043.

Myhre, G., Shindell, D., Bréon, F.M., et al. (2013). Anthropogenic and Natural Radiative Forcing, Climate Change 2013: The Physical Science Basis. Contribution of Working Group I to the Fifth Assessment Report of the Intergovernmental Panel on Climate Change, pp. 659–740.

Nakagahra, M., Okuno, K., and Vaughan, D. (1997). Rice genetic resources: history, conservation, investigative characterization and use in Japan. *Oryza Mol. Plant* 35: 69–77.

Nelson, G.C., Rosegrant, M.W., Koo, J. et al. (2009). Climate change impacts on agriculture and costs of adaptation. *Intl. Food Policy Res.* 30.

Noya, I., González-García, S., Bacenetti, J. et al. (2018). Environmental impacts of the cultivation-phase associated with agricultural crops for feed production. *J. Clean. Prod.* 172: 3721–3733.

Oehl, F., Sieverding, E., Ineichen, K. et al. (2003). Impact of land use intensity on the species diversity of arbuscular mycorrhizal fungi in agroecosystems of Central Europe. *Appl. Environ. Microbiol.* 69 (5): 2816–2824.

Oliveira, T.C., Cabral, J.S.R., Santana, L.R. et al. (2022). The arbuscular mycorrhizal fungus Rhizophagus clarus improves physiological tolerance to drought stress in soybean plants. *Sci. Rep.* 12 (1): 1–15.

Ortiz, R., Braun, H.J., Crossa, J. et al. (2008). Wheat genetic resources enhancement by the International Maize and Wheat Improvement Center (CIMMYT). *Genet. Resour. Crop. Evol.* 55 (7): 1095–1140.

Oruru, M.B. and Njeru, E.M. (2016). Upscaling arbuscular mycorrhizal symbiosis and related agroecosystems services in smallholder farming systems. *BioMed Res. Intl.* 2016: 4376240.

Oshunsanya, S.O., Nwosu, N.J., and Li, Y. (2019). Abiotic stress in agricultural crops under climatic conditions. In: *Sustainable Agriculture, Forest and Environmental Management* (ed. M. Jhariya, A. Banerjee, R. Meena, and D. Yadav), 71–100. Singapore: Springer.

Pandey, G., Yadav, C.B., Sahu, P.P. et al. (2017). Salinity induced differential methylation patterns in contrasting cultivars of foxtail millet (*Setaria italica* L.). *Plant Cell Rep.* 36 (5): 759–772.

Pareek, A., Dhankher, O.P., and Foyer, C.H. (2020). Mitigating the impact of climate change on plant productivity and ecosystem sustainability. *J. Exp. Bot.* 71 (2): 451–456.

Parihar, M., Rakshit, A., Rana, K. et al. (2020). The effect of arbuscular mycorrhiza l fungi inoculation in mitigating salt stress of Pea (*Pisum sativum* L.). *Commun. Soil Sci. Plant Anal.* 51: 1545–1559.

Parniske, M. (2008). Arbuscular mycorrhiza: the mother of plant root endosymbioses. *Nat. Rev. Microbiol.* 6 (10): 763–775.

Pasbani, B., Salimi, A., Aliasgharzad, N., and Hajiboland, R. (2020). Colonization with arbuscular mycorrhizal fungi mitigates cold stress through improvement of antioxidant defense and accumulation of protecting molecules in eggplants. *Sci. Hortic.* 272: 109575.

Paustian, K., Larson, E., Kent, J. et al. (2019). Soil C sequestration as a biological negative emission strategy. *Front. Clim.* 8: https://doi.org/10.3389/fclim.2019.0000.

Popkin, B.M. (1998). The nutrition transition and its health implications in lower-income countries. *Public Health Nutr.* 1 (1): 5–21.

Porcel, R., Aroca, R., and Ruiz-Lozano, J. (2012). Salinity stress alleviation using arbuscular mycorrhizal fungi. A review. *Agron. Sustain. Dev.* 32: 181–200.

Porcel, R., Redondogómez, S., Mateosnaranjo, E. et al. (2015). Arbuscular mycorrhizal symbiosis ameliorates the optimum quantum yield of photosystem II and reduces non-photochemical quenching in rice plants subjected to salt stress. *J. Plant Physiol.* 185: 75–83.

Prasad, S.C. (2016). Innovating at the margins: the system of rice intensification in India and transformative social innovation. *Ecol. Soc.* 21 (4): 7.

Prasad, P.V.V., Pisipati, S.R., Momčilović, I., and Ristic, Z. (2011). Independent and combined effects of high temperature and drought stress during grain filling on plant yield and chloroplast EF-Tu expression in spring wheat. *J. Agron. Crop Sci.* 197 (6): 430–441.

Rahman, M.S., Di, L., Eugene, G.Y. et al. (2018). Impact of climate change on soil salinity: a remote sensing based investigation in Coastal Bangladesh. *7th International Conference on Agro-geoinformatics (Agro-geoinformatics)*, Hangzhou, China (6–9 August 2018), 1–5.

Rashidi, M. and Keshavarzpour, F. (2007). Effect of different tillage methods on grain yield and yield components of maize (*Zea mays* L.). *Int. J. Agric. Biol.* 9 (2): 274–277.

Rasmussen, K.J. (1999). Impact of ploughless soil tillage on yield and soil quality: a Scandinavian review. *Soil Tillage Res.* 53 (1): 3–14.

Rasmussen, P.U., Hugerth, L.W., Blanchet, F.G. et al. (2018). Multiscale patterns and drivers of arbuscular mycorrhizal fungal communities in the roots and root-associated soil of a wild perennial herb. *New Phytol.* 220 (4): 1248–1261.

Reay, D.S., Davidson, E.A., Smith, K.A. et al. (2012). Global agriculture and nitrous oxide emissions. *Nat. Clim. Chang.* 2 (6): 410–416.

Reva, M., Cano, C., Herrera, M.A., and Bago, A. (2021). Arbuscular mycorrhizal inoculation enhances endurance to severe heat stress in three horticultural crops. *HortScience* 56 (4): 396–406.

Rillig, M.C. and Mummey, D.L. (2006). Mycorrhizas and soil structure. *New Phytol.* 171 (1): 41–53.

Rillig, M.C., Wright, S.F., Nichols, K.A. et al. (2001). Large contribution of arbuscular mycorrhizal fungi to soil carbon pools in tropical forest soils. *Plant Soil* 233 (2): 167–177.

Rillig, M.C., Sosa-Hernández, M.A., Roy, J. et al. (2016). Towards an integrated mycorrhizal technology: harnessing mycorrhiza for sustainable intensification in agriculture. *Front. Plant Sci.* 7: 1625.

Ruiz-Lozano, J.M. (2003). Arbuscular mycorrhizal symbiosis and alleviation of osmotic stress. New perspectives for molecular studies. *Mycorrhiza* 13 (6): 309–317.

Ruiz-Lozano, J.M., Aroca, R., Zamarreño, Á.M. et al. (2016). Arbuscular mycorrhizal symbiosis induces strigolactone biosynthesis under drought and improves drought tolerance in lettuce and tomato. *Plant Cell Environ.* 39 (2): 441–452.

Ruiz-Vera, U.M., Siebers, M., Gray, S.B. et al. (2013). Global warming can negate the expected CO_2 stimulation in photosynthesis and productivity for soybean grown in the Midwestern United States. *Plant Physiol.* 162 (1): 410–423.

Rymen, B., Fiorani, F., Kartal, F. et al. (2007). Cold nights impair leaf growth and cell cycle progression in maize through transcriptional changes of cell cycle genes. *Plant Physiol.* 143 (3): 1429–1438.

Sairam, R.K. and Tyagi, A. (2004). Physiology and molecular biology of salinity stress tolerance in plants. *Curr. Sci.* 86: 407–421.

Sakata, T. and Higashitani, A. (2008). Male sterility accompanied with abnormal anther development in plants–genes and environmental stresses with special reference to high temperature injury. *Int. J. Plant Dev. Biol.* 2 (1): 42–51.

Salim, B.B.M. and Abou El-Yazied, A. (2015). Effect of mycorrhiza on growth, biochemical constituents and yield of snap bean plants under water deficit conditions. *J. Hortic. Sci. Ornam. Plants* 7 (3): 131–140.

Sanghera, G.S., Wani, S.H., Hussain, W., and Singh, N.B. (2011). Engineering cold stress tolerance in crop plants. *Curr. Genomics* 12 (1): 30.

Santander, C., Ruiz, A., Garcia, S. et al. (2019). Efficiency of two arbuscular mycorrhizal fungi inocula to improve saline stress tolerance in lettuce plants by changes of antioxidant defense mechanisms. *J. Sci. Food Agric.* 100: 1577–1587.

Sarkar, A. (2015). Modern agriculture and food and nutrition insecurity: paradox in India. *Public Health* 129 (9): 1291–1293.

Schnepf, A., Jones, D., and Roose, T. (2011). Modelling nutrient uptake by individual hyphae of arbuscular mycorrhizal fungi: temporal and spatial scales for an experimental design. *Bull. Math. Biol.* 73 (9): 2175–2200.

Schoumans, O.F. (2015). *Phosphorus Leaching from Soils: Process Description, Risk Assessment and Mitigation. Wageningen University and Research.*

Schüßler, A., Schwarzott, D., and Walker, C. (2001). A new fungal phylum, the Glomeromycota: phylogeny and evolution. *Mycol. Res.* 105 (12): 1413–1421.

Selmar, D. and Kleinwächter, M. (2013). Stress enhances the synthesis of secondary plant products: the impact of stress-related over-reduction on the accumulation of natural products. *Plant Cell Physiol.* 54 (6): 817–826.

Sharif, M., Ahmad, E., Sarir, M.S. et al. (2011). Response of different crops to arbuscular mycorrhiza fungal inoculation in phosphorus-deficient soil. *Commun. Soil Sci. Plant Anal.* 42 (19): 2299–2309.

Sheng, M., Tang, M., Zhang, F., and Huang, Y. (2011). Influence of arbuscular mycorrhiza on organic solutes in maize leaves under salt stress. *Mycorrhiza* 21: 423–430.

Shi, G., Yao, B., Liu, Y. et al. (2017). The phylogenetic structure of AMF communities shifts in response to gradient warming with and without winter grazing on the Qinghai–Tibet Plateau. *Appl. Soil Ecol.* 121: 31–40.

Shimono, H., Okada, M., Kanda, E., and Arakawa, I. (2007). Low temperature-induced sterility in rice: evidence for the effects of temperature before panicle initiation. *Field Crop Res.* 101 (2): 221–231.

Shirani Bidabadi, S. and Mehralian, M. (2020). Arbuscular mycorrhizal fungi inoculation to enhance chilling stress tolerance of watermelon. *Gesunde Pflanz.* 72 (2): 171–179.

Singh, R.P., Brennan, J.P., Farrell, T. et al. (2005). Economic analysis of breeding for improved cold tolerance in rice in Australia. *Aust. Agribus. Rev.* 13: 1–9.

Sivakumar, K., Kumaresan, G., and Sugapriya, N. (2019). Arbuscular Mycorrhizal Fungi (AMF) and phosphate solubilizing bacteria (PSB) on tolerance of tomato under salt stress. *J. Pharmacogn. Phytochem.* 8: 4717–4721.

Skiba, U.M. and Rees, R.M. (2014). Nitrous oxide, climate change and agriculture. *CAB Rev.* 9 (010): 1–7.

Smith, S.E. and Read, D.J. (2008). *Mycorrhizal Symbiosis*. San Diego, CA: Academic Press.

Smith, F.A. and Smith, S.E. (2011). What is the significance of the arbuscular mycorrhizal colonisation of many economically important crop plants? *Plant Soil* 348 (1): 63–79.

Smith, S.E., Smith, F.A., and Jakobsen, I. (2003). Mycorrhizal fungi can dominate phosphate supply to plants irrespective of growth responses. *Plant Physiol.* 133 (1): 16–20.

Smith, S.E., Anderson, I.C., and Smith, F.A. (2018). Mycorrhizal associations and phosphorus acquisition: from cells to ecosystems. In: *Annual Plant Reviews*, vol. 48 (ed. W.C. Plaxton and H. Lambers), 409–439. Hoboken, NJ: *Wiley*.

Sosa-Hernández, M.A., Leifheit, E.F., Ingraffia, R., and Rillig, M. (2019). Subsoil arbuscular mycorrhizal fungi for sustainability and climate-smart agriculture: a solution right under our feet? *Front. Microbiol.* 10: 744.

Sowiński, P., Rudzińska-Langwald, A., Adamczyk, J. et al. (2005). Recovery of maize seedling growth, development and photosynthetic efficiency after initial growth at low temperature. *J. Plant Physiol.* 162 (1): 67–80.

Storer, K., Coggan, A., Ineson, P., and Hodge, A. (2018). Arbuscular mycorrhizal fungi reduce nitrous oxide emissions from N_2O hotspots. *New Phytol.* 220 (4): 1285–1295.

Strom, N., Hu, W., Haarith, D. et al. (2020). Interactions between soil properties, fungal communities, the soybean cyst nematode, and crop yield under continuous corn and soybean monoculture. *Appl. Soil Ecol.* 147: 103388.

Sun, X., Su, Y., Zhang, Y. et al. (2013). Diversity of arbuscular mycorrhizal fungal spore communities and its relations to plants under increased temperature and precipitation in a natural grassland. *Chin. Sci. Bull.* 58 (3): 4109–4119.

Thakur, P. and Nayyar, H. (2013). Facing the cold stress by plants in the changing environment: sensing, signaling, and defending mechanisms. In: *Plant Acclimation to Environmental Stress* (ed. N. Tuteja and S.S. Gill), 29–69. New York, NY: Springer.

Thakur, P., Kumar, S., Malik, J.A. et al. (2010). Cold stress effects on reproductive development in grain crops: an overview. *Environ. Exp. Bot.* 67 (3): 429–443.

Theocharis, A., Clément, C., and Barka, E.A. (2012). Physiological and molecular changes in plants grown at low temperatures. *Planta* 235 (6): 1091–1105.

Thirkell, T.J., Charters, M.D., Elliott, A.J. et al. (2017). Are mycorrhizal fungi our sustainable saviours? Considerations for achieving food security. *J. Ecol.* 105: 921–929.

Tian, H., Drijber, R.A., Zhang, J.L. et al. (2004). Different effects of arbuscular mycorrhizal fungal isolates from saline or non-saline soil on salinity tolerance of Plants. *Appl. Soil Ecol.* 26: 143–148.

Tilman, D., Fargione, J., Wolff, B. et al. (2001). Forecasting agriculturally driven global environmental change. *Science* 292 (5515): 281–284.

Tilman, D., Balzer, C., Hill, J., and Befort, B.L. (2011). Global food demand and the sustainable intensification of agriculture. *Proc. Natl. Acad. Sci.* 108 (50): 20260–20264.

Treseder, K.K., Egerton-Warburton, L.M., Allen, M.F. et al. (2003). Alteration of soil carbon pools and communities of mycorrhizal fungi in chaparral exposed to elevated carbon dioxide. *Ecosystems* 6 (8): 786–796.

Trnka, M., Rötter, R.P., Ruiz-Ramos, M. et al. (2014). Adverse weather conditions for European wheat production will become more frequent with climate change. *Nat. Clim. Chang.* 4 (7): 637–643.

Uemura, M., Joseph, R.A., and Steponkus, P.L. (1995). Cold acclimation of *Arabidopsis thaliana* (effect on plasma membrane lipid composition and freeze-induced lesions). *Plant Physiol.* 109 (1): 15–30.

UN DESA (2015). United Nations, Department of Economic and Social Affairs, Population Division (UN DESA), world population prospects: the 2015 revision, key findings and advance tables. Working Paper No. ESA/P/WP. 241, United Nations, Department of Economics and Social Affairs, New York.

United Nations Framework Convention on Climate Change (2015). Adoption of the Paris Agreement: draft decision -/CP.21: proposal / by the President. Conference of the Parties to the United Nations Framework Convention on Climate Change. President. Geneva : UN, 12 Dec 2015.

Van Der Heijden, M.G., Verkade, S., and de Bruin, S.J. (2008). Mycorrhizal fungi reduce the negative effects of nitrogen enrichment on plant community structure in dune grassland. *Global Change Biol.* 14 (11): 2626–2635.

Vance, C.P., Uhde-Stone, C., and Allan, D.L. (2003). Phosphorus acquisition and use: critical adaptations by plants for securing a nonrenewable resource. *New Phytol.* 157 (3): 423–447.

Verbruggen, E. and Toby Kiers, E. (2010). Evolutionary ecology of mycorrhizal functional diversity in agricultural systems. *Evol. Appl.* 3 (5–6): 547–560.

Várallyay, G. (2010). The impact of climate change on soils and on their water management. *Agron. Res.* 8 (2): 385–396.

Vermeulen, S.J., Aggarwal, P.K., Ainslie, A. et al. (2012). Options for support to agriculture and food security under climate change. *Environ. Sci. Pol.* 15 (1): 136–144.

Vollenweider, P. and Günthardt-Goerg, M.S. (2005). Diagnosis of abiotic and biotic stress factors using the visible symptoms in foliage. *Environ. Pollut.* 137 (3): 455–465.

Wahid, A. (2007). Physiological implications of metabolite biosynthesis for net assimilation and heat-stress tolerance of sugarcane (*Saccharum officinarum*) sprouts. *J. Plant Res.* 120 (2): 219–228.

Wang, Y. and Lambers, H. (2020). Root-released organic anions in response to low phosphorus availability: recent progress, challenges and future perspectives. *Plant Soil* 447 (1): 135–156.

Wang, D., Hejazi, M., Cai, X., and Valocchi, A.J. (2011). Climate change impact on meteorological, agricultural, and hydrological drought in central Illinois. *Water Resour. Res.* 47 (9): https://doi.org/10.1029/2010WR009845.

Wang, Z.G., Bi, Y.L., Jiang, B. et al. (2016). Arbuscular mycorrhizal fungi enhance soil carbon sequestration in the coalfields, northwest China. *Sci. Rep.* 6 (1): 1–11.

Wasaya, A., Tahir, M., Manaf, A. et al. (2011). Improving maize productivity through tillage and nitrogen management. *Afr. J. Biotechnol.* 10 (82): 19025–19034.

Weis, E. and Berry, J.A. (1988). Plants and high temperature stress. *Symp. Soc. Exp. Biol.* 42: 329–346.

WHO (2013). *Global Health Observatory*. WHO.

Wik, M., Pingali, P. and Brocai, S. (2008). *Global Agricultural Performance: Past Trends and Future Prospects*. Washington, DC: World Bank.

Williams, A., Manoharan, L., Rosenstock, N.P. et al. (2017). Long-term agricultural fertilization alters arbuscular mycorrhizal fungal community composition and barley (*Hordeum vulgare*) mycorrhizal carbon and phosphorus exchange. *New Phytol.* 213: 874–885.

Wolf, J., Johnson, N.C., Rowland, D.L. et al. (2003). Elevated CO_2 and plant species richness impact arbuscular mycorrhizal fungal spore communities. *New Phytol.* 157: 579–588.

Wu, J., Han, Z., Xu, Y. et al. (2020). Changes in extreme climate events in china under 1.5 °C–4 °C global warming targets: projections using an ensemble of regional climate model simulations. *J. Geophys. Res. Atmos.* 125: e2019JD03105.

Xiao, L., Liu, L., Asseng, S. et al. (2018). Estimating spring frost and its impact on yield across winter wheat in China. *Agric. For. Meteorol.* 260: 154–164.

Xie, W., Hao, Z., Zhou, X. et al. (2018). Arbuscular mycorrhiza facilitates the accumulation of glycyrrhizin and liquiritin in *Glycyrrhiza uralensis* under drought stress. *Mycorrhiza* 28 (3): 285–300.

Xin, Z. and Browse, J. (2000). Cold comfort farm: the acclimation of plants to freezing temperatures. *Plant Cell Environ.* 23 (9): 893–902.

Yadav, S.K. (2010). Cold stress tolerance mechanisms in plants. A review. *Agron. Sustain. Dev.* 30: 515–527.

Yang, W., Zheng, Y., Gao, C. et al. (2013). The arbuscular mycorrhizal fungal community response to warming and grazing differs between soil and roots on the Qinghai-Tibetan plateau. *PLoS One* 8: e76447.

Yeo, A. (1998). Predicting the interaction between the effects of salinity and climate change on crop plants. *Sci. Hortic.* 78 (1–4): 159–174.

Yoshida, R., Kanno, A., Sato, T., and Kameya, T. (1996). Cool-temperature-induced chlorosis in rice plants (I. Relationship between the induction and a disturbance of etioplast development). *Plant Physiol.* 110 (3): 997–1005.

Yue, Y., Zhou, Y., Wang, J., and Ye, X. (2016). Assessing wheat frost risk with the support of GIS: an approach coupling a growing season meteorological index and a hybrid fuzzy neural network model. *Sustainability* 8: 1308.

Zhang, Q., Chen, Q., Wang, S. et al. (2014). Rice and cold stress: methods for its evaluation and summary of cold tolerance-related quantitative trait loci. *Rice* 7 (1): 1–12.

Zheng, B., Chapman, S.C., Christopher, J.T. et al. (2015). Frost trends and their estimated impact on yield in the Australian wheatbelt. *J. Exp. Bot.* 66: 3611–3623.

Zhu, X., Song, F., and Xu, H. (2010). Influence of arbuscular mycorrhiza on lipid peroxidation and antioxidant enzyme activity of maize plants under temperature stress. *Mycorrhiza* 20 (5): 325–332.

Zhu, X.C., Song, F.B., Liu, S.Q., and Liu, T.D. (2011). Effects of arbuscular mycorrhizal fungus on photosynthesis and water status of maize under high temperature stress. *Plant Soil* 346 (1): 189–199.

Zhu, G., Wang, S., Wang, W. et al. (2013). Hotspots of anaerobic ammonium oxidation at land–freshwater interfaces. *Nat. Geosci.* 6 (2): 103–107.

Zhu, X., Song, F., and Liu, F. (2017). Arbuscular mycorrhizal fungi and tolerance of temperature stress in plants. In: *Arbuscular Mycorrhizas and Stress Tolerance of Plants* (ed. Q.-S. Wu), 163–194. Singapore: *Springer*.

Zhu, J., Li, M., and Whelan, M. (2018). Phosphorus activators contribute to legacy phosphorus availability in agricultural soils: a review. *Sci. Total Environ.* 612: 522–537.

Part 4

Plant Stress Under Climate Change: Molecular Insights

15

Plant Stress and Climate Change

Molecular Insight

Anamika Roy[1], *Mamun Mandal*[1], *Ganesh Kumar Agrawal*[2,3], *Randeep Rakwal*[2,3,4], *and Abhijit Sarkar*[1]

[1] *Laboratory of Applied Stress Biology, Department of Botany, University of Gour Banga, Malda, West Bengal, India*
[2] *Department of Education, Global Research Arch for Developing Education (GRADE) Academy Pvt. Ltd., Birgunj, Nepal*
[3] *Department of Biotechnology, Research Laboratory for Biotechnology and Biochemistry (RLABB), Kathmandu, Nepal*
[4] *Department of Health and Sport Science, Faculty of Health and Sport Sciences, University of Tsukuba, Tsukuba, Japan*

15.1 Introduction

Stress can be any unfavorable environmental condition that has the potential to impede agricultural output and plant growth. Some of major abiotic stress factors are drought and flood, low, and high temperature, salinity, ultraviolet (UV) radiation, and heavy metal stress. These stresses negatively affect plant growth and development (Levitt 1980; Wang et al. 2004; Wania et al. 2016). Environmental stress factors can change speedily or gradually. So, reorganization of these stress factors and give responses against them in appropriate time is very crucial for plants for their survival (Takahashi and Shinozaki 2019).

Under this abiotic stress, some cellular adaptations at physiological and biochemical levels and molecular adaptations are observed in plants. It finally brings changes in plant morphology, which helps plants decrease the damaging effects caused by abiotic factors (Wang et al. 2020b). Generally, plants have two strategies against stress. Either they avoid the stress (stress avoidance) or they tolerate the stress (stress tolerance). Stress avoidance is a physiologically nonactive stage, like mature seed, whereas stress tolerance is an active mechanism where adjustments are made in response to stress, known as acclimation. Desert plants like ephemerals can escape the stress (Zhu 2002). Acclimation is a time-dependent process (Levitt 1980; Larcher 2003). Acclimations bring changes in plant proteomes, transcriptomes, and metabolomes (Table 15.1). Reports prove that changes in transcriptome levels often do not affect plant proteins. So, plant proteins are highly important for their crucial role against plant stress (Gygi et al. 1999; Bogeat-Triboulot et al. 2007).

Intergovernmental Panel on Climate Change (IPCC) reported that plants are exposed to stress as a result of climate changes (Andy 2016). Climate changes increase the levels of both carbon dioxide and temperature (Hirayama and Shinozaki 2010). Deforestation and the use of fuel increased the carbon dioxide concentration from 280 to 400 μmol^{-1}. It is also predicted that the elevated level of CO_2 will become 800 μmol^{-1} at the end of this century. This elevated level of carbon dioxide is the main cause of the greenhouse effect as well as global warming (Vaughan et al. 2018).

Plant stress proteomics is a dynamic process that mainly deals with plant proteins that are formed during stress and also depends on the stress intensity and duration. Mainly, there are four stages in plant stress response. First, an initial alarm phase; then an acclimation phase; then maintenance phase and exhaustion phase; and lastly recovery phase. The alarm phase is the very initial phase when stress is introduced for the first time and stress tolerance decreases. During the acclimation phase, stress tolerance increased. In the maintenance phase, stress tolerance levels remain constant. Lastly, an exhaustion phase is noticed, where stress tolerance levels severely decrease. After plant stress has ceased, a recovery phase can be observed (Larcher 2003). Stresses stimulate different responses in plants, like changes in growth, membrane fluidity, altered gene expression, cellular metabolism, protein accumulation and their cellular localization, and protein function (Verma et al. 2013). This chapter therefore aimed to shed light on climate change and plant responses to different stress factors (drought and flooding, high and low temperatures, salinity, ultraviolet radiation, and heavy metal stress), their management strategies, and the regulatory pathways.

Global Climate Change and Plant Stress Management, First Edition. Edited by Mohammad Wahid Ansari, Anil Kumar Singh, and Narendra Tuteja.

Table 15.1 Molecular responses of important crops under diverse climate factors.

Stress	Plant	Level	Response	Method	References
Drought	*Sorghum bicolor*	Proteomic	Starch biosynthesis, RUBISCO activation, abscisic acid signaling, HSPs, phospholipid metabolism, ROS scavenging, epicuticular wax production	LC-MS/MS analysis	Ogden et al. (2020)
	Chickpea	Proteomic	34 differentially expressed proteins interrelated to photosynthesis, protein synthesis and degradation, stress response, gene transcription and signaling	2D-E with (LC-MS/MS)	Vessal et al. (2020)
Waterlogging	*Manihot* sp. (cassava)	Transcriptome	MYB, AP2/ERF, WRKY, NAC	qRT-PCR RNA sequencing	Cao et al. (2022)
	Triticum sp. (Wheat)	Transcriptome	AP2/ERF, WRKY, MYB	Transcriptome sequencing (Illumina NovaSeq^{TM} 6000)	Shen et al. (2020)
	Helianthus sp. (sunflower), *Elaeis* sp. (oil palm)	Metabolomic	Upregulation of GABA, polyols, organic acid, down-TCA cycle, N_2 metabolism	LC-MS/MS	Cui et al. (2019b)
	Medicago trancatula	Metabolomic	Accumulation of GABA, glutamine, spermidine, arginine	GC-MS	Lothier et al. (2020)
	Hordeum sp. (barley)	Genomic	Upregulation of gene of trichome birefringence, α/β-hydrolases, serine carboxypeptidase, auxin-responsive protein	RNA-sequencing	Borrego-Benjumea et al. (2020)
	Cucumis sp. (cucumber)	Proteomic	Pyruvate decarboxylase, alcohol dehydrogenases, 1-amino cyclopropane-1-carboxylic acid oxidases, peroxidases	iTRAQ (isobaric tags for relative and absolute quantification) followed by qRT-PCR	Xu et al. (2016)
	Vitis sp. (grapevine)	Proteomic	Upregulation of proline-rich protein kinase, PERK15 isoform X2, ATP-dependent zinc metalloprotease, polyphenol oxidase, glycine-rich protein	Tandem mass tag (TMT)	Wang et al. (2021)
High temperature	*Monochamus alternatus*	Transcriptome	Upregulation of several DEGs	RNA sequencing	Li et al. (2020a)
	Oryza sp.	Transcriptome	Heat shock transcription factors (HSF) and non-HSF, DREB, NAC, OsBBX22, OsBBX24, MYB	Microarray analysis	Zhang et al. (2013)
	Solanum sp. (tomato)	Genomic	HSP gene	MACE	Fragkostefanakis et al. (2015)
	Arabidopsis sp.	Genomic	Induce male meiotic genome restitution	Radial microtubule array (RMA)	De Storme and Geelen (2020)
	Lepidium sp. (maca)	Proteomic	HSP protein, ROS-scavenging enzyme	TMT	Wang et al. (2020c)
	Solanum sp. (tomato)	Proteomic	Upregulation of proteins involve in photosynthesis, defense, protein folding and metabolism	MALDI-TOF	Sang et al. (2017)
	Arabidopsis sp.	Metabolomic	Accumulation of beta-alanine, naringenin, glycine betaine, amino acids and peptides	GC-MS	Wang et al. (2020a)

Table 15.1 (Continued)

Stress	Plant	Level	Response	Method	References
	Pinus radiata	Metabolomic	Phenylalanine, dihydromyricetin, hexadecanoic acid, cytokinin, flavonoid, terpenoid	GC-MS and LC-Orbitap-MS	Escandon et al. (2018)
Low temperature	*Eukalyptus globules*	Genomic	Upregulation of EglWRKY	RNA sequencing	Aguayo et al. (2019)
	Brassica rapa	Genomic	Genes of MAPK pathway, ribosome biogenesis, signal transduction, basal transcription factor	qRT-PCR	Ma and Jeffrey (2019)
	Medicago falcate	Transcriptome	WRKY, NAC, MYB, bHLH, ERF	SMRT-sequencing	Cui et al. (2019a)
	Solanum sp. (potato)	Proteome	Upregulation of defense protein, ribosome formation, protein folding and degradation Downregulation of chlorophyll protein	LC-ESI-MS/MS	Li et al. (2021)
	Alium sativum	Proteome	Antioxidative enzymes, enzymes of carbohydrate and nucleotide metabolism, protein folding, splicing factors	LC-ESI-MS/MS	Dufoo-Hurtado et al. (2015)
	Solanum sp. (potato)	Metabolome	Upregulation of proline, monoaldehyde, soluble protein	LC-ESI-MS/MS	Li et al. (2021)
Salinity	*Gossypium hirsutum*	Proteomic	Proteins of cytoskeleton, membrane transport, signal transduction, stress response, antioxidative protein	iTRAQ	Li et al. (2015)
	Hibiscus hamabo	Transcriptomic	WRKY, bHLH	RNA sequencing	Ni et al. (2022)
	Triticum aestivum	Transcriptomic	Transcription factors like WD40, MYB-HB, bZIP, WRKY	qRT-PCR	Goyal et al. (2016)
	Lactuca sativa	Metabolomic	Secondary metabolites, proline, soluble sugar, ascorbic acid	Colorimetric analysis and spectrometric analysis	Zhang et al. (2021)
UV radiation	*Populus cathayana*	Proteomic	Protein involve in photosynthesis, transcription, translation and post-translational modification, redox system, carbohydrate and energy metabolism	iTRAQ analysis	Zhang et al. (2017)
	Oryza sp.	Proteomic	Carbohydrate and energy metabolism, translation and transcription, antioxidative enzyme, N_2 metabolism, fatty acid biosynthesis	MALDI-TOF-MS	Kumar et al. (2016)
	Nymphoides humboldtiana	Metabolomic	Secondary metabolite production (flavonoid), antioxidant activity	—	Nocchi et al. (2020)
Heavy metal	*Arabidopsis* sp.	Genomic	Heavy metal associated protein (HMP) genes	Pfam database, followed by phylogenetic and gene duplication analysis	Li et al. (2020b)
Air pollution	*Oryza sativa* L.	Proteomic	PR (pathogen related) proteins, ATP-dependent CLP protease, superoxide dismutase, ascorbate peroxidase(s), calcium-binding protein	2-DE protein profiles	Agrawal et al. (2002)

(*Continued*)

Table 15.1 (Continued)

Stress	Plant	Level	Response	Method	References
Climate changes	Ozone-tolerant *Arabidopsis* sp.	Transcriptomic	Ethylene signaling, jasmonic acid signaling	Expressed sequence tags (EST)	Tamaoki et al. (2003)
	Glycine sp. (soybean)	Genomic	Upregulation of photosynthetic gene, respiration gene and hypoxia responsive gene	Genome analysis and FACE	Leakey et al. (2009)
	Glycine sp. (soybean)	Proteomic	Enzymes related to carbohydrate and energy metabolism, dark reaction of photosynthesis	Microarray	Leakey et al. (2009)
	Arabidopsis sp.	Proteomic	Upregulation of dark reaction enzymes of photosynthesis	Gene chip and FACE	Markelz et al. (2014)
	Arabidopsis sp.	Transcriptomic	Transcripts of Glycolysis, TCA cycle, electron transport system	Microarray	Markelz et al. (2014)
	Arabidopsis sp.	Metabolomic	Carbohydrate, sugar alcohol, maltose, galactose	Principal component analysis (PCA) of metabolome	Noguchi et al. (2015) and Kaplan et al. (2012)
	Festuca sp. *Raphanus sativus*	Metabolomic	Accumulation of monosaccharide, sucrose, hexose	Biometric analysis	Urbonaviciute et al. (2006) and Noguchi et al. (2015)

LC-ESI-MS/MS, liquid chromatography-electrospray ionization-tandem mass spectrometry; FACE, free-air CO_2 enrichment.

15.2 Different Stress Factors and Climate Changes Effects in Plants

15.2.1 Water Stress

Water stress can be further divided into drought (water scarcity) and flooding (excess water availability).

15.2.1.1 Drought

Drought means decreased water potential in the soil. It causes cellular dehydration. High temperatures and low humidity in the soil and atmosphere are the main causes of drought. An imbalance between evapotranspiration flux and water intake may result in a drought-like condition (Lipiec et al. 2013). Two thirds of global land areas and 15% of rural land areas are affected by drought (Cramer et al. 2011). According to Anjum et al. (2011), plants are exposed to drought when water uptake by the root is limited or water loss through transpiration is very high. Poor germination and damaged seedling establishment are main effects of drought (Kaya et al. 2006; Farooq et al. 2009). Drought also limits mitosis and cell elongation that causes reduction in cell growth (Hussain et al. 2008). Drought also causes reduction in cell growth due to the absence of turgor (Taiz and Zeiger 2006). Drought alters the enzymatic activity of RUBISCO large unit and FBP aldolase, which results in the formation of new homeostasis (Hajheidari et al. 2007).

15.2.1.2 Flooding or Waterlogging

Flooding occurs when water is present in excess amounts in the soil. Plants cannot absorb this excess amount of water, so traces of water are present in the soil surface. This also leads to unavailability of oxygen (Kosova et al. 2011). Continuous rainfall in poorly drained areas is one of the main reasons for flooding. Flooding can be divided into two types waterlogging, when only plant roots are covered by water, and submergence, when aerial plant tissue completely covered by water. Both conditions result in an anaerobic condition, also known as hypoxia, where oxygen level ≤21% (Sasidharan et al. 2018). Rice is a flood-tolerant species, whereas wheat, barley, and maize are flood-sensitive species (Komatsu et al. 2013). On a global scale, during 2006–2016, floods caused two-thirds of the damage and also caused losses in crop production (Food and Agriculture Organization of the United Nations [FAO] 2017). Waterlogging negatively affects plant growth and development, photosynthetic rates, and chlorophyll reduction (Kuai et al. 2014; Yan et al. 2018).

15.2.2 Temperature Stress

15.2.2.1 High Temperature Stress

It is predicted that temperature increased by 0.2 °C $decade^{-1}$, it leads to temperature level higher (2–11 °C) than the current level by the end of 2100 (Lee 2007). Heat stress affects plant growth, development, and crop yields and also generates reactive oxygen species (ROS) (Hasanuzzaman et al. 2012, 2013). Heat stress could also destabilize and melt DNA, RNA, and proteins (Sharma et al. 2017). Elevated temperatures cause scorching of leaves and twigs, leaf abscission, senescence, and inhibition of root and shoot growth (Vollenweider and Gunthardt-Goerg 2005). High temperatures also affect the shoot net assimilation rates and total dry weight, which result in a reduction in plant growth (Wahid et al. 2007). It alters the composition of the cytoskeleton, induces the accumulation of different eukaryotic transcriptional factors, which finally result in programmed cell death (Majoul et al. 2004; Zhang et al. 2010).

15.2.2.2 Low Temperature Stress

Low temperature stress can be of two types: chilling (temperature ≤20 °C) and freezing (temperature ≤0 °C) (Chinnusamy et al. 2007). Cold stress could decrease membrane fluidity, disrupt ion and electron flow, and other metabolic activities in plants. Cold stress also causes freezing of cell, cell membrane destruction, and finally, cell death (Kidokoro et al. 2017). The low temperature also generates reactive oxygen species (ROS) (Rezaie et al. 2020). Low temperature stress causes reduction in the activity of photosystems I and II, Calvin cycle related enzymatic activity (Hajihashemi et al. 2018). Both heat and cold temperature stress create an imbalance between photosynthetic electron transport and carbon assimilation processes, which leads to photoinhibition and thermal energy dissipation (Kosova et al. 2015).

15.2.3 Salinity Stress

Soil salinity is one of the major worldwide problems (Evelin et al. 2019). Salinity stress is caused by the accumulation of excess amount of salt ions (mainly Na^+ and Cl^-) in the soil. The soil salinity is developed due to accumulation of salt by primary (natural) and secondary (anthropogenic) processes. Weathering from parent rock and deposition from seawater are primary processes, whereas poor rainage and continuous irrigation are secondary processes (Evelin et al. 2019). This excess salt concentration causes reduction in water potential of the soil (Sakamoto et al. 2008). High salt stress results in ion toxicity, oxidative stress, genotoxicity, altered metabolic pathways, and reduction in both cell division and elongation. It also alters protein and lipid composition of plant cell plasma membrane, generates ion imbalance, and hampers plant growth and development (Fujii and Zhu 2009; Lopez-Perez et al. 2009; Rodriguez-Milla and Salinas 2009). Salt stress decreases agricultural productivity and crop yields (Rahman et al. 2019). Globally, almost 800 million ha are affected by salinity (FAO 2017).

15.2.4 Ultraviolet (UV) Radiation Stress

Ozone layer is like a sunshade that protects life forms on Earth from ultraviolet radiation. Depletion of this ozone layer can cause UV damage by releasing chlorofluorocarbons (CFCs) (Moan 2011). UV radiation includes UV-A (315–400 nm), UV-B (280–315 nm), and UV-C (100–280 nm). Among them, UV-B is the most potential stress factor (Singh et al. 2006). UV effects on organisms depend on the intensity of radiation and the developmental stages of organisms (Moreira-Rodriguez et al. 2017). UV radiation can affect plants both positively and negatively. Positively, it induces crop production and enhances defense mechanism against pathogens (Wargent and Jordan 2013). But it also causes DNA damage, the formation of reactive oxygen species (ROS) and alteration of cellular processes. These damages directly or indirectly inhibit several cellular processes like photosynthesis, respiration, growth, and development. UV radiation also hampers photosynthetic electron transport and results in increased activity of NADPH-oxidases and peroxidases as a result of ROS generation (Jenkins 2009; Li et al. 2013).

15.2.5 Heavy Metal Stress

Heavy metal stress is developed due to metal ion penetration into the plant cell, which results in ion toxicity. Generally, heavy metal binds with proteins and alters their conformation as well as functions (Ahsan et al. 2009; Hossain and Komatsu 2013). Some metals and metalloids play very crucial roles (components of chloroplast and mitochondrial electron transport systems, transcription factors, etc.) in plant growth and development. But increasing concentration of these metals in the soil can inhibit plant growth and development (Kosova et al. 2011). Toxic metals like cadmium (Cd), copper (Cu), chromium (Cr), lead (Pb), zinc (Zn) as well as hazardous metals like arsenic (As), boron (B) all are considered as heavy metals. These heavy metals negatively affect plant growth and development (Hossain et al. 2012a). Besides affecting plant growth and development, heavy metals also induce misfolding in both newly synthesized as well as newly synthesized proteins (Amm et al. 2014; Zhou et al. 2016).

15.2.6 Air Pollution Stress

Air pollution is a major problem nowadays. Air pollutants adversely affect plant growth, development, and yield (Gostin 2016). Urbanization, industrialization, and deforestation are some of the major causes of air pollution in developing countries like India. In India, Delhi is in the worst conditions, though CNG-driven public transports are used there (Saxena et al. 2012). Air pollutants like sulfur dioxide and flyash can reduce photosynthetic pigments: chlorophyll and carotenoids. These photopigments play a crucial role in plant photosynthesis, so a reduction in their concentration can negatively affect plant photosynthesis. Environmental stress also inactivates RUBISCO (Govindaraju et al. 2011). Plants leaves are surrounded by several pollutants like O_3 and NO_x, which modify net carbon fixation. Sometimes these pollutants accumulate in the soil and thus disturbing the function of plant roots (mineral and water uptake). Air pollutants can also imbalance plant communities, species interactions, plant growth, development, and economic yield (Weber et al. 2002).

15.2.7 Climate Change

Deforestation, industrialization, increasing livestock farming, more nitrogen-containing fertilizer use are some of the human causes that are responsible for climate change. Atmospheric CO_2 levels will be increased from 730 to 1000 ppm by the end of the century (Intergovernmental Panel on Climate Change [IPCC] 2007, 2014). Elevated temperatures, drought, flooding, rising sea levels, increased carbon dioxide and greenhouse gases concentration all are byproduct of climate change. Increasing level of carbon dioxide and temperature can directly affect the growth and development of plants both positively and negatively (Kukal and Irmak 2018). Positive effects of elevated carbon dioxide is that it helps increasing the rate of plant photosynthesis which further results in greater productivity, growth, yield (Ainsworth and Long 2005; Van der Kooi et al. 2016). CO_2 affects the quality of crops negatively, as it reduces nutrients, protein concentration, vitamins, and some macro- and micro-elements (Hogy et al. 2013; Myers et al. 2014; Fernando et al. 2015; Broberg et al. 2017).

15.3 Plant Responses Against Stress

These stresses not only bring changes physiologically but also at proteome and transcriptome level (Li et al. 2008; Shulaev et al. 2008; Chae et al. 2009). Mark Wilkins first coined the term "proteome" in 1994, which is a combination of protein and genome. Mainly, proteome means the entire complement or database set of proteins. Plant proteins help alter physiological traits in response to stress. So, plant proteins play very crucial role against stress (Jorrin-Novo et al. 2009; Kosova et al. 2011). Transcriptome refers to the set of all RNA transcripts of a cell including both coding and non-coding RNA (Wang et al. 2020b).

15.3.1 Water Stress Responses

15.3.1.1 Drought Responses

Stomatal closure is the first physiological response to drought. Accumulation of osmolytes (proline, glycine-betaine, polyamines, and polyols) and hydrophilic proteins (LEA proteins) can be observed at the molecular level (Kosova et al. 2018). In *Arabidopsis thaliana*, phytohormone signaling regulates responses against drought (Gupta et al. 2020). Abscisic acid (ABA) is a key phytohormone that regulates drought responses by closing stomata and stress-responsive gene expression (Addicott et al. 1968; Cutler et al. 2010). Plants first sense the water deficit signal, then transmit it from root to leaves, and regulate drought in both ABA-dependent and independent manner (Takahashi et al. 2018). According to recent studies, drought stress responses are also regulated by some hormones like peptides (Takahashi et al. 2019). Calcium waves, hydraulic signals, and mRNAs also play crucial roles in drought stress responses (Notaguchi and Okamoto 2015; Choi et al. 2017; Kudla et al. 2018). Increased expression of HSP 70 also observed (Kosova et al. 2011). Some key genes enhance drought tolerance, like DRO1 in rice and ERECTA in *Arabidopsis*. Besides that ethylene response factors (ERF), DREB, ZFP, and MYB also regulate drought responses in wheat, maize, and rice (Kulkarni et al. 2017).

Genomic Responses Drought stress causes the production of ROS, which function as a major signal for the activation of adaptation plant defense. ROS are detoxified in two ways: enzymatically and non-enzymatically. Drought-induced accumulation of osmoprotectants also boosts the activity of a number of pathways. During drought conditions, the biosynthesis of SOD, POD, TSP, and CAT, as well as polyols/sugars, glycine betaine, and dehydrins, is considerably increased (Singh et al. 2022). Through transcriptional regulation, nitric oxide (NO) works as a secondary signaling molecule in ROS scavenging, further activating antioxidant defense enzymes like POD, CAT, SOD, TSP. Drought tolerance is a complicated and multidimensional issue mediated by multiple complex systems, as evidenced by the cross-talk between signaling molecules and regulatory genes including MYB, DEREB/ERF, WRKY, NOS, and NAC (Singh et al. 2022).

Proteomic Responses The photosynthetic electron transport chain was severely hampered by drought stress, and the additional excitation energy was channeled into the generation of ROS. Numerous ROS scavengers were produced under drought conditions to mitigate the negative impacts of such ROS. Dehydrins, dehydro ascorbate reductase, γ-glutamyl cysteine synthetase, quinine reductase, and glutathione S-transferases are among the enzymes found in plants, as founded proteomic investigations (Mishra and Das 2003). Furthermore, in drought circumstances, plant roots have higher quantities of molecular chaperones including heat shock proteins. By inhibiting aggregation and helping in the non-active proteins refolding, those proteins may perform a cytoprotective effect within plants root. Numerous proteomic investigations revealed that roots respond to drought stress utilizing processes that are identical to those seen in damaged cells. Drought-tolerant types of rapeseed seedlings, for example, exhibited higher amounts of proteosomal factors (Mohammadi et al. 2012). Likewise, dehydration was discovered to stimulate many proteolytic enzymes that aid in the breakdown of permanently degraded proteins in plants. Leucine aminopeptidases, cysteine proteases, ubiquitin family proteases, and multi-catalytic endopeptidases are only a few examples (Ghosh and Xu 2014). In addition, defense-related proteins and protease inhibitors, including pentatrico peptide repeat (PPR) protein, phloem serpin-1, nucleotide binding domain-leucine rich repeat protein (NB-LPR), Knotted 1 (kn1), and jasmonate-inducible proteins, were abundant in plants. These observations clearly imply that protease inhibitors and defense-related proteins that govern programmed cell death (PCD) are also implicated in the drought stress response (Ghosh and Xu 2014).

Transcriptomic Responses In drought-induced abiotic stress responses, transcription factors are critical components of the transcriptional regulatory network. DREB or CBF, MYB, bZIP, NAC, and MYC families' members have been identified as having ABA-independent and ABA-dependent functions in the regulation of gene expression in drought responses (You et al. 2019). The identified ABA receptors and other fundamental signaling molecules have properly explained the ABA signaling pathway, which is crucial to drought tolerance responses in plants (Zhu 2016). The ABA receptors PYR/PYL/RCAR block the phosphatase activity of Group A PP2Cs, leading to activation of SnRK2s, which phosphorylates or activates downstream substrates including bZIP transcriptional regulators and ion channels (Ma et al. 2016). RNA-seq is widely utilized to uncover the mechanistic mechanisms of drought adaptations in so many plant species, due to the advent of high-throughput sequencing technologies (Egea et al. 2018). In most investigations, transcriptome analysis demonstrated that water scarcity condition causes alterations in photosynthesis, hormone signal transduction, amino acid metabolism, carbohydrate metabolism, secondary metabolites, and fatty acid metabolism (You et al. 2019).

Metabolomic Responses The sugar (such as fructan and trehalose) and amino acids (such as proline) accumulations that function as osmoprotectants during drought conditions are an evident example of metabolites associated with plant responses to stress (You et al. 2019). Modifications in primary and secondary metabolites have recently been linked to drought responses in various plant species, which were confirmed by metabolic analysis in previous studies. In *Arabidopsis* plant, many studies have previously reported that the levels of mainly lipids (including the acylated steryl glycosides and glycosyl inositol phosphoceramides), amino acids (proline, tryptophan, glutamine, aspartate, alanine, isoleucine, ornithine, valine, and leucine), flavonoids (cyaniding and quercetin), and TCA cycle intermediated (*cis*-aconitate, 2-oxoglutarate, and succinate) were tremendously enhanced during the water scarcity conditions (Tarazona et al. 2015; Pires et al. 2016). Likewise, water deficit caused significant alterations in the levels of metabolites especially in the case of organic acids in crop plants such as maize (Obata et al. 2015), rice (Ma et al. 2016), and barley (Chmielewska et al. 2016).

15.3.1.2 Waterlogging Responses

To cope with waterlogging stress, plants modify their morphological structure, endogenous hormone biosynthesis, signaling process, and metabolic pathway (Pan et al. 2021).

Genomic Responses Genes involved in antioxidant system, glycolysis, amino acid metabolism, and fermentation pathways are regularly upregulated. Photosynthetic genes and genes involved in starch, sucrose metabolism, N_2 metabolism are downregulated, whereas stress-tolerance genes, endogenous hormone producing genes, and ethylene-binding factors are upregulated. Ethylene increases shoot length during waterlogged conditions (Wang et al. 2020b). Prolonged waterlogging stress (hypoxia condition) affects plant growth, development, and metabolism negatively; as a response, "core hypoxia responsive genes" are induced. Such genetic responses lead to reorganization of root metabolism (Hsu and Shih 2013).

Proteomic Responses Glycolysis enzymes are upregulated, resulting in increased carbohydrate metabolism. Waterlogging also results in an anaerobic, or lack of oxygen, condition. So, fermentation enzymes and cell wall modifying enzymes are also increased in response to waterlogging (Kosova et al. 2018). Proteome changes and elevated hydrogen peroxide levels are observed in plants (Ahsan et al. 2007; Alam et al. 2010). Proteins involved in photosynthesis, signal transduction, programmed cell death (PCD), stress and defense mechanisms, mRNA processing are upregulated; in contrast, activity of large subunit of RUBISCO is downregulated (Kosova et al. 2011).

Transcriptomic Responses Eukaryotic transcription factor-EIF-5A-1, apoptosis antagonizing transcription factor (AATF) become more abundant to increase oxygen supply in tissue by forming aerenchyma (Kosova et al. 2011). Initial phase of flooding alters splicing factors U3, snoRNP, NOP1/NOP56 complex that alters expression of mRNA, ribosome, and protein in soybean (Yin and Komatsu 2015, 2016, 2017). Transcription factors such as MYBs, AP2/ERFs, WRKYs, and NACs are identified as a response to waterlogging in cassava. Transcription factors like AP2/ERF (Sub1A, HRE1, HRE2, Snorkel1, Snorkel2, RAP2.2) enhance hypoxia tolerance; WRKY and NAC transcriptionally regulate waterlogging genes (Van Veen et al. 2016; Cao et al. 2022).

Metabolomic Responses Previous metabolomic studies in *Arabidopsis* showed a slight difference in shoot and root responses under waterlogging stress. Root responded more strongly as compared to shoot. GABA (γ-aminobutyrate), alanine, and lactate are accumulated in root; while shoot faces less hypoxic condition, only alanine is produced due to oxygen generation by photosynthesis (Mustroph et al. 2014). Increase in GABA, polyols (glycerol and inositol), and organic acids (citrate, aconitate, and malate) and decrease in many amino acids, TCA cycle, and nitrogen metabolism were reported in sunflower and oil palm (Cui et al. 2019b). GC-MS analysis in *Medicago* showed that GABA, polyols, and polyamines like spermidine levels are increased, and raffinose and xylose levels are decreased significantly (Lothier et al. 2020). Prolonged waterlogging induces accumulation of toxic metabolites like ethanol, hydrogen peroxide, and lactic acid, which finally leads to cell senescence and plant death (Zhang et al. 2017).

15.3.2 Temperature Stress Responses

15.3.2.1 High Temperature Stress Responses

Plants develop different strategies against high temperatures.

Genomic Responses Different genes linked to both stress recognition and stress response are mainly regulated at transcriptional and post-transcriptional level. Altered gene expression by transcriptomic approach has been studied in rice (Mangrauthia et al. 2017), wheat (Qin et al. 2008), tomato (Bita et al. 2011), barley (Mangelsen et al. 2011), maize (Shi et al. 2017), soybean (Gillman et al. 2019), brassica (Dong et al. 2015). These heat responsive genes are mainly involved in translation, transcription, post-translational modification, primary and secondary metabolism (Takahashi and Shinozaki 2019).

Proteomic Responses High temperature not only increases protein mis-folding but also denatures intracellular proteins and membrane complexes (Kosova et al. 2011). Several proteins with chaperone (chaperonin 60β subunit CPN60-β, HSP90, chaperonin CPN10, chloroplast chaperonin) levels are increased; five different subfamilies of HSPs: HSP110, HSP90, HSP70, HSP60, and small HSPs (sHSP) are produced; a balanced level of HSP70 is maintained throughout the stress (Baniwal et al. 2004). Upregulation of HSP100, HSP70, and sHSP is observed in rice (Lee 2007), wheat (Majoul et al. 2004), and heat-tolerant poplar (Ferreira et al. 2006). Enzymes related to glycolysis and Krebs cycle (pyruvate dehydrogenase, transketolase, and thiamine biosynthesis) are upregulated, which finally increases the activity of RUBISCO activase and pentose phosphate pathway. Ribulose-1,5-bis-phosphate (RuBP), the substrate of RUBISCO, is regenerated by the pentose phosphate pathway (Baniwal et al. 2004).

Transcriptomic Responses Eukaryotic translation initiation factors (eIF4F, eIF5A-3) are continuously accumulated leading to programmed cell death (PCD) under a long-term heat treatment (Majoul et al. 2004; Zhang et al. 2010). In *Arabidopsis*, HSFA1 and dehydration-responsive element binding proteins (DREBs) are upregulated (Holladay et al. 2004; Wang et al. 2020a). Transcription factors like ERF, bHLH, and NAC are continuously reported. These transcripts are mainly expressed during light responses, biotic and abiotic stress responses, and development (Hwang et al. 2019). In heat-stressed transgenic *A. thaliana*, CsCML45 gene is regularly overexpressed, which enhances the heat tolerance level of the plant (Wang et al. 2019). Differentially expressed genes (DEGs) enhances expression of ROS-scavenging enzyme to protect from high oxidative stress during high temperature by Li et al. in *Monochamus alternates* (2020a) (Bakku et al. 2020).

Metabolomic Responses Plant produced antioxidants, hormones (ABA, salicylic acid, and ethylene), secondary metabolite (phenolics, anthocyanin, flavonoids, and flavons), osmoprotectants, polyamines, and different regulatory pathways (like violaxanthin cycle) are ongoing to protect themselves from stress (Nahar et al. 2015). Many osmolytes (glycine betaine, soluble sugars) are accumulated; these osmolytes help to maintain protein stability and membrane bilayer (Sung et al. 2003; Sihag et al. 2023). Carbohydrates and proline play important roles in heat stress. Proline content is increased by almost 6.03 log fold level. Different carbohydrate conjugates like galactinol, raffinose, and melibiose also serve as osmoprotectants during heat stress (Ren et al. 2019).

15.3.2.2 Low Temperature Stress Responses

Genomic Responses Cold stress enhances expression of multiple EglWRKY genes. Almost 11 EglWRKY genes were reported in *Eukalyptus globulus* plants (Aguayo et al. 2019). Knockdown of SIF3HL through RNA interference makes plants sensitive to cold stress. SIICE1 induces expression of SICBF1 and SIDRCi7, which help to tolerate chilling stress in tomato (Miura 2012). This SICBF1 of tomato, when induced in *Arabidopsis*, becomes cold tolerant (Zhang 2004). Cold stress induces CBF family genes in plants, which encode transcription factors like CBF1, CBF2, and CBF3 (also known as DREB1B, DREB1C, and DREB1A). This CBF gene regulates almost 100 downstream genes; among them, dehydration responsive element (DRE) is one of the major genes that act in cold responsiveness (Yamaguchishinozaki and Shinozaki 1994).

Proteomic Responses Besides high temperature, low temperature also induces mis-folding in newly formed as well as existed protein. Cold stress decreased photosynthesis rate (Kosmala et al. 2009); to cope with this, RUBISCO subunits and RUBISCO activase levels were increased (Hashimoto and Komatsu 2007; Gao et al. 2009). In cold stress, HSPs play a very crucial role. Expression of some HSPs, like HSP70, is continuously upregulated, while expression of HSP90 is downregulated (Bae et al. 2003; Taylor et al. 2005). ROS-scavenging enzymes, Cu–Zn superoxide dismutase, enzymes involved in ascorbate glutathione, glycolysis, and carbohydrate metabolism levels are continuously maintained throughout the stress (Kawamura and Uemura 2003; Degand et al. 2009; Kosmala et al. 2009). Riskie-Fe-S protein of cytochrome b_6/f complex becomes abundant in *Arabidopsis* in response to cold (Amme et al. 2006; Gao et al. 2009). The RNA-binding protein CP29's level is increased, which mainly involves in mRNA processing (Kawamura and Uemura 2003; Amme et al. 2006).

Transcriptomic Responses Along with proteome, transcriptome changes are observed in rice seedlings (Yan et al. 2006). Expression of transcription factor MeTCP4 is increased in cold stress. MeTCP4 not only enhance cold stress tolerance ability of plants but also decrease level of ROS (Cheng et al. 2019). Cold stress increases expression of the AP2 domain protein CBF, which finally activates downstream cold responsive genes (Wanqian et al. 2018). This CBF is regulated by the ICE (bHLH transcription factors) by sumonylation and poly-ubiquitinization (Zhang et al. 2016). Transcription factors belonging to MYB (AtMYB2), bZIP (OBF4), and bHLH similar to MYC groups are upregulated in *Arabidopsis* and CRT/DRE binding factor 1 is upregulated in poplar leaves (Renaut et al. 2004). Besides that, upregulation of some transcription factors involved in stress-responsive gene expression and RNA splicing factors (U2 snRNP-A′) is reported for *A. thaliana* by Bae et al. (2003).

Metabolomic Responses Cold stress causes metabolomic imbalances (Kosova et al. 2008, 2011). Hydrophilic protein LEA, especially LEAII (dehydrin) is increased during cold stress, which serves as high molecular osmoprotectants (Gao et al. 2009; Tunnacliffe and Wise 2007; Reiland et al. 2009). Besides that, monosaccharides and oligosaccharides are accumulated in the cell cytoplasm, which also serve as osmoprotectants (Kosova et al. 2011). *S*-Adenosyl methionine (SAM), which is a precursor of ethylene and polyamines, is also shooting up (Cui et al. 2005; Amme et al. 2006; Yan et al. 2006). Low molecular osmolytes (glycine betaine, proline, raffinose-derived oligosaccharides) and ROS-scavenging systems (catalase, peroxidase, superoxide dismutase, lignin and ascorbic acid) are also reported in *Arabidopsis* like annual plants (Ding et al. 2018; Yao et al. 2020). Suberin and lignin also accumulate to enhance cold tolerance of plants (Jian et al. 2020).

15.3.3 Salinity Stress Responses

Plant responses to salinity can be divided into two main phases. First phase takes place within minutes to days, where ion-independent growth reduction can be observed. Stomata are closed, and expansion of cells is inhibited (Munns and Passioura 1984; Rajendran et al. 2008). Second phase observed within days to weeks, which reduces metabolic activity that finally leads to reduction in plant growth and development, early senescence, and cell death (Munns and Tester 2008; Roy et al. 2014).

15.3.3.1 Genomic Responses

Salt stress induces the biosynthesis of endogenous hormones-producing genes such as ethylene, salicylic acid, ABA-producing gene (Zhu 2002). Salt stress can be tolerated by both ABA-dependent and ABA-independent pathways (Kumar et al. 2013). Salt stress responsive genes can be divided into two major categories: early- and late-induced gene. Several transcription factors belong to early-induced gene, as they are expressed in early, whereas late-induced gene (COR, RD, membrane stabilizing proteins) are expressed slowly (Rao et al. 2015). To cope with salt stress, *Arabidopsis* contains only subclass 1, Na^+ selective, AtHKT1, while monocotyledons contain multiple HKT isoforms (Berthomieu et al. 2003; Sunarpi et al. 2005). Besides that, upregulation of certain genes, like C2C2-CO-like WRKY, (members of the MYB and NAC families) AKT1, KAT1, and SOS1, is reported in salt-tolerant varieties (Li et al. 2019).

15.3.3.2 Proteomic Responses

High salinity stress induces an increased level of different proteins in the roots of different species like receptor protein kinase (RTK) in rice and small GTP binding protein in wheat, *Arabidopsis*, and rice (Jiang et al. 2007; Wang et al. 2008; Cheng et al. 2009; Zhang et al. 2009). Ca^{+2}-related proteins calmodulin and calreticulin are also observed in plant salinity responses (Li et al. 2010; Zörb et al. 2010). 14-3-3 family proteins like GF14a and GF14b are also observed, which regulate H^+-ATPase activity and initiate salt response (Malakshah et al. 2007). Increasing levels of proteins that are involved in membrane stabilization, ion homeostasis, and signal transduction are observed. Leucine-rich-repeat type receptor-like protein kinase, OsRPK1 is also observed. Proteins involved in glycolysis, carbohydrate metabolism, nucleic acid metabolism, and fatty acid metabolism as well as ROX-scavenging enzyme (APX, DHAR, and SOD) are upregulated in glycophytes like rice (Dooki et al. 2006; Cheng et al. 2009), wheat (Caruso et al. 2008; Wang et al. 2008), and also in *Arabidopsis* (Ndimba et al. 2005). Both enzymatic and non-enzymatic antioxidant systems protect plants from oxidative stress caused by salinity (Kaur et al. 2019). Pathogenesis-related (PR) proteins were also observed (Dani et al. 2005). Protective proteins like HSP-70 and other chaperones are also upregulated (Kosova et al. 2011).

15.3.3.3 Transcriptomic Responses

Upregulation of several transcription and translation factors like transcription factor BTF3 from NAC family, DEAD box RNA helicase involved in modulation of stress-inducible CBF transcriptional activators, and eukaryotic initiation translation factor eIF5A, was reported by Wang et al. (2008) in wheat. MdMYB46 induces secondary cell wall formation and deposition of lignin; thus, MdMYB46 enhances salt tolerance in plants (Chen et al. 2019). Transcription factors like bZIP,

WRKY, AP2, NAC, C2H2 zinc finger gene, MYB, and DREB bind to promoter region of different stress-responsive genes in cis-acting manner and regulates different stress responsive genes by either upregulating or downregulating (Gupta and Huang 2014).

15.3.3.4 Metabolomic Responses

Salt stress initiates accumulation of amino acids like proline, serine, threonine, tryptophan, histidine, lysine, and ornithine. Besides that, secondary metabolites like quinic acid and pipecolic acid are also observed. Accumulation of raffinose and its precursor galactinol was reported by Joshi et al. (2021). Hydrophilic LEA protein is accumulated in the cell cytoplasm in response to salinity. Sequestration of inorganic ions in vacuole is observed (Flowers 2004; Bartels and Sunkar 2005). Compatible solutes, uncharged, polar, soluble, low molecular weight compounds are accumulated under salt stress. These compounds include glycine betaine, raffinose-derived oligosaccharides, derivative sugars, and polyols, which maintain osmotic balance within the cell by water influx but do not involve in cellular metabolism (Hasegawa et al. 2000).

15.3.4 Ultraviolet (UV) Radiation Stress

Under UV-radiation stress, plants adapt different mechanisms to reduce the damaging effect of oxidative stress (Singh et al. 2014).

15.3.4.1 Genomic Responses

UV stress targets different genes such as chalcone synthase (CHS), chalcone isomerase (CHI), flavonol synthase (FLS), di-hydro flavonol 4-reductase (DFR), and leucoanthocyanidin diozygenase (LDOX). These genes involve in regulating the flavonoid biosynthetic pathway (Tilbrook et al. 2013). UV resistance locus 8 (UVR8) is a key regulator of different genes involved in UV damage (Favory et al. 2009). UVR8 regulates photomorphogenic response by regulating different genes such as inhibition of hypocotyl gene and DNA repair gene (Rizzini et al. 2011). UVR8 encodes a 440-amino-acid, seven-bladed β-propeller protein (Christie et al. 2012).

15.3.4.2 Proteomic Responses

To deal with oxidative stress, plants evolve their ROS-scavenging machinery (Singh et al. 2010; Tripathi et al. 2010). Besides that, they also produce enzymatic antioxidant molecules (SOD, catalase, ascorbate peroxidase, glutathione reductase, monodehydroascorbate reductase, isoflavone reductase, etc.). These enzymes help to reduce the damaging effects of oxidative stress on plants (Kosova et al. 2011). Activity of SOD enzymes increase in *Arabidopsis*, rice, pea, and wheat but not in barley and soybean. Besides SOD, ascorbate peroxidase level also increases in *Arabidopsis* (Rao et al. 1996). Proteins like HSP70 and disulfide isomerase are also involved in regulating oxidative stress (Bohler et al. 2007). Different plant defense protein levels are continuously enhanced under UV stress (Agrawal et al. 2009). RUBISCO and proteins related to photosynthesis are drastically reduced. In contrast, enzymes related to Calvin cycle and carbohydrate metabolism are upregulated (Bohler et al. 2007).

15.3.4.3 Transcriptomic Responses

UV ray causes DNA damages that negatively affect transcription; to protect from this damage, multiple DNA repair mechanisms (nucleotide excision repair, photolyase repair) are ongoing in plants (Ries et al. 2000). UV stress induces alterations in transcriptome and transcriptional factors in multiple plant species including *Arabidopsis*. Transcription factors like AP-1 and P53 are activated by UV signaling (Brown and Jenkins 2008). Multiple DEGs are reported in *Arabidopsis* under UV stress. These DEGs are involved in stress response, defense signaling, ROS metabolism, antioxidation, and detoxification, and it enhances stress tolerance capability (Yoon et al. 2016).

15.3.4.4 Metabolomic Responses

Under UV stress, plants accumulate different secondary metabolites (Nybakken et al. 2012). Besides enzymatic antioxidant molecule, plants also produce non-enzymatic antioxidant molecules like tocopherol, carotenoid, and phenolics (Kosova et al. 2011). These antioxidant compounds not only protect plants from oxidative stress but also enhance growth and development of plants (de Pinto and de Gara 2004). A balanced level of soluble sugar such as sucrose is maintained throughout the stress in both plants and cyanobacteria (Kolman et al. 2015; Shen et al. 2018). Proline, mannitol, and glycine betaine serve as osmoprotectants that maintain plasma membrane integrity, while stabilizing photosynthetic apparatus and protecting from oxidative damages (Chen and Murata 2008; Szabados and Savoure 2010).

15.3.5 Heavy Metal Stress Responses

Plants adapt different strategies to cope with different heavy metal (HM) stress. Plants restrict the movement or long distance transport of HMs by binding them to cell wall or cellular exudates – this is the first line of defense (Manara 2012; Hasan et al. 2015).

15.3.5.1 Genomic Responses

Modification of gene expression is very fast strategy in response to HM stress. HM stress is very specific to the metal that is present in toxic concentration in nearby environment (Verkleij et al. 1991). Genes related to oxidative stress (GSH), chemical stress, stress responsive production, and secondary metabolite production are continuously upregulated, whereas genes related to photosynthesis and various development processes are downregulated (Thakur et al. 2019). Expression of dehydration genes is reported in barley by Tamas et al. (2010). Previous studies reported the presence of Nramp gene, which encodes the protein involved in metal ion translocation (Belouchi et al. 1997). HMT1 genes encode ATP binding cassette (ABC) which helps in sequestration of ions into the vacuole; this HMT1 transporter was first identified in *Schizosaccharomyces pombe* (Kuriakose and Prasad 2008).

15.3.5.2 Proteomic Responses

Increased concentration of heavy metal significantly reduces the activity of RUBISCO in rice; in contrast, ROS-scavenging enzyme, mainly SOD, and pathogenesis-related protein (OsPR5 and OsPR10) are increased (Hajduch et al. 2021). When level of HMs is increased, plant induces complex strategies for storage and detoxification by using phytochelatin and metallothionein. Phytochelatins are cysteine-rich oligomers that help in trafficking and sequestration of HMs into the vacuole by vacuole transporters (Zhao and Chengcai 2011; Pochodylo and Aristilde 2017). Increased level of phytochelatin not always increases HMs tolerance response. In comparison to wild plants, *Arabidopsis* increased level (2.1 folds) of phytochelatins makes plants sensitive to Cd and Zn (Lee et al. 2003). Metallothionein is a cysteine-rich biomolecule that involves in trafficking, detoxification, and sequestration of HMs and also protects plants from oxidative stress caused by HMs (Zhao and Chengcai 2011; Hossain et al. 2012b; Ansarypour and Shahpiri 2017). Besides phytochelatin and metalothione, HSPs (HSP70, HSP90) play a very crucial role in plant stress responses. They help in maintaining cell environment as well as preventing protein accumulation and degraded mis-folded proteins (Park and Seo 2015).

15.3.5.3 Transcriptomic Responses

HM stress activates different stress-responsive genes, which further activate multiple signaling cascades. These stress-responsive genes can be further divided into two major categories: regulatory genes and functional genes; regulatory genes encode transcription factors, whereas functional genes encode metabolite-encoding genes (Tran et al. 2010). HM stress triggers mitogen-activating protein kinase (MAPK) signaling cascade; this MAPK further switches on downstream transcription factors such as MYB, DREB, bZIP, AP2, and ERF (Li et al. 2016). Different MYBs such as MYB4, MYB28, MYB43, MYB48, MYB72, and MYB124 are continuously upregulated in Cd and Zn metal stresses in Arabidopsis (Van de Mortel et al. 2008). WRKY transcription factors mainly bind to W box of promoter region in many stress-responsive gene (Opdenakker et al. 2012).

15.3.5.4 Metabolomic Responses

HM stress triggers accumulation of low molecular osmolytes like proline and sugar derivatives. These osmolytes help to maintain osmotic balance at the cellular level and increase HM tolerance (Song et al. 2020). Amino acids, mainly proline and histidine, enhance metal tolerance by chelating metal ions (Sharma and Dietz 2006). Besides amino acids, an amino acid derivative (aminocarboxylate) plays an important role in stress tolerance by chelating and transporting of metal ions (Kerkeb and Kramer 2003). Biosynthesis of phenolic compounds is also increased under HM stress; these compounds bind to metal ions, especially iron and copper, by their hydroxyl and carboxyl groups, and help to chelate the metal ions (Jun et al. 2003). Previous studies reported that expression of some organic acids such as malate, oxalate, and citrate was also increased, which involved metal transportation and ion sequestration into the vacuole (Rauser 1999).

15.3.6 Air Pollution Stress Responses

Plants adapt different mechanisms to deal with air pollution. Changes in the structure of the outermost cuticle are the first line of defense (Gostin 2016). They increase humidity through transpiration, absorb carbon dioxide, and release oxygen through photosynthesis. Plants adapt several mechanisms to detoxify air pollutants, like they can also immobilize the absorbing pollutant (phytostabilization) or simply absorb the pollutant into leaves and roots (phytoaccumulation) (Brilli et al. 2018).

15.3.6.1 Genomic Responses

Ozone stress activates several genes that are involved in secondary metabolite production, stress response, transcription and post-transcription, DNA repair, and RNA processing (Cho et al. 2008). Upregulation of PAL, SOD, Glutathione *S*-transferase (GST), and cytosolic copper/zinc (Cu/Zn) genes have been studied previously. Ethylene-related genes are upregulated, as they promote early senescence and cell death; in contrast, auxin-related genes are downregulated as they are responsible for cell elongation and growth. Except ethylene, some of the senescence-associated genes (SAG) like cysteine protease and glutamine synthase are also expressed (Jones and Herman 1993; Wolfgang et al. 2002; Gupta et al. 2005). Besides that, genes involved in lipid metabolism, amino acid and protein metabolism, carbohydrate metabolism, glycolysis, PPP, and cell wall metabolism are triggered (Gupta et al. 2005).

15.3.6.2 Proteomic Responses

The secondary pollutant ozone induces oxidative stress (Sarkar and Agrawal 2010a, b; Sarkar et al. 2010, 2015); to deal with this oxidative stress, plants activate multiple ROS-scavenging enzymes such as SOD, catalase, ascorbate peroxidase, glutathione reductase, and isoflavone reductase (Renaut et al. 2009; Mandal et al. 2022). Ascorbic acid (AA) is one of the major antioxidants that helps reduce damaging effects of ROS (Pasqualini et al. 2002). Chaperones, HSP70, and disulfide isomerase are also involved in regulating oxidative stress. Enzymes related to glycolysis, carbohydrate metabolism, stress responsive proteins (hairpin-binding protein, germin protein), cellular signaling proteins, and PR proteins are continuously upregulated; in contrast, large and small subunits of RUBISCO and enzymes related to respiration and the dark and light reactions of photosynthesis are downregulated (Kosova et al. 2011).

15.3.6.3 Transcriptomic Responses

Ozone induces transcription of defense-related gene (Iriti and Faoro 2009; Cho et al. 2013). Phenylalanine ammonia lyase (PAL) mRNA induction is also increased, which finally results in the production of defensive compounds under ozone stress (Toumainen et al. 1996). Multiple transcription factors, such as AP_2/ERF, WRKY, MYB, bHLH, and WD40, that enhance stress tolerance and regulate expression of anthocyanin and proanthocyanin; these anthocyanin and proanthocyanin help to reduce oxidative stress (Allan et al. 2008). The mRNA of photosynthetic proteins and ubiquitination proteins (F box protein and E3 ligase) are continuously upregulated (Gupta et al. 2005).

15.3.6.4 Metabolomic Responses

Air pollutants, especially ozone, alter metabolic activities and induce accumulation of secondary metabolites, low molecular osmolytes, derivative sugars, and polyamines. These secondary metabolites not only help to overcome abiotic stress but also involve in plant defense. Phenolics, one of the major secondary compounds, enhance stress tolerance, whereas anthocyanin and proanthocyanin serve as antioxidants under ozone stress (Iriti and Faoro 2009). Not only secondary metabolites but also primary metabolites like amino acids accumulate in response to the stress (Sarkar and Singh 2015). Expression of BVOC was also reported under both biotic and abiotic stress by Blande et al. (2014).

15.3.7 Climate Change Responses

Climate change mainly associated with heat stress and an elevated CO_2 level. Increasing levels of CO_2 induce leaf thickening in plants (Urton 2018). According to Gray and Brady (2016), plants change their flowering time and enhance expansion of root biomass in response to climate change, especially high temperature and elevated level of CO_2. In *A. thaliana*, earlier shift from vegetative to reproductive stages is reported (Balasubramanian et al. 2006). In contrast, larger leaves and extra leaf nodes are found in soybean (Dermody et al. 2006).

15.3.7.1 Genomic Responses

Climate changes upregulate the expression of photosynthetic gene and respiration gene. Genes related to hypoxia responses are expressed (Leaky et al. 2009). Genes for HIC (3-ketoacyl coenzyme A synthase) not only are important for cell wall wax biosynthesis but also regulate stomatal development in response to elevated levels of CO_2 (Gray et al. 2000).

15.3.7.2 Proteomic Responses

In soybean, enzymes involved in carbohydrate metabolism have an electron transport system that is continuously upregulated, and the rate of dark reaction increases (Leaky et al. 2009). Increasing rate of dark reaction is also noticed in *Arabidopsis* (Markelz et al. 2014). Elevated level of CO_2 ensures the availability of CO_2 for RUBISCO; as a result, photorespiration is inhibited (Drake et al. 1997).

15.3.7.3 Transcriptomic Responses

Elevated level of CO_2 alters the expression of several transcription factors, especially ERF and WRKY (Zhu et al. 2019). In low oxygen conditions, transcription factor like $AtMYB_2$ is also expressed in *Arabidopsis*, which further enhances $AtADH_1$ expression (Abe et al. 2003). Heat shock factor $HsfA_2$ is also expressed, which further activates the downstream ADH gene to increase hypoxia tolerance (Banti et al. 2010). Some transcription factors *trans*-activate hypoxia-responsive genes to cope with the stress (Zhu et al. 2019).

15.3.7.4 Metabolomic Responses

Both high temperature and elevated level of CO_2 induce secondary metabolite production. Phenolic compounds are produced under elevated level of CO_2, whereas biogenic volatile organic compounds (BVOCs) are produced under high temperature stress (Ramakrishna and Ravishankar 2011). Generally, synthesis of BVOC by non-mevalonate pathway is directly related with short-term environmental stress, or more specifically, short-term heat shock in plants (Li and Sharkey 2013). Though the production of BVOC is an energetically expensive process, it has the ability to protect plants from environmental stress (Wahid et al. 2012).

15.4 Conclusion

The massive amounts of data generated through omics approaches, including transcriptomics, proteomics, and metabolomics, permit plant scientists may unite and incorporate alternative methodologies to generate insights on gene function, genetic alterations, and plants' capacity to cope with different environmental stresses. Despite the fact that technologies are continually developing, the observed global molecular alterations have always been part of the overall scheme. The capacity to correlate and analyze the massive amounts of data generated by research initiatives will be the next difficult step toward a complete knowledge of plant responses to stress. Furthermore, it is important to keep in mind that the ultimate goal of this abiotic stress research is to discover how the plant reacts to the stressful factor in order to develop appropriate solutions to the problem.

References

Abe, H., Urao, T., Ito, T. et al. (2003). Arabidopsis AtMYC2 (bHLH) and AtMYB2 (MYB) function as transcriptional activators in abscisic acid signaling. *Plant Cell* 15 (1): 63–78.

Addicott, F.T., Lyon, J.L., Ohkuma, K. et al. (1968). Abscisic acid: a new name for abscisin 2 (dormin). *Science* 159: 3822.

Agrawal, G.K., Rakwal, R., Yonekura, M. et al. (2002). Proteome analysis of differentially displayed proteins as a tool for investigating ozone stress in rice (*Oryza sativa* L.) seedlings. *Proteomics* 2 (8): 947–959.

Agrawal, S.B., Singh, S., and Agrawal, M. (2009). Ultraviolet-B induce changes in gene expression and antioxidants in plants. *Adv. Bot. Res.* 52: 47–86.

Aguayo, P., Lagos, C., Conejera, D. et al. (2019). Transcriptome-wide identification of WRKY family genes and their expression under cold acclimation in *Eucalyptus globulus*. *Trees (Berl)* 33: 1313–1327.

Ahsan, N., Lee, D.G., Lee, S.H. et al. (2007). A comparative proteomic analysis of tomato leaves in response to waterlogging stress. *Physiol. Plant.* 131: 555–570.

Ahsan, N., Renaut, J., and Komatsu, S. (2009). Recent developments in the application of proteomics to the analysis of plant responses to heavy metals. *Proteomics* 9: 2602–2621.

Ainsworth, E.A. and Long, S.P. (2005). What have we learned from 15 years of free air CO_2 enrichment (FACE)? A meta-analytic review of the responses of photosynthesis, canopy properties and plant production to rising CO_2. *New Phytol.* 165: 351–372.

Alam, I., Lee, D.G., Kim, K.H. et al. (2010). Proteome analysis of soybean roots under waterlogging stress at an early vegetative stage. *J. Biosci.* 35: 49–62.

Allan, A.C., Hellens, R.P., and Laing, W.A. (2008). MYB transcription factors that colour our fruit. *Trends Plant Sci.* 13: 99–102.

Amm, I., Sommer, T., and Wolf, D.H. (2014). Protein quality control and elimination of protein waste: the role of the ubiquitin proteasome system. *Biochim. Biophys. AcaBBA. Molecul. Cell Res.* 1843: 182–196.

Amme, S., Matros, A., Schlesier, B., and Mock, H.P. (2006). Proteome analysis of cold stress response in *Arabidopsis thaliana* using DIGE-technology. *J. Exp. Bot.* 57: 1537–1546.

Andy, P. (2016). Abiotic stress tolerance in plants. *Plant Sci.* 7: 1–9.

Anjum, S.A., Wang, L.C., Farooq, M. et al. (2011). Brassinolide application improves the drought tolerance in maize through modulation of enzymatic antioxidants and leaf gas exchange. *J. Agron. Crop Sci.* 197: 177–185.

Ansarypour, Z. and Shahpiri, A. (2017). Heterologous expression of a rice metallothionein isoform (OsMTI-1b) in *Saccharomyces cerevisiae* enhances cadmium, hydrogen peroxide and ethanol tolerance. *Braz. J. Microbiol.* 48: 537–543.

Bae, M.S., Cho, E.J., Choi, E.Y., and Park, O.K. (2003). Analysis of the *Arabidopsis* nuclear proteome and its response to cold stress. *Plant J.* 36: 652–663.

Bakku, R.K., Rakwal, R., Shibato, J. et al. (2020). Transcriptomics of mature rice (*Oryza Sativa* L. Koshihikari) seed under hot conditions by DNA microarray analyses. *Atmosphere* 11: 528.

Balasubramanian, S., Sureshkumar, S., Lempe, J., and Weigel, D. (2006). Potent induction of *Arabidopsis thaliana* flowering by elevated growth temperature. *PLoS Genet.* 2: 980–989.

Baniwal, S.K., Bharti, K., Chan, K.Y. et al. (2004). Heat stress response in plants: a complex game with chaperones and more than twenty heat stress transcription factors. *J. Biosci.* 29: 471–487.

Banti, V., Mafessoni, F., Loreti, E. et al. (2010). The heat-inducible transcription factor HsfA2 enhances anoxia tolerance in *Arabidopsis*. *Plant Physiol.* 152 (3): 1471–1483.

Bartels, D. and Sunkar, R. (2005). Drought and salt tolerance in plants. *Crit. Rev. Plant Sci.* 24: 23–58.

Belouchi, A., Kwan, T., and Gros, P. (1997). Cloning and characterization of the OsNramp family from *Oryza sativa*, a new family of membrane proteins possibly implicated in the transport of metal ions. *Plant Mol. Biol.* 33: 1085–1092.

Berthomieu, P., Conejero, G., Nublat, A. et al. (2003). Functional analysis of AtHKT1 in *Arabidopsis* shows that Na^+ recirculation by the phloem is crucial for salt tolerance. *EMBO J.* 22: 2004–2914.

Bita, C.E., Zenoni, S., Vriezen, W.H. et al. (2011). Temperature stress differentially modulates transcription in meiotic anthers of heat-tolerant and heat-sensitive tomato plants. *BMC Genomics* 12: 384.

Blande, J.D., Holopainen, J.K., and Niinemets, U. (2014). Plant volatiles in polluted atmospheres: stress responses and signal degradation. *Plant Cell Environ.* 37: 1892–1904.

Bogeat-Triboulot, M.B., Brosché, M., Renaut, J. et al. (2007). Gradual soil water depletion results in reversible changes of gene expression, protein profiles, ecophysiology, and growth performance in *Populus euphratica*, a poplar growing in arid regions. *Plant Physiol.* 143 (2): 876–892.

Bohler, S., Bagard, M., Oufir, M. et al. (2007). A DIGE analysis of developing poplar leaves subjected to ozone reveals major changes in carbon metabolism. *Proteomics* 7: 1584–1599.

Borrego-Benjumea, A., Carter, A., Tucker, J.R. et al. (2020). Genome-wide analysis of gene expression provides new insights into waterlogging responses in barley (*Hordeum vulgare* L.). *Plants (Basel)* 9 (2): 240.

Brilli, F., Fares, S., Ghirardo, A. et al. (2018). Plants for sustainable improvement of indoor air quality. *Trends Plant Sci.* 23 (6): 507–512.

Broberg, M., Hogy, P., and Pleijel, H. (2017). CO_2 induced changes in wheat grain composition: meta-analysis and response function. *Agronomy* 7: 32.

Brown, B.A. and Jenkins, G.I. (2008). UV-B signaling pathways with different fluence-rate response profiles are distinguished in mature *Arabidopsis* leaf tissue by requirement for UVR8, HY5, and HYH. *Plant Physiol.* 146: 576–588.

Cao, M., Zheng, L., Li, J. et al. (2022). Transcriptomic profiling suggests candidate molecular responses to waterlogging in cassava. *PLoS One* 17 (1): e0261086.

Caruso, G., Cavaliere, C., Guarino, C. et al. (2008). Identification of changes in *Triticum durum* L. leaf proteome in response to salt stress by two-dimensional electrophoresis and MALDI-TOF mass spectrometry. *Anal. Bioanal. Chem.* 391: 381–390.

Chae, L., Sudat, S., Dudoit, S. et al. (2009). Diverse transcriptional programs associated with environmental stress and hormones in the *Arabidopsis* receptor like kinase gene family. *Mol. Plant* 2: 84–107.

Chen, T.H. and Murata, N. (2008). Glycinebetaine: an effective protectant against abiotic stress in plants. *Trends Plant Sci.* 13: 499–505.

Chen, K., Song, M., Guo, Y. et al. (2019). MdMYB46 could enhance salt and osmotic stress tolerance in apple by directly activating stress responsive signals. *Plant Biotechnol. J.* 17 (12): 2341–2355.

Cheng, Y., Qi, Y., Zhu, Q. et al. (2009). New changes in the plasma-membrane associated proteome of rice root sunder salt stress. *Proteomics* 9: 3100–3114.

Cheng, Z., Lei, N., Li, S. et al. (2019). The regulatory effects of MeTCP4 on cold stress tolerance in *Arabidopsis thaliana*: a transcriptome analysis. *Plant Physiol. Biochem.* 138: 9–16.

Chinnusamy, V., Zhu, J., and Zhu, J.K. (2007). Cold stress regulation of gene expression in plants. *Trends Plant Sci.* 12: 444–451.

Chmielewska, K., Rodziewicz, P., Swarcewicz, B. et al. (2016). Analysis of drought-induced proteomic and metabolomic changes in barley (*Hordeum vulgare* L.) leaves and roots unravels some aspects of biochemical mechanisms involved in drought tolerance. *Front. Plant Sci.* 7: 1108.

Cho, K., Shibato, J., Agrawal, G.K. et al. (2008). Integrated transcriptomics, proteomics and metabolomics analysis to survey ozone responses in the leaves of rice seedling. *J. Proteome Res.* 7: 2980–2998.

Cho, K., Shibato, J., Kubo, A. et al. (2013). Genome-wide mapping of the ozone-responsive transcriptomes in rice panicle and seed tissues reveals novel insight into their regulatory events. *Biotechnol. Lett.* 35: 647–656.

Choi, W.G., Miller, G., Wallace, I. et al. (2017). Orchestrating rapid long-distance signaling in plants with CA^{+2}, ROS and electric signals. *Plant J.* 90: 698–707.

Christie, J.M., Arvai, A.S., Baxter, K.J. et al. (2012). Plant UVR8 photoreceptor senses UV-B by tryptophan-mediated disruption of cross-dimer salt bridges. *Science* 335: 1492–1496.

Cramer, G.R., Urano, K., Delrot, S. et al. (2011). Effects of abiotic stress on plants: a system biology perspective. *BMC Plant Biol.* 11: 163.

Cui, S., Huang, F., Wang, J. et al. (2005). A proteomic analysis of cold stress responses in rice seedlings. *Proteomics* 5: 3162–3172.

Cui, G., Chai, H., Yin, H. et al. (2019a). Full-length transcriptome sequencing reveals the low-temperature-tolerance mechanism of *Medicago falcata* roots. *BMC Plant Biol.* 19: 575.

Cui, J., Abadie, C., Carroll, A. et al. (2019b). Responses to K deficiency and waterlogging interact via respiratory and nitrogen metabolism. *Plant Cell Environ.* 42: 647–658.

Cutler, S.R., Rodriguez, P.L., Finkelstein, R.R., and Abrams, S.R. (2010). Abscisic acid: emergence of a core signaling network. *Annu. Rev. Plant Biol.* 61: 651–679.

Dani, V., Simon, W.J., Duranti, M., and Croy, R.R.D. (2005). Changes in the tobacco leaf apoplast proteome in response to salt stress. *Proteomics* 5: 737–745.

De Storme, N. and Geelen, D. (2020). High temperatures alter cross-over distribution and induce male meiotic restitution in *Arabidopsis thaliana*. *Commun. Biol.* 3: 187.

Degand, H., Faber, A.M., Dauchot, N. et al. (2009). Proteomic analysis of chicory root identifies proteins typically involved in cold acclimation. *Proteomics* 9: 2903–2907.

Dermody, O., Long, S.P., and Delucia, E.H. (2006). How does elevated CO_2 or ozone affect the leaf area index of soybean when applied independently? *New Phytol.* 169: 145–155.

Ding, Y., Lv, J., Shi, Y. et al. (2018). EGR2 phosphatase regulates OST1 kinase activity and freezing tolerance in *Arabidopsis*. *EMBO J.* 38 (1): e99819.

Dong, X., Yi, H., Lee, J. et al. (2015). Global gene-expression analysis to identify differentially expressed genes critical for the heat stress response in *Brassica rapa*. *PLoS One* 10: e0130451.

Dooki, A.D., Mayer-Posner, F.J., Askari, H. et al. (2006). Proteomic responses of rice young panicles to salinity. *Proteomics* 6: 6498–6507.

Drake, B.G., Gonzalez-Meler, M.A., and Long, S.P. (1997). More efficient plants: a consequence of rising atmospheric CO_2? *Annu. Rev. Plant Physiol. Plant Mol. Biol.* 48: 609–639.

Dufoo-Hurtado, M.D., Huerta-Ocampo, J.A., Barrera-Pacheco, A. et al. (2015). Low temperature conditioning of garlic (*Allium sativum L.*) "seed" cloves induces alterations in sprouts proteome. *Front. Plant Sci.* 6: 332.

Egea, I., Albaladejo, I., Meco, V. et al. (2018). The drought-tolerant *Solanum pennellii* regulates leaf water loss and induces genes involved in amino acid and ethylene/jasmonate metabolism under dehydration. *Sci. Rep.* 8: 2791.

Escandon, M., Meijón, M., Valledor, L. et al. (2018). Metabolome Integrated Analysis of High-Temperature Response in *Pinus radiata*. *Front. Plant Sci.* 9: 485.

Evelin, H., Devi, T.S., Gupta, S., and Kapoor, R. (2019). Mitigation of salinity stress in plants by arbuscular mycorrhizal symbiosid: current understanding and new challenges. *Front. Plant Sci.* 10: 470.

Farooq, M., Wahid, A., Kobayashi, N. et al. (2009). Improving water relation and gas exchange with brassinosteroids in rice under drought stress. *J. Agron. Crop Sci.* 195: 262–269.

Favory, J.J., Stec, A., Gruber, H. et al. (2009). Interaction of COP1 and UVR8 regulates UV-B induced photomorphogenesis and stress acclimation in *Arabidopsis*. *EMBO J.* 28: 591–601.

Fernando, N., Panozzo, J., Tausz, M. et al. (2015). Rising CO_2 concentration altered wheat grain proteome and flour rheological characteristics. *Food Chem.* 170: 448–454.

Ferreira, S., Hjerno, K., Larsen, M. et al. (2006). Proteome profiling of *Populus euphratica* Oliv. upon heat stress. *Ann. Bot.* 98: 361–377.

Flowers, T.J. (2004). Improving crop salt tolerance. *J. Exp. Bot.* 55: 307–319.

Food and Agriculture Organization of the United Nations [FAO] (2017). *The Impact of Disasters and Crises on Agriculture and Food Security*. Rome: FAO.

Fragkostefanakis, S., Simm, S., Paul, P. et al. (2015). Chaperone network composition in *Solanum lycopersicum* explored by transcriptome profiling and microarray meta-analysis. *Plant Cell Environ.* 38: 693–709.

Fujii, H. and Zhu, J.-K. (2009). An autophosphorylation site of the protein kinase SOS2 is important for salt tolerance in *Arabidopsis*. *Mol. Plant.* 2: 183–190.

Gao, F., Zhou, Y., Zhu, W. et al. (2009). Proteomic analysis of cold stress-responsive proteins in *Thellungiella rosette* leaves. *Planta* 230: 1033–1046.

Ghosh, D. and Xu, J. (2014). Abiotic stress responses in plant roots: a proteomics perspective. *Front. Plant Sci.* 5: 6.

Gillman, J.D., Biever, J.J., Ye, S. et al. (2019). A seed germination transcriptomic study contrasting two soybean genotypes that differ in terms of their tolerance to the deleterious impacts of elevated temperatures during seed fill. *BMC Res. Notes* 12: 522.

Gostin, I. (2016). Air pollution stress and plant response. *Environ. Sci.* 99–117.

Govindaraju, M., Ganeshkumar, R.S., Suganthi, P. et al. (2011). *Impact Assessment of Air Pollution Stress on Plant Species Through Biochemical Estimations*. World Academy of Science, Engineering and Technology.

Goyal, E., Amit, S.K., Singh, R.S. et al. (2016). Transcriptome profiling of the salt-stress response in *Triticum aestivum* cv. Kharchia Local. *Sci. Rep.* 6: 27752.

Gray, S.B. and Brady, S.M. (2016). Plant developmental responses to climate change. *Dev. Biol.* 419: 64–77.

Gray, J.E., Holroyd, F.M., Sijmons, P.C. et al. (2000). The HIC signaling pathway links CO_2 perception to stomatal development. *Nature* 408: 713–716.

Gupta, B. and Huang, B. (2014). Mechanism of salinity tolerance in plants: physiological, biochemical, and molecular characterization. *Int. J. Gen. Syst.* 2014: 701596.

Gupta, P., Duplesis, S., White, H. et al. (2005). Gene expression patterns of trembling aspen trees following long term exposure to interacting elevated CO_2 and tropospheric O_3. *New Phytol.* 167: 129–141.

Gupta, A., Rico-Medina, A., and Cano-Delgado, A.I. (2020). The physiology of plant responses to drought. *Science* 368: 266–269.

Gygi, S.P., Rochon, Y., Franza, B.R., and Aebersold, R. (1999). Correlation between protein and mRNA abundance in yeast. *Mol. Cell Bio.* 19 (3): 1720–1730.

Hajduch, M., Rakwal, R., Agrawal, G.K. et al. (2021). High-resolution two-dimensional electrophoresis separation of proteins from metal-stressed rice (*Oryza sativa* L.) leaves: drastic reductions/fragmentation of ribulose-1,5-bisphosphate carboxylase/oxygenase and induction of stress-related proteins. *Electrophoresis* 22: 2824–2831.

Hajheidari, M., Eivazi, A., Buchanan, B.B. et al. (2007). Proteomics uncovers a role for redox in drought tolerance in wheat. *J. Proteome Res.* 6: 1451–1460.

Hajihashemi, S., Noedoost, F., Genus, J.M.C. et al. (2018). Effect of cold stress on photosynthetic traits, carbohydrates, morphology and anatomy in nine cultivars of *Stevia rebaudiana*. *Front. Plant Sci.* 9: 1430.

Hasan, M., Ahammed, G.J., Yin, L. et al. (2015). Melatonin mitigates cadmium phytotoxicity through modulation of phytochelatins biosynthesis, vacuolar sequestration, and antioxidant potential in *Solanum lycopersicum* L. *Front. Plant Sci.* 6: 601.

Hasanuzzaman, M., Hossain, M.A., da Silva, J.A.T., and Fujita, M. (2012). Plant responses and tolerance to abiotic oxidative stress: antioxidant defenses is a key factor. In: *Crop stress and Its Management: Perspectives and Strategies* (ed. B. Venkateswarlu, A. Shanker, C. Shanker, and M. Maheswari), 261–315. Dordrecht: Springer.

Hasanuzzaman, M., Nahar, K., and Fujita, M. (2013). Extreme temperatures, oxidative stress and antioxidant defense in plants. In: *Abiotic Stress: Plant Responses and Application in Agriculture* (ed. K. Vahdati and C. Leslie), 169–205. INTECH.

Hasegawa, P.M., Bressan, R.A., Zhu, J.K., and Bohnert, H.J. (2000). Plant cellular and molecular responses to high salinity. *Annu. Rev. Plant Biol.* 51: 463–499.

Hashimoto, M. and Komatsu, S. (2007). Proteomic analysis of rice seedlings during cold stress. *Proteomics* 7: 1293–1302.

Hirayama, T. and Shinozaki, K. (2010). Research on plant abiotic responses in the post-genome era: past, present and future. *Plant J.* 61: 1041–1052.

Hogy, P., Brunnbauer, M., Koehler, P. et al. (2013). Grain quality characteristics of spring wheat as affected by free air CO_2 enrichment. *Environ. Exp. Bot* 88: 11–18.

Holladay, J.D., Jones, E.O., Dagle, R.A. et al. (2004). High efficiency and low carbon monoxide micro-scale methanol processors. *J. Power Sources* 131: 69–72.

Hossain, Z. and Komatsu, S. (2013). Contribution of proteomic studies towards understanding plant heavy metal stress response. *Front. Plant Sci.* 3: 310.

Hossain, M.A., Piyatida, P., da Teixeira, Silva, J.A., and Fujita, M. (2012a). Molecular mechanism of heavy metal toxicity and tolerance in plants: central role of glutathione in detoxification of reactive oxygen species and methyl glyoxal and in heavy metal chelation. *J. Bot.* 2012: 1–37.

Hossain, Z., Nouri, M.Z., and Komatsu, S. (2012b). Plant cell organelle proteomics in response to abiotic stress. *J. Proteome Res.* 11: 37–48.

Hsu, F.C. and Shih, M.C. (2013). Plant defense after flooding. *Plant Signal. Behav.* 8: 2699–2713.

Hussain, M., Malik, M.A., Farooq, M. et al. (2008). Improving drought tolerance by exogenous application of glycine betaine and salicylic acid in sunflower. *J. Agron. Crop Sci.* 194: 193–199.

Hwang, J.E., Kim, Y.J., Jeong, D.Y. et al. (2019). Transcriptome analysis of Korean fir (*Abies koreana*) in response to elevated carbon dioxide and high temperature. *Plant Biotechnol. Rep.* 13: 603–612.

IPCC (2007). *Climate Change 2007: Synthesis Report. Contribution of Working Groups I, II and III to the Fourth Assessment Report of the Intergovernmental Panel on Climate Change* [Core Writing Team, Pachauri, R.K and Reisinger, A. (eds.)], 104. Geneva, Switzerland: IPCC.

IPCC (2014). *Climate Change 2014: Synthesis Report. Contribution of Working Groups I, II and III to the Fifth Assessment Report of the Intergovernmental Panel on Climate Change* [Core Writing Team, R.K. Pachauri and L.A. Meyer (eds.)], 151. Geneva, Switzerland: IPCC.

Iriti, M. and Faoro, F. (2009). Chemical diversity and defence metabolism: how plants cope with pathogens and ozone pollution. *Intl. J. Mol. Sci.* 10 (8): 3371–3399.

Jenkins, G.I. (2009). Signal transduction in response to UV-B radiation. *Annu. Rev. Plant Biol.* 60: 407–431.

Jian, H., Xie, L., Wang, Y. et al. (2020). Characterization of cold stress responses in different rapeseed ecotypes based on metabolomics and transcriptomics analyses. *PeerJ* 8: e8704.

Jiang, Y., Yang, B., Harris, N.S., and Deyholos, M.K. (2007). Comparative proteomic analysis of NaCl stress-responsive proteins in *Arabidopsis* roots. *J. Exp. Bot.* 58: 3591–3607.

Jones, A.M. and Herman, E.M. (1993). KDEL-containing auxin-binding protein is secreted to the plasma membrane and cell wall. *Plant Physiol.* 101: 595–606.

Jorrin-Novo, J.V., Maldonado, A.M., Echevarria-Zomeno, S. et al. (2009). Plant proteomics update (2007–2008): second-generation proteomic techniques, an appropriate experimental design, and data analysis to fulfill MIAPE standards, increase plant proteome coverage, and expand biological knowledge. *J. Proteomics* 72: 285–314.

Joshi, J., Hasnain, G., Logue, T. et al. (2021). A core metabolome response of maize leaves subjected to long-duration abiotic stresses. *Metabolites* 11: 797.

Jun, M., Fu, H.Y., Hong, J. et al. (2003). Comparison of antioxidant activities of isoflavones from kudzu root (*Puerarialobata Ohwi*). *J. Food Sci.* 68: 2117–2122.

Kaplan, F., Zhao, W., Richards, J.T. et al. (2012). Transcriptional and metabolic insights into the differential physiological response of *Arabidopsis* to optimal and supraoptimal atmospheric CO_2. *PLoS One* 7: e43583.

Kaur, P., Handa, N., Kumar, V. et al. (2019). Cross talk among reactive oxygen, nitrogen and sulfur during abiotic stress in plants. In: *Reactive Oxygen, Nitrogen and Sulfur During Abiotic Stress in Plants: Production, Metabolism, Signaling and Defense Mechanism* (ed. M. Hasanuzzaman, V. Fotopoulos, K. Nahar, and M. Fujita), 857–871. Hoboken, NJ: Wiley-Blackwell.

Kawamura, Y. and Uemura, M. (2003). Mass spectrometric approach for identifying putative plasma membrane proteins of *Arabidopsis* leaves associated with cold acclimation. *Plant J.* 36: 141–154.

Kaya, M.A., Okcub, G., Ataka, M. et al. (2006). Seed treatments to overcome salt and drought stress during germination in sunflower (*Helianthus annus* L.). *Eur. J. Agron.* 24: 291–295.

Kerkeb, L. and Kramer, U. (2003). The role of free histidine in xylem loading of nickel in *Alyssum lesbiacum* and *Brassica juncea*. *Plant Physiol.* 131: 716–724.

Kidokoro, S., Yoneda, K., Takasaki, H. et al. (2017). Different cold-signaling pathways function in the responses to rapid and gradual decreases in temperature. *Plant Cell* 29: 760–774.

Kolman, M., Nishi, C., Perez-Cenci, M., and Salerno, G. (2015). Sucrose in cyanobacteria: from a salt-response molecule to play a key role in nitrogen fixation. *Life* 5: 102–126.

Komatsu, S., Shirasaka, N., and Sakata, K. (2013). "Omics" techniques for identifying flooding-response mechanisms in soybean. *J. Proteomics* 93: 169–178.

Kosmala, A., Bocian, A., Rapacz, M. et al. (2009). Identification of leaf proteins differentially accumulated during cold acclimation between *Festuca pratensis* plants with distinct levels of frost tolerance. *J. Exp. Bot.* 60: 3595–3560.

Kosova, K., Prasil, I.T., and Vitamvas, P. (2008). The relationship between vernalization-and photoperiodically-regulated genes and the development of frost tolerance in wheat and barley. *Biol. Plant.* 52: 601–615.

Kosova, K., Vitamvas, P., Prasil, I.T., and Renaut, J. (2011). Plant proteome changes under abiotic stress-contribution to proteomic studies to understanding plant stress response. *J. Proteomics* 74: 1301–1322.

Kosova, K., Vitamvas, P., Urban, M.O. et al. (2015). Biological networks underlying abiotic stress tolerance in temperate crops – a proteomic perspective. *Int. J. Mol. Sci.* 16: 20913–20942.

Kosova, K., Vitamvas, P., Urban, M.O. et al. (2018). Plant abiotic stress proteomics: the major factors determining alterations in cellular proteome. *Front. Plant Sci.* 9: 122.

Kuai, J., Liu, Z., Wang, Y. et al. (2014). Waterlogging during flowering and boll forming stages affects sucrose metabolism in the leaves subtending the cotton boll and its relationship with boll weight. *Plant Sci.* 223: 79–98.

Kudla, J., Becker, D., Grill, E. et al. (2018). Advances and current challenges in calcium signaling. *New Phytol.* 218: 414–431.

Kukal, M.S. and Irmak, S. (2018). Climate-driven crop yield and yield variability and climate change impacts on the U.S. great plains agricultural production. *Sci. Rep.* 8: 3450.

Kulkarni, M., Soolanayakanahally, R., Ogawa, S. et al. (2017). Drought response in wheat: key genes and regulatory mechanisms controlling root system architecture and transpiration efficiency. *Front. Chem.* 5: 106.

Kumar, K., Kumar, M., Kim, S.R. et al. (2013). Insights into genomics of salt stress response in rice. *Rice* 6: 27.

Kumar, J., Babele, P.K., Singh, D., and Kumar, A. (2016). UV-B radiation stress causes alteration in whole cell protein profile and expression of certain genes in the rice phyllosphereic bacterium *Enterobacter clocae*. *Front. Microbiol.* 7: 1440.

Kuriakose, S.V. and Prasad, M.N.V. (2008). Cadmium stress affects seed germination and seedling growth in *Sorghum bicolor* (L.) Moench by changing the activities of hydrolyzing enzymes. *Plant Growth Regul.* 54: 143–156.

Larcher, W. (2003). *Physiological Plant Ecology*, 4e. Berlin, Heidelberg: Springer.

Leaky, A.D.B., Xu, F., Gillespie, K.M. et al. (2009). Genomic basis for stimulated respiration of plants growing under elevated carbon dioxide. *Proc. Natl. Acad. Sci. U.S.A.* 106: 3597–3602.

Lee, H. (2007). Intergovernmental Panel on Climate Change (IPCC) Contribution of Working Group I to the Fourth Assessment Report of the Intergovernmental Panel on Climate Change. Cambridge University.

Lee, S., Moon, J.S., Ko, T.S. et al. (2003). Overexpression of *Arabidopsis* phytochelatin synthase paradoxically leads to hypersensitivity to cadmium stress. *Plant Physiol.* 131: 656–663.

Levitt, J. (1980). *Responses of Plants to Environmental Stresses: Chilling, Freezing and High Temperature Stresses*. New York, NY: Academic Publisher.

Li, Z. and Sharkey, T.D. (2013). Molecular and pathway controls on biogenic volatile organic compound emissions. In: *Biology, Controls and Models of Tree Volatile Organic Compounds* (ed. Ü. Niinemets and R.K. Monson), 119–151. London: Springer.

Li, P., Ainsworth, E.A., Leakey, A.D.B. et al. (2008). *Arabidopsis* transcript and metabolite profiles: ecotype-specific responses to open-air elevated [CO_2]. *Plant Cell Environ.* 31: 1673–1687.

Li, X.J., Yang, M.F., Chen, H. et al. (2010). Abscisic acid pretreatment enhances salt tolerance of rice seedlings: proteomic evidence. *Biochim. Biophys. Acta* 1804: 929–940.

Li, J., Yang, L., Jin, D. et al. (2013). UV-B induced photomorphogenesis in *Arabidopsis*. *Protein Cell.* 4: 485–492.

Li, W., Zhao, F., Fang, W. et al. (2015). Identification of early salt stress responsive proteins in seedling roots of upland cotton (*Gossypium hirsutum* L.) employing iTRAQ-based proteomic technique. *Front. Plant Sci.* 6: 732.

Li, S., Gao, J., Yin, K. et al. (2016). MYB75 phosphorylation by MPK4 is required for light-induced anthocyanin accumulation in *Arabidopsis*. *Plant Cell.* 16: 130.

Li, J., Gao, Z., Zhou, L. et al. (2019). Comparative transcriptome analysis reveals Kþ transporter gene contributing to salt tolerance in eggplant. *BMC Plant Biol.* 19 (01): 67–85.

Li, H., Zhao, X., Qiao, H. et al. (2020a). Comparative transcriptome analysis of the heat stress response in *Monochamus alternatus* hope (Coleoptera: Cerambycidae). *Front Physiol.* 10: 1568.

Li, J., Zhang, M., Sun, J. et al. (2020b). Heavy metal stress associated proteins in rice and *Arabidopsis*: genome wide identification, phylogenetics, duplication, and expression profiles analyses. *Front. Genet.* 11: 477.

Li, H., Luo, W., Ji, R. et al. (2021). A comparative proteomic study of cold responses in potato leaves. *Heliyon* 7: e06002.

Lipiec, J., Doussan, C., Nosalewicz, A., and Kondracka, K. (2013). Effect of drought and heat stresses on plant growth and yield: a review. *Int. Agrophys.* 27: 463–477.

Lopez-Perez, L., Martínez-Ballesta, M.D.C., Maurel, C., and Carvajal, M. (2009). Changes in plasma membrane lipids, aquaporins and proton pump of broccoli roots, as an adaptation mechanism to salinity. *Phytochemistry* 70: 492–500.

Lothier, J., Diab, H., Cukier, C. et al. (2020). Metabolic responses to waterlogging differ between roots and shoots and reflect phloem transport alteration in *Medicago truncatula*. *Plants (Basel)* 9 (10): 1373.

Ma, L. and Jeffrey, A. (2019). Transcriptome analysis reveals key cold-stress-responsive genes in winter rapeseed (*Brassica rapa* L.). *Intl. J. Mol. Sci.* 5: 1071.

Ma, X., Xia, H., Liu, Y. et al. (2016). Transcriptomic and metabolomic studies disclose key metabolism pathways contributing to well-maintained photosynthesis under the drought and the consequent drought-tolerance in rice. *Front. Plant Sci.* 7: 1886.

Majoul, T., Bancel, E., Triboi, E. et al. (2004). Proteomic analysis of the effect of heat stress on hexaploid wheat grain: characterization of heat-responsive proteins from non-prolamins fraction. *Proteomics* 4: 505–513.

Malakshah, S.N., Rezaei, M.H., Heidari, M., and Salekdeh, G.H. (2007). Proteomics reveals new salt responsive proteins associated with rice plasma membrane. *Biosci. Biotechnol. Biochem.* 71: 2144–2154.

Manara, A. (2012). Plant responses to heavy metal toxicity. In: *Plants and Heavy Metals* (ed. A. Furini), 27–53. Springer.

Mandal, M., Sarkar, M., Khan, A. et al. (2022). Reactive Oxygen Species (ROS) and Reactive Nitrogen Species (RNS) in plants–maintenance of structural individuality and functional blend. *Adv. Redox Res.* 5: 100039.

Mangelsen, E., Kilian, J., Harter, K. et al. (2011). Transcriptome analysis of high-temperature stress in developing barley caryopses: early stress responses and effects on storage compound biosynthesis. *Mol. Plant.* 4: 97–115.

Mangrauthia, S.K., Bhogireddy, S., Agarwal, S. et al. (2017). Genome-wide changes in microRNA expression during short and prolonged heat stress and recovery in contrasting rice cultivars. *J. Exp. Bot.* 68: 2399–2412.

Markelz, R.J.C., Lai, L.X., Vosseler, L.N., and Leaky, A.D.B. (2014). Transcriptional reprogramming and stimulation of leaf respiration by elevated CO_2 concentration is diminished, but not eliminated, under limiting nitrogen supply. *Plant Cell Environ.* 37: 2542–2552.

Mishra, S. and Das, A.B. (2003). Effect of NaCl on leaf salt secretion and antioxidative enzyme level in roots of a mangrove, *Aegiceras corniculatum*. *Indian J. Exp. Biol.* 41: 160–166.

Miura, K. (2012). SlICE1 encoding a MYC-type transcription factor controls cold tolerance in tomato, *Solanum lycopersicum*. *Plant Biotechnol.* 29: 253–260.

Moan, J. (2011). Visible light and UV radiation. In: *Radiation at Home, Outdoors and in the Workplaces* (ed. L.T. Dauer), 69–85. Scandinavian Science Publisher.

Mohammadi, P.P., Moieni, A., and Komatsu, S. (2012). Comparative proteome analysis of drought-sensitive and drought-tolerant rape seed roots and their hybrid F1 line under drought stress. *Amino Acids* 43: 2137–2152.

Moreira-Rodriguez, M., Nair, V., Benavides, J. et al. (2017). UVA, UVB light and methyl jasmonate, alone or combined redirect the biosynthesis of glucosinolates, phenolics, carotenoids and chlorophyll in *Broccoli sprouts*. *Int. J. Mol. Sci.* 18: 2330.

Munns, R. and Passioura, J.B. (1984). Whole plant responses to salinity. *Aust. J. Plant Physiol.* 13: 143–160.

Munns, R. and Tester, M. (2008). Mechanisms of salinity tolerance. *Annu. Rev. Plant Biol.* 59: 651–681.

Mustroph, A., Barding-Jr, G.A., Kaiser, K.A. et al. (2014). Characterization of distinct root and shoot responses to low-oxygen stress in *Arabidopsis* with a focus on primary C-and N-metabolism. *Plant Cell Environ.* 37: 2366–2380.

Myers, S.S., Zanobetti, A., Kloog, I. et al. (2014). Increasing CO_2 threatens human nutrition. *Nature* 510: 139–142.

Nahar, K., Hasanuzzaman, M., Ahamed, K.U. et al. (2015). Plant responses and tolerance to high temperature stress: role of exogenous phytoprotectants. In: *Crop Production and Global Environment Issues* (ed. K.R. Hakeem). Cham: Springer.

Ndimba, B.K., Chivasa, S., Simon, W.J., and Slabas, A.R. (2005). Identification of *Arabidopsis* salt and osmotic stress responsive proteins using two-dimensional difference gel electrophoresis and mass spectrometry. *Proteomics* 5: 4185–4196.

Ni, L., Wang, Z., Liu, X. et al. (2022). Transcriptome analysis of salt stress in *Hibiscus hamabo* Sieb. et Zucc based on Pacbio full-length transcriptome sequencing. *Int. J. Mol. Sci.* 23: 138.

Nocchi, N., Duarte, H., Pereira, R.C., and Konno, T.U.P. (2020). Effects of UV-B radiation on secondary metabolite production, antioxidant activity, photosynthesis and herbivory interaction in *Nymphoides humbodtiana* (Menyanthaceae). *J. Photchem. Photobiol. B* 212: 112021.

Noguchi, K., Watanabe, C.K., and Terashima, I. (2015). Effect of elevated atmospheric CO_2 on primary metabolite levels in *Arabidopsis thaliana* Col-o leaves: an examination of metabolome data. *Plant Cell Physiol.* 56: 2069–2078.

Notaguchi, M. and Okamoto, S. (2015). Dynamics of long-distance signaling via plant vascular tissues. *Front. Plant Sci.* 6: 161.

Nybakken, L., Horkka, R., and Julkunen-Tiitto, R. (2012). Combined enhancements of temperature and UVB influence growth and phenolics in clones of the sexually dimorphic *Salix myrsinifolia*. *Physiol. Plant.* 145: 551–564.

Obata, T., Witt, S., Lisec, J. et al. (2015). Metabolite profiles of maize leaves in drought, heat, and combined stress field trials reveal the relationship between metabolism and grain yield. *Plant Physiol.* 169: 2665–2683.

Ogden, A.J., Abdali, S., Engbrecht, K.M. et al. (2020). Distinct preflowering drought tolerance strategies of *Sorghum bicolor* genotype RTx430 revealed by subcellular protein profiling. *Int. J. Mol. Sci.* 21 (24): 9706.

Opdenakker, K., Remans, T., Vangronsveld, J., and Cuypers, A. (2012). Mitogen-activated protein (MAP) kinases in plant metal stress: regulation and responses in comparison to other biotic and abiotic stresses. *Int. J. Mol. Sci.* 13: 7828–7853.

Pan, J., Sharif, R., Xu, X., and Chen, X. (2021). Mechanisms of waterlogging tolerance in plants: research progress and prospects. *Front. Plant Sci.* 11: 627331.

Park, C.J. and Seo, Y.S. (2015). Heat shock proteins: a review of the molecular chaperones for plant immunity. *Plant Pathol. J.* 31: 323–333.

Pasqualini, S., Della Torre, G., Ferranti, F. et al. (2002). Salicylic acid modulates ozone-induced hypersensitive cell death in tobacco plants. *Physiol. Plant.* 115: 204–212.

de Pinto, M.C. and de Gara, L. (2004). Changes in the ascorbate metabolism of apoplastic and symplastic spaces are associated with cell differentiation. *J. Exp. Bot.* 55: 2559–2569.

Pires, M.V., Pereira Junior, A.A., Medeiros, D.B. et al. (2016). The influence of alternative pathways of respiration that utilize branched-chain amino acids following water shortage in *Arabidopsis*. *Plant Cell Environ.* 39: 1304–1319.

Pochodylo, A.L. and Aristilde, L. (2017). Molecular dynamics of stability and structures in phytochelatin complexes with Zn, Cu, Fe, Mg, and Ca: implications for metal detoxification. *Environ. Chem. Lett.* 15: 1–6.

Qin, D., Wu, H., Peng, H. et al. (2008). Heat stress-responsive transcriptome analysis in heat susceptible and tolerant wheat (*Triticum aestivum* L.) by using wheat genome array. *BMC Genomics* 9: 432.

Rahman, M., Ijaj, M., Qamar, S. et al. (2019). Abiotic stress signaling in rice crop. In: *Advances in Rice Research for Abiotic Stress Tolerance* (ed. M. Hasanuzzaman, M. Fujita, K. Nahar, and J.K. Biswas), 551–569. Woodhead Publishing.

Rajendran, K., Tester, M., and Roy, S.J. (2008). Quantifying the main three components of salinity tolerance in cereals. *Plant Cell Environ.* 32: 237–249.

Ramakrishna, A. and Ravishankar, G.A. (2011). Influence of abiotic stress signals on secondary metabolites in plants. *Plant Signal Behav.* 6: 1720–1731.

Rao, M.V., Paliyath, G., and Ormrod, D.P. (1996). Ultraviolet-B and ozone induced biochemical changes in antioxidant enzymes of *Arabidopsis thaliana*. *Plant Physiol.* 110: 125–136.

Rao, A.Q., Din, S., Akhtar, S. et al. (2015). Genomics of salinity tolerance in plants. *Plant Genomics* https://doi.org/10.5772/63361.

Rauser, W.E. (1999). Structure and function of metal chelators produced by plants: the case for organic acids, amino acids, phytin, metallothioneins. *Cell Biochem. Biophys.* 31: 19–48.

Reiland, S., Messerli, G., Baerenfaller, K. et al. (2009). Large-scale *Arabidopsis* phosphoproteome profiling reveals novel chloroplast kinase substrates and phosphorylation networks. *Plant Physiol.* 150: 889–903.

Ren, S., Ma, K., Lu, Z. et al. (2019). Transcriptomic and metabolomic analysis of the heat-stress response of *Populus tomentosa* Carr. *Forests* 10: 383.

Renaut, J., Lutts, S., Hoffmann, L., and Hausman, J.F. (2004). Responses of poplar to chilling temperatures: proteomics and physiological aspects. *Plant Biol.* 6: 81–90.

Renaut, J., Bohler, S., Hausman, J.F. et al. (2009). The impact of atmospheric composition on plants: a case study of ozone and poplar. *Mass Spectrom. Rev.* 28: 495–516.

Rezaie, R., Abdollahi Mandoulakani, B., and Fattahi, M. (2020). Cold stress changes antioxidant defense system, phenylpropanoid contents and expression of genes involved in their biosynthesis in *Ocimum basilicum* L. *Sci. Rep.* 10 (1): 5290.

Ries, G., Heller, W., Puchta, H. et al. (2000). Elevated UV-B radiation reduces genome stability in plants. *Nature* 406: 98–101.

Rizzini, L., Favory, J.J., Cloix, C. et al. (2011). Perception of UV-B by the *Arabidopsis* UVR8 protein. *Science* 332: 103–106.

Rodriguez-Milla, M.A. and Salinas, J. (2009). Prefoldins 3 and 5 play an essential role in *Arabidopsis* tolerance to salt stress. *Mol. Plant.* 2: 526–534.

Roy, S.J., Negrao, S., and Tester, M. (2014). Salt resistant crop plants. *Curr. Opin. Biotechnol.* 26: 115–124.

Sakamoto, H., Matsuda, O., and Iba, K. (2008). ITN1, a novel gene encoding an ankyrin repeat protein that affects the ABA-mediated production of reactive oxygen species and is involved in salt-stress tolerance in *Arabidopsis thaliana*. *Plant J.* 56 (3): 411–422.

Sang, Q., Shan, X., An, Y. et al. (2017). Proteomic analysis reveals the positive effect of exogenous spermidine in tomato seedlings' response to high-temperature stress. *Front. Plant Sci.* 8: 120.

Sarkar, A. and Agrawal, S.B. (2010a). Elevated ozone and two modern wheat cultivars: an assessment of dose dependent sensitivity with respect to growth, reproductive and yield parameters. *Environ. Exp. Bot.* 69 (3): 328–337.

Sarkar, A. and Agrawal, S.B. (2010b). Identification of ozone stress in Indian rice through foliar injury and differential protein profile. *Environ. Monit. Assess.* 161: 205–215.

Sarkar, A., Rakwal, R., Agrawal, S.B. et al. (2010). Investigating the impact of elevated levels of ozone on tropical wheat using integrated phenotypical, physiological, biochemical, and proteomics approaches. *J. Proteome Res.* 9 (9): 4565–4864.

Sarkar, A., Singh, A.A., Agrawal, S.B. et al. (2015). Cultivar specific variations in antioxidative defense system, genome and proteome of two tropical rice cultivars against ambient and elevated ozone. *Ecotoxicol. Environ. Saf.* 115: 101–111.

Sasidharan, R., Hartman, S., Liu, Z. et al. (2018). Signal dynamics and interactions during flooding stress. *Plant Physiol.* 176: 1106–1117.

Saxena, P., Bharadwaj, R., and Ghosh, C. (2012). Status of air pollutants after implementation of CNG in Delhi. *Curr. World Environ.* 7 (1): 109–115.

Sharma, S.S. and Dietz, K.J. (2006). The significance of amino acids and amino acid derived molecules in plant responses and adaptation to heavy metal stress. *J. Exp. Bot.* 57: 711–726.

Sharma, R., Sharma, P., Kumar, S. et al. (2017). Heat treatment affects regeneration, protein expression and genetic makeup of *Vigna aconitifolia* (Jacq.) Marechal. *Ann. Agrar. Sci.* 16: 116–220.

Shen, S.G., Jia, S.R., Yan, R.R. et al. (2018). The physiological responses of terrestrial cyanobacterium: *Nostoc flagelliforme* to different intensities of ultraviolet-B radiation. *RSC Adv.* 8: 21065–21074.

Shen, C., Yuan, J., Qiao, H. et al. (2020). Transcriptomic and anatomic profiling reveal the germination process of different wheat varieties in response to waterlogging stress. *BMC Genet.* 21: 93.

Shi, J., Yan, B., Lou, X. et al. (2017). Comparative transcriptome analysis reveals the transcriptional alterations in heat-resistant and heat-sensitive sweet maize (*Zea mays* L.) varieties under heat stress. *BMC Plant Biol.* 17: 26.

Shulaev, V., Cortes, D., Miller, G., and Mittler, R. (2008). Metabolomics for plant stress response. *Physiol. Plant.* 132: 199–208.

Sihag, P., Kumar, U., Sagwal, V. et al. (2023). Effect of terminal heat stress on osmolyte accumulation and gene expression during grain filling in bread wheat (*Triticum aestivum* L.). *Plant Genome* e20307. https://doi.org/10.1002/tpg2.20307.

Singh, S.S., Kumar, P., and Rai, A.K. (2006). Ultraviolet radiation stress: molecular and physiological adaptation in trees. In: *Abiotic Stress Tolerance in Plants* (ed. A.K. Rai and T. Takabe), 91–110. Dordrecht: Springer.

Singh, A., Sarkar, A., Singh, S. et al. (2010). Investigation of supplemental ultraviolet-B-induced changes in antioxidative defense system and leaf proteome in radish (*Raphanus sativus* L. cv Truthful): an insight to plant response under high oxidative stress. *Protoplasma* 245: 75–83.

Singh, S., Sarkar, A., Agrawal, S. et al. (2014). Impact of ambient and supplemental ultraviolet-B stress on kidney bean plants: an insight into oxidative stress management. *Protoplasma* 251: 1395–1405.

Singh, P.K., Indoliya, Y., Agrawal, L. et al. (2022). Genomic and proteomic responses to drought stress and biotechnological interventions for enhanced drought tolerance in plants. *Curr. Plant Biol.* 29: 100239.

Song, X., Zhang, C., Chen, W. et al. (2020). Growth responses and physiological and biochemical changes in five ornamental plants grown in urban lead-contaminated soils. *Plant Environ. Int.* 1: 29–47.

Sunarpi, Horiew, T., Motoda, J. et al. (2005). Enhanced salt tolerance mediated by AtHKT1 transporter induced Na unloading from xylem vessels to xylem parenchyma cells. *Plant J.* 44: 928–938.

Sung, D.Y., Kaplan, F., Lee, K.J., and Guy, C.L. (2003). Acquired tolerance to temperature extremes. *Trends Plants Sci.* 8: 179–187.

Szabados, L. and Savoure, A. (2010). Proline: a multifunctional amino acid. *Trends Plant Sci.* 15: 89–97.

Taiz, L. and Zeiger, E. (2006). *Plant Physiology*, 4e. Sunderland, MA: Sinauer Associates Inc. Publishers.

Takahashi, F. and Shinozaki, K. (2019). Long-distance signaling in plant stress response. *Curr. Opin. Plant Biol.* 47: 106–111.

Takahashi, F., Kuromori, T., Sato, H., and Shinozaki, K. (2018). Regulatory gene networks in drought stress responses and resistance in plants. *Adv. Exp. Med. Biol.* 1081: 189–214.

Takahashi, F., Hanada, K., Kondo, T., and Shinozaki, K. (2019). Hormone like peptides and small coding genes in plant stress signaling and development. *Curr. Opin. Plant Biol.* 51: 88–95.

Tamaoki, M., Nakajima, N., Kubo, A. et al. (2003). Transcriptome analysis of O_3-exposed *Arabidopsis* reveals that multiple signal pathways act mutually antagonistically to induce gene expression. *Plant Mol. Biol.* 53: 443–456.

Tamas, L., Mistril, I., Huttova, J. et al. (2010). Role of reactive oxygen species-generating enzymes and hydrogen peroxide during cadmium, mercury, and osmotic stress in barley root tip. *Planta* 231: 221–231.

Tarazona, P., Feussner, K., and Feussner, I. (2015). An enhanced plant lipidomics method based on multiplexed liquid chromatography-mass spectrometry reveals additional insights into cold- and drought-induced membrane remodeling. *Plant J.* 84: 621–633.

Taylor, N.L., Heazlewood, J.L., Day, D.A., and Millar, A.H. (2005). Differential impact of environmental stresses on the pea mitochondrial proteome. *Mol. Cell Proteomics* 4: 1122–1133.

Thakur, S., Choudhary, S., and Bhardwaj, P. (2019). Comparative transcriptome profiling under cadmium stress reveals the uptake and tolerance mechanism in *Brassica juncea*. *J. Plant Growth Regul.* 38 (03): 1141–1152.

Tilbrook, K., Arongaus, A.B., Binkert, M. et al. (2013). The UVR8 UV-B photoreceptor: perception, signaling and response. *Arabidopsis Book* 11: e0164.

Toumainen, J., Pellinen, R., Roy, S. et al. (1996). Ozone affect birch (*Betula pendula* Roth) phenylpropanoid, polyamine and reactive oxygen detoxifying pathways at biochemical and gene expression levels. *J. Plant Physiol.* 148: 179–188.

Tran, L.S.P., Nishiyama, R., Yamaguchi-Shinozaki, K., and Shinozaki, K. (2010). Potential utilization of NAC transcription factors to enhance abiotic stress tolerance in plants by biotechnological approach. *GM Crops.* 1: 32–39.

Tripathi, R., Sarkar, A., Pandey Rai, S., and Agrawal, S.B. (2010). Supplemental ultraviolet-B and ozone: impact on antioxidants, proteome and genome of linseed (*Linum usitatissimum* L. cv. Padmini). *Plant Biol.* 13 (1): 93–104.

Tunnacliffe, A. and Wise, M.J. (2007). The continuing conundrum of the LEA proteins. *Naturwissenschaften* 94: 791–812.

Urbonaviciute, A., Samuoliene, G., Sakalauskaite, J. et al. (2006). The effect of elevated CO_2 concentrations on leaf carbohydrate, chlorophyll contents and photosynthesis on radish. *Pol. J. Environ. Stud.* 15: 921–925.

Urton, J. (2018). High levels cause plants to thicken their leaves which could worsen climate change effects, researcher say. *UW news* (1 October 2018).

Van de Mortel, J.E., Schat, H., Moerland, P.D. et al. (2008). Expression differences for genes involved in lignin, glutathione and sulphate metabolism in response to cadmium in *Arabidopsis thaliana* and the related Zn/Cd-hyper accumulator *Thlaspi caerulescens*. *Plant Cell Environ.* 31: 301–324.

Van der Kooi, C.J., Reich, M., De Kok, L.J., and Tausz, M. (2016). Growth and yield stimulation under elevated CO_2 and drought: a meta-analysis on crops. *Environ. Exp. Bot.* 122: 150–157.

Van Veen, H., Vashisht, D., Akman, M. et al. (2016). Transcriptomes of eight *Arabidopsis thaliana* accessions reveal core conserved genotype and organ specific responses to flooding stress. *Plant Physiol.* 172: 668–689.

Vaughan, M.M., Block, A., Christensen, S.A. et al. (2018). The effects of climate associated abiotic stresses on maize phytochemical defense. *Phytochem. Rev.* 17: 37–49.

Verkleij, J.A.C., Lolkema, P.C., De Neeling, A.L., and Harmens, H. (1991). Heavy metal resistance in higher plants: biochemical and genetic aspects. In: *Ecological Responses to Environmental Stress* (ed. J. Rozema and J.A.C. Verkleij), 8–19. Dordrecht: Springer.

Verma, S., Nizam, S., and Verma, P.K. (2013). Biotic and abiotic stress signaling in plants. *Stress Signal. Plants Genomics Proteomics Perspect.* 1: 25–49.

Vessal, S., Arefian, M., and Siddique, K.H. (2020). Proteomic responses to progressive dehydration stress in leaves of chickpea seedlings. *BMC Genomics* 21 (1): 1–15.

Vollenweider, P. and Gunthardt-Goerg, M.S. (2005). Diagnosis of abiotic and biotic stress factors using the visible symptoms in foliage. *Environ. Pollut.* 137: 455–465.

Wahid, A., Gelani, S., Ashraf, M., and Foolad, M. (2007). Heat tolerance in plants: an overview. *Environ. Exp. Bot.* 61: 199–223.

Wahid, A., Farooq, M., Hussain, I. et al. (2012). Responses and management of heat stress in plants. In: *Environmental Adaptations and Stress Tolerance of Plants in the Era of Climate Change* (ed. P. Ahmad and M.N.V. Prasad), 135–157. New York, Dordrecht, Heidelberg, London: Springer.

Wang, W., Vinocur, B., Shoseyov, O., and Altman, A. (2004). Role of plant heat-shock proteins and molecular chaperones in the abiotic stress response. *Trends Plant Sci.* 9: 244–252.

Wang, M.C., Peng, Z.Y., Li, C.L. et al. (2008). Proteomic analysis on a high salt tolerance introgression strain of *Triticum aestivum/Thinopyrum ponticum*. *Proteomics* 8: 1470–1489.

Wang, M., Zhang, X., Li, Q. et al. (2019). Comparative transcriptome analysis to elucidate the enhanced thermotolerance of tea plants (*Camellia sinensis*) treated with exogenous calcium. *Planta.* 249 (03): 775–786.

Wang, L., Ma, K.B., Lu, Z.G. et al. (2020a). Differential physiological, transcriptomic and metabolomic responses of *Arabidopsis* leaves under prolonged warming and heat shock. *BMC Plant Biol.* 20: 86.

Wang, X., Li, N., Li, W. et al. (2020b). Advances in transcriptomics in the response to stress in plants. *Global Med Genet.* 7: 30–34.

Wang, Z.Q., Zhao, Q.M., Zhong, X. et al. (2020c). Comparative analysis of maca (*Lepidium meyenii*) proteome profiles reveals insights into response mechanisms of herbal plants to high-temperature stress. *BMC Plant Biol.* 20: 431.

Wang, X., Yan, L., Wang, B. et al. (2021). Comparative proteomic analysis of grapevine rootstock in response to waterlogging stress. *Front. Plant Sci.* 12: 749184.

Wania, S.H., Kumar, V., Shriram, V., and Sah, S.K. (2016). Phytohormones and their metabolic engineering for abiotic stress tolerance in crop plants. *Crop J.* 4: 162–176.

Wanqian, F., Jing, L., Sixin, L., and Shanjun, W. (2018). A DREB1 gene from zoysiagrass enhances *Arabidopsis* tolerance to temperature stresses without growth inhibition. *Plant Sci.* 278: 20–31.

Wargent, J.J. and Jordan, B.R. (2013). From ozone depletion to agriculture: understanding the role of UV radiation in sustainable crop production. *New Phytol.* 197: 1058–1076.

Weber, J., Tingey, D., and Andersen, C. (2002). *Plant Response to Air Pollution*. Washington: U.S. Environmental Protection Agency.

Wolfgang, M., Barry, C.S., Tauriainen, A.A. et al. (2002). Ethylene synthesis regulated by biphasic induction of 1-aminocyclopropane-1-carboxylic acid synthase and 1-aminocyclopropane-1-carboxylic acid oxidase genes is required for hydrogen peroxide accumulation and cell death in ozone exposed tomato. *Plant Physiol.* 130: 1918–1926.

Xu, X., Ji, J., Ma, X. et al. (2016). Comparative proteomic analysis provides insight into the key proteins involved in cucumber (*Cucumis sativus* L.) adventitious root emergence under waterlogging stress. *Front. Plant Sci.* 7: 1515.

Yamaguchishinozaki, K. and Shinozaki, K. (1994). A novel *cis*-acting element in an *Arabidopsis* gene is involved in responsiveness to drought, low-temperature, or high-salt stress. *Plant Cell.* 6: 251–264.

Yan, S.P., Zhang, Q.Y., Tang, Z.C. et al. (2006). Comparative proteomic analysis provides new insights into chilling stress responses in rice. *Mol. Cell Proteomics* 5: 484–496.

Yan, K., Zhao, S., Cui, M. et al. (2018). Vulnerability of photosynthesis and photosystem I in Jerusalem artichoke (*Helianthus tuberosus* L.) exposed to waterlogging. *Plant Physiol. Biochem.* 125: 239–246.

Yao, L., Hao, X., Cao, H. et al. (2020). ABA-dependent bZIP transcription factor, CsbZIP18, from *Camellia sinensis* negatively regulates freezing tolerance in *Arabidopsis*. *Plant Cell Rep.* 39: 553–565.

Yin, X. and Komatsu, S. (2015). Quantitative proteomics of nuclear phosphoproteins in the root tip of soybean during the initial stages of flooding stress. *J. Proteomics* 119: 183–195.

Yin, X. and Komatsu, S. (2016). Nuclear proteomics reveals the role of protein synthesis and chromatin structure in root tip of soybean during the initial stage of flooding stress. *J. Proteome Res.* 15: 2283–2298.

Yin, X. and Komatsu, S. (2017). Comprehensive analysis of response and tolerant mechanisms in early stage soybean at initial-flooding stress. *J. Proteomics.* 169: 225–232.

Yoon, M.Y., Kim, M.Y., Shim, S. et al. (2016). Transcriptomic profiling of soybean in response to high-intensity UV-B irradiation reveals stress defense signaling. *Front. Plant Sci.* 7: 1917.

You, J., Zhang, Y., Liu, A. et al. (2019). Transcriptomic and metabolomic profiling of drought-tolerant and susceptible sesame genotypes in response to drought stress. *BMC Plant Biol.* 19 (1): 1–16.

Zhang, X. (2004). Freezing sensitive tomato has a functional CBF regulon that differs from that of freezing tolerant *Arabidopsis*. *Plant J.* 39: 905–919.

Zhang, L., Tian, L.H., Zhao, J.F. et al. (2009). Identification of an apoplastic protein involved in the initial phase of salt stress response in rice root by two-dimensional electrophoresis. *Plant Physiol.* 149: 916–928.

Zhang, M.H., Li, G.W., Huang, W. et al. (2010). Proteomic study of *Carissa spinarum* in response to combined heat and drought stress. *Proteomics* 10: 3117–3129.

Zhang, X., Rerksiri, W., Liu, A. et al. (2013). Transcriptome profile reveals heat response mechanism at molecular and metabolic levels in rice flag leaf. *Gene* 530: 185–192.

Zhang, R.F., Guo, Y., Li, Y.Y. et al. (2016). Functional identification of MdSIZ1 as a SUMO E3 ligase in apple. *J. Plant Physiol.* 198: 69–80.

Zhang, P., Lyu, D., Jia, L. et al. (2017). Physiological and *de novo* transcriptome analysis of the fermentation mechanism of *Cerasus sachalinensis* roots in response to short-term waterlogging. *BMC Genomics* 18: 649.

Zhang, G., Wang, Y., Wu, K. et al. (2021). Exogenous application of chitosan alleviate salinity stress in lettuce (*Lactuca sativa* L.). *Horticulturae* 7: 342.

Zhao, Y. and Chengcai, C. (2011). Towards understanding plant response to heavy metal stress. In: *Abiotic Stress in Plants – Mechanisms and Adaptations* (ed. A. Shanker and B. Venkateswarlu), 59–78. Rijeka: Tech Europe.

Zhou, S., Okekeogbu, I., Sangireddy, S. et al. (2016). Proteome modification in tomato plants upon long-term aluminum treatment. *J. Proteom. Res.* 15: 1670–1684.

Zhu, J.K. (2002). Salt and drought stress signal transduction in plants. *Annu. Rev. Plant Biol.* 53: 247–273.

Zhu, J.K. (2016). Abiotic stress signaling and responses in plants. *Cell* 167: 313–324.

Zhu, Q., Gong, Z., Huang, J. et al. (2019). High-CO_2/hypoxia-responsive transcription factors DkERF24 and DkWRKY1 interact and activate *DkPDC2* promoter. *Plant Physiol.* 180 (1): 621–633.

Zörb, C., Schmitt, S., and Mühling, K.H. (2010). Proteomic changes in maize roots after short-term adjustment to saline growth conditions. *Proteomics* 10: 4441–4449.

16

Developing Stress-Tolerant Plants

Role of Small GTP Binding Proteins (RAB and RAN)

Manas K. Tripathy[1] *and Sudhir K. Sopory*[2]

[1] *Division of Plant and Microbial Biotechnology, Institute of Life Sciences, Bhubaneswar, Odisha, India*
[2] *Department of Plant Molecular Biology, International Centre for Genetic Engineering and Biotechnology, New Delhi, India*

16.1 Introduction

In addition to changes caused by natural processes such as variations in the solar cycle, human activity is considered as a major factor in climate change through burning fossil fuels like coal, oil, and gas. This has led to changes in the earth's temperature, which is estimated to rise to 2.7 °C at the end of this century. As the earth is an interconnected system, changes in one factor bring about changes in all other factors. The rise in temperature brings intense droughts, water scarcity, severe fires, rising sea levels, flooding, melting polar ice, catastrophic storms, and declining biodiversity (United Nations Climate Action 2021). Of the many effects, climate change, in terms of increase in carbon dioxide (CO_2), methane (CH_4), and ozone (O_3) and variations in temperature will have direct impact on crop productivity (IPCC 2014). In some cases, higher concentrations of CO_2 will improve net photosynthesis rates and increase water use efficiency (Long et al. 2004; Long and Ort 2010; Deryng et al. 2016). This positive effect is higher in C_3 plants such as wheat, rice, and soybean, due to the limited photosynthetic output of photorespiratory carbon losses. Nevertheless, in the long term, the constant increment in CO_2 concentration will have a negative impact in the climate, thus counterbalancing the increase in crop yield (Specht et al. 1999; Long et al. 2004; Dong et al. 2018; Senapati et al. 2019; Wei et al. 2019). On the other hand, increases in O_3 levels have significant negative effect on crop production. O_3 is a reactive oxidant that damages plant tissues, brings leaf injuries that lead to decreased photosynthesis, accelerated senescence, and cell death (Vandermeiren et al. 2009), though O_3 sensitivity between species varies to a great extent (Mills et al. 2007).

Climate change also brings changes in the rainfall patterns. Heavy rainfall leads to flood, whereas less rainfall causes drought. With intense rainfall, the disease load also increases. All these conditions affect crop growth and development and ultimately hamper crop yield (Lobell and Burke 2008; Urban et al. 2015; Sharif et al. 2017). Moreover, in future, there is a risk of frequent floods in coastal areas due to increasing sea level caused by strong waves and storms (Vitousek et al. 2017). The high salinity of seawater will become an additional problem for coastline crops due to osmotic and anionic stress. Drought stress affects vegetative stages like reduction in kernel number and biomass and reproductive stages such as pollen sterility and ovarian abortion, which lead to loss of crop productivity (Boyer and Westgate 2004). During drought, plants use less water, which brings stomatal closure and lower CO_2 intake. Thus, reduction of crop biomass decreases the rate of photosynthesis (Garofalo et al. 2019). Water scarcity caused by drought leads to salinity stress. Reduction of soil water potential and ion toxicity affects plant growth and development. Soil salinity reduces yield in tolerant crops such as cotton, barley, and sugar beet as well as in salinity-sensitive crops like sweet potato, wheat, or maize (Zörb et al. 2019).

Climate change due to global warming leads to increasing heat waves, decrease in winter season, less duration but heavy rainfall, increase in droughts, and other extreme weather, which adversely affect agricultural production (Calleja-Cabrera et al. 2020). Plants' response to high temperatures under extreme conditions is characterized by the heat shock response. However, crop growth and yield are affected by small changes in ambient growth temperature. Although more detailed studies have reported on how plants tolerate extreme heat, less is understood on how crops adapt to moderately increased or warmer temperatures (Quint et al. 2016; Vu et al. 2019). Different prediction models suggest that frequent increases in temperature bring heavy losses in crop yield at medium latitudes (Liu and Howell 2016), whereas extreme latitudes with

Global Climate Change and Plant Stress Management, First Edition. Edited by Mohammad Wahid Ansari, Anil Kumar Singh, and Narendra Tuteja.

less fertile soil are becoming more appropriate for agriculture (Long and Ort 2010; Lobell et al. 2011; Iizumi and Ramankutty 2015; Sharif et al. 2017). Thus, warmer temperature could expand the areas potentially suitable for cropping, increase the length of growing period, and increase the crop yields (How to Feed the World in 2050, Rome; FAO 2009). However, higher temperatures shorten the growth season, allowing the crops a shorter time for photosynthesis, even in well-irrigated and tolerant crops. Moreover, during this shorter growth period, heat stress directly affects photosynthetic rate. As a result, crops have less biomass to face the anthesis and the consequent grain filling. Warmer environments also affect post-anthesis stages, reducing grain growth and increasing fruit senescence. Additionally, the increase in temperature promotes a higher evapotranspiration rate that, ultimately, reduces soil moisture and the available water needed for grain filling. When plants face extreme temperatures for a short duration, these processes are even more severely affected (Asseng et al. 2011, 2015, 2019; Liu et al. 2014, 2019; Lesk et al. 2016). Accordingly, reports suggest that in wheat, rice, and sorghum, heat causes loss of grain yield by shortening the growth period, affecting spikelet's development, reducing grains per spike and grain size (Prasad et al. 2006; Jagadish et al. 2010; Fahad et al. 2017). Similarly, in oilseed rape, *Brassica rapa* and *Brassica juncea* yield losses due to decrease in siliques number per plant as well as defects in pod formation (Angadi et al. 2000; Morrison and Stewart 2002). High temperatures also lead to a decrease in crop quality by changing seed composition. Thus, in cereals and oilseed crops, heat stress reduces the oil, starch, and protein contents of seeds (Jagadish et al. 2015; Fahad et al. 2017). In wheat, increased temperatures reduce the levels of valuable protein while accumulating proline and soluble carbohydrates (Qaseem et al. 2019). On the other hand, higher temperatures also reduce oilseed rape seeds quality by reducing the amounts of oil and increasing the levels of proteins and glucosinolates (Aksouh et al. 2001). In rice, high temperatures during ripening led to the deterioration of grain quality including starch accumulation (Morita et al. 2016; Chen et al. 2017). In brief, crops are substantially but heterogeneously affected by temperature variability (Thiault et al. 2019). As a solution, we need to evaluate and understand further the changes that crops undergo under the future climatic scenario.

In order to sustain crop productivity under changing climate for the ever-increasing population, there is an urgent need to maintain and improve crops that can overcome climatic constraints (Bailey-Serres et al. 2019; Shan-e-Ali Zaidi et al. 2019). To tackle the problem, it is important to understand the molecular basis of plant physiology and metabolism and discover tools and techniques to develop climate-resilient plants with respect to specific regions as per future predictions.

In this chapter, we are focusing on small GTP-binding proteins as one of the groups of candidate genes to be utilized for tailoring plants to acquire stress tolerance to different abiotic and biotic factors caused by climate change.

16.2 A Brief Overview of GTP-Binding Proteins

GTP (guanosine triphosphate)-binding proteins or G-proteins are signal-transducing proteins that transmit signals outside the cell and bring about changes inside cells. G-proteins function as a molecular switches that become active by binding GTP and inactive when they are bound by GDP. G-proteins can be divided into two classes viz heterotrimeric G-protein complexes and monomeric small GTP-binding proteins. The heterotrimeric G-proteins is made up of alpha (α), beta (β), and gamma (γ) subunits. Additionally, the beta and gamma subunits can form a stable dimeric complex called as beta-gamma complex. Heterotrimeric G-proteins are located within the cell and activated by G protein-coupled receptors (GPCRs), which span the cell membrane (Neves et al. 2002; McCudden et al. 2005; Milligan and Kostenis 2006; Oldham and Hamm 2008; Urano et al. 2013; Maruta et al. 2021). However, in this chapter, we are discussing only monomeric small GTPases, their structure, and function in plants.

16.3 Small GTP-Binding Proteins

In recent years, remarkable progress has been made in understanding the endomembrane system in plants. However, it is still not clear how small GTP-binding proteins (small GTPases) participate in the signaling steps that regulate developmental processes, control hormone biosynthesis or activity, and mediate biotic and abiotic stress tolerance. Like heterotrimeric G-proteins, small GTPases also work as protein switches that are turned on and off by the nucleotide GTP. When GTP is bound, the G-proteins and GTPases are "on" and activate downstream signals (Figure 16.1). Small GTPases present in all eukaryotes, from yeast to humans, and these constitute a superfamily consisting of more than 100 members. This superfamily is structurally classified into at least five families: the Ras, Rho, Rab, Sar1/Arf, and Ran families (Bischoff et al. 1999;

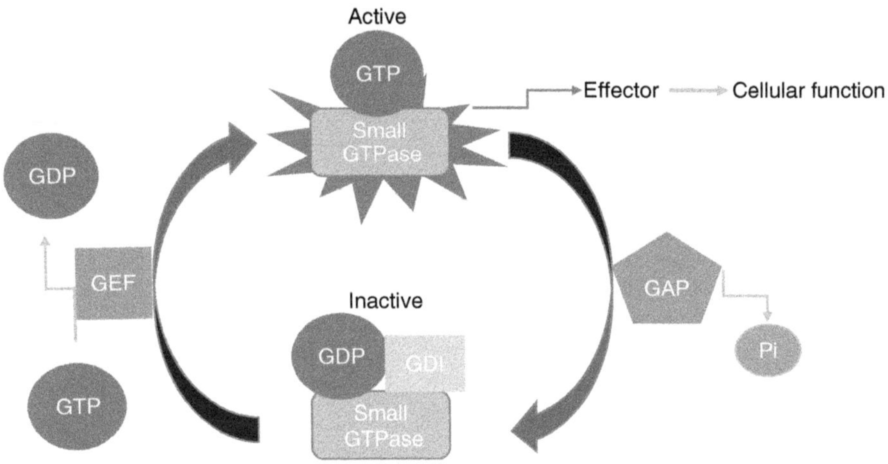

Figure 16.1 Small GTPase regulatory cycle. Small GTPases cycle between an active GTP-bound and an inactive GDP-bound state. GEFs activate GTPase, which later interact with specific effectors to perform cellular functions. GAPs stimulate the GTPase activity of proteins and catalyze the inactivation of their regulatory activity.

Takai et al. 2001; Agarwal et al. 2009; Tripathy et al. 2021). However, Ras homologs are absent in plants. In this chapter, we further focus our presentation on only two main small GTPases, RAB and RAN, and discuss their role in plant abiotic and biotic stress responses.

16.3.1 RAB

RAB (Ras-related proteins in brain) GTPases are important proteins that regulate intracellular vesicular transport, which consequently influences many cellular functions and responses. In both plants and animals, it helps in trafficking of proteins between different organelles via endocytotic and secretory pathways. RAB proteins consist of multiple conserved domains (Figure 16.2), including four guanine nucleotide-binding domains (G1, G3, G4, and G5), an effector-binding domain (G2), a membrane attachment domain (C-residues), and YRG domain (Takai et al. 2001; Tripathy et al. 2021). The G1 domain is involved in Mg^+ and phosphate binding. The G3 domain is the site where DTAG sequence interacts with the γ-phosphate of GTP. The G4 domain is the region that determines guanine specificity, and the G5 domain is the site where ETSA sequence interacts with the D residue in the NKXD sequence. The effector region G2 specifies RAB function, and its TIGADF motif interacts with specific GTPase-activating proteins (GAPs) (Haubruck et al. 1990; Zerial and Stenmark 1993). The role of YRG is not well established. In comparison to yeast, RABs are more diverse in plants and mammals. The RAB GTPases make up the largest subfamily of the RAS superfamily and are divided into eight sub-families based on their localization and/or function in membrane trafficking, such as RAB1, RAB2, RAB5, RAB6, RAB7, RAB8, RAB11, and RAB18. Reports suggested, multiple copies of RABs are present in plants and other organisms; 52 in *Oryza sativa*, 57 in *Arabidopsis thaliana*, 67 in *Populus trichocarpa*, 11 in *Saccharomyces cerevisiae*, 7 in *Schizosaccharomyces pombe*, 31 in *Caenorhabditis elegans*, 33 in *Drosophila melanogaster*, and 65 in *Homo sapiens* (Agarwal et al. 2009; Tripathy et al. 2021).

16.3.1.1 Role of RAB's in Plant

In animals, RAB family has been well studied, whereas less work has been reported in plants. Previous reports suggest RABs play important role in plant growth and development, autophagy, abiotic stress, plant-microbe interaction, and biotic stresses (Agarwal et al. 2009; Kwon et al. 2013; Tripathy et al. 2021; Wang et al. 2021). Among these functions, in this chapter we have elaborated on RABs role in abiotic stress, plant–microbe interaction, and biotic stresses.

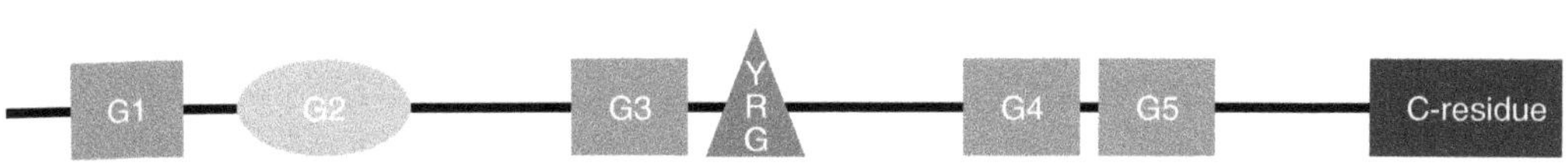

Figure 16.2 Schematic representation of RAB protein domain architecture. Different conserved motifs of GTP binding and hydrolysis are represented by multiple color polygons. G1 (GDSGVGKT), G3 (WDTAGQ), G4 (GNKXD), and G5 (ETSAK) are guanine nucleotide-binding domains, and G2 (YKATIGADF) is an effector-binding domain. YRG domain function is unknown.

Abiotic Stress Several reports suggest RAB protein plays important role during abiotic stress tolerance. In a desiccation-tolerant grass, *Sporoboluss tapfianus,* under dehydration conditions, a RAB2 homolog was induced. During drying and rehydration, the accumulation of *SsRab2* transcript indicates that activation of the endomembrane system may be necessary for both a protection- and/or repair-based mechanism of desiccation tolerance. The accumulation of RAB2 protein is necessary and acts as an integral part but not an exclusive component during desiccation tolerance (O'Mahony and Oliver 1999).

Similarly, overexpression of plant-unique RAB5 protein RABF1 (ARA6) confers increased tolerance to salinity stress. It may be due to RABF1 associated to membrane and plays a role in recycling and degradation processes during senescence and stress responses. The response to salt stress might be linked to close association of RabF1 with the plasma membrane (PM), since it contains several transporters involved in signaling in response to salt stress (Yin et al. 2017). In mango tree (*Mangifera indica*), *MiRAB5*, a RAB5 gene, is ubiquitously expressed in various tissues. In response to various abiotic stresses such as cold, salinity, and PEG treatments, its expression is upregulated, indicating that it participates in the signaling response to these stresses (Liu et al. 2014).

In rice, OsRAB7 is localized to the vacuolar membrane, suggesting its involvement in vesicular transport to the vacuole and a role in vacuole biogenesis. Under various environmental stimuli such as cold, NaCl, dehydration, and ABA, the OsRAB7 was differentially expressed, suggesting its role in stress tolerance (Nahm et al. 2003). The transgenic plants overexpressing AtRAB7 (*AtRabG3e*) showed increased tolerance to salt and osmotic stresses and reduced accumulation of reactive oxygen species (ROS) during salt stress. This finding suggests profound effects of membrane endocytosis and stress tolerance in plants (Mazel et al. 2004). Tobacco transgenic plants constitutively overexpressing *Pennisetum gluacum* RAB7 (PgRAB7) showed enhanced tolerance to both salinity and dehydration stress. This finding suggested that *PgRab7* is a potent gene involved in abiotic stress adaptation. The tolerance of transgenics, which may be either due to increase in vesicle transport or due to protective or repair-based mechanism of the endomembrane system as RAB7, is involved in vacuolar transport inside cell (Agarwal et al. 2008). Tobacco plants overexpressing the RAB7 from a phreatophyte *Prosopis juliflora* (PjRAB7) showed enhanced tolerance to salt stress. During salt stress, the PjRAB7 transgenic plants were found to accumulate more sodium than control plants. This indicates PjRAB7 may have a role in sodium sequestration into the vacuole, which leads to stress tolerance (George and Parida 2011). Rice plants overexpressing a vesicle trafficking gene, *OsRab7* showed enhanced tolerance to salt stress with an increased number of vesicles in the root tip of transgenic plants. This finding further supports the hypothesis that during salt stress tolerance of OsRAB7, overexpressing plants use vacuolar trafficking as one of the mechanisms to survive high salt concentration (Peng et al. 2014). Consistent with the above finding, transgenic rice plants constitutively overexpressing PgRAB7 also showed tolerance to both drought and salinity stress. Compared to wild-type rice plants, the transgenic plants overexpressing PgRAB7 under NaCl stress showed higher seed germination, survival, and better growth. The transgenic plants have better performance due to controlled entry of Na^+ ions into the cell, selective exclusion of Na^+ ions, and/or Na^+ ions compartmentalization in the vacuole. During salt stress, the structural integrity of the photosynthetic apparatus and thus overall photosynthetic activity was intact and active due to the upregulation of WD domain protein and chlorophyll A–B binding protein. There was a better ROS scavenging mechanism in transgenic lines, possibly through improving antioxidant defense machinery (Tripathy et al. 2017). A recently published report suggests that overexpression of the *OsRab7* gene also improved drought and heat tolerance, and increased grain yield in transgenic rice by modulating osmolytes, antioxidants, and the expression of abiotic stress-responsive genes (El-Esawi and Alayafi 2019). Above all, these findings add stronger support for the importance of RAB7 in abiotic stress tolerance.

In *Arabidopsis*, a chloroplast localized novel RAB GTPase protein, CPRABA5e, plays a role in stress responses, development, and thylakoid biogenesis processes. Similar to the cytosolic RAB protein, which functions in vesicle transport, this CPRABA5e protein functions in transport to and from thylakoids (Karim et al. 2014). A study in Arabidopsis demonstrated that RABA1 GTPases (a RAB11 member) function in transport between the trans-Golgi network and the PM and is involved in salinity stress tolerance (Asaoka et al. 2013). In *P. trichocarpa,* overexpression of the constitutive activation mutant PtRABE1b (Q74L) showed a larger root number, better root growth, and salt tolerance. In the overexpression transgenic lines, the transcript level of trafficking-related genes (Got1-like, KEU, p24, PHF1, SEC14, SKD1, and SYP61), a stress-related transcription factor (bZIP60), an autophagy-related gene (ATG18a-1), and development-related genes (*EPC1*) were up-regulated (Zhang et al. 2018). In response to heat, drought, and salt stress conditions, transcriptional profiles of different tissues at different developmental stages of *Solanum lycopersicum* showed that *RabGAP* genes associated with endocytic and pre-vacuolar trafficking are up-regulated in roots subjected to salt stress, suggesting that these genes play an important role during salt stress in tomato (Madrid-Espinoza et al. 2019).

Transcript expression profile study of chickpea (*Cicer arietinum L.*) found that the *CaRabC* (*Rab18*) gene sub-family has five isoforms such as *CaRabC-1a*, *-1b*, *-1c*, *-2a*, and *-2a** and the isoforms were differentially expressed under salinity, rapid dehydration, and drought stresses. Under control and salinity stress, the isoform *CaRabC-1c* is expressed at a higher level in leaves, whereas in rapidly dehydrated leaves the isoform *CaRabC-1a* is highly expressed. This study explored a possible link between sequence variants and the level of abiotic stress tolerance in chickpea (Khassanova et al. 2019). Whether the stress response is through endocytosis and Na^+ compartmentalization needs to be addressed, even though there is increasing evidence suggesting the role of endocytosis under salinity and osmotic stress (Martín-Davison et al. 2017). There is a report on Na^+ ion compartmentalization in the vacuole (Baral et al. 2015), and this was regarded as one of the mechanisms of RAB tolerance to salinity and drought stress. In chickpea, intracellular trafficking *CaRab-GTP* gene expression patterns under salt stress suggest that a coordinated network operates during intracellular trafficking pathways. Under salt stress, the different gene clades showed variable responses. Out of six *CaRab* gene clades, *CaRabB*, *-D*, and *-E* show similar expression profiles. These three genes form the major trafficking route through the endosomal system, and the expression of CaRABA2 in response to salt was strongly correlated with Na^+ accumulation in leaves. This finding suggests salinity tolerance was not correlated with either expression pattern of any of the six *CaRab* gene clades or Na^+ accumulation in the leaves. In chickpea, only Na^+ accumulation may not be the sole cause for salinity tolerance, but rather only one out of multiple different mechanisms, and many more genes are likely to be involved. This report puts light on the role of *CaRab* gene family in the response of chickpea plants to salinity (Sweetman et al. 2020). RABA member-*RabA2b* transcription was upregulated during several abiotic stresses. The *RabA2b* promoter activity revealed that it contains binding motifs for several ABA-responsive transcription factors. The *RabA2b* overexpressed Arabidopsis transgenic plants showed tolerance to drought stress. PM proteome study showed overexpression of *RabA2b* brings about changes in the PM proteome. In transgenic lines, stress-coping proteins as well as cell wall/cuticle modifiers are increased. Compared to wild type plants, the cuticle permeability of transgenic leaves was significantly reduced, suggesting that it plays a role in its drought tolerance properties (Ambastha et al. 2021). Overall these findings suggest that a vesicle trafficking protein, RAB, have an important role in plant tolerance to abiotic stresses like salinity, osmotic, and cold and heat stress.

Plant–Microbe Interaction and Biotic Stress Multiple studies confirmed that during symbiosis and nodule formation in legumes, vesicular trafficking plays an important role. The peri-bacteroid membrane (PBM) surrounding rhizobium in nodules originates from the fusion and expansion of newly synthesized vesicles with the PM. During functional characterization of RAB GTPases in PBM biogenesis and nodule development, several genes encoding RAB GTPases were isolated from legumes, including sRAB1 and sRAB7 from *Glycine max* and vRAB7 from *Vigna aconitifolia* (Cheon et al. 1993). During nodule formation, an increase in the transcript abundance of *sRab1* and *vRab7* genes was observed, and nodule development was affected by antisense suppression of these genes. Antisense suppression of sRAB1 reduced the number of bacteroids in the cytoplasm and vacuoles, while antisense suppression of vRAB7 led to the accumulation of late endosomal structures, including multivesicular bodies in the perinuclear region. These results indicate RAB's critical roles in symbiosis and nodule formation.

In legumes, root nodule cells, Rhizobium bacteria form N_2-fixing organelles called symbiosomes. The bacteria enter the cells through an endocytosis-like process. Using membrane identity markers such as regulatory small GTPases and SNARE proteins in *Medicago truncatula*, symbiosomes contain MtRAB7 in the early stages of development. However, the symbiosomes do not acquire vacuolar SNAREs (SYP22 and VTI11) until the onset of their senescence. By contrast, symbiosomes acquire the PM-located SNARE SYP132 from the start of symbiosome formation throughout their development. Symbiosomes appear to acquire a unique MtSYP132 and MtRAB7 for the endosome stage (Limpens et al. 2009).

RAB's role in biotic stress responses is also well documented. Worldwide, a plant-damaging oomycete, *Phytophthora brassicae*, causes significant crop losses. It suppresses the disease resistance of host cells by producing virulence factors, known as RxLR effector proteins. In the host cell, the conserved RxLR effector interacts with host RABA-type GTPases to block vesicle-mediated secretion of antimicrobial proteins such as PATHOGENESIS-RELATED PROTEIN 1 (PR-1) and DEFENSIN (PDF1.2), which inhibits pathogen growth in extracellular space. This finding suggested the important role of RABs in plant-pathogen interactions during host defenses (Brandizzi 2018; Tomczynska et al. 2018). Downregulation of RABE did not affect plant susceptibility to the pathogenic bacteria *Pseudomonas syringae*. Whereas the plants defenses conferring resistance to *P. syringae* infection are enhanced by constitutive expression of RABE1d-Q74L (Speth et al. 2009). In plants, RAB-SNARE interactions mediate basal defense responses by triggering microbial origin elicitor molecules called microbe-associated molecular patterns (MAMPs). In Arabidopsis, a member of the SNARE superfamily, syntaxin-PEN1, is needed for elicitor-mediated induction of lipopolysaccharides (LPS) as part of plant defenses (Erbs et al. 2010).

Using virus-induced gene silencing, knocking down TaRAB7 (wheat RAB7 homologue) expression enhanced the susceptibility of wheat cv. Suwon 11 to an avirulent race, CYR23. These findings indicate that TaRAB7 plays an important role in the early stage of wheat-stripe rust fungus interaction and increases the defense of wheat plants against the fungus *Pucinia striiformis* f. sp. *Tritici* (Liu et al. 2012). Recently, it was also reported that in the fungus *Arthrobotrys oligospora* two RAB GTPases (AoRAB-7A and AoRAB-2) are involved in the regulation of mycelial growth, conidiation, trap formation, stress resistance, and pathogenicity in the nematode (Yang et al. 2018).

16.3.2 RAN

RAN (Ras-related nuclear protein) is a small GTP-binding protein and well conserved in plants and animals. RAN is mainly localized inside the nucleus. It is essential for nuclear transport by trafficking of proteins and RNA in and out of the nucleus, nuclear assembly, and mRNA processing. The RAN proteins also regulate nucleocytoplasmic transport during the G1, S, and G2 phases of the cell cycle and microtubule organization during the M phase (Xu and Cai 2014). Like other small GTPases, RAN functions by binding and hydrolyzing GTP. RAN can switch between a GDP- and GTP-bound state, and the transition from RAN-GDP to RAN-GTP can occur by nucleotide exchange. However, RAN has very low intrinsic nucleotide exchange and GTP hydrolysis. It is stimulated by the GTPase-activating protein (RAN-GAP1) and the guanine nucleotide exchange factor RCC1 (regulator of chromosome condensation 1). After RAN-GAP and RAN-BP1 (RAN binding protein 1) released from the nucleus to the cytoplasm, it hydrolyzes RAN-GTP (Kim and Huh 2013; Xu and Cai 2014; Xu et al. 2016). In contrast to other small signaling GTPases like RAS and RHO, RAN itself is not membrane-bound. It is unique in that the sequestering of its accessory proteins provides the spatial information for its respective activities (Agarwal et al. 2009).

As a member of the RAS superfamily, RAN is structurally composed of a fold known as the G-domain (GTP-binding domain), a central six- stranded β-sheet that is surrounded by α-helices (de Boor et al. 2015). Structurally, plant RAN protein N- and C-terminal sequences are different from those of humans, *C. elegans* and yeast. The N-terminal sequence is highly conserved among plant RAN proteins. Similar to known animal RAN proteins, plant RAN proteins had highly conserved RAN-GAP effector-binding motif (KKYEPTIGVEV) and GTP-binding sites (G1–G4). A stretch of five Asp residues present in the acidic domain, which is a unique characteristic of plant RAN proteins. In *AtRan4*, the acidic tail is absent, and it was suggested that *AtRan4* might have a different function. The G-domain is conserved across RAS superfamily. It consists of five motifs (G-boxes), namely G1 (GXXXXGK[S/T], where X is any amino acid), G2 (XTX), G3 (DXXG), G4 (N/T)(K/Q)xD, and G5 ((T/G/C)(C/S)AK) (Figure 16.3). Among these, G1 motif, also called as the P-loop, binds to β-phosphates and Mg^{2+} ion. The conserved threonine (T) residue of G2 contacts γ-phosphate and Mg^{2+} ion. G3-motif is involved in hydrolysis. The G4 and G5 recognize guanine base and confer nucleotide specificity (Toma-Fukai and Shimizu 2019; Choudhury et al. 2021).

16.3.2.1 Role of RAN in Plants

In animal cells, stress treatments cause RAN re-localization and slowing of nuclear transport, but in plant cells, the role of RAN protein exposed to stress is still not well understood. RAN in plants is a multigene protein family, while there is a single *Ran* gene in the genome of mammalian and *C. elegans* (Kim and Huh 2013). In Arabidopsis, RAN is encoded by a family of four genes and in rice, three genes (Xu and Cai 2014). The physiological functions of the *Ran* genes in monocot and dicot plants are not always identical. In plants, RAN functions in the signaling pathways in response to changing environmental stimuli is poorly defined. A very few reports regarding RAN or its specific binding proteins involved in abiotic and biotic stress response are available.

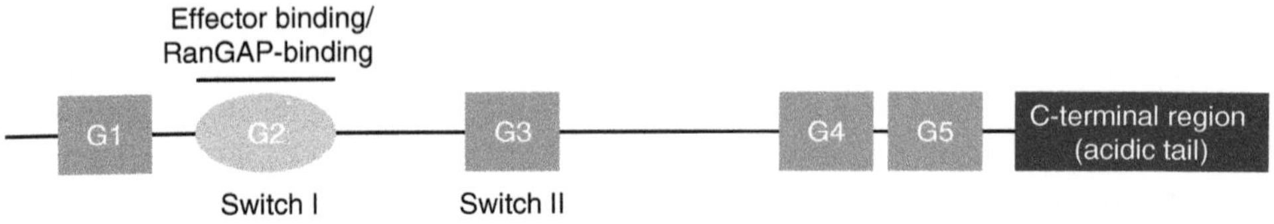

Figure 16.3 Schematic representation of RAN protein domain architecture. Different colored polygons represent conserved motifs of GTP binding and hydrolysis. G1 (GxxxxGK[S/T]), G2 (XTX), G3 (DXXG), G4 (N/T)(K/Q)xD, and G5 ((T/G/C)(C/S)AK). "x" is any amino acid. G1 motif (P-loop), binds to β-phosphates and Mg^{2+} ion. The conserved threonine (T) residue of G2 contacts γ-phosphate and Mg^{2+} ion. G3 motif is involved in hydrolysis. The G4 and G5 recognize the guanine base and confer nucleotide specificity.

Abiotic Stress Based on amino acid similarity, in Arabidopsis, four RAN-GTPases (AtRAN1, AtRAN2, AtRAN3, and AtRAN4) have been identified (Haizel et al. 1997; Vernoud et al. 2003). RAN-GTPases interact directly with AtKPNB1 (a homolog of human importin β1; KPNB1), which is involved in abscisic acid response and drought tolerance. Several studies have shown that cell division is closely related to stress tolerance. Xu et al. (2016) reported that under cold stress, AtRAN1 expression was up-regulated. This study suggests that AtRAN1 may function as a regulator in the cold signaling pathway in Arabidopsis. Additionally, Arabidopsis plants overexpressing AtRAN1 showed increased freezing tolerance after 4 °C acclimation. Also, the *atran1* and *atran3* mutants and wild-type plants have different levels of freezing tolerance. This finding suggests that AtRAN1 positively regulates freezing tolerance by maintaining the cell cycle in Arabidopsis. Another report suggests that OsRAN1 is ubiquitously expressed in rice tissues, with the highest expression in the spike. Under cold stress and indoleacetic acid treatment, the level of OsRAN1 mRNA was greatly increased, whereas the transcript level decreased with the addition of salt and polyethylene glycol (PEG). Further, OsRAN1 overexpression in Arabidopsis increased tiller number, and altered root development. OsRAN1 overexpression in rice improves cold tolerance. The levels of cellular free proline and sugar levels were highly increased in transgenic plants under cold stress. OsRAN1 maintained cell division and cell cycle progression and also promoted the formation of an intact nuclear envelope. This finding suggested that OsRAN1 may have an important and conserved role in cold stress signaling in plants (Xu and Cai 2014). As in its animal counterparts, OsRAN2 is mainly localized in the nucleus, with some in the cytoplasm. It is predicted that OsRAN2 might be involved in nucleo-cytoplasmic trafficking similar to that of animals, in addition to roles in the mitotic process. OsRAN2 transcript level was reduced by salt, osmotic, and exogenous abscisic acid (ABA) treatments. Overexpression of OsRAN2 in rice resulted in enhanced sensitivity to salinity, osmotic stress, and ABA. Seedlings of *A. thaliana* transgenic plants overexpressing OsRAN2 were overly sensitive to salinity stress and exogenous ABA treatment. This finding suggests that RAN, in part, influence the abiotic stress response. RAN is a required factor in importin β-mediated nucleo-cytoplasmic transport. SENSITIVE TO ABA AND DROUGHT2 (SAD2) encodes an importin β, which is a transporter receptor in nucleo-cytoplasmic trafficking. Mutation in SAD2 in Arabidopsis caused hypersensitivity to ABA, salt, and PEG treatments. Thus, it is predicted that OsRAN2' roles in abiotic stress might be somehow related to importin-β roles in abiotic stresses (Zang et al. 2010). In rice under cold treatment, transcript level of OsRAN2 was increased, but not during salt and drought stress. During cold stress, the OsRAN2 was involved in the regulation of cell division. The transgenic rice plants overexpressing OsRAN2 enhanced cold tolerance, whereas knockdown rice mutant lines showed an aberrant organization of spindles during mitosis and stunted growth during development. Under cold stress, the OsRAN2 overexpressed transgenic rice plants maintained cell division, with decreased proportion of cells with intranuclear tubulin with normal nuclear envelope. This finding suggests that during cold resistance in rice, OsRAN2 maintained cell division through promoting the normal export of intranuclear tubulin at the end of mitosis (Chen et al. 2011). Transgenic tobacco overexpressing *Lepidium latifolium* RAN (*LlaRan*) showed hypersensitivity to cold stress. In transgenic tobacco, at the cellular level, during cold stress, the membrane stability was declined drastically which may be due to the failure in enhancing tolerance to wounding stress. Also, ROS scavenging system was compromised (Sinha et al. 2014). Characterization of two sweet potato plants (*Ipomoea batatas* cv. Yulmi) *Ran* genes *IbRan1* and *IbRan2* suggests that under chilling stress (15 °C) transcription machinery of these two genes did respond differently in roots, but that their expression pattern is similar in petioles and leaves. This finding indicates that both the genes were responsible for the signaling transduction of chilling stress and that there may be functional redundancy in RAN family members against chilling stress (Kim and Huh 2013). Tissue-specific expression study of *MsRan* genes suggests that MsRAN transcripts including its alternatively spliced transcripts were differentially regulated in various tissues, and they were involved in responses to low temperature stress (Zhang et al. 2017).

During drought stress, plants first response is the closure of stomata to prevent excessive water loss by transpiration. In the guard cell, the AtRAN1 expression was higher, which suggests AtRAN1 active involvement in the control of the stomatal movement. *atran1* mutant plants have compromised chitin-induced stomata closer activity. Furthermore, compared to wild-type plants, *atran1* mutant plants were hypersensitive to drought stress. This finding suggest that AtRAN1 plays role in plants survival during drought stress tolerance (Song et al. 2021). Genome-wide analysis of RAN GTPases in the wheat (*Triticum aestivum*) genome led to identification of two *RAN* genes (*TaRAN1* and *TaRAN2*). Under different abiotic stresses (drought, cold, and salt), *RAN* gene expression may be species-specific. In wheat, after initial upregulation and downregulation in drought and cold stress, respectively, expression of both the *RAN* genes is restored to control levels, while in salt stress, expression of these genes is significantly downregulated in the shoot. This finding suggests differential regulation of *TaRAN* genes under developmental stages and abiotic stresses (Choudhury et al. 2021). Overall, RAN proteins regulate plant growth and development through their cellular processes and may contribute differentially to the abiotic stress response in plants.

Biotic Stress The small GTPase AtRAN1, which is highly expressed in Arabidopsis guard cells, shows its involvement in the control of the stomatal movement. Chitin, a fungal, microbial-associated molecular pattern, induces stomatal closure. *atran1* mutant plants also exhibited impaired chitin-induced stomatal closure. Thus, under chitin treatment, the stomatal closure was less sensitive in the *atran1* mutant than in wild-type plants. Upon chitin treatment, *atran1* mutant plants contained more *P. syringae* pv. tomato (Pst) DC3000 biomass than wild-type plants. Due to compromised stomatal closure in *atran1* mutant plants, which provides a free passage through which more bacteria move into the stoma and get into the leaf apoplast to cause disease. As chitin is an important component of the fungal cell wall, enhanced susceptibility to *Sclerotinia sclerotiorum* upon mutation of AtRAN1 is not surprising. This finding suggests that AtRAN1 may also be involved in the response to bacterial pathogen-associated molecular patterns (PAMPs); this hypothesis needs to be tested (Song et al. 2021). Recent reports also support the finding that overexpression of RAN1 in Arabidopsis leads to elongation growth and enhanced disease resistance against Pst DC3000 by promoting gibberellin (GA) and salicylic acid (SA) signaling, respectively. Overexpression of the constitutively active form, mutant RAN_{Q72L}, inhibits elongation growth and the defense response by blocking GA and SA signaling. Furthermore, AtRAN1 overexpression stimulates GA signaling by promoting DELLA protein degradation in the nucleus and increases SA signaling by promoting NPR1 nuclear localization. On the other hand, overexpression of $AtRAN1_{Q72L}$ prevents the degradation of DELLA protein and the localization of NPR1 in the nucleus, suggesting that AtRAN1-mediated nucleo-cytoplasmic transport plays a unique role in both GA and SA signaling pathways. Thus, this finding provides genetic and molecular evidence for further understanding of AtRAN1-mediated elongation growth and defense responses in *A. thaliana* (Xu et al. 2021). In *Nicotiana benthamiana*, NbRanBP1-1 (Ran-binding protein1-1) a regulator of nuclear export of RNA and proteins was identified as an essential gene for the resistance to potato blight pathogen *Phytophthora infestans*. In response to elicitor treatment, the NbRanBP1-1-silenced plants showed delayed accumulation of capsidiol, a sesquiterpenoid phytoalexin, which leads to reduced resistance to *P. infestans*. In NbRanBP1-1-silenced plants, mRNA was abnormally accumulated, suggesting NbRanBP1-1 participation in the nuclear export of mRNA. In NbRanBP1-1-silenced plants, elicitor-induced expression of defense genes, NbEAS and NbWIPK, was not affected in the early stage of defense induction, but the accumulation of NbWIPK protein was reduced. Nuclear export of the small G-protein NbRan1a was activated during the induction of plant defense, whereas this process was compromised in NbRanBP1-1-silenced plants. Gene silencing of the nuclear pore proteins, Nup75 and Nup160, also brings abnormal nuclear accumulation of mRNA, defects in the nuclear export of NbRan1a, and reduced production of capsidiol, resulting in decreased resistance to *P. infestans*. Overall, these findings suggest that nuclear export of NbRan is a key event for defense induction in *N. benthamiana*, and both RanBP1-1 and nucleoporins play important roles in the process (Mizuno et al. 2019).

16.4 Conclusions

Climate change, as reflected in an increase in the global temperature, unusual precipitation patterns resulting in floods and drought, rising sea levels, melting polar ice, will have severe impact on biodiversity, plant adaptation, and agricultural productivity. To tackle the adverse effects of stress, both abiotic and biotic, and simultaneously feed the ever-growing world population, the development of climate-resilient plant varieties and/or improvement of new trait(s) is inevitable. Tailoring of better crop plants through molecular approach is one of the breeding opportunities to develop crop plants tolerant to abiotic and biotic stresses. In recent years, remarkable progress has been made in our understanding of the genomic basis of tolerance, and a number of genes and QTL have been identified. Among these, the role of GTP-binding proteins in regulating the endomembrane system during stress is also gaining importance. However, it is still not clear how small GTPases participate in the signaling steps that regulate developmental processes, control hormone biosynthesis or activity, and mediate biotic and abiotic stress tolerance. Like heterotrimeric G-proteins, small GTPases also work as protein switches that are turned on and off by the nucleotide GTP. Small GTPases are present in most eukaryotes and constitute a superfamily of more than 100 members. RAB GTPases are important proteins and regulate intracellular vesicular transport, which consequently influences many cellular functions and responses. Several reports suggest its role in abiotic stress tolerance like salinity, drought, cold. The tolerance may be due to an increase in vesicle transport or due to a protective or repair-based mechanism of the endomembrane system, as RAB protein is involved in vacuolar transport inside the cell. RAB's role in biotic stress resistance and plant–microbe interaction was also reported. But the molecular mechanisms are poorly understood. RAN, also a GTP-binding protein, functions in the nucleus-cytoplasmic transport of RNA and proteins. RAN's role in abiotic stress such as cold is well documented. However, very few reports suggest RAN's role in biotic stress. It is suggested that small GTPase can be used as candidate genes/protein for developing crop plants that can withstand growing abiotic and biotic stress situations arising due to climate change; this has been demonstrated in many plants now. In future, further detailed study is needed to understand the molecular mechanism as to how small GTPase helps in stress resistance and signaling pathways, more so to understand the interlinking relationship with other signaling pathways in plants.

Acknowledgments

MKT acknowledges support from CSIR, Government of India, in the form of Senior Research Associateship (Scientist Pool Scheme) Fellowship, and SKS acknowledges support from SERB, Government of India, in the form of Distinguished Fellowship.

References

Agarwal, P.K., Agarwal, P., Jain, P. et al. (2008). Constitutive overexpression of a stress-inducible small GTP binding protein PgRab7 from *Pennisetum glaucum* enhances abiotic stress tolerance in transgenic tobacco. *Plant Cell Rep.* 27: 105–115.

Agarwal, P., Reddy, M.K., Sopory, S.K., and Agarwal, P.K. (2009). Plant Rabs: characterization, functional diversity, and role in stress tolerance. *Plant Mol. Biol. Report.* 27: 417–430.

Aksouh, N.M., Jacobs, B.C., Stoddard, F.L., and Mailer, R.J. (2001). Response of canola to dierent heat stresses. *Aust. J. Agric. Res.* 52: 817–824.

Ambastha, V., Matityahu, I., Tidhar, D., and Leshem, Y. (2021). RabA2b overexpression alters the plasma-membrane proteome and improves drought tolerance in *Arabidopsis*. *Front. Plant Sci.* 12: 738694.

Angadi, S.V., Cutforth, H.W., Miller, P.R. et al. (2000). Response of three *Brassica* species to high temperature stress during reproductive growth. *Can. J. Plant Sci.* 80: 693–701.

Asaoka, R., Uemura, T., Ito, J. et al. (2013). Arabidopsis RABA1 GTPases are involved in transport between the trans-Golgi network and the plasma membrane, and are required for salinity stress tolerance. *Plant J.* 73: 240–249.

Asseng, S., Foster, I., and Turner, N.C. (2011). The impact of temperature variability on wheat yields. *Glob. Chang. Biol.* 17: 997–1012.

Asseng, S., Ewert, F., Martre, P. et al. (2015). Rising temperatures reduce global wheat production. *Nat. Clim. Chang.* 5: 143–147.

Asseng, S., Martre, P., Maiorano, A. et al. (2019). Climate change impact and adaptation for wheat protein. *Glob. Chang. Biol.* 25: 155–173.

Bailey-Serres, J., Parker, J.E., Ainsworth, E.A. et al. (2019). Genetic strategies for improving crop yields. *Nature* 575: 109–118.

Baral, A., Shruthi, K.S., and Mathew, M.K. (2015). Vesicular trafficking and salinity responses in plants. *IUBMB Life* 67: 677–686.

Bischoff, F., Molendijk, A., Rajendrakumar, C.S., and Palme, K. (1999). GTP-binding proteins in plants. *Cell. Mol. Life Sci.* 55: 233–256.

de Boor, S., Knyphausen, P., Kuhlmann, N. et al. (2015). Small GTP-binding protein Ran is regulated by posttranslational lysine acetylation. *Proc. Natl. Acad. Sci. U.S.A.* 112: E3679–E3688.

Boyer, J.S. and Westgate, M. (2004). Grain yields with limited water. *J. Exp. Bot.* 55: 2385–2394.

Brandizzi, F. (2018). Rabs: targets in the plant-pathogen battlefield. *Plant J.* 95: 185–186.

Calleja-Cabrera, J., Boter, M., Oñate-Sánchez, L., and Pernas, M. (2020). Root growth adaptation to climate change in crops. *Front. Plant Sci.* 11: 544.

Chen, N., Xu, Y., Wang, X. et al. (2011). OsRAN2, essential for mitosis, enhances cold tolerance in rice by promoting export of intranuclear tubulin and maintaining cell division under cold stress. *Plant Cell Environ.* 34: 52–64.

Chen, J., Tang, L., Shi, P. et al. (2017). Eects of short- term high temperature on grain quality and starch granules of rice (*Oryza sativa* L.) at post-anthesis stage. *Protoplasma* 254: 935–943.

Cheon, C.I., Lee, N.G., Siddique, A.B. et al. (1993). Roles of plant homologs of Rab1p and Rab7p in the biogenesis of the peribacteroid membrane, a subcellular compartment formed *de novo* during root nodule symbiosis. *EMBO J.* 12: 4125–4135.

Choudhury, S., Mansi, Muthusamy, S.K. et al. (2021). Genome-wide identification of Ran GTPase family genes from wheat (*T. aestivum*) and their expression profile during developmental stages and abiotic stress conditions. *Funct. Integr. Genomics* 21: 239–250.

Deryng, D., Elliott, J., Folberth, C. et al. (2016). Regional disparities in the beneficial effects of rising CO_2 concentrations on crop water productivity. *Nat. Clim. Chang.* 6: 786–790.

Dong, J., Gruda, N., Lam, S.K. et al. (2018). Effects of elevated CO_2 on nutritional quality of vegetables: a review. *Front. Plant Sci.* 9: 924.

El-Esawi, M.A. and Alayafi, A.A. (2019). Overexpression of rice Rab7 gene improves drought and heat tolerance and increases grain yield in rice (*Oryza sativa L.*). *Genes (Basel)* 10: E56.

Erbs, G., Molinaro, A., Dow, J.M., and Newman, M.A. (2010). Lipopolysaccharides and plant innate immunity. *Subcell. Biochem.* 53: 387–403.

Fahad, S., Bajwa, A.A., Nazir, U. et al. (2017). Crop production under drought and heat stress: plant responses and management options. *Front. Plant Sci.* 8: 1147.

Food and Agriculture Organization of the United Nations (FAO) (2009). *Global Agriculture Towards 2050*. Rome: FAO.

Garofalo, P., Ventrella, D., Kersebaum, K.C. et al. (2019). Water footprint of winter wheat under climate change: trends and uncertainties associated to the ensemble of crop models. *Sci. Total Environ.* 658: 1186–1208.

George, S. and Parida, A. (2011). Over-expression of a Rab family GTPasefrom phreatophyte *Prosopis juliflora* confers tolerance to salt stress on transgenic tobacco. *Mol. Biol. Rep.* 38: 1669–1674.

Haizel, T., Merkle, T., Pay, A. et al. (1997). Characterization of proteins that interact with the GTP-bound form of the regulatory GTPase Ran in *Arabidopsis*. *Plant J.* 11: 93–103.

Haubruck, H., Engelke, U., Mertins, P., and Gallwitz, D. (1990). Structural and functional analysis of ypt2, an essential Ras-related gene in the fission yeast *Schizosaccharomyces pombe* encoding a Sec4 protein homologue. *EMBO J.* 9: 1957–1962.

Iizumi, T. and Ramankutty, N. (2015). How do weather and climate influence cropping area and intensity? *Glob. Food Sec.* 4: 46–50.

IPCC (2014). Climate change 2014: synthesis report. In: *Contribution of Working Groups I, II and III to the Fifth Assessment Report of the Intergovernmental Panel on Climate Change* (ed. R.K. Pachauri and L.A. Meyer). Geneva: IPCC.

Jagadish, S.V.K., Muthurajan, R., Oane, R. et al. (2010). Physiological and proteomic approaches to address heat tolerance during anthesis in rice (Oryza sativa L.). *J. Exp. Bot.* 61: 143–156.

Jagadish, S.V.K., Murty, M.V.R., and Quick, W.P. (2015). Rice responses to rising temperatures - challenges, perspectives and future directions. *Plant. Cell Environ.* 38: 1686–1698.

Karim, S., Alezzawi, M., Garcia-Petit, C. et al. (2014). A novel chloroplast localized Rab GTPase protein CPRabA5e is involved in stress, development, thylakoid biogenesis and vesicle transport in *Arabidopsis*. *Plant Mol. Biol.* 84: 675–692.

Khassanova, G., Kurishbayev, A., Jatayev, S. et al. (2019). Intracellular vesicle trafficking genes, *RabC*-GTP, are highly expressed under salinity and rapid dehydration but down-regulated by drought in leaves of Chickpea (*Cicer arietinum L.*). *Front. Genet.* 10: 40.

Kim, Y.-H. and Huh, G.H. (2013). Members of the ran family of stress-inducible small GTP-binding proteins are differentially regulated in sweet potato plants. *J. Plant Biotechnol.* 40: 9–17.

Kwon, S.I., Cho, H.J., Kim, S.R., and Park, O.K. (2013). The Rab GTPase RabG3b positively regulates autophagy and immunity-associated hypersensitive cell death in *Arabidopsis*. *Plant Physiol.* 161: 1722–1736.

Lesk, C., Rowhani, P., and Ramankutty, N. (2016). Influence of extreme weather disasters on global crop production. *Nature* 529: 84–87.

Limpens, E., Ivanov, S., van Esse, W. et al. (2009). Medicago N_2-fixing symbiosomes acquire the endocytic identity marker Rab7 but delay the acquisition of vacuolar identity. *Plant Cell* 21: 2811–2828.

Liu, J. and Howell, S.H. (2016). Managing the protein folding demands in the endoplasmic reticulum of plants. *New Phytol.* 211: 418–428.

Liu, F., Guo, J., Bai, P. et al. (2012). Wheat TaRab7 GTPase is part of the signaling pathway in responses to stripe rust and abiotic stimuli. *PLoS One* 7: e37146.

Liu, Z.L., Luo, C., Dong, L. et al. (2014). Molecular characterization and expression analysis of a GTP-binding protein (MiRab5) in *Mangifera indica*. *Gene* 540: 86–91.

Liu, B., Martre, P., Ewert, F. et al. (2019). Global wheat production with 1.5 and 2.0C above pre-industrial warming. *Glob. Chang. Biol.* 25: 1428–1444.

Lobell, D.B. and Burke, M.B. (2008). Why are agricultural impacts of climate change so uncertain? The importance of temperature relative to precipitation. *Environ. Res. Lett.* 3: 7.

Lobell, D.B., Schlenker, W., and Costa-Roberts, J. (2011). Climate trends and global crop production since 1980. *Science* 333: 616–620.

Long, S.P. and Ort, D.R. (2010). More than taking the heat: crops and global change. *Curr. Opin. Plant Biol.* 13: 240–247.

Long, S.P., Ainsworth, E.A., Rogers, A., and Ort, D.R. (2004). Rising atmospheric carbon dioxide: plants FACE the future. *Annu. Rev. Plant Biol.* 55: 591–628.

Madrid-Espinoza, J., Salinas-Cornejo, J., and Ruiz-Lara, S. (2019). The Rab-GAP gene family in tomato (*Solanum lycopersicum*) and wild relatives: identification, interaction networks, and transcriptional analysis during plant development and in response to salt stress. *Genes (Basel)* 10: E638.

Martín-Davison, A.S., Pérez-Díaz, R., Soto, F. et al. (2017). Involvement of SchRabGDI1 from *Solanum chilense* in endocytic trafficking and tolerance to salt stress. *Plant Sci.* 263: 1–11.

Maruta, N., Trusov, Y., Jones, A.M., and Botella, J.R. (2021). Heterotrimeric G proteins in plants: canonical and atypical Gα subunits. *Int. J. Mol. Sci.* 22: 11841.

Mazel, A., Leshem, Y., Tiwari, B.S., and Levine, A. (2004). Induction of salt and osmotic stress tolerance by overexpression of an intracellular vesicle trafficking protein AtRab7 (AtRabG3e). *Plant Physiol.* 134: 118–128.

McCudden, C.R., Hains, M.D., Kimple, R.J. et al. (2005). G-protein signaling: back to the future. *Cell. Mol. Life Sci.* 62: 551–577.

Milligan, G. and Kostenis, E. (2006). Heterotrimeric G-proteins: a short history. *Br. J. Pharmacol.* 147: S46–S55.

Mills, G., Hayes, F., Jones, M.L., and Cinderby, S. (2007). Identifying ozone-sensitive communities of (semi-) natural vegetation suitable for mapping exceedance of critical levels. *Environ. Pollut.* 146: 736–743.

Mizuno, Y., Ohtsu, M., Shibata, Y. et al. (2019). *Nicotiana benthamiana* RanBP1-1 is involved in the induction of disease resistance via regulation of nuclear-cytoplasmic transport of small GTPase Ran. *Front. Plant Sci.* 10: 222.

Morita, S., Wada, H., and Matsue, Y. (2016). Countermeasures for heat damage in rice grain quality under climate change. *Plant Prod. Sci.* 19: 1–11.

Morrison, M.J. and Stewart, D.W. (2002). Heat stress during flowering in summer Brassica. *Crop Sci.* 42: 797–803.

Nahm, M.Y., Kim, S.W., Yun, D. et al. (2003). Molecular and biochemical analyses of OsRab7, a rice Rab7 homolog. *Plant Cell Physiol.* 44: 1341–1349.

Neves, S.R., Ram, P.T., and Iyengar, R. (2002). G protein pathways. *Science* 296: 1636–1639.

O'Mahony, P.J. and Oliver, M.J. (1999). Characterization of a desiccation responsive small GTP-binding protein (Rab2) from the desiccation-tolerant grass *Sporobolus stapfianus*. *Plant Mol. Biol.* 39: 809–821.

Oldham, W.M. and Hamm, H.E. (2008). Heterotrimeric G protein activation by G-protein-coupled receptors. *Nat. Rev. Mol. Cell. Biol.* 9: 60–71.

Peng, X., Ding, X., Chang, T. et al. (2014). Overexpression of a vesicle trafficking gene, OsRab7, enhances salt tolerance in rice. *Sci. World J.* 2014: 483526.

Prasad, P.V.V., Boote, K.J., Allen, L.H. et al. (2006). Species, ecotype and cultivar differences in spikelet fertility and harvest index of rice in response to high temperature stress. *F. Crop. Res.* 95: 398–411.

Qaseem, M.F., Qureshi, R., and Shaheen, H. (2019). Effects of pre-anthesis drought, heat and their combination on the growth, yield and physiology of diverse wheat (*Triticum aestivum* L.) genotypes varying in sensitivity to heat and drought stress. *Sci. Rep.* 9: 477.

Quint, M., Delker, C., Franklin, K.A. et al. (2016). Molecular and genetic control of plant thermomorphogenesis. *Nat. Plants* 2: 15190.

Senapati, N., Brown, H.E., and Semenov, M.A. (2019). Raising genetic yield potential in high productive countries: designing wheat ideotypes under climate change. *Agric. For. Meteorol.* 271: 33–45.

Shan-e-Ali Zaidi, S., Vanderschuren, H., Qaim, M. et al. (2019). New plant breeding technologies for food security. *Science* 363: 1390–1391.

Sharif, B., Makowski, D., Plauborg, F., and Olesen, J.E. (2017). Comparison of regression techniques to predict response of oilseed rape yield to variation in climatic conditions in Denmark. *Eur. J. Agron.* 82: 11–20.

Sinha, V.B., Grover, A., Singh, S. et al. (2014). Overexpression of Ran gene from *Lepidium latifolium* L. (*LlaRan*) renders transgenic tobacco plants hypersensitive to cold stress. *Mol. Bio. Rep.* 41: 5989–5996.

Song, Z., Zhang, C., Chen, L. et al. (2021). The *Arabidopsis* small G-protein AtRAN1 is a positive regulator in chitin-induced stomatal closure and disease resistance. *Mol. Plant Pathol.* 22: 92–107.

Specht, J.E., Hume, D.J., and Kumudini, S.V. (1999). Soybean yield potential – a genetic and physiological perspective. *Crop Sci.* 39: 1560–1570.

Speth, E.B., Imboden, L., Hauck, P., and He, S.Y. (2009). Subcellular localization and functional analysis of the *Arabidopsis* GTPase RabE. *Plant Physiol.* 149: 1824–1837.

Sweetman, C., Khassanova, G., Miller, T.K. et al. (2020). Salt-induced expression of intracellular vesicle trafficking genes, CaRab-GTP, and their association with Na^+ accumulation in leaves of chickpea (*Cicer arietinum* L.). *BMC Plant Biol.* 20: 183.

Takai, Y., Sasaki, T., and Matozaki, T. (2001). Small GTP-binding proteins. *Physiol. Rev.* 81: 153–208.

Thiault, L., Mora, C., Cinner, J.E. et al. (2019). Escaping the perfect storm of simultaneous climate change impacts on agriculture and marine fisheries. *Science* 5: eaaw9976.

Toma-Fukai, S. and Shimizu, T. (2019). Structural insights into the regulation mechanism of small GTPases by GEFs. *Molecules* 24: 3308.

Tomczynska, I., Stumpe, M., and Mauch, F. (2018). A conserved RxL reffector interacts with host RABA-type GTPases to inhibit vesicle-mediated secretion of antimicrobial proteins. *Plant J.* 95: 187–203.

Tripathy, M.K., Tiwari, B.S., Reddy, M.K. et al. (2017). Ectopic expression of PgRab7 in rice plants (*Oryza sativa* L.) results in differential tolerance at the vegetative and seed setting stage during salinity and drought stress. *Protoplasma* 254: 109–124.

Tripathy, M.K., Deswal, R., and Sopory, S.K. (2021). Plant RABs: role in development and in abiotic and biotic stress responses. *Curr. Genomics* 22: 26–40.

United Nations Climate Action (2021). What is climate change. https://www.un.org/en/climatechange/what-is-climate-change.

Urano, D., Chen, J.G., Botella, J.R., and Jones, A.M. (2013). Heterotrimeric G protein signalling in the plant kingdom. *Open Biol.* 3: 120186.

Urban, D.W., Roberts, M.J., Schlenker, W., and Lobell, D.B. (2015). The effects of extremely wet planting conditions on maize and soybean yields. *Clim. Change* 130: 247–260.

Vandermeiren, K., Harmens, H., Mills, G., and De Temmerman, L. (2009). Impacts of ground-level ozone on crop production in a changing climate. In: *Climate Change and Crops. Environmental Science and Engineering* (ed. S.N. Singh), 213–243. (Berlin: Springer).

Vernoud, V., Horton, A.C., Yang, Z.B., and Nielsen, E. (2003). Analysis of the small GTPase gene superfamily of *Arabidopsis*. *Plant Physiol.* 131: 1191–1208.

Vitousek, S., Barnard, P.L., Fletcher, C.H. et al. (2017). Doubling of coastal flooding frequency within decades due to sea-level rise. *Sci. Rep.* 7: 1362.

Vu, L.D., Xu, X., Gevaert, K., and De Smet, I. (2019). Developmental plasticity at high temperature. *Plant Physiol.* 181: 399–411.

Wang, P., Wang, T., Han, J. et al. (2021). Plant autophagy: an intricate process controlled by various signaling pathways. *Front. Plant Sci.* 12: 754982.

Wei, P.S., Chiu, H.H., Hsieh, Y.C. et al. (2019). Absorption coefficient of water vapor across atmospheric troposphere layer. *Heliyon* 5: e01145.

Xu, P. and Cai, W. (2014). RAN1 is involved in plant cold resistance and development in rice (*Oryza sativa*). *J. Exp. Bot.* 65: 3277–3287.

Xu, P., Zang, A., Chen, H., and Cai, W. (2016). The small G protein AtRAN1 regulates vegetative growth and stress tolerance in *Arabidopsis thaliana*. *PLoS One* 11: e0154787.

Xu, P., Ma, W., Liu, J. et al. (2021). Overexpression of a small GTP-binding protein Ran1 in *Arabidopsis* leads to promoted elongation growth and enhanced disease resistance against *P. syringae* DC3000. *Plant J.* 108: 977–991.

Yang, X., Ma, N., Yang, L. et al. (2018). Two Rab GTPases play different roles in conidiation, trap formation, stress resistance, and virulence in the nematode-trapping fungus *Arthrobotrys oligospora*. *Appl. Microbiol. Biotechnol.* 102: 4601–4613.

Yin, C., Karim, S., Zhang, H., and Aronsson, H. (2017). Arabidopsis RabF1 (ARA6) is involved in salt stress and dark-induced senescence (DIS). *Int. J. Mol. Sci.* 18: 309.

Zang, A., Xu, X., Neill, S., and Cai, W. (2010). Overexpression of OsRAN2 in rice and *Arabidopsis* renders transgenic plants hypersensitive to salinity and osmotic stress. *J. Exp. Bot.* 61: 777–789.

Zerial, M. and Stenmark, H. (1993). RabGTPases in vesicular transport. *Curr. Opin. Cell Biol.* 5: 613–620.

Zhang, Y.L., Fang, Z.Z., and Lai, Z.X. (2017). Identification of alternatively spliced *MsRan* transcripts involved in low temperature response in *Musa* spp. *Biol. Plant.* 61: 483–493.

Zhang, J., Li, Y., Liu, B. et al. (2018). Characterization of the Populus Rab family genes and the function of PtRabE1b in salt tolerance. *BMC Plant Biol.* 18: 124.

Zörb, C., Geilfus, C.M., and Dietz, K.J. (2019). Salinity and crop yield. *Plant Biol.* 21: 31–38.

17

Biotechnological Strategies to Generate Climate-Smart Crops

Recent Advances and Way Forward

Jyoti Maurya, Roshan Kumar Singh, and Manoj Prasad

National Institute of Plant Genome Research (NIPGR), Jawaharlal Nehru University, New Delhi, India

17.1 Introduction

Greenhouse effect is crucial to sustain life on earth as it maintains the average global temperature at 14 °C which otherwise is −19 °C. Anthropogenic activities such as industrialization, deforestation, urbanization, and exploitation of fossil fuel reservoirs are responsible for increase in greenhouse effect due to excessive emission of greenhouse gases (GHGs) and is leading to increased average temperature of globe. Excessive greenhouse effect is a result of overproduction of several GHGs such as carbon dioxide (CO_2), methane (CH_4), nitrous oxide (N_2O), chlorides and fluorides of carbon (CFC and CCl_4), and water vapors (Cassia et al. 2018; Chaudhry and Sidhu 2022). It is important to point out here that the concentration of atmospheric CO_2 has jumped from 390 to 409 ppm of CH_4 from 1803 to 1866 ppb during 2011–2019 (IPCC 2021). Enhanced GHGs are responsible for the increase in global temperature and changing climatic conditions. As per the Intergovernmental Panel on Climate Change (IPCC), there was an increase of 1.09 °C in global mean surface temperatures (GMST) during 2011–2020. According to the IPCC report on climate change 2021, the key indicators of climate system, such as atmosphere, oceans, cryosphere, and biosphere, are changing at unprecedented rates since the last several decades, at least past 2000 years (IPCC 2021).

Climate change majorly include increase in global temperature and unpredicted precipitation patterns leading to frequent heat, drought, salinity, and flood incidents in agricultural lands across the globe (Zandalinas et al. 2021). Since plants are unable to relocate themselves toward optimum conditions, their growth and development are retarded under these stresses (Zandalinas et al. 2021). These unfavorable environments have been drastically impacted the crop productivity of both developing and developed countries ever since and are expected to increase in coming years (Lobell and Gourdji 2012). Also, the population around the globe is expected to increase up to nine billion, which will lead to an increase in food demands of around 85% by 2050 (FAOSTAT 2017). Crop losses due to changing climatic conditions and increasing food demands by ever-increasing global population are major causes of growing food insecurity (Ehrlich et al. 1993; Gregory et al. 2005). To address these problems, a shift toward generation of climate-smart crops that are more tolerant of adverse climatic conditions is an urgent need to ensure food security.

Climate change affects water availability, soil fertility, environmental pollution, and other environmental factors that are crucial to agricultural crop productivity (Gornall et al. 2010). Abiotic stresses induce several morpho-physiological, physiological, and molecular alterations in plants, such as reduced plant growth, decreased photosynthetic rate, accumulation of reactive oxygen species (ROS), stress metabolites, and hence reduced crop yield (Bita and Gerats 2013; Hasanuzzaman et al. 2013). In response, these plants have also devised mechanisms to reduce its adverse impacts through physiological adjustments such as partial closure of stomatal aperture to reduce water loss, activation of antioxidative pathway to scavenge excessive ROS, and expression of stress-responsive genes leading to stress tolerance (Dos Reis et al. 2012; Pereira 2016; Zandalinas et al. 2022). The magnitude of stress tolerance in plants depends on its inherent ability for morpho-physiological and biochemical adaptation owing to its genetic architecture. For instance, significant yields of major cereal crops such as wheat, rice, and maize are found to be affected by a slight increase in temperature (Chuang et al. 2017). On the other hand, several underutilized cereal crops, including foxtail millet, pearl millet, and sorghum, can grow on marginal lands and

Global Climate Change and Plant Stress Management, First Edition. Edited by Mohammad Wahid Ansari, Anil Kumar Singh, and Narendra Tuteja.

sustain irregular precipitation and increasing temperatures up to a certain extent. Moreover, millets are nutritionally superior and have higher water and nitrogen use efficiency as compared to major cereal crops (Halli et al. 2016; Muthamilarasan et al. 2016). Although millets are known for their tolerance against multiple abiotic stresses, it has been estimated that the yield of sorghum (a millet) will be reduced by up to 32% by 2080 due to climate change (Soora et al. 2013). Therefore, to generate climate-smart crops, it is important to understand and harness the innate capability of multiple stress tolerance of plants with a special attention toward millet crops (Gupta et al. 2017; Shivhare and Lata 2017).

Both breeding and biotechnological strategies are being applied to generate climate-resilient crops. Breeding led to crop improvement through creating novel genetic variation via identification, selection, and domestication of stress-tolerant phenotypes (Caligari 2001). On the other end, biotechnological approach involves identification and introgression of stress-responsive genes for enhanced tolerance. Biotechnological approaches have several advantages over breeding as it is less time consuming, is simple, and allows to cross interspecies barriers. Due to these advantages, nowadays biotechnological strategies such as gene expression modulation through overexpression or downregulation and gene editing through CRISPR-Cas 9 system are serving as crucial approaches toward generating stress tolerance in crop plants (Jaganathan et al. 2018; Baillo et al. 2019; Mandal et al. 2021).

On the basis of this background, this chapter briefs about the effect of climate change on crop growth, development, and yield, as well as plant responses and adaptations during changing environments. This chapter also briefs about different strategies implied to combat the climate change, elaborates the integrative-omics, and details biotechnological strategies to generate the stress-tolerant crops.

17.2 Climate Change and Crop Yield

The altered physiology of plant systems due to unwanted climatic stress is reflected in plant performance and yield. Effect of changing climate on crop productivity has been estimated that to be 15–35% of crop yield loss for every 3–4 °C rise in temperature (www.fao.org). Delincé and Witzke (2015) claim that there will be 2–15% loss in agricultural production at the global level by 2050 because of climate change. Although it is very difficult to perform a precise assessment of abiotic stresses on crop productivity due to climate change, it is evident from different surveys done across 26 countries in the European region (Olesen et al. 2011). There are low-temperature environments and short time period available for cropping in northern regions, whereas higher temperatures and low rainfall are limiting factors in southern regions (Olesen et al. 2011).

Among the various abiotic stresses that plants encounter due to changing climate, high temperature, and water deficit are well reported to have severe impacts on yield (Chuang et al. 2017; Dietz et al. 2021). Using different analytical methods, it has been found that every 1 °C rise in mean temperature of globe would cause global yield reduction of wheat up to 6%, rice up to 3.2%, maize up to 7.4%, and soybean up to 3.1% (Chuang et al. 2017). The yield of two maize cultivars, Şafak and Ant-İ90, has been reduced from 9.35 to 3.62 tons per hectare ($t\,ha^{-1}$) and 7.4 to 1.4 $t\,ha^{-1}$, respectively, due to severe drought stress in subtropical climate of Turkey (Aydinsakir et al. 2013). Similar pattern of yield reduction due to drought in different global climatic regions has been observed; as in cereals (wheat, oat, rye, and soybean), vegetables (tomato, potato, and common bean), fruits (grape, apple, banana, strawberry, orange, and mango), and others as reviewed by Dietz et al. (2021). The combined effect of temperature and drought stress on crop yields is found to be more devastating as compared to their individual effects (Matiu et al. 2017). In wheat, it has been observed that the weight of a single grain has been reduced by 13–27% because of drought and by 43–83% because of combined heat and drought stress (Mahrookashani et al. 2017). However, heat stress alone has no effect on single grain weight, but it leads to reduction in grain number of up to 14–28% and yield of up to 16–25% (Mahrookashani et al. 2017).

Both water deficits and extreme temperatures impede plant growth and development at all stages, but their effects during various reproductive stages on stigma, anther, pollen development, and fertilization are more critical (Sinha et al. 2021). Elevated temperature is the cause of male sterility in cereal crops. For instance, when wheat plants grown at 20 °C were subjected to 30 °C during meiotic phase initiation in pollen mother cells for a short duration, this caused abnormal male gametophyte development and reduction in seed setting (Saini and Aspinall 1982). Evaluation of the combined effects of drought, heat, and higher CO_2 levels on the biomass accumulation yield of several wheat genotypes has been done during post-anthesis. It has been observed that there is 45–50% loss in plant biomass and grain yield (Dias de Oliveira et al. 2013). Another study suggests that high temperature stress in wheat during anthesis (grain filling) results in 29–44% decline in grain yield per plant as compared to optimum temperature conditions (Djanaguiraman et al. 2020). The yield of legume crops has also been compromised in response to drought during reproduction and grain filling stages, as reviewed by Farooq et al. (2017). All these studies, along with continuous reports coming every day, give a view of the changing climate and its impact on crop yields.

17.3 Effect of Climate Change on Crop Morpho-physiology, and Molecular Level

Plants need optimum environmental conditions to perform maximally. Increased CO_2 levels are directly associated with gaseous exchange, photosynthesis, and carbon partitioning, and other developmental processes in terms of growth and yield (Katul et al. 2010; Gray and Brady 2016; Thompson et al. 2017). An increase in CO_2 level imparts both positive and negative effects; for example, in C_3 plants, increased CO_2 availability increases the affinity of RUBISCO toward CO_2, despite the presence of oxygen, which reduces the likelihood of photorespiration-mediated energy loss and increases carbon assimilation (Tomimatsu et al. 2014; Hay et al. 2017). In foxtail millet, a C_4 model crop, it has been observed that increased CO_2 level results in enhanced biomass and yield due to an increase in tiller number, seed number, improved water use, and photosynthetic efficiency (Li et al. 2019a). On the other hand, higher levels of CO_2 exposure for prolonged periods are reported to affect source-sink balance, the electron transport chain (ETC), decrease in RUBISCO content, and hence plant productivity (Gamage et al. 2018). Furthermore, elevated CO_2 champers shortening of root and shoot growth of herbaceous plants and decreased nutritional content such as iron (Fe), sulfur (S), zinc (Zn), and nitrogen (N) of non-legumes (Cohen et al. 2018; Loladze et al. 2019).

Increase in temperature beyond optimum causes heat injury and disturbs intermolecular interactions essential for normal physiological functions of a plant (Bita and Gerats 2013; Hasanuzzaman et al. 2013). Heat stress affects plants at every developmental stage from germination to reproduction. High temperature leads to decrease in percentage of seed germination and its viability, reduced radicle and plumule growth (Dürr et al. 2018; Yuan and Wen 2018). Increased temperature lowers the stability of cellular structures, membranes, and proteins and impairs metabolic functions, and this leads to ROS generation in plants (He et al. 2017; Kumar et al. 2018). Plants enable their thermosensory mechanisms to sense heat stress through calcium-mediated signaling, cytoskeletal reorganization, and lipid signaling pathways (Carmody et al. 2016). Heat stress causes reduced levels of photosynthetic pigments and antioxidant activity to affect photosynthetic processes (Carmody et al. 2016; Chalanika De Silva and Asaeda 2017). Recently, it was reported that FaHSP17.8-CII is involved in retaining heat stress induced memory via epigenetic modifications and leads to a decrease in ROS accumulation and chloroplast ultrastructure damage, improved chlorophyll content, and photosystem II activity (Bi et al. 2021a).

The frequency of drought stress has increased over the past few decades due to climate change. It is estimated that by 2050, drought stress will affect 50% of crops globally (Vurukonda et al. 2016). Roots are the initial organs to perceive water deficit and coordinate with above ground parts; this causes reduction in turgor pressure and partial closure of stomatal aperture to decrease transpiration and ensure water availability (Comstock 2002; Mahmood et al. 2019). Moreover, the nutrient uptake is done by the roots along with water; hence, water deficit conditions hinder nutrient uptake (Rouphael et al. 2012; Bista et al. 2018). Low cellular water content leads to increase in solute concentration, causing damage to critical enzymes and proteins involved in physiological processes (Zlatev and Lidon 2012). Turgor pressure and water potential are responsible for maintaining turgidity and optimum solution-solute conditions of a plant cell. Water deficit conditions cause change in these parameters resulting in altered plants morpho-physiology. Decreased turgidity of guard cells leads to partial stomatal closure to prevent water loss (Huber et al. 2019; Knipfer et al. 2020). Due to the partial closure of stomatal opening, the entry of CO_2 is limited inside, which leads to increased photorespiration, ROS generation, and decreased photosynthetic rate (Noctor et al. 2014). The increased ROS generation during water deficit conditions causes reduction in chlorophyll levels and hinders photosynthetic operations in crops (Sohrabi et al. 2012; Meher et al. 2018). This is associated with low turgor pressure, low rate of leaf expansion, decrease in C-assimilation, damage to photosystems, and electron transport chain (Zargar et al. 2017).

Excess concentrations of salts like chlorides of sodium and calcium, calcium carbonate are major reasons for soil salinity, and its harmful effect on plants is termed as salinity stress. Soil salinity is increasing as there is increase in transpiration and water evaporation due to enhanced temperatures and decreased rainfall (Mukhopadhyay et al. 2021; Sahab et al. 2021). On the basis of their response to salinity stress, plants are classified as glycophytes (salt-sensitive) and halophytes (salt-tolerant) in which most of them are glycophytes (van Zelm et al. 2020; Zhao et al. 2020). Immediate effect of salinity stress induces osmotic stress in plants. But if the stress persists for longer durations, it leads to ion toxicity (Parihar et al. 2015). Salinity stress alters the water status to cause dehydration and affects cellular metabolism. Seed germination is hampered by salinity as the germination percentage of tomato plants was found to be reduced under salt stress (Singh et al. 2012). Formation of ROS during salinity stress leads to reduction in stomatal conductance and lower chlorophyll content, hence reduced photosynthetic rates and cell death (Silveira and Carvalho 2016; Hameed et al. 2021; Pompelli et al. 2021). Moreover, excess salinity causes complete destruction of chlorophyll molecules (chlorosis) and premature cell death (necrosis) due to increased ROS accumulation (Okon 2019). ROS also leads to degradation of macromolecules such as carbohydrates, proteins and lipids and accumulation of by-products generated from these (Kumar et al. 2021b; Nefissi Ouertani et al. 2021).

17.4 Plant Responses to Stress Conditions

The inherent capability of plant adaptations toward changing climate serves as the basics to develop the stress tolerant/resistant crops against stress environments. Therefore, it is a prerequisite to understand the physiological and genetic basis of adaptations used by plants. Phytohormones regulate plant growth and development under normal as well as under different stress conditions by regulating many physiological, biochemical, and signaling pathways. Several phytohormones, including abscisic acid (ABA), salicylic acid (SA), jasmonic acid (JA), ethylene (ET), and brassinosteroids (BR), are induced during abiotic stress conditions (Eyidogan et al. 2012; Kumar et al. 2016). Among these, ABA, SA, JA, and ET are well studied in case of abiotic stress response (Verslues and Zhu 2005; Tuteja 2007; Yuan and Lin 2008; Kazan 2015; Per et al. 2018). Basal levels of ABA play an important role in regulating biogenesis of chloroplasts, stomatal movements, cell cycle, and central metabolism of cells (Brookbank et al. 2021). Under drought and salt stress, ABA biosynthesis increases and leads to stomatal closure to reduce water loss (Ma and Qin 2014). Moreover, ABA also induces expression of different genes to achieve homeostasis; for example, a cotton NAC transcription factor, *GhNAC4*, is induced by abiotic stress, and exogenous ABA applications have shown increased abiotic stress tolerance in tobacco plants upon overexpression (Trishla and Kirti 2021). SA is known to be a crucial regulator of flowering, senescence, nitrogen metabolism, photosynthesis, and water status in plants during abiotic stresses (Yuan and Lin 2008; Hasanuzzaman et al. 2017). SA is reported to provide tolerance against high temperatures, drought, and salinity stresses (Miura and Tada 2014; Hasanuzzaman et al. 2017). Furthermore, SA is shown to regulate redox mediated pathways through regulation of the *PR1*, *GRXC9*, and *ORA59* genes in Arabidopsis (Herrera-Vásquez et al. 2015). Similarly, JA is also involved in the regulation of different physiological and molecular events in response to abiotic stress (Wang et al. 2020a). Raza et al. (2021a) have advocated the role of JA as a part of signal transduction and antioxidant defense pathways under abiotic stress environments. ET regulates plant growth and development through induction of members of transcription factor (TF) family, namely ethylene responsive factors (ERFs) (Klay et al. 2018). It has been shown that different sets of ERFs are associated with heat, cold, salt, drought, and flooding conditions in tomatoes (Klay et al. 2018).

Heat shock proteins (HSPs) are molecular chaperons, and in plants, they are categorized into five classes based on their size: small HSP (sHSP), HSP60, HSP70, HSP90, and HSP100 (Al-Whaibi 2011; Singh 2019). HSPs are known to play a crucial role in response to abiotic stresses such as heat, drought, and salinity in crop plants (Aneja et al. 2015; Khan and Shahwar 2020). The role of HSPs is well documented during heat stress as compared to other stresses. Heat stress lead to induction of basic heat stress response and induce the expression of heat shock factors (HSFs) and HSPs (Qu et al. 2013). HSPs help to maintain protein structure and, hence, its function at high temperatures and redox balance in plants (Qu et al. 2013; Khan and Shahwar 2020). When temperature exceeds more than 30 °C, crop plants such as wheat, soybean, maize accumulate HSPs to cope heat stress (Yadav et al. 2020). In naturally abiotic stress tolerant plant, foxtail millet, a total of 113 HSP genes have been identified, and SisHSP27 was found to express highly differential during various abiotic stresses (Singh et al. 2016). A genome-wide study led to the identification of 24 HSFs and 117 HSPs from eggplant, and their expression was found to be increased during heat stress (Gong et al. 2021). Fragkostefanakis et al. (2015) outlined the prospects of thermotolerance by engineering HSFs and HSPs. sHSPs are reported to play an important role in heat, drought, and salinity stress tolerance in crop plants (Liu et al. 2021b; Wu et al. 2022). Overexpression of *OsHSP18.6* imparts drought tolerance to rice (Wang et al. 2015). An sHSP from tomato, *SlHSP17.1*, provides cold stress tolerance by reducing cold stress induced oxidative stress (Zhang et al. 2020).

Osmoprotactants or osmolytes are inert organic solutes that help in maintaining osmotic balance by imparting overall negative osmotic potential to allow optimal water movement throughout the cell. During abiotic stress conditions, osmolytes such as sugars, sugar alcohols, amino acids, and amines are accumulated to protect biological macromolecules from oxidative damage (Pattnaik et al. 2021). Osmolytes like sucrose, trehalose, proline, and glycine betaine are known to accumulate under drought and salinity stress and help to maintain redox homeostasis and plant water status (Singh et al. 2015). Proline accumulation is a prevalent abiotic stress mediated physiological response. For instance, overexpression of a brassica gene, *BrCIPK1*, in rice led to enhanced tolerance toward cold, drought, and salinity stress through activating pyrroline-5-carboxylate synthase enzyme of the proline biosynthesis pathway (Abdula et al. 2016). Another study suggests over-accumulation of proline protects photosynthetic enzymes and shows increased antioxidant enzyme activity in response to increased salt concentration in transgenic sorghum plants overexpressing a proline biosynthesis gene *P5CSF129A* (Surender Reddy et al. 2015).

Peroxides and superoxides are produced and accumulate under stress conditions to cause redox imbalance and impaired cellular functions. Therefore, plants have developed antioxidant mechanisms to scavenge and inhibit overproduction of ROS as an adaptive strategy for stress tolerance (Sharma et al. 2012; Das and Roychoudhury 2014). Plants synthesize non-enzymatic antioxidants such as phenols, alkaloids, flavonoids, ascorbate, glutathione, and enzymes with antioxidant properties such as catalases, peroxidases, and superoxide dismutase to check the ROS levels under stress environments (Sharma et al. 2012; Das and Roychoudhury 2014). Both non-enzymatic and enzymatic antioxidants work in a coordinated fashion and lead to ultimate conversion of ROS into H_2O and O_2 molecules (Huang et al. 2019). Studies have shown the role of antioxidant activity of these compounds during salinity stress; for example, overexpression of a sweet potato NAC TF, *IbNAC7*, in Arabidopsis showed increased salt tolerance due to increased catalase activity and proline accumulation as compared with wild type plants (Meng et al. 2020). Similarly, drought stress also leads to ROS generation, and there are several reports that advocate involvement of antioxidative mechanisms as an adaptive response (Verma et al. 2019; Wang et al. 2019c).

17.5 Strategies to Combat Climate Change

17.5.1 Cultural and Conventional Methods

Different cultural methods practiced by farmers to combat changing environmental conditions include various irrigation techniques, changes in seed sowing and harvesting time, giving preferences to crop varieties having a shorter life span, and crop rotation schemes. Altogether, these crop management strategies are being used to achieve crop resilience to some extent (Battisti et al. 2018; Duku et al. 2018; Teixeira et al. 2018; Deligios et al. 2019). Application of fertilizers are important as they provide required nutrients to crop plants to obtain better yields during stress conditions (Henderson et al. 2018). Introducing crop varieties that are tolerant to abiotic stress conditions such as high temperatures, drought, and salinity in agricultural systems is needed and being implemented.

Conventional methods imply plant breeding for development and selection of stress-tolerant cultivars. Plant breeders take advantage of the changed genetic constitution of a population over time, measured through genetic variance analysis in terms of polymorphism, inbreeding, assessment, assortment, and recombination (Saravanan and Pachiyappan 2013; Priyanka et al. 2017). It involves collection of phenotypic variability along with desired traits across the landraces and followed by selection of superior parents for desired traits to the generated pure lines for cross-hybridization and screening of hybrid progeny showing better adaptability and improved yield.

Although cultural and conventional methods have been practiced for a long time to combat environmental changes, the time constraints associated with breeding techniques have become challenging. In continuation to this, recently the concept of speed breeding has been introduced where plants are grown in fully controlled environmental conditions to obtain crop plants such as wheat, barley, chickpea, and pea up to six generation per year (Watson et al. 2018). Speed breeding resulted in enhanced crop yields and accelerated scientific research by reducing its generation time. Hence, inclusion of speed breeding technology along with different crop breeding technologies supported by advanced multi-omic approaches has immense potential in crop improvement.

17.5.2 Multi-omics Approach

When plants are subjected to varying environmental cues, it leads to changes in protein and other metabolite levels and hence the morpho-physiological response of plants, due to altered gene expression. These changes are crucial for plants adaptation and survival. Integrated omics and systems biology approaches such as genomics, transcriptomics, proteomics, metabolomics, ionomics, miRNAome, phenome, and others led to developing an understanding of biological systems as a whole so that it can be implied to improve crop performance (Figure 17.1).

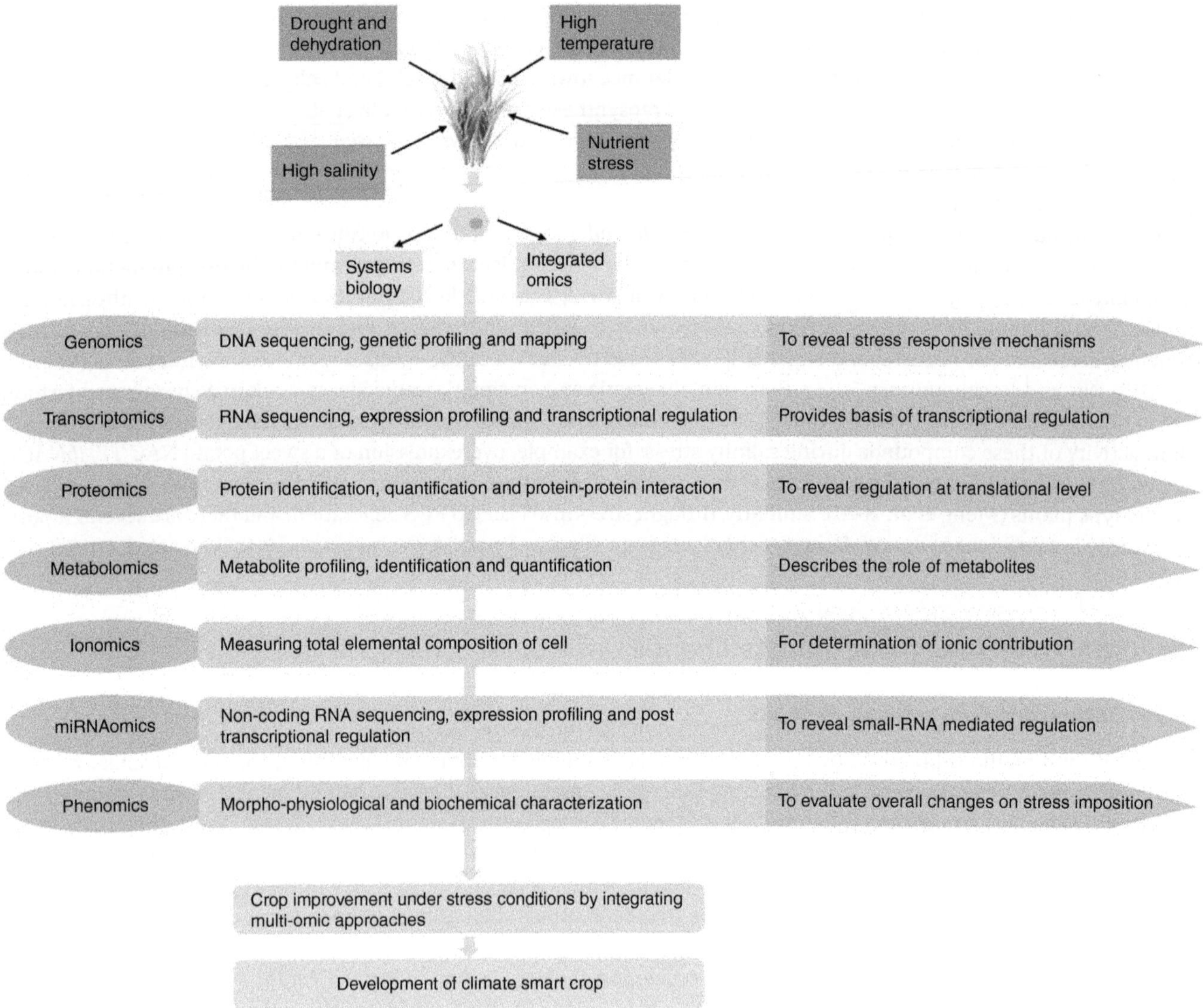

Figure 17.1 Integration of multi-omics based approaches to achieve stress tolerance against changing climate and to generate climate-smart crops.

Genomic approaches include sequencing of whole genome, genetic profile, mapping, and characterizing structural and functional genomics. Introduction of high-throughput DNA sequencing led to identification of high degree of DNA polymorphism, and the most popular are single nucleotide polymorphisms (SNPs) (D'Agostino and Tripodi 2017). SNPs along with other DNA markers help in marker-assisted selection of valuable stress resistant/tolerant traits and lead to accelerated molecular plant breeding through marker-assisted breeding approach (Collard and Mackill 2008). High-throughput sequencing coupled with phenotypic studies leads to identification of quantitative trait loci (QTLs). QTLs are often associated with one or more gene functions under stressed conditions. For example, rice QTL that affects a trait, cold tolerance at the bud burst stage (CTB), was mapped through genome-wide association studies (GWAS) (Yang et al. 2020a). Also, it has been shown that this QTL is associated with the expression of rice gene *OsRab11,* involved in providing cold tolerance through modulation of the abscisic acid pathway (Wang et al. 2021). Another study suggests that drought tolerance in soybean was associated with multiple QTL that govern the expression of an array of genes responsible for drought tolerance (Wang et al. 2020b). GWAS is used to link allelic variants with specific traits among different crop cultivars and has played a crucial role in identifying the genetic basis of biotic and abiotic tolerance/resistance during changing climatic events (Mousavi-Derazmahalleh et al. 2019; Khan et al. 2021). A GWAS study coupled with the reverse genetic approach carried out on 300 Arabidopsis accessions led to identification of genomic regions responsible for abscisic acid accumulation under drought stress (Kalladan et al. 2017). Similar studies have been carried out to reveal genes associated with proline accumulation during drought (Verslues et al. 2014). Also, GWAS has made a way to understand the basis of salinity, heat, and cold resistance in a wide range of crop species (Kumar et al. 2015; Chen et al. 2017; Chopra et al. 2017; Lafarge et al. 2017). Therefore, GWAS provides an understanding of the genetic basis of stress tolerance and enables breeders to screen the elite germplasm for crop improvement.

Transcriptomic studies are performed to assess the changes in mRNA levels in different plant tissues under stress conditions. It involves RNA sequencing, expression profiling, and pinpointing crucial transcripts of a pathway and hence gives an insight about transcriptional regulation mechanisms of stress tolerance. Transcriptome studies during biotic and abiotic stress conditions have been done in different crop species and suggest altered transcript levels of many functional genes such as various transcription factors, kinases, phosphatases, and others (Bashir et al. 2019; Wang et al. 2019b; Bala et al. 2020; Kang et al. 2020; Liu et al. 2021a). For instance, when wheat plants are treated with ABA and gibberellic acid (GA), they become more susceptible to fusarium head blight disease. So, to gain a molecular insight, transcriptome profiling of wheat revealed differential expression of more than 30,000 gene transcripts related to defense responses, molecular transport, carbohydrate, and lipid metabolism. Another transcriptome study has been done in pepper during heat, cold, salinity, and osmotic stresses, which suggested differential expression of genes (Kang et al. 2020). These differentially expressed genes were found to perform a role in cellular response to osmotic stress, starvation, immune response, hypersensitive salinity response, and oxidative and organellar stress responses in pepper (Kang et al. 2020). Therefore, transcriptomic studies provide a basis to understand the role of a gene transcript and its associated pathway during stressed conditions.

Both the genomic and transcriptomic studies have helped to understand the multigenecity of plants genome and provided a global picture of plant responses during stress conditions. These strategies altogether shifted our approach more toward focusing the multicomponent interactions and co-expression circuits from the single gene approach (Bohnert et al. 2006; Weston et al. 2008). However, study focusing only on genes may not give the exact idea about a cell's response during a particular time and stress due to regulatory checkpoints at translational and post-translational levels. Proteomics is an emerging approach leading the way toward rapid identification and assessment of novel proteins of stress tolerance and resistance. A wide range of proteomic studies have been done to elucidate abiotic stress mediated protein changes in major crop plants (Barkla et al. 2013). Mostly, these protein changes have been shown to be associated with metabolic pathway proteins such as glycolysis and citric acid cycle enzymes, stress-mediated ROS scavenging pathway proteins such as ascorbate peroxidase, superoxide dismutase, and glutathione reductase, and enzymes involved in protein synthesis, processing, and degradation (Barkla et al. 2013). For instance, an integrative study using transcriptomics and proteomics in cucumber seed post-germination events suggests that drought and salt stress led to differential expression of 4197 mRNAs, 36 microRNAs, and 768 proteins when compared to control conditions and are correlated with proline, polyamine, ethylene, and nitric oxide metabolism (Du et al. 2021). Another proteomic study performed in wheat endosperm during drought, salinity, and submergence stress delineates differentially expressed proteins related to amino and organic acid metabolisms (Yan et al. 2021). These findings provide preliminary insights about how stress affects plant cell proteomes, and further studies are needed to comprehend the mechanism and application of changing plant proteomes and their response to their environment.

Other omics-led approaches becoming popular during recent years include metabolomics for metabolite profiling, identification, and quantification to address whether metabolites are responsible for a particular phenotypic trait (Ghatak et al. 2018). Several studies highlight the role and applications of metabolomics in plants during various abiotic stresses (Nakabayashi and Saito 2015; Ghatak et al. 2018; Rastogi et al. 2019; Zhao et al. 2019b; Feng et al. 2020; Kumar et al. 2021a). Ionomics is about the measurement of total elemental composition of a plant (Salt et al. 2008; Huang and Salt 2016). Recently, more progress has been made in plant ionomics and has delineated the role of gene and gene networks that control plant ionomes and their applications in response to plant stress, along with proteomic, transcriptomic, and metabolomics approaches (Sanchez et al. 2011; Kumari et al. 2015; Ali et al. 2021; Peng et al. 2021). Further, miRNAs and long-noncoding RNAs are well known key regulators of biotic and abiotic stress in crop plants (Noman et al. 2017; Zhang et al. 2019; Ali et al. 2020; Sharma et al. 2020; Singh et al. 2021). To emphasize more on these regulatory roles, miRNAome is an omics approach that applies sequencing, expression profiling, and post-transcriptional regulation of noncoding RNAs has been done. A study shows comparison of miRNA profiling of roots of salt-tolerant rice varieties; Pokkali rice and salt-sensitive cultivar suggest the basis of salt adaptability in Pokkali rice through miRNA-mediated genetic regulation of cellular metabolism (Goswami et al. 2020). Phenomics is morphological, physiological, and biochemical characterization of stress-induced changes to provide a picture of plants adaptive strategies during changing environmental conditions (Houle et al. 2010; Zhao et al. 2019a; Deery and Jones 2021). Further advances and implications of integrated omic approaches in combination with biotechnological approaches will potentiate the development of climate-smart crops.

17.5.3 Biotechnological Approaches

Interventions of biotechnology in crop improvement against climate change have been involved generation of genetically modified (GM) crops, also referred as transgenic crops. GM crops have their genes altered by using different genetic engineering techniques. Genetic engineering techniques have an upper hand over breeding techniques. Whereas breeding techniques facilitate transfer of genes among closely associated species, laborious and time-consuming genetic engineering techniques have been shown to overcome these barriers. The prerequisite to imply the genetic engineering technique for any trait improvement is that one must have a trait-/stress-responsive gene to start with. The screening of desired trait-responsive genes is done through different omic-based approaches discussed in Sections 17.5.1 and 17.5.2. After identification, the expression and regulation of gene of interest can be engineered to obtain the desired phenotype. For example, if a gene is acting as a positive regulator for plant growth and development during stressed conditions, its expression can be specifically enhanced for desired phenotype. Similarly, if a gene is acting as a negative regulator during stressed conditions, it is needed to downregulate the expression of that gene to enable the plants to overcome the stress consequences. Both gene silencing and overexpression approaches along with a newly revolutionary genetic engineering tool, genome editing, are being extensively used in crop improvement and are discussed in further sections. Figure 17.2 summarizes the basics of different biotechnological approaches that are being implemented in crop improvement.

17.5.3.1 Combating Climate Change Through Overexpression of Candidate Gene(s)

Once identified for stress tolerant/resistant traits in a species, genes are cloned into expression vectors and then transferred through various genetic engineering techniques such as particle bombardment, electroporation, and *Agrobacterium*-mediated gene transfer for successful incorporation of gene in the genome of target species. These transformations are either transient (the incorporated gene will express for a short duration) or stable (incorporation of gene in plant genome). *Agrobacterium*-mediated gene transfer method provides stable integration of transgene to the host genome and is widely used in the generation of genetically modified plants (Singh et al. 2016). One of the popular examples of this approach is the expression of a bacterial gene, *Cry* indifferent crop species such as cotton and brinjal to achieve pest resistance against

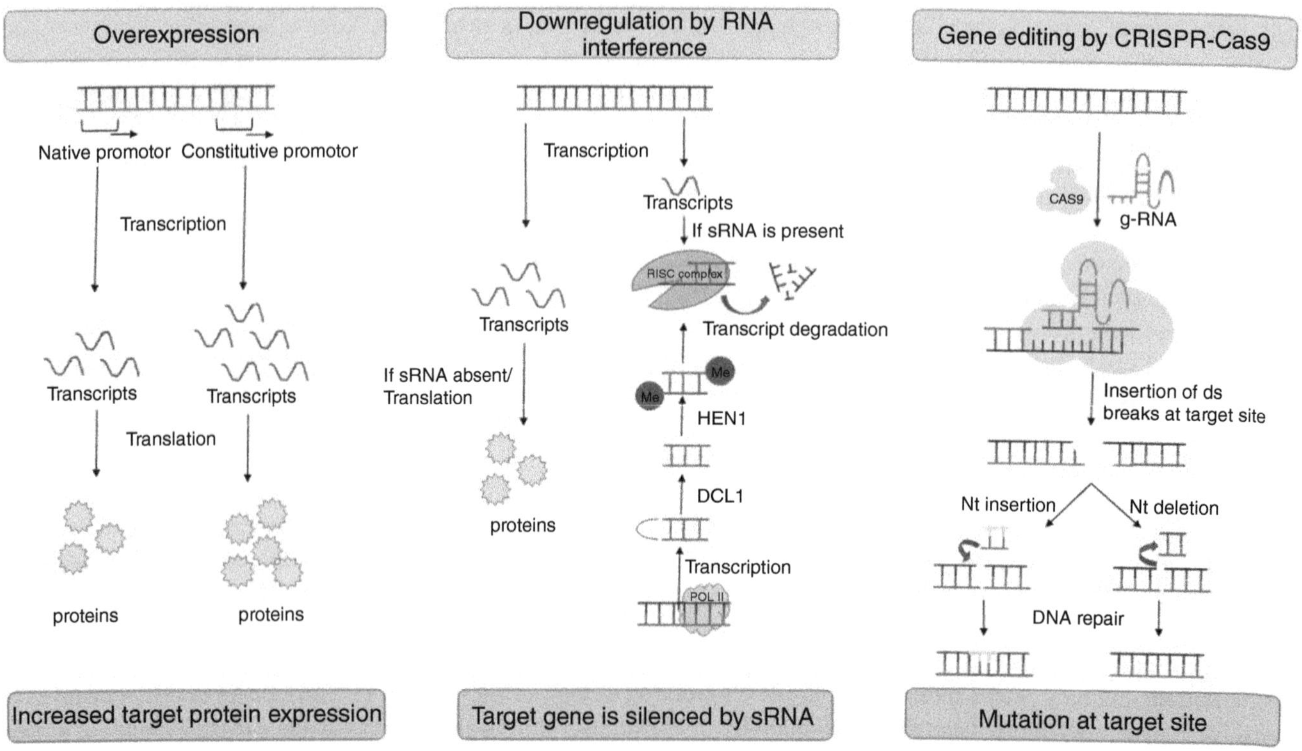

Figure 17.2 Biotechnological approaches for crop improvement. Overexpression of the candidate gene under constitutive promotor leads to increase in transcription of target gene, which can be translated into protein. RNA interference (RNAi) causes downregulation of the target gene. Gene editing through CRISPR-Cas9 leads to nucleotide insertions or deletions at the target site, which result in mutations. *Source:* Adopted from Singh et al. (2020).

lepidopterans and decrease yield loss in these crops. Cotton cultivation in India has been drastically increased after BT cotton introduction (Venugopalan et al. 2009). Similarly, a vast number of genes showing stress tolerance properties from different crops have been expressed in target species to achieve the desired stress response (reviewed by Singh et al. 2021).

Genes from diverse families have been expressed in many crops to attain biotic and abiotic stress tolerance. Transcription factor (TF) encoding genes are most prominently used as they act as nodal expression regulators for a number of genes underlying different pathways. For instance, AP2/ERFBP are plant-specific TF families and constitute subfamilies, namely Apetala2 (AP2), dehydration stress responsive binding proteins (DREB), ERF, and are related to AB3/VP1 (RAV). Among these, DREB and ERF are widely studied for their specific roles during plant stress (Sharoni et al. 2011). Two members of DREB subfamily DREB1 and DREB2 have been thoroughly shown to play a role in dehydration and high and low temperature stress in Arabidopsis (Liu et al. 1998). Further, elucidation and application of abiotic stress tolerance of DREB subfamily members have been extended to other crop species from vegetables (tomato and potato), major cereals (wheat, maize, and rice), millets to crops such as cotton and tobacco (Zhao et al. 2010; Lata et al. 2011; Bouaziz et al. 2012, 2013; Li et al. 2014, 2018b; Akbudak et al. 2018; El-Esawi and Alayafi 2019; Wang et al. 2019a; Mao et al. 2020; Niu et al. 2020; Yang et al. 2020b). Overexpression of the soybean transcription factor *GmMYB4* is shown to improve high density yield and drought tolerance by regulating plant architecture by decreasing the levels of brassinosteroids (Chen et al. 2021). Similarly, basic leucine zippers (bZIP), WRKY, NAC and other TF families have been expressed in several crop species (Shao et al. 2015; Jiang et al. 2017; Bi et al. 2021b;Baillo et al. 2019). These studies collectively suggest that these TFs are crucial for increased tolerance to plant biotic and abiotic stresses, which could be reflected in terms of plant survival and yield performance in adverse environments.

In addition to generating genetically modified plants, the overexpression approach also helps in elucidating the function of novel gene (s) during stress conditions through evaluating morpho-physiological response of plants. A sweet potato TF encoding gene *IbMYB116* when overexpressed in Arabidopsis, increased level of methyl jasmonate content and decreased level of hydrogen peroxide molecules were recorded during drought stress (Zhou et al. 2019). This study revealed that *IbMYB116* might activate a reactive oxygen species scavenging system by modulating the jasmonic acid signaling pathway and impart drought tolerance in transgenic Arabidopsis. A large number of novel stress-responsive genes from different crops, such as *GmCAMTA12* from soybean, *VvNAC17* from grape, and *RSZ21b* from cassava, have been functionally characterized by generating Arabidopsis overexpression lines (Noman et al. 2019; Ju et al. 2020; Chen et al. 2022). Along with overexpression Arabidopsis lines, the T-DNA insertion lines of *Arabidopsis thaliana* have also been widely used to elucidate the function of novel genes (Lu et al. 2014; Qu and Qin 2014; Liu et al. 2021c). Sequence information access through public domain has accelerated the large-scale identification and functional characterization of novel genes in response to stresses and is being used to overexpression target crop plants to attain desired traits.

17.5.3.2 Small RNA-Mediated Gene Silencing Approach

RNA interference (RNAi) evolved in eukaryotes as a protective cell mechanism against foreign nucleic acid molecules and is also widely known for its crucial role in maintaining genomic stability, epigenetic modifications, transcriptional regulation, and translational regulation of genes (Wassenegger 2005; Laudadio et al. 2019; Hung and Slotkin 2021). In plants, gene silencing was discovered by chance while trying to deepen the color of petunia flowers through expression of chalcone synthase (CHS), a gene responsible for pigment production. Napoli et al. (1990), observed that instead of pigment enhancement in petunia flowers, a variegated color phenotype is obtained. It was also showed that the level of CHS transcript levels was reduced up to 50-folds in transgenic plants without altering endogenous CHS transcript levels as compared to wild type and concluded that there is an unclear molecular mechanism involved in this reversible co-suppression of homologous genes in trans. Later, similar phenomenon was also observed in a nematode *Caenorhabditis elegans* that after injecting a dsRNA the levels of endogenous gene homologous to dsRNA were reduced (Timmons and Fire 1998). Hence, the unclear molecular mechanism of "Napoli et al. (1990)" was described as RNAi by "Timmons and Fire." Further, the RNAi phenomenon has emerged as a promising biotechnological tool to silence target genes and has wide applications in crop improvement (Jagtap et al. 2011). Plants have endogenous machinery of small RNAs (sRNA) to regulate the gene expression, which consist of short interfering RNAs (siRNA) and microRNAs (miRNA). These sRNAs are noncoding, are 21–28 nucleotides in length, and are formed by cleavage of dsRNA fragments. Although both siRNA and miRNA are similar in their function, including regulation of their target gene through transcriptional gene silencing (TGS)/post-transcriptional gene silencing (PTGS), they differ from each other in terms of their biogenesis and processing. While miRNAs are derived from the genomic region of DNA, siRNAs can be obtained through the cleavage of dsRNA into shorter fragments.

miRNAs are encoded by MIR genes and are transcribed with the help of RNA polymerase II (RNA pol II) to make hairpin loop like structure; pri-miRNA. pri-miRNAs are further processed to pre-miRNA and then to mature mi-RNA containing a guide strand (complementary to the target mRNA) and remaining passenger strand through the action of dicer like protein (DCL). Furthermore, HUA ENHANCER1 (HEN1) protein assists in the 3′ methylation of mature miRNA to preserve it against degradation. Mature miRNA then moves into the cytoplasm and forms RNA induced silencing complex (RISC) with Argonaut (AGO) and other effector proteins. When guide RNA strand complements with target mRNA as RISC complex, the nuclease activity of AGO proteins leads to its degradation or translational inhibition in cases of imperfect complementarity (Prasad et al. 2019). In contrast to miRNAs, which arise from genomic regions, siRNAs can be formed due to cleavage of endogenous dsRNA or can be of exogenous origin such as virus, transgene, or transposons. To generate siRNAs, DCL cuts these dsRNAs into 21–24 nucleotide long fragments with 2 nucleotide overhangs at both 3′-OH and 5′-PO_4 of RNA duplex (Prasad et al. 2019). Further siRNA induced silencing complex (siRISC) has been formed, which led to transcriptional and post-transcriptional gene silencing in a similar fashion as described for miRNA.

RNAi technology has been utilized to understand the basis of abiotic stress tolerance by developing genetically engineered plants in response to high temperature, drought, dehydration, salinity, flood, and heavy metal toxicity. Glutaredoxins (GRXs) are known to perform a crucial role in redox-signaling pathways. RNAi-mediated silencing of the rice *OsGRX17* resulted in improved drought tolerance compared with wild type (Hu et al. 2017). Similarly, RNAi-based silencing of a rice cytokinin oxidase, *OsCKX2*, specific to the inflorescence meristem, is performed to decipher its role in maintaining rice yield performance under salinity stress conditions. Silenced plants have been observed with better vegetative growth, improved photosynthetic efficiency, and relative water content, decreased electrolyte leakage under NaCl-mediated salinity stress (Joshi et al. 2018). Dehydrins are an important class of proteins known to prevent osmotic damage in plants. Silencing of a wheat dehydrin, *WZY2*, results in reduced activity of oxidative stress associated enzymes, lower relative water content, and increased malonaldidehyde (MDA) content, which leads to increased osmotic stress intolerance in wheat (Yu et al. 2019b).This suggests that *WZY2* gene has a crucial role in maintaining osmotic balance in wheat. GRAS TFs are important regulators of plant growth and development under abiotic stress conditions. Silencing of a tomato GRAS TF, *SlGRAS10*, leads to reduced plant height and leaf size, short internodes, and increased flavonoid content during salt and drought stress in transgenic tomato. Also, transgenic tomato plants have shown increased expression of genes that encode enzymes involved in the scavenging of ROS such as catalases, peroxidases, and superoxide dismutases (Habib et al. 2021).

Different classes of miRNAs have been shown to regulate gene expression during plant abiotic stresses. In maize, *zma-miR160*, *zma-miR390*, *zma-miR393*, and *zma-miR2275*, *zma-miR166*, and *zma-miR395* are shown to contribute to drought tolerance in the tolerant HKI-1532 as compared to the sensitive variety V-372. These miRNAs target TFs NAC, HD-Zip III, and GAMYB and reduce their transcript levels more in V-372 as compared to HKI-1532. Since HKI-1532 has high m-RNA activity due to lower miRNA-mediated downregulation, it performs better under drought conditions (Aravind et al. 2017). Another miRNA of maize, *ZmmiR190*, and its target *ZmCRP04* have shown to enhance drought tolerance in Arabidopsis and rice (Chai et al. 2021). In apple, the role of *miR156/SPL* module has been shown to regulate salt stress tolerance (Ma et al. 2021).

17.5.3.3 Gene Editing Through Clustered Regularly Interspaced Short Palindromic Repeats (CRISPR) Approach

Prior to the invention of CRISPR, gene editing methods were based on DNA recognition and binding properties of specified proteins such as meganucleases, Transcription Activator Like Effector Nucleases (TALENs), and Zinc Finger Nucleases (ZFNs) (Gaj et al. 2013). These proteins are differentiated into a DNA recognition domain derived from TALEs or ZF TFs, and an endonuclease domain composed of type II restriction enzymes. The recognition domain can be customized to recognize the target DNA sequence, and the endonuclease domain inserts a double stranded break (DSB) at the targeted site (Carroll 2011; Hensel and Kumlehn 2019). While these tools are widely used, certain drawbacks associated with them are summarized by Gardiner et al. (2022). Since last decade, CRISPR and CRISPR-associated proteins (Cas) have emerged as simple, efficient, and multifaceted genome editing tools. CRISPR is composed of short palindromic DNA sequences and is spaced by nonrepetitive spacer sequences. Findings suggest that these spacer sequences are of foreign origin, such as viral or plasmid, and are associated with prokaryotic adaptive immunity (Sorek et al. 2013). CRISPR-Cas system uses noncoding RNAs to identify its target DNA sequence, in contrast to TALEs or ZFs, which rely on amino acid sequences for target specificity. It is comprised of a complementary RNA sequence against target DNA and is termed as spacer or CRISPR RNA (crRNA) and another RNA sequence, trans-activating crRNA (tracrRNA). The TracrRNA remains attached to the Cas protein and pairs with crRNA to form guide RNA (gRNA). Casproteins and gRNA form a complex that identifies specific target DNA sequence with the help of protospacer adjacent motif (PAM) sequences to create a DSB (Amitai and Sorek 2016). DSBs are further repaired through existing DNA repair pathways, homology-directed repair, and non-homologous end joining. Inactive Cas (dCas) are being used nowadays to achieve a range of targeted effects with associated effector domains, for example Suntag and MS2 systems (Pflueger et al. 2018; Lee et al. 2019). The recent advances in CRISPR-Cas system have eliminated the need for PAM sequences to delimit its targeting scope (Ren et al. 2021a).

Continuously emerging CRISPR editing tools are being utilized in crop improvement and to develop new varieties that are more resilient to stressed environments (Bao et al. 2019; Debbarma et al. 2019; Li et al. 2021; Sami et al. 2021). Auxin-Regulated Genes (ARGOS) are known to regulate auxin-mediated lateral organ size and are a component of a negative feedback loop that desensitizes ethylene responses of Arabidopsis (Hu et al. 2003; Rai et al. 2015). CRISPR-Cas tool was used to modify the promoter of maize ARGOS genes and replace it with the native promoter. ARGOS gene variants with modified promoters showed an increase in the level of their transcripts as compared to wild types. Field experiments suggested that, as compared with wild-type maize plants having native promoters, ARGOS gene variants showed improved grain yield under drought stress conditions (Shi et al. 2017). Lateral organ boundary (LBD) family of TFs are plant-specific TFS and have well described roles in lateral organ development and other functions (Xu et al. 2016). A tomato LBD TF encoding gene, *SlLBD40* was found to be deferentially upregulated when tomato plants were subjected to PEG, methyl jasmonate, and salt treatments. These observations suggested that *SlLBD40* might act as an important regulator of the jasmonic acid (JA) pathway and in plants water retention capacity during dehydration and salt stress. A CRISPR-Cas9 directed knockout of *SlLBD40* in tomato plants showed less wilting when compared to *SlLBD40* overexpressing plants (Liu et al. 2020). In an attempt to develop drought and salt-tolerant rice varieties, the CRISPR-Cas9 system was used to generate mutants in drought and salt tolerance (*OsDST*) gene of *indica* rice cultivar, MTU1010. It has been found that *dst* mutants had broader leaves with decreased stomatal density, resulting in an increase in water retention capacity during dehydration stress. Also, *dst* mutants show high levels of salt tolerance and osmotic stress tolerance up to certain extent (Santosh Kumar et al. 2020). In addition to these, there are numerous studies suggesting CRISPR-Cas9 mediated editing of genes for improved drought and salt tolerance, for example, ABA receptors of rice; *OsPYL9*, pathogenesis-related gene; *SlNPR1* in tomato (Li et al. 2019b; Usman et al. 2020; Chennakesavulu et al. 2021).

CRISPR-Cas9 mediated generation of single, double, and triple mutants of Arabidopsis CBF (C-repeat Binding Factor) genes *CBF1, CBF2*, and *CBF3* encoding (CBF) TFs has been done to elucidate the basis of freezing tolerance in plants. Triple mutants showed extreme freezing sensitivity after cold acclimation (Jia et al. 2016; Zhao et al. 2016). Similarly, a tomato CBF mutant, *slcbf1*, also showed decreased chilling tolerance with higher malonaldehyde levels and elevated electrolyte leakage (Li et al. 2018a). These findings suggest that *CBFs* are crucial for freezing tolerance. Another application of CRIPR-Cas9 technology has been made to obtain mutants of rice *PRP* (proline-rich protein) gene, *OsPRP1, to* decipher its function. *osprp1* mutants displayed decreased levels of proline accumulation and antioxidant enzyme activities, abscisic acid, ascorbic acid, and chlorophyll and hence were more cold sensitive as compared to wild type plants (Nawaz et al. 2019). Mitogen-activated protein kinases (MAPKs) are well-conserved protein kinases and are extensively studied in response to plant biotic and abiotic stress (He et al. 2020). A tomato MAPK, *SlMAPK3*, shows decrease in relative transcript levels during heat stress. Mutants of the *SlMAPK3* gene raised using CRISPR-Cas9 tool were showing improvements in water retention capacity, reduced cell membrane damage, and decrease in ROS species. Furthermore, the expression of HSFs (Heat Shock Factors) and HSPs (Heat Shock Proteins) was found to be increased in *SlMAPK3* mutants as compared to wild types (Yu et al. 2019a). Functional characterization of rice NAC TFs, ONAC127, and ONAC127 in response to heat stress has been done using mutants generated through CRISPR-Cas9 system. CRISPR knockout rice plants have incomplete grain filling and collapsed grains, and this effect appeared more severe in response to heat stress (Ren et al. 2021b).

17.6 Conclusion and Way Forward

Rapidly changing global climatic conditions have direct consequences on crop growth, development, and yield, which is reflected in terms of increasing food insecurity. Climate change has caused an increase in heat, drought, salinity, dehydration, and flood events more often than before. These events affect crop productivity either individually or, more likely, in combination. Stress conditions lead to reduced plant performance as in case; heat stress causes heat injury (Bita and Gerats 2013; Hasanuzzaman et al. 2013), ROS generation causes reduction in photosynthetic rate due to photo-oxidative damage of crucial proteins and enzymes (Carmody et al. 2016; Chalanika De Silva and Asaeda 2017). Plants have devised several mechanisms to combat the adverse effects of climate change such as induction of signaling mechanisms through phytohormones, heat shock proteins, osmolytes accumulation, and antioxidant activities (Sharma et al. 2012; Das and Roychoudhury 2014). To understand these adaptive strategies and utilize them to develop crops that are more resilient and are sustainable during ever changing climatic conditions, attempts have been made through breeding, integrated multi-omics, and biotechnological strategies. Systems biology and integrated omics approaches are functions from the very beginning, from the identification of a stress-responsive gene to its characterization and its implication in stress tolerance (Cramer et al. 2011; Raza et al. 2021b). Conventional biotechnological approaches, including overexpression and RNAi-mediated silencing techniques, helped us to understand the role of novel stress-tolerant genes and also to develop genetically modified crops with improved yields under stress conditions. CRISPR-Cas9 tool is a new domain of biotechnological approach, facilitating a simple, efficient, and multifaceted editing of genes for increased stress tolerance and understanding the role of stress responsive novel genes. Recent developments in CRISPR-based editing tools are more exciting owing to their increased efficiency to achieve a range of targeted effects (Ren et al. 2021a). Further understanding of stress responsive genes and their implications by using biotechnological approaches to make climate-smart crops will lead to agricultural sustainability and food security in the future.

Acknowledgments

Authors' work in this area is supported by J. C. Bose National Fellowship Grant of Department of Science and Technology (File No.: JCB/2018/000001). Jyoti Maurya acknowledges the Council of Scientific and Industrial Research, Government of India for the research fellowship. Roshan Kumar Singh acknowledges the DBT Multi-institutional project entitled "Germplasm Characterization and Trait Discovery in Wheat Using Genomics Approaches and Its Integration for Improving Climate Resilience, Productivity and Nutritional Quality" under mission program of "Characterisation of Genetic Resources," [BT/Ag/Network/Wheat/2019-20] for the research grant. The authors are also thankful to DBT-eLibrary Consortium (DeLCON) for providing access to the e-resources.

Declaration of Interest Statement

No conflict of interest declared.

References

Abdula, S.E., Lee, H.-J., Ryu, H. et al. (2016). Overexpression of BrCIPK1 gene enhances Abiotic stress tolerance by increasing proline biosynthesis in rice. *Plant Mol. Biol. Report.* 34: 501–511.

Akbudak, M.A., Filiz, E., and Kontbay, K. (2018). DREB2 (dehydration-responsive element-binding protein 2) type transcription factor in sorghum (*Sorghum bicolor*): genome-wide identification, characterization and expression profiles under cadmium and salt stresses. *3 Biotech* 8: 426.

Ali, M., Javaid, A., Naqvi, S.H. et al. (2020). Biotic stress triggered small RNA and RNAi defense response in plants. *Mol. Biol. Rep.* 47: 5511–5522.

Ali, S., Tyagi, A., and Bae, H. (2021). Ionomic approaches for discovery of novel stress-resilient genes in plants. *Int. J. Mol. Sci.* 22: 7182.

Al-Whaibi, M.H. (2011). Plant heat-shock proteins: a mini review. *J. King Saud Univ. Sci.* 23: 139–150.

Amitai, G. and Sorek, R. (2016). CRISPR–Cas adaptation: insights into the mechanism of action. *Nat. Rev. Microbiol.* 14: 67–76.

Aneja, B., Yadav, N.R., Kumar, N., and Yadav, R.C. (2015). Hsp transcript induction is correlated with physiological changes under drought stress in Indian mustard. *Physiol. Mol. Biol. Plants* 21: 305–316.

Aravind, J., Rinku, S., Pooja, B. et al. (2017). Identification, characterization, and functional validation of drought-responsive microRNAs in subtropical maize inbreds. *Front. Plant Sci.* 8: 941.

Aydinsakir, K., Erdal, S., Buyuktas, D. et al. (2013). The influence of regular deficit irrigation applications on water use, yield, and quality components of two corn (*Zea mays* L.) genotypes. *Agric. Water Manag.* 128: 65–71.

Baillo, E.H., Kimotho, R.N., Zhang, Z., and Xu, P. (2019). Transcription factors associated with abiotic and biotic stress tolerance and their potential for crops improvement. *Genes (Basel)* 10, 771.

Bala, M., Sinha, R., Mallick, M.A. et al. (2020). Methods of gene expression profiling to understand abiotic stress perception and response in legume crops. *Methods Mol. Biol.* 2107: 99–126.

Bao, A., Burritt, D.J., Chen, H. et al. (2019). The CRISPR/Cas9 system and its applications in crop genome editing. *Crit. Rev. Biotechnol.* 39: 321–336.

Barkla, B.J., Vera-Estrella, R., and Pantoja, O. (2013). Progress and challenges for abiotic stress proteomics of crop plants. *Proteomics* 13: 1801–1815.

Bashir, K., Matsui, A., Rasheed, S., and Seki, M. (2019). Recent advances in the characterization of plant transcriptomes in response to drought, salinity, heat, and cold stress. *F1000Research* 8: F1000.

Battisti, R., Sentelhas, P.C., Parker, P.S. et al. (2018). Assessment of crop-management strategies to improve soybean resilience to climate change in Southern Brazil. *Crop Pasture Sci.* 69: 154–162.

Bi, A., Wang, T., Wang, G. et al. (2021a). Stress memory gene FaHSP17.8-CII controls thermotolerance via remodeling PSII and ROS signaling in tall fescue. *Plant Physiol.* 187: 1163–1176.

Bi, C., Yu, Y., Dong, C. et al. (2021b). The bZIP transcription factor TabZIP15 improves salt stress tolerance in wheat. *Plant Biotechnol. J.* 19: 209–211.

Bista, D.R., Heckathorn, S.A., Jayawardena, D.M. et al. (2018). Effects of drought on nutrient uptake and the levels of nutrient-uptake proteins in roots of drought-sensitive and-tolerant grasses. *Plants (Basel)* 7: 28.

Bita, C.E. and Gerats, T. (2013). Plant tolerance to high temperature in a changing environment: scientific fundamentals and production of heat stress-tolerant crops. *Front. Plant Sci.* 4: 273.

Bohnert, H.J., Gong, Q., Li, P., and Ma, S. (2006). Unraveling abiotic stress tolerance mechanisms – getting genomics going. *Curr. Opin. Plant Biol.* 9: 180–188.

Bouaziz, D., Pirrello, J., Ben Amor, H. et al. (2012). Ectopic expression of dehydration responsive element binding proteins (StDREB2) confers higher tolerance to salt stress in potato. *Plant Physiol. Biochem. PPB* 60: 98–108.

Bouaziz, D., Pirrello, J., Charfeddine, M. et al. (2013). Overexpression of StDREB1 transcription factor increases tolerance to salt in transgenic potato plants. *Mol. Biotechnol.* 54: 803–817.

Brookbank, B.P., Patel, J., Gazzarrini, S., and Nambara, E. (2021). Role of Basal ABA in plant growth and development. *Genes (Basel)* 12: 1936.

Caligari, P. (2001). *Plant Breeding and Crop Improvement*. eLS.

Carmody, M., Waszczak, C., Idänheimo, N. et al. (2016). ROS signalling in a destabilised world: a molecular understanding of climate change. *J. Plant Physiol.* 203: 69–83.

Carroll, D. (2011). Genome engineering with zinc-finger nucleases. *Genetics* 188: 773–782.

Cassia, R., Nocioni, M., Correa-Aragunde, N., and Lamattina, L. (2018). Climate change and the impact of greenhouse gasses: CO_2 and NO, friends and foes of plant oxidative stress. *Front. Plant Sci.* 9: 273.

Chai, W., Song, N., Su, A. et al. (2021). ZmmiR190 and its target regulate plant responses to drought stress through an ABA-dependent pathway. *Plant Sci.* 312: 111034.

Chalanika De Silva, H.C. and Asaeda, T. (2017). Effects of heat stress on growth, photosynthetic pigments, oxidative damage and competitive capacity of three submerged macrophytes. *J. Plant Interact.* 12: 228–236.

Chaudhry, S. and Sidhu, G.P.S. (2022). Climate change regulated abiotic stress mechanisms in plants: a comprehensive review. *Plant Cell Rep.* 41: 1–31.

Chen, J., Chopra, R., Hayes, C. et al. (2017). Genome-wide association study of developing leaves' heat tolerance during vegetative growth stages in a sorghum association panel. *Plant Genome* 10: https://doi.org/10.3389/fpls.2018.01272.

Chen, L., Yang, H., Fang, Y. et al. (2021). Overexpression of GmMYB14 improves high-density yield and drought tolerance of soybean through regulating plant architecture mediated by the brassinosteroid pathway. *Plant Biotechnol. J.* 19: 702–716.

Chen, Y., Weng, X., Zhou, X. et al. (2022). Overexpression of cassava RSZ21b enhances drought tolerance in Arabidopsis. *J. Plant Physiol.* 268: 153574.

Chennakesavulu, K., Singh, H., Trivedi, P.K. et al. (2021). State-of-the-art in CRISPR technology and engineering drought, salinity, and thermo-tolerant crop plants. *Plant Cell Rep.* 41: 815–831. https://doi.org/10.1007/s00299-021-02681-w.

Chopra, R., Burow, G., Burke, J.J. et al. (2017). Genome-wide association analysis of seedling traits in diverse *Sorghum germplasm* under thermal stress. *BMC Plant Biol.* 17: 12.

Chuang, Z., Bing, L., Shilong, P. et al. (2017). Temperature increase reduces global yields of major crops in four independent estimates. *Proc. Natl. Acad. Sci. U. S. A.* 114: 9326–9331.

Cohen, I., Rapaport, T., Berger, R.T., and Rachmilevitch, S. (2018). The effects of elevated CO_2 and nitrogen nutrition on root dynamics. *Plant Sci.* 272: 294–300.

Collard, B.C.Y. and Mackill, D.J. (2008). Marker-assisted selection: an approach for precision plant breeding in the twenty-first century. *Philos. Trans. R. Soc. Lond. Ser. B Biol. Sci.* 363: 557–572.

Comstock, J.P. (2002). Hydraulic and chemical signalling in the control of stomatal conductance and transpiration. *J. Exp. Bot.* 53: 195–200.

Cramer, G.R., Urano, K., Delrot, S. et al. (2011). Effects of abiotic stress on plants: a systems biology perspective. *BMC Plant Biol.* 11: 163.

D'Agostino, N. and Tripodi, P. (2017). NGS-based genotyping, high-throughput phenotyping and genome-wide association studies laid the foundations for next-generation breeding in horticultural crops. *Diversity* 9: 38.

Das, K. and Roychoudhury, A. (2014). Reactive oxygen species (ROS) and response of antioxidants as ROS-scavengers during environmental stress in plants. *Front. Environ. Sci.* 2: 53.

Debbarma, J., Sarki, Y.N., Saikia, B. et al. (2019). Ethylene response factor (ERF) family proteins in abiotic stresses and CRISPR-Cas9 genome editing of ERFs for multiple abiotic stress tolerance in crop plants: a review. *Mol. Biotechnol.* 61: 153–172.

Deery, D.M. and Jones, H.G. (2021). Field phenomics: will it enable crop improvement? *Plant Phenomics* 2021: 9871989.

Deligios, P.A., Chergia, A.P., Sanna, G. et al. (2019). Climate change adaptation and water saving by innovative irrigation management applied on open field globe artichoke. *Sci. Total Environ.* 649: 461–472.

Delincé, J. and Witzke, H. (2015). Economic impacts of climate change on agriculture: the AgMIP approach. *J. Appl. Remote. Sens.* 1: 97099–97099.

Dias de Oliveira, E., Bramley, H., Siddique, K.H.M. et al. (2013). Can elevated CO_2 combined with high temperature ameliorate the effect of terminal drought in wheat? *Funct. Plant Biol.* 40: 160–171.

Dietz, K.-J., Zörb, C., and Geilfus, C.-M. (2021). Drought and crop yield. *Plant Biol. (Stuttg.)* 23: 881–893.

Djanaguiraman, M., Narayanan, S., Erdayani, E., and Prasad, P.V.V. (2020). Effects of high temperature stress during anthesis and grain filling periods on photosynthesis, lipids and grain yield in wheat. *BMC Plant Biol.* 20: 268.

Dos Reis, S.P., Lima, A.M., and De Souza, C.R.B. (2012). Recent molecular advances on downstream plant responses to abiotic stress. *Int. J. Mol. Sci.* 13: 8628–8647.

Du, C., Li, H., Liu, C., and Fan, H. (2021). Understanding of the postgerminative development response to salinity and drought stresses in cucumber seeds by integrated proteomics and transcriptomics analysis. *J. Proteome* 232: 104062.

Duku, C., Zwart, S.J., and Hein, L. (2018). Impacts of climate change on cropping patterns in a tropical, sub-humid watershed. *PLoS One* 13: e0192642.

Dürr, C., Brunel-Muguet, S., Girousse, C. et al. (2018). Changes in seed composition and germination of wheat (*Triticum aestivum*) and pea (*Pisum sativum*) when exposed to high temperatures during grain filling and maturation. *Crop Pasture Sci.* 69: 374–386.

Ehrlich, P.R., Ehrlich, A.H., and Daily, G.C. (1993). Food security, population and environment. *Popul. Dev. Rev.* 19: 1–32.

El-Esawi, M.A. and Alayafi, A.A. (2019). Overexpression of StDREB2 transcription factor enhances drought stress tolerance in cotton (*Gossypium barbadense* L.). *Genes (Basel)* 10: 142.

Eyidogan, F., Oz, M.T., Yucel, M., and Oktem, H.A. (2012). Signal transduction of phytohormones under abiotic stresses. In: *Phytohormones and Abiotic Stress Tolerance in Plants* (ed. N.A. Khan), 1–48. Springer.

FAOSTAT (2017). http://www.fao.org/faostat/en/#home

Farooq, M., Gogoi, N., Barthakur, S. et al. (2017). Drought stress in grain legumes during reproduction and grain filling. *J. Agron. Crop Sci.* 203: 81–102.

Feng, Z., Ding, C., Li, W. et al. (2020). Applications of metabolomics in the research of soybean plant under abiotic stress. *Food Chem.* 310: 125914.

Fragkostefanakis, S., Roth, S., Schleiff, E., and Scharf, K.-D. (2015). Prospects of engineering thermotolerance in crops through modulation of heat stress transcription factor and heat shock protein networks. *Plant Cell Environ.* 38: 1881–1895.

Gaj, T., Gersbach, C.A., and Barbas, C.F. 3rd (2013). ZFN, TALEN, and CRISPR/Cas-based methods for genome engineering. *Trends Biotechnol.* 31: 397–405.

Gamage, D., Thompson, M., Sutherland, M. et al. (2018). New insights into the cellular mechanisms of plant growth at elevated atmospheric carbon dioxide concentrations. *Plant Cell Environ.* 41: 1233–1246.

Gardiner, J., Ghoshal, B., Wang, M., and Jacobsen, S.E. (2022). CRISPR-CAS mediated transcriptional control and epi-mutagenesis. *Plant Physiol.* 188: 1811–1824. https://doi.org/10.1093/plphys/kiac033.

Ghatak, A., Chaturvedi, P., and Weckwerth, W. (2018). Metabolomics in plant stress physiology. *Adv. Biochem. Eng. Biotechnol.* 164: 187–236.

Gong, C., Pang, Q., Li, Z. et al. (2021). Genome-wide identification and characterization of Hsf and Hsp gene families and gene expression analysis under heat stress in eggplant (*Solanum melongema* L.). *Horticulturae* 7: 149.

Gornall, J., Betts, R., Burke, E. et al. (2010). Implications of climate change for agricultural productivity in the early twenty-first century. *Philos. Trans. R. Soc. Lond. Ser. B Biol. Sci.* 365: 2973–2989.

Goswami, K., Mittal, D., Gautam, B. et al. (2020). Mapping the salt stress-induced changes in the root miRNome in Pokkali rice. *Biomolecules* 10: 498.

Gray, S.B. and Brady, S.M. (2016). Plant developmental responses to climate change. *Dev. Biol.* 419: 64–77.

Gregory, P.J., Ingram, J.S.I., and Brklacich, M. (2005). Climate change and food security. *Philos. Trans. R. Soc. London Ser. B Biol. Sci.* 360: 2139–2148.

Gupta, S.M., Arora, S., Mirza, N. et al. (2017). Finger millet: a 'certain' crop for an 'uncertain' future and a solution to food insecurity and hidden hunger under stressful environments. *Front. Plant Sci.* 8: 643.

Habib, S., Lwin, Y.Y., and Li, N. (2021). Down-regulation of SlGRAS10 in tomato confers abiotic stress tolerance. *Genes (Basel)* 12: 623.

Halli, H., Angadi, S., and Patil, R. (2016). Water and nutrient use efficiency in agriculture and the role of cereals: a review. *J. Farm. Sci.* 29: 299–306.

Hameed, A., Ahmed, M.Z., Hussain, T. et al. (2021). Effects of salinity stress on chloroplast structure and function. *Cells* 10: 2023.

Hasanuzzaman, M., Nahar, K., Alam, M.M. et al. (2013). Physiological, biochemical, and molecular mechanisms of heat stress tolerance in plants. *Int. J. Mol. Sci.* 14: 9643–9684.

Hasanuzzaman, M., Nahar, K., Bhuiyan, T.F. et al. (2017). Salicylic acid: an all-rounder in regulating abiotic stress responses in plants. In: *Phytohormones-Signaling Mechanisms and Crosstalk in Plant Development and Stress Responses* (ed. M. El-Esawi), 31–75. London: IntechOpen.

Hay, W.T., Bihmidine, S., Mutlu, N. et al. (2017). Enhancing soybean photosynthetic CO_2 assimilation using a cyanobacterial membrane protein, ictB. *J. Plant Physiol.* 212: 58–68.

He, B., Huang, H., Guo, T. et al. (2017). Physiological responses of *Scaevola aemula* seedlings under high temperature stress. *South Afr. J. Bot.* 112: 203–209.

He, X., Wang, C., Wang, H. et al. (2020). The function of MAPK cascades in response to various stresses in horticultural plants. *Front. Plant Sci.* 11: 952.

Henderson, B., Cacho, O., Thornton, P. et al. (2018). The economic potential of residue management and fertilizer use to address climate change impacts on mixed smallholder farmers in Burkina Faso. *Agric. Syst.* 167: 195–205.

Hensel, G. and Kumlehn, J. (2019). Genome engineering using TALENs. *Methods Mol. Biol.* 1900: 195–215.

Herrera-Vásquez, A., Salinas, P., and Holuigue, L. (2015). Salicylic acid and reactive oxygen species interplay in the transcriptional control of defense genes expression. *Front. Plant Sci.* 6: 171.

Houle, D., Govindaraju, D.R., and Omholt, S. (2010). Phenomics: the next challenge. *Nat. Rev. Genet.* 11: 855–866.

Hu, Y., Xie, Q., and Chua, N.-H. (2003). The Arabidopsis auxin-inducible gene ARGOS controls lateral organ size. *Plant Cell* 15: 1951–1961.

Hu, Y., Wu, Q., Peng, Z. et al. (2017). Silencing of OsGRXS17 in rice improves drought stress tolerance by modulating ROS accumulation and stomatal closure. *Sci. Rep.* 7: 15950.

Huang, X.-Y. and Salt, D.E. (2016). Plant ionomics: from elemental profiling to environmental adaptation. *Mol. Plant* 9: 787–797.

Huang, H., Ullah, F., Zhou, D.-X. et al. (2019). Mechanisms of ROS regulation of plant development and stress responses. *Front. Plant Sci.* 10: 1–10.

Huber, A.E., Melcher, P.J., Piñeros, M.A. et al. (2019). Signal coordination before, during and after stomatal closure in response to drought stress. *New Phytol.* 224: 675–688.

Hung, Y.-H. and Slotkin, R.K. (2021). The initiation of RNA interference (RNAi) in plants. *Curr. Opin. Plant Biol.* 61: 102014.

IPCC, 2021: Climate Change 2021. Cambridge University Press, Cambridge, United Kingdom and New York, NY, USA, 2391pp doi:https://doi.org/10.1017/9781009157896.

Jaganathan, D., Ramasamy, K., Sellamuthu, G. et al. (2018). CRISPR for crop improvement: an update review. *Front. Plant Sci.* 9: 985.

Jagtap, U.B., Gurav, R.G., and Bapat, V.A. (2011). Role of RNA interference in plant improvement. *Naturwissenschaften* 98: 473–492.

Jia, Y., Ding, Y., Shi, Y. et al. (2016). The cbfs triple mutants reveal the essential functions of CBFs in cold acclimation and allow the definition of CBF regulons in Arabidopsis. *New Phytol.* 212: 345–353.

Jiang, J., Ma, S., Ye, N. et al. (2017). WRKY transcription factors in plant responses to stresses. *J. Integr. Plant Biol.* 59: 86–101.

Joshi, R., Sahoo, K.K., Tripathi, A.K. et al. (2018). Knockdown of an inflorescence meristem-specific cytokinin oxidase – OsCKX2 in rice reduces yield penalty under salinity stress condition. *Plant Cell Environ.* 41: 936–946.

Ju, Y.-L., Yue, X.-F., Min, Z. et al. (2020). VvNAC17, a novel stress-responsive grapevine (*Vitis vinifera* L.) NAC transcription factor, increases sensitivity to abscisic acid and enhances salinity, freezing, and drought tolerance in transgenic Arabidopsis. *Plant. Physiol. Biochem. PPB* 146: 98–111.

Kalladan, R., Lasky, J.R., Chang, T.Z. et al. (2017). Natural variation identifies genes affecting droughtinduced abscisic acid accumulation in *Arabidopsis thaliana*. *Proc. Natl. Acad. Sci. U. S. A.* 114: 11536–11541.

Kang, W.-H., Sim, Y.M., Koo, N. et al. (2020). Transcriptome profiling of abiotic responses to heat, cold, salt, and osmotic stress of *Capsicum annuum* L. *Sci. Data* 7: 17.

Katul, G., Manzoni, S., Palmroth, S., and Oren, R. (2010). A stomatal optimization theory to describe the effects of atmospheric CO_2 on leaf photosynthesis and transpiration. *Ann. Bot.* 105: 431–442.

Kazan, K. (2015). Diverse roles of jasmonates and ethylene in abiotic stress tolerance. *Trends Plant Sci.* 20: 219–229.

Khan, Z. and Shahwar, D. (2020). Role of heat shock proteins (HSPs) and heat stress tolerance in crop plants. In: *Sustainable Agriculture in the Era of Climate Change* (ed. R. Roychowdhury, S. Choudhury, M. Hasanuzzaman, and S. Srivastava), 211–234. Springer.

Khan, S.U., Saeed, S., Khan, M.H.U. et al. (2021). Advances and challenges for QTL analysis and GWAS in the plant-breeding of high-yielding: a focus on rapeseed. *Biomolecules* 11: 1516.

Klay, I., Gouia, S., Liu, M. et al. (2018). Ethylene response factors (ERF) are differentially regulated by different abiotic stress types in tomato plants. *Plant Sci.* 274: 137–145.

Knipfer, T., Bambach, N., Hernandez, M.I. et al. (2020). Predicting stomatal closure and turgor loss in woody plants using predawn and midday water potential. *Plant Physiol.* 184: 881–894.

Kumar, V., Singh, A., Mithra, S.V.A. et al. (2015). Genome-wide association mapping of salinity tolerance in rice (*Oryza sativa*). *DNA Res.* 22: 133–145.

Kumar, V., Sah, S.K., Khare, T. et al. (2016). Engineering phytohormones for abiotic stress tolerance in crop plants. In: *Plant Hormones Under Challenging Environmental Factors* (ed. G.J. Ahammed and J.-Q. Yu), 247–266. Springer.

Kumar, J., Basu, P.S., Gupta, S. et al. (2018). Physiological and molecular characterisation for high temperature stress in *Lens culinaris*. *Funct. Plant Biol.* 45: 474–487.

Kumar, M., Kumar Patel, M., Kumar, N. et al. (2021a). Metabolomics and molecular approaches reveal drought stress tolerance in plants. *Int. J. Mol. Sci.* 22: 9108.

Kumar, S., Li, G., Yang, J. et al. (2021b). Effect of salt stress on growth, physiological parameters, and ionic concentration of water dropwort (*Oenanthe javanica*) cultivars. *Front. Plant Sci.* 12: 660409.

Kumari, A., Das, P., Parida, A.K., and Agarwal, P.K. (2015). Proteomics, metabolomics, and ionomics perspectives of salinity tolerance in halophytes. *Front. Plant Sci.* 6: 537.

Lafarge, T., Bueno, C., Frouin, J. et al. (2017). Genome-wide association analysis for heat tolerance at flowering detected a large set of genes involved in adaptation to thermal and other stresses. *PLoS One* 12: e0171254.

Lata, C., Bhutty, S., Bahadur, R.P. et al. (2011). Association of an SNP in a novel DREB2-like gene SiDREB2 with stress tolerance in foxtail millet [*Setaria italica* (L.)]. *J. Exp. Bot.* 62: 3387–3401.

Laudadio, I., Carissimi, C., and Fulci, V. (2019). How RNAi machinery enters the world of telomerase. *Cell Cycle* 18: 1056–1067.

Lee, J.E., Neumann, M., Duro, D.I., and Schmid, M. (2019). CRISPR-based tools for targeted transcriptional and epigenetic regulation in plants. *PLoS One* 14: e0222778.

Li, C., Yue, J., Wu, X. et al. (2014). An ABA-responsive DRE-binding protein gene from Setaria italica, SiARDP, the target gene of SiAREB, plays a critical role under drought stress. *J. Exp. Bot.* 65: 5415–5427.

Li, R., Zhang, L., Wang, L. et al. (2018a). Reduction of tomato-plant chilling tolerance by CRISPR-Cas9-mediated SlCBF1 mutagenesis. *J. Agric. Food Chem.* 66: 9042–9051.

Li, S., Zhao, Q., Zhu, D., and Yu, J. (2018b). A DREB-like transcription factor from maize (*Zea mays*), ZmDREB4.1, plays a negative role in plant growth and development. *Front. Plant Sci.* 9: 395.

Li, P., Li, B., Seneweera, S. et al. (2019a). Photosynthesis and yield response to elevated CO_2, C_4 plant foxtail millet behaves similarly to C_3 species. *Plant Sci.* 285: 239–247.

Li, R., Liu, C., Zhao, R. et al. (2019b). CRISPR/Cas9-Mediated SlNPR1 mutagenesis reduces tomato plant drought tolerance. *BMC Plant Biol.* 19: 38.

Li, C., Brant, E., Budak, H., and Zhang, B. (2021). CRISPR/Cas: a Nobel Prize award-winning precise genome editing technology for gene therapy and crop improvement. *J. Zhejiang Univ. Sci. B* 22: 253–284.

Liu, Q., Kasuga, M., Sakuma, Y. et al. (1998). Two transcription factors, DREB1 and DREB2, with an EREBP/AP2 DNA binding domain separate two cellular signal transduction pathways in drought- and low-temperature-responsive gene expression, respectively, in Arabidopsis. *Plant Cell* 10: 1391–1406.

Liu, L., Zhang, J., Xu, J. et al. (2020). CRISPR/Cas9 targeted mutagenesis of SlLBD40, a lateral organ boundaries domain transcription factor, enhances drought tolerance in tomato. *Plant Sci.* 301: 110683.

Liu, H., Able, A.J., and Able, J.A. (2021a). Small RNA, transcriptome and degradome analysis of the transgenerational heat stress response network in durum wheat. *Int. J. Mol. Sci.* 22.

Liu, Y.-L., Liu, S., Xiao, J.-J. et al. (2021b). CaHSP18. 1a, a small heat shock protein from pepper (*Capsicum annuum* L.), positively responds to heat, drought, and salt tolerance. *Horticulturae* 7: 117.

Liu, Y., Chen, X., Xue, S. et al. (2021c). SET DOMAIN GROUP 721 protein functions in saline-alkaline stress tolerance in the model rice variety Kitaake. *Plant Biotechnol. J.* 19: 2576–2588.

Lobell, D.B. and Gourdji, S.M. (2012). The influence of climate change on global crop productivity. *Plant Physiol.* 160: 1686–1697.

Loladze, I., Nolan, J.M., Ziska, L.H., and Knobbe, A.R. (2019). Rising atmospheric CO_2 lowers concentrations of plant carotenoids essential to human health: a meta-analysis. *Mol. Nutr. Food Res.* 63: e1801047.

Lu, G., Wang, X., Liu, J. et al. (2014). Application of T-DNA activation tagging to identify glutamate receptor-like genes that enhance drought tolerance in plants. *Plant Cell Rep.* 33: 617–631.

Ma, Y. and Qin, F. (2014). ABA regulation of plant responses to drought and salt stresses. In: *Abscisic Acid: Metabolism, Transport and Signaling* (ed. D.-P. Zhang), 315–336. Dordrecht: Springer.

Ma, Y., Xue, H., Zhang, F. et al. (2021). The miR156/SPL module regulates apple salt stress tolerance by activating MdWRKY100 expression. *Plant Biotechnol. J.* 19: 311–323.

Mahmood, T., Khalid, S., Abdullah, M. et al. (2019). Insights into drought stress signaling in plants and the molecular genetic basis of cotton drought tolerance. *Cells* 9: 105.

Mahrookashani, A., Siebert, S., Hüging, H., and Ewert, F. (2017). Independent and combined effects of high temperature and drought stress around anthesis on wheat. *J. Agron. Crop Sci.* 203: 453–463.

Mandal, K., Boro, P., and Chattopadhyay, S. (2021). Micro-RNA based gene regulation: a potential way for crop improvements. *Plant Gene* 27: 100312.

Mao, L., Deng, M., Jiang, S. et al. (2020). Characterization of the DREBA4-type transcription factor (SlDREBA4), which contributes to heat tolerance in tomatoes. *Front. Plant Sci.* 11: 554520.

Matiu, M., Ankerst, D.P., and Menzel, A. (2017). Interactions between temperature and drought in global and regional crop yield variability during 1961–2014. *PLoS One* 12: e0178339.

Meher, S.P., Ashok Reddy, K., and Manohar Rao, D. (2018). Effect of PEG-6000 imposed drought stress on RNA content, relative water content (RWC), and chlorophyll content in peanut leaves and roots. *Saudi J. Biol. Sci.* 25: 285–289.

Meng, X., Liu, S., Dong, T. et al. (2020). Comparative transcriptome and proteome analysis of salt-tolerant and salt-sensitive sweet potato and overexpression of IbNAC7 confers salt tolerance in Arabidopsis. *Front. Plant Sci.* 11: 572540.

Miura, K. and Tada, Y. (2014). Regulation of water, salinity, and cold stress responses by salicylic acid. *Front. Plant Sci.* 5: 4.

Mousavi-Derazmahalleh, M., Bayer, P.E., Hane, J.K. et al. (2019). Adapting legume crops to climate change using genomic approaches. *Plant Cell Environ.* 42: 6–19.

Mukhopadhyay, R., Sarkar, B., Jat, H.S. et al. (2021). Soil salinity under climate change: challenges for sustainable agriculture and food security. *J. Environ. Manag.* 280: 111736.

Muthamilarasan, M., Dhaka, A., Yadav, R., and Prasad, M. (2016). Exploration of millet models for developing nutrient rich graminaceous crops. *Plant Sci.* 242: 89–97.

Nakabayashi, R. and Saito, K. (2015). Integrated metabolomics for abiotic stress responses in plants. *Curr. Opin. Plant Biol.* 24: 10–16.

Napoli, C., Lemieux, C., and Jorgensen, R. (1990). Introduction of a chimeric chalcone synthase gene into petunia results in reversible co-suppression of homologous genes *in trans*. *Plant Cell* 2: 279–289.

Nawaz, G., Han, Y., Usman, B. et al. (2019). Knockout of OsPRP1, a gene encoding proline-rich protein, confers enhanced cold sensitivity in rice (*Oryza sativa* L.) at the seedling stage. *3 Biotech* 9: 254.

Nefissi Ouertani, R., Abid, G., Karmous, C. et al. (2021). Evaluating the contribution of osmotic and oxidative stress components on barley growth under salt stress. *AoB Plants* 13: plab034.

Niu, X., Luo, T., Zhao, H. et al. (2020). Identification of wheat DREB genes and functional characterization of TaDREB3 in response to abiotic stresses. *Gene* 740: 144514.

Noctor, G., Mhamdi, A., and Foyer, C.H. (2014). The roles of reactive oxygen metabolism in drought: not so cut and dried. *Plant Physiol.* 164: 1636–1648.

Noman, A., Fahad, S., Aqeel, M. et al. (2017). miRNAs: Major modulators for crop growth and development under abiotic stresses. *Biotechnol. Lett.* 39: 685–700.

Noman, M., Jameel, A., Qiang, W.-D. et al. (2019). Overexpression of GmCAMTA12 enhanced drought tolerance in Arabidopsis and Soybean. *Int. J. Mol. Sci.* 20: 4849.

Okon, O. (2019). Effect of salinity on physiological processes in plants. In: *Microorganisms in Saline Environments: Strategies and Functions* (ed. B. Giri and A. Varma), 237–262.

Olesen, J.E., Trnka, M., Kersebaum, K.C. et al. (2011). Impacts and adaptation of European crop production systems to climate change. *Eur. J. Agron.* 34: 96–112.

Parihar, P., Singh, S., Singh, R. et al. (2015). Effect of salinity stress on plants and its tolerance strategies: a review. *Environ. Sci. Pollut. Res. Int.* 22: 4056–4075.

Pattnaik, D., Dash, D., Mishra, A. et al. (2021). Emerging roles of osmoprotectants in alleviating abiotic stress response under changing climatic conditions. In: *Climate Impacts on Sustainable Natural Resource Management* (ed. P. Kumar, R.K. Singh, M. Kumar, et al.), 303–324. Wiley.

Peng, C.-Y., Xu, X.-F., Zhu, H.-Y. et al. (2021). Metabolics and ionomics responses of tea leaves (*Camellia sinensis* (L.) *O. Kuntze*) to fluoride stress. *Plant Physiol. Biochem. PPB* 158: 65–75.

Per, T.S., Khan, M.I.R., Anjum, N.A. et al. (2018). Jasmonates in plants under abiotic stresses: crosstalk with other phytohormones matters. *Environ. Exp. Bot.* 145: 104–120.

Pereira, A. (2016). Plant abiotic stress challenges from the changing environment. *Front. Plant Sci.* 7: 1123.

Pflueger, C., Tan, D., Swain, T. et al. (2018). A modular dCas9-SunTag DNMT3A epigenome editing system overcomes pervasive off-target activity of direct fusion dCas9-DNMT3A constructs. *Genome Res.* 28: 1193–1206.

Pompelli, M.F., Ferreira, P.P.B., Chaves, A.R.M. et al. (2021). Physiological, metabolic, and stomatal adjustments in response to salt stress in *Jatropha curcas*. *Plant Physiol. Biochem. PPB* 168: 116–127.

Prasad, A., Sharma, N., Muthamilarasan, M. et al. (2019). Recent advances in small RNA mediated plant-virus interactions. *Crit. Rev. Biotechnol.* 39: 587–601.

Priyanka, D.V., Reddy, M.T., Begum, H. et al. (2017). Genetic divergence analysis of inbred lines of okra [*Abelmoschus esculentus* (L.) Moench]. *Int. J. Curr. Microbiol. App. Sci.* 6: 379–388.

Qu, L.-J. and Qin, G. (2014). Generation and characterization of Arabidopsis T-DNA insertion mutants. *Methods Mol. Biol.* 1062: 241–258.

Qu, A.-L., Ding, Y.-F., Jiang, Q., and Zhu, C. (2013). Molecular mechanisms of the plant heat stress response. *Biochem. Biophys. Res. Commun.* 432: 203–207.

Rai, M.I., Wang, X., Thibault, D.M. et al. (2015). The ARGOS gene family functions in a negative feedback loop to desensitize plants to ethylene. *BMC Plant Biol.* 15: 157.

Rastogi, S., Shah, S., Kumar, R. et al. (2019). Ocimum metabolomics in response to abiotic stresses: cold, flood, drought and salinity. *PLoS One* 14: e0210903.

Raza, A., Charagh, S., Zahid, Z. et al. (2021a). Jasmonic acid: a key frontier in conferring abiotic stress tolerance in plants. *Plant Cell Rep.* 40: 1513–1541.

Raza, A., Razzaq, A., Mehmood, S.S. et al. (2021b). Omics: The way forward to enhance abiotic stress tolerance in *Brassica napus* L. *GM Crops Food* 12: 251–281.

Ren, Q., Sretenovic, S., Liu, S. et al. (2021a). PAM-less plant genome editing using a CRISPR-SpRY toolbox. *Nat. Plants* 7: 25–33.

Ren, Y., Huang, Z., Jiang, H. et al. (2021b). A heat stress responsive NAC transcription factor heterodimer plays key roles in rice grain filling. *J. Exp. Bot.* 72: 2947–2964.

Rouphael, Y., Cardarelli, M., Schwarz, D. et al. (2012). Effects of drought on nutrient uptake and assimilation in vegetable crops. In: *Plant Responses to Drought Stress* (ed. R. Aroca), 171–195. Springer.

Sahab, S., Suhani, I., Srivastava, V. et al. (2021). Potential risk assessment of soil salinity to agroecosystem sustainability: current status and management strategies. *Sci. Total Environ.* 764: 144164.

Saini, H.S. and Aspinall, D. (1982). Abnormal sporogenesis in wheat (*Triticum aestivum* L.) induced by short periods of high temperature. *Ann. Bot.* 49: 835–846.

Salt, D.E., Baxter, I., and Lahner, B. (2008). Ionomics and the study of the plant ionome. *Annu. Rev. Plant Biol.* 59: 709–733.

Sami, A., Xue, Z., Tazein, S. et al. (2021). CRISPR-Cas9-based genetic engineering for crop improvement under drought stress. *Bioengineered* 12: 5814–5829.

Sanchez, D.H., Pieckenstain, F.L., Escaray, F. et al. (2011). Comparative ionomics and metabolomics in extremophile and glycophytic Lotus species under salt stress challenge the metabolic pre-adaptation hypothesis. *Plant Cell Environ.* 34: 605–617.

Santosh Kumar, V.V., Verma, R.K., Yadav, S.K. et al. (2020). CRISPR-Cas9 mediated genome editing of drought and salt tolerance (OsDST) gene in indica mega rice cultivar MTU1010. *Physiol. Mol. Biol. Plants* 26: 1099–1110.

Saravanan, K. and Pachiyappan, R. (2013). Genetic divergence analysis in rice bean [*Vigna umbellate* (L.)]. *Int. J. Agric. Sci.* 8: 165–167.

Shao, H., Wang, H., and Tang, X. (2015). NAC transcription factors in plant multiple abiotic stress responses: progress and prospects. *Front. Plant Sci.* 6: 902.

Sharma, P., Jha, A.B., Dubey, R.S., and Pessarakli, M. (2012). Reactive oxygen species, oxidative damage, and antioxidative defense mechanism in plants under stressful conditions. *J. Bot.* 2012: 26.

Sharma, R., Upadhyay, S., Bhat, B. et al. (2020). Abiotic stress induced miRNA-TF-gene regulatory network: a structural perspective. *Genomics* 112: 412–422.

Sharoni, A.M., Nuruzzaman, M., Satoh, K. et al. (2011). Gene structures, classification and expression models of the AP2/EREBP transcription factor family in rice. *Plant Cell Physiol.* 52: 344–360.

Shi, J., Gao, H., Wang, H. et al. (2017). ARGOS8 variants generated by CRISPR-Cas9 improve maize grain yield under field drought stress conditions. *Plant Biotechnol. J.* 15: 207–216.

Shivhare, R. and Lata, C. (2017). Exploration of genetic and genomic resources for abiotic and biotic stress tolerance in pearl millet. *Front. Plant Sci.* 7: 2069.

Silveira, J.A.G. and Carvalho, F.E.L. (2016). Proteomics, photosynthesis and salt resistance in crops: an integrative view. *J. Proteome* 143: 24–35.

Singh, R. (2019). Plant molecular chaperones: structural organization and their roles in abiotic stress tolerance. In: *Molecular Plant Abiotic Stress: Biology and Biotechnology* (ed. A. Roychoudhury and D.K. Tripathi), 221. Hoboken, NJ: Wiley.

Singh, J., Sastry, E.V.D., and Singh, V. (2012). Effect of salinity on tomato (*Lycopersicon esculentum* Mill.) during seed germination stage. *Physiol. Mol. Biol. Plants* 18: 45–50.

Singh, M., Kumar, J., Singh, S. et al. (2015). Roles of osmoprotectants in improving salinity and drought tolerance in plants: a review. *Rev. Environ. Sci. Bio/Technol.* 14: 407–426.

Singh, R.K., Jaishankar, J., Muthamilarasan, M. et al. (2016). Genome-wide analysis of heat shock proteins in C_4 model, foxtail millet identifies potential candidates for crop improvement under abiotic stress. *Sci. Rep.* 6: 32641.

Singh, R.K., Prasad, A., Muthamilarasan, M. et al. (2020). Breeding and biotechnological interventions for trait improvement: status and prospects. *Planta* 252: 54.

Singh, R.K., Prasad, A., Maurya, J., and Prasad, M. (2021). Regulation of small RNA-mediated high temperature stress responses in crop plants. *Plant Cell Rep.* 41: 765–773. https://doi.org/10.1007/s00299-021-02745-x.

Sinha, R., Fritschi, F.B., Zandalinas, S.I., and Mittler, R. (2021). The impact of stress combination on reproductive processes in crops. *Plant Sci.* 311: 111007.

Sohrabi, Y., Heidari, G., Weisany, W. et al. (2012). Changes of antioxidative enzymes, lipid peroxidation and chlorophyll content in chickpea types colonized by different Glomus species under drought stress. *Symbiosis* 56: 18.

Soora, N.K., Aggarwal, P.K., Saxena, R. et al. (2013). An assessment of regional vulnerability of rice to climate change in India. *Clim. Chang.* 118: 683–699.

Sorek, R., Lawrence, C.M., and Wiedenheft, B. (2013). CRISPR-mediated adaptive immune systems in bacteria and archaea. *Annu. Rev. Biochem.* 82: 237–266.

Surender Reddy, P., Jogeswar, G., Rasineni, G.K. et al. (2015). Proline over-accumulation alleviates salt stress and protects photosynthetic and antioxidant enzyme activities in transgenic sorghum [*Sorghum bicolor* (L.) Moench]. *Plant Physiol. Biochem.* 94: 104–113.

Teixeira, E.I., de Ruiter, J., Ausseil, A.-G. et al. (2018). Adapting crop rotations to climate change in regional impact modelling assessments. *Sci. Total Environ.* 616–617: 785–795.

Thompson, M., Gamage, D., Hirotsu, N. et al. (2017). Effects of elevated carbon dioxide on photosynthesis and carbon partitioning: a perspective on root sugar sensing and hormonal crosstalk. *Front. Physiol.* 8: 578.

Timmons, L. and Fire, A. (1998). Specific interference by ingested dsRNA. *Nature* 395: 854.

Tomimatsu, H., Iio, A., Adachi, M. et al. (2014). High CO_2 concentration increases relative leaf carbon gain under dynamic light in *Dipterocarpus sublamellatus* seedlings in a tropical rain forest, Malaysia. *Tree Physiol.* 34: 944–954.

Trishla, V.S. and Kirti, P.B. (2021). Structure-function relationship of *Gossypium hirsutum* NAC transcription factor, GhNAC4 with regard to ABA and abiotic stress responses. *Plant Sci.* 302: 110718.

Tuteja, N. (2007). Abscisic acid and abiotic stress signaling. *Plant Signal. Behav.* 2: 135–138.

Usman, B., Nawaz, G., Zhao, N. et al. (2020). Precise editing of the OsPYL9 gene by RNA-guided Cas9 nuclease confers enhanced drought tolerance and grain yield in rice (*Oryza sativa* L.) by regulating circadian rhythm and abiotic stress responsive proteins. *Int. J. Mol. Sci.* 21: E7854.

Venugopalan, M.V., Sankaranarayanan, K., Blaise, D. et al. (2009). Bt cotton (*Gossypium* sp.) in India and its agronomic requirements–a review. *Indian J. Agron.* 54: 343–360.

Verma, G., Srivastava, D., Tiwari, P., and Chakrabarty, D. (2019). ROS modulation in crop plants under drought stress. In: *Reactive Oxygen, Nitrogen and Sulfur Species in Plants: Production, Metabolism, Signaling and Defense Mechanisms* (ed. M. Hasanuzzaman, V. Fotopoulos, K. Nahar, and M. Fujita), 311–336. Wiley.

Verslues, P.E. and Zhu, J.-K. (2005). Before and beyond ABA: upstream sensing and internal signals that determine ABA accumulation and response under abiotic stress. *Biochem. Soc. Trans.* 33: 375–379.

Verslues, P.E., Lasky, J.R., Juenger, T.E. et al. (2014). Genome-wide association mapping combined with reverse genetics identifies new effectors of low water potential-induced proline accumulation in Arabidopsis. *Plant Physiol.* 164: 144–159.

Vurukonda, S.S.K.P., Vardharajula, S., Shrivastava, M., and SkZ, A. (2016). Enhancement of drought stress tolerance in crops by plant growth promoting rhizobacteria. *Microbiol. Res.* 184: 13–24.

Wang, A., Yu, X., Mao, Y. et al. (2015). Overexpression of a small heat-shock-protein gene enhances tolerance to abiotic stresses in rice. *Plant Breed.* 134: 384–393.

Wang, G., Xu, X., Wang, H. et al. (2019a). A tomato transcription factor, SlDREB3 enhances the tolerance to chilling in transgenic tomato. *Plant Physiol. Biochem. PPB* 142: 254–262.

Wang, J., Sun, H., Sheng, J. et al. (2019b). Transcriptome, physiological and biochemical analysis of *Triarrhena sacchariflora* in response to flooding stress. *BMC Genet.* 20: 88.

Wang, X., Liu, H., Yu, F. et al. (2019c). Differential activity of the antioxidant defence system and alterations in the accumulation of osmolyte and reactive oxygen species under drought stress and recovery in rice (*Oryza sativa* L.) tillering. *Sci. Rep.* 9: 8543.

Wang, J., Song, L., Gong, X. et al. (2020a). Functions of jasmonic acid in plant regulation and response to abiotic stress. *Int. J. Mol. Sci.* 21: 1446.

Wang, W., Zhou, B., He, J. et al. (2020b). Comprehensive identification of drought tolerance QTL-allele and candidate gene systems in Chinese cultivated soybean population. *Int. J. Mol. Sci.* 21: 4830.

Wang, D., Liu, Z., Xiao, Y. et al. (2021). Association mapping and functional analysis of rice cold tolerance QTLs at the bud burst stage. *Rice* 14: 98.

Wassenegger, M. (2005). The role of the RNAi machinery in heterochromatin formation. *Cell* 122: 13–16.

Watson, A., Ghosh, S., Williams, M.J. et al. (2018). Speed breeding is a powerful tool to accelerate crop research and breeding. *Nat. Plants* 4: 23–29.

Weston, D.J., Gunter, L.E., Rogers, A., and Wullschleger, S.D. (2008). Connecting genes, coexpression modules, and molecular signatures to environmental stress phenotypes in plants. *BMC Syst. Biol.* 2: 16.

Wu, J., Gao, T., Hu, J. et al. (2022). Research advances in function and regulation mechanisms of plant small heat shock proteins (sHSPs) under environmental stresses. *Sci. Total Environ.* 825: 154054.

Xu, C., Luo, F., and Hochholdinger, F. (2016). LOB domain proteins: beyond lateral organ boundaries. *Trends Plant Sci.* 21: 159–167.

Yadav, A., Singh, J., Ranjan, K. et al. (2020). Heat shock proteins: master players for heat-stress tolerance in plants during climate change. In: *Heat Stress Tolerance in Plants: Physiological, Molecular and Genetic Perspectives* (ed. S.H. Wani and V. Kumar), 189–211. Hoboken, NJ: Wiley.

Yan, M., Zheng, L., Li, B. et al. (2021). Comparative proteomics reveals new insights into the endosperm responses to drought, salinity and submergence in germinating wheat seeds. *Plant Mol. Biol.* 105: 287–302.

Yang, J., Li, D., Liu, H. et al. (2020a). Identification of QTLs involved in cold tolerance during the germination and bud stages of rice (*Oryza sativa* L.) via a high-density genetic map. *Breed. Sci.* 70: 292–302.

Yang, Y., Al-Baidhani, H.H.J., Harris, J. et al. (2020b). DREB/CBF expression in wheat and barley using the stress-inducible promoters of HD-Zip I genes: impact on plant development, stress tolerance and yield. *Plant Biotechnol. J.* 18: 829–844.

Yu, W., Wang, L., Zhao, R. et al. (2019a). Knockout of SlMAPK3 enhances tolerance to heat stress involving ROS homeostasis in tomato plants. *BMC Plant Biol.* 19: 354.

Yu, Z., Wang, X., Mu, X., and Zhang, L. (2019b). RNAi mediated silencing of dehydrin gene WZY2 confers osmotic stress intolerance in transgenic wheat. *Funct. Plant Biol.* 46: 877–884.

Yuan, S. and Lin, H.-H. (2008). Minireview: role of salicylic acid in plant abiotic stress. *Z. Naturforsch. C J. Biosci.* 63: 313–320.

Yuan, X. and Wen, B. (2018). Seed germination response to high temperature and water stress in three invasive Asteraceae weeds from Xishuangbanna, SW China. *PLoS One* 13: e0191710.

Zandalinas, S.I., Fritschi, F.B., and Mittler, R. (2021). Global warming, climate change, and environmental pollution: recipe for a multifactorial stress combination disaster. *Trends Plant Sci.* 26: 588–599.

Zandalinas, S.I., Balfagón, D., Gómez-Cadenas, A., and Mittler, R. (2022). Plant responses to climate change: metabolic changes under combined abiotic stresses. *J. Exp. Bot.* 73: 3339–3354. https://doi.org/10.1093/jxb/erac073.

Zargar, S.M., Gupta, N., Nazir, M. et al. (2017). Impact of drought on photosynthesis: molecular perspective. *Plant Gene* 11: 154–159.

van Zelm, E., Zhang, Y., and Testerink, C. (2020). Salt tolerance mechanisms of plants. *Annu. Rev. Plant Biol.* 71: 403–433.

Zhang, X., Wang, W., Zhu, W. et al. (2019). Mechanisms and functions of long non-coding RNAs at multiple regulatory levels. *Int. J. Mol. Sci.* 20: 5573.

Zhang, N., Zhao, H., Shi, J. et al. (2020). Functional characterization of class I SlHSP17.7 gene responsible for tomato cold-stress tolerance. *Plant Sci.* 298: 110568.

Zhao, L., Hu, Y., Chong, K., and Wang, T. (2010). ARAG1, an ABA-responsive DREB gene, plays a role in seed germination and drought tolerance of rice. *Ann. Bot.* 105: 401–409.

Zhao, C., Zhang, Z., Xie, S. et al. (2016). Mutational evidence for the critical role of CBF transcription factors in cold acclimation in Arabidopsis. *Plant Physiol.* 171: 2744–2759.

Zhao, C., Zhang, Y., Du, J. et al. (2019a). Crop phenomics: current status and perspectives. *Front. Plant Sci.* 10: 714.

Zhao, Y., Xing, H., Li, X. et al. (2019b). Physiological and metabolomics analyses reveal the roles of fulvic acid in enhancing the production of astaxanthin and lipids in *Haematococcus pluvialis* under abiotic stress conditions. *J. Agric. Food Chem.* 67: 12599–12609.

Zhao, C., Zhang, H., Song, C. et al. (2020). Mechanisms of plant responses and adaptation to soil salinity. *Innovation (New York)* 1: 100017.

Zhou, Y., Zhu, H., He, S. et al. (2019). A novel sweetpotato transcription factor gene IbMYB116 enhances drought tolerance in transgenic Arabidopsis. *Front. Plant Sci.* 10: 1025.

Zlatev, Z. and Lidon, F.C. (2012). An overview on drought induced changes in plant growth, water relations and photosynthesis. *Emirates J. Food Agric.* 24: 57–72.

18

Receptor-Like Kinases and ROS Signaling

Critical Arms of Plant Response to Stress

Samir Sharma

Department of Biochemistry, University of Lucknow, Lucknow, India

18.1 Preamble

Climate change has occurred with increasing rapidity recently, not giving life forms on Earth time to adapt. While climate change is primarily manifest as a rise in temperature, the sequelae of this rise are changes in wind currents and rain patterns, causing seasons to change at unusual times of the year. All these changes upset plant–environment coordination. These effect plant life deleteriously and bring about a marked reduction in yield and, at times, failure to reproduce or even survive. What is even worse is that when plants are compromised by climate change induced abiotic stresses, they become more susceptible to bacterial and fungal infections, and biotic stress further compounds abiotic stress. Stresses drastically reduce productivity and, at times, threaten the very existence of plant types endemic to a particular region. For this reason, stresses and plants' response to them have become a subject of intense scientific interest and activity. The present chapter endeavors to give a composite view of the molecules and mechanisms used by plants to sense stress and the response this sensing elicits.

Phosphorylation is the most common way in which all organisms control metabolic activity. Metabolites, as well as the enzymes and effectors that control them, frequently undergo reversible phosphorylation. Kinases of various descriptions and classes phosphorylate metabolites or proteins, usually in a highly regulated manner. Conversely, the phosphatases reverse these phosphorylation events and are generally constitutively active and relatively infrequently regulated. Separately, receptor–ligand interactions form the backbone of intracellular and intercellular communication in all organisms. A combination of protein kinases and receptors has slowly evolved over time. Natural selection has strongly favored the evolution of protein molecules with the characteristics of receptors and kinases as "receptor kinases." On the other hand, reactive oxygen species (ROS) are the inevitable outcome of reactions or pathways involving electron transport. Their inherent ability to react with and covalently modify proteins and nucleic acids and generate volatile lipid signals makes ROS critical signaling entities if their concentration remains within their safe, dynamic range. Receptor kinases abound in the microbial, animal, and plant world and have vital roles in cellular communication. Much as the simplistic view of these receptor kinases would lead us to believe, it is hard to classify all kinases as receptor kinases, despite their roles in sensing and responding to cellular conditions through phosphorylating activity. Receptor-like cytoplasmic kinases (RLCKs) tend to act like signaling buffers, parsing signals received from the changing concentration of small molecules and stabilizing them as relatively longer-lasting phosphorylation states. They lack an extracellular domain and act through a system of kinase relays to affect signal transduction and changes in the physiological state. Abiotic stresses directly or indirectly interfere with membrane integrity and molecular function, causing significant disruption of redox processes and the generation of ROS. Very recently, there has been a large body of information that implicates both ROS and protein kinase function in cellular regulation ranging from altered metabolic flux to cell proliferation or cell death.

Global Climate Change and Plant Stress Management, First Edition. Edited by Mohammad Wahid Ansari, Anil Kumar Singh, and Narendra Tuteja.

18.2 Climate Change: The Agent of Stress

The fact that the climate of Earth has changed significantly is as clearly visible as the reasons of this change are obscure (Zandalinas et al. 2021). There are two principal lines of thought. One of these squarely blames greenhouse gas emissions for global warming and the associated geochemical changes. The other argues that climate change is a part of the long-term changes, associated with Earth's continuing changing relationship with the sun, resulting in such things as the Milankovitch Cycles (Campisano 2012). Regardless of the cause bringing about these changes, it is a fact that our planet is experiencing change in climate, mainly by way of a slow but steady rise in temperature, commonly referred to as global warming. This gradual rise in temperature triggers a domino effect, causing floods, tropical storms, and long periods of drought. In future, warming could possibly change Earth's rotation due to increased volume of seas caused by reallocation of water distribution because of glacier melting. Catastrophic as these changes may seem in the long run, on a much shorter timescale, they are causing small but steady and significant changes to plant life also. Plants have evolved over a period of almost three billion years. Their inherent genetic variability as well as cross-pollination have led to great diversification and adaptation. However, climate changes occurring with great rapidity over the past century have upset the delicate equilibrium that plants have attained with their surroundings (Chaudhry and Sidhu 2022). Significant climatic variation has already been witnessed all over the world, and even more severe changes are predicted in the coming years. Thus, the challenge is two-fold. First, to increase productivity to cater to the express needs of an ever-increasing population, and second, to accomplish this under the stress conditions brought about by climate change. The toughest thing about dealing with climate change is that it cannot be combated with agents that may suffice against biotic stresses.

18.3 Abiotic Stress: A Severe Threat by Itself and a Window of Opportunity for Biotic Stress Agents

Abiotic stresses, which include drought, salinity, heat, cold, nutrient deficiencies, and the presence of heavy metal, affect crops negatively and are of major concern to us due to the drastic yield reductions they bring about (Jewell et al. 2010; Ma et al. 2021). Being sessile, plants can either adapt to stresses and tide them over or succumb. In every species, there occur some cultivars that appear to be less susceptible to stresses than others (Cramer et al. 2011). Still others adopt a slightly different strategy and become tolerant to these stresses that lower the yield somewhat but allow the tolerant plants to grow. The major components of this adaptive process are extensive metabolic and biochemical modifications that lead to altered physiology and, at times, even morphological adaptation (Janicka-Russak et al. 2008). There is a fundamental difference between plants' responses to biotic stresses and abiotic stresses. Plants have an enemy that can be counterattacked and even subdued, as far as biotic stresses are concerned. This is just not possible in the case of abiotic stresses. The only way out for plants is alterations in their metabolism and, at times, even in their morphology. However, unlike biotic stresses, abiotic stresses may be temporary, and non-restrictive conditions may be restored within a reasonable span of time. Adaptations to abiotic stresses are marked by a one-way interaction between plant and stressor, unlike biotic stress, in which this interaction is two-way, with both organisms reacting to each other. If the quantum of abiotic stress is too high or prolonged, the strategies of adaptation employed by plants are not likely to succeed, negatively impacting survival or reproduction and transition into the next generation. While certain habitats may be poor in nutrients or rich in toxic elements, they rarely are the sites of lush vegetation. Gradual decline in nutrients or increase in heavy metals give plants an opportunity to respond. There are, however, situations where a primary stress conditions causes the emergence of a secondary one. A prime example is drought, which is usually associated with intense light stress. Nutrient uptake is always carried out from a solution of these elements in the soil, and poor water availability makes it difficult for the plant to import nutrient elements. This adds nutrient stress to drought and light stress and compounds the whole stress situation. Biotic stresses find a brief mention in the present chapter because a plant stressed by abiotic agents is easy prey for biotic stresses that would otherwise have failed to breach its passive defenses.

Like any other stimulus, stress needs to be perceived, interpreted as a signal, and transduced into a chemical signal that can elicit a physiological and, quite often, a morphogenetic response. This requires a system of sensors and transducers and a response system that can react to the stress by altering the physiology of the plant, perhaps also employing morphogenesis as a tool in this response. The present chapter will briefly deal with the general description of the response system that reacts to stress, both abiotic and biotic, and in brief the response to the stress condition.

18.4 Plant Receptor-Like Kinases (RLKs)

Transmembrane receptors abound in all organisms. However, there is a considerable difference between animals and plants when it comes to receptor-linked kinases. The common factor in signaling systems across all species is their adherence to the "two-component system" of signaling (Urao et al. 2000). Perceiving and responding to environmental stimuli and growth regulators are perhaps the most important roles of two-component systems in plants. In plants, almost all major signaling pathways involve transmembrane signaling. The perception of several signaling entities such as peptides, plant hormones, nucleotides, oligo and polysaccharides, ion species, and other small molecule effectors and the subsequent signal transduction inside the target cell is mediated by cell surface receptors. Signal transduction results in altering the physiological response or bringing about developmental changes in the target tissue and, ultimately, the plant. The signaling molecules and ions mentioned here connect biotic and abiotic stimuli to plant responses and ensure survival and evolutionary success. Post-translational modification of the phosphorylation status of a downstream signaling partner alters its reactivity in a signaling pathway. The signaling pathways are reversibly activated or inactivated because of this phosphorylation (Lehti-Shiu et al. 2012). The eukaryotic protein kinases (ePKs), not all of which are found in plants, responsible for this are broadly divided into several families of kinases. The more significant of these are the mitogen-activated protein kinases (MAPKs), receptor-like kinases (RLKs), glycogen synthase kinases (GSKs), NIMA-related kinases (NEKs), calcium-dependent protein kinases (CDPKs). This list, though far from complete, includes most of the ePKs currently known. Every class in the list is characterized by specific structural and functional attributes (Lehti-Shiu and Shiu 2012).

The plant RLKs are majorly implicated in plant–microbe relationships ranging from the establishment of symbiosis to adverse relationships like plant–pathogen interactions (Busch and Benfey 2010; Van Norman et al. 2011). In this way, the membrane serves to separate the external and internal milieu, while the RLKs serve to convey extracellular messages to elicit a physiological response Zhu 2016; Zipfel and Oldroyd 2017). Apart from plant–microbe interactions and perceiving molecular messages (Couto and Zipfel 2016; Jones and Dangl 2006), RLKs control intercellular communication and regulate events such as fertilization, self-incompatibility, meristem maintenance, cell differentiation, and expansion. RLKs and receptor-like proteins (RLPs) without kinase activity, located on the cell surface of plant cells, form most receptors (Tang et al. 2017).

The RLKs follow a relatively conserved architectural pattern in possessing an extracellular or "Ecto" domain, responsible for ligand-binding, linked to the single-pass transmembrane (TM) domain. The kinase domain of the RLK projects into the cytoplasmic domain and is responsible for relaying the extracellular signal intracellularly. Recent studies regarding RLK structure have addressed the "Juxtamembrane" or JM domain (Aifa et al. 2005; Thiel and Carpenter 2007), which is sandwiched between the transmembrane and the kinase domains. This domain was hard to observe in crystals and was assigned the relatively mundane role of acting as a linker between the TM and kinase domains. However, studies in animal systems have revealed that these domains control activities like dimerization and activation of kinase activity, calmodulin binding (Hedger et al. 2015; Martín-Nieto and Villalobo 1998) and are found to contain trafficking signal motifs (Lemmon and Schlessinger 2010). It is important to clarify that the JM domains have different roles in regulating kinase activation states in different receptors. In several receptors, they are known to cause autoinhibition of the kinase domain, while in others they cause activation (Oh et al. 2012). It is vital to remember that the classification of a putative kinase as an RLK is based on sequence homology and structural similarity. Some of these RLKs have been proven to possess ligand-binding induced kinase activity and have therefore been confirmed as Receptor Kinases (RKs). At present, the number of kinases with clearly defined ligand-induced kinase activity (RKs) is relatively small with respect to RLKs (Dievart et al. 2020). The structure of a typical animal Receptor Tyrosine Kinase (RTK) along with the function associated with various domains is shown in Figure 18.1.

RLKs follow the structural principles and functional relations of domains found in mammalian RTKs. Like animal RTKs, the juxtamembrane domains of RLKs do have important roles in regulating kinase activity (Griffith et al. 2004; Brewer et al. 2009). The RLPs differ from the RLKs in possessing a ligand binding domain, a transmembrane domain, or

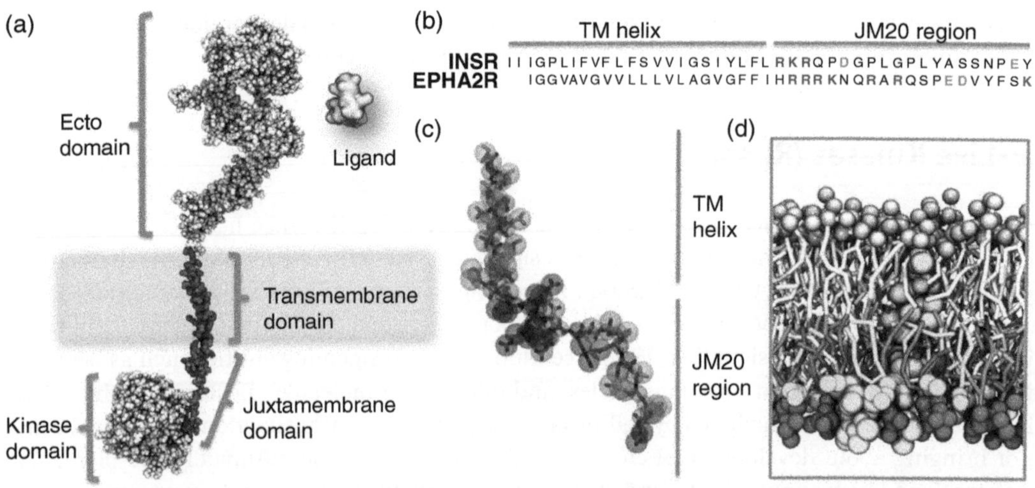

Figure 18.1 RTK architecture includes a JM region enriched in basic residues. (a) Schematic representation of a generic RTK model (generated using MODELLER based on PDB id: 2GS6, 3GOP, 2JIU 2M20, and 3NJP) with an extracellular ligand binding ectodomain, a single transmembrane spanning helix, a basic JM region, and a kinase domain. (b) Sequence of INSR and the EphA2 receptor with charged residues within the JM region. (c) Protein model of the TM-JM region of the EphA2 receptor with CG beads overlaid onto an atomistic model. (d) Cross-section of the same EphA2 receptor TM-JM model embedded in a lipid bilayer containing 20% PS and 20% PIP_2 within the inner leaflet (Figure courtesy Hedger et al. 2015).

a glycosylphosphatidylinositol anchor but a complete absence of the JM and kinase domains. Close to or more than a thousand RLKs have been reported from the Arabidopsis and Rice genomes, while more than a hundred RLPs have been reported from the same plants. The reason why the present review chooses to dwell upon the RLKs in general and their JM domain more specifically is that the JM domain is considered a point of interaction between the RLKs and their interacting RLCKs. In fact, as will be considered in this chapter, RLCKs interact with the cytosolic kinase or juxtamembrane domains of the RLKs and act as transducers for relaying the signal further downstream. They also act to amplify the ambit of phosphorylation-mediated control of metabolism by interacting with a far greater number of substrates than the RLKs can directly modify. Some of the RLKs found in plants are given in Table 18.1, while domain structure of *Arabidopsis* RLKs is shown in Figure 18.2.

These are "dual-specificity kinases" that possess conserved motifs of both Ser/Thr kinases as well as Tyr kinases and are therefore efficient at phosphorylating serine/threonine as well as tyrosine residues (Oh et al. 2009; Madsen et al. 2011). Histidine-Arginine-Aspartate (HRD), Valine-Alanine-Isoleucine-Lysine (VAIK), and Aspartate-Phenylalanine-Glycine (DFG) are three conserved motifs that are discovered to be present in their cytoplasmic domains and that allow them to be classified as members of the kinase family. RLKs in both plants and animals have similar downstream targets, such as ROS and MAPKs, aside from their similar functions. Furthermore, the desensitization pathways involve ubiquitination and endocytosis and are common between animal and plant RLKs. Interestingly, RLKs would have been universal in their distributions, were it not for the fungi that do not possess these kinases (Turra et al. 2014; Liang and Zhou 2018).

Table 18.1 Some representative receptor like kinases found in plants (Jose et al. 2020).

	Name	Description
1	NFR1	Nod factor receptor 1
2	BRI	Brassinosteroid insensitive 1
3	PRK1	Pollen-expressed receptor kinase 1
4	SERK1	Somatic embryogenesis receptor kinase 1
5	HAESA	Bak1-like kinase 1

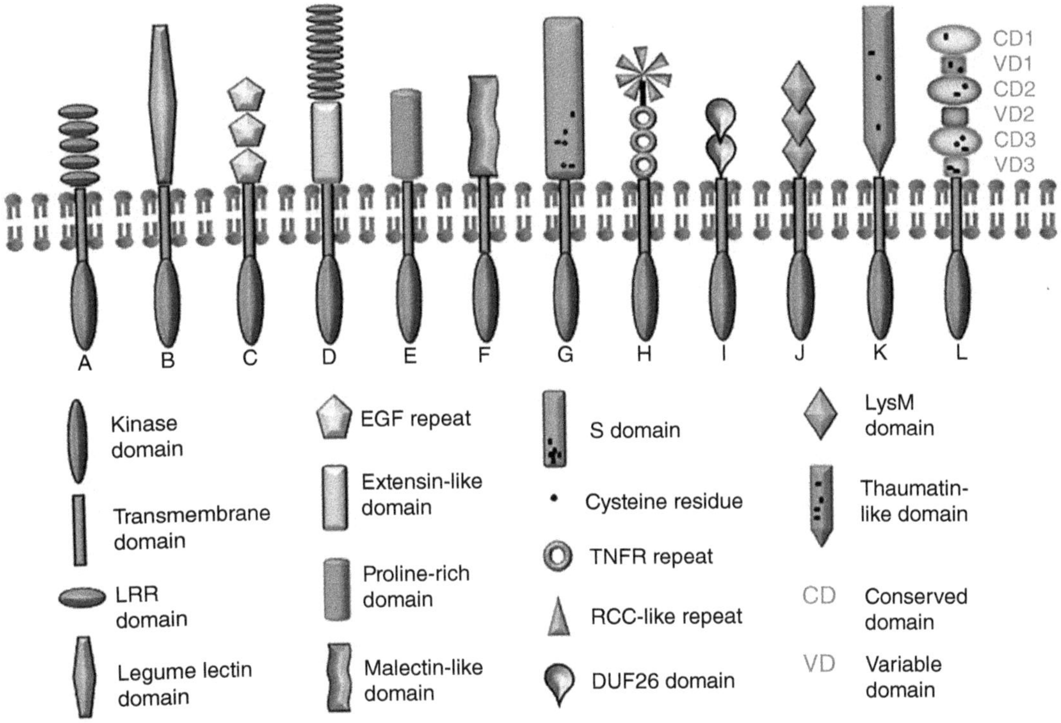

Figure 18.2 Domain architecture of Arabidopsis RLKs. (a) SERK (LRR), (b) LecRK1 (Lectin), (c) WAK1 (WAK), (d) LRX1 (Extensin + LRR), (e) PERK4 (PERK), (f) FER (CrRLK), (g) AtS1 (S-domain), (h) ACR4 (CR-like), (i) CRK (DUF26), (j) AtCERK1 (LysM), (k) PR5K (Thaumatin), (l) LRK10L1.2 (LRK). RCC, regulator of chromosome condensation (Jose et al. 2020).

18.5 Receptor-Like Cytosolic Kinases

RLCKs are a superfamily comprising several structurally similar proteins with kinase activity (Liang and Zhou 2018). These RLCKs form an interface between development- and defense-related signals and the cellular response to these signals (Busch and Benfey 2010). Among the development-related signals are the hardwired triggers, hormonal influences, and light-triggered responses, while defense-related responses are related to pathogen attack-related signals (Goff and Ramonell 2007). RLCKs do not possess an extracellular ligand-binding domain. They constitute a significant class of signaling proteins regulating the plant's physiological responses to stress and signaling molecules. They engage in crosstalk with the plant defense or immune receptor kinases (RKs) and participate in multiple signaling schemes to coordinate defense against pathogenic attacks. They also regulate growth, embryonic patterning, pollen tube guidance, abscission, and response to abiotic stresses along with RLKs. Significant among these are the RLKs that respond to Brassinosteroids and signaling peptides. Apart from the RLKs, which are the primary regulators of RLCKs, several other interacting proteins modify RLCK action significantly. Put together, the system comprising RLKs, RLCKs, and accessory proteins represents a system of phosphorylation relays that serves to regulate plant responses to signals of internal and external origin (Van Norman et al. 2011). However, it is pertinent to caution the reader to note the distinction between RLKs and RLCKs. The terms will be used frequently and are not interchangeable. The RLCKs constitute a robust signaling system in plants and are usually interacting partners in a complex signaling network.

The extracellular and transmembrane domains that define the RLK family of proteins are absent from the RLCKs (Shiu et al. 2004). RLCKs are distinguished by a serine/threonine protein kinase domain that exhibits as many as 11 highly conserved kinase subdomains numbered I to XI. I addition, they have short N- and C-terminal regions (Hanks and Hunter 1995). Studies (Shiu and Bleecker 2001; Vij et al. 2008) have established 147 RLCKs in *Arabidopsis* and as many as 379 RLCK members in rice. Structural analysis predicts RLCKs to be anchored to the plasma membrane via post-translational N-terminal myristoylation/palmitoylation motifs (Veronese et al. 2006; Tang et al. 2008) that guide these proteins to be myristoylated/palmitoylated via the N-terminal Gly residue (Podell and Gribskov 2004). Myristoyl/palmitoyl anchoring of these RLCKs to the inner leaflet of the plasma membrane enables size economy in these proteins and allows them to be accessed by other PM-resident RLCKs and also numerous other membrane and cytosolic proteins (Figure 18.3).

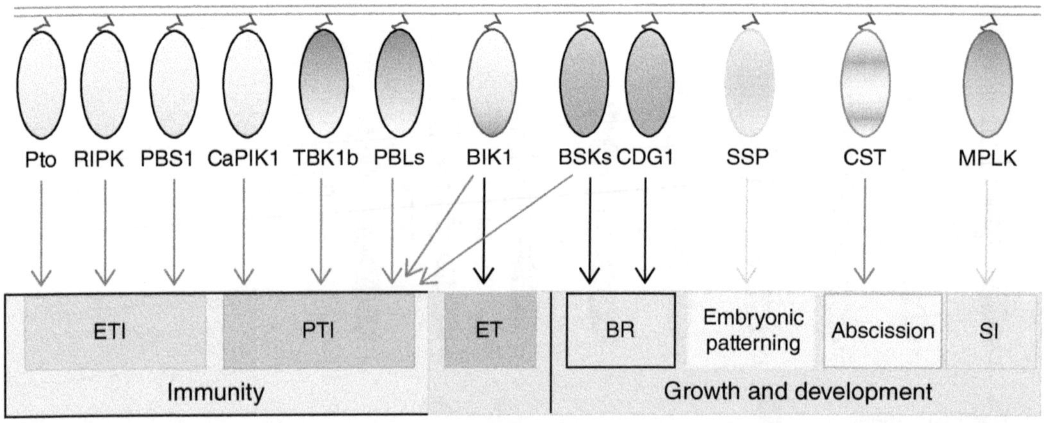

Figure 18.3 RLCKs that are potentially anchored to the plasma membrane through N-terminal putative myristoylation and/or palmitoylation motifs. Pto, RPM1-induced protein kinase (RIPK), and PBS1 are involved in effector-triggered immunity (ETI); capsicum annuum receptor-like cytoplasmic protein kinase 1 (CaPIK1), tomato BIK1 ortholog tomato protein kinase 1b (TBK1b), and PBLs are involved in PAMP-triggered immunity (PTI); BIK1 is involved in PTI and ET signaling; brassinosteroid signaling kinases (BSKs) are involved in PTI; and BR signaling; constitutive differential growth 1 (CDG1) is involved in BR signaling; short suspensor (SSP) is involved in embryonic patterning; cast away (CST) is involved in abscission; and M locus protein kinase (MPLK) is involved in self-incompatibility (SI) (Lin et al. 2013).

RLCKs are often found to associate with RLKs to relay the signals originating in extracellular cues perceived by RLKs (Yamaguchi et al. 2013). It is apparent that these cytosolic kinases are crowded in close proximity to RLKs and other signaling elements and serve to translate, modulate, potentiate, or perhaps even attenuate their actions so that the most appropriate cytosolic (physiological) response can be ensured in response to an extracellular stimulus. These post-translational modifications to the cytosolic kinases also ensure a high rate of mobility to support the interactions with their upstream signaling partners (Salaun et al. 2010).

18.6 Why Are Receptor-Like Cytosolic Kinases Needed?

RLP-type receptors lack a cytosolic domain and are functionally reminiscent of G-protein coupled receptors, where the two-box model of the receptor input and signaling output (transducer) domains is divided into two sets of proteins, represented by the GPCRs and their cognate heterotrimeric G-proteins. Lately, there have been changes in the way we view these GPCRs as switches that exist in the two alternative states of on and off. Recent structural studies of GPCR-transducer complexes have forced us to view the GPCR-G protein combine as multistate allosteric microprocessors, in which biased ligands transmit distinct structural information, which is then converted to distinct physiological responses (Smith et al. 2018). By extension, since certain RLP-type receptors lack a cytoplasmic catalytic domain, the mind wanders towards a GPCR-like situation where the receptor and transducer are two different molecules and are likely to be present in close proximity to facilitate the rapid transfer of the transduced signal to the cytosol and elicit a physiological response. In most other cases, the huge diversity of RLKs that possess cytoplasmic kinase domains requires their actions to be relayed via *trans*-phosphorylation events involving RLCKs (Lin et al. 2013). The serine/threonine protein kinase domain of RLCKs has 11 highly conserved kinase subdomains (I to XI), as well as brief N- and C-terminal region. (Hanks and Hunter 1995). These characteristics enable the RLCKs to serve as an apt foil to the RLKs and as signaling partners to the RLPs that do not have kinase activity of their own.

The sessile nature of plants has forced several adaptations on them. Plants with these adaptations have passed through the brutal filter of natural selection and have reached the stage they are in today. Broadly, these can be classified as development and defense. Unlike animals, plants do follow a genetically imprinted or hardwired pattern of development, but the extent of that development fits in a very broad window. For example, the branching pattern, arrangement of leaves in a plant are quite well defined, but the number, length, and spread of branches are not. Thus, while qualitatively, the traits are genetically imprinted, their quantitative expression is quite variable. These differences are governed to a very large extent by the environmental conditions prevailing in the region where the plant is growing and are dependent strictly on the perception of environmental conditions by the plant and its reaction to those conditions. For this, it needs environment sensors and the downstream signaling cascades capable of converting the sensory perception into a physiological, and ultimately a developmental response. This process and the components responsible for it help the plant optimize its development to best suit the prevailing environmental conditions. These adaptations are seen as extent and pattern of root and shoot development, ratio of lateral branching to apical growth, and overall canopy development. Summarily, the collection of these developmental responses can be termed as plants response to its abiotic environment and the genetic cues hard-wired into its genome. Secondly, plants are deeply concerned with defending themselves against pathogens and other organisms that might physically harm them. This involves a set of responses no less demanding of a system of sensors and responders than developmental cues, and it enables the plant to survive amidst pathogens and other harmful beings that cohabit its space at different times. Appropriate and timely responses to pathogens is the only other set of characteristics apart from developmental cues that have allowed plants to pass through the filter of natural selection and diversify as they have don't though almost a billion years from when the first multicellular plant arose. While this system of plant defense, frequently referred to as its immune system bears little resemblance to is counterpart in animals, it definitely is a very strong defense and complements the physical and mechanical barriers of plant cells. Together, the so called "innate immune system" of plants, is effectively used to repel or even exterminate completely the microbial threats encountered by plants (Lin et al. 2013). The relay and amplification system comprising RLKs and RLCKs is vital to the developmental as well as defense responses exhibited by the plants and will be discussed separately.

18.7 Receptor-Like Cytosolic Kinases in Plant Defense

As mentioned before, plants do not possess a sophisticated adaptive immune system, but they do possess a sensory system for the perception of adversity or stress. In this way, their defense system resembles the innate immunity system of higher animals. Since stress presents itself in many forms, with each form having numerous biological manifestations, having numerous sensory-response systems is a tremendous advantage. It is possibly why plants possess a huge number of genes representing the similarly massive family of RLKs and their corresponding RLCK counterparts. Microbial invasions represent a major and perhaps the commonest part of biotic stress. As microbes invade and multiply in an attempt to cause a successful invasion of plants, they generate characteristic molecules during the process of division. Every pathogen is associated with a combination of these molecules that is a signature of the presence of that pathogen or class of pathogens and are termed microbe- or pathogen-associated molecular patterns (MAMPs or PAMPs) (Boller and Felix 2009). The huge diversity of RLKs helps the MAMPs be detected precisely. This detection and the processes downstream of it have been naturally selected during a process of co-evolution of the host and its pathogen, which leads to a very interesting variation in this interaction. MAMP recognition leads to conformational change in the RLKs, followed frequently by homo- or hetero-dimerization mediated activation in a process somewhat similar to mammalian receptor tyrosine kinases (Shiu and Bleecker 2001; Antolin-Llovera et al. 2012).

The chain of plant defense reactions triggered after perceiving MAMPs or PAMPs is termed PAMP-triggered immunity (PTI). As innate immunity would function in higher animals, PTI comprises a set of physiological responses that form the foundations of plant defense and frequently manages to avert a broad spectrum of microbial invasions. Elicitation of PTI is manifested through a major change in transcriptional programming, production of ROS, and reinforcing of the cell wall through the synthesis of callose and stomatal closure to block pathogen entry (Boller and Felix 2009). The process of co-evolution has equipped the pathogens with various virulence mechanisms to dampen the host immune systems or interfere with host physiological and cellular responses. These include injecting a repertoire of virulence effector proteins into host cells through the type III secretion system (TTSS) of Gram-negative bacteria. These proteins successfully prevent the immune system from responding effectively (Shan et al. 2007; Feng and Zhou 2012). Certain plants manage to directly or indirectly counter the injected virulence effector proteins through the action of evolutionarily acquired polymorphic intracellular receptors termed as disease resistance proteins (R proteins). This second line of plant inducible defense is defined as effector-triggered immunity (ETI) (Spoel and Dong 2012). In this context, different RLCKs are known to be involved in this plant-microbe contest for superiority. In various plant species, including Arabidopsis, rice, and tomato, RLCKs are implicated in plant immunological responses (Veronese et al. 2006; AbuQamar et al. 2008; Vij et al. 2008; Ao et al. 2014). These investigations offer proof that BIK1 plays a crucial part in plant innate immune responses. Some illustrative examples of the role of RLCKs in plant defense are listed in Table 18.2.

Table 18.2 Some representative receptor-like cytosolic kinases in plants and their functions.

	Species	Receptor-like cytosolic kinase	Description	Reference
1.	Tomato	Tomato protein kinase 1b (TPK1b)	Plasma membrane localized RLCK. Provides resistance against pathogen *B. cinerea* and chewing insect tobacco hornworm	AbuQamar et al. (2008)
2.	Rice	OsRLCK185	Phosphorylated by PAMP-receptor OsCERK1. Downregulation causes MAP kinase activation and downregulation of chitin-induced genes PBZ1 and PAL2. OsRLCK185/OsCERK1 interaction suppressed by blight-causing *Xanthomonas oryzae* effector Xoo1488, suggesting essentiality of PAMP-RLCK for OTI	Yamaguchi et al. (2013), Wang et al. (2017)
3.	Rice	OsRLCK176	OsRLCK176/OsCERK1 interaction required for chitin-induced ROS production	Ao et al. (2014)
4.	Maize	ZmBLK1	Putative ortholog of tomato TPK1b. Overexpression causes enhanced resistance to Goss' wilt.	Li et al. (2022)
5.	Arabidopsis	FLS2	Associates with BAK1 on sensing a 22 AA residue polypeptide of bacterial flagellin and triggers downstream signaling resulting in transcriptional reprogramming.	Gomez-Gomez and Boller (2000)

Structural and phylogenetic analysis of RLCKs in Arabidopsis and Rice indicates the majority of RLCKs related to immunity to belong to the RLCK VII subfamily (Shiu et al. 2004; Rao et al. 2018). This is a clear example of the highly conserved structures of the RLCK VII subfamily members being related to their function in conferring immunity against various pathogens and insects. Nothing can underline the importance of RLCKs in plant defense than the fact that more than 2.5% of the Arabidopsis genome codes only for RLKs, including receptor like cytosolic kinases. Thus, a close coordination between the RLKs, RLCKs, and downstream signaling elements brings about the transcriptional reprogramming and the physiological response needed for disease resistance.

Membrane-bound RLKs and RLCKs sit at the crossroads of plant hormone signaling, triggering of MAP kinase pathways, regulation of Ca^{2+} channels, and a global triggering of stress response and adaptation. Mitogen activated protein kinases (MAP kinases) are a group of Ser/Thr kinases, first isolated and characterized in vertebrates, other animals (Pelech and Sanghera 1992; Ruderman 1993; Jonak et al. 1994) and later in plants also. They have been named so, because they were found to link the presence of a mitogenic signal to cell-proliferation. However, later they were found to control a large segment of plant responses to the environment, including stress responses. MAP kinase activity increases dramatically after certain extracellular responses are received. They are therefore aptly named extracellular signal-regulated kinases (ERKs). One of the most important triggers of MAP kinase cascade activation in plants is the activation of RLKs and RLCKs (Figure 18.4).

18.8 Receptor-Like Cytosolic Kinases in Plant Development

The attainment of plant form, completely in consonance with the environment, is the key to success in the brutal process of natural selection. While every organism is genetically hardwired to express its physical and biochemical characteristics, the extent of expression of these characteristics is determined by the environment of the organism (Buchanan et al. 2015). The perception of environmental factors, their subsequent integration, and the physiological response they elicit depend on a massive network of sensors and the signaling pathways downstream of sensory perception. Apart from its genetic information, plant development is driven almost completely by abiotic environmental factors like light, temperature, availability of water. Between sensing the nature of the environment and responding to it physiologically, lies a huge network of signaling molecules. Depict it as we might, with the help of diagrams and data, the sensing-response cycle in any organism is a completely dynamic process, sensitive to the continuous change in environmental conditions. As discussed in a previous section, reversible phosphorylation of substrates is an extremely successful method for physiological alteration and changes in gene expression brought about by kinases. The RLCKs, discussed previously in the context of response to

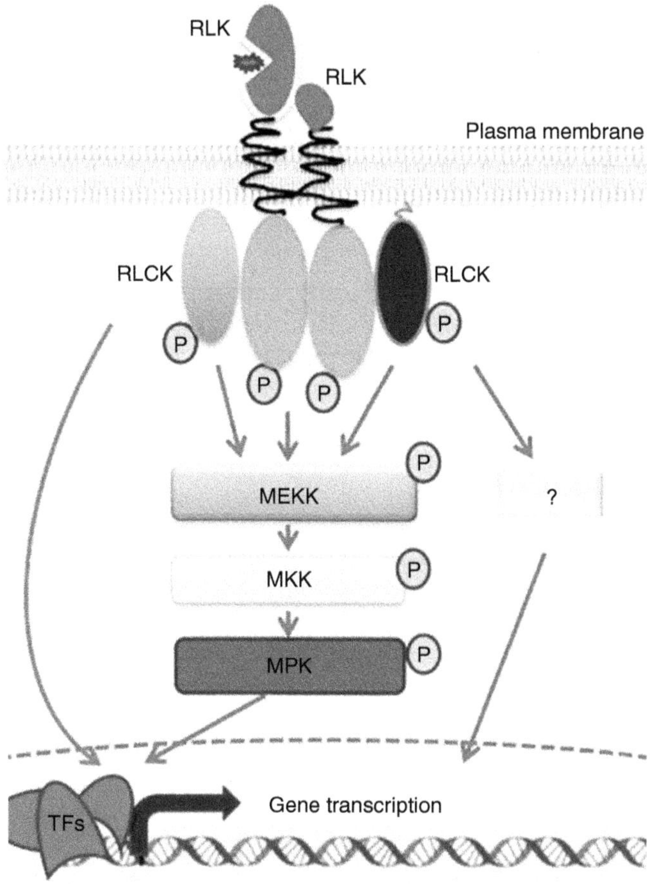

Figure 18.4 A model of receptor-like cytoplasmic kinase (RLCK) functions in different biological processes through association and transphosphorylation with RLK receptor complexes: RLCK/RLK complexes could activate mitogen-activated protein kinase (MAPK) cascade (MEKK-MKK-MPK) and/or MAPK-independent signaling to regulate gene expression via transcription factors (TFs) or other physiological responses (Lin et al. 2013).

plant pathogens, are also vital to the sensing of the environment. Apart from transient phosphorylation, plant hormones like auxin, cytokinin (CK), gibberellins (GA), abscisic acid (ABA), ethylene (ET), and brassinosteroids (BRs) form the backbone of this signaling network (Vanstraelen and Benkova 2012). SA and jasmonic acid (JA) are concerned more significantly with plant defense, and their action is vital to plant protection. The action of plant hormones, as in the case of their animal counterparts, is a complex area. Rarely can their effects be viewed as standalone. The extent of crosstalk between the signaling networks of different plant hormones as well as shared downstream signaling effectors make it a very interesting, albeit very complex, area of study (Robert-Seilaniantz et al. 2011). RLCKs have recently been shown to be intimately involved in these signaling networks that are vital to plant development in response to environmental cues. A recent genome-wide transcriptional profiling of rice RLCK genes revealed differential, growth stage linked expression of about 100 out of 379 OsRLCKs, strongly implicating them in diverse plant developmental processes (Vij et al. 2008). Similarly, multiple RLCKs have been reported to function in BR-, ET-, and SA-mediated signaling pathways, modulating various plant growth and development processes.

Plants start life as a single cell called a zygote. The Arabidopsis zygote elongates about threefold before dividing into a small apical and a large basal cell. The apical cell divides and forms the spherical pre-embryo, while the basal cell produces the filamentous suspensor. The subfamily II RLCK, short suspensor (SSP), has been shown to play important roles in the embryonic patterning process (Bayer et al. 2009). The ssp mutant suppresses zygote elongation and suspensor formation. Interestingly, ssp-mediated embryonic phenotypes possess a unique parent-of-origin effect. It has been proposed that SSP mRNA produced in sperm cells remains untranslated and is delivered to the zygote, where SSP proteins accumulate transiently upon fertilization. Genetic analyses indicate that a MAPK cascade constituting of YDA (a MAPK kinase)-MPK3/6 (MAPKs) and putative transcription factor Grounded (GRD) act downstream of SSP. Transient accumulation of SSP protein triggers activation of YDA signaling (Lukowitz et al. 2004; Bayer et al. 2009). The study provides an example that RLCKs function genetically upstream of the MAPK cascade in transducing RLK signaling (Figure 18.4). However, it remains unknown whether SSP directly activates YDA by phosphorylation or via other mechanisms.

18.9 Reactive Oxygen Species: Dual Role in Plants and Links to Receptor-Like Protein Kinases

Hardwired genetic programs and environmental stimuli play equally important roles in plant morphogenesis and help shape the plant morphologically and physiologically in accordance with its environment. While genetic imprinting provides a general framework for responses, environmental stimuli, including stress, guide the plant within that genetically imprinted framework. Generally speaking, environmental influences bring about adaptations that are more usually quantitative, rather than qualitative, with exceptions. All higher plants follow an aerobic lifestyle, and therefore oxygen forms a central component of energy generating pathways and certain morphogenetic pathways. ROS, namely singlet oxygen, superoxide, hydrogen peroxide, hydroxyl radical, etc., form the bulk of ROS and are the inevitable by-products of aerobic metabolism (Sandalio and Romero-Puertas 2015; Dietz et al. 2016; Huang et al. 2018). Due to their inherent reactivity, ROS cause covalent modifications to proteins, lipids, carbohydrates, and nucleic acids, among other biomolecules. If allowed uncontrolled, these modifications can be highly damaging and result in plant death. However, a comprehensive system of enzymatic and non-enzymatic antioxidants manages ROS levels to a comfortable level without eliminating them. The low, completely tolerable levels at which ROS fluctuate within an organism have been favored by natural selection as a marker of metabolic status, and an elaborate system of signaling has evolved that senses ROS levels and adjusts cellular metabolism accordingly. Thus, ROS can be dangerous if their transient concentration exceeds a certain value, can be used as agents of death in the event of apoptosis, and are consistently used as signaling entities within safe limits of their transient concentration. This area has been covered in several excellent reviews (Apel and Hirt 2004; Gill and Tuteja 2010; Mansoor et al. 2022).

RLKs have been considered the central regulator of plant responses to abiotic stress, biotic stress, and development under unstressed conditions. They are supposed to integrate messages from the external environment as well as from internal cues in the form of hormones, metabolite gradients, and the interpretation of light quality, quantity, and duration (Marshall et al. 2012).

ROS are generated in the apoplast, intracellularly in chloroplasts, mitochondria, peroxisomes, and cytosolic redox reactions characterized by large difference in redox potential (Figure 18.5). They serve as signals if maintained in a restricted concentration range. Every ROS has distinct chemical and physical properties and can affect the cellular redox status drastically (Joo et al. 2005; Vahisalu et al. 2010) and even that of distant cells (Gilroy et al. 2016). Generation of ROS in the

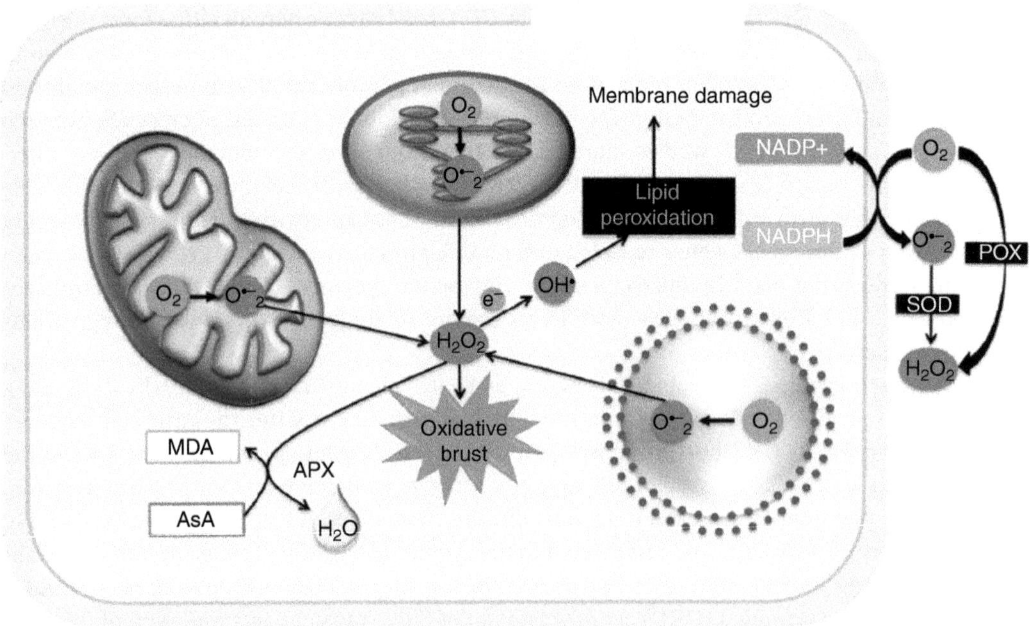

Figure 18.5 Generation of ROS intracellularly and in the apoplast: the processes involved in the production and control of ROS in different organelles within a plant system. The above is observed during stress response in plants. All these organelles are collectively responsible at maintaining ROS homeostasis in the cell (Nadarajah 2020).

apoplastic space is a consequence of cell wall resident peroxidase activity and of a group of plasma membrane bound NADPH oxidases (Kärkönen and Kuchitsu 2015). Although the mechanism of plants' response to ROS generation and the specificity associated with ROS-induced signaling are unclear, transcriptomic evidence clearly indicates the elicitation of responses specific to the class, source, and localization of the ROS (Vaahtera et al. 2014; Willems et al. 2016). At any given point in time, ROS relative concentrations are arrived at after cross-talk between Ca^{2+} flux, ABA, and various other hormones and signaling molecules. Activation of gene cascades is transduced by ROS waves, and these result in mounting physiological and possibly morphological changes that combat abiotic stress as well as keep ROS in check (Nadarajah 2020). Perhaps the most significant gene cascade activated during ROS signaling is that of mitogen-activated protein kinases (Nadarajah and Sidek 2010). These elaborate cascades lead to gene activation. As an example, MPK6 in *Arabidopsis* is activated under conditions of drought and repressed as soon as hydration is restored (Jagodzik et al. 2018).

18.10 Conclusion

As functionally diverse as the population of RLKs might be, they cannot control polygenic activities like cell proliferation and differentiation, unless the signaling pathway downstream of RLK activation controls several protein substrates in a concerted manner and amplifies the signal at the same time. This is accomplished by MAP kinase cascade activation. MAP kinase activation can be brought about by various intracellular factors such as metabolic status and extracellular factors like plant hormones. ROS levels have been naturally selected as one of the best markers of metabolic status throughout animal and plant species, and the interpretation of their fluctuating levels has been linked to MAP kinase activation via different mechanisms. Levels of hormones like ethylene and ABA change in response to stress and are perceived by RLKs and RLCKs via an elaborate sensing system described earlier. These serve to activate MAP kinase cascades and regulate the plant response to stress extremely effectively.

References

AbuQamar, S., Chai, M.F., Luo, H. et al. (2008). Tomato protein kinase 1b mediates signaling of plant responses to necrotrophic fungi and insect herbivory. *Plant Cell.* 20: 1964–1983.

Aifa, S., Nordvall, G., Lundstrom, I. et al. (2005). A basic peptide within the juxtamembrane region is required for EGF receptor dimerization. *Exp. Cell Res.* 302: 108–114.

Antolin-Llovera, M., Ried, M.K., Binder, A., and Parniske, M. (2012). Receptory kinase signaling pathways in plant-microbe interactions. *Annu. Rev. Phytopathol.* 50: 451–473.

Ao, Y., Li, Z., Feng, D. et al. (2014). OsCERK1 and OsRLCK176 play important roles in peptidoglycan and chitin signaling in rice innate immunity. *Plant J.* 80: 1072–1084.

Apel, K. and Hirt, H. (2004). Reactive oxygen species: metabolism, oxidative stress, and signal transduction. *Annu. Rev. Plant Biol.* 55 (1): 373–399.

Bayer, M., Navi, T., Giglione, C. et al. (2009). Paternal control of embryonic patterning in Arabidopsis thaliana. *Science* 332: 1485–1488.

Boller, T. and Felix, G. (2009). A renaissance of elicitors: perception of microbe-associated molecular patterns and danger signals by pattern-recognition receptors. *Annu. Rev. Plant Biol.* 60: 379–406.

Brewer, M.R., Choi, S.H., Alvarado, D. et al. (2009). The juxtamembrane region of the EGF receptor functions as an activation domain. *Mol. Cell* 34: 641–651.

Buchanan, B.B., Gruissem, W., and Jones, R.L. (ed.) (2015). *Biochemistry and Molecular Biology of Plants*, 2e. Wiley-Blackwell. ISBN: 9780470714218.

Busch, W. and Benfey, P.N. (2010). Information processing without brains- the power of intercellular regulators in plants. *Development* 137: 1215–1226.

Campisano, C.J. (2012). Milankovitch cycles, paleoclimatic change, and hominin evolution. *Nat. Educ. Knowl.* 4: 5.

Chaudhry, S. and Sidhu, G.P.S. (2022). Climate change regulated abiotic stress mechanisms in plants: a comprehensive review. *Plant Cell Rep.* 41 (1): 1–31.

Couto, D. and Zipfel, C. (2016). Regulation of pattern recognition receptor signalling in plants. *Nat. Rev. Immunol.* 16: 537–552.

Cramer, G.R., Urano, K., Delrot, S. et al. (2011). Effects of abiotic stress on plants: a systems biology perspective. *BMC Plant Biol.* 11: 1–14.

Dietz, K.J., Turkan, I., and Krieger-Liszkay, A. (2016). Redox- and reactive oxygen species-dependent signaling into and out of the photosynthesizing chloroplast. *Plant Physiol.* 171: 1541–1550.

Dievart, A., Perin, C., and Ranvez, V. (2020). Origin and diversity of plant receptor-like kinases. *Annu. Rev. Plant Biol.* 71: 131–156.

Feng, F. and Zhou, J.M. (2012). Plant-bacterial pathogen interactions mediated by type III effectors. *Curr. Opin. Plant Biol.* 15: 469–476.

Gill, S.S. and Tuteja, N. (2010). Reactive oxygen species and antioxidant machinery in abiotic stress tolerance in crop plants. *Plant Physiol. Biochem.* 48: 909–930.

Gilroy, S., Białasek, M., Suzuki, N. et al. (2016). ROS, calcium, and electric signals: key mediators of rapid systemic signaling in plants. *Plant Physiol.* 171: 1606–1615.

Goff, K.E. and Ramonell, K.M. (2007). The role and regulation of receptor-like kinases in plant defence. *Gene Regul. Syst. Biol.* 1: 167–175.

Gomez-Gomez, L. and Boller, T. (2000). An LRR receptor-like kinase involved in the perception of the bacterial elicitor flagellin in *Arabidopsis*. *Mol. Cell.* 5: 1003–1011.

Griffith, J., Black, J., Faerman, C. et al. (2004). The structural basis for autoinhibition of FLT3 by the juxtamembrane domain. *Mol. Cell* 13: 169–178.

Hanks, S.K. and Hunter, T. (1995). The eukaryotic protein kinase superfamily: kinase (catalytic) domain structure and classification. *FASEB J* 9: 576–596.

Hedger, G., Sansom, M.S.P., and Koldsø, H. (2015). The juxtamembrane regions of human receptor tyrosine kinases exhibit conserved interaction sites with anionic lipids. *Sci. Rep.* 5: 9198.

Huang, L., Jia, J., Zhao, X. et al. (2018). The ascorbate peroxidase APX1 is a direct target of a zinc finger transcription factor ZFP36 and a late embryogenesis abundant protein OsLEA5 interacts with ZFP36 to co-regulate OsAPX1 in seed germination in rice. *Biochem. Biophys. Res. Commun.* 495: 339–345.

Jagodzik, P., Tajdel-Zielińska, M., Cieśla, A. et al. (2018). Mitogen-activated protein kinase cascades in plant hormone signaling. *Front. Plant Sci.* 9: 1387. https://doi.org/10.3389/fpls.2018.01387.

Janicka-Russak, M., Kabała, K., Burzyński, M., and Kłobus, G. (2008). Response of plasma membrane H+-ATPase to heavy metal stress in *Cucumis sativus* roots. *J. Exp. Bot.* 59: 3721–3728.

Jewell, M.C., Campbell, B.C., and Godwin, I.D. (2010). Transgenic plants for abiotic stress resistance. In: *Transgenic Crop Plants* (ed. C. Kole, C.H. Michler, A.G. Abbott, and T.C. Hall), 67–132. Berlin/Heidelberg: Springer.

Jonak, C., Heberle-Bors, E., and Hirt, H. (1994). MAP Kinases: universal multipurpose signaling tools. *Plant Mol. Biol.* 24: 407–416.

Jones, J.D.G. and Dangl, J.L. (2006). The plant immune system. *Nature* 444: 323–329.

Joo, J.H., Wang, S., Chen, J.G. et al. (2005). Different signaling and cell death roles of heterotrimeric G Protein α and β subunits in the *Arabidopsis* oxidative stress response to ozone. *Plant Cell* 17: 957–970.

Jose, J., Ghantasala, S., and Roy, C.S. (2020). Arabidopsis transmembrane receptor-like kinases (RLKs): a bridge between extracellular signal and intracellular regulatory machinery. *Int. J. Mol. Sci.* 21: 4000. https://doi.org/10.3390/ijms21114000.

Kärkönen, A. and Kuchitsu, K. (2015). Reactive oxygen species in cell wall metabolism and development in plants. *Phytochemistry* 112: 22–32.

Lehti-Shiu, M.D. and Shiu, S.H. (2012). Diversity, classification and function of the plant protein kinase superfamily. *Philos. Trans. R. Soc. B Biol. Sci.* 367: 2619–2639.

Lehti-Shiu, M.D., Zou, C., and Shiu, S.H. (2012). Origin, Diversity, Expansion History, and Functional Evolution of the Plant Receptor-Like Kinase/*Pelle* Family. In: *Receptor-like Kinases in Plants. Signaling and Communication in Plants*, vol. 13 (ed. F. Tax and B. Kemmerling). Berlin, Heidelberg: Springer.

Lemmon, M.A. and Schlessinger, J. (2010). Cell signaling by receptor tyrosine kinases. *Cell* 141: 1117–1134.

Li, W., Liao, C.J., Bluhm, B.H. et al. (2022). A Maize (*Zea mays* L.) BIK1-like receptor-like cytoplasmic kinase contributes to disease resistance. *Plant Mol. Biol. Rep.* 40: 28–42.

Liang, X. and Zhou, J.M. (2018). Receptor-like cytoplasmic kinases: central players in plant receptor kinase–mediated signaling. *Annu. Rev. Plant Biol.* 69: 267–299.

Lin, W., Ma, X., Shan, L., and He, P. (2013). Big roles of small kinases: the complex functions of receptor-like cytoplasmic kinases in plant immunity and development. *J. Int. Plant Biol.* 55: 1188–1197.

Lukowitz, W., Roeder, A., Parmenter, D., and Somerville, C. (2004). A MAPKK kinase gene regulates extra-embryonic cell fate in arabidopsis. *Cell* 116: 109–119.

R. solanacearum. Furthermore, the resistance was compromised when the homogenate was applied to *jai1-1* mutants with defective signaling.

Generally, SA is assumed to be involved with the induction of defense responses related to biotrophic and hemi-biotrophic pathogens along with the establishment of SAR, while JA signaling is assumed to be involved with the defense induction with leaf-chewing and necrotrophic microbes (Glazebrook 2005). However, recent evidences have highlighted the potentiality of JA toward both the necrotrophic and some biotrophic fungal infections mainly the powdery and downy mildews (Trouvelot et al. 2008; Antico et al. 2012; Guerreiro et al. 2016). Infection with necrotrophic fungi triggers the JA-dependent signaling pathway, initiating an effective defense response as compared to SA-mediated defense (Spoel et al. 2007). SAR is a broad-spectrum disease resistance and can be induced by the exogenous application of SA, JA, jasmonic methyl ester, etc. (Métraux et al. 2002). Induction of SAR in plants results in the accumulation of PR proteins (Ryals et al. 1996). Phytopathogens triggers the plant defense responses through upregulation of the plant defensing gene like *PDF1.2* along with *PR-3* and *PR-4* genes via both SA and JA dependent pathways (Thomma et al. 1998). Hydrolytic enzymes like β-1,3-glucanases (PR-2) that play an important role in catalyzing the hydrolysis of β-1,3-glucan, a major constituent of the fungal cell wall (Wessels 1993), which results in growth inhibition of the pathogen. Induction of β-1,3-glucanases was observed through western blot analysis of wheat leaves that were treated with JA, which effectively reduced the disease incidence caused by *Stagonospora nodorum* (Jayaraj et al. 2004).

19.2.3 Ethylene (ET)

Ethylene is a volatile molecule with the chemical formula C_2H_2 that plays an important role in plant growth, development, and survival, which includes germination, root/shoot development, abscission in fruits and leaves, flowering, and senescence (Abeles et al. 2012; McManus 2012). The amino acid methionine is required for ET synthesis, which is acted upon by the enzyme S-AdoMet synthase (ADS) to form *S*-adenosyl-methionine (AdoMet). The enzyme 1-aminocyclopropane-1-ca rboxylate synthase (ACS) converts AdoMet to 5′-methylthioadenosine (MTA). MTA then enters Yang cycle and is converted back to methionine and the precursor of ET, 1-aminocyclopropane-1-carboxylic acid (ACC) which is then oxidized to ET by the enzyme ACC oxidase. ET is distributed from the point of synthesis through diffusion across membranes into nearby cells and does not require transporters. Nevertheless, 1-aminocyclopropane-1-carboxylic acid (ACC), an intermediate precursor of ethylene, requires transporter (Van de Poel and Van Der Straeten 2014). Plants that are under biotic stress show increased production of ET, which through a series of events regulates defense-related genes mainly, the ethylene-responsive transcription factors (Broekaert et al. 2006).

According to van Loon et al. (2006), ET acts as an important modulator of responses in plants toward pathogens. Exogenous application of ET has shown gene regulatory activity including the defense related genes in *Arabidopsis thaliana* (Van Zhong and Burns 2003). As per Nascimento et al. (2018), both ET as well as its precursor ACC show active participation in plant defense through modulation of plant immune responses. However, in some plant species, ET shows negative results in terms of disease resistance. For example, a tomato mutant, Never ripe (Nr), with impaired ethylene perception displayed significant disease reduction as compared to wild-type tomato plants against both fungal and bacterial pathogens (Lund et al. 1998). Nevertheless, positive effects of ET in disease resistance have been reported on the other hand. Penmetsa et al. (2008) reported the involvement of EIN2 protein, which is a positive ET signaling regulator in *Medicago truncatula* defense against *Phytophthora medicaginis*. Similarly, reduced hyphal growth was observed in excised soybean hypocotyls that were treated with ACC with increased levels of β-1,3-endoglucanase (Yoshikawa et al. 1990). Blast causing *Magnaporthe grisea* resistant cultivars of rice displays enhancement in ET synthesis followed by elevated levels of ACC and ACC oxidase activity as compared to wild genotypes. Furthermore, exogenous ethephon induces resistance against blast, whereas inhibition of ET leads to expanding lesions in place of HR lesions (Iwai et al. 2006). Overexpression of ACS2 genes in transgenic rice provides increased resistance toward necrotrophic and hemi-biotrophic fungal phytopathogens like *Magnaporthe oryzae* and *Rhizoctonia solani* through enhanced ET production (Helliwell et al. 2013).

19.2.4 Abscisic Acid (ABA)

ABA is a sesquiterpene containing 15 carbon atoms, synthesized from large carotenoids precursor via plastids through 2-*C*-methyl-D-erythritol 4-phosphate (MEP) pathway in plants. ABA plays a crucial role in plant growth and development including germination, dormancy, and seed development. It also confers a diversified role in plant processes including adaptive responses through gene expression regulation against various environmental stresses. The hormone induces varying changes in plant physiological processes that help in overcoming various stresses. Efforts have been made in the past decade to understand the role of ABA in plant defense responses. According to Ton et al. (2009), ABA functions as the cost-efficient strategy at the early stages of pathogen colonization, promoting early defense barriers while preventing the activation of SA- and JA- dependent defense responses.

The phytohormone ABA plays a crucial role in alleviation of number of abiotic stresses including drought, salinity along with dormancy and maturation of seeds (Uno et al. 2000; Marion-Poll and Leung 2006). Despite the fact ABA is a crucial molecule for growth, development, and stress regulation, some phytopathogens are successful in suppressing the SAR by induced overproduction of ABA through their evolved mechanisms. Applications of exogenous ABA have shown to increase disease severity in plants like *Hordeum vulgare* (Ulferts et al. 2015) and *Beta vulgaris* (Schmidt et al. 2008). SA functions as an important signaling molecule for the activation of SAR against pathogen proliferation (Sillero et al. 2012). Investigations have revealed negative regulation of ABA in disease resistance by negatively modulating the SA-dependent pathway (Audenaert et al. 2002) through ABA-SA antagonism, resulting in reduced expression of *PR* genes like *PR1* and *PR10* (Xiong and Yang 2003). However, some studies have also shown the potential role of ABA in regulating plant defense responses. For example, exogenous application of ABA provided resistance to *Xylella fastidiosa* by inducing the production of phenolic compounds in *Vitis vinifera* (Meyer and Kirkpatrick 2011). Investigation have also shown resistant roles of ABA against necrotrophic fungi like *Alternaria brassicicola* in *constitutive disease susceptibility2-1D* (*cds2-1D*) line of *Arabidopsis*, while the mutant *aba3-1* showed susceptibility toward the pathogen (Fan et al. 2009). Microscopic analysis in ABA-treated plants have shown ABA-induced resistance toward invading pathogen through restriction of growth in the mesophyll. In addition, ABA-induced resistance is independent of SA-, JA-, or callose-dependent resistance mechanisms but utilizes another functional protein known as Gα-protein in *Oryza sativa* against *Cochliobolus miyabeanus* (De Vleesschauwer et al. 2010).

Phytopathogens generally enter the plant system directly by dissolving plant cell walls through the production of cell wall degrading enzymes or by utilizing the natural openings present on the plant body (Melotto et al. 2008). Stomata are natural pores that facilitate gaseous exchange between the plant body and the environment. However, these openings also act as an important gateway for the entry of pathogens inside the plant tissues (Melotto et al. 2006). The phytohormone ABA plays an important role in regulating plant defense responses by regulating stomatal activity. The abilities of pathogenic bacteria to induce stomatal closure suggest that plants can sense some of the conserved bacterial molecules, such as pathogen-associated molecular pattern (PAMPs), which stimulate the innate immunity in plants (Gómez-Gómez and Boller 2002). Inability of PAMPs like flg22 and LPS to induce stomatal closure in *aba3-1* mutant plants indicates the important role of ABA in stomatal closure against pathogen invasion (Melotto et al. 2006). For example, enhanced resistance through stomatal closure towards *P. syringae* pv. *tomato* DC3000 was observed in *Arabidopsis* with two overexpressed ABA receptors, *RCAR4/PYL10-OX* and *RCAR5/PYL11-OX* (Lim and Lee 2015).

Transcriptome profiling showed upregulation of several genes related to various transcription factors, heat shock proteins, pathogen resistance, SA, JA, and ethylene signaling pathways that have been reported with the application of ABA (Wang et al. 2013). According to Cheng et al. (2016), ABA may upregulate the expression of various genes related to stress through miRNA-mediated posttranscriptional regulation, thereby displaying its potential in pathogen resistance. ABA treatment induces *ABA-RESPONSIVE1* (*ABR1*) which is essential for cell death priming and for the production of ROS. *ABR1* silencing compromised hypersensitive response (HR) while enhancing the levels of ABA and bacterial growth in *C. annuum* plants (Choi and Hwang 2011). Furthermore, heterologous transgenic expression of *ABR1* enhanced resistance, while the ortholog mutant, *abr1,* was susceptible toward *P. syringae* pv. *tomato* and *Hyaloperonospora arabidopsidis* in *A. thaliana*.

Plasmodesmata are the junction between two adjacent cells through which transfer of various cell materials occurs based on their sizes. The pore size of plasmodesmata is regulated by the controlled deposition of callose, a polysaccharide belonging to the class of β-1,3-glucan (Li et al. 2012). In *Arabidopsis*, callose is encoded by *callose synthase* (*CalS*) genes like *CalS10* (Ellinger and Voigt 2014) and is involved in plant defense responses toward phytopathogens, with their deposition on cell walls and plasmodesmata providing barriers restricting the proliferation of pathogens (Luna et al. 2011; Ellinger and Voigt 2014). Callose deposition is mainly controlled by the enzyme β-1,3-glucanases by hydrolyzing the 1,3-β-D-glucosidic linkages, generating β-1,3-glucan units as the end products (De Storme and Geelen 2014). Viral infections are known to induce such hydrolytic enzymes that hydrolyze the callose, leading to changes in plasmodesmata pore size and viral proliferation (Oide et al. 2013). The ability of ABA in suppressing the enzymatic activity of PR2 (β-1,3-glucanase) permits more callose deposition at the sites of plasmodesmata, thereby restricting viral proliferation (Rezzonico et al. 1998; Heinlein 2015).

19.3 Phytohormone Mediated Cross-Talk in Plant Defense Under Biotic Stress

Plants under stress are often challenged for the efficient utilization of their limited resources in order to adapt and survive the biotic environment in cost-efficient manner (Pieterse et al. 2012). Therefore, the antagonistic interaction between SA and JA appears to play an important role in optimizing the immune responses toward specific pathogens. Plant defense against biotrophic pathogens is usually mediated through SA signaling, while on the other hand, JA mediates defense against necrotrophic pathogens and herbivores (Thaler et al. 2012). Suppression of JA-dependent responses as seen after the infection with SA-inducing biotrophic pathogen shows the investment of SA-dependent resources over the JA-dependent (Spoel et al. 2007). However, investigations have also shown negative regulatory roles of JA on SA responsive genes. According to Koo et al. (2007), JA induces *AtBSMT1*, which converts SA to MeSA, resulting in lower SA levels.

WRKY family of plant-specific transcriptional regulatory factors have shown their involvement in disease resistance (De-slandes et al. 2002). According to Li et al. (2004), SA activates the expression of *WRKY70* that requires functional regulatory protein NONEXPRESSER OF PR GENES 1 (NPR1) for full-scale induction. WRK70 binds to the promoter region of *PR1* and promotes *PR1* expression. JA, on the other hand, works oppositely to SA by suppressing the activity of *WRKY70* (Li et al. 2004). *Arabidopsis* mutant *npr1* with reduced SA signal transduction showed the requirement of regulatory protein NPR1 for the SA-JA antagonism in cytosol (Spoel et al. 2003). NPR1 homologs NPR3 and NPR4 act as a Cullin 3 ubiquitin E3 (CUL3) ligase adapter protein, mediating the degradation of NPR1, and are dependent on the SA levels. For example, during higher SA levels, NPR3 interacts with NPR1, while the lower levels of SA mediate NPR4-mediated degradation of NPR1 (Fu et al. 2012). Suppression of *WRKY70* results in the activation of JA-responsive/CORONATINE INSENSITIVE 1 (COI1) dependent genes (Li et al. 2004). *NahG* plants that were unable to accumulate SA show higher levels of JA leading to the expression of JA-responsive genes like *LOX2* (*LIPOXYGENASE2*), *PDF1.2*, encoding plant defensin, and *VSP* encoding for vegetative storage protein against *P. syringae* pv. *tomato* DC3000 in *Arabidopsis* (Spoel et al. 2003). Among the responsive genes in *Arabidopsis*, *PDF1.2* and *VSP2* show higher sensitivity toward the suppression by SA, and the suppression is independent of the types of JA-responsive gene elicitors (Koornneef and Pieterse 2008; Leon-Reyes et al. 2010).

MAP kinase 4 (MPK4) activity negatively regulates the activity of SA signaling and thereby SAR, while it is required for the induction of JA associated genes. For example, *Arabidopsis mpk4* knockout mutant shows higher accumulation of SA and constitutive expressions of *PR* genes, while JA-dependent *PDF1.2* and *THI2.1* genes are blocked (Petersen et al. 2000). Defense regulators EDS1 and PAD4 effect SA–ET/JA signal antagonism by exerting positive regulation toward SA accumulation and PR expression while repressing the activity of PDF1.2 downstream of MPK4, as it is evident from the studies of Brodersen et al. (2006). They further added that these defense regulators, EDS1 and PAD4, are negatively regulated by MPK4.

JA dependent signaling can occur in two distinct pathways. Firstly, MYC branch mainly regulated by MYC2 occurs through the co-regulation of JA and ABA that induces downstream genes like VSP2 and LOX2 (Lorenzo et al. 2004; Vos et al. 2013). Secondly, ERF pathway occurs with the help of transcription factors like EIN3, EIL1, and ERF, which are co-regulated by phytohormones like JA and ET that results in the activation of downstream genes like PDF1.2 (Zhu et al. 2011; Pieterse et al. 2012). JA signaling induces MCY2 that activates the NAC transcriptional factors, which further inhibits the expression of ICS1 (Zheng et al. 2012). Thus, JA signaling inhibits the accumulation of SA by inhibiting its biosynthesis.

During necrotrophic infection, ET and JA work in cooperation with one another to present plant defense responses. It is evident from the work of Berrocal-Lobo et al. (2002) that in *Arabidopsis*, Ethylene Response Factor1 (ERF1) regulates ethylene defense responses toward the necrotrophic pathogens *B. cinerea* and *Plectosphaerella cucumerina*. Simultaneous activation of ERF1 by ET and JA generates transcriptional factors that regulate the genes associated with defense response (Lorenzo et al. 2004). JASMONATE ZIM-DOMAIN (JAZ) family interacts with various phytohormone signaling pathways and plays a crucial role in the regulation of defense responses, reviewed in Kazan and Manners (2012). At lower levels of JA, JAZ proteins in association with their co-repressor TOPLESS (TPL) with the help of the adapter protein NOVEL INTERACTOR OF JAZ (NINJA) or histone deacetylase (HDA6) and suppress the gene expression through transcriptional repression (Song et al. 2014). According to Zhu et al. (2011), two ET transcription factors of *Arabidopsis* ETHYLENE INSENSITIVE 3 (EIN3) and EIN3-LIKE 1 (EIL1) converge signals from ET and JA and regulate the expression of genes including defense response genes. JAZ proteins with its co-repressor HDA6 modify histone deacetylation and suppress the EIN3/EIL1-dependent transcription. Furthermore, they also reported that the action JA itself encourages the removal of JAZ from EIN3/EIL1. When JA (JA-Ile) associates with the COI1 F-box protein, JAZ repressors undergo ubiquitination and are directed for degradation in S-phase kinase-associated protein1/Cullin1/F-box protein COI1 (SCF^{COI1}) dependent manner, thereby re-establishing the activity of EIN3/EIL1 and various JA responses (Howe 2010; Song et al. 2014).

Signaling between ABA and JA occurs directly through the interaction of the ABA receptor PYL6 with the MYC2 transcription factor, where ABA enhances the interaction between PYL6 and MYC2. In the absence of ABA, weak interaction between PYL6 and MYC2 exists, which allows the binding of transcription factors to JAZ6 promoter. At higher levels of ABA, strong PYL6 and MYC2 interaction occurs, which leads to the expression of JAZ8 (Aleman et al. 2016). Thus, receptor of ABA, PYRABACTIN RESISTANCE LIKE (PYL) forms a complex with JAZ, activating MYC2 transcriptional activity (Aleman et al. 2016). MYC2 together with MED25 subunit then induces the expression of the JA-related responsive gene *VSP2* in response to wounding (Wang et al. 2019). E3 RING ligase KEEP ON GOING (KEG), repressor of ABSCISIC ACID INSENSITIVE5 (ABI5) transcription factor involved in ABA signaling shows positive regulatory effect on JAZ12, where JAZ12 negatively regulates the function of MYC2. Furthermore, both JA and ABA applications lead to the degradation of JAZ12 (Pauwels et al. 2015). JA/ET-independent suppression of SAR induction with ABA treatment show antagonistic relationship between SA and ABA (Yasuda et al. 2008). Stomatal defense is another important defense mechanism that restricts pathogen entry, and the response is known to be associated with ABA. However, Wu et al. (2018) report the simultaneous activity of ABA and SA regulated pathways that induce stomatal closure in *N. benthamiana* infected with *Bacillus amyloliquefaciens* FZB42. Zeng and He (2010) reported the involvement of NPR1 in stomatal defense acting upstream of ABA but downstream of SA signaling. Thus, the interplay between different plant hormones, especially SA, JA, ET and ABA, coordinates to generate an efficient and specific effect against the plant pathogen.

References

Abeles, F.B., Morgan, P.W., and Saltveit, M.E. Jr. (2012). *Ethylene in Plant Biology*. New York: Academic Press.

Aleman, F., Yazaki, J., Lee, M. et al. (2016). An ABA-increased interaction of the PYL6 ABA receptor with MYC2 transcription factor: a putative link of ABA and JA signaling. *Sci. Rep.* 6: 1–10.

AL-Saleh, M.A. (2011). Pathogenic variability among five bacterial isolates of *Xanthomonas campestris* pv. *vesicatoria*, causing spot disease on tomato and their response to salicylic acid. *J. Saudi Soc. Agric. Sci.* 10: 47–51.

An, C. and Mou, Z. (2011). Salicylic acid and its function in plant immunity F. *J. Integr. Plant Biol.* 53: 412–428.

Anand, A., Uppalapati, S.R., Ryu, C.M. et al. (2008). Salicylic acid and systemic acquired resistance play a role in attenuating crown gall disease caused by *Agrobacterium tumefaciens*. *Plant Physiol.* 146: 703–715.

Antico, C.J., Colon, C., Banks, T., and Ramonell, K.M. (2012). Insights into the role of jasmonic acid-mediated defenses against necrotrophic and biotrophic fungal pathogens. *Front. Biol.* 7: 48–56.

Audenaert, K., De Meyer, G.B., and Höfte, M.M. (2002). Abscisic acid determines basal susceptibility of tomato to *Botrytis cinerea* and suppresses salicylic acid-dependent signaling mechanisms. *Plant Physiol.* 128: 491–501.

Bajwa, A.A., Farooq, M., Al-Sadi, A.M. et al. (2020). Impact of climate change on biology and management of wheat pests. *Crop Prot.* 137: 105304.

Berrocal-Lobo, M., Molina, A., and Solano, R. (2002). Constitutive expression of ETHYLENE-RESPONSE-FACTOR1 in *Arabidopsis* confers resistance to several necrotrophic fungi. *Plant J.* 29: 23–32.

Brodersen, P., Petersen, M., Bjørn Nielsen, H. et al. (2006). *Arabidopsis* MAP kinase 4 regulates salicylic acid-and jasmonic acid/ethylene-dependent responses via EDS1 and PAD4. *Plant J.* 47: 532–546.

Broekaert, W.F., Delauré, S.L., De Bolle, M.F., and Cammue, B.P. (2006). The role of ethylene in host-pathogen interactions. *Annu. Rev. Phytopathol.* 44: 393–416.

Chaloner, T.M., Gurr, S.J., and Bebber, D.P. (2021). Plant pathogen infection risk tracks global crop yields under climate change. *Nat. Clim. Chang.* 11: 710–715.

Chandrasekhar, B., Umesha, S., and Kumar, H.N. (2017). Proteomic analysis of salicylic acid enhanced disease resistance in bacterial wilt affected chilli (*Capsicum annuum*) crop. *Physiol. Mol. Plant Pathol.* 98: 85–96.

Cheng, H.Y., Wang, Y., Tao, X. et al. (2016). Genomic profiling of exogenous abscisic acid-responsive microRNAs in tomato (*Solanum lycopersicum*). *BMC Genomics* 17: 1–13.

Choi, D.S. and Hwang, B.K. (2011). Proteomics and functional analyses of pepper abscisic acid–responsive 1 (ABR1), which is involved in cell death and defense signaling. *Plant Cell* 23: 823–842.

De-slandes, L., Olivier, J., Theulières, T. et al. (2002). Resistance to *Ralstonia solanacearum* in *Arabidopsis thaliana* is conferred by the recessive *RRS1-R* gene, a member of a novel family of resistance genes. *Proc. Natl. Acad. Sci.* 99: 2404–2409.

De Storme, N. and Geelen, D. (2014). Callose homeostasis at plasmodesmata: molecular regulators and developmental relevance. *Front. Plant Sci.* 5: 138.

De Vleesschauwer, D., Yang, Y., Vera Cruz, C., and Höfte, M. (2010). Abscisic acid-induced resistance against the brown spot pathogen *Cochliobolus miyabeanus* in rice involves MAP kinase-mediated repression of ethylene signaling. *Plant Physiol.* 152: 2036–2052.

Delgado-Baquerizo, M., Guerra, C.A., Cano-Díaz, C. et al. (2020). The proportion of soil-borne pathogens increases with warming at the global scale. *Nat. Clim. Chang.* 10: 550–554.

Dempsey, D.M.A., Vlot, A.C., Wildermuth, M.C., and Klessig, D.F. (2011). Salicylic acid biosynthesis and metabolism. *Arabidopsis Book* 9: e0156.

Dezar, C.A., Giacomelli, J.I., Manavella, P.A. et al. (2011). HAHB10, a sunflower HD-Zip II transcription factor, participates in the induction of flowering and in the control of phytohormone-mediated responses to biotic stress. *J. Exp. Bot.* 62: 1061–1076.

Dudney, J., Willing, C.E., Das, A.J. et al. (2021). Nonlinear shifts in infectious rust disease due to climate change. *Nat. Commun.* 12: 1–13.

Durrant, W.E. and Dong, X. (2004). Systemic acquired resistance. *Annu. Rev. Phytopathol.* 42: 185–209.

Ellinger, D. and Voigt, C.A. (2014). Callose biosynthesis in *Arabidopsis* with a focus on pathogen response: what we have learned within the last decade. *Ann. Bot.* 114: 1349–1358.

Fan, J., Hill, L., Crooks, C. et al. (2009). Abscisic acid has a key role in modulating diverse plant-pathogen interactions. *Plant Physiol.* 150: 1750–1761.

Fu, Z.Q., Yan, S., Saleh, A. et al. (2012). NPR3 and NPR4 are receptors for the immune signal salicylic acid in plants. *Nature* 486: 228–232.

Ghini, R., Bettiol, W., and Hamada, E. (2011). Diseases in tropical and plantation crops as affected by climate changes: current knowledge and perspectives. *Plant Pathol.* 60: 122–132.

Glazebrook, J. (2005). Contrasting mechanisms of defense against biotrophic and necrotrophic pathogens. *Annu. Rev. Phytopathol.* 43: 205–227.

Gómez-Gómez, L. and Boller, T. (2002). Flagellin perception: a paradigm for innate immunity. *Trends Plant Sci.* 7: 251–256.

Guerreiro, A., Figueiredo, J., Sousa Silva, M., and Figueiredo, A. (2016). Linking jasmonic acid to grapevine resistance against the biotrophicoomycete *Plasmopara viticola*. *Front. Plant Sci.* 7: 565.

Hase, S., Takahashi, S., Takenaka, S. et al. (2008). Involvement of jasmonic acid signalling in bacterial wilt disease resistance induced by biocontrol agent *Pythium oligandrum* in tomato. *Plant Pathol.* 57: 870–876.

Heinlein, M. (2015). Plasmodesmata: channels for viruses on the move. *Methods Mol. Biol.* 1217: 25–52.

Helliwell, E.E., Wang, Q., and Yang, Y. (2013). Transgenic rice with inducible ethylene production exhibits broad-spectrum disease resistance to the fungal pathogens *Magnaporthe oryzae* and *Rhizoctonia solani*. *Plant Biotechnol. J.* 11: 33–42.

Howe, G.A. (2010). Ubiquitin ligase-coupled receptors extend their reach to jasmonate. *Plant Physiol.* 154: 471–474.

Iwai, T., Miyasaka, A., Seo, S., and Ohashi, Y. (2006). Contribution of ethylene biosynthesis for resistance to blast fungus infection in young rice plants. *Plant physiol.* 142: 1202–1215.

Jayaraj, J., Muthukrishnan, S., Liang, G.H., and Velazhahan, R. (2004). Jasmonic acid and salicylic acid induce accumulation of β-1,3-glucanase and thaumatin-like proteins in wheat and enhance resistance against *Stagonospora nodorum*. *Biol. Plant.* 48: 425–430.

Jia, C., Zhang, L., Liu, L. et al. (2013). Multiple phytohormone signalling pathways modulate susceptibility of tomato plants to *Alternaria alternata* f. sp. *lycopersici*. *J. Exp. Bot.* 64: 637–650.

Jones, J.D. and Dangl, J.L. (2006). The plant immune system. *Nature* 444: 323–329.

Kachroo, P., Yoshioka, K., Shah, J. et al. (2000). Resistance to turnip crinkle virus in *Arabidopsis* is regulated by two host genes and is salicylic acid dependent but NPR1, ethylene, and jasmonate independent. *Plant Cell* 12: 677–690.

Kazan, K. and Manners, J.M. (2012). JAZ repressors and the orchestration of phytohormone crosstalk. *Trends Plant Sci.* 17: 22–31.

Khan, N., Bano, A., Ali, S., and Babar, M. (2020). Crosstalk amongst phytohormones from planta and PGPR under biotic and abiotic stresses. *Plant Growth Regul.* 90: 189–203.

Koo, Y.J., Kim, M., Kim, E.H. et al. (2007). Overexpression of salicylic acid carboxyl methyltransferase reduces salicylic acid-mediated pathogen resistance in *Arabidopsis thaliana*. *Plant Mol. Biol.* 64: 1–15.

Koornneef, A. and Pieterse, C.M. (2008). Cross talk in defense signaling. *Plant Physiol.* 146: 839–844.

Kouzai, Y., Kimura, M., Watanabe, M. et al. (2018). Salicylic acid-dependent immunity contributes to resistance against *Rhizoctonia solani*, a necrotrophic fungal agent of sheath blight, in rice and *Brachypodium distachyon*. *New Phytol.* 217: 771–783.

Leon-Reyes, A., Van der Does, D., De Lange, E.S. et al. (2010). Salicylate-mediated suppression of jasmonate-responsive gene expression in *Arabidopsis* is targeted downstream of the jasmonate biosynthesis pathway. *Planta* 232: 1423–1432.

Li, J., Brader, G., and Palva, E.T. (2004). The WRKY70 transcription factor: a node of convergence for jasmonate-mediated and salicylate-mediated signals in plant defense. *Plant Cell* 16: 319–331.

Li, W., Zhao, Y., Liu, C. et al. (2012). Callose deposition at plasmodesmata is a critical factor in restricting the cell-to-cell movement of Soybean mosaic virus. *Plant Cell Rep.* 31: 905–916.

Lim, C.W. and Lee, S.C. (2015). *Arabidopsis* abscisic acid receptors play an important role in disease resistance. *Plant Mol. Biol.* 88: 313–324.

van Loon, L.C., Geraats, B.P., and Linthorst, H.J. (2006). Ethylene as a modulator of disease resistance in plants. *Trends Plant Sci.* 11: 184–191.

López, M.A., Bannenberg, G., and Castresana, C. (2008). Controlling hormone signaling is a plant and pathogen challenge for growth and survival. *Curr. Opin. Plant Biol.* 11: 420–427.

Lorenzo, O., Chico, J.M., Saénchez-Serrano, J.J., and Solano, R. (2004). JASMONATE-INSENSITIVE1 encodes a MYC transcription factor essential to discriminate between different jasmonate-regulated defense responses in *Arabidopsis*. *Plant Cell* 16: 1938–1950.

Love, A.J., Yun, B.W., Laval, V. et al. (2005). Cauliflower mosaic virus, a compatible pathogen of *Arabidopsis*, engages three distinct defense-signaling pathways and activates rapid systemic generation of reactive oxygen species. *Plant Physiol.* 139: 935–948.

Luck, J., Spackman, M., Freeman, A. et al. (2011). Climate change and diseases of food crops. *Plant Pathol.* 60: 113–121.

Luna, E., Pastor, V., Robert, J. et al. (2011). Callose deposition: a multifaceted plant defense response. *Mol. Plant Microbe Interact.* 24: 183–193.

Lund, S.T., Stall, R.E., and Klee, H.J. (1998). Ethylene regulates the susceptible response to pathogen infection in tomato. *Plant Cell* 10: 371–382.

Maksimov, I., Troshina, N., Surina, O., and Cherepanova, E. (2014). Salicylic acid increases the defense reaction against bunt and smut pathogens in wheat calli. *J. Plant Interact.* 9: 306–314.

Marion-Poll, A. and Leung, J. (2006). Abscisic acid synthesis, metabolism and signal transduction. In: *Plant Hormone Signaling, Annual Plant Reviews* (ed. P. Hedden and S.G. Thomas), 1–36. Oxford: Blackwell Publishing.

Mathur, P., Singh, V.P., and Kapoor, R. (2018). Interactive effects of CO 2 concentrations and Alternaria brassicae (Berk.) Sacc. infection on defense signalling in Brassica juncea (L.) Czern. & Coss. *Eur. J. Plant Pathol.* 151: 413–425.

McManus, M.T. (2012). *Annual Plant Reviews Vol. 44: The Plant Hormone Ethylene*. Oxford: Wiley-Blackwell.

Melotto, M., Underwood, W., Koczan, J. et al. (2006). Plant stomata function in innate immunity against bacterial invasion. *Cell* 126: 969–980.

Melotto, M., Underwood, W., and He, S.Y. (2008). Role of stomata in plant innate immunity and foliar bacterial diseases. *Annu. Rev. Phytopathol.* 46: 101–122.

Métraux, J.P., Nawrath, C., and Genoud, T. (2002). Systemic acquired resistance. *Euphytica* 124: 237–243.

Meyer, M.M. and Kirkpatrick, B.C. (2011). Exogenous applications of abscisic acid increase curing of Pierce's disease-affected grapevines growing in pots. *Plant Dis.* 95: 173–177.

Nascimento, F.X., Rossi, M.J., and Glick, B.R. (2018). Ethylene and 1-aminocyclopropane-1-carboxylate (ACC) in plant–bacterial interactions. *Front. Plant Sci.* 9: 114.

Oide, S., Bejai, S., Staal, J. et al. (2013). A novel role of PR 2 in abscisic acid (ABA) mediated, pathogen-induced callose deposition in *Arabidopsis thaliana*. *New Phytol.* 200: 1187–1199.

Pandey, P., Irulappan, V., Bagavathiannan, M.V., and Senthil-Kumar, M. (2017). Impact of combined abiotic and biotic stresses on plant growth and avenues for crop improvement by exploiting physio-morphological traits. *Front. Plant Sci.* 8: 537.

Pauwels, L., Ritter, A., Goossens, J. et al. (2015). The ring e3 ligase keep on going modulates jasmonate zim-domain12 stability. *Plant Physiol.* 169: 1405–1417.

Penmetsa, R.V., Uribe, P., Anderson, J. et al. (2008). The *Medicago truncatula* ortholog of *Arabidopsis* EIN2, sickle, is a negative regulator of symbiotic and pathogenic microbial associations. *Plant J.* 55: 580–595.

Petersen, M., Brodersen, P., Naested, H. et al. (2000). *Arabidopsis* MAP kinase 4 negatively regulates systemic acquired resistance. *Cell* 103: 1111–1120.

Pieterse, C.M., Van der Does, D., Zamioudis, C. et al. (2012). Hormonal modulation of plant immunity. *Annu. Rev. Cell Dev. Biol.* 28: 489–521.

Qi, P.F., Johnston, A., Balcerzak, M. et al. (2012). Effect of salicylic acid on *Fusarium graminearum*, the major causal agent of fusarium head blight in wheat. *Fungal Biol.* 116: 413–426.

Rezzonico, E., Flury, N., MeinsJr, F., and Beffa, R. (1998). Transcriptional down-regulation by abscisic acid of pathogenesis-related β-1,3-glucanase genes in tobacco cell cultures. *Plant Physiol.* 117: 585–592.

Ryals, J.A., Neuenschwander, U.H., Willits, M.G. et al. (1996). Systemic acquired resistance. *Plant Cell* 8: 1809.

Sabbagh, E., Sabbagh, S.K., Panjehkeh, N., and Bolok-Yazdi, H.R. (2018). Jasmonic acid induced systemic resistance in infected cucumber by *Pythium aphanidermatum*. *J. Agric. Sci.* 24: 143–152.

Schmidt, K., Pflugmacher, M., Klages, S. et al. (2008). Accumulation of the hormone abscisic acid (ABA) at the infection site of the fungus *Cercospora beticola* supports the role of ABA as a repressor of plant defence in sugar beet. *Mol. Plant Pathol.* 9: 661–673.

Shigenaga, A.M. and Argueso, C.T. (2016). No hormone to rule them all: interactions of plant hormones during the responses of plants to pathogens. *Semin. Cell Dev. Biol.* 56: 174–189.

Sillero, J.C., Rojas-Molina, M.M., Ávila, C.M., and Rubiales, D. (2012). Induction of systemic acquired resistance against rust, ascochyta blight and broomrape in faba bean by exogenous application of salicylic acid and benzothiadiazole. *Crop Prot.* 34: 65–69.

Song, S., Qi, T., Wasternack, C., and Xie, D. (2014). Jasmonatesignaling and crosstalk with gibberellin and ethylene. *Curr. Opin. Plant Biol.* 21: 112–119.

Spoel, S.H., Koornneef, A., Claessens, S.M. et al. (2003). NPR1 modulates cross-talk between salicylate-and jasmonate-dependent defense pathways through a novel function in the cytosol. *Plant Cell* 15: 760–770.

Spoel, S.H., Johnson, J.S., and Dong, X. (2007). Regulation of tradeoffs between plant defenses against pathogens with different lifestyles. *Proc. Natl. Acad. Sci. U. S. A.* 104: 18842–18847.

Thaler, J.S., Humphrey, P.T., and Whiteman, N.K. (2012). Evolution of jasmonate and salicylate signal crosstalk. *Trends Plant Sci.* 17: 260–270.

Thomma, B.P., Eggermont, K., Penninckx, I.A. et al. (1998). Separate jasmonate-dependent and salicylate-dependent defense-response pathways in *Arabidopsis* are essential for resistance to distinct microbial pathogens. *Proc. Natl. Acad. Sci. U. S. A.* 95: 15107–15111.

Ton, J., Flors, V., and Mauch-Mani, B. (2009). The multifaceted role of ABA in disease resistance. *Trends Plant Sci.* 14: 310–317.

Trouvelot, S., Varnier, A.L., Allegre, M. et al. (2008). A β-1,3 glucan sulfate induces resistance in grapevine against *Plasmopara viticola* through priming of defense responses, including HR-like cell death. *Mol. Plant Microbe Interact.* 21: 232–243.

Ulferts, S., Delventhal, R., Splivallo, R. et al. (2015). Abscisic acid negatively interferes with basal defence of barley against *Magnaporthe oryzae*. *BMC Plant Biol.* 15: 1–13.

Uno, Y., Furihata, T., Abe, H. et al. (2000). *Arabidopsis* basic leucine zipper transcription factors involved in an abscisic acid-dependent signal transduction pathway under drought and high-salinity conditions. *Proc. Natl. Acad. Sci. U. S. A.* 97: 11632–11637.

Van de Poel, B. and Van Der Straeten, D. (2014). 1-Aminocyclopropane-1-carboxylic acid (ACC) in plants: more than just the precursor of ethylene! *Front. Plant Sci.* 5: 640.

Van Zhong, G. and Burns, J.K. (2003). Profiling ethylene-regulated gene expression in *Arabidopsis thaliana* by microarray analysis. *Plant Mol. Biol.* 53: 117–131.

Vlot, A.C., Dempsey, D.M.A., and Klessig, D.F. (2009). Salicylic acid, a multifaceted hormone to combat disease. *Annu. Rev. Phytopathol.* 47: 177–206.

Vos, I.A., Verhage, A., Schuurink, R.C. et al. (2013). Onset of herbivore-induced resistance in systemic tissue primed for jasmonate-dependent defenses is activated by abscisic acid. *Front. Plant Sci.* 4: 539.

Wang, Y., Tao, X., Tang, X.M. et al. (2013). Comparative transcriptome analysis of tomato (*Solanum lycopersicum*) in response to exogenous abscisic acid. *BMC Genomics* 14: 1–14.

Wang, H., Li, S., Li, Y.A. et al. (2019). MED25 connects enhancer–promoter looping and MYC2-dependent activation of jasmonate signalling. *Nat. Plants* 5: 616–625.

Wasternack, C. and Kombrink, E. (2010). Jasmonates: structural requirements for lipid-derived signals active in plant stress responses and development. *ACS Chem. Biol.* 5: 63–77.

Wessels, J.G. (1993). Tansley Review No. 45 Wall growth, protein excretion and morphogenesis in fungi. *New Phytol.* 123: 397–413.

Wu, L., Li, X., Ma, L. et al. (2018). Acetoin and 2,3-butanediol from *Bacillus amyloliquefaciens* induce stomatal closure in *Arabidopsis thaliana* and *Nicotiana benthamiana*. *J. Exp. Bot.* 69: 5625–5635.

Xiong, L. and Yang, Y. (2003). Disease resistance and abiotic stress tolerance in rice are inversely modulated by an abscisic acid–inducible mitogen-activated protein kinase. *Plant Cell* 15: 745–759.

Yasuda, M., Ishikawa, A., Jikumaru, Y. et al. (2008). Antagonistic interaction between systemic acquired resistance and the abscisic acid–mediated abiotic stress response in *Arabidopsis*. *Plant Cell* 20: 1678–1692.

Yoshikawa, M., Takeuchi, Y., and Horino, O. (1990). A mechanism for ethylene-induced disease resistance in soybean: enhanced synthesis of an elicitor-releasing factor, β-1,3-endoglucanase. *Physiol. Mol. Plant Pathol.* 37: 367–376.

Zeng, W. and He, S.Y. (2010). A prominent role of the flagellin receptor FLAGELLIN-SENSING2 in mediating stomatal response to *Pseudomonas syringae* pv. *tomato* DC3000 in *Arabidopsis*. *Plant Physiol.* 153: 1188–1198.

Zhang, Y., Shi, X., Li, B. et al. (2016). Salicylic acid confers enhanced resistance to *Glomerella* leaf spot in apple. *Plant Physiol. Biochem.* 106: 64–72.

Zheng, X.Y., Spivey, N.W., Zeng, W. et al. (2012). Coronatine promotes *Pseudomonas syringae* virulence in plants by activating a signaling cascade that inhibits salicylic acid accumulation. *Cell Host Microbe* 11: 587–596.

Zhu, Z., An, F., Feng, Y. et al. (2011). Derepression of ethylene-stabilized transcription factors (EIN3/EIL1) mediates jasmonate and ethylene signaling synergy in *Arabidopsis*. *Proc. Natl. Acad. Sci. U. S. A.* 108: 12539–12544.

20

Recent Perspectives of Drought Tolerance Traits

Physiology and Biochemistry

Priya Yadav[1], Mohammad Wahid Ansari[1], Narendra Tuteja[2], and Moaed Al Meselmani[3]

[1] *Department of Botany, Zakir Husain Delhi College, University of Delhi, New Delhi, India*
[2] *Plant Molecular Biology Group, International Centre for Genetic Engineering and Biotechnology, New Delhi, India*
[3] *School of Biosciences, Grantham Centre, The University of Sheffield, Sheffield, England*

20.1 Introduction

Climate change and global warming cause environmental stress, which affects plant growth and production. Environmental stresses are of two types: biotic and abiotic stresses. Salinity stress, water stress, temperature stress, pH stress, and heavy metal stress are referred to as abiotic stress (Wang et al. 2003; Wania et al. 2016). These stresses affect plants at every developmental stage from germination to maturity. Drought is a type of water stress in which soil faces a lower level of water concentration. Drought is a meteorological phenomenon in which a spell of precipitation is insufficient. Water deficiency results in harmful impacts on plant growth and production, especially at reproductive phase, as reported in legumes (Golldack et al. 2014; Farooq et al. 2016). Harmful effects caused by drought stress include disturbance in water uptake due to low diffusion pressure and high diffusion pressure deficit, loss of turgor pressure, cellular dehydration, disturbed cell elongation and division, reduced metabolism rate, increased oxidative damage, and decreased yield. Also, it affects the activity of enzymes and ion balance (Anjum et al. 2017; Hussain et al. 2018). These effects depend on the duration, severity of stress, and the sensitivity of plants (Zhu 2002). Drought stress depends on various factors such as insufficient rainfall, water holding capacity of the soil, high light intensity and temperature, and evapotranspiration. Drought is of two types: physiological drought and physical drought. Physiological drought, also called pseudo-drought, is a condition where plants are not able to absorb water due to high salt concentration in the soil, whereas soil contains sufficient amounts of water. Physical drought is a real drought; this is a condition in which the water level in the soil is very low (Passioura and Angus 2010; Salehi-Lisar and Bakhshayeshan-Agdam 2020). Not only water uptake, nutrient uptake is also affected during drought stress. Nutrients are a basic need for plant growth, and a lack of nutrients negatively affects plant growth and development (Da Silva et al. 2011). Major sources of drought stress include abnormal rainfall, global warming, and shifts in monsoon pattern (Seleiman et al. 2020). Global warming results in increased temperature, which enhances the water deficiency in plants as this increased temperature increases the evapotranspiration rate in plants and decreases the internal water (Sultan et al. 2019). Due to global warming, carbon dioxide levels increased in the environment, which results in enhanced photosynthetic rate (Brown et al. 2018). Some of the reasons behind disturbed patterns of rainfall or less rainfall are urbanization, industrialization, and urbanization (Fatima et al. 2020). Rainfall dependent Crop land faces more drought conditions. Less or no rainfall largely affects plant production (Konapala et al. 2020). Plants develop different resistance mechanisms to manage drought stress. Improves morphological, physiological, and biochemical traits, cellular signaling pathway, physical adaptation (Osakabe et al. 2014; Yadav et al. 2020). Cuticular wax accumulation is found as a potent defense against drought stress in many plants (Borisjuk et al. 2014). Some acclimatization strategies plants follow are (i) drought escape plants complete their life cycle before onset of drought, (ii) drought avoidance (formation of thick critical wax deposition and modifications of leaf into spines), and (iii) drought tolerance. These mechanisms made plants survive and adapt during drought stress (Levitt 1980). This chapter provides information about drought stress responses and recent advances in tolerance mechanisms involving physiological and biochemical traits.

Global Climate Change and Plant Stress Management, First Edition. Edited by Mohammad Wahid Ansari, Anil Kumar Singh, and Narendra Tuteja.

20.2 Effects and Response During Drought Stress on Physiological and Biochemical Traits of Plants

Plant growth and development are affected by soil water levels that are higher or lower than optimal. Proper water level and nutrient availability are necessary for plant growth (Elemike et al. 2019). Water deficiency in soil affects the uptake of nutrients and water, which results in leaf yellowing, dropping and senescence, education, flower sagging, short plant height, leaf rolling, overall plant wilting (Corso et al. 2020). In comparison to cell division, cell enlargement gets affected more during drought stress (Humplík et al. 2017). During drought stress, the metabolic process of plants gets affected, such as respiration and photosynthesis (Birami et al. 2018). Drought stress affects seed germination in a negative manner. During germination, absorption of water by seed is the first step. Seeds imbibe water and swell. Chemical composition of the seed defines the amount of water to be absorbed (Rahmani et al. 2006). If water is not available in sufficient amounts for imbibition, germination will be reduced or stopped completely (Leila et al. 2007). The impact of drought stress during the vegetative stage is less important as compared to the reproductive stage as plant received overall production after reproduction. Whereas at the vegetative stage, drought also shows negative effects, reducing leaf and stem development, disrupting physiological processes, and changing metabolism (Kabiri 2010). Root length is used to assess drought tolerance (Ali et al. 2011), as during drought stress root length and diameter reduce (Bibi et al. 2010). Water usage efficiency during drought stress is an important physiological aspect to understand tolerance. It was found that tolerant species show better water usage efficiency, low transpiration, and less water consumption as compared to susceptible cultivars (Abbate et al. 2004). Physiological functions show changes during drought stress like relative water content, transpiration rate, stomatal conductance, flowering time, chlorophyll content, and plant architecture (Din et al. 2011). Photosynthetic rate is affected by the content of photosynthetic pigment and stress. During drought stress, plants show early stomatal closure and halt carbon dioxide diffusion, which limits photosynthesis and carbon dioxide assimilation. Hence, plant dry weight is reduced (Pandey and Shukla 2015). Drought stress also affects chlorophyll content; tolerant species are found to have more chlorophyll compared to susceptible species (Dawood et al. 2019). Rubisco enzyme activity decreases, chloroplast photooxidation occurs, chlorophyllase activity enhances, and chlorophyll synthesis reduces (Kabiri et al. 2014). Relative water content, water loss rate, and transpiration rate regulate dehydration in plants. During drought stress RWC, water loss rate and transpiration rate reduce; leaf surface area exposure reduces leaf rolling, which decreases transpiration rate (Lonbani and Arzani 2011). Epicuticular waxes and organic compounds of cuticle and trichome density reduce water loss and protect plants against drought stress (Akrami and Yousefi 2015). During drought stress, root length and volume reduce, which results in reduced uptake of macronutrients (N, P, and K) as their availability during drought conditions is reduced (Noman et al. 2018). As water level decreases in soil, phosphorus mobility also decreases, which results in reduced uptake of phosphorus (Faye et al. 2006). Noman et al. (2018) report reductions in calcium, phosphorus, and potassium uptake in wheat plants. Not only macronutrients, plants also face micronutrient deficiency during drought stress. Nutrient deficiency causes chlorosis and necrosis in plants (Havlin 1999) (Table 20.1). Biochemical effects involve decrease in ATP production, a reduction in activity of Rubisco, phosphoenolpyruvate carboxylase, and NADP-dependent glyceraldehyde phosphate dehydrogenase enzyme, and an increase in activity of proline dehydrogenase, trehalose phosphate synthase. Accumulation of osmoprotectant, abscisic acid (ABA), and ethylene helps in maintaining tissue water, stomatal closing, leaf senescence, and fruit ripening, respectively (Ozturk et al. 2021; Zhang et al. 2021). Drought stress affects the activity of enzymes involved in the synthesis and transport of sugars and amino acids from source to sink like glutamate pyruvic aminotransferase (GPT), soluble starch synthase (SSS), granule-bound starch synthase (GBSS), sucrose synthase (SS), and glutamine synthetase (GS) (Saeidi and Abdoli 2015; Singh et al. 2018).

Resistance mechanism through which plants manage to cope up with drought stress includes basically three types, as explained in Figure 20.1. Plant production gets adversely affected by harmful effects of drought stress. Plants show different adaptive mechanisms to protect themselves from drought stress (Batool et al. 2020). Escape mechanism is one of the adaptive or resistant mechanisms that plants show against drought stress. Plants shorten their life cycle, and their seasonal growth and reproduction occur before the onset of drought (Alvarez et al. 2018). Plants' growth period reduces; early following occur and completes their life cycle early to protect themselves from drought (Blum 2011). Drought avoidance is another adaptive mechanism in which plants use different strategies to avoid the effects of drought stress, such as closing their stomata to reduce stomatal transpiration rate and increase water uptake (Dobra et al. 2010). Deposition of wax and formation of cuticles are good defense against drought stress (Boulard et al. 2017). Tolerance response mechanism works basically at the biochemical and physiological levels. Metabolic and biochemical dynamics change to make plants adaptive against drought stress. Solute accumulation occurs for osmotic adjustment, antioxidant defense enzymes activate, and photosynthesis and respiration processes modify (Lopez-Galiano et al. 2019; Zhang et al. 2019).

Table 20.1 Drought-mediated physiological and biochemical consequences and plant response to mitigate its various negative effects.

Drought stress			
Various adverse effects		Plant response	
Physiological	**Biochemical**	**Physiological**	**Biochemical**
Decrease in stomatal closure and CO_2 intake	Overproduction of ROS	Decrease in stomatal closure and leaf expansion	Increase in ROS scavenging species production (SOD, CAT, GR, and APX)
Decrease in photosynthesis and respiration	Protein and DNA damage	Decrease in transpiration	Production of ABA and jasmonates
Disturbance in cell wall integrity and increase in electrolyte leakage	Oxidative damage	Osmotic adjustment, accumulation of sugar, amino acid, and alkaloids	Changes in carbohydrate metabolism
Disturbance in the electron transport chain	Reduction in chlorophyll content	Maintenance of turgor pressure	Accumulation of polyamines
Capillary movement inhibited and translocation affected	Photochemical efficiency affected	Improved water usages efficiency	Increase in antioxidant enzyme activity
Decrease in transportation rate	Rubisco activity reduced	Improved root growth and increased nutrient uptake	Upregulation of hormonal signaling
Reduction in relative water content	Antioxidant defense system affected	Ethylene production Increase in fruit-ripening stimulation	
Increase in internal temperature	Carbohydrate productivity reduced	Leaf senescence altered by ethylene	
Reduction in macro- and micronutrients uptake Seed germination inhibited and increase in epinasty	Down regulation of primary metabolites	Production of abscisic acid (ABA), resulting in leaf senescence and stomatal closing	

20.3 Recent Advances in Drought Stress Tolerance

There are several methods used to alleviate the negative effects of drought stress. Plant growth regulators play an important role in improving the adverse effects of drought stress. Gibberellic acid (GA) improves the internal water balance, photosynthesis, respiration, and stomatal conductance during drought stress reported in wheat and maize; enhances the synthesis of APX and CAT; and increases the overall yield of plants (Javid et al. 2011). Salicylic acid application in the case of wheat plants increases the accumulation of proline content, which indirectly helps in drought stress tolerance. Salicylic acid treatment improves the growth and yield of plants under stress (Miura and Tada 2014). ABA and jasmonic acid's exogenous application found beneficial during stress conditions. ABA promotes stomatal closure, which reduces transpiration, enhancing water conservation strategies. The improved SOD and CAT activity lead to enhanced proline content that improves crop productivity (Ahmad et al. 2018). Increased ethylene concentration stimulates leaf abscission and decreases water loss (Sewelam et al. 2014). Brassinolide activates the BES1/BZR1 transcription factor to enhance plant defense and regulate ROS (Im-Kim et al. 2013). Auxin cytokinin ratio favors more and more root growth; increases water absorption and activities of SOD, peroxidase, and catalase; and improves plant survival in drought stress (Jan et al. 2019). Hydrogel is a type of water-absorbing polymer that helps in locking soil moisture. Hydrogel helps plants in semiarid and arid conditions by reducing the need for regular irrigation. By maintaining the soil moisture, hydrogel protects plants from wilting and increases the survival time of plants during drought stress. (Tu et al. 2018). Soil with hydrogel shows less water loss through evaporation as compared to non-hydrogel-containing soil. Hydrogel also decreases the hydraulic conductance of soil (Saha et al. 2020). Seed priming approach for drought stress tolerance is the best short-term approach. This process mainly helps at the time of germination of seed. Primed seeds show enhanced germination rates (Nawaz et al. 2013). Calcium chloride osmopriming in wheat plants helps to accumulate osmolytes. These osmolytes lighten up the osmotic stress

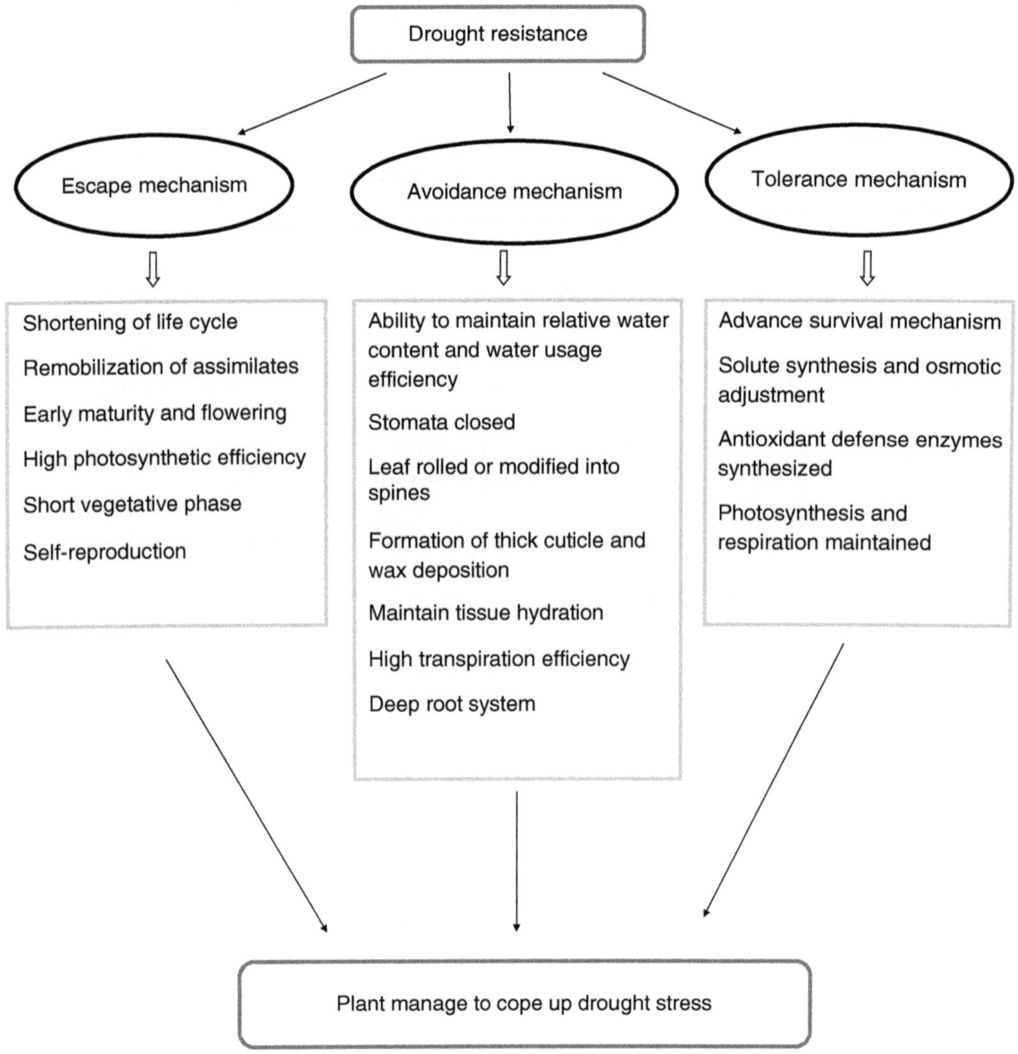

Figure 20.1 Drought resistance mechanisms shown by plants.

caused by drought. Also, maintain relative water content and improve the grain yield of plants (Tabassum et al. 2018). Molecular priming has been found to be beneficial in inducing tolerance against stress. It mainly works against oxidative stress (Kerchev et al. 2020). SNP osmopriming enhances the antioxidant enzymes production, maintains relative water content, and increases the solute accumulation. It improves photosynthesis and maintains membrane stability (Farooq et al. 2009). Drought stress tolerance using nanoparticles (NPs) is a novel approach that shows good results to a greater extent. Nanoparticles of zinc, iron, silver, and silicon were used, but the exact mechanism of action is still unknown. Some studies reported that these nanoparticles induce drought tolerance by regulating water channels, seed germination, nutrient uptake, photosynthetic rate, etc. (Khodakovskaya et al. 2011; Li et al. 2020). Ze et al. (2011) reported that TiO_2 NPs enhanced the photosynthetic rate during drought stress by improving light absorption and light distribution between photosystems I and II. ZnO NPs were also found beneficial in the case of maize during drought stress. ZnO NPs regulate stomatal movement, improve water content, promote photosynthesis, and also stimulate some enzymes that help in metabolism (Sun et al. 2021). Polyamines not only influence growth and development but also play an important role in stress mitigation. It was reported that the endogenous level of polyamines was higher in plants under stress compared to non-stressed plants. Overexpression of polyamines improves abiotic stress tolerance (Kasukabe et al. 2004). Yang et al. (2007) reported that rice produces polyamines in response to drought stress. Exogenous application of *spermine* improves the antioxidant defense mechanism, oxidative stress, and glyoxalase system and helps in mitigating drought stress (Hasan et al. 2021). Polyamines in *Agrostis stolonifera* create tolerance against drought stress by regulating nitrogen metabolism, tricarboxylic acid cycle (TCA), glutamate decarboxylase, and α-ketoglutarase and declining nitrite nitrogen content; nitrate reductase, glutamine synthetase, glutamate synthetase, and GABA aminotransferase also maintain higher photosynthetic rate, cell membrane stability, water usage efficiency (Tan et al. 2022). Antioxidant defense system of plants includes enzymatic and non-enzymatic components and plays an important role in stress mitigation. The reactive oxygen species in plants are removed by antioxidant enzymes, which provide a most efficient defense against osmotic stress (Farooq et al. 2009). During water deficit condition, expression of antioxidant genes increases (Ratnayaka et al. 2003).

20.4 Arbuscular Mycorrhizal Fungi (AMF) and Plant Growth-Promoting Rhizobacteria (PGPRs) in Drought Stress Tolerance

Arbuscular mycorrhizal fungi (AMF) associated plants experience increased water and nutrient uptake. AMF hyphae help in water absorption and maintaining water content in symbiotic plants (Huang et al. 2017). AMF also increases the production of enzymatic and non-enzymatic antioxidant, non-enzymatic antioxidants include glutathione (GSH), ascorbate (ASC), carotenoid, flavonoid, and tocopherol. Enzymatic antioxidants include ascorbate peroxidases (APX), superoxide dismutase (SOD), catalase (CAT), glutathione reductase (GR), guaiacol peroxidase (G-POD), and glutathione peroxidase (GPX); they help in ROS scavenging (Bahmani et al. 2018). AMF in symbiotic association regulates enzymatic antioxidant transcription (Abbaspour et al. 2012). AMF increases the accumulation of proline in plants. Proline acts as an osmoprotectant and protects many proteins from denaturation (Goicoechea et al. 1998). Also, soluble sugar accumulation takes place, which helps in adjusting osmotic stress in plants. AMF also helps in maintaining photosynthetic apparatus and improves phenotypic plasticity (Bakr et al. 2018). AM symbiosis helps in water uptake and transport. Mycorrhizal roots with *Rhizophagus irregularis* mycelia show overexpression of gintAQPF1 and gintAQPF2 aquaporins, which improve water uptake during drought stress (Li et al. 2013). Drought stress tolerance is also achieved by the application of nanoparticles. Nanoparticles improve productivity and enhance the survival of plants during stress conditions (Jsarotia et al. 2018). Titanium dioxide nanoparticles were found beneficial in improving seed germination in wheat varieties (Shafea et al. 2017). Zinc and copper nanoparticles improve relative water content, increase antioxidant enzymes, and decrease MDA (Seleiman et al. 2020). Zinc oxide nanoparticles improve physiological as well as biochemical traits of plants, like photosynthesis pigment level and leaf water content (Taran et al. 2017) (Table 20.2). Another side of the story must be investigated to determine whether these nanoparticles that induce stress tolerance have any negative impact on grain quality, grain weight, and flowering time (Abdelsalam et al. 2018). Osmoprotectant signaling plays an important role in drought stress tolerance. Osmolytes accumulate and maintain the physiological processes and improve plant growth and development (Brito et al. 2019). Exogenous application of proline, mannitol, and trehalose, glycine betaine helps plants maintain osmotic balance, antioxidant enzyme activity, photosynthetic efficiency, and plant growth (Zouari et al. 2019). Fructan accumulation reported in sugar beet helps in osmotic adjustment, ROS scavenging mechanisms, and stabilization of the membrane during drought stress (Pilon-Smits et al. 1999). PGPRs play an important role in drought stress tolerance. PGPRs present symbiotically and non-symbiotically in rhizosphere and provide many growth-promoting benefits (Kabiraj et al. 2020). PGPRs improve micronutrient availability, are involved in nitrogen fixation, release phytohormones, improve root length and number of root hairs by colonizing roots, hence increasing water uptake, and maintain soil composition (Kabiraj et al. 2020; Pathania et al. 2020). *Aospirillum* sp. increase proline content enhanced seedling growth, plant biomass, and seed germination in maize (Garcia et al. 2017). Niu et al. (2018) reported that *Pseudomonas fluorescens* enhanced seed germination and plant growth in foxtail millet (Table 20.2).

20.5 Genomic Level Approach in Drought Stress Tolerance

Environmental interaction and genotype together involved in drought stress (Fleury et al. 2010). Drought tolerance related traits need to be controlled at the genetic level for better tolerance. QTL hotspot detection, nested associated mapping, AB-QTL are some mapping techniques used to target traits that play a role in drought tolerance (Yuliasti and Reflinur 2015). Drought stress tolerance traits were found in wheat (Kollers et al. 2013) and rice (Huang et al. 2010) using QTL mapping. Molecular approaches and traditional breeding in combination are good strategies to enhance stress tolerance (Bhatnagar-Mathur et al. 2008). Using conventional breeding methods and molecular approaches, drought-tolerant transgenic plants can be made with better water usage efficiency, water uptake quality, and stomatal conductance (Oliveira et al. 2020).

Table 20.2 Drought stress resistance induction using a different approach in plants.

Strategies	Crops	Effects	References
Phytohormones			
Auxin	Potato	ROS produce and maintain oxidative stress and water loss reduction	Im-Kim et al. (2013)
GA (gibberellic acid)	Tomato	Reduce transpiration, reduce stomatal pores, and improve growth	Nir et al. (2014)
ABA (abscisic acid)	Rice	Increase SOD, increase osmoprotectant, reduce stomata and leaf area, reduce transpiration, and improve water usage efficiency	Yu et al. (2013)
GA (gibberellic acid)	Rice	Increase APX and CAT, improve photosynthesis and root length increases, and improve proline content	Wu et al. (2016)
Osmoprotectant			
Glycine betaine	Rice, barley	Improve osmotic disturbance and enhance photosynthetic efficiency	Ashraf and Foolad (2007)
Mannitol D-Ononitol	Maize	Free radical scavenging, improved water usage efficiency, and osmotic adjustment	Kaya et al. (2013)
Nanoparticles			
Iron	Sunflower	Improve yield and oil content	Davar et al. (2014)
Zinc oxide	Soybean	Improve seed germination and growth	Sedghi et al. (2013)
Silicon	Tomato, rice	Increase CAT, GR, and SOD activity and increase ascorbate content	Shi et al. (2014) and Ma et al. (2006)
Potassium	Rice, wheat	Increase osmotic protectant, improve growth, and increase chlorophyll content and proline content	Zain et al. (2014) and Jatav et al. (2021)
Microbes			
Piriformospora indica	*Arabidopsis*	Decrease saliyclic acid and ABA content and enhance jasmonic acid level	Sherameti et al. (2008)
Trichoderma harzianum	Rice	Delay drought stress and improve root growth	Shukla et al. (2012)
Funneliformis mosseae, *Rhizophagus irregularis*	*Pelargonium graveolens*	Increase oil content and yield, decrease MDA, and increase SOD, CAT, and APX	Amiri et al. (2015)
Rhizophagus irregularis	*Zea mays*	Improve plant growth, decrease MDA content, increase proline, and improve relative water content	Zhao et al. (2015)
Glomus intraradices *Bacillus* spp. *Pseudomonas azotoformans*	*Calotropis Procera* Maize Wheat	Sodium, potassium content increases, CAT, SOD, APX activity increases improved nutrient uptake, reduced ascorbate peroxidase and glutathione reductase activities, increased proline content. Improved seed germination, increased root length and shoot length	Bahmani et al. (2018), Silva et al. (2020), and Ansari et al. (2021)

Genes involved in drought stress tolerance include TaNAC69 gene in wheat, which increases tolerance to drought stress (Budak et al. 2013); NF-YB2 gene in maize, which increases photosynthetic rate and improves crop yield during drought stress (Vandenbroucke and Metzlaff 2013); HSP70-1 in tobacco plant, which enhances tolerance mechanism (Cho and Hong 2006); AP37 and OSNAC10 in rice, which increase yield and tolerance during drought stress (Oh et al. 2013). Signal peptides control various growth and development processes, and recent studies have reported that these peptides are involved in abiotic stress responses. These signaling peptides are small, 5–20 amino acid or 40–100 amino acids, in length. They are produced by direct translation of open reading frame or by proteolytic processing of precursor proteins. Under drought stress, CLE25 increases ABA biosynthesis, which regulates stomatal closure (Tan et al. 2022).

20.6 Conclusion

Environmental stresses create global food security threats nowadays. Losses in agricultural productivity due to these environmental stresses are a serious problem. These stresses affect plants right from morphology to physiology, biochemical, and metabolic processes, which results in a direct negative effect on crop yield. Plants adopt some mechanisms to withstand these stresses. Drought, being a major environmental stress, harms a wide range of crop production. Plants try to minimize the harmful effects caused by drought stress by using response mechanisms like avoidance, tolerance, and escape. Changes at structural level also found leaf rolling, stomatal closure, root length increment. At physiological level, plants show increased water usage efficiency, decrease transpiration, regulations of photosynthetic efficiency, respiration, and osmotic regulations. Involvement of microbes, nanoparticles, phytohormones, and priming mechanisms was also found beneficial in ameliorating the adverse effects of drought stress. Other methods involve breeding strategies, management practices, and molecular and genomic perspectives. Future research should focus toward multidisciplinary approaches to understand and alleviate drought stress tolerance.

References

Abbaspour, H., Saeidi-Sar, S., Afshari, H., and Abdel-Wahhab, M.A. (2012). Tolerance of mycorrhiza infected pistachio (*Pistacia vera* L.) seedling to drought stress under glasshouse conditions. *J. Plant Physiol.* 169: 704–709.

Abbate, P.E., Dardanelli, J.L., Cantarero, M.G. et al. (2004). Climatic and water availability effects on water-use efficiency in wheat. *Crop Sci.* 44: 474–483.

Abdelsalam, N.R., Abdel-Megeed, A., Ali, H.M. et al. (2018). Genotoxicity effects of silver nanoparticles on wheat (*Triticum aestivum* L.) root tip cells. *Ecotoxicol. Environ. Saf.* 155: 76–85.

Ahmad, I., Kamran, M., Ali, S. et al. (2018). Seed filling in maize and hormones crosstalk regulated by exogenous application of uniconazole in semiarid regions. *Environ. Sci. Pollut. Res.* 25: 33225–33239.

Akrami, M. and Yousefi, Z. (2015). Biological control of Fusarium wilt of tomato (*Solanum lycopersicum*) by *Trichoderma* spp. as antagonist fungi. *Biol. Forum Int. J.* 7: 887–892.

Ali, M.A., Abbas, A., Awan, S.I. et al. (2011). Correlated response of various morpho-physiological characters with grain yield in sorghum landraces at different growth phases. *J. Anim. Plant Sci.* 21: 671–679.

Álvarez, S., Rodríguez, P., Broetto, F., and Sánchez-Blanco, M.J. (2018). Long term responses and adaptive strategies of *Pistacia lentiscus* under moderate and severe deficit irrigation and salinity: osmotic and elastic adjustment, growth, ion uptake and photosynthetic activity. *Agric. Water Manag.* 202: 253–262.

Amiri, R., Nikbakht, A., and Etemadi, N. (2015). Alleviation of drought stress on rose geranium [*Pelargonium graveolens* (L.) Herit.] in terms of antioxidant activity and secondary metabolites by mycorrhizal inoculation. *Sci. Hortic.* 197: 373–380.

Anjum, S.A., Ashraf, U., Zohaib, A. et al. (2017). Growth and developmental responses of crop plants under drought stress: a review. *Zemdirbyste-Agriculture* 104: 267–276.

Ansari, F.A., Jabeen, M., and Ahmad, I. (2021). *Pseudomonas azotoformans* FAP5, a novel biofilm forming PGPR strain, alleviates drought stress in wheat plant. *Int. J. Environ. Sci. Technol.* 18: 3855–3870.

Ashraf, M. and Foolad, M.R. (2007). Roles of glycine betaine and proline in improving plant abiotic stress resistance. *Environ. Exp. Bot.* 59: 206–216.

Bahmani, M., Naghdi, R., and Kartoolinejad, D. (2018). Milkweed seedlings tolerance against water stress: comparison of inoculations with *Rhizophagus irregularis* and *Pseudomonas putida*. *Environ. Technol.* 10: 111–121.

Bakr, J., Pék, Z., Helyes, L., and Posta, K. (2018). Mycorrhizal inoculation alleviates water deficit impact on field-grown processing tomato. *Pol. J. Environ. Stud.* 27: 1949–1958.

Batool, T., Ali, S., Seleiman, M.F. et al. (2020). Plant growth promoting rhizobacteria alleviates drought stress in potato in response to suppressive oxidative stress and antioxidant enzymes activities. *Sci. Rep.* 10: 16975.

Bhatnagar-Mathur, P., Vadez, V., and Sharma, K.K. (2008). Transgenic approaches for abiotic stress tolerance in plants: retrospect and prospects. *Plant Cell Rep.* 27: 411–424.

Bibi, A., Sadaqat, H.A., Akram, H.M., and Mohammed, M.I. (2010). Physiological markers for screening sorghum (*Sorghum bicolor*) germplasm under water stress condition. *Int. J. Agric. Biol.* 12: 451–455.

Birami, B., Gattmann, M., Heyer, A.G. et al. (2018). Heat waves alter carbon allocation and increase mortality of Aleppo pine under dry conditions. *Front. For. Glob. Chang. Front.* 1: 8.

Blum, A. (ed.) (2011). Plant water relations, plant stress and plant production. In: *Plant Breeding for Water-Limited Environments*, 11–52. Berlin/Heidelberg: Springer.

Borisjuk, N., Hrmova, M., and Lopato, S. (2014). Transcriptional regulation of cuticle biosynthesis. *Biotechnol. Adv.* 32: 526–540.

Boulard, T., Roy, J.C., Pouillard, J.B. et al. (2017). Modeling of micrometeorology, canopy transpiration and photosynthesis in a closed greenhouse using computational fluid dynamics. *Biosyst. Eng.* 158: 110–133.

Brito, C., Dinis, L.T., Moutinho-Pereira, J., and Correia, C.M. (2019). Drought stress effects and olive tree acclimation under a changing climate. *Plants* 8: 232.

Brown, S., Nicholls, R.J., Lázár, A.N. et al. (2018). What are the implications of sea-level rise for a 1.5, 2 and 3 °C rise in global mean temperatures in the Ganges-Brahmaputra-Meghna and other vulnerable deltas? *Regul. Environ. Chang.* 18: 1829–1842.

Budak, H., Kantar, M., and Yucebilgili Kurtoglu, K. (2013). Drought tolerance in modern and wild wheat. *Sci. World J.* 4: 66–75.

Cho, E.K. and Hong, C.B. (2006). Over-expression of tobacco NtHSP70-1 contributes to drought-stress tolerance in plants. *Plant Cell Rep.* 25: 349–358.

Corso, D., Delzon, S., Lamarque, L.J. et al. (2020). Neither xylem collapse, cavitation, or changing leaf conductance drive stomatal closure in wheat. *Plant Cell Environ.* 43: 854–865.

Da Silva, E.C., Nogueira, R., da Silva, M.A., and de Albuquerque, M.B. (2011). Drought stress and plant nutrition. *Plant Stress* 5: 32–41.

Davar, F., Zareii, A.R., and Amir, H. (2014). Evaluation the effect of water stress and foliar application of Fe nanoparticles on yield, yield components and oil percentage of safflower (*Carthamus tinctorious* L.). *Int. J. Adv. Biol. Biomed. Res.* 2: 1150–1159.

Dawood, M.F.A., Abeed, A.H.A., and Aldaby, E.E.S. (2019). Titanium dioxide nanoparticles model growth kinetic traits of some wheat cultivars under different water regimes. *Indian J. Plant Physiol.* 24: 129–140.

Din, J., Khan, S.U., Ali, I., and Gurmani, A.R. (2011). Physiological and agronomic response of canola varieties to drought stress. *J. Anim. Plant Sci.* 21: 78–82.

Dobra, J., Motyka, V., Dobrev, P. et al. (2010). Comparison of hormonal responses to heat, drought and combined stress in tobacco plants with elevated proline content. *J. Plant Physiol.* 167: 1360–1370.

Elemike, E.E., Uzoh, I.M., Onwudiwe, D.C., and Babalola, O.O. (2019). The role of nanotechnology in the fortification of plant nutrients and improvement of crop production. *Appl. Sci.* 9: 499.

Farooq, M., Basra, S.M.A., Wahid, A., and Rehman, H. (2009). Exogenously applied nitric oxide enhances the drought tolerance in fine grain aromatic rice (*Oryza sativa* L.). *J. Agron. Crop Sci.* 195: 254–261.

Farooq, M., Gogoi, N., Barthakur, S. et al. (2016). Drought stress in grain legumes during reproduction and grain filling. *J. Agron. Crop Sci.* 203: 81–102.

Fatima, A., Farid, M., Safdar, K. et al. (2020). Loss of AgroBiodiversity and productivity due to climate change in continent Asia: a review. In: *Plant Ecophysiology and Adaptation Under Climate Change: Mechanisms and Perspectives* (ed. M. Hasanuzzaman), 51–71. Berlin/Heidelberg: Springer.

Faye, I., Diouf, O., Guisse, A. et al. (2006). Characterizing root responses to low phosphorus in pearl millet [*Pennisetum glaucum* (L.) R. Br.]. *Agron. J.* 98: 1187–1194.

Fleury, D., Jefferies, S., Kuchel, H., and Langridge, P. (2010). Genetic and genomic tools to improve drought tolerance in wheat. *J. Exp. Bot.* 61: 3211–3222.

García, J.E., Maroniche, G., Creus, C. et al. (2017). in vitro PGPR properties and osmotic tolerance of different Azospirillum native strains and their effects on growth of maize under drought stress. *Microbiol. Res.* 202: 21–29.

Goicoechea, N., Szalai, G., Antolín, M.C. et al. (1998). Influence of arbuscular mycorrhizae and Rhizobium on free polyamines and proline levels in water-stressed alfalfa. *J. Plant Physiol.* 153: 706–711.

Golldack, D., Li, C., Mohan, H., and Probst, N. (2014). Tolerance to drought and salt stress in plants: unraveling the signaling networks. *Front. Plant Sci.* 5: 151.

Hasan, M.M., Skalicky, M., Jahan, M.S. et al. (2021). Spermine: its emerging role in regulating drought stress responses in plants. *Cells* 10 (2): 261.

Havlin, J. (1999). *Soil Fertility and Fertilizers: An Introduction to Nutrient Management*, 6e. Upper Saddle River, NJ: Prentice Hall.

Huang, X., Wei, X., Sang, T. et al. (2010). Genome-wide association studies of 14 agronomic traits in rice landraces. *Nat. Genet.* 42: 961–967.

Huang, Y.M., Zou, Y.N., and Wu, Q.S. (2017). Alleviation of drought stress by mycorrhizas is related to increased root H_2O_2 efflux in trifoliate orange. *Sci. Rep.* 7: 42335.

Humplík, J.F., Bergougnoux, V., and Van Volkenburgh, E. (2017). To stimulate or inhibit? That is the question for the function of abscisic acid. *Trends Plant Sci.* 22: 830–841.

Hussain, M., Farooq, S., Hasan, W. et al. (2018). Drought stress in sunflower: physiological effects and its management through breeding and agronomic alternatives. *Agric. Water Manag. J.* 201: 152–166.

Im-Kim, J., Baek, D., and Park, H.C. (2013). Overexpression of Arabidopsis YUCCA6 in potato results in high-auxin developmental phenotypes and enhanced resistance to water deficit. *Mol. Plant* 6: 337–349.

Jan, S., Abbas, N., Ashraf, M., and Ahmad, P. (2019). Roles of potential plant hormones and transcription factors in controlling leaf senescence and drought tolerance. *Protoplasma* 256: 313–329.

Jatav, K.S., Agarwal, R.M., Singh, R.P., and Shrivastava, M. (2021). Growth and yield responses of wheat (*Triticum aestivum* L.) to suboptimal water supply and different potassium doses. *J. Funct. Environ. Bot.* 2: 39–51.

Javid, M.G., Sorooshzadeh, A., Moradi, F. et al. (2011). The role of phytohormones in alleviating salt stress in crop plants. *Aust. J. Crop Sci.* 5: 726.

Jsarotia, P., Kashyap, P.L., Bhardwaj, A.K. et al. (2018). Nanotechnology scope and applications for wheat production and quality enhancement: a review of recent advances. *Wheat Barley Res.* 10: 137–150.

Kabiraj, A., Majhi, K., Halder, U. et al. (2020). Role of plant growth-promoting rhizobacteria (PGPR) for crop stress management. In: *Sustainable Agriculture in the Era of Climate Change* (ed. R. Roychowdhury, S. Choudhury, M. Hasanuzzaman, and S. Srivastava), 367–389. Cham, Switzerland: Springer.

Kabiri, R. (2010). Effect of salicylic acid to reduce oxidative stress caused by drought in the hydroponic cultivation of *Nigella sativa*. MA thesis. Kerman University Shahid Bahonar.

Kabiri, R., Nasibi, F., and Farahbakhsh, H. (2014). Effect of exogenous salicylic acid on some physiological parameters and alleviation of drought stress in *Nigella sativa* plant under hydroponic culture. *Plant Prot.* 50: 43–51.

Kasukabe, Y., He, L., Nada, K. et al. (2004). Overexpression of spermidine synthase enhances tolerance to multiple environmental stresses and up-regulates the expression of various stress-regulated genes in transgenic Arabidopsis thaliana. *Plant Cell Physiol.* 45 (6): 712–722.

Kaya, C., Sonmez, O., Aydemir, S. et al. (2013). Exogenous application of mannitol and thiourea regulates plant growth and oxidative stress responses in salt-stressed maize (*Zea mays* L.). *J. Plant Interact.* 3: 234–344.

Kerchev, P., van der Meer, T., Sujeeth, N. et al. (2020). Molecular priming as an approach to induce tolerance against abiotic and oxidative stresses in crop plants. *Biotechnol. Adv.* 1: 107503.

Khodakovskaya, M.V., de Silva, K., Nedosekin, D.A. et al. (2011). Complex genetic, photothermal, and photoacoustic analysis of nanoparticle-plant interactions. *Proc. Natl. Acad. Sci.* 108: 1028–1033.

Kollers, S., Rodemann, B., Ling, J. et al. (2013). Whole genome association mapping of Fusarium head blight resistance in European winter wheat (*Triticum aestivum* L.). *PLoS One* 22: 57500.

Konapala, G., Mishra, A.K., Wada, Y., and Mann, M.E. (2020). Climate change will affect global water availability through compounding changes in seasonal precipitation and evaporation. *Nat. Commun.* 11: 1–10.

Leila, R. (2007). Response of Tunisian autochthonous pearl millet to drought stress induced by polyethylene glycol 6000. *Afr. J. Biotechnol.* 6: 1102–1105.

Levitt, J. (ed.) (1980). Chilling, freezing, and high temperature stresses. In: *Responses of Plants to Environmental Stress*, vol. 2, 23–445. New York: Academic Press.

Li, T., Hu, Y.J., Hao, Z.P. et al. (2013). First cloning and characterization of two functional aquaporin genes from an arbuscular mycorrhizal fungus *Glomus intraradices*. *New Phytol.* 197: 617–630.

Li, H., Guo, Q. et al. (2020). Application of *Streptomyces pactum* Act12 enhances drought resistance in wheat. *J. Plant Growth Regul.* 39: 122–132.

Lonbani, M. and Arzani, A. (2011). Morpho-physiological traits associated with terminal drought-stress tolerance in triticale and wheat. *Agron. Res.* 9: 315–329.

López-Galiano, M.J., García-Robles, I., González-Hernández, A.I. et al. (2019). Expression of miR159 is altered in tomato plants undergoing drought stress. *Plants* 8: 201.

Ma, J.F., Tamai, K., Yamaji, N. et al. (2006). A silicon transporter in rice. *Nature* 440: 688–691.

Miura, K. and Tada, Y. (2014). Regulation of water, salinity, and cold stress responses by salicylic acid. *Front. Plant Sci.* 5: 4.

Nawaz, J., Hussain, M., Jabbar, A. et al. (2013). Seed priming a technique. *Int. J. Agric. Crop Sci.* 6: 1373.

Nir, I.D., Moshelion, M., and Weiss, D. (2014). The Arabidopsis GIBBERELLIN METHYL TRANSFERASE 1 suppresses gibberellin activity, reduces whole-plant transpiration and promotes drought tolerance in transgenic tomato. *Plant Cell Environ.* 37: 113–123.

Niu, X., Song, L., Xiao, Y., and Ge, W. (2018). Drought-tolerant plant growth-promoting rhizobacteria associated with foxtail millet in a semi-arid agroecosystem and their potential in alleviating drought stress. *Front. Microbiol.* 8: 2580.

Noman, A., Ali, Q., Naseem, J. et al. (2018). Sugar beet extract acts as a natural bio-stimulant for physio-biochemical attributes in water stressed wheat (*Triticum aestivum* L.). *Acta Physiol. Plant.* 40: 110.

Oh, S.J., Kim, Y.S., Kwon, C.W. et al. (2013). Overexpression of the transcription factor AP37 in rice improves grain yield under drought conditions. *Plant Physiol.* 150: 1368–1379.

Oliveira, I.C.M., Guilhen, J.H.S., de Oliveira Ribeiro, P.C. et al. (2020). Genotype-by environment interaction and yield stability analysis of biomass sorghum hybrids using factor analytic models and environmental covariates. *Field Crop Res.* 257: 107929.

Osakabe, Y., Osakabe, K., Shinozaki, K., and Tran, L.S.P. (2014). Response of plants to water stress. *Front. Plant Sci.* 5: 86.

Ozturk, M., TurkyilmazUnal, B., García-Caparros, P. et al. (2021). Osmoregulation and its actions during the drought stress in plants. *Physiol. Plant.* 172: 1321–1335.

Pandey, V. and Shukla, A. (2015). Acclimation and tolerance strategies of rice under drought stress. *Rice Sci.* 22: 147–161.

Passioura, J.B. and Angus, J.F. (2010). Improving productivity of crops in water-limited environments. In: *Advances in Agronomy*, vol. 106 (ed. D.L. Sparks), 37–75. Cambridge, MA: Academic Press.

Pathania, P., Rajta, A., Singh, P.C., and Bhatia, R. (2020). Role of plant growth-promoting bacteria in sustainable agriculture. *Biocatal. Agric. Biotechnol.* 30: 101842.

Pilon-Smits, E.A.H., Terry, N., Sears, T., and Van-Dun, K. (1999). Enhanced drought resistance in fructan-producing sugar beet. *Plant Physiol. Biochem.* 37: 313–317.

Rahmani, N. (2006). Effect of irrigation and nitrogen application on the quantity and quality of medicinal plant Marigold (*Calendula officinalis* L.). Master Thesis Agronomy. University of Takestan Branch.

Ratnayaka, H.H., Molin, W.T., and Sterling, T.M. (2003). Physiological and antioxidant responses of cotton and spurred anoda under interference and mild drought. *J. Exp. Bot.* 54: 2293–2305.

Saeidi, M. and Abdoli, M. (2015). Effect of drought stress during grain filling on yield and its components, gas exchange variables, and some physiological traits of wheat cultivars. *J. Agric. Sci. Technol.* 17: 885–898.

Saha, A., Sekharan, S., and Manna, U. (2020). Superabsorbent hydrogel (SAH) as a soil amendment for drought management: a review. *Soil Tillage Res.* 204: 104736.

Salehi-Lisar, S.Y. and Bakhshayeshan-Agdam, H. (2020). Agronomic crop responses and tolerance to drought stress. In: *Agronomic Crops*, 63–91. Berlin/Heidelberg: Springer.

Sedghi, M., Hadi, M., and Toluie, S.G. (2013). Effect of nano zinc oxide on the germination parameters of soybean seeds under drought stress. *Ann. WUT Ser. Biol.* XVI: 73–78.

Seleiman, M.F., Almutairi, K.F., Alotaibi, M. et al. (2020). Nano fertilization as an emerging fertilization technique: why modern agriculture can benefit from its use? *Plants* 10: 2.

Sewelam, N., Oshima, Y., Mitsuda, N., and Ohme-Takagi, M. (2014). A step towards understanding plant responses to multiple environmental stresses: a genome-wide study. *Plant Cell Environ.* 37: 2024–2035.

Shafea, A., Dawood, M., and Zidan, M. (2017). Wheat seedlings traits as affected by soaking at titanium dioxide nanoparticles. *Environ. Earth Ecol.* 1: 102–111.

Sherameti, I., Tripathi, S., Varma, A., and Oelmuller, R. (2008). The root-colonizing endophyte *Pirifomospora indica* confers drought tolerance in Arabidopsis by stimulating the expression of drought stress–related genes in leaves. *Mol. Plant Microbe Interact.* 21: 799–807.

Shi, Y., Zhang, Y., Yao, H. et al. (2014). Silicon improves seed germination and alleviates oxidative stress of bud seedlings in tomato under water deficit stress. *Plant Physiol. Biochem.* 78: 27–36.

Shukla, N., Awasthi, R.P., Rawat, L., and Kumar, J. (2012). Biochemical and physiological responses of rice (*Oryza sativa* L.) as influenced by *Trichoderma harzianum* under drought stress. *Plant Physiol. Biochem.* 54: 78–88.

Silva, R., Filgueiras, L., Santos, B. et al. (2020). *Gluconacetobacter diazotrophicus* changes the molecular mechanisms of root development in *Oryza sativa* L. Growing under water stress. *Int. J. Mol. Sci.* 21: 333.

Singh, S., Prasad, S., Yadav, V. et al. (2018). Effect of drought stress on yield and yield components of rice (*Oryza sativa* L.) genotypes. *Int. J. Curr. Microbiol. Appl. Sci.* 7: 2752–2759.

Sultan, B., Defrance, D., and Iizumi, T. (2019). Evidence of crop production losses in West Africa due to historical global warming in two crop models. *Sci. Rep.* 9: 1–15.

Sun, L., Song, F., Zhu, X. et al. (2021). Nano-ZnO alleviates drought stress via modulating the plant water use and carbohydrate metabolism in maize. *Arch. Agron. Soil Sci.* 67 (2): 245–259.

Tabassum, T., Farooq, M., Ahmad, R. et al. (2018). Terminal drought and seed priming improves drought tolerance in wheat. *Physiol. Mol. Biol. Plants* 24: 845–856.

Tan, M., Hassan, M.J., Peng, Y. et al. (2022). Polyamines metabolism interacts with γ-aminobutyric acid, proline and nitrogen metabolisms to affect drought tolerance of creeping bentgrass. *Int. J. Mol. Sci.* 23 (5): 2779.

Taran, N., Storozhenko, V., Svietlova, N. et al. (2017). Effect of zinc and copper nanoparticles on drought resistance of wheat seedlings. *Nanoscale Res. Lett.* 12: 60.

Tu, Y., Wang, R., Zhang, Y., and Wang, J. (2018). Progress and expectation of atmospheric water harvesting. *Joule* 2: 1452–1475.

Vandenbroucke, K. and Metzlaff, M. (2013). Abiotic stress tolerant crops: genes, pathways and bottlenecks. In: *Sustainable Food Production* (ed. R.A. Meyers), 1–17. Berlin/Heidelberg: Springer.

Wang, W., Vinocur, B., and Altman, A. (2003). Plant responses to drought, salinity and extreme temperatures: towards genetic engineering for stress tolerance. *Planta* 28: 1–14.

Wania, S.H., Kumar, V., Shriram, V., and Sah, S.K. (2016). Phytohormones and their metabolic engineering for abiotic stress tolerance in crop plants. *Crop J.* 4: 162–176.

Wu, J., Kim, S.G., Kang, K.Y. et al. (2016). Overexpression of a pathogenesis related protein 10 enhances biotic and abiotic stress tolerance in rice. *Plant Pathol. J.* 32: 552–562.

Yadav, S., Modi, P., Dave, A. et al. (2020). Effect of abiotic stress on crops. In: *Sustainable Crop Production* (ed. M. Hasanuzzaman, M. Filho, M. Fujita, and T. Nogueira). Rijeka, Croatia: IntechOpen.

Yang, J., Zhang, J., Liu, K. et al. (2007). Involvement of polyamines in the drought resistance of rice. *J. Exp. Bot.* 58 (6): 1545–1555.

Yu, L., Chen, X., Wang, Z. et al. (2013). Arabidopsis enhanced drought tolerance1/HOMEODOMAIN GLABROUS11 confers drought tolerance in transgenic rice without yield penalty. *Plant Physiol.* 162: 1378–1391.

Yuliasti, Y. and Reflinur, R. (2015). Evaluation of mungbean mutant lines to drought stress and their genetic relationships using SSR markers. *Atom Indones. J.* 41: 161–167.

Zain, N.A.M., Ismail, M.R., Puteh, A. et al. (2014). Drought tolerance and ion accumulation of rice following application of additional potassium fertilizer. *Commun. Soil Sci. Plant Anal.* 45: 2502–2514.

Ze, Y., Liu, C., Wang, L. et al. (2011). The regulation of TiO 2 nanoparticles on the expression of light-harvesting complex II and photosynthesis of chloroplasts of Arabidopsis thaliana. *Biol. Trace Element Res.* 143: 1131–1141.

Zhang, F., Wang, P., Zou, Y.N. et al. (2019). Effects of mycorrhizal fungi on root-hair growth and hormone levels of taproot and lateral roots in trifoliate orange under drought stress. *Arch. Agron. Soil Sci.* 65: 1316–1330.

Zhang, A., Liu, M., Gu, W. et al. (2021). Effect of drought on photosynthesis, total antioxidant capacity, bioactive component accumulation, and the transcriptome of *Atractylodes lancea*. *BMC Plant Biol.* 21: 1–14.

Zhao, R., Guo, W., Bi, N. et al. (2015). Arbuscular mycorrhizal fungi affect the growth, nutrient uptake and water status of maize (*Zea mays* L.) grown in two types of coal mine spoils under drought stress. *Appl. Soil Ecol.* 88: 41–49.

Zhu, J.K. (2002). Salt and drought stress signal transduction in plants. *Annu. Rev. Plant Biol.* 53: 247–273.

Zouari, M., Hassena, A.B., Trabelsi, L. et al. (2019). Exogenous proline-mediated abiotic stress tolerance in plants: possible mechanisms. In: *Osmoprotectant-Mediated Abiotic Stress Tolerance in Plants* (ed. M.A. Hossain, V. Kumar, D.J. Burritt, et al.), 99–121. Berlin/Heidelberg: Springer.

21

Understanding the Role of Key Transcription Factors in Regulating Salinity Tolerance in Plants

Sahana Basu and Gautam Kumar

Department of Life Science, Central University of South Bihar, Gaya, Bihar, India

21.1 Introduction

Salinity is one of the major abiotic stresses frequently affecting more than 20% of the total agricultural lands throughout the world (Shrivastava and Kumar 2015). Salinity is predominantly employed by excess NaCl concentration in the soil. Poor irrigation, defective drainage, and leaching are the major factors for increasing soil salinity. Every year about 10 million ha agricultural land all over the world is destroyed by excessive salt accumulation. It is therefore a serious threat to combat the situation for sustaining the food security.

Salt stress frequently affects seed germination, plant growth, physiological activities, metabolism, and reproduction, consequently posing threat to the sustainable agricultural productivity (Basu et al. 2022). It decreases the plants' water uptake ability, thereby inhibiting plant growth. Salinity-induced inhibition of cell divisions along with increased cell death suppress the plant growth (Basu et al. 2021). Metabolic disorders and limited nutrient uptake are also responsible for the hindrance of plant growth. Salinity stress also interferes with the major physiological processes in plants (Kumar et al. 2022). High Na^+ concentration in plant cells impedes the K^+ uptake, thereby impairing the cellular ionic homeostasis. Additionally, Na^+ toxicity in plant cells causes early leaf senescence and necrosis (Park et al. 2016). Deteriorated photosynthesis and stomatal conductance decrease the biomass and productivity of plants (Basu et al. 2017). Salinity stress in plant cells causes the overproduction of reactive oxygen species (ROS) leading to oxidative stress (Miller et al. 2010). Plants develop improved antioxidant defense system to maintain redox homeostasis through ROS scavenging. Salinity tolerance in plants is associated with the regulation of cytosolic Na^+ concentration, thereby preventing ionic and oxidative damage. Therefore, understanding the ionic and redox homeostasis mechanisms is essential for enhancing salinity tolerance in plants.

21.2 Transcription Factors Conferring Salinity Tolerance

Transcription factors (TFs) being the key regulators of numerous stress-responsive genes act as significant candidates facilitating salinity tolerance in plants (Figure 21.1, Singh et al. 2002). The TFs activate or repress the expression of specific sets of stress-responsive genes by binding to their promoters. Previous studies have characterized several plant TF members of APETALA2/ethylene responsive factor (AP2/ERF), WRKY, nuclear factor-Y (NF-Y), NAM (no apical meristem, Petunia), ATAF1–2 (Arabidopsis thaliana activating factor), and CUC2 (cup-shaped cotyledon, Arabidopsis) (NAC), basic helix-loop-helix (bHLH), v-Myb myeloblastosis viral oncogene homolog (MYB), and basic leucine zipper (bZIP) in response to salinity stress, which are listed in Table 21.1.

21.2.1 APETALA2/Ethylene Responsive Factor

APETALA2/ethylene responsive factor (AP2/ERF) TF family is a large family of TFs in plants that are the key regulators of different abiotic stresses, including salinity (Fu et al. 2021). This TF family was first characterized by Jofuku et al. (1994) in *Arabidopsis* and named as APETALA2 after the founding member.

Global Climate Change and Plant Stress Management, First Edition. Edited by Mohammad Wahid Ansari, Anil Kumar Singh, and Narendra Tuteja.

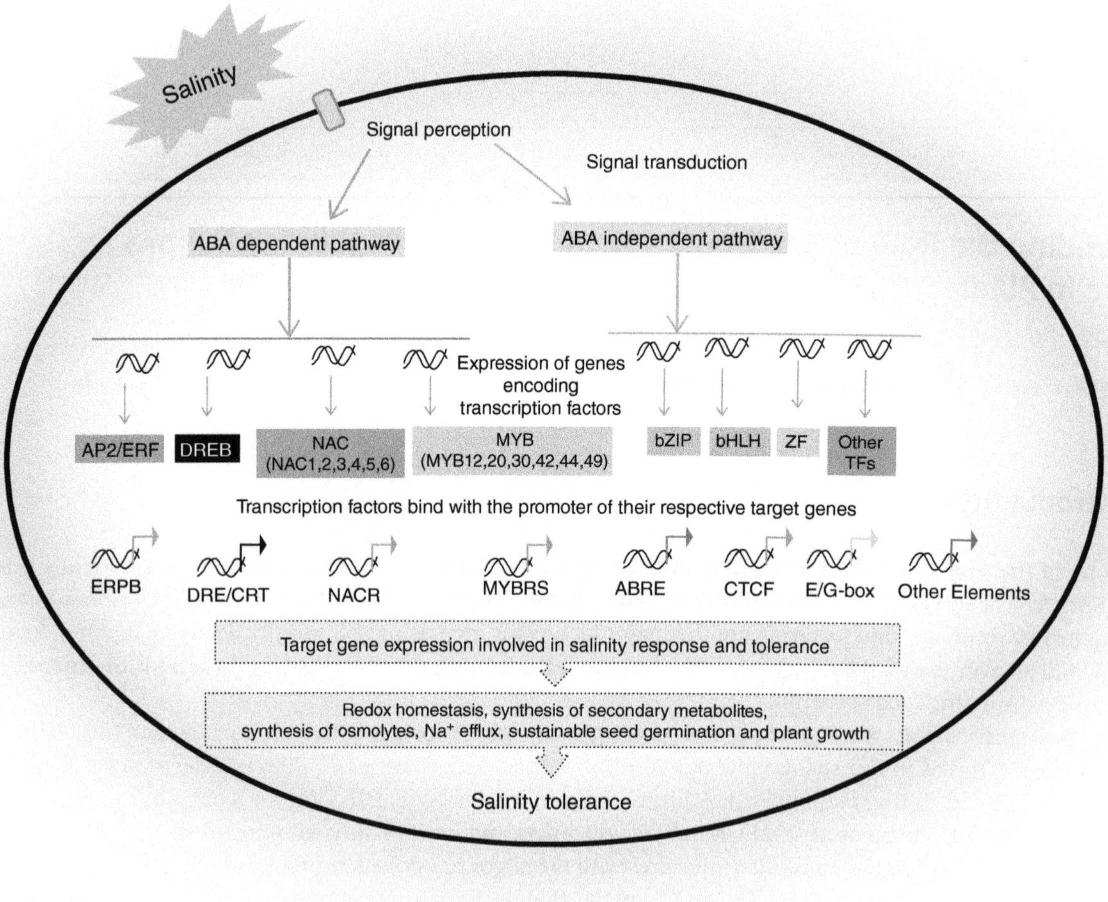

Figure 21.1 Salinity tolerance in plants mediated by major transcription factors (TFs). Plant hormone abscisic acid (ABA) plays an important role under abiotic stresses. Adverse environmental conditions increase the ABA levels in plants thereby improving their survivability. Exogenous ABA often stimulates a number of salinity-responsive genes. However, several salinity-induced genes have been found that are unresponsive to exogenous ABA treatment indicating the presence of ABA-dependent and ABA-independent regulatory systems in plants. Among the major TFs, AP2/ERF (including DREB subfamily), MYB, and NAC have been revealed to be driven by the ABA-dependent pathway. In contrast, bHLH, bZIP, and other TFs have been found to follow the ABA-independent pathway. The TFs bind to the promoter regions of their target genes, thereby conferring salinity tolerance in plants.

Table 21.1 Major TFs characterized in response to salinity stress.

Gene family	Gene name	Studied plant	Salinity stress applied	References
ERF	*SlERF*	*Solanum lycopersicum* L. cv. Pusa Ruby	200 mM NaCl	Sharma et al. (2010)
bHLH	*BdbHLH*	*Brachypodium distachyon*	200 mM NaCl	Niu et al. (2017)
	MdbHLH	*Malus × domestica* Borkh.	200 mM NaCl	Mao et al. (2017)
	PvbHLH	*Phaseolus vulgaris* L. cv. Ispir	150 mM NaCl	Kavas et al. (2016)
bZIP	*bZIP57, bZIP59*	*Suaeda fruticosa*	900 mM NaCl	Diray-Arce et al. (2019)
CAMTA	*CAMTA12*			
MYB	*MYB13, MYB88, MYB72*			
M-Type	*M-Type26, M-Type28*			

21.2.1.1 Structure of AP2/ERF Transcription Factors

The AP2/ERFs are characterized by 40–70 conserved amino acids containing DNA-binding domain (DBD), termed as APETALA2 (AP2)/ethylene responsive element binding factor (EREB; Dietz et al. 2010). The AP2/ERF TFs have been extensively studied in plants. Genome-wide identification and characterization of the AP2/ERF family TFs have been conducted in different plant species, including rice (Sharoni et al. 2011), sorghum (Yan et al. 2013), castor bean (Xu et al. 2013), cucumber (Hu and Liu 2011), poplar (Zhuang et al. 2008), and soybean (Zhang et al. 2008).

cis-Regulatory elements present in the promoter of the target genes are the main site for strong binding of the AP2/ERF proteins. The GCC-box (5′-AGCCGCC-3′) and DRE/CRT (dehydration responsive element/C-repeat, 5′-RCCGCC-3′) are the main DB elements (Gu et al. 2017). Most of the AP2/ERF proteins bind to the promoter containing GCC-box but the degree of activation of promoter differs between the members. The members of group C are strong activators, groups G and H are neutral activators, and groups A, B, and E are weak activators. In disparity members of F group are repressors (Pirrello et al. 2012). Additionally, elements originated from the GCC-box and DRE/CRT belong to the *cis*-regulatory elements that regulate the target gene expression under various environmental stresses (Mizoi et al. 2012). The ERF protein can also bind to the CE1 (coupling element, 5′-CACCG-3′; Wu et al. 2008) and VWRE (vascular wounding responsive element, 5′-GAAAAGAAAATTTC-3′; Sasaki et al. 2007) in tobacco. The ERF proteins have also been revealed to directly interact with promoters containing a non-GCC element (Chakravarthy et al. 2003).

21.2.1.2 Classification of AP2/ERF Transcription Factors

Based on the number of AP2-DBD, the AP2/ERF family has been broadly classified into four subfamilies, viz., Apetala2 (AP2), ethylene responsive factors (ERF; subgroup V-X), dehydration-responsive element binding proteins (DREB; subgroup A1–A6), and related to ABA-insensitive3(ABI3)/viviparous1 (RAV) (Figure 21.2, Wang et al. 2015). The AP2 subfamily containing two AP2/ERF domains regulates the development of plants (Chandler 2018). The ERF subfamily containing one AP2/ERF domain constitutes the largest group of the AP2/ERF family. This subfamily was first recognized as the binding factors facilitating the ethylene response that recognizes the GCC-box thereby, regulating the plants' responses to stresses (Fujimoto et al. 2000). The dehydration-responsive element or C-repeat (DRE/CRT) are recognized by AP2/ERF domain containing DREB subfamily and consequently regulating the plants' responses to abiotic stresses

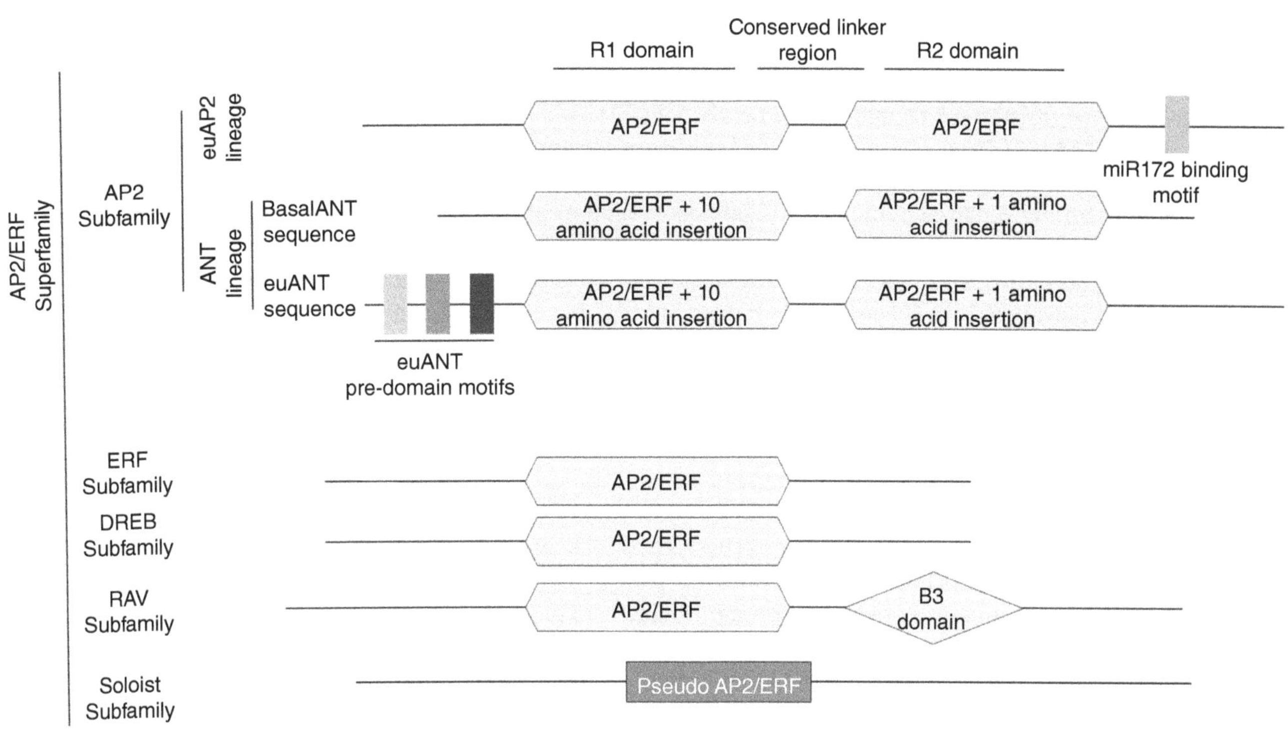

Figure 21.2 Schematic representation of different members of the APETALA 2 (AP2)/ethylene response factor (ERF) transcription factor superfamily. The AP2 subfamily contains two AP2/ERF domains. The ERF and DREB subfamilies contain one AP2/ERF domain. The RAV subfamily contains one AP2/ERF domain and a B3 domain. The Soloist subfamily contains one AP2 domain (AP2-like domain).

(Nguyen et al. 2019). The RAV subfamily contains one AP2/ERF domain and a B3 domain that differs from the other subfamilies and may inhibit transcription (Kagaya et al. 1999). The Soloist subfamily is a small group of TFs with an incredibly altered gene structure and single AP2 domain (AP2-like domain) (Li et al. 2017).

21.2.1.3 Role of AP2/ERF Transcription Factors in Salinity Tolerance

The AP2/ERFs expression is strongly regulated by different abiotic stresses, including salinity. However, under normal conditions most of the AP2/ERFs exhibit low expressions though phytohormones and stress stimuli at different growth stages can induce or repress their expressions (Li et al. 2017). Candidate genes encoding several TFs have been cloned and overexpressed in different plant species raising transgenics (Table 21.2).

A recent study has revealed the *ZmEREB20* isolated from *Zea mays* to confer salinity tolerance in *Arabidopsis thaliana*. The transgenic plants showed increased root hair growth, elevated ROS scavenging, increased expression of auxin-related genes, ion transporter genes, and root hair growth genes under salinity stress (Fu et al. 2021). Another AP2/ERF gene *OsSTAP1* isolated from *Oryza sativa* has been shown to confer salinity tolerance in *Nicotiana tabacum*. The *OsSTAP1* overexpressing plants showed lower Na^+/K^+ ratios in shoots, increased antioxidative enzyme (superoxide dismutase, SOD; peroxidase, POD; catalase, CAT) activities, and upregulation of stress-responsive genes following salinity stress (Wang et al. 2020). Overexpression of *GmDREB6* in *Glycine max* has been found to enhance salinity tolerance with increased proline content and higher expression of stress-responsive genes (Nguyen et al. 2019). On the other hand, two ERF genes *JcERF1* (Wang et al. 2015) and *JcERF2* (Yang et al. 2014) isolated from *Jatropha curcas* has been revealed to confer salinity tolerance in *N. tabacum* by increasing germination rate and root length, enhanced expression of stress-related genes.

21.2.2 WRKY

One of the large families of regulatory proteins are WRKY proteins in plants (Bakshi and Oelmuller 2014). A large number of studies have revealed that WRKY proteins play a key role in the complex signaling processes during stress responses in plants (Rushton et al. 2010). These proteins also have been found to regulate the developmental and physiological processes in plants (Besseau et al. 2012). These TFs usually act as activators or repressors of the stress-responsive genes.

21.2.2.1 Structure of WRKY Transcription Factors

The WRKY TFs contain a 60 amino acid sequence, whose conserved region is termed as the WRKY domain comprising of a highly conserved N-terminal heptapeptide WRKYGQK and a C-terminal zinc finger-like motif Cys(2)-His(2) or Cys(2)-HisCys (Rushton et al. 2010). This TF binds to the *cis*-element TTGAC(C/T) W-box located in the promoter of their target genes. The WRKY TFs have been extensively studied in plants. Genome-wide identification and characterization of the AP2/ERF family TFs have been executed in different plant species, including *Arabidopsis* containing 74 members, rice with 104 members, soybean and Poplar containing >100 members, sorghum containing 68 members (Pandey and Somssich 2009), and barley with 45 members (Mangelsen et al. 2008). The WRKY TFs have also been reported in bryophyte (*Physcomitrella patens* containing 38 members), pteridophyte (*Sellaginella moellendorffii* containing 35 members; Bakshi and Oelmuller 2014), and gymnosperm (*Pinus monticola* containing 80 members; Liu and Ekramoddoullah 2009).

21.2.2.2 Classification of WRKY Transcription Factors

The WRKY proteins identified till date contain one or two WRKY domains (Eulgem et al. 2000). The WRKY proteins are split into three main groups (I–III) based on the number of WRKY domains and the pattern of the zinc finger-like motif. Group I WRKY proteins contain two WRKY domains, whereas group II proteins are mostly composed of one WRKY domain (Figure 21.3). However, these two groups possess the identical pattern of zinc ligands (C–X4–5–C–X22–23–H–X1–H).The WRKY domain of group III WRKY proteins has various zinc finger motif sequences. This group contains a C2–HC motif (C–X7–C–X23–H–X1–C) instead of a C2–H2 pattern. Group II proteins are divided into five different subgroups, viz. IIa, IIb, IIc, IId, and IIe, based on the primary amino acid sequence. The C-terminal WRKY domain consists of a four-stranded antiparallel β-sheet with the zinc coordinating Cys/His residues, developing a zinc-binding pocket. The β1-strand of the β-sheet containing the WRKYGQK motif interacts with the major groove of DNA (Figure 21.4, Yamasaki et al. 2012). The WRKY proteins regulate the stress responses in plants through binding with the W-box elements [(T)TGAC(C/T)] in the promoters of their target gene, thereby upregulating or downregulating their transcription.

Table 21.2 Transgenic plants overexpressing major TF genes for enhanced salinity tolerance.

Gene family	Gene name	Source plants	Transgenic plants	Salinity tolerance conferred	Functions	References
AP2/ERF	*ZmEREB20*	*Zea mays*	*Arabidopsis thaliana* L. cv. Col-0	150 mM NaCl	Increased root hair growth, elevated ROS scavenging, increased expression of auxin-related genes, ion transporter genes, and root hair growth genes	Fu et al. (2021)
	OsSTAP1	*Oryza sativa*	*Nicotiana tabacum* L.	150 mM NaCl	Higher SOD, POD, and CAT activities, lower Na^+/K^+ ratios in shoots, increased antioxidative enzymes (SOD, POD, and CAT) activities, upregulation of stress-responsive genes, including other ERF- and peroxidase-encoding genes	Wang et al. (2020)
	GmDREB6	*Glycine max*	*G. max* L. cv. DT84	300 mM NaCl	Increased proline content, higher expression of stress-responsive genes	Nguyen et al. (2019)
	JcDREB2	*Jatropha curcas*	*O. sativa* cv. Zhonghua 11	150 mM NaCl	Higher survival rate under salinity stress, increased proline content, lower relative electrolyte leakage (REL) and MDA content, enhanced SOD and CAT activities, increased expressions of salt-tolerance-related genes	Tang et al. (2017)
	VrDREB2A	*Vigna radiata* L.	*A. thaliana* L. cv. Col-0	210 mM NaCl	Higher germination rate under salinity stress	Chen et al. (2016)
				150 mM NaCl	Increased root elongation, increased survival rate, lower REL, expression of salinity-responsive genes	
	JcERF2	*J. curcas*	*N. tabacum* L.	200 mM NaCl	Increased germination rate and root length, enhanced expression of stress-related genes	Wang et al. (2015)
	JcERF1	*J. curcas*	*N. tabacum* cv. Nc89	300 mM NaCl	Increased root length, decreased MDA content, increased soluble carbohydrates and free proline contents, higher expression of stress-responsive genes	Yang et al. (2014)
	GmDREB1	*G. max*	*Triticum aestivum* L. cv. Jimai 19	300 mM NaCl	Longer coleoptiles and radicles and a greater radicle number; greater root length, fresh weight, and tiller number per plant; improved yield-related traits; higher levels of proline and betaine and lower levels of MDA and REL; upregulated osmotic- and oxidative-stress-related proteins	Jiang et al. (2014)
	AtDREB2C	*A. thaliana*	*A. thaliana* L. cv. Col-0	300 mM NaCl	Increased chlorophyll content and relative ion leakage, increased fresh weight of roots, enhanced primary root length, number of lateral root (LR), branch points, and total LR length	Song et al. (2014)
			Brassica napus L. cv. Youngsan	400 mM NaCl	Decreased Na^+ content, increased relative water content (RWC) of leaves	
	OsDREB2A	*O. sativa*	*G. max* cv. Huachun 3	300 mM NaCl	Increased proline and soluble sugar contents, upregulated expression of stress-responsive genes	Zhang et al. (2013)

(*Continued*)

Table 21.2 (Continued)

Gene family	Gene name	Source plants	Transgenic plants	Salinity tolerance conferred	Functions	References
	OsDREB1D	*O. sativa*	*A. thaliana* L. cv. Col-0	100 mM NaCl	Increased seed germination, higher expression of stress-responsive genes	Zhang et al. (2009)
	AtDREB1A	*A. thaliana*	*N. tabacum* L. cv. NC89	300 mM NaCl	Higher root length and chlorophyll content, elevated levels of soluble sugars, increased expression of osmotic stress and sucrose biosynthesis related genes	Cong et al. (2008)
	GmDREB2	*G. max*	*A. thaliana* L. cv. Col-0	200 mM NaCl	Higher level of free proline, increased expression of stress-responsive genes	Chen et al. (2007)
	JcDREB	*J. curcas*	*A. thaliana* L. cv. Col-0	200 mM NaCl	Increased chlorophyll content, higher expression of stress-responsive genes	Tang et al. (2011)
WRKY	*MfWRKY70*	*Myrothamnus flabellifolia*	*A. thaliana* L. cv. Col-0	300 mM NaCl	Promotion of root growth and water Retention; enhanced SOD, POD, and CAT activities; maintenance of membrane-lipid stability; higher expression of stress-associated genes	Xiang et al. (2021)
	DnWRKY11	*Dendrobium nobile*	*N. tabacum* cv. Huangmiaoyu	200 mM NaCl	Higher germination rate; longer root length; higher fresh weight; higher SOD, POD, and CAT activities; lower MDA content; higher expression of stress-related genes	Xu et al. (2014)
	GhWRKY17	*Gossypium hirsutum* L.	*N. benthamiana*	200 mM NaCl	Increased germination rate and root elongation; increased chlorophyll content; decreased H_2O_2 and $^{\cdot}O_2^-$ accumulation; decreased MDA content and electrolyte leakage; increased proline content; enhanced SOD, POD, CAT, and APX activities; higher expression of stress-responsive genes	Yan et al. (2014)
	DgWRKY3	*Dendranthema grandiflorum*	*N. tabacum*	150 mM NaCl	Increased proline content, reduced H_2O_2 and MDA accumulation, higher antioxidant enzyme (SOD, POD, CAT, and APX) activities, increased antioxidants (AsA, GSH) accumulation, enhanced expression of osmotic adjustment and membrane protection-related genes	Liu et al. (2013)
bHLH	*CgbHLH001*	*Chenopodium glaucum*	*N. benthamiana*	200 mM NaCl	Higher expressions of stress-responsive genes	Wang et al. (2017)
	TabHLH39	*T. aestivum*	*A. thaliana* L. cv. Col-0	250 mM NaCl	Increased root length, increased soluble sugar and proline contents, decreased electrolyte leakage, higher expressions of stress-responsive genes	Zhai et al. (2016)
	AtbHLH112	*A. thaliana*	*A. thaliana* L. cv. Col-0	150 mM NaCl	Higher germination rate, increased root growth and fresh weight, lower electrolyte leakage, higher proline content, decreased ROS in guard cells and leaves, increased SOD and POD activities, higher expressions of stress-responsive genes	Liu et al. (2015)
	OrbHLH001	*Oryza rufipogon*	*A. thaliana* L. cv. Col-0	300 mM NaCl	Higher survival rate under salinity stress, increased root length, enhanced expression of salinity-responsive genes	Li et al. (2010)

MYB	*OsMYB6*	*O. sativa*	*O. sativa* cv. ZH_{11}	150 mM NaCl	Higher proline content, higher CAT and SOD activities, lower relative electrolyte leakage (REL) and MDA content, higher expression of stress-responsive genes	Tang et al. (2019)
	OsMYB48-1	*O. sativa*	*O. sativa* L. cv. Nipponbare	150 mM NaCl	Reduced rate of water loss, lower MDA content, higher proline content	Xiong et al. (2014)
	OsMYB3R-2	*O. sativa*	*A. thaliana* L. cv. Col-0	250 mM NaCl	Higher germination rate, longer root length, higher expression of stress-related genes	Dai et al. (2007)
NAC	*OsNAC022*	*O. sativa*	*N. benthamiana*	150 mM NaCl	Increased shoot and root length and number of lateral roots, decreased Na^+ contents in roots and shoots, higher expression of stress-responsive genes	Hong et al. (2016)
	OsSNAC1	*O. sativa*	*G. hirsutum* L. cv. YZ1	250 mM NaCl	Increased root development, increased biomass, sustainable photosynthesis rate, enhanced proline content, decreased MDA content	Liu et al. (2014)
	OsSNAC1	*O. sativa*	*O. sativa* L. cv. Nipponbare	200 mM NaCl	Increased fresh weight, higher expression of stress-responsive genes	Hu et al. (2006)
NF-Y	*Cdt-NF-YC1*	*Cynodon dactylon*	*O. sativa* L. cv. Zhonghua 11	200 mM NaCl	Higher relative water content (RWC) and chlorophyll content	Chen et al. (2015)
	PwNF-YB3	*Picea wilsonii Mast.*	*A. thaliana* L. cv. Col-0	150 mM NaCl	Accelerated onset of flowering, vigorous seed germination, increased root length and RWC, higher expression of stress-responsive genes	Zhang et al. (2015)
	PwHAP5	*P. wilsonii*	*A. thaliana* L. cv. Col-0	150 mM NaCl	Higher germination rate, increased primary root growth, enhanced expression of stress-responsive genes	Li et al. (2013a)
	AtNFYA1	*A. thaliana*	*A. thaliana* L. cv. Col-0	125 mM NaCl	Higher germination rate, enhanced seedling survivability and fresh weight, higher expression of stress-responsive genes	Li et al. (2013b)
	GmNFYA3	*G. max*	*A. thaliana* L. cv. Col-0	300 mM NaCl	Higher germination rate, increased chlorophyll content, enhanced expression of stress-responsive genes	Ni et al. (2013)
bZIP	*TabZIP15*	*T. aestivum*	*T. aestivum* cv. Kenong 199	300 mM NaCl	Increased plant height, root length, and fresh weight; sustainable spike length and grain number spike^{-1}; decreased H_2O_2 and MDA content; enhanced expression of salt stress-responsive genes	Bi et al. (2021)
	GhABF2	*G. hirsutum*	*A. thaliana* L. cv. Col-0 *G. hirsutum* L. cv. Sumian-12	200 mM NaCl	Increased proline content, increased SOD and CAT activities, increased fiber yield	Liang et al. (2016)
	GmbZIP44 *GmbZIP6* *GmbZIP78*	*G. max*	*A. thaliana* L. cv. Col-0	150 mM NaCl	Higher germination rate and survivability, increased plant height, enhanced expression of stress-responsive genes	Liao et al. (2008)
HD-ZIP	*AtEDT1/ HDG11*	*G. hirsutum*	*G. hirsutum* L. cv. R15 *G. hirsutum* L. cv. WC *Populus tomentosa* Carr.	0.6% NaCl	Well-developed root system, increased proline and soluble sugar content, decreased MDA content, enhanced SOD and CAT activities	Yu et al. (2015)
MADS-box	*OsMADS57*	*O. sativa*	*A. thaliana* L. cv. Col-0 *O. sativa* L. cv. Nipponbare	200 mM NaCl	Higher germination rate, longer root length, better growth status, increased SOD and POD activities, reduced MDA content, improved expression of stress-related genes	Wu et al. (2021)

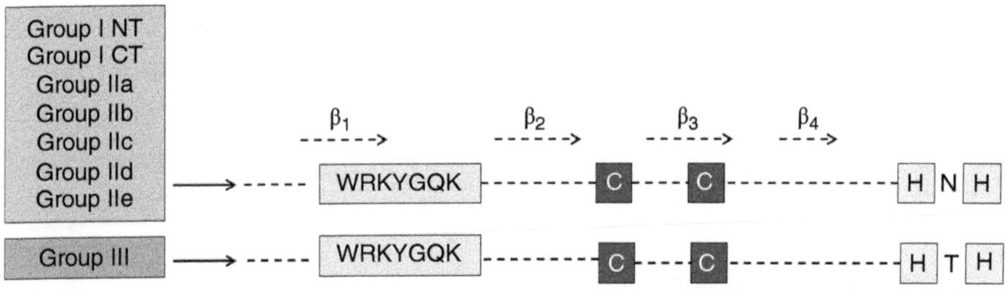

Figure 21.3 The consensus WRKY domain of three major WRKY groups in plants. The WRKY motif, the cysteines (C), and histidines (H) forming the zinc finger are shown in boxes. The four β-strands are shown with dotted arrows. "Group I CT" and "Group I NT" symbolize the N-terminal and C-terminal WRKY domains of group I WRKY proteins.

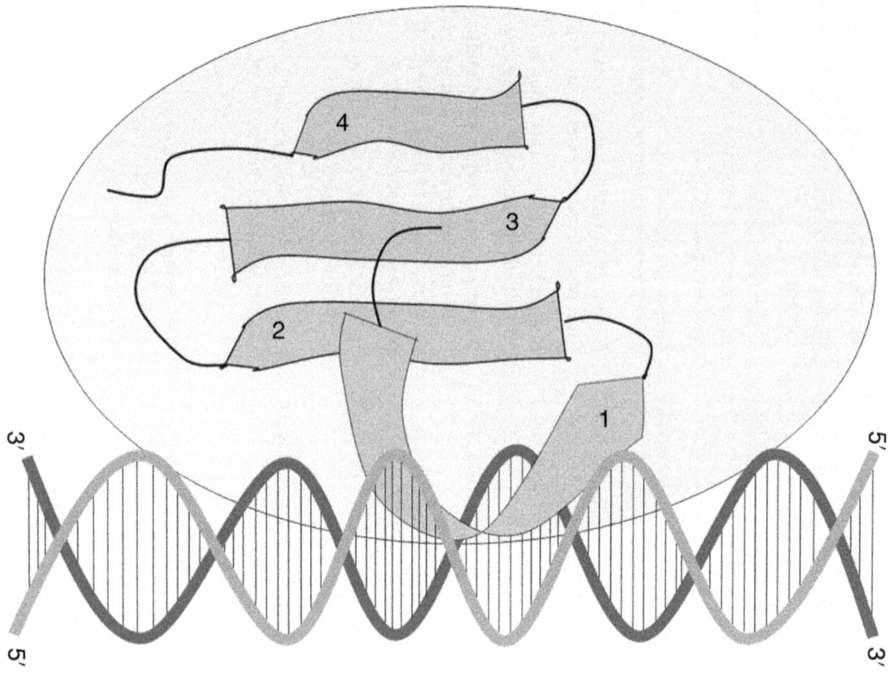

Figure 21.4 Diagram of a C-terminal WRKY domain interacting with the DNA. The domain consists of a four-stranded antiparallel β-sheet (1–4). The DNA recognition takes place along the major grove by the β1-strand containing the WRKYQK motif.

21.2.2.3 Role of WRKY Transcription Factors in Salinity Tolerance

The WRKY TFs have been shown to play a major role in salinity tolerance in several plant species (Bakshi and Oelmuller 2014). In a recent study, *MfWRKY70* isolated from *Myrothamnus flabellifolia* has been revealed to confer salinity tolerance, thereby promoting root growth and water retention capacity in *A. thaliana* (Xiang et al. 2021). The study also showed the *MfWRKY70* overexpressed plants to maintain membrane-lipid stability mediated by enhanced antioxidant enzyme (SOD, POD, and CAT) activities. The transgenic plants also exhibited higher expressions of the candidate genes associated with the salinity tolerance. Similarly, overexpression of *DnWRKY11* isolated from *Dendrobium nobile* showed salinity tolerance in *N. tabacum* (Xu et al. 2014). The transgenic plants had higher germination rate, longer root length, and higher fresh weight under salinity stress conditions. The plants also displayed higher SOD, POD, and CAT activities with lower malondealdehyde (MDA) content and higher expression of stress-related genes under NaCl-mediated salt stress. Yan et al. (2014) showed the overexpression of *GhWRKY17* isolated from *Gossypium hirsutum* to confer salinity tolerance in *Nicotiana benthamiana*. The transgenic plants showed increased germination rate, root elongation, and improved chlorophyll content under salinity stress. The *GhWRKY17* overexpressed plants also showed efficient ROS scavenging with decreased MDA content facilitated by enhanced SOD, POD, CAT, and APX activities. Increased proline content in transgenic plants also provided improved osmotic adjustment under salinity stress. Overexpression of the *DgWRKY3* isolated from *Dendranthema grandiflorum* provided enhanced salinity tolerance in *N. tabacum* associated with increased proline content, reduced ROS and MDA contents, and higher accumulation of enzymatic and nonenzymatic antioxidants (Liu et al. 2013). Jiang and Deyholos (2009) have also revealed the overexpression of WRKY25 and WRKY33 in *A. thaliana* to increase salinity tolerance and ABA sensitivity.

21.2.3 Basic Helix-Loop-Helix

Basic helix-loop-helix (bHLH) TF family comprising a highly conserved basic/helix-loop-helix motif is the second largest TF gene family in plants (Hao et al. 2021). The bHLH TFs regulate the expression of numerous genes involved in specific and overlapping regulatory pathways. These TFs regulate the target gene expression involved in biosynthesis and metabolism of different phytohormones, including ABA (Abe et al. 2003), ethylene, gibberellin (GA), brassinosteroid (BR; Friedrichsen et al. 2002), and jasmonic acid (JA).

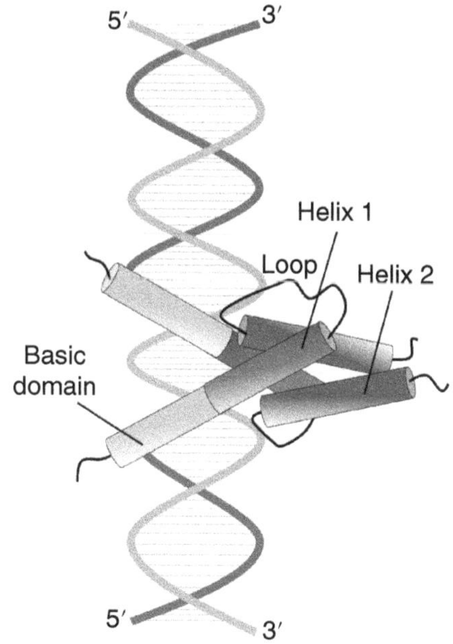

Figure 21.5 Diagram of a bHLH transcription factor interacting with the DNA. The bHLH TF consists of two conserved motifs, a basic DNA binding domain, and a helix-loop-helix (HLH) region. These proteins containing bHLH domain act as homo- or heterodimers to regulate the target gene expression.

21.2.3.1 Structure of bHLH Transcription Factors

The members of this family contain two conserved motifs having about 50–60 amino acids with different functions, a basic DNA binding region and a helix-loop-helix (HLH) region (Figure 21.5). The basic amino acid region containing around 15 amino acids, including approximately 6 basic residues, is located at the N-terminus of the bHLH domain. This region recognizes and precisely binds the DNA motif to the promoter of the target gene (Heim et al. 2003). The helix-loop-helix region having around 40–50 amino acids, including hydrophobic residue containing two amphipathic alpha helices connected by a hydrophobic ring, is located at the C-terminus of the domain. This region is responsible for the dimerization of domain allowing the formation of homo- or hetero-dimeric complexes for promoting protein–protein interactions and regulating transcription of target gene (Feller et al. 2011). Highly conserved regions include Leu23/64, Leu/Iso54, and the Val61, while the Leu23 and Leu52 are responsible for the dimerization. The bHLH protein binds to the core DNA sequence motif known as E-box (5′-CANNTG-3′) in animals, identical to the G-box (5′-CACGTG-3′) motif in plants (Sun et al. 2018).

21.2.3.2 Classification of bHLH Transcription Factors

The bHLH has not been widely explored in plants. However, the phylogenetic tree of bHLH was first constructed in *A. thaliana*. The AtbHLH family containing 162 bHLH members has been divided into 12 subfamilies (Bailey et al. 2003). In rice, 167 bHLH members have been identified in the OsbHLH family, which has been divided into 22 subfamilies (Li et al. 2006). Genome-wide identification and characterization of the bHLH family TFs have been accomplished in different plant species, including apple (Mao et al. 2017), *Brachypodium* (Niu et al. 2017), common bean (Kavas et al. 2016), and poplar (Zhao et al. 2018) containing 188, 146, 155, and 202 bHLH genes, respectively (Table 21.1).

Based on the expression patterns the bHLH TFs are categorized into two classes. Class A comprising E12, E47 (E2a/TCF3 isoforms), and HEB (human E-box binding factor) is ubiquitously expressed. Class B including MyoD, Neurog1, and Mash1 is tissue-specifically expressed. Proteins belong to Class B form homo- and/or hetero-dimers with Class A or other Class B proteins through their HLH domains and bind to the genomic DNA with their basic amino acid stretch. This recognizes the consensus sequence termed an E-box (CANNTG) residing in the *cis*-regulatory elements of the downstream target genes.

21.2.3.3 Role of bHLH Transcription Factors in Salinity Tolerance

A number of studies have revealed the involvement of bHLH TFs in conferring salinity tolerance in plants (Table 21.2). Zhai et al. (2016) has shown the overexpression of *TebHLH39* isolated from wheat to improve salinity tolerance in transgenic *Arabidopsis*. Following the study, the transgenic seedlings exhibited higher survival rate than that of the wild type. The transgenic seedlings also had increased soluble sugar and proline contents under salinity stress. Additionally, the transgenic plants also maintained lower electrolyte leakage under salinity stress as compared to the wild type plants. The *TebHLH39* overexpressed plants also displayed higher expressions of stress-responsive genes, consequently having higher level of tolerance under 250 mM NaCl-induced salt stress. Wang et al. (2017) has shown *CgbHLH001* to confer salinity tolerance in *Chenopodium glaucum*. Another study by Liu et al. (2015) showed overexpression of *AtbHLH112* in *A. thaliana* conferred salinity tolerance in transgenic plants. Following salinity stress, the transgenic plants showed higher germination rate, increased root growth, and fresh weight. The transgenic plants also exhibited improved osmotic adjustment through increased proline level. The transgenic plants had increased antioxidant enzyme activities (SOD and POD) that led to lower accumulation of ROS in leaves and guard cells, thereby decreasing the electrolyte leakage as compared to the wild type plants. The AtbHLH112 overexpressed plants also showed higher expressions of stress-responsive genes that provided salinity tolerance. Li et al. (2010) has revealed *OrbHLH001* isolated from *Oryza rufipogon* to enhance salinity tolerance in transgenic *A. thaliana* L. The study also showed the transgenic plants to possess higher survival rate, increased root length, and higher expression of salinity-responsive genes under 300 mM NaCl-induced salinity stress.

21.2.4 v-Myb Myeloblastosis Viral Oncogene Homolog

The v-Myb myeloblastosis viral oncogene homolog (MYB) TFs belong to one of the largest gene families in plants playing important roles in various biological processes, including cell morphology construction, plant growth, physiological activities, metabolism, and responses to abiotic stresses (Cao et al. 2020).

21.2.4.1 Structure of MYB Transcription Factors

The MYB TFs are characterized by a highly conserved DBD (MYB domain), usually consisting of one to four imperfect repeats in plants (Martin and Paz-Ares 1997). Each MYB repeat encodes three α-helices containing approximately 50–53 amino acids. The second and third helices among the three helices form a helix-turn-helix (HTH) structure through which it can interact and combine with the major groove of the target DNA, thereby regulating the expression of the target gene (Cao et al. 2020). Individual MYB repeat domain comprises three tryptophan residues separated by 18–19 amino acids developing a secondary structure.

21.2.4.2 Classification of MYB Transcription Factors

Based on the structure of the DBD, the MYB TFs in plants are classified into four subfamilies, viz. 1R-MYB, 2R-MYB, 3R-MYB, and 4R-MYB (Figure 21.6, Martin and Paz-Ares 1997). The 1R-MYB containing one MYB domain is the second-largest subfamily of the MYB TFs that maintains the chromosomal structure by regulating transcription in plants. This subfamily is divided into different subgroups, including CCA1-like, CPC-like, I-box-binding-like, TRF-like, and TBP-like MYB-related proteins. Several studies have identified 68, 64, and 42 1R-MYB members in *Arabidopsis*, rice, and *Dendrobium candidum*, respectively (Cao et al. 2020). The 2R-MYB containing two MYB domains is the most extensively occurred protein that regulates cell differentiation and the transcription of secondary metabolic pathways. This subfamily has been believed to be evolved from the R1-MYB through duplication or from the 3R-MYB through loss of R1 (Rosinski and Atchley 1998). The 2R-MYB is grouped into 28 subgroups based on their C-terminal MYB domain localized conserved amino acid sequence motifs. In rice, *Arabidopsis*, *D. candidum*, and *J. curcas*, respectively, 90, 120, 117, and 123 members of 2R-MYB have been reported so far. The 3R-MYB containing three MYB domains is the evolutionarily conserved group in plants that contributes in cell cycle control and regulates cell differentiation. There are five 3R-MYB members in both rice and *Arabidopsis*, whereas *D. candidum* and *J. curcas* contain four members. The 4R-MYB containing four R1/R2 domains is the smallest subfamily of MYB TFs in plants. Till date only two 4R-MYB members have been identified in *Arabidopsis* and *D. candidum*. However, rice and *J. curcas* contains one 4R-MYB TF.

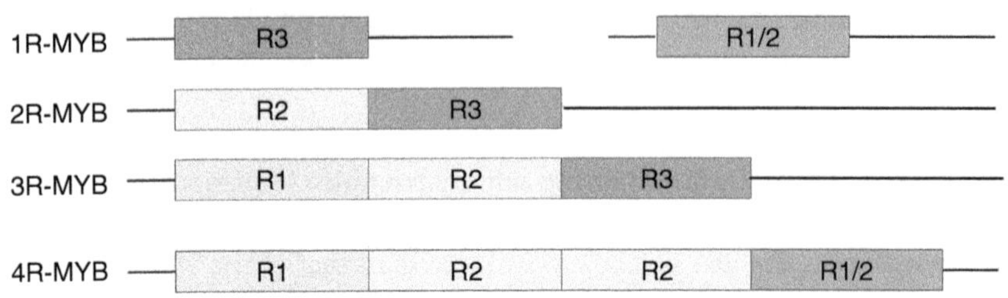

Figure 21.6 Schematic representation of different members of MYB TFs with one to four MYB domain repeats in plants.

21.2.4.3 Role of MYB Transcription Factors in Salinity Tolerance

Role of MYB TFs in enhancing salinity tolerance in plants is well documented. Tang et al. (2019) has shown the overexpression of *OsMYB6* to improve salinity tolerance in rice. The *OsMYB6* overexpressing plants exhibited higher proline content and enhanced antioxidant enzyme activities with lower relative electrolyte leakage (REL) and MDA content under salinity stress. The transgenic plants also displayed higher expression of the stress-responsive genes following NaCl treatment. Overexpression of the *OsMYB48-1* has also been reported to confer salinity tolerance in rice with reduced rate of water loss, lower MDA content, and higher proline content under 150 mM NaCl treatment (Xiong et al. 2014). Another study by Dai et al. (2007) showed the overexpression of *OsMYB3R-2* in *A. thaliana* to improve seed germination and root length. The *OsMYB3R-2* overexpressing plants also showed higher expression of the stress-responsive genes following salinity stress.

21.2.5 NAM (for no apical meristem), ATAF1 and −2, and CUC2 (for cup-shaped cotyledon)

NAM (for no apical meristem), ATAF1 and −2, and CUC2 (for cup-shaped cotyledon) proteins establish one of the largest families of plant-specific TFs. The NAC gene family name has been derived from three TFs: NAM (no apical meristem, Petunia), ATAF1-2 (*A. thaliana* activating factor), and CUC2 (cup-shaped cotyledon, *Arabidopsis*) sharing the same DBD (Souer et al. 1996; Aida et al. 1997). Genome-wide identification and characterization of the NAC family TFs have been accomplished in different plant species, including *Arabidopsis* containing 117, rice with 151, citrus containing 26, soybean and tobacco with 152, poplar with 163, and grapes with 79 NAC genes (So and Lee 2019).

21.2.5.1 Structure of NAC Transcription Factors

A typical NAC TF is comprised of N-terminal about 150 amino acids containing conserved NAC domain and a more divergent transcription regulatory (TR) region at the C-terminus (Figure 21.7, Olsen et al. 2005). The NAC domain consists of five subdomains (A–E) representing DB and protein–protein interaction motifs (Jensen et al. 2010). The A subdomain is responsible for the dimerization of the TF, whereas the highly conserved, positive charge containing C and D subdomains bind to the specific *cis*-regulatory DNA sequences. The diverse B and E subdomains are accountable for the functional diversity of the NAC gene (Puranik et al. 2012).

21.2.5.2 Classification of NAC Transcription Factors

Grounded on the motifs rice NAC proteins (OsNAC) are classified into 15 types (A–O). The A–E types contain five motifs in DBD, whereas F–O types containing distinct motifs are present in NAC-like proteins (Fang et al. 2008). Thus variations in the motifs represent the miscellaneous functions of the NAC TFs. The NAC domain regulates the protein binding, thereby determining the fate and functions of the NAC proteins. In some NACs, a highly hydrophobic region in a negative regulatory domain (NRD) forms a part of the D subunit suppressing the transcriptional activity of the NAC TFs.

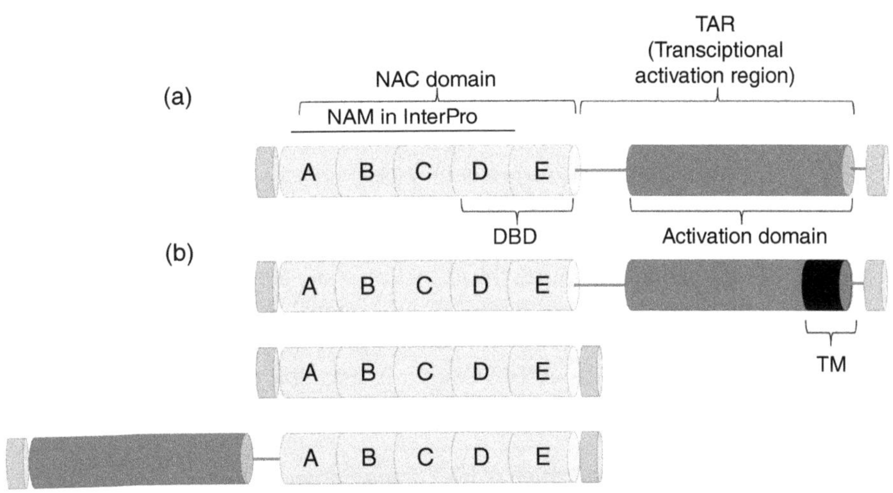

Figure 21.7 Schematic diagram of structures of NAC proteins. (a) The N-terminal region contains highly conserved NAC domain comprising five subdomains (A–E). The C-terminal region contains highly divergent transcriptional activation domain. (b) Structural modifications of NAC protein. The C-terminal region of some NAC proteins contains transmembrane domains (TM).

The NAC-repressor domain inhibits the activity of the members of other TF families (e.g. AP227 and WRKY). The TR domain in the C-terminus acts as an activator or repressor. This occasionally displays protein-binding activity. In *Arabidopsis*, the highly divergent TR domain contains 13 common motifs in 12 subgroups out of total 18 subgroups of the NAC TF (Ooka et al. 2003). Additionally, some NACs contain an α-helical transmembrane motif (NTLs) for plasma membrane or endoplasmic reticulum membrane anchoring of the C-terminal region. Several studies have identified 18 NTLs in *Arabidopsis*, 7 in maize, and 11 in soybean that play key regulatory roles in response to abiotic stresses (Seo et al. 2008).

21.2.5.3 Role of NAC Transcription Factors in Salinity Tolerance

Several studies have shown the NAC TFs to confer salinity tolerance in several plant species. Hong et al. (2016) have revealed *OsNAC022* isolated from *O. sativa* to provide salinity tolerance in *N. benthamiana*. The *OsNAC022* overexpressing plant exhibited increased shoot and root length with increased number of lateral roots under salinity stress. The transgenic plants also exhibited decreased level of Na^+ accumulation in roots and shoots with higher expression of stress-responsive genes following salinity stress. Overexpression of *OsSNAC1* in *G. hirsutum* showed increased root development and biomass under salinity stress. The transgenic plants also displayed sustainable photosynthesis rate, enhanced proline content, and decreased MDA content under salinity stress (Liu et al. 2014). Further, overexpression of *OsSNAC1* in rice showed increase in fresh weight accompanied by higher expression of stress-responsive genes under salinity stress (Hu et al. 2006).

21.2.6 Nuclear Factor-Y

Nuclear factor-Y (NF-Y) is a ubiquitous and highly conserved TF found in eukaryotes, including plants (Li et al. 1992). The NF-Y is also known as heme-associated protein (HAP) or CCAAT-binding factor (CBF). Several studies have identified a large number of NF-Y subunits in different plant species, including *Arabidopsis*, rice, soybean, canola, poplar, and barley (Chen et al. 2015). The NF-Y has been found to frequently regulate the development in plants, including root growth, flowering time, embryogenesis, fatty acid biosynthesis, and chloroplast biogenesis (Miyoshi et al. 2003; Wei et al. 2010).

21.2.6.1 Structure of NF-Y Transcription Factors

The NF-YA subunit is localized to the nucleus that binds to the CCAAT *cis*-element in the promoter regions of the target genes (Petroni et al. 2012). The core regions of NF-YA subunits contain two conserved α-helix domains, namely A1 and A2 (Figure 21.8). The A1 α-helix contains 20 amino acids and is located in the N-terminal of the core region that facilitates the interaction with NF-YB and NF-YC subunits (Zhao et al. 2017). The A2 α-helix contains 21 amino acids and is located in the C-terminal that provides sequence specificity for recognition and binding of the CCAAT *cis*-element. The NF-YB and NF-YC subunits contain highly conserved histone fold domain (HFD) contributing in the protein-DNA and protein–protein interactions (Laloum et al. 2013). The HFD of NF-YB and NF-YC contains three α-helices (α1, α2, and α3) separated by two loops (Mantovani 1999). The α1 helix makes essential stabilizing contacts with the DNA backbone flanking the central CCAAT pentamer. The NF-YB and NF-YC form a heterodimer via α2 and α3 of their HFDs. The conserved domain of NF-YB and NF-YC also contains a fourth helix (αC) (Laloum et al. 2013). Translocation of NF-YB from the cytoplasm to the nucleus requires the formation of NF-YB/YC heterodimers due to the absence of a nuclear localization signal (NLS) in the NF-YB family members, contrasting with that of the NF-YA and NF-YC subunits. The HFDs of NF-YB and NF-YC subunits assemble on a head-to-tail manner forming a dimer to interact with the NF-YA subunits.

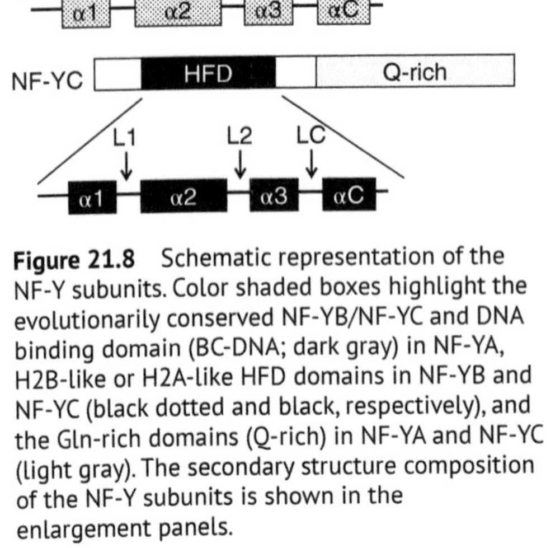

Figure 21.8 Schematic representation of the NF-Y subunits. Color shaded boxes highlight the evolutionarily conserved NF-YB/NF-YC and DNA binding domain (BC-DNA; dark gray) in NF-YA, H2B-like or H2A-like HFD domains in NF-YB and NF-YC (black dotted and black, respectively), and the Gln-rich domains (Q-rich) in NF-YA and NF-YC (light gray). The secondary structure composition of the NF-Y subunits is shown in the enlargement panels.

21.2.6.2 Classification of NF-Y Transcription Factors

The NF-Y TF in plants includes three discrete subunits, viz., NF-YA/HAP2, NF-YB/HAP3, and NF-YC/HAP5 encoded by a large family of genes (Mantovani 1999). The NF-Y shows different combinations of the subunits for the transcriptional control (Hackenberg et al. 2012). In *Arabidopsis* NF-YA, NF-YB, and NF-YC are encoded by 10, 13, and 13 genes, respectively (Petroni et al. 2012). In rice 10, 11, and 7 (Thirumurugan et al. 2008) and in canola 14, 14, and 5 (Liang et al. 2014) NF-YA, NF-YB, and NF-YC encoding genes have been reported.

21.2.6.3 Role of NF-Y Transcription Factors in Salinity Tolerance

NF-Y is also involved in salinity tolerance in plants. The *Cdt-NF-YC1* isolated from *Cynodon dactylon* has been shown to confer salinity tolerance in rice (Chen et al. 2015). The *Cdt-NF-YC1* overexpressing plants maintained higher relative water content (RWC) and chlorophyll content following 200 mM NaCl treatment. Zhang et al. (2015) has revealed overexpression of *PwNF-YB3* isolated from *Picea wilsonii* to confer salinity tolerance in *A. thaliana*. The study showed transgenic plants to possess accelerated onset of flowering, vigorous germination rate, increased root length, and maintained RWC under salinity stress. Another study by Li et al. (2013a) also showed *PwHAP5* (*Pw NF-YC*) to provide salinity tolerance in *A. thaliana* with higher germination rate, increased primary root growth, and enhanced expression of stress-responsive genes. Overexpression of *AtNFYA1* has been found to increase germination rate, enhance seedling survivability and fresh weight, and intensify expression of stress-responsive genes in *A. thaliana* under salinity stress (Li et al. 2013b). Overexpression of *GmNFYA3* isolated from *G. max* has been shown to confer salinity tolerance in *A. thaliana* with increased germination rate, higher chlorophyll content, and enhanced expression of stress-responsive genes under salinity stress (Ni et al. 2013).

21.2.7 Basic Leucine Zipper

Basic leucine zipper (bZIP) is one of the largest TF families in plants. The bZIP TF family members play critical roles in various biological processes, including floral initiation, seed maturation, and embryogenesis (Alonso et al. 2009). These proteins have also been reported to function as vital components in response to different abiotic stresses. Genome-wide analyses have identified members of bZIP gene family in various plant species, including 75 in *Arabidopsis*, 92 in sorghum, 89 in barley, 89 in rice, 69 in tomato, 160 in soybean, 55 in grapevine, 247 in canola, 125 in maize, and 96 in *Brachypodium distachyon* (Wang et al. 2018).

21.2.7.1 Structure of bZIP Transcription Factors

The bZIP gene family is characterized by 60–80 amino acid containing highly conserved bZIP domain (Nijhawan et al. 2008). This domain comprises two functional regions, a highly conserved basic region and a more diversified leucine zipper region (Figure 21.9). The basic region consisting of an invariant N-x7-R/K motif with approximately 16 amino acid residues is located at the N-terminus of the bZIP domain. This region is responsible for DB and nuclear localization. The leucine zipper includes a heptad repeat of leucine or other hydrophobic amino acids located accurately nine amino acids toward the C-terminus. The zipper is involved in specific recognition and dimerization. The bZIP TFs recognize the *cis*-acting DNA elements containing ACGT core motif, including A-box (TACGTA), C-box (GACGTC), and G-box (CACGTG) (Glover and Harrison 1995). However, some bZIP proteins (*GmbZIP44* from *G. max*) identify and bind to non-palindromic sequences (Liao et al. 2008). The bZIPs recognize the DNA with the two continuous α-helices wrapped around their C-terminal

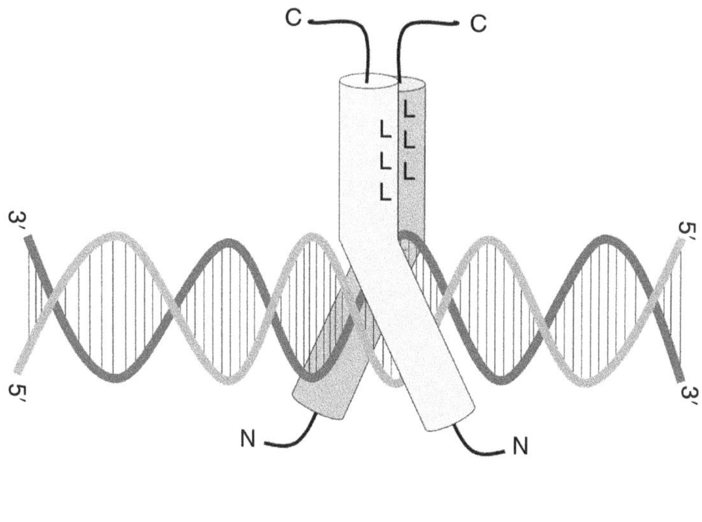

Figure 21.9 Schematic diagram of a bZIP dimer bound to the DNA. Two proteins form a Y-shape structure holding a DNA molecule. The major groove is contacted by both bZIPs via their DNA-binding domains. L represents the leucines forming the interface in the bZIP dimer.

leucine zipper motifs and slightly disintegrate on their N-terminus, forming a Y-shaped structure that holds the DNA duplex (Figure 21.9, Glover and Harrison 1995). In this complex, each basic region contacts the DNA along the major groove on opposite sides of the double helix so that each monomer binds one-half of the DNA target sequence binding the whole sequence with the bZIP dimer.

21.2.7.2 Classification of bZIP Transcription Factors

Based on the sequence similarities and functional features, the bZIP family is subdivided into 10 groups, named A–I, and S in *Arabidopsis* (Nijhawan et al. 2008). The bZIP members classified in different groups can be combined through heterodimerization to form specific bZIP pairs with discrete functionalities.

21.2.7.3 Role of bZIP Transcription Factors in Salinity Tolerance

The bZIP TFs have also been revealed to confer salinity tolerance in different plant species. A recent study has shown the overexpression of *TabZIP15* to enhance salinity tolerance in *Triticum aestivum* (Bi et al. 2021). The transgenic plants showed increased plant height, root length, and fresh weight with sustainable spike length and grain number spike^{-1} following salinity stress. The *TabZIP15* overexpressing plants also displayed decreased ROS and MDA accumulation with enhanced expression of salt stress-responsive genes under salinity stress. The overexpression of *GhABF2* isolated from *G. hirsutum* has been found to confer salinity tolerance in *G. hirsutum* and *A. thaliana*. The transgenic plants showed increased osmotic adjustment and redox homeostasis with increased fiber yield under salinity stress (Liang et al. 2016). Another study showed the overexpression of GmbZIP44, GmbZIP6, and GmbZIP78 isolated from *G. max* to enhance salinity tolerance in *Arabidopsis* by improving germination rate, survivability, and plant height under salinity stress (Liao et al. 2008).

21.3 Conclusion

Salinity tolerance in plants is a complicated molecular mechanism coordinating numerous gene functions. Although substantial research progresses have been made to our understanding of the molecular mechanism of salinity tolerance, further studies are necessary to increase our understanding in this particular direction. Intricate analyses of the candidate genes encoding TFs may facilitate in enhancing salinity tolerance in plants. Development of biotechnological approaches may help in understanding the molecular background of salinity tolerance in plants.

References

Abe, H., Urao, T., Ito, T. et al. (2003). Arabidopsis *AtMYC2* (bHLH) and *AtMYB2* (MYB) function as transcriptional activators in abscisic acid signaling. *Plant Cell* 15 (1): 63–78.

Aida, M., Ishida, T., Fukaki, H. et al. (1997). Genes involved in organ separation in Arabidopsis: an analysis of the cup-shaped cotyledon mutant. *Plant Cell* 9: 841–857.

Alonso, R., Onate-Sanchez, L., Weltmeier, F. et al. (2009). A pivotal role of the basic leucine zipper transcription factor bZIP53 in the regulation of Arabidopsis seed maturation gene expression based on heterodimerization and protein complex formation. *Plant Cell* 21: 1747–1761.

Bailey, P.C., Martin, C., Toledo-Otriz, G. et al. (2003). Update on the basic helix-loop-helix transcription factor gene family in *Arabidopsis thaliana*. *Plant Cell* 15 (11): 2497–2501.

Bakshi, M. and Oelmuller, R. (2014). WRKY transcription factors: Jack of many trades in plants. *Plant Signal. Behav.* 9 (2): e27700.

Basu, S., Giri, R.K., Benazir, I. et al. (2017). Comprehensive physiological analyses and reactive oxygen species profiling in drought tolerant rice genotypes under salinity stress. *Physiol. Mol. Biol. Plants* 23: 837–850.

Basu, S., Kumar, A., Benazir, I., and Kumar, G. (2021). Reassessing the role of ion homeostasis for improving salinity tolerance in crop plants. *Physiol. Plant.* 171 (4): 502–519.

Basu, S., Sarkar, S., Prasad, R., and Kumar, G. (2022). Current approaches for alleviating abiotic stress tolerance in crops: role of beneficial microorganisms. In: *Environmental and Microbial Biotechnology* (ed. R. Prasad and S.-H. Zhang). Singapore: Springer.

Besseau, S., Li, J., and Palva, E.T. (2012). WRKY54 and WRKY70 co-operate as negative regulators of leaf senescence in *Arabidopsis thaliana*. *J. Exp. Bot.* 63: 2667–2679.

Bi, C., Yu, Y., Dong, C. et al. (2021). The bZIP transcription factor *TabZIP15* improves salt stress tolerance in wheat. *Plant Biotechnol. J.* 19 (2): 209–211.

Cao, Y., Li, K., Li, Y. et al. (2020). MYB transcription factors as regulators of secondary metabolism in plants. *Biology* 9 (3): 61.

Chakravarthy, S., Tuori, R.P., D'Ascenzo, M.D. et al. (2003). The tomato transcription factor Pti4 regulates defense-related gene expression via GCC box and non-GCC box cis elements. *Plant Cell* 15: 3033–3050.

Chandler, J.W. (2018). Class VIIIb APETALA2 ethylene response factors in plant development. *Trends Plant Sci.* 23: 151–162.

Chen, M., Wang, Q.-Y., Cheng, X.-G. et al. (2007). *GmDREB2*, a soybean DRE-binding transcription factor, conferred drought and high-salt tolerance in transgenic plants. *Biochem. Biophys. Res. Commun.* 353: 299–305.

Chen, M., Zhao, Y., Zhuo, C. et al. (2015). Overexpression of a *NF-YC* transcription factor from bermudagrass confers tolerance to drought and salinity in transgenic rice. *Plant Biotechnol. J.* 13 (4): 482–491.

Chen, H., Liu, L., Wang, L. et al. (2016). VrDREB2A, a DREB-binding transcription factor from *Vigna radiata*, increased drought and high-salt tolerance in transgenic *Arabidopsis thaliana*. *J. Plant Res.* 129: 263–273.

Cong, L., Zheng, H.C., Zhang, Y.X., and Yao, T. (2008). *Arabidopsis DREB1A* confers high salinity tolerance and regulates the expression of GA dioxygenases in tobacco. *Plant Sci.* 174: 156–164.

Dai, X., Xu, Y., Ma, Q. et al. (2007). Overexpression of an R1R2R3 MYB gene, *OsMYB3R-2*, increases tolerance to freezing, drought, and salt stress in transgenic Arabidopsis. *Plant Physiol.* 143: 1739–1751.

Dietz, K.J., Vogel, M.O., and Viehhauser, A. (2010). AP2/EREBP transcription factors are part of gene regulatory networks and integrate metabolic, hormonal and environmental signals in stress acclimation and retrograde signalling. *Protoplasma* 245: 3–14.

Diray-Arce, J., Knowles, A., Suvorov, A. et al. (2019). Identification and evolutionary characterization of salt-responsive transcription factors in the succulent halophyte *Suaeda fruticosa*. *PLoS One* 14 (9): e0222940.

Eulgem, T., Rushton, P.J., Robatzek, S., and Somssich, I.E. (2000). The WRKY superfamily of plant transcription factors. *Trends Plant Sci.* 5 (5): 199–206.

Fang, Y., You, J., Xie, K. et al. (2008). Systematic sequence analysis and identification of tissue-specific or stress-responsive genes of NAC transcription factor family in rice. *Mol. Genet. Genom.* 280: 547–563.

Feller, A., Machemer, K., Braun, E.L., and Grotewold, E. (2011). Evolutionary and comparative analysis of MYB and bHLH plant transcription factors. *Plant J.* 66: 94–116.

Friedrichsen, D.M., Nemhauser, J., Muramitsu, T. et al. (2002). Three redundant brassinosteroid early response genes encode putative bHLH transcription factors required for normal growth. *Genetics* 162 (3): 1445–1456.

Fu, J., Zhu, C., Wang, C. et al. (2021). Maize transcription factor ZmEREB20 enhanced salt tolerance in transgenic Arabidopsis. *Plant Physiol. Biochem.* 159: 257–267.

Fujimoto, S.Y., Ohta, M., Usui, A. et al. (2000). Arabidopsis ethylene-responsive element binding factors act as transcriptional activators or repressors of GCC box-mediated gene expression. *Plant Cell* 12: 393–404.

Glover, J.N. and Harrison, S.C. (1995). Crystal structure of the heterodimeric bZIP transcription factor c-Fos-c-Jun bound to DNA. *Nature* 373: 257–261.

Gu, C., Guo, Z.H., Hao, P.P. et al. (2017). Multiple regulatory roles of AP2/ERF transcription factor in angiosperm. *Bot. Stud.* 58 (1): 6.

Hackenberg, D., Keetman, U., and Grimm, B. (2012). Homologous NF-YC2 subunit from Arabidopsis and tobacco is activated by photooxidative stress and induces flowering. *Int. J. Mol. Sci.* 13: 3458–3477.

Hao, Y., Zong, X., Ren, P. et al. (2021). Basic helix-loop-helix (bHLH) transcription factors regulate a wide range of functions in Arabidopsis. *Int. J. Mol. Sci.* 22: 7152.

Heim, M.A., Jakoby, M., Werber, M. et al. (2003). The basic Helix-Loop-Helix transcription factor family in plants: a genome-wide study of protein structure and functional diversity. *Mol. Biol. Evol.* 20 (5): 735–747.

Hong, Y., Zhang, H., Huang, L. et al. (2016). Overexpression of a stress-responsive NAC transcription factor gene *ONAC022* improves drought and salt tolerance in rice. *Front. Plant Sci.* 7: 4.

Hu, L. and Liu, S. (2011). Genome-wide identification and phylogenetic analysis of the ERF gene family in cucumbers. *Genet. Mol. Biol.* 34 (4): 624–633.

Hu, H., Dai, M., Yao, J. et al. (2006). Overexpressing a NAM, ATAF, and CUC (NAC) transcription factor enhances drought resistance and salt tolerance in rice. *Proc. Natl. Acad. Sci. U.S.A.* 103 (35): 12987–12992.

Jensen, M.K., Kjaersgaard, T., Nielsen, M.M. et al. (2010). The *Arabidopsis thaliana* NAC transcription factor family: structure-function relationships and determinants of ANAC019 stress signaling. *Biochem. J.* 426: 183–196.

Jiang, Y. and Deyholos, M.K. (2009). Functional characterization of Arabidopsis NaCl-inducible WRKY25 and WRKY33 transcription factors in abiotic stresses. *Plant Mol. Biol.* 69: 91–105.

Jiang, Q., Hu, Z., Zhang, H., and Ma, Y. (2014). Overexpression of *GmDREB1* improves salt tolerance in transgenic wheat and leaf protein response to high salinity. *Crop J.* 2 (2-3): 120–131.

Jofuku, K.D., Denboer, B.G.W., van Montagu, M., and Okamuro, J.K. (1994). Control of *Arabidopsis* flower and seed development by the homeotic gene APETALA2. *Plant Cell* 6: 1211–1225.

Kagaya, Y., Ohmiya, K., and Hattori, T. (1999). RAV1, a novel DNA-binding protein, binds to bipartite recognition sequence through two distinct DNA-binding domains uniquely found in higher plants. *Nucleic Acids Res.* 27: 470–478.

Kavas, M., Baloglu, M.C., Atabay, E.S. et al. (2016). Genome-wide characterization and expression analysis of common bean bHLH transcription factors in response to excess salt concentration. *Mol. Genet. Genomics* 291: 129–143.

Kumar, G., Basu, S., Singla-Pareek, S.L., and Pareek, A. (2022). Unraveling the contribution of *OsSOS2* in conferring salinity and drought tolerance in a high-yielding rice. *Physiol. Plant.* 174 (1): e13638.

Laloum, T., De Mita, S., Gamas, P. et al. (2013). CCAAT-box binding transcription factors in plants: Y so many? *Trends Plant Sci.* 18 (3): 157–166.

Li, X.Y., Mantovani, R., Hooft van Huijsduijnen, R. et al. (1992). Evolutionary variation of the CCAAT-binding transcription factor NF-Y. *Nucleic Acids Res.* 20: 1087–1091.

Li, X., Duan, X., Jiang, H. et al. (2006). Genome-wide analysis of basic/helix-loop-helix transcription factor family in rice and Arabidopsis. *Plant Physiol.* 141 (4): 1167–1184.

Li, F., Guo, S., Zhao, Y. et al. (2010). Overexpression of a homopeptide repeat-containing bHLH protein gene (*OrbHLH001*) from Dongxiang Wild Rice confers freezing and salt tolerance in transgenic Arabidopsis. *Plant Cell Rep.* 29: 977–986.

Li, L.L., Yu, Y.L., Wei, J. et al. (2013a). Homologous HAP5 subunit from *Picea wilsonii* improved tolerance to salt and decreased sensitivity to ABA in transformed Arabidopsis. *Planta* 238: 345–356.

Li, Y.J., Fang, Y., Fu, Y.R. et al. (2013b). NFYA1 is involved in regulation of postgermination growth arrest under salt stress in Arabidopsis. *PLoS One* 8: e61289.

Li, H., Wang, Y., Wu, M. et al. (2017). Genome-Wide Identification of AP2/ERF transcription factors in cauliflower and expression profiling of the ERF family under salt and drought stresses. *Front. Plant Sci.* 8: 946.

Liang, M., Yin, X., Lin, Z. et al. (2014). Identification and characterization of NF-Y transcription factor families in Canola (*Brassica napus* L.). *Planta* 239: 107–126.

Liang, C., Meng, Z., Meng, Z. et al. (2016). GhABF2, a bZIP transcription factor, confers drought and salinity tolerance in cotton (*Gossypium hirsutum* L.). *Sci. Rep.* 6: 35040.

Liao, Y., Zou, H.F., Wei, W. et al. (2008). Soybean *GmbZIP44*, *GmbZIP62* and *GmbZIP78* genes function as negative regulator of ABA signaling and confer salt and freezing tolerance in transgenic Arabidopsis. *Planta* 228: 225–240.

Liu, J.J. and Ekramoddoullah, A.K. (2009). Identification and characterization of the WRKY transcription factor family in *Pinus monticola*. *Genome* 52: 77–88.

Liu, Q.L., Zhong, M., Li, S. et al. (2013). Overexpression of a chrysanthemum transcription factor gene, DgWRKY3, intobacco enhances tolerance to salt stress. *Plant Physiol. Biochem.* 69: 27–33.

Liu, G., Li, X., Jin, S. et al. (2014). Overexpression of rice NAC gene *SNAC1* improves drought and salt tolerance by enhancing root development and reducing transpiration rate in transgenic cotton. *PLoS One* 9 (1): e86895.

Liu, Y., Ji, X., Nie, X. et al. (2015). Arabidopsis AtbHLH112 regulates the expression of genes involved in abiotic stress tolerance by binding to their E-box and GCGbox motifs. *New Phytol.* 207 (3): 692–709.

Mangelsen, E., Kilian, J., Berendzen, K.W. et al. (2008). Phylogenetic and comparative gene expression analysis of barley (*Hordeum vulgare*) WRKY transcription factor family reveals putatively retained functions between monocots and dicots. *BMC Genomics* 9: 194.

Mantovani, R. (1999). The molecular biology of the CCAAT-binding factor NF-Y. *Gene* 239: 15–27.

Mao, K., Dong, Q., Li, C. et al. (2017). Genome wide identification and characterization of apple bHLH transcription factors and expression analysis in response to drought and salt stress. *Front. Plant Sci.* 8: 480.

Martin, C. and Paz-Ares, J. (1997). MYB transcription factors in plants. *Trends Genet.* 13: 67–73.

Miller, G., Suzuki, N., Ciftci-Yilmaz, S., and Mittler, R. (2010). Reactive oxygen species homeostasis and signalling during drought and salinity stresses. *Plant Cell Environ.* 33: 453–467.

Miyoshi, K., Ito, Y., Serizawa, A., and Kurata, N. (2003). *OsHAP3* genes regulate chloroplast biogenesis in rice. *Plant J.* 36: 532–540.

Mizoi, J., Shinozaki, K., and Yamaguchi-Shinozaki, K. (2012). AP2/ERF family transcription factors in plant abiotic stress responses. *Bba-Gene Regul. Mech.* 1819: 86–96.

Nguyen, Q.H., Vu, L.T.K., Nguyen, L.T.N. et al. (2019). Overexpression of the *GmDREB6* gene enhances proline accumulation and salt tolerance in genetically modified soybean plants. *Sci. Rep.* 9: 19663.

Ni, Z., Hu, Z., Jiang, Q., and Zhang, H. (2013). *GmNFYA3*, a target gene of miR169, is a positive regulator of plant tolerance to drought stress. *Plant Mol. Biol.* 82: 113–129.

Nijhawan, A., Jain, M., Tyagi, A.K., and Khurana, J.P. (2008). Genomic survey and gene expression analysis of the basic leucine zipper transcription factor family in rice. *Plant Physiol.* 146: 333.

Niu, X., Guan, Y., Chen, S., and Li, H. (2017). Genome-wide analysis of basic helix-loop-helix (bHLH) transcription factors in *Brachypodium distachyon*. *BMC Genomics* 18: 619.

Olsen, A.N., Ernst, H.A., Leggio, L.L., and Skriver, K. (2005). NAC transcription factors: structurally distinct, functionally diverse. *Trends Plant Sci.* 10: 79–87.

Ooka, H., Satoh, K., Doi, K. et al. (2003). Comprehensive analysis of NAC family genes in *Oryza sativa* and *Arabidopsis thaliana*. *DNA Res.* 10 (6): 239–247.

Pandey, S.P. and Somssich, I.E. (2009). The role of WRKY transcription factors in plant immunity. *Plant Physiol.* 150: 1648–1655.

Park, H.J., Kim, W.-Y., and Yun, D.-J. (2016). A new insight of salt stress signaling in plant. *Mol. Cells* 39: 447–459.

Petroni, K., Kumimoto, R.W., Gnesutta, N. et al. (2012). The promiscuous life of plant NUCLEAR FACTOR Y transcription factors. *Plant Cell* 24: 4777–4792.

Pirrello, J., Prasad, B.C., Zhang, W. et al. (2012). Functional analysis and binding affinity of tomato ethylene response factors provide insight on the molecular bases of plant differential responses to ethylene. *BMC Plant Biol.* 12: 190.

Puranik, S., Sahu, P.P., Srivastava, P.S., and Prasad, M. (2012). NAC proteins: regulation and role in stress tolerance. *Trends Plant. Sci.* 17: 369–381.

Rosinski, J.A. and Atchley, W.R.J. (1998). Molecular evolution of the MYB family of transcription factors: evidence for polyphyletic origin. *J. Mol. Evol.* 46: 74–83.

Rushton, P.J., Somssich, I.E., Ringler, P., and Shen, Q.J. (2010). WRKY transcription factors. *Trends Plant Sci.* 15: 247–258.

Sasaki, K., Mitsuhara, I., Seo, S. et al. (2007). Two novel AP2/ERF domain proteins interact with *cis*-element VWRE for wound-induced expression of the Tobacco tpoxN1 gene. *Plant J.* 50: 1079–1092.

Seo, P.J., Kim, S.G., and Park, C.M. (2008). Membrane-bound transcription factors in plants. *Trends Plant. Sci.* 13: 550–556.

Sharma, M.K., Kumar, R., Solanke, A.U. et al. (2010). Identification, phylogeny, and transcript profiling of ERF family genes during development and abiotic stress treatments in tomato. *Mol. Genet. Genomics* 284: 455–475.

Sharoni, A.M., Nuruzzaman, M., Satoh, K. et al. (2011). Gene structures, classification and expression models of the AP2/EREBP transcription factor family in rice. *Plant Cell Physiol.* 52 (2): 344–360.

Shrivastava, P. and Kumar, R. (2015). Soil salinity: a serious environmental issue and plant growth promoting bacteria as one of the tools for its alleviation. *Saudi J. Biol. Sci.* 22 (2): 123–131.

Singh, K., Foley, R.C., and Onate-Sanchez, L. (2002). Transcription factors in plant defense and stress responses. *Curr. Opin. Plant Biol.* 5 (5): 430–436.

So, H.-A. and Lee, J.-H. (2019). NAC transcription factors from soybean (*Glycine max* L.) differentially regulated by abiotic stress. *J. Plant Biol.* 62: 147–160.

Song, C., Je, J., Hong, J.K., and Lim, C.O. (2014). Ectopic expression of an *Arabidopsis* dehydration-responsive element-binding factor *DREB2C* improves salt stress tolerance in crucifers. *Plant Cell Rep.* 33: 1239–1254.

Souer, E., van Houwelingen, A., Kloos, D. et al. (1996). The no apical meristem gene of Petunia is required for pattern formation in embryos and flowers and is expressed at meristem and primordia boundaries. *Cell* 85: 159–170.

Sun, X., Wang, Y., and Sui, N. (2018). Transcriptional regulation of bHLH during plant response to stress. *Biochem. Biophys. Res. Commun.* 503 (2): 397–401.

Tang, M., Liu, X., Deng, H., and Shen, S. (2011). Over-expression of *JcDREB*, a putative AP2/EREBP domain-containing transcription factor gene in woody biodiesel plant *Jatropha curcas*, enhances salt and freezing tolerance in transgenic *Arabidopsis thaliana*. *Plant Sci.* 181 (6): 623–631.

Tang, Y., Liu, K., Zhang, J. et al. (2017). *JcDREB2*, a physic nut AP2/ERF gene, alters plant growth and salinity stress responses in transgenic rice. *Front. Plant Sci.* 8: 306.

Tang, Y., Bao, X., Zhi, Y. et al. (2019). Overexpression of a MYB family gene, *OsMYB6*, increases drought and salinity stress tolerance in transgenic rice. *Front. Plant Sci.* 10: 168.

Thirumurugan, T., Ito, Y., Kubo, T. et al. (2008). Identification, characterization and interaction of HAP family genes in rice. *Mol. Genet. Genomics* 279: 279–289.

Wang, X., Han, H., Yan, J. et al. (2015). A new AP2/ERF transcription factor from the oil plant *Jatropha curcas* confers salt and drought tolerance to transgenic tobacco. *Appl. Biochem. Biotechnol.* 176: 582–597.

Wang, J., Cheng, G., Wang, C. et al. (2017). The bHLH transcription factor CgbHLH001 is a potential interaction partner of CDPK in halophyte *Chenopodium glaucum*. *Sci. Rep.* 7 (1): 8441.

Wang, Y., Zhang, Y., Zhou, R. et al. (2018). Identification and characterization of the bZIP transcription factor family and its expression in response to abiotic stresses in sesame. *PLoS One* 13 (7): e0200850.

Wang, Y., Wang, J., Zhao, X. et al. (2020). Overexpression of the transcription factor gene *OsSTAP1* increases salt tolerance in rice. *Rice* 13: 50.

Wei, X., Xu, J., Guo, H. et al. (2010). DTH8 suppresses flowering in rice, influencing plant height and yield potential simultaneously. *Plant Physiol.* 153: 1747–1758.

Wu, L.J., Zhang, Z.J., Zhang, H.W. et al. (2008). Transcriptional modulation of ethylene response factor protein JERF3 in the oxidative stress response enhances tolerance of tobacco seedlings to salt, drought, and freezing. *Plant Physiol.* 148: 1953–1963.

Wu, J., Yu, C., Huang, L., and Gan, Y. (2021). A rice transcription factor, OsMADS57, positively regulates high salinity tolerance in transgenic *Arabidopsis thaliana* and *Oryza sativa* plants. *Physiol. Plant.* 173 (3): 1120–1135.

Xiang, X.-Y., Chen, J., Xu, W.-X. et al. (2021). Dehydration-induced WRKY transcriptional factor *MfWRKY70* of *Myrothamnus flabellifolia* enhanced drought and salinity tolerance in Arabidopsis. *Biomolecules* 11: 327.

Xiong, H., Li, J., Liu, P. et al. (2014). Overexpression of *OsMYB48-1*, a novel MYB-related transcription factor, enhances drought and salinity tolerance in rice. *PLoS One* 9 (3): e92913.

Xu, W., Li, F., Ling, L., and Liu, A. (2013). Genome-wide survey and expression profiles of the AP2/ERF family in castor bean (*Ricinus communis* L.). *BMC Genomics* 14: 785.

Xu, X.-B., Pan, Y.-Y., Wang, C.-L. et al. (2014). Overexpression of *DnWRKY11* enhanced salt and drought stress tolerance of transgenic tobacco. *Biologia* 69: 994–1000.

Yamasaki, K., Kigawa, T., Watanabe, S. et al. (2012). Structural basis for sequence-specific DNA recognition by an Arabidopsis WRKY transcription factor. *J. Biol. Chem.* 287: 7683–7691.

Yan, H.W., Hong, L., Zhou, Y.Q. et al. (2013). A genome-wide analysis of the ERF gene family in sorghum. *Genet. Mol. Res.* 12 (2): 2038–2055.

Yan, H., Jia, H., Chen, X. et al. (2014). The cotton WRKY transcription factor *GhWRKY17* functions in drought and salt stress in transgenic *Nicotiana benthamiana* through ABA signalling and the modulation of reactive oxygen species production. *Plant Cell Physiol.* 55 (12): 2060–2076.

Yang, H., Yu, C., Yan, J. et al. (2014). Overexpression of the *Jatropha curcas JcERF1* gene coding an AP2/ERF-type transcription factor increases tolerance to salt in transgenic tobacco. *Biochemistry* 79 (11): 1226–1236.

Yu, L.-H., Wu, S.-J., Peng, Y.-S. et al. (2015). *Arabidopsis EDT1/HDG11* improves drought and salt tolerance in cotton and poplar and increases cotton yield in the field. *Physiol. Plant.* 14 (1): 72–84.

Zhai, Y., Zhang, L., Xia, C. et al. (2016). The wheat transcription factor, TabHLH39, improves tolerance to multiple abiotic stressors in transgenic plants. *Biochem. Biophys. Res. Commun.* 473 (4): 1321–1327.

Zhang, G., Chen, M., Chen, X. et al. (2008). Phylogeny, gene structures, and expression patterns of the ERF gene family in soybean (*Glycine max* L.). *J. Exp. Bot.* 59 (15): 4095–4107.

Zhang, Y., Chen, C., Jin, X.-F. et al. (2009). Expression of a rice DREB1 gene, *OsDREB1D*, enhances cold and high-salt tolerance in transgenic Arabidopsis. *BMB Rep.* 42 (8): 486–492.

Zhang, X.-X., Tang, Y.-J., Ma, Q.-B. et al. (2013). *OsDREB2A*, a rice transcription factor, significantly affects salt tolerance in transgenic soybean. *PLoS One* 8 (12): e83011.

Zhang, T., Zhang, D., Liu, Y. et al. (2015). Overexpression of a NF-YB3 transcription factor from *Picea wilsonii* confers tolerance to salinity and drought stress in transformed *Arabidopsis thaliana*. *Plant Physiol. Biochem.* 94: 153–164.

Zhao, H., Wu, D., Kong, F. et al. (2017). The *Arabidopsis thaliana* nuclear factor Y transcription factors. *Front. Plant Sci.* 7: 2045.

Zhao, K., Li, S., Yao, W. et al. (2018). Characterization of the basic helixeloopehelix gene family and its tissue-differential expression in response to salt stress in poplar. *Peer J.* 6: e4502.

Zhuang, J., Cai, B., Peng, R.H. et al. (2008). Genome-wide analysis of the AP2/ERF gene family in *Populus trichocarpa*. *Biochem. Biophys. Res. Commun.* 371 (3): 468–474.

Part 5

Stress Management Strategies for Sustainable Agriculture

22

Seed Quality Assessment and Improvement Between Advancing Agriculture and Changing Environments

Andrea Pagano, Paola Pagano, Conrado Dueñas, Adriano Griffo, Shraddha Shridhar Gaonkar, Francesca Messina, Alma Balestrazzi, and Anca Macovei

Department of Biology and Biotechnology, University of Pavia, Pavia, Italy

22.1 Introduction: A Seed's Viewpoint on Climate Change

Through the voice of the United Nations, the international community has established seventeen Sustainable Development Goals to coordinate the current strive toward a more equitable and sustainable society by 2030. The second goal plans to eliminate hunger by promoting sustainable agriculture, supporting small farmers, and proposing market-effective technologies (Shelar et al. 2021). In this sense, sustainable agriculture is gaining importance in today's food production landscape because it considers the farmer's profitability along with its environmental impact (Gold and Gates 2017). The availability of high-quality seeds represents a crucial element for the vast majority of farming systems, standing at the crossroads between human development and environmental sustainability in these decades that are witnessing the severe effects of anthropogenic climate change.

To meet the nutritional requirements of the growing population, grain yields per land unit need to double within the next century, and since some of the most widely consumed crops might have already reached their maximal productivity, a deeper understanding of seed ecophysiology and further developments in seed technology will be necessary to increase grain production (Long and Ort 2010).

The consequences of climate change on seed ecophysiology are variegated and species-dependent. Seeds perceive increasing carbon dioxide concentrations, extreme temperature variations, and insufficient or excessive precipitations as environmental stresses that affect different stages, including maturation, dormancy, germination, and seedling establishment. In other words, from the point of view of seed producers, both pre-harvest phases (development on the mother plant) and post-harvest phases (processing, storage) can be affected by environmental variables, reducing seed quality (Parmesan and Hanley 2015; Lamichaney and Maity 2021).

The concentration of atmospheric carbon dioxide has increased approximately 25% since 1959 and is predicted to surpass 550 ppm during the current century from the current 410 ppm. Beside the effects of carbon dioxide as a greenhouse gas and the central role of carbon metabolism in plant physiology, the accumulation of atmospheric carbon dioxide may directly affect seed development by altering carbon/nitrogen ratios in the mother plant, thus making nitrogen availability a further limiting factor called into question by climate change (Ziska et al. 2012; Lamichaney and Maity 2021).

As the main consequence of greenhouse emissions, rising temperatures threaten seed production from cold-adapted crops, such as cool season grain legumes (lentils, faba beans, chickpeas, field peas, etc.) (Kumar et al. 2021), but also wheat and rice (Langridge and Reynolds 2021; Heredia et al. 2022). Heat stress directly affects dormancy regulation, seed longevity, germination, and seedling development by accelerating the deterioration of biological macromolecules and, indirectly, by triggering the accumulation of reactive oxygen species (ROS), which in turn causes oxidative stress and alters hormonal signaling (Finch-Savage and Footitt 2017; Zhou et al. 2020; Farooq et al. 2021). On the other hand, low humidity and low temperatures promote dormancy maintenance as well as seed longevity under storage (Toh et al. 2008).

Water is one of the major factors influencing plant growth and the most relevant one when considering the drought risks and desertification trends exacerbated by climate change. Many aspects of seed physiology reflect adaptive responses to

Global Climate Change and Plant Stress Management, First Edition. Edited by Mohammad Wahid Ansari, Anil Kumar Singh, and Narendra Tuteja.

variable water availability, including the capacity of orthodox seeds to withstand extensive dehydration and survive in dormant states until suitable conditions for germination occur. Also, water uptake during imbibition triggers the reactivation of pre-germinative metabolism at the beginning of the germination process (Paparella et al. 2015; Penfield, 2017; Farrant and Hilhorst 2021). Moreover, desiccation-tolerant seeds represent a promising model to understand the molecular bases of drought stress response and design biotechnological approaches and breeding programs to improve it also in vegetative tissues beyond the scope of seed industry (Farrant and Hilhorst 2021).

Conversely, the culture requirements of certain crops expose seed germination and seedling establishment to the anoxic condition due to severe flooding. This is the case of rice, whose seedlings require transferring from nurseries to fields, whereas direct sowing would require uniform germination, improved root plasticity, and resistance to prolonged anaerobic conditions (Heredia et al. 2022). On the other hand, unlike many other widespread crops, rice cultivation itself contributes to greenhouse gas emissions because of the methane produced in flooded paddy fields (Heredia et al. 2022), constituting a further example of how systemic views are needed to meet agricultural requirements with environmentally sustainable solutions.

"Diversity," constitutes a keyword to plan solutions for complex issues, including the role of seed quality in farming systems facing climate change, considering different markets and the requirements of each species. Uniformity is useful for farming organization but represents a risk factor under fluctuating conditions, whereas diversity would ensure higher degrees of adaptability. In the seed industry, this is true at many levels, since coordinated germination is good for timely harvests but dangerous when the weather is voluble in the same way in which, macroscopically, the lack of genetic diversity can be convenient for many farming systems yet exposing essential crops to the effects of climate change.

Seed dormancy itself is a crucial adaptive mechanism to ensure the persistence of seed populations under unfavorable periods and the survival of plant populations in diverse environments (Forbis et al. 2002). Variations in germination rates constitute an evolutionary advantage in wild seed populations, as the temporal dispersal of germination minimizes the competition for resources and the exposure to stress events. In contrast, variability in germination rates is disadvantageous in crops, where uniformity is preferable to plan sowing and harvesting (Bewley and Black 2013). Furthermore, the rising temperatures and water deficiency caused by climate change could anticipate, delay, or prevent regeneration from dormant seed populations (Kim and Han 2018). For example, with shorter winters, seeds may remain partly dormant during spring and need a longer time to germinate (Walck et al. 1999). Moreover, climate change is a threat for the existing crop species due to the temporal mismatch of growth periods and seasonal cycles, which can affect the behavior of pollinators and facilitate the spreading of seed-borne diseases, whereas the need for seed storage affects dormancy regulation and seed longevity (Anderson et al. 2004, 2020; Maity and Pramanik 2013).

Agrobiodiversity constitutes another aspect to be considered when the seed industry requires new solutions to tackle exposure to environmental stresses. In the last century, three-quarters of genetic diversity in agriculture has been lost as high-performing and genetically uniform varieties appeared on the market. Today, less than 200 edible plant species are cultivated worldwide, and a dozen of them cover 75% of the global food demand (FAO 1999).

With these premises, seeds are crucial for crop diversity. Seed banks ensure long-term preservation of genetic diversity, and the efforts to enrich germplasm collections and make them available to farmers are more and more necessary along with targeted studies to improve seed longevity and storability (Westengen et al. 2020; Zhou et al. 2020). In this scenario, the valorization of local varieties, neglected species, and crop wild relatives provides new solutions to improve stress tolerance by promoting genetic variability. Operatively, screening and phenotyping under different conditions have been proven useful for the selection of traits involved in stress resistance as genetic resources for breeders and biotechnologists (Muñoz et al. 2017; Forti et al. 2021; Langridge and Reynolds 2021).

From an economic perspective, while discussing seed technology, the capabilities of developed and developing economies need to be considered. Risk assessment studies indicated the regions more prone to food insecurity related to climate change are located in Africa, not because of a higher exposure to the consequences of climate change but because of social and economic issues (Parry et al. 2005). Consequently, new developments and proposed solutions in seed technology need to be contextualized according to their target farming system. In this sense, sustainable, versatile, and resource-effective approaches, such as seed priming, are gaining popularity to improve yields while coping with the consequences of climate change, especially in developing countries (Devika et al. 2021).

Needless to say, safeguarding biological diversity is a major imperative not only when applied to agriculture and human development, but also concerning the preservation of wild environments and ecological niches threatened by climate change (Kupika and Nhamo 2016; Vázquez-Ramírez and Venn 2021). This chapter provides an overview of the current

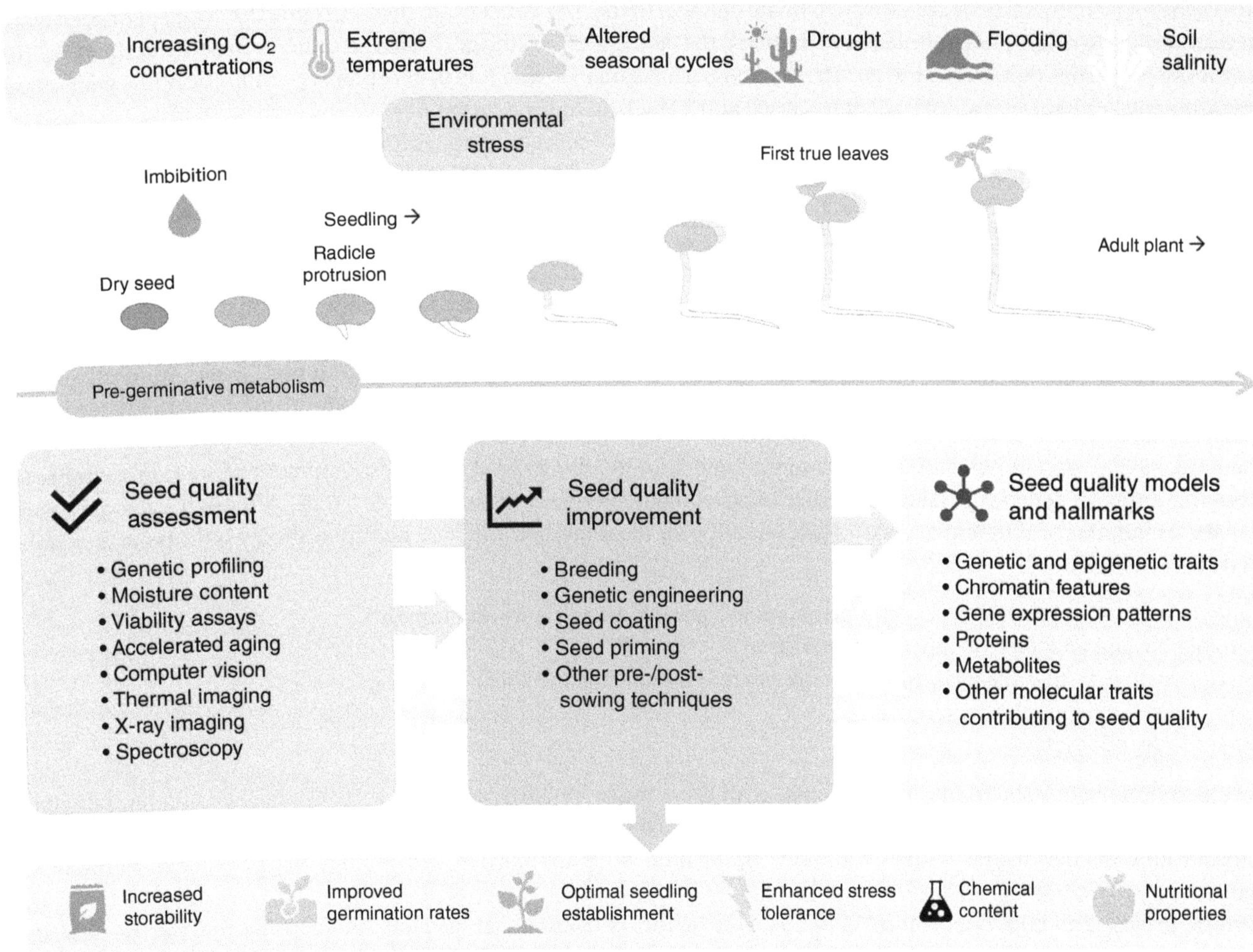

Figure 22.1 Outline of the main sources of environmental stress related to climate change to be considered for their effects on pre-germinative metabolism and an overview of the integrative approach between seed quality assessment and improvement and the search for reproducible models and hallmarks of seed quality, along with some of the most relevant traits to be considered for targeted enhancements.

state of seed technology in relation with the stress conditions caused by climate change, summarizing the techniques that can be applied to seed testing, the approaches to improve seed quality, and the current knowledge about the molecular mechanisms of stress response in the context of pre-germinative metabolism (Figure 22.1).

22.2 Assessing Seed Quality: Invasive and Non-invasive Techniques for Grain Testing

In the complexity of the social and environmental landscape outlined so far, what defines seed quality? The concept of seed quality is extremely broad and includes multiple aspects. Seed companies, breeders, and biotechnologists have identified several traits of interest that are useful to optimize various phases of seed production, including harvesting, storage, sowing, germination, and post-germination. These include germination percentage and speed, homogeneous seedling emergence and development, longevity, storability, resistance to mechanical damage, absence of biological and chemical contaminants, and tolerance to biotic and abiotic stress. Other features, including chemical composition and nutritional properties, might be also held into account for certain grains destined for human or livestock consumption (Finch-Savage and Bassel 2016; Rahman and Cho 2016).

The whole complex of these traits constitutes seed quality. These features can be defined quantitatively, species by species, through specific parameters, evaluated by official seed testing institutions, and improved according to the requirements of farming systems adapting to climate change. In this last concern, to further refine the concept of seed quality in relationship to diverse environmental conditions, the concept of "seed vigor" was introduced, broadly defined as "the sum of those properties that determine the activity and performance of seed lots of acceptable germination in a wide range of environments" (Ventura et al. 2012; Finch-Savage and Bassel 2016; ISTA 2018).

An early, reliable, and non-invasive assessment of the quality of a seed lot constitutes a precious tool to optimize seed production and safety controls, with practical and economic benefits (Huang et al. 2015). Currently, standard tests approved by the International Seed Testing Association (ISTA), such as the germination test, allow one to observe the behavior of seeds in the post-sowing phase. However, these procedures present issues related to the direct use of the seeds for testing and the time required to obtain relevant information. Furthermore, despite its practicality, the use of a single test does not allow the analysis of multiple factors that may directly contribute to seed quality or be indirectly predictive for its evaluation, so the obtained information is inevitably partial. Beside the classical seed testing parameters relying on germination performances, new technologies are emerging as predictive tools for seed quality, integrating high-throughput phenotyping, advanced digital imaging, and modeling with the detection of compounds related to quality and stress tolerance (Feng et al. 2019; ElMasry et al. 2020).

The need for precise and reliable elements to evaluate seed quality has led to the development of new procedures to analyze the different genetic, physical, and physiological characteristics of this complex trait. Although each approach allows the evaluation of different aspects of seed quality, an ideal method for seed quality testing would possess certain crucial characteristics:

- highly informative (different types of information are provided accurately and rapidly);
- reproducible and adaptable to different situations;
- non-destructive, so that the analyzed seeds can be used for sowing or further analyses;
- no specialized personnel is required;
- economically affordable.

Unfortunately, none of the current methods possess all these major benefits, and all require established and reproducible correlations between the measured parameters and the traits directly related to seed quality. Subsequently, it is necessary to integrate different existing techniques while developing innovative approaches, with studies on a variety of experimental systems.

Genetic purity is an important feature to evaluate the quality of commercial seed lots, as it certifies that the tested seeds belong to a certain variety with its specific qualities, with practical consequences for its farming procedures and legal implications for its commercialization. In this context, recombinant DNA technology and the continuous search for new molecular markers associated with seed quality have brought profound changes to the methods of determining genetic purity. In particular, the analysis of certain genetic polymorphisms using PCR technology has proved to be a more sensitive and accurate method than the characterization of proteins by HPLC and electrophoresis (McDonald 1998; Crockett et al. 2002). Random Amplification of Polymorphic DNA-PCR (RAPD-PCR) is based on the use of arbitrarily designed oligonucleotides (random primers) for the amplification of DNA fragments (RAPD markers), which, overall, generate a genotype-specific profile. This procedure has been widely used for genotypic analysis on different seed plants (Joel et al. 1996; Fu et al. 2003). Similarly, Simple Sequence Repeat PCR (SSR-PCR), Inter Simple Sequence Repeat PCR (ISSR-PCR), and Sequence-Related Amplified polymorphism PCR (SRAP-PCR) allow the evaluation of seed genetic purity through the amplification of specific markers (Liu et al. 2007). Genome Wide Association Studies (GWAS) and Quantitative Trait Loci (QTL) mapping can also be applied to identify genetic hallmarks helpful for breeding, especially considering the complex multigenic nature of many traits associated with stress tolerance (Macovei et al. 2017; Choudhary et al. 2019).

Moisture content is another factor that determines seed quality, since it can reveal the presence of physical damage and give indications regarding the harvesting period and the general health state of the grains. Moreover, moisture content is one of the key determinants of seed ripeness and storability, since dry and cold conditions are most suitable for the long-term conservation of mature orthodox seeds (Mersal et al. 2006; Elias et al. 2012; Kim 2018). Several methods are available to evaluate seed moisture content. One of the most common protocols consists of weighting the seeds before and after drying at high (>80 °C) temperatures. Other methods rely on moisture extraction by organic solvents followed by the chromatographic analysis of the obtained extract, while others allow the indirect quantitation of moisture content through the evaluation of the physical and/or chemical characteristics of the seeds. Although the methods that involve a direct determination of the seed moisture content are the most accurate, their use on a large scale is time-consuming (Grabe 1989).

Chemical approaches to the evaluation of seed viability are available, including the Tetrazolium Test. It relies on TTC (2,3,5-triphenyl tetrazolium chloride), a colorless dye that can penetrate inside the seed cells. At the cellular level, TTC molecules are reduced by the dehydrogenases involved in metabolic activities, such as the malate dehydrogenase enzyme. This reaction results in the formation of formazan, a non-diffusible molecule that acquires a strong red coloration visible to the naked eye. In other words, living, metabolically active, cells convert the starting dye and take on a red color, unlike dead cells, thus providing an indication about the seed viability state. Compatibly with the chromatic and morphological features of the target seed lot, this approach can also allow a visual screening for viability. The impossibility of using the analyzed seeds for cultivation constitutes one of the major limitations of the tetrazolium test. Moreover, it requires expert personnel for the visual assessments (Soares et al. 2016; França-Neto and Krzyzanowski 2019).

Accelerated aging is a popular approach to determine seed quality by predicting the germination capacity of seeds that have been exposed to environmental conditions that are detrimental for their viability, usually high temperatures and high humidity, that speed up seed spoilage by simulating deterioration under controlled conditions. In fact, low-quality seeds tend to deteriorate more quickly compared to high-quality seeds when subjected to particular temperature and humidity conditions. These tests are reproducible and adaptable and represent the standard test for the evaluation of viability in numerous species (Demir et al. 2004; Dutra and Vieira 2006).

Up until now, the most widespread and well-known techniques for seed quality evaluation have been described, highlighting the different aspects of seed quality that can be evaluated. However, all have a major limitation: the seeds are damaged or altered in the process and, consequently, cannot be used for sowing or further tests. This may represent an obstacle if the availability of seeds is low or their cost is high, or if multiple analyses are planned on the same material to help with experimental reproducibility and resource optimization.

Fortunately, non-destructive approaches have been developed in recent years, allowing for a more rapid and accurate assessment of seed quality. Specifically, optical analysis techniques that have already been applied in the medical, diagnostic, and food sectors, have also found their place in the agricultural sector for seed quality assessment. An interesting aspect of these new approaches is certainly the ability to collect information on new and multiple levels with single and non-invasive assays. For example, it is possible to determine the main physical characteristics and chemical composition, the presence or absence of damage or parasite infections, and other valuable information at the same time. In this sense, the use of optical methods for seed quality evaluation represents a time-saving and cost-effective alternative compared to chromatographic and electrophoresis techniques. Despite these advantages, the required equipment is still expensive and often requires specialized training, concerning both the sensors for data acquisition and the high-performance computers for data processing (Xia et al. 2019).

Computer vision (or machine vision) is a method based on artificial intelligence that simulates human vision to rapidly collect and interpret high-resolution images of an object (in this case, a seed). For example, this technology has been used to predict the germination potential of pepper (*Capsicum* spp.) seeds through the evaluation of their morphological features (Tu et al. 2018). Besides providing a basic physical characterization of a seed lot (e.g. shape, size, color), it is also able to detect the presence of morphological anomalies caused by damage or infections (Patel et al. 2012; Lin et al. 2019). Unfortunately, machine vision is not generally conceived to provide information about chemical composition, so its contribution to the evaluation of seed health state is partial.

Thermal imaging allows image acquisition in the infrared wavelengths, thus displaying the temperature gradients that characterize the observed object without physical contact and without using external light sources. The retrieved information can provide a good estimate of seed viability and germinative potential. A limitation of thermal imaging is the excessive sensitivity to the surrounding environment that impairs the accuracy of the analyses if a suitable space is not available (ElMasry et al. 2020). Nonetheless, thermal imaging has been used, for example, to screen viable, aged and non-viable pea (*Pisum sativum* L.) seeds by exploiting contrastive differences in temperatures (Men et al. 2017).

X-ray imaging, despite being a widely used technology in the medical field, is also popular for the analysis of food and seed quality. Its principles are relatively simple: the sample (e.g. a single seed) is hit by short-wavelength electromagnetic radiations that undergo attenuation phenomena determined by its density. A detector measures the attenuated radiation, and a software displays the differences in density of the sample as changes in color (or contrast) of two-dimensional images. With each new measurement, radiation can be administered from different angles, generating a set of images that can be combined into a detailed representation of the external and internal structure of a seed, along with eventual morphological evidence of damages or infections. Currently, X-ray imaging is continuously optimized to allow a rapid and reliable estimation of seed quality in different experimental systems (Medeiros et al. 2020; Bianchini et al. 2021) and has been used, for example, to analyze the internal morphology of Brassicaceae seeds and bell pepper (*Capsicum annuum*) embryos (Gagliardi and Marcos-Filho 2011; Musaev et al. 2021).

The spectroscopic techniques applied to seed quality evaluation deserve a separate discussion. Unlike many other non-invasive approaches, they are able to extract accurate data concerning the chemical composition of the analyzed samples. Despite the associated costs, this is perhaps one of the deepest levels of analysis to consider for a non-destructive assessment of seed quality.

Near-infrared spectroscopy (NIRS) technologies, for example, are widely used in agriculture and industry. Like other spectroscopic approaches, they analyze the sample through electromagnetic radiation that, as the name suggests, belongs to the near-infrared (780–2500 nm), whose penetrative capacity causes the vibration of the atoms within the sample. These vibrations are not free, as the movement of the atoms is constrained by the chemical bonds determining their mutual distance. The result is a highly specific emission spectrum, whose detection and interpretation are able to reveal the chemical composition of the object (Font et al. 2006; Huang et al. 2015; Al-Amery et al. 2018). For example, NIR technology has been used to accurately estimate the protein and oil content in pea seeds (Hacisalihoglu et al. 2020). This short explanation certainly does not convey the complexity of the NIR technology, but it serves to broadly understand how it is possible to gather useful information to estimate the quality of a seed lot.

Raman spectroscopy also exploits the atom vibrations induced by radiation to obtain a highly specific emission spectrum applicable to the estimation of seed quality (Ambrose et al. 2016a, 2016b). It is named after the physicist who first described the underlying phenomenon, known as "Raman Scattering." When a light beam interacts with an object, most of the photons spread with the same initial energy (Rayleigh scattering), while a small amount of photons diffuses with different energy (Raman diffusion). The resulting Raman spectrum, typically induced by a monochromatic laser beam and possibly integrated with other spectroscopic approaches, allows an accurate and non-destructive evaluation of the chemical composition of a given sample (Rahman and Cho 2016). Raman spectroscopy has been successfully applied to study the chemical composition of cocoa seeds and their extracts (Edwards et al. 2005).

Hyperspectral imaging allows the acquisition of both spectral and spatial information, thus allowing the simultaneous assessment of chemical composition and physical features (Feng et al. 2019). Explaining this complex technique simplistically, in hyperspectral imaging, the sample is hit by electromagnetic radiation at different wavelengths (λ) and, for each wavelength, the detector collects images of the given spatial coordinates of the object (x, y). The resulting data are called hypercubes and are three-dimensional (x, y, λ), consisting of the coordinates of the pixels (x, y) detected at a specific wavelength (λ). The possibility of obtaining different types of information on the quality of the seed is certainly the strong point of this technology. For example, it has been utilized to distinguish corn seeds treated at high temperatures from untreated seeds by analyzing their protein, carbohydrate, and moisture contents (Ambrose et al. 2016a, 2016b).

Ultimately, the non-invasive methods that have been outlined certainly represent a faster, deeper, and more precise way to retrieve information for seed quality evaluation, but, on the other hand, the associated costs and the required expertise are still prohibitive for large-scale screening of seed lots. Therefore, the future goal is to optimize these methods to reduce their costs and promote synergies for their diffusion.

22.3 Improving Seed Quality: Optimizing Priming Techniques to Face the Challenges of Climate Changes

As exposed earlier, seed quality mainly relies on, but is not limited to, viability and longevity, high germination rates and consistency, seed and seedling health, and other parameters according to the requirements of crop species, seed producers, and markets (Sadeghi et al. 2011; Maity and Pramanik 2013; Do Espirito Santo Pereira et al. 2021). In interaction with a seed's genetic background, each of these features is influenced by environmental variables occurring during maturation on the mother plant, as well as during harvesting, processing, storage, sowing, and germination. Adverse conditions affecting any of these stages can impair seed quality, with substantial implications on agricultural productivity, supply chains, and food security (DaMatta et al. 2010; Maity and Pramanik 2013). Germination is a particularly sensitive stage, with young seedlings exposed to numerous stress conditions that require optimal seed vigor to ensure the development of uniform and healthy seedlings (Yari et al. 2010; Paparella et al. 2015; Do Espirito Santo Pereira et al. 2021).

The improvement of seed quality is an ancient issue that has integrated artificial selection of useful traits as well as direct pre-sowing treatments. A diverse panel of approaches has been established to improve seed quality, mediating between the ancient necessity of higher yields and the current need for environmental sustainability. The most utilized include intervention on whole varieties in the case of selective breeding and genetic engineering as well as targeted treatments directly on seeds, as in the case of coating and priming (Wimalasekera 2015; Chakraborti et al. 2022).

Conventional breeding has played a significant role in the improvement of stress response in plants. Despite their general acceptance by the public opinion, these techniques tend to require time-consuming programs and a large amount of trained manpower, although marker-assisted breeding programs considerably improve their global efficiency at producing lines with traits of interest, also concerning biotic and abiotic stress resistance (Macovei et al. 2017; Choudhary et al. 2019; Farrant and Hilhorst 2021).

Genetic engineering, including gene silencing, gene mutation, and gene knock-in and knock-out, is fast developing and very promising in terms of efficiency and results, but, among its drawbacks, public and regulators raise concerns about its environmental safety and pose restrictions to its large-scale use. Globally, their intrinsic costs, the required biosafety testing, and the heterogeneity of their legal status still represent substantial impairments for their diffusion (van Acker et al. 2017; Johnson and Puthur 2021).

Seed coating consists of covering the seed with a protective layer of various materials according to a wide range of aims, including physical protection against mechanical damage, adding nutritional supplements for seed germination, facilitating water absorption or retention, easing seed handling, and administration of pesticides or beneficial microorganisms, thus enhancing seed resistance against biotic and abiotic stress. Being a well-established and versatile approach for the enhancement of seed vigor, costs and biosafety concerns can vary according to the composition of the protective layer (Ehsanfar and Modarres-Sanavy 2005; Pedrini et al. 2017; Ma 2019; Rocha et al. 2019).

Among the pre-sowing techniques for crop improvement and stress management, seed priming is among the most adaptable, cost-effective, and environmentally friendly ones, reflecting the main principles of sustainable agriculture (Afzal et al. 2016). Seed priming includes a broad family of well-established approaches that, as a general principle, rely on incomplete imbibition steps to improve germination rates and synchronization, seedling vigor, and growth in a wide range of environments (Paparella et al. 2015).

Several attempts to enhance germination through pre-sowing treatments are reported throughout history, notably by Theophrastus (371–287 BCE), Gaius Plinius Secundus (23–79 CE), Oliver de Serres (1539–1619), and Charles Darwin (1809–1882). Such approaches evolved into the modern seed priming, along with its physiological explanations: the controlled imbibition-dehydration cycle activates pre-germinative metabolism along with seed stress response mechanisms, thus preparing the primed seeds for a faster and more uniform germination and a more efficient stress response (Varier et al. 2010; Paparella et al. 2015). Seed priming often involves the administration of specific agents during the imbibition phase to modulate water uptake or trigger specific responses to enhance stress tolerance in seeds and seedlings, resulting in improved vigor under a variety of environments (Varier et al. 2010; Jisha et al. 2012; Paparella et al. 2015). For example, the rice genotype CO 51 was subjected to artificially elevated carbon dioxide levels. Seed priming treatments with salicylic acid (SA), citric acid (CA), ascorbic acid (AsA), and distilled water (hydropriming) were effective at improving germination rates and mitigating the oxidative stress caused by the increased carbon dioxide levels (Nedunchezhiyan et al. 2019). Moreover, "priming memory" established during seed priming can last and be beneficial for stress tolerance also in later developmental stages (Wojtyla et al. 2016).

Seed priming protocols can be classified according to the priming agent that is administered, which can be water, an osmotic agent, an antioxidant compound, a phytohormone, or another chemical, as well as a physical treatment such as temperature or a magnetic field that can induce positive effects on features of interest.

Hydropriming. This priming technique is the earliest one historically and the simplest one operatively. Seeds are soaked in water to activate pre-germinative metabolism. The patterns of imbibition and dehydration can vary according to the target crop, as well as the outputs in terms of germination performances. For example, crop wild relatives (CWR) *Solanum villosum* Miller (hairy nightshade) subjected to two hours of hydropriming displayed synchronized germination and accumulation of phenolic compounds, known to be involved in abiotic stress response (Forti et al. 2021). Different positive effects were obtained for other crops, including quinoa (*Chenopodium quinoa*), in which hydropriming improved seedling development and photosynthetic pigment content along with significantly enhancing resistance to drought stress (Nadali et al. 2021).

Osmopriming. This priming method uses osmotic agents to prime the seeds. Some of the commonly used osmotic agents are polyethylene glycol (PEG), sorbitol, and mannitol. Inorganic salts like NaCl, KCl, KNO_3, $CaCl_2$, K_3PO_4, and $MgSO_4$ (Lutts et al. 2016). Sugarcane (*Saccharum officinarum* L.) seeds that were primed with NaCl (100 mM) helped in overcoming the negative effects induced by salt and drought stress. The osmoprimed plants exhibited lower leaf senescence induced by salt and dehydration (Patade et al. 2009). Priming of rice seeds (*Oryza sativa*) with 5% PEG and 5% NaCl solution proved to be beneficial for the seeds to thrive under drought conditions. Increased glutathione peroxidase (GPX) activity associated with antioxidant activity and lower malondialdehyde (MDA) levels associated with lower lipid peroxidation seemed to be correlated with drought resistance in primed seeds (Goswami et al. 2013). Sunflowers (*Helianthus annuus* L.) under drought stress that were osmoprimed with 10% PEG for 16 hours showed increased germination capacity along with increased soluble sugars and chlorophyll contents. This indicates that osmopriming of sunflower seeds improved their stress response toward drought (Bourioug et al. 2020).

Hormopriming. These approaches to priming rely on the administration of solutions containing phytohormones to modulate specific metabolic pathways relevant for seed germination or stress response. The most used phytohormones in hormopriming include salicylic acid (SA), abscisic acid, auxins, gibberellins, kinetin, polyamines, and ethylene (Sytar et al. 2018). For example, exogenous application of ethephon (ETP, donor source of ethylene) was beneficial at mitigating the effects of waterlogging stress in soybean seeds (*Glycine max* L.) by activating transcriptional programs and metabolic pathways involved into abiotic stress response (Kim et al. 2018). While *Medicago truncatula* seeds treated with kinetin show anticipated and coordinated germination along with the accumulation of antioxidant metabolites and transcripts (Araújo et al. 2019; Pagano et al. 2022).

Biopriming. Beneficial microorganisms can be utilized as priming agents to enhance stress response and interact with plant growth regulators (Singh et al. 2015). Rice (*O. sativa* L.) seeds bioprimed with *Trichoderma harzianum* showed enhanced plant growth under drought stress, with transcriptomic analyses highlighting the increased transcription of genes related to antioxidant mechanism, photosynthesis, and hormonal signaling (Bashyal et al. 2021). Similar effects were observed in tomato seeds (*Solanum lycopersicum*) bioprimed with *Pseudomonas putida* strain SAESo11 showing increased drought stress tolerance when compared to unprimed seeds. Further analyses demonstrated that *P. putida* SAESo11 strain stimulated nutrient and water uptake and promoted photosynthetic activity. Moreover, this *P. putida* strain was isolated from harsh environments, suggesting its possible ecophysiological implications for promoting plant stress tolerance (Papadopoulou et al. 2022).

Nanopriming. In this recently introduced family of priming methods, seeds are soaked in solutions containing nanoparticles (NPs) that, despite certain biosafety concerns dye to toxic side effects, are reported to improve germination, stress response, and antioxidant activities (Shelar et al. 2021). This is the case with manganese nanoparticles, which have been reported to improve response to abiotic stress with better toxicological profiles compared to bulk or ionic counterparts (Ye et al. 2019). Similarly, *Brassica napus* seeds primed with selenium and zinc oxide nanoparticles (SeNPs and ZnONPs) displayed improved germination rates under salt stress. Concomitantly, changes in antioxidant enzyme activity and phytohormone signaling were observed (El-Badri et al. 2021). The positive effects of nanopriming were reported also for wheat seeds (*Triticum aestivum*), which displayed enhanced drought stress response after priming with silicon oxide (SiO_2) nanoparticles (Rai-Kalal et al. 2021).

Beside the aforementioned priming approaches, many others are continuously being established and optimized, also including physical treatments. Among them, thermopriming consists of pre-sowing at specific temperatures to reduce thermo-inhibition, with positive effects demonstrated for crops adapted to warm environments (Huang et al. 2002). Conversely, treatments such as cold stratification and cold exposure of seeds are able to improve seed germination and stress resilience in adult plants (Ding et al. 2020).

Other physical techniques have been successfully utilized for seed invigoration, including magnetic fields, microwaves, and electromagnetic radiations (UV rays, γ-rays, and X-rays) that have been shown to enhance the response to different kinds of abiotic stresses in many model and crop species. Methods of physical invigoration need careful optimization of the administered doses and the evaluation of the changes induced in seed morphology, gene expression, and metabolite accumulation. Nonetheless, they are considered among the most eco-friendly, cost-effective, and non-invasive priming techniques (Araújo et al. 2016).

Despite its versatility to improve stress tolerance during the germination phase, seed priming presents certain critical aspects to be considered in order to optimize a given priming protocol on a new crop variety or seed lot according to the requirements of the farmers. Specifically, the effectiveness of each method depends on treatment duration, water potential, and many other elements that determine the physiological state of the average seed in a seed lot (Varier et al. 2010). For example, priming protocols are carried out without allowing the completion of the germination process or the depletion of the compounds that ensure seed desiccation tolerance. Otherwise, excessively prolonged imbibition results in impaired priming efficiency or loss of seed viability, a condition known as "overpriming" (Pagano et al. 2022). Subsequently, the optimization of the imbibition-dehydration cycle represents a critical variable in the design of priming protocols. Among the side effects of seed priming, an impaired seed longevity is observed in many experimental systems in terms of a shorter shelf-life and reduced storability due to the activation of pre-germinative metabolism, which implies the partial depletion of the compounds that prevent deterioration in dry quiescent seeds.

Currently, the optimization of priming protocols requires time-consuming empirical assessments, and, to tackle its critical aspects and side effects, a detailed knowledge of pre-germinative metabolism would be precious to drive innovation in seed priming technology.

22.4 Understanding Seed Quality: Molecular Hallmarks and Experimental Models for Future Perspectives in Seed Technology

The integration of theory, experimental evidence, and observations often struggles to provide reliable models to explain and predict plant responses to variable environments (Parmesan and Hanley 2015). This applies also to seed technology, where detailed models of pre-germinative metabolism in diverse experimental systems would help the development of more efficient approaches to test and enhance seed quality and vigor.

The complexity of seed metabolism requires multilevel and integrative approaches to be investigated and understood. Recent advancements in this direction have been made possible by forward and reverse genetics as well as high throughput analyses able to provide a comprehensive picture of the genome, transcriptome, proteome, or metabolome of a given experimental system, including more specific levels of analysis like epigenome, lipidome, or ionome. Hence, these technologies are globally referred to as "-omic" (Ahuja et al. 2010). To integrate these data, exploratory data analyses and system biology offer a promising perspective to identify the traits and pathways contributing to seed phenotypic plasticity in relation to environmental variables from a translational perspective, from model plants to crop and wild species (Macovei et al. 2017).

The processes constituting seed metabolism can be grouped into three broad phases, initiated and governed by water uptake, each with its metabolic profiles at the cellular level (Varier et al. 2010). The first phase involves rapid watering due to low water potential of the seeds. Simultaneously, proteins are synthesized from pre-existing mRNA, while DNA and mitochondria are repaired from damage incurred during seed maturation, storage, or early stress events (Rajjou et al. 2004). A decrease in water uptake occurs during the second phase, in which the majority of physiological activities are initiated. This includes protein translation from newly transcribed mRNA and synthesis of new mitochondria, reservoir mobilization. Rapid water uptake increases again during the third phase, characterized by the completion of the germination process with radicle protrusion and the beginning of post-germinative metabolism (Wojtyla et al. 2016; Macovei et al. 2017).

Climate changes exacerbate a combination of environmental stress factors, such as carbon dioxide concentration increases, higher temperatures, drought, and salinity, that affect plant growth and pose rising threats to agriculture systems, especially in a time of significant growth of the global population (Ahuja et al. 2010). Traditional and high-throughput approaches have allowed the identification and characterization of the pathways linked to stress perception and response in both model plants and crop species. Understanding how plants face abiotic and biotic stress is especially important at the level of seed germination and seedling establishment, crucial phases for the plant life cycle and, subsequently, for agriculture (Gorai and Neffati 2007; Ahuja et al. 2010; Macovei et al. 2017).

As sessile organisms, plants have evolved an array of cellular, physiological, and morphological defense responses to avoid or tolerate biotic and abiotic stresses (Wojtyla et al. 2016; Macovei et al. 2017; He et al. 2018).

Temperature is considered among the most critical environmental factors affecting plant growth. Despite the extensive number of studies on the effects of global warming on plants, few attempts have been made to estimate its effect on seed germination (Hedhly et al. 2009). Variations in the optimal temperatures inhibit germination and promote irreversible damage (Gorai et al. 2006). Altered germination and seed vigor due to high temperature have been linked with reduced thermo-stability of the plasma membrane and membrane fluidity, with consequently delayed calcium signaling cascade, kinases, and heat shock elements recruitment and activation (Begcy et al. 2018). Beside species-specific variability, the effects of high temperatures on seed physiology are influenced by other co-occurring environmental parameters. For instance, elevated temperatures associated with high humidity accelerate seed deterioration during storage, while high temperatures following fertilization are known to impair grain germination and seedling development (Debouza et al. 2021).

Water loss, exacerbated by heat events, affects seeds and results in mechanical stress in the cells by destabilizing cellular structures and turgor pressure. These changes are detected by a complex signaling system and transmitted to the nucleus, where they activate transcriptional programs involved in the heat shock response. At the cellular level, protective mechanisms are activated, involving cell wall plasticity, membrane stabilization, DNA damage response (DDR), and chromatin condensation (Chen and Arora 2013; Dekkers et al. 2015).

Stabilizing metabolites. Late embryogenesis abundant (LEA) proteins, heat shock proteins (HSP), and RFO (non-reducing oligosaccharides of the raffinose group) play crucial roles in seed desiccation tolerance, which in orthodox seeds is acquired during late seed maturation and is progressively lost during radicle protrusion (Smolikova et al. 2021). During the last phases of seed maturation, LEA proteins acquire the hydroxylated residues necessary for protein–protein interactions, forming hydrogen bonds and stabilizing water molecules in their interactions with protein structures. HSP proteins, on the other hand, act as molecular chaperons, maintaining correct protein folding and limiting the formation of aggregates (Hoekstra et al. 2001; Gaff and Melvin 2013). Desiccation tolerance is an adaptive trait of orthodox seeds and is fundamental for seed priming technology, which is based on rehydration-dehydration cycles. Subsequently, these molecular dynamics can represent a useful guide for the identification of molecular hallmarks to monitor and optimize seed priming protocols (Paparella et al. 2015).

Reactive oxygen species (ROS) and antioxidant mechanisms. One of the most pervasive elements of the seed stress response is represented by ROS, including, among the most relevant ones, the peroxide anion ($O^{2-}\bullet$), hydrogen peroxide (H_2O_2), and hydroxyl radical ($\bullet OH$). The production of these highly reactive chemical species is related to many metabolic pathways that are activated along with pre-germinative metabolism and seed reservoir mobilization under physiological conditions, including peroxisomal reactions electron transport chains in mitochondria and plastids as a byproduct of aerobic metabolism. Under stress conditions, ROS levels tend to increase due to leakage from damaged membranes. An excessive amount of ROS causes oxidative damage to biological structures and macromolecules, including proteins, lipids, and nucleic acids, and represents a significant cause of seed quality deterioration (Bailly et al. 2008; Singh et al. 2016; Kurek et al. 2019).

On the other hand, a baseline ROS level is necessary to mediate numerous molecular, biochemical, and physiological processes in seeds, including cell wall plasticity during radicle elongation, endosperm weakening, dormancy release, and modulation of hormonal and stress signaling. Because of this double nature, it is important that the intracellular ROS levels are maintained within certain physiological ranges known as "oxidative window of germination," although no universally established thresholds clearly differentiate the negative and positive function of ROS, that highly depends on the experimental system and its physiological conditions (Bailly et al. 2008; Kurek et al. 2019; Do Espirito Santo Pereira et al. 2021).

To maintain intracellular redox state and to cope oxidative damage, active and passive mechanisms coexist in seeds, relying on antioxidant enzymes and non-enzymatic antioxidant molecules. Among the enzymatic routes of ROS detoxification, ascorbate peroxidase (APX), catalase (CAT), superoxide dismutase (SOD), dehydroascorbate reductase (DHAR), glutathione peroxidase (GPX), glutathione reductase (GSSGR), and monodehydroascorbate reductase (MDHAR) can be cited (Oracz et al. 2007). Dehydrated and metabolically dormant seeds cannot rely on antioxidant enzymes, whose functioning requires a sufficiently hydrated environment. Instead, dry seeds rely on their endowment in antioxidant molecules such as glutathione, ascorbate, tocopherol, quinones and flavonoids that accumulate during seed maturation (Hoekstra et al. 2001).

The pathways that maintain ROS levels within safe ranges constitute a tightly regulated system. For example, in *Arabidopsis thaliana*, a relevant subnetwork of 152 genes is directly involved in ROS-scavenging or ROS-producing proteins (Mittler 2017).

DNA damage response (DDR). Seed protective mechanisms relying on osmoprotectants and antioxidants allow cellular compartments a wide range of resistance to heat and drought but are not sufficient to counteract DNA damage. Abiotic stress can cause damage at the DNA level, which can result in aborted germination (Debouza et al. 2021), and DNA is particularly sensitive to oxidative damage, associated with desaturation of deoxyribose or covalent modifications of bases (Waterworth et al. 2019).

As a result, seeds have developed powerful DNA repair mechanisms to cope with single- or double-strand breaks and other types of injuries. Many players are involved in the plant DDR to preserve the integrity and the correct functioning of the genome. This signaling network is an ensemble of distinct but often overlapping subnetworks, where different components fulfil distinct functions in precise spatial and temporal scenarios (Arcas et al. 2014). Most studies about plant DDR have been carried out in the model plant *A. thaliana* (Manova and Gruszka 2015), although the overall organization of the DDR signaling cascade appears to be conserved between plants and animals, involving numerous actors among DNA damage sensors, signal transducers, mediators, and effectors (Yoshiyama et al. 2013; Nisa et al. 2019).

In mammals, DDR activation depends on two protein kinases, Ataxia Telangiectasia Mutated (ATM) and ATM and Rad3-related (ATR) (Yoshiyama et al. 2013). The homologs of ATR and ATM were first isolated in *A. thaliana* in the early 2000s, based on their sequence conservation with their counterparts in animals and yeast. Garcia et al. (2003) and

Culligan et al. (2004), respectively, demonstrated that the atm mutant is sensitive to DSB-inducing agents and the atr mutant is sensitive to replication stress-inducing agents. ATR and ATM initiate a phosphorylation-mediated signal transduction cascade that leads to the regulation of downstream processes.

DNA lesions are detected by DDR sensors, which recognize single- or double-strand breaks in DNA.

The MRE11/RAD50/NBS1 (MRN) complex acts as DSB sensors in the ATM pathway, recruiting ATM at the DSB sites and interacting with NBS1 (Gallego and White 2001; Rupnik et al. 2010). In the ATR pathway, replication protein A (RPA) binds to single-stranded DNA (ssDNA) generated at the site of stalled replication forks (Garcia-Maya and Buck 1998; Zou and Elledge 2003).

Mediator proteins work to coordinate the temporal-spatial regulation of the various factors in the DDR, promote their activation, recruit additional substrates, and control their association with damaged DNA (Yoshiyama et al. 2013).

Regarding the effectors of DDR in plants, no plant homologs of p53 have been identified in any of the model plants (Yoshiyama et al. 2013). In plants, the ATR and ATM branches of DDR signaling converge on Suppressor Of Gamma-response 1 (SOG1), a transcription factor of the NAC (NAM, ATAF1/2, and CUC2) family. SOG1, the central regulator of plant DDR, controls the expression of DNA repair genes and cell cycle regulators (Nisa et al. 2019). Recent genome-wide analyses of SOG1 targets confirmed the central role of SOG1 in the early transcriptional response to DSBs, placing it at the top of the regulatory DDR network (Bourbousse et al. 2018; Ogita et al. 2018).

These molecular mechanisms are crucial for seed viability to maintain seed genome stability, counteracting the DNA lesions that accumulate during storage, as well as the damages resulting from environmental stress occurring during germination (Waterworth et al. 2019).

Regulation. A tuned network of regulative pathways coordinates pre-germinative metabolism and ensures prompt responses to different environmental conditions, operating at the transcriptional, post-transcriptional, translational, and post-translational levels and modulated by epigenetic signatures and chromatin condensation state. Above them, hormonal signaling coordinates stress responses on a higher scale, in interconnection with the regulation of dormancy release and germination. Abscisic acid (ABA) is central in seed metabolism, inhibiting germination and promoting stress responses, counteracting the effect of ethylene and gibberellins (GAs) that promote germination. The role of ABA/GA ratio in germination and stress response is reportedly conserved. Several ABA-responsive elements, transcription factors, protein kinases, and phosphatases are involved in seed stress responses, and ABA accumulation has been observed during seed maturation and dormancy establishment, as well as under a variety of unfavorable conditions, including drought, elevated temperatures, salinity, and carbon dioxide concentration (Toh et al. 2008; Maity and Pramanik 2013; Wojtyla et al. 2016). These evolutionary conserved hormonal dynamics, integrated with species-typical responses, are also useful for the design of hormopriming protocols (Paparella et al. 2015).

22.5 Conclusive Remarks

The interaction of the environmental and biological factors governing seed quality and vigor is complex, with a high degree of inter- and intra-specific variability that causes seed technology to rely on empirical validation rather than predictive models. Advancements in high throughput analyses are strengthening the current understanding of seed ecophysiology and pre-germinative metabolism from a translational perspective, from model species to crop and wild varieties, under a variety of environmental conditions. These developments will provide precious resources to corroborate empirical data and to develop precise and versatile protocols for seed quality conservation, assessment, and improvement, finding economically and environmentally sustainable solutions to face the present and future challenges of climate change.

References

van Acker, R., Rahman, M.M., and Cici, S.Z.H. (2017). *Pros and Cons of GMO Crop Farming*. Oxford: Research Encyclopedia of Environmental Science.

Afzal, I., Rehman, H.U., Naveed, M. et al. (2016). Recent advances in seed enhancements. In: *New Challenges in Seed Biology – Basic and Translational Research Driving Seed Technology* (ed. S. Araújo and A. Balestrazzi), 47–74. IntechOpen. Chapter 2.

Ahuja, I., de Vos, R.C., Bones, A.M., and Hall, R.D. (2010). Plant molecular stress responses face climate change. *Trends Plant Sci.* 15 (12): 664–674.

Al-Amery, M., Geneve, R.L., Sanches, M.F. et al. (2018). Near-infrared spectroscopy used to predict soybean seed germination and vigor. *Seed Sci. Res.* 28 (3): 245–252.

Ambrose, A., Kandpal, L.M., Kim, M.S. et al. (2016a). High speed measurement of corn seed viability using hyperspectral imaging. *Infrared Phys. Technol.* 75: 173–179.

Ambrose, A., Lohumi, S., Lee, W.H., and Cho, B.K. (2016b). Comparative nondestructive measurement of corn seed viability using Fourier transform near-infrared (FT-NIR) and Raman spectroscopy. *Sens. Actuators B Chem.* 224: 500–506.

Anderson, P.K., Cunningham, A.A., Patel, N.G. et al. (2004). Emerging infectious diseases of plants: pathogen pollution, climate change and agrotechnology drivers. *Trends Ecol. Evol.* 19 (10): 535–544.

Anderson, R., Bayer, P.E., and Edwards, D. (2020). Climate change and the need for agricultural adaptation. *Curr. Opin. Plant Biol.* 56: 197–202.

Araújo, S., Paparella, S., Dondi, D. et al. (2016). Physical methods for seed invigoration: advantages and challenges in seed technology. *Front. Plant Sci.* 7: 646.

Araújo, S., Pagano, A., Dondi, D. et al. (2019). Metabolic signatures of germination triggered by kinetin in *Medicago truncatula*. *Sci. Rep.* 9 (1): 1–13.

Arcas, A., Fernández-Capetillo, O., Cases, I., and Rojas, A.M. (2014). Emergence and evolutionary analysis of the human DDR network: implications in comparative genomics and downstream analyses. *Mol. Biol. Evol.* 31 (4): 940–961.

Bailly, C., El-Maarouf-Bouteau, H., and Corbineau, F. (2008). From intracellular signaling networks to cell death: the dual role of reactive oxygen species in seed physiology. *C. R. Biol.* 331 (10): 806–814.

Bashyal, B.M., Parmar, P., Zaidi, N.W., and Aggarwal, R. (2021). Molecular programming of drought-challenged *Trichoderma harzianum*-bioprimedrice (*Oryza sativa* L.). *Front. Microbiol.* 12: 573.

Begcy, K., Sandhu, J., and Walia, H. (2018). Transient heat stress during early seed development primes germination and seedling establishment in rice. *Front. Plant Sci.* 9: 1768.

Bewley, J.D. and Black, M. (2013). *Seeds: Physiology of Development and Germination*. Springer Science & Business Media.

Bianchini, V.D.J.M., Mascarin, G.M., Silva, L.C.A.S. et al. (2021). Multispectral and X-ray images for characterization of *Jatropha curcas* L. seed quality. *Plant Methods* 17 (1): 1–13.

Bourbousse, C., Vegesna, N., and Law, J.A. (2018). SOG1 activator and MYB3R repressors regulate a complex DNA damage network in Arabidopsis. *Proc. Natl. Acad. Sci. U. S. A.* 115: E12453–E12462.

Bourioug, M., Ezzaza, K., Bouabid, R. et al. (2020). Influence of hydro- and osmo-priming on sunflower seeds to break dormancy and improve crop performance under water stress. *Environ. Sci. Pollut. Res.* 27 (12): 13215–13226.

Chakraborti, S., Bera, K., Sadhukhan, S., and Dutta, P. (2022). Bio-priming of seeds: plant stress management and its underlying cellular, biochemical and molecular mechanisms. *Plant Stress* 3: 100052.

Chen, K. and Arora, R. (2013). Priming memory invokes seed stress-tolerance. *Environ. Exp. Bot.* 94: 33–45.

Choudhary, M., Wani, S.H., Kumar, P. et al. (2019). QTLian breeding for climate resilience in cereals: progress and prospects. *Funct. Integr. Genom.* 19 (5): 685–701.

Crockett, P.A., Singh, M.B., Lee, C.K., and Bhalla, P.L. (2002). Genetic purity analysis of hybrid broccoli (*Brassica oleracea* var. *italica*) seeds using RAPD PCR. *Aust. J. Agric. Res.* 53 (1): 51–54.

Culligan, K., Tissier, A., and Britt, A. (2004). ATR regulates a G2-phase cell-cycle checkpoint in *Arabidopsis thaliana*. *Plant Cell* 16: 1091–1104.

DaMatta, F.M., Grandis, A., Arenque, B.C., and Buckeridge, M.S. (2010). Impacts of climate changes on crop physiology and food quality. *Int. Food Res. J.* 43 (7): 1814–1823.

Debouza, N.E., BabuThruppoyil, S., Gopi, K. et al. (2021). Plant and seed germination responses to global change, with a focus on CO_2: a review. *One Ecosyst.* 6: e74260.

Dekkers, B.J., Costa, M.C., Maia, J. et al. (2015). Acquisition and loss of desiccation tolerance in seeds: from experimental model to biological relevance. *Planta* 241 (3): 563–577.

Demir, I., Ozden, Y.S., and Yilmaz, K. (2004). Accelerated ageing test of aubergine, cucumber and melon seeds in relation to time and temperature variables. *Seed Sci. Technol.* 32 (3): 851–855.

Devika, O.S., Singh, S., Sarkar, D. et al. (2021). Seed priming: a potential supplement in integrated resource management under fragile intensive ecosystems. *Front. Sustain. Food Syst.* 5: 654001.

Ding, Y., Shi, Y., and Yang, S. (2020). Molecular regulation of plant responses to environmental temperatures. *Mol. Plant.* 13 (4): 544–564.

Do Espirito Santo Pereira, A., Caixeta Oliveira, H., Fernandes Fraceto, L., and Santaella, C. (2021). Nanotechnology potential in seed priming for sustainable agriculture. *Nanomaterials* 11 (2): 267.

Dutra, A.S. and Vieira, R.D. (2006). Accelerated ageing test to evaluate seed vigor in pumpkin and zucchini seeds. *Seed Sci. Technol.* 34 (1): 209–214.

Edwards, H.G., Villar, S.E.J., de Oliveira, L.F.C., and Le Hyaric, M. (2005). Analytical Raman spectroscopic study of cacao seeds and their chemical extracts. *Anal. Chim. Acta* 538 (1–2): 175–180.

Ehsanfar, S. and Modarres-Sanavy, S.A. (2005). Crop protection by seed coating. *Commun. Agric. Appl. Biol.* 70 (3): 225–229.

El-Badri, A.M., Batool, M., Wang, C. et al. (2021). Selenium and zinc oxide nanoparticles modulate the molecular and morpho-physiological processes during seed germination of *Brassica napus* under salt stress. *Ecotoxicol. Environ. Saf.* 225: 112695.

Elias, S.G., Copeland, L.O., McDonald, M.B., and Baalbaki, R.Z. (2012). *Seed Testing: Principles and Practices*. Michigan State University Press.

ElMasry, G., ElGamal, R., Mandour, N. et al. (2020). Emerging thermal imaging techniques for seed quality evaluation: principles and applications. *Int. Food Res. J.* 131: 109025.

FAO (1999). *Women: Users, Preservers and Managers of Agrobiodiversity*. Rome: FAO. https://www.fao.org/3/y5609e/y5609e02.htm (accessed 26 February 2023).

Farooq, M.A., Zhang, X., Zafar, M.M. et al. (2021). Roles of reactive oxygen species and mitochondria in seed germination. *Front. Plant Sci.* 12: 781734.

Farrant, J.M. and Hilhorst, H. (2021). Crops for dry environments. *Curr. Opin. Biotechnol.* 74: 84–91.

Feng, L., Zhu, S., Liu, F. et al. (2019). Hyperspectral imaging for seed quality and safety inspection: a review. *Plant Methods* 15: 91.

Finch-Savage, W.E. and Bassel, G.W. (2016). Seed vigor and crop establishment: extending performance beyond adaptation. *J. Exp. Bot.* 67 (3): 567–591.

Finch-Savage, W.E. and Footitt, S. (2017). Seed dormancy cycling and the regulation of dormancy mechanisms to time germination in variable field environments. *J. Exp. Bot.* 68 (4): 843–856.

Font, R., del Río-Celestino, M., and de Haro-Bailón, A. (2006). The use of near-infrared spectroscopy (NIRS) in the study of seed quality components in plant breeding programs. *Ind. Crops Prod.* 24 (3): 307–313.

Forbis, T.A., Floyd, S.K., and de Queiroz, A. (2002). The evolution of embryo size in angiosperms and other seed plants: implications for the evolution of seed dormancy. *Evolution* 56 (11): 2112–2125.

Forti, C., Ottobrino, V., Doria, E. et al. (2021). Hydropriming applied on fast germinating *Solanum villosum* Miller seeds: impact on pre-germinative metabolism. *Front. Plant Sci.* 12: 639336.

França-Neto, J.D.B. and Krzyzanowski, F.C. (2019). Tetrazolium: an important test for physiological seed quality evaluation. *J. Seed Sci.* 41: 359–366.

Fu, Y.B., Guerin, S., Peterson, G.W. et al. (2003). RAPD analysis of genetic variability of regenerated seeds in the Canadian flax cultivar CDC Normandy. *Seed Sci. Technol.* 31 (1): 207–211.

Gaff, D. and Melvin, O. (2013). The evolution of desiccation tolerance in angiosperm plants: a rare yet common phenomenon. *Funct. Plant Biol.* 40 (4): 315–328.

Gagliardi, B. and Marcos-Filho, J. (2011). Relationship between germination and bell pepper seed structure assessed by the X-ray test. *Sci. Agric.* 68 (4): 411–416.

Gallego, M.E. and White, C.I. (2001). Rad50 function is essential for telomere maintenance in Arabidopsis. *Proc. Natl. Acad. Sci. U. S. A.* 98: 1711–1716.

Garcia, V., Bruchet, H., Camescasse, D. et al. (2003). AtATM is essential for meiosis and the somatic response to DNA damage in plants. *Plant Cell* 15 (1): 119–132.

Garcia-Maya, M.M. and Buck, K.W. (1998). Purification and properties of a DNA primase from *Nicotiana tabacum*. *Planta* 204 (1): 93–99.

Gold, M.V. and Gates, J.P. (2017). Tracing the evolution of organic/sustainable agriculture: a selected and annotated bibliography. In: *Sustainable Development of Organic Agriculture* (ed. K. Etingoff), 209–314. Apple Academic Press.

Gorai, M. and Neffati, M. (2007). Germination responses of *Reaumuria vermiculata* to salinity and temperature. *Ann. Appl. Biol.* 151 (1): 53–59.

Gorai, M., Vadel, A.M., and Neffati, M. (2006). Seed germination characteristics of *Phragmites communis*: effects of temperature and salinity. *Belg. J. Bot.* 139 (1): 78–86.

Goswami, A., Banerjee, R., and Raha, S. (2013). Drought resistance in rice seedlings conferred by seed priming: role of the anti-oxidant defense mechanisms. *Protoplasma* 250 (5): 1115–1129.

Grabe, D.F. (1989). Measurement of seed moisture. In: *Seed Moisture*, vol. 14 (ed. P.C. Stanwood and M.B. McDonald), 69–92. CSSA.

Hacisalihoglu, G., Freeman, J., Armstrong, P.R. et al. (2020). Protein, weight, and oil prediction by single-seed near-infrared spectroscopy for selection of seed quality and yield traits in pea (*Pisum sativum*). *J. Sci. Food Agric.* 100 (8): 3488–3497.

He, M., He, C.Q., and Ding, N.Z. (2018). Abiotic stresses: general defenses of land plants and chances for engineering multistress tolerance. *Front. Plant Sci.* 9: 1771.

Hedhly, A., Hormaza, J.I., and Herrero, M. (2009). Global warming and sexual plant reproduction. *Trends Plant Sci.* 14 (1): 30–36.

Heredia, M.C., Kant, J., Prodhan, M.A. et al. (2022). Breeding rice for a changing climate by improving adaptations to water saving technologies. *Theor. Appl. Genet.* 135 (1): 17–33.

Hoekstra, F.A., Golovina, E.A., and Buitink, J. (2001). Mechanisms of plant desiccation tolerance. *Trends Plant Sci.* 6: 431–438.

Huang, Y.M., Wang, H.H., and Chen, K.H. (2002). Application of seed priming treatments in spinach (*Spinacia oleracea* L.) production. *J. Chin. Soc. Hortic. Sci.* 48: 117–123.

Huang, M., Wang, Q.G., Zhu, Q.B. et al. (2015). Review of seed quality and safety tests using optical sensing technologies. *Seed Sci. Technol.* 43 (3): 337–366.

ISTA (International Seed Testing Association) (2018). *International Rules for Seed Testing*. Basserdorf, Switzerland: ISTA.

Jisha, K.C., Vijayakumari, K., and Puthur, J.T. (2012). Seed priming for abiotic stress tolerance: an overview. *Acta Physiol. Plant.* 35 (5): 1381–1396.

Joel, D.M., Portnoy, V., and Katzir, N. (1996). Identification of single tiny seeds of Orobanche using RAPD analysis. *Plant Mol. Biol. Rep.* 14 (3): 243–248.

Johnson, R. and Puthur, J.T. (2021). Seed priming as a cost effective technique for developing plants with cross tolerance to salinity stress. *Plant Physiol. Biochem.* 162: 247–257.

Kim, D.H. (2018). Extending *Populus* seed longevity by controlling seed moisture content and temperature. *PLoS One* 13 (8): e0203080.

Kim, D.H. and Han, S.H. (2018). Direct effects on seed germination of 17 tree species under elevated temperature and CO_2 conditions. *Open Life Sci.* 13 (1): 137–148.

Kim, Y., Seo, C.W., Khan, A.L. et al. (2018). Exo-ethylene application mitigates waterlogging stress in soybean (*Glycine max* L.). *BMC Plant Biol.* 18 (1): 1–16.

Kumar, J., Mir, R.R., Shafi, S. et al. (2021). Genomics associated interventions for heat stress tolerance in cool season adapted grain legumes. *Int. J. Mol. Sci.* 23 (1): 399.

Kupika, O.L. and Nhamo, G. (2016). Mainstreaming biodiversity and wildlife management into climate change policy frameworks in selected east and southern African countries. *Jamba* 8 (3): 254.

Kurek, K., Plitta-Michalak, B., and Ratajczak, E. (2019). Reactive oxygen species as potential drivers of the seed aging process. *Plants* 8 (6): 174.

Lamichaney, A. and Maity, A. (2021). Implications of rising atmospheric carbon dioxide concentration on seed quality. *Int. J. Biometeorol.* 65 (6): 805–812.

Langridge, P. and Reynolds, M. (2021). Breeding for drought and heat tolerance in wheat. *Theor. Appl. Genet.* 134 (6): 1753–1769.

Lin, P., Xiaoli, L., Li, D. et al. (2019). Rapidly and exactly determining postharvest dry soybean seed quality based on machine vision technology. *Sci. Rep.* 9 (1): 1–11.

Liu, L., Liu, G., Gong, Y. et al. (2007). Evaluation of genetic purity of F1 hybrid seeds in cabbage with RAPD, ISSR, SRAP, and SSR markers. *HortScience* 42 (3): 724–727.

Long, S.P. and Ort, D.R. (2010). More than taking the heat: crops and global change. *Curr. Opin. Plant Biol.* 13 (3): 241–248.

Lutts, S., Benincasa, P., Wojtyla, L. et al. (2016). Seed priming: new comprehensive approaches for an old empirical technique. In: *New Challenges in Seed Biology – Basic and Translational Research Driving Seed Technology* (ed. S. Araújo and A. Balestrazzi), 1–46. IntechOpen. Chapter 1.

Ma, Y. (2019). Seed coating with beneficial microorganisms for precision agriculture. *Biotechnol. Adv.* 37 (7): 107423.

Macovei, A., Pagano, A., Leonetti, P. et al. (2017). Systems biology and genome-wide approaches to unveil the molecular players involved in the pre-germinative metabolism: implications on seed technology traits. *Plant Cell Rep.* 36 (5): 669–688.

Maity, A. and Pramanik, P. (2013). Climate change and seed quality: an alarming issue in crop husbandry. *Curr. Sci.* 105 (10): 1336–1338.

Manova, V. and Gruszka, D. (2015). DNA damage and repair in plants – from models to crops. *Front. Plant Sci.* 6: 885.

McDonald, M.B. (1998). Seed quality assessment. *Seed Sci. Res.* 8 (2): 265–276.

Medeiros, A.D.D., Martins, M.S., Silva, L.J.D. et al. (2020). X-ray imaging and digital processing application in non-destructive assessing of melon seed quality. *J. Seed Sci.* 42: e202042005.

Men, S., Yan, L., Liu, J. et al. (2017). A classification method for seed viability assessment with infrared thermography. *Sensors* 17 (4): 845.

Mersal, E.F., El-Emam, A.A., and Selim, A.H. (2006). Effect of storage period, seed moisture content and insecticides treatment on wheat, *Triticum aestivum* L. seed quality. *Ann. Agric. Sci. Moshtohor* 44 (1): 111–124.

Mittler, R. (2017). ROS are good. *Trends Plant Sci.* 22 (1): 11–19.

Muñoz, N., Liu, A., Kan, L. et al. (2017). Potential uses of wild germplasms of grain legumes for crop improvement. *Int. J. Mol. Sci.* 18 (2): 328.

Musaev, F., Priyatkin, N., Potrakhov, N. et al. (2021). Assessment of Brassicaceae seeds quality by X-ray analysis. *Horticulturae* 8 (1): 29.

Nadali, F., Asghari, H.R., Abbasdokht, H. et al. (2021). Improved quinoa growth, physiological response, and yield by hydropriming under drought stress conditions. *GesundePflanzen* 73 (1): 53–66.

Nedunchezhiyan, V., Velusamy, M., and Subburamu, K. (2019). Seed priming to mitigate the impact of elevated carbon dioxide associated temperature stress on germination in rice (*Oryza sativa* L.). *Arch. Agron. Soil Sci.* 66 (1): 83–95.

Nisa, M.U., Huang, Y., Benhamed, M., and Raynaud, C. (2019). The plant DNA damage response: signaling pathways leading to growth inhibition and putative role in response to stress conditions. *Front. Plant Sci.* 10: 653.

Ogita, N., Okushima, Y., Tokizawa, M. et al. (2018). Identifying the target genes of SUPPRESSOR OF GAMMA RESPONSE 1, a master transcription factor controlling DNA damage response in Arabidopsis. *Plant J.* 94 (3): 439–453.

Oracz, K., El-MaaroufBouteau, H., Farrant, J.M. et al. (2007). ROS production and protein oxidation as a novel mechanism for seed dormancy alleviation. *Plant J.* 50 (3): 452–465.

Pagano, A., Folini, G., Pagano, P. et al. (2022). ROS accumulation as a hallmark of dehydration stress in primed and overprimed *Medicago truncatula* seeds. *Agronomy* 12: 268.

Papadopoulou, A., Matsi, T., Kamou, N. et al. (2022). Decoding the potential of a new *Pseudomonas putida* strain for inducing drought tolerance of tomato (*Solanum lycopersicum*) plants through seed biopriming. *J. Plant Physiol.* 271: 153658.

Paparella, S., Araújo, S.S., Rossi, G. et al. (2015). Seed priming: state of the art and new perspectives. *Plant Cell Rep.* 34 (8): 1281–1293.

Parmesan, C. and Hanley, M.E. (2015). Plants and climate change: complexities and surprises. *Ann. Bot.* 116 (6): 849–864.

Parry, M., Rosenzweig, C., and Livermore, M. (2005). Climate change, global food supply and risk of hunger. *Philos. Trans. R. Soc. Lond. B Biol. Sci.* 360 (1463): 2125–2138.

Patade, V.Y., Bhargava, S., and Suprasanna, P. (2009). Halopriming imparts tolerance to salt and PEG induced drought stress in sugarcane. *Agric. Ecosyst. Environ.* 134 (1–2): 24–28.

Patel, K.K., Kar, A., Jha, S.N., and Khan, M.A. (2012). Machine vision system: a tool for quality inspection of food and agricultural products. *J. Food Sci. Technol.* 49 (2): 123–141.

Pedrini, S., Merritt, D.J., Stevens, J., and Dixon, K. (2017). Seed coating: science or marketing spin? *Trends Plant Sci.* 22 (2): 106–116.

Penfield, S. (2017). Seed dormancy and germination. *Curr. Biol.* 27 (17): R874–R878.

Rahman, A. and Cho, B.K. (2016). Assessment of seed quality using non-destructive measurement techniques: a review. *Seed Sci. Res.* 26 (4): 285–305.

Rai-Kalal, P., Tomar, R.S., Jajoo, A. et al. (2021). Seed nanopriming by silicon oxide improves drought stress alleviation potential in wheat plants. *Funct. Plant Biol.* 48 (9): 905–915.

Rajjou, L., Gallardo, K., Debeaujon, I. et al. (2004). The effect of α-amanitin on the Arabidopsis seed proteome highlights the distinct roles of stored and neosynthesized mRNAs during germination. *Plant Physiol.* 134 (4): 1598–1613.

Rocha, I., Ma, Y., Souza-Alonso, P. et al. (2019). Seed coating: a tool for delivering beneficial microbes to agricultural crops. *Front. Plant Sci.* 10: 1357.

Rupnik, A., Lowndes, N.F., and Grenon, M. (2010). MRN and the race to the break. *Chromosoma* 119 (2): 115–135.

Sadeghi, H., Khazaei, F., Yari, L., and Sheidaei, S. (2011). Effect of seed osmopriming on seed germination behavior and vigor of soybean (*Glycine max* L.). *ARPN J. Agric. Biol. Sci.* 6 (1): 39–43.

Shelar, A., Singh, A.V., Maharjan, R.S. et al. (2021). Sustainable agriculture through multidisciplinary seed nanopriming: prospects of opportunities and challenges. *Cells* 10 (9): 2428.

Singh, H.B., Bisen, K., Keswani, C. et al. (2015). Unrealized potential of seed biopriming for versatile agriculture. In: *Nutrient Use Efficiency: From Basics to Advance* (ed. A. Rakshit, A. Sen, and H.B. Singh), 193–206. India: Springer.

Singh, R., Singh, S., Parihar, P. et al. (2016). Reactive oxygen species (ROS): beneficial companions of plants' developmental processes. *Front. Plant Sci.* 7: 1299.

Smolikova, G., Leonova, T., Vashurina, N. et al. (2021). Desiccation tolerance as the basis of long-term seed viability. *Int. J. Mol. Sci.* 22 (1): 1–24.

Soares, V.N., Elias, S.G., Gadotti, G.I. et al. (2016). Can the tetrazolium test be used as an alternative to the germination test in determining seed viability of grass species? *Crop Sci.* 56 (2): 707–715.

Sytar, O., Kumari, P., Yadav, S. et al. (2018). Phytohormone priming: regulator for heavy metal stress in plants. *J. Plant Growth Regul.* 38 (2): 739–752.

Toh, S., Imamura, A., Watanabe, A. et al. (2008). High temperature-induced abscisic acid biosynthesis and its role in the inhibition of gibberellin action in Arabidopsis seeds. *Plant Physiol.* 146 (3): 1368–1385.

Tu, K.L., Li, L.J., Yang, L.M. et al. (2018). Selection for high quality pepper seeds by machine vision and classifiers. *J. Integr. Agric.* 17 (9): 1999–2006.

Varier, A., Vari, A.K., and Dadlani, M. (2010). The subcellular basis of seed priming. *Curr. Sci.* 99 (4): 450–456.

Vázquez-Ramírez, J. and Venn, S.E. (2021). Seeds and seedlings in a changing world: a systematic review and meta-analysis from high altitude and high latitude ecosystems. *Plants* 10 (4): 768.

Ventura, L., Donà, M., Macovei, A. et al. (2012). Understanding the molecular pathways associated with seed vigor. *Plant Physiol. Biochem.* 60: 196–206.

Walck, J., Baskin, J., and Baskin, C. (1999). Effects of competition from introduced plants on establishment, survival, growth and reproduction of the rare plant *Solidago shortii* (Asteraceae). *Biol. Conserv.* 88 (2): 213–219.

Waterworth, W.M., Bray, C.M., and West, C.E. (2019). Seeds and the art of genome maintenance. *Front. Plant Sci.* 10: 706.

Westengen, O.T., Lusty, C., Yazbek, M. et al. (2020). Safeguarding a global seed heritage from Syria to Svalbard. *Nat. Plants* 6 (11): 1311–1317.

Wimalasekera, R. (2015). Role of Seed Quality in Improving Crop Yields. In: *Crop Production and Global Environmental Issues* (ed. K.R. Hakeem), 153–168. Cham: Springer.

Wojtyla, Ł., Lechowska, K., Kubala, S., and Garnczarska, M. (2016). Molecular processes induced in primed seeds-increasing the potential to stabilize crop yields under drought conditions. *J. Plant Physiol.* 203: 116–126.

Xia, Y., Xu, Y., Li, J. et al. (2019). Recent advances in emerging techniques for non-destructive detection of seed viability: a review. *Artif. Intell. Agric.* 1: 35–47.

Yari, L., Aghaalikani, M., and Khazaei, F. (2010). Effect of seed priming duration and temperature on seed germination behavior of bread wheat (*Triticum aestivum* L.). *ARPN J. Agric. Biol. Sci.* 5 (1): 1–6.

Ye, Y., Medina-Velo, I.A., Cota-Ruiz, K. et al. (2019). Can abiotic stresses in plants be alleviated by manganese nanoparticles or compounds? *Ecotoxicol. Environ. Saf.* 184: 109671.

Yoshiyama, K.O., Sakaguchi, K., and Kimura, S. (2013). DNA damage response in plants: conserved and variable response compared to animals. *Biology* 2 (4): 1338–1356.

Zhou, W., Chen, F., Luo, X. et al. (2020). A matter of life and death: molecular, physiological, and environmental regulation of seed longevity. *Plant Cell Environ.* 43 (2): 293–302.

Ziska, L.H., Bunce, J.A., Shimono, H. et al. (2012). Food security and climate change: on the potential to adapt global crop production by active selection to rising atmospheric carbon dioxide. *Proc. Biol. Sci.* 279 (1745): 4097–4105.

Zou, L. and Elledge, S.J. (2003). Sensing DNA damage through ATRIP recognition of RPA-ssDNA complexes. *Science* 300 (5625): 1542–1548.

23

CRISPR/Cas9 Genome Editing and Plant Stress Management

Isorchand Chongtham[1] and Priya Yadav[2]

[1] *Department of Molecular Biotechnology and Health Sciences, Molecular Biotechnology Center, University of Turin, Turin, Italy*
[2] *Department of Botany, Zakir Husain Delhi College, University of Delhi, New Delhi, India*

23.1 Introduction

Observable traits of an organism, be it a unicellular or as complex as ourselves, are indubitably linked to the genetic materials that code for them. When we have the mastery to manipulate the specific responsible region of the gene, we can outdo certain maladies that we are facing right now. In general, gene editing is the process of modifying and manipulating a specific target region of a genome, its contexts like epigenetics, or its outputs. Genes of interest or regulatory element of a pathway can be manipulated according to our technological capability through deletion, insertion, and modification of the DNA sequence. In particular, gene editing can correct a plant's susceptibility to diseases, dependency on specific climatic condition, overreaction to stresses, or produce a completely new generation with profitable and favorable phenotypes. For this, related pathway genes, crucial transcripts of a metabolism, or genes involved in the translation of important biomolecules involved in the plant stress management pathways are targeted.

However, unlike animal cells, many plant cells are intrinsically pluripotent, and this is a major advantage of applying gene editing for generating a new genome of the plant (Paszkowski et al. 1988).With billions of DNA in eukaryotes, there sure are many double-stranded DNA breaks (DSBs), and it was discovered that organisms can repair them mainly in two ways viz nonhomologous DNA end joining (NHEJ) and homologous recombination (HR) (Lieber 2010). In comparison of the two, HR is regarded to be of lower error rate, while NHEJ is more error-prone (Bétermier et al. 2014). Early studies unearthed the mechanism of homologous recombination of a specific sequence of DNA to the excised DNA, and proved to be an important advancement in genome editing as it raised the utility of recombining another new fragment of DNA (Capecchi 1989). Earlier before CRISPR/Cas, even with such a precision of homologous recombination, the technique was extremely difficult to use as the frequency of HR was as low as 1 in 10^6–10^9 cells (Capecchi 1989). To succeed in such targeted recombination technique of gene editing, a method to excise a region of DNA where recombination must be introduced has become quintessential. Many attempts, including the use of oligonucleotide and chemical recognition of DNA sequences, were experimented but were of no considerable applicability. But still, it established the prospective concept of using homologous recombination as a tool for gene editing (Doudna and Charpentier 2014). At the recognition site of E-DreI (Engineered I-DmoI/I-CreI), an artificially engineered endonuclease with the use of fusing domains of I-DmoI and I-CreI homing endonucleases, researchers incorporated sequences into the intron (Chevalier et al. 2002; Jacquier and Dujon 1985). Also, zinc finger (ZF) proteins along with FokI endonuclease can cleave DNA at specific sites (Kim et al. 1996). Further, the utilities of ZF-based nucleases cleaving a targeted chromosome site and direct HR were found out (Bibikova et al. 2001, 2002).

Till now, four important DNA binding proteins for site specific introduction of DNA DSBs are produced, namely homing endonucleases (/meganucleases) based (Smith et al. 2006), ZF nucleases (Miller et al. 2007; Urnov et al. 2005), transcription activator-like effectors (TALEs) (Boch et al. 2009; Christian et al. 2010; Miller et al. 2011; Moscou and Bogdanove 2009), and CRISPR/Cas9 (Cong et al. 2013; Mali et al. 2013a). It is hard to correspond the protein residues of meganuclease with their target DNA sequence (Hsu et al. 2014). For targeting specific sequences, ZF nuclease and TALENs, depends on protein–DNA interaction (Hassan et al. 2021). However, CRISPR/Cas uses RNA–DNA interaction for the cleavage of the

Global Climate Change and Plant Stress Management, First Edition. Edited by Mohammad Wahid Ansari, Anil Kumar Singh, and Narendra Tuteja.

specific site by the endonuclease Cas9 (Hassan et al. 2021). Recently developed CRISPR/Cas9 genome editing technology has gained more popularity and versatility (Doudna and Charpentier 2014). In this chapter, different novel methods of CRISPR/Cas9 genome editing application in plant stress management will be studied.

23.2 CRISPR/Cas9

23.2.1 CRISPR Cas System

Since its discovery in *Escherichia coli (*Ishino et al. 1987*)*, Clustered Regularly Interspaced Palindromic Repeats (CRISPR) have also been found in bacteria and archaea (Mojica et al. 2000). It was also established that their genomes contain CRISPR/Cas (CRISPR-associated protein) loci, which are responsible for their adaptive immunity against viruses (Makarova et al. 2013). These loci have clustered Cas genes, the distinctive CRISPR array (repeat sequences/direct repeats) intercalated by spacers (variable sequences), which correspond to protospacers (foreign nucleic acid sequences) (see Figure 23.1, Hsu et al. 2014; van der Oost et al. 2014).

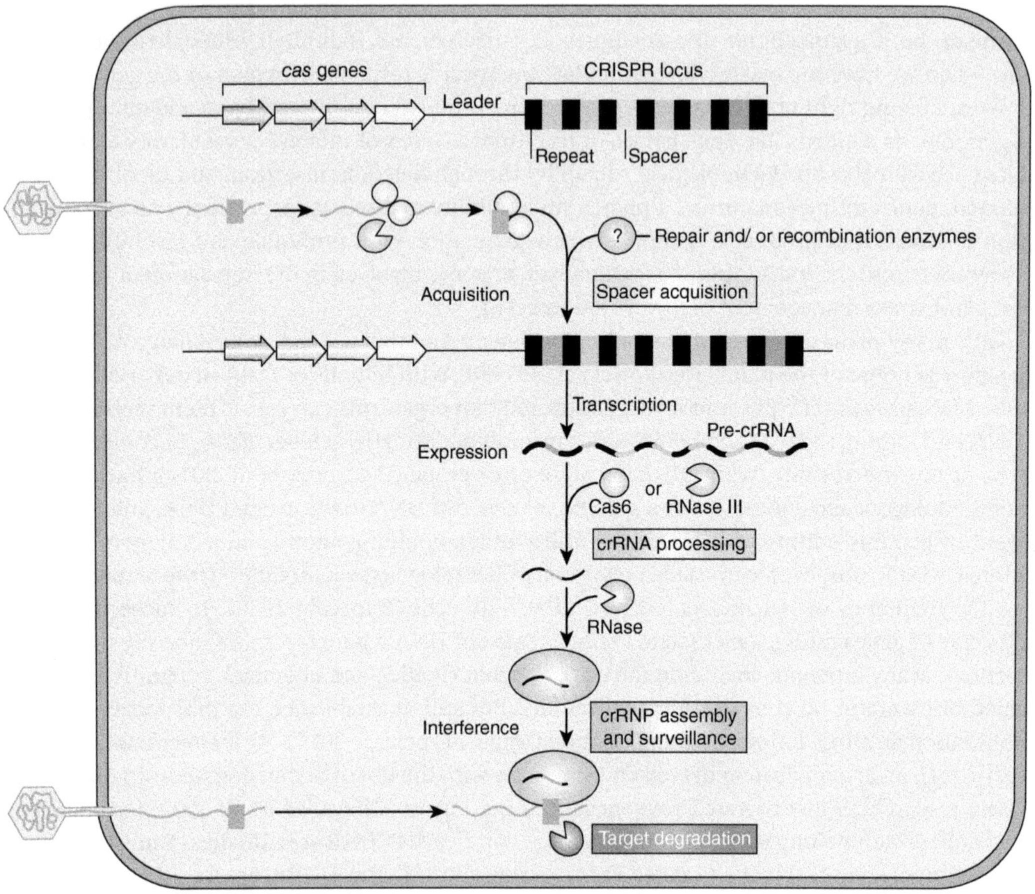

Figure 23.1 Overview of the CRISPR/Cas system. Adaptive immunity by CRISPR/Cas (clustered regularly interspaced short palindromic repeats–CRISPR-associated proteins) systems is mediated by CRISPR RNAs (crRNAs) and Cas proteins, which form multicomponent CRISPR ribonucleoprotein (crRNP) complexes. The first stage is known as acquisition, which occurs following the entry of an invading mobile genetic element (in this case, a viral genome). The invading DNA is fragmented, and a new protospacer is selected, processed, and integrated as a new spacer at the leader end of the CRISPR array. During the second stage, which is known as expression, the CRISPR locus is transcribed and the pre-crRNA is processed into small crRNAs by CRISPR-associated (Cas6) and/or housekeeping ribonucleases (such as RNase III). The mature crRNAs and Cas proteins assemble to form a crRNP complex. During the final stage of interference, the crRNP scans invading DNA for a complementary nucleic acid target, and on successful recognition, the target is eventually degraded by Cas nucleases.

Adaptive immunity of the CRISPR/Cas system occurs through acquisition of foreign DNA sequence as spacer in the CRISPR array, expression of the CRISPR RNA as precursor CRISPR RNA (pre-crRNA), maturation to functional crRNA through crRNA processing, and interference by the complex of crRNAs and Cas proteins at the complementary of the spacer in the invading nucleic acid (van der Oost et al. 2014). Here, the Cas genes are translated into protein products of Cas like Cas9 or Cas6 that are important in the later immunological function of the CRISPR/Cas system. There are three classes of CRISPR/Cas system: type I, type II, and type III (Haft et al. 2005; Makarova et al. 2011b). In comparison, type I and type III have more Cas proteins than type II (Brouns et al. 2008; Hale et al. 2009). Type II system uses only Cas9 enzyme or protein for RNA-guided DNA recognition and cleavage, which due to its simplicity is easy to pragmatize in genome editing (Gasiunas et al. 2012; Jinek et al. 2012).

23.2.2 CRISPR Cas9

In *Streptococcus pyogenes*, the maturation of crRNA requires a **trans-activating crRNA (tracrRNA),** which is trans-encoded upstream of CRISPR/Cas locus (Deltcheva et al. 2011). The duplex of tracrRNA hybridized with crRNA (tracrRNA:crRNA) along with Cas9 and endogenous RNase III enable the maturation of the crRNA (Deltcheva et al. 2011) (see Figure 23.2, Doudna and Charpentier 2014). The duplex (**also called single-guide RNA/sgRNA**) also facilitates DNA cleavage by acting as a guide to the specific site through complementarity (Brocken et al. 2018; Jinek et al. 2012). Cas9 (formerly COG3513, Csx12, Cas5, or Csn1) has two distinctive signature nuclease domains: **HNH** (Bolotin et al. 2005; Haft et al. 2005; Makarova et al. 2006) and **RuvC-like** (Makarova et al. 2006). It was observed in *Streptococcus thermophilus* that both HNH and RuvC domains were needed to have interference of the plasmid transformation by the CRISPR/Cas system (Sapranauskas et al. 2011). For the cleavage of the target, the Cas9 recognizes **Protospacer adjacent motif (PAM)** (Jiang and Doudna 2017), a short sequence of usually 2–6 bp downstream of 3–4 nucleotides from the cut site of the target sequence (Gasiunas et al. 2012; Jinek et al. 2012) (see Figure 23.3, Doudna and Charpentier 2014). PAM acts as an additional determinant in target recognition (Jinek et al. 2012). In the initial DNA binding PAM is necessary, without which the guide RNA cannot be recognized by Cas9 – in the absence of PAM, even target sequences fully complementary to the guide RNA sequence are not recognized by Cas9 (Sternberg et al. 2014). Then, the DSB of the foreign DNA by Cas9 nuclease will

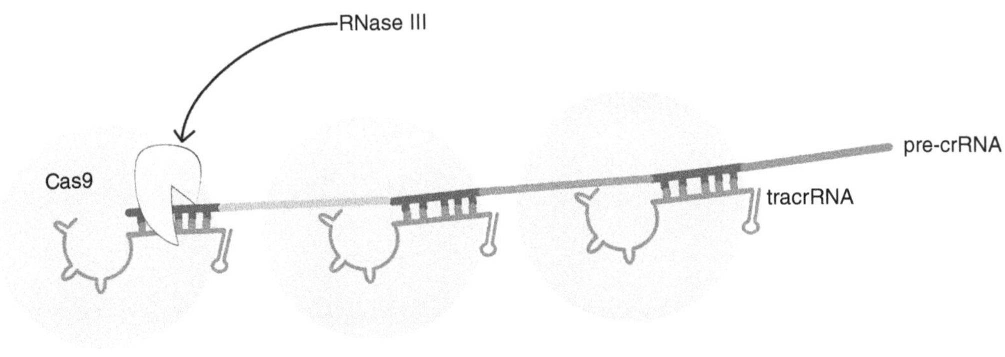

Figure 23.2 tracrRNA:crRNA co-maturation and Cas9 co-complex formation. The natural pathway of antiviral defense involves association of Cas9 with antirepeat-repeat RNA (tracrRNA:crRNA) duplexes, RNA co-processing by ribonuclease III, further trimming, R-loop formation, and target DNA cleavage.

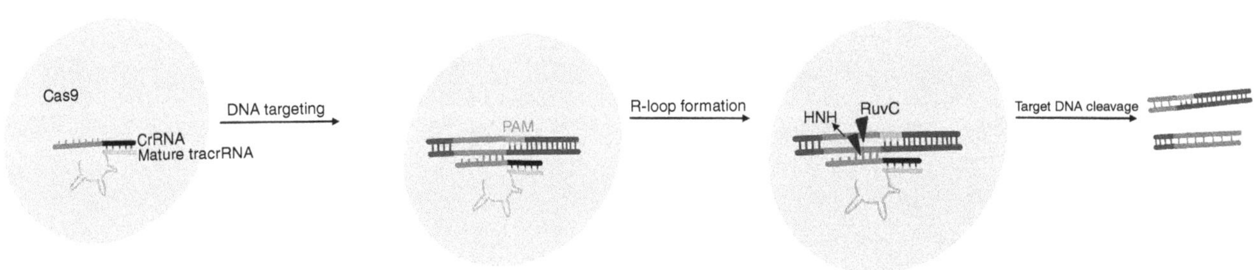

Figure 23.3 RNA-guided cleavage of target DNA. Details of duplex tracrRNA:crRNA cleaving DNA.

be introduced: one in the complementary of 20 nucleotide sequence with the crRNA by the HNH domain, while the RuvC-like domain will facilitate cleavage of the opposite strand (Gasiunas et al. 2012; Jinek et al. 2012) (see Figure 23.3, Doudna and Charpentier 2014). If both HNH and RuvC-like domains are mutated (Asp10→Ala, His840→Ala respectively), dead Cas9 (dCas9) is generated, which is used as a novel tool apart from the original genome editing as transcriptional blockage (Bikard et al. 2013; Qi et al. 2013), gene expression modulation (Cheng et al. 2013; Gilbert et al. 2013), epigenetic editing (Hilton et al. 2015; Kearns et al. 2015), and staining of chromosomal regions for live cell imaging (Anton et al. 2014; Chen et al. 2013; Ma et al. 2015a). However, if either HNH or RuvC-like domain is mutated, the protein variant will act as nickase and cleave the single strand of the DNA (Gasiunas et al. 2012; Jinek et al. 2012).

In genome editing, the Cas9 from *S. pyogenes* (spCas9) is the most widely used one (Hassan et al. 2021). The PAM sequence for SpCas9 is 5′-NGG-3′ (N = A, T, C, G) (Hassan et al. 2021). The sgRNA can be designed to recognize any targeted sequence of DNA of 20 nucleotides (guide sequence) as long as it is adjacent to the PAM and has that double-stranded structure at 3′ to bind to Cas9 (Jinek et al. 2012). This simplicity of Cas9 to only engineer the desired target sequence proves to be advantageous over other system of genome editing (Doudna and Charpentier 2014). However, the nucleotide sequence region close to the PAM site of about ~8–12 bp called seed region is also important for target recognition, and mismatches in this region might affect the gene editing (Hassan et al. 2021).

23.2.3 CRISPR/Cas9 Mechanism

The SpCas9 in complex with sgRNA and target DNA (forming heteroduplex) has a helical recognition (REC) lobe and a nuclease (NUC) lobe consisting of the HNH domain, assembled RuvC subdomains, and a PAM-interacting (PI) C-terminal region (see Figure 23.4, Nishimasu et al. 2014). Studies suggest that at the apo state of the SpCas9, the enzyme is autoinhibited by virtue of the conformation of RuvC-like domain blocking the HNH domain active site and positioning away from the REC lobe (Jinek et al. 2014; Nishimasu et al. 2014).

When Cas9 binds to the gRNA, the conformation of the complex changes, and there is more rearrangement with the association of foreign DNA. This creates a central channel between the two structural lobes that will accommodate the bound target DNA and guide RNA as well as the RNA duplex of crRNA and tracrRNA (Doudna and Charpentier 2014; Jinek et al. 2014; Nishimasu et al. 2014). An α helix rich in arginine, bridge helix (BH), hinges REC and NUC lobes of the Cas9 associating the 3′ seed sequence of the RNA–DNA heteroduplex (Cong et al. 2013; Fu et al. 2013; Hsu et al. 2013; Jinek et al. 2012; Mali et al. 2013b; Nishimasu et al. 2014; Pattanayak et al. 2013). Thus, the conformational rearrangement of Cas9 might contribute to the mechanism of introduction of DSB by the CRISPR/Cas9 system (Doudna and Charpentier 2014). Crystal structure of complexed Cas9 with gRNA and a partially dsDNA target illustrated how the PAM interacts with the

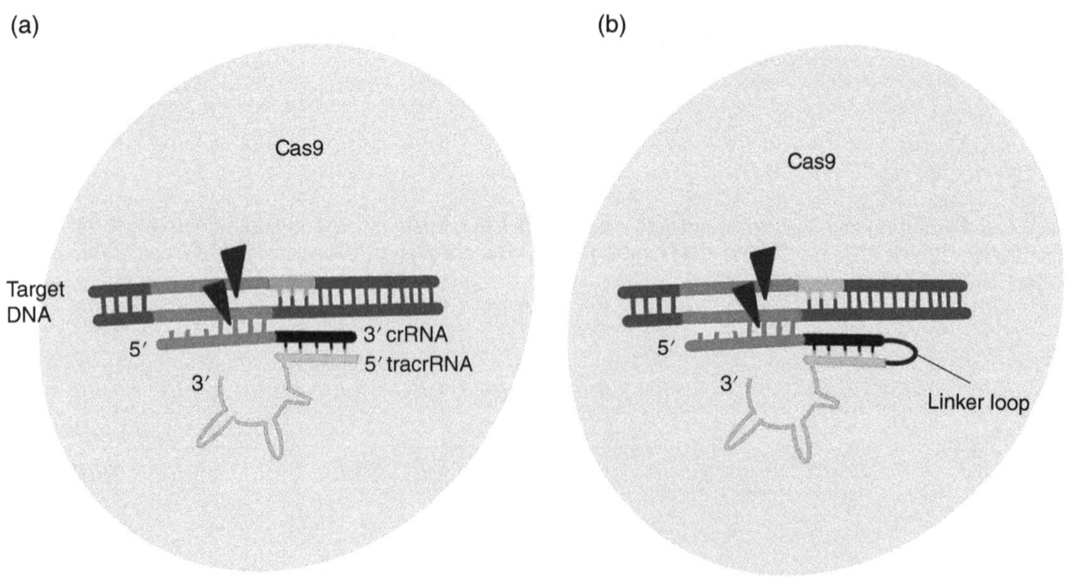

Figure 23.4 RNA–DNA heteroduplex formation. **RNA–DNA heteroduplex formation.** (a) Cas9 programmed by crRNA:tracrRNA duplex. (b) Cas9 programmed by single guide RNA.

arginine motifs of the C-terminal domain of the Cas9 on the non-complementary strand within the major groove of the duplexed PAM (Anders et al. 2014). The interaction between the minor groove of PAM and the phosphodiester group at +1 in the complexed DNA forms an R-loop just next to PAM (Anders et al. 2014). PAM is shown to be responsible for R-loop stability and association rates (Szczelkun et al. 2014). It has been proposed that DNA melting is initiated by PAM recognition with R-loop formation expanding toward the distal protospacer end and concomitant RNA strand invasion and RNA–DNA hybrid formation (Anders et al. 2014; Sternberg et al. 2014; Szczelkun et al. 2014). Active Cas9 is very specific about its target, and the generation of off-target DSBs are seldom observed (Sternberg et al. 2014). Cas proteins, gRNAs, and gene regulatory elements (GREs) for expressing Cas proteins and gRNAs are important for designing the CRISPR genome editing construct (Hassan et al. 2021).

23.2.4 CRISPR/Cas9 Types of Gene Editing

Type II CRISPRs, which till now are only found in bacterial strains (Chylinski et al. 2013), can be sub-grouped into three subtypes viz IIA, IIB, and IIC, based on the CRISPR loci (Chylinski et al. 2013; Makarova et al. 2011a) (see Figure 23.5, Hsu et al. 2014). However, all these three subtypes share a very similar domain architecture of HNH, RuvC-like, and REC domains, with the arginine-rich BH (Chylinski et al. 2013, 2014; Fonfara et al. 2014; Makarova et al. 2011a). Presently, there are three types of CRISPR systems, i.e. CRISPR Cas that uses the Cas(9) nuclease activity to introduce DSBs, the base editors, and the prime editors for the transitioning bases (Gao 2021; Zhu et al. 2020). The base editors and the prime editors would not require the DSB for genome editing (Zhu et al. 2020).

The base editors can be cytosine base editing, adenine base editing, or both, which in this case called dual base editing (Zhu et al. 2020). It uses the Cas9 nickase (nCas9), which results from the deactivation of the RuvC-like domain by D10A mutation. This nCas9 is fused with protein deaminase (either cytidine deaminase or adenosine deaminase) and a uracil DNA glycosylase inhibitor (UGI) if the editing to be carried out is cytidine (Zhu et al. 2020). Cytosine Base Editor edits C:G bp into T:A bp in the specific target DNA of sgRNA (Jin et al. 2020; Komor et al. 2016; Molla and Yang 2019; Ren et al. 2018; Shimatani et al. 2017; Zong et al. 2017, 2018). The deaminase converts the cytidines that are not complexed along with the UGI + nCas9 to uridines, i.e. those that are part of the R-loop generated. The nick, when introduced, initiates the DNA mismatch repair pathway and repairs the U:G mismatch to U:A, and a DNA replication generates the desired base transition (Komor et al. 2016; Zhu et al. 2020) (see Figure 23.6(a), Zhu et al. 2020). While in the case of adenosine base editing, adenosine deaminase as an effector, fused with nCas9 (D10A), deaminates adenosines to inosines (Gaudelli et al. 2017; Hua et al. 2020; Kang et al. 2018; Li et al. 2018a; Molla and Yang 2019; Zhu et al. 2020); it is then repaired and replicated (see Figure 23.6(a), Zhu et al. 2020). In dual base editing, a fusion of an adenosine deaminase (ecTadA-ecTadA), a UGI, cytidine deaminase (APOBEC3A), and nCas9 (D10A) called saturated targeted endogenous mutagenesis editor (STEME) is used (Li et al. 2020) (see Figure 23.6(a), Zhu et al. 2020).

APOBEC–Cas9 fusion-induced deletion (AFID) system has been developed for specific deletions within the protospacer (Wang et al. 2020). It is composed of a cytidinedeaminase, Cas9, UDG (uracil DNA glycosylase), and apyrimidinic

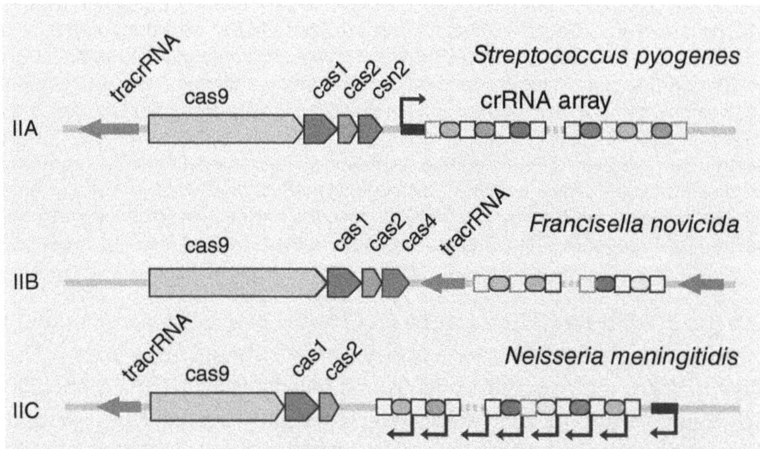

Figure 23.5 Some orthologs of Crispr loci of type II. Canonical CRISPR locus organization from type II CRISPR systems, which can be classified into IIA–IIC based on their Cas gene clusters. Whereas type IIC CRISPR loci contain the minimal set of cas9, cas1, and cas2, IIA and IIB retain their signature csn2 and cas4 genes, respectively.

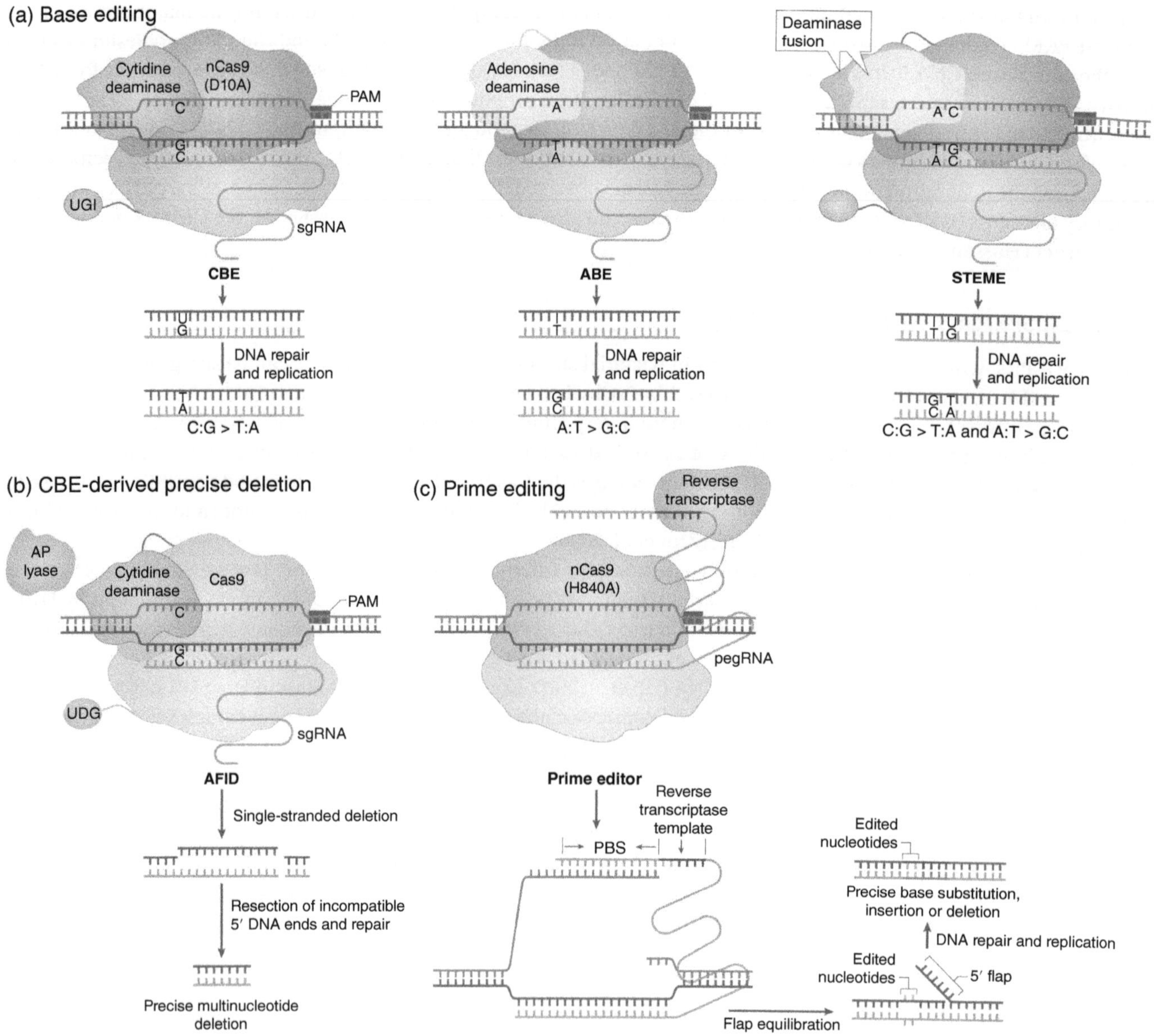

Figure 23.6 Deaminase-mediated and reverse transcriptase-mediated precise genome editing technologies in plants. (a) **Base editing technologies in plants.** Base editors comprise Cas9 nickase (nCas9; bearing a D10A mutation) fused with two proteins: a deaminase and an uracil DNA glycosylase inhibitor (UGI). The cytosine base editor (CBE) generates C:G > T:A base substitutions, whereas the adenine base editor ABE generates A: T > G:C base substitutions. A fusion of nCas9 with these two base editors, referred to as a STEME, generates simultaneous C:G > T:A and A:T > G:C dual base substitutions using one single guide RNA (sgRNA). (b) **The APOBEC-Cas9 fusion-induced deletion system (AFID).** Cytidine deaminase converts cytidine to uridine, and then a uracil DNA glycosylase (UDG), which is fused to the cytidine deaminase-Cas9 complex, excises the uridine and generates an apyrimidinic (AP) site, which is nicked by the separate yet co-expressed AP lyase. Following base excision repair, a precise deletion is generated between the site of the Cas9-induced DNA double-strand break and the deaminated cytidine. (c) **Prime editing technology.** The prime editor tool is composed of a fusion of nCas9 with reverse transcriptase and a prime editing guide RNA (pegRNA). The pegRNA carries the desired mutations (green pegs) at the 3′ end of the reverse transcriptase template. The primer-binding site (PBS) binds to the nicked DNA strand, thereby priming reverse transcription of the template into the desired DNA sequence. The edited nucleotides are then inserted into the target site in a precise way. PAM, protospacer adjacent motif.

site lyase (AP lyase). This is based on cytosine base editors except that instead of inhibiting the activity of UDG, UDG is overexpressed to obtain a base excision repair (Wang et al. 2020). The uridines are excised creating AP (apyrimidinic) sites that are then nicked by AP lyases (Wang et al. 2020). A specific and precise deletion should occur between the deaminated cytidine that was nicked eventually by the AP lyase and the DSB generated by the Cas9 activity (Wang et al. 2020; Zhu et al. 2020) (see Figure 23.6(b), Zhu et al. 2020).

Prime editing, which is a very recent tool tested in human cells, is more precise and can substitute 12 kinds of bases, insert up to 44 bp, delete up to 80 bp, and combine all these gene edits (Anzalone et al. 2019; Lin et al. 2020; Xu et al. 2020a, b; Zhu et al. 2020). Prime editor utilize either nCas9 (H840A – mutation HNH domain) or dCas9 conjugated to a reverse transcriptase (RT) from the Moloney murine leukemia virus (M-MLV) that has a primer binding site at the 3′ end of the sgRNA and a template for the RT (Anzalone et al. 2019; Zhu et al. 2020). RT template has the desired sequence to be mutated, nCas9 paired with the primer binding site (PBS) ssDNA strand priming the RT and integrating the genetic information from the RT template into the genome (Anzalone et al. 2019; Zhu et al. 2020) (see Figure 23.6(c), Zhu et al. 2020). The edited and the unedited 3′ flaps are equilibrated, ligated, and repaired generating the precise edition of the sequence (Zhu et al. 2020). This method, unlike the others, is not limited by the complementarity of the PAM (Zhu et al. 2020). This system has been developed and tested in rice and wheat (Lin et al. 2020; Xu et al. 2020a, b).

23.3 Construct of the CRISPR/Cas9

23.3.1 The gRNA

Using the CRISPR/Cas system, different results for gene product can be obtained, and they are summarized in Figure 23.7 (Hassan et al. 2021). It is therefore pertinent to choose the right target site and the design of an effective gRNA against that target (Hassan et al. 2021). Depending on the purpose of the final edit, gRNAs are chosen accordingly. For the generation of knockouts due to premature termination of codon (PTC) by NHEJ, targets located near the earlier exons are preferred, while exons very close to intron-exon or ATG should be avoided as non-sense mediated decay will destroy the PTC when PTC-bearing transcript is located near the intron-exon junction (≥50–55 nt) (Hassan et al. 2021; Popp and Maquat 2016). In base and prime editors, gRNA should be within the editing window (Hassan et al. 2021). Especially in prime editing, it is favorable for the pegRNA to include a PBS of 13 nt and an RT of 10–16 nt, without an adjacent C at the 3′-end of the gRNA (Hassan et al. 2021). About 30 web-based tools are available for designing the specific gRNAs (Gerashchenkov et al. 2020; Hanna and Doench 2020; Hassan et al. 2021). CRISPy Web is a handy website for designing a gRNA with a provided genome (Blin et al. 2016). And for the downstream process involving U6 and U3 promoters, expression of gRNA is enhanced if “G” or “A” is the first base of the RNAs, respectively (Hassan et al. 2021).

23.3.2 The Choice of Gene Regulatory Elements (GREs)

GREs are the promoters, enhancers, insulators, and silencers that regulate the expression of the gene. The GRE must be tailored in such a way, that it favors our need. Strong and constitutive (rather than inducible) promoters have been widely used; as the more the expression of Cas proteins and gRNAs, the more efficient is the genome editing (Feng et al. 2018; Hassan et al. 2021; Johnson et al. 2015; Li et al. 2013; Mikami et al. 2015a, b; Osakabe et al. 2016; Yamamoto et al. 2019). Examples include promoters extracted from plant microbes like CaMV, NOS or housekeeping genes like UBIQUITIN, ACTIN (Hassan et al. 2021). Instead, in the egg cells of *Arabidopsis* (*Arabidopsis thaliana*), tissue-specific, egg cell, ovule, or meiotic cell-specific promoter YAO or EC1 is a better choice for genome editing in germline cells over the strong promoters (Castel et al. 2019; Feng et al. 2018; Hu et al. 2018; Kishi-Kaboshi et al. 2017; Ma et al. 2015b; Ordon et al. 2020; Shockey 2020; Xie et al. 2015; Yamamoto et al. 2019).

Terminators too affect the efficiency of the CRISPR/Cas system through the stability of Cas and gRNA transcripts (Castel et al. 2019; Wang et al. 2015). It is commendable to use a strong terminator for the Cas9 expression cassette (to minimize RNA Pol II read through) and to construct gRNA and Cas9 expression cassettes in opposite directions (Castel et al. 2019; Hassan et al. 2021). Until now, studies have concluded that the best terminator for Cas9 expression cassette in plants is rbcS-E9 from *Pisum sativum* (Castel et al. 2019; Ordon et al. 2020; Wang et al. 2015).

23.3.3 Multiplex CRISPR

Multiple gRNAs are used to target multiple sites for editing simultaneously the multiple targeted regions of the genome due to the need to rewire complex pathways and promoter editing for introducing quantitative trait or directed evolution (Hassan et al. 2021). Albeit limitations, this can be achieved by designing stacking gRNAs transcription units driven by U6 or U3 promoters together in one construct (Lowder et al. 2015; Ma et al. 2015b). This may cause variation in gRNA transcripts

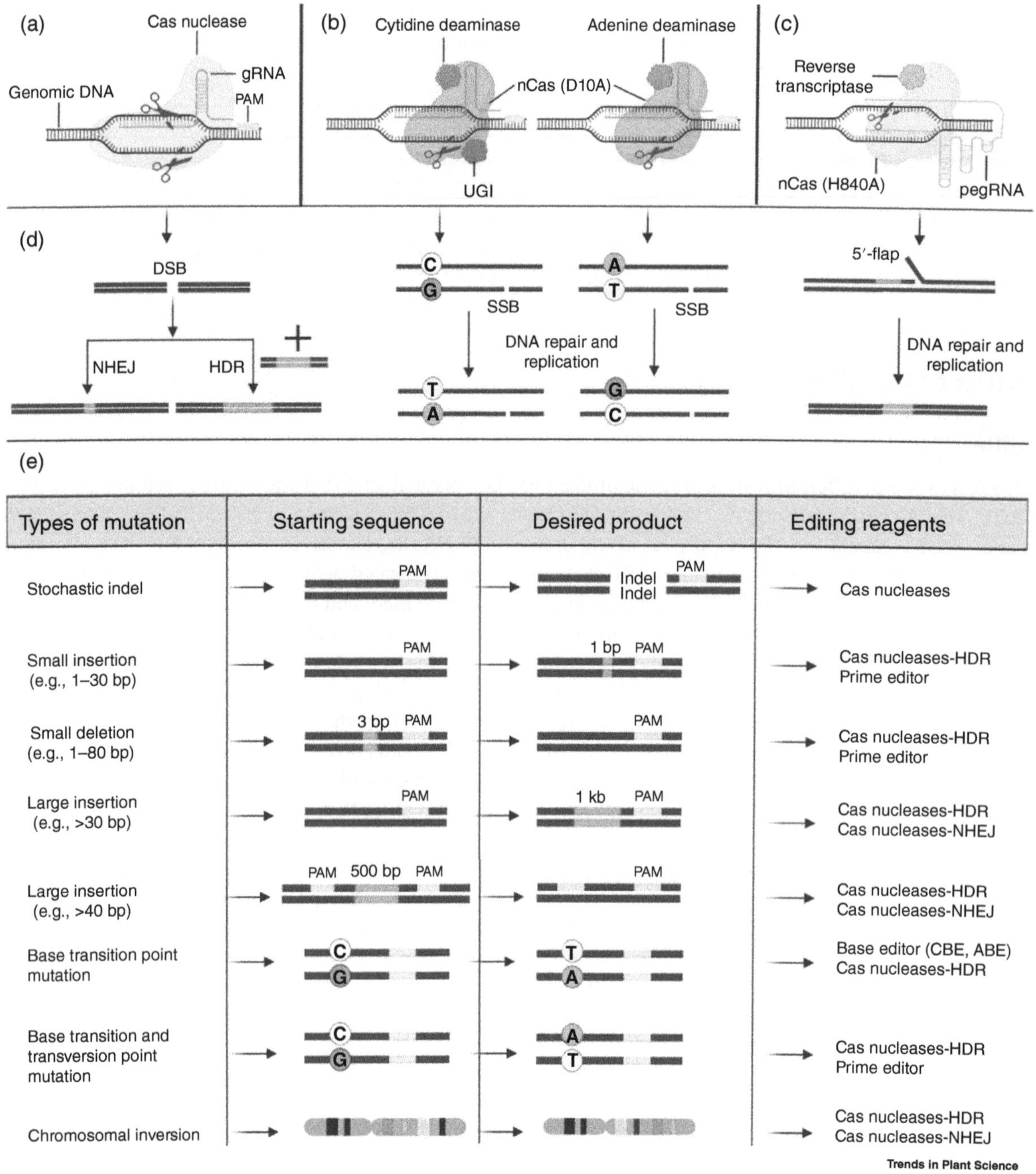

Figure 23.7 Different types of genetic modifications generated by CRISPR-based genome editors. (a) CRISPR/Cas nucleases. (b) Base editors. (c) Prime editor. (d) Editing mechanisms. (e) Different types of editing outcomes generated by various genome editors. ABE, adenine base editor; CBE, cytidine base editor; CRISPR, clustered regularly interspaced short palindromic repeats; DSB, double-strand break; gRNA, guide RNA; HDR, homology-dependent repair; nCas, Cas nickase; NHEJ, non-homologous end-joining; PAM, protospacer adjacent motif; pegRNA, prime editing guide RNA; SSB, single-strand break; UGI, uracil DNA glycosylase inhibitor.

(due to repeated use of U6 or U3 promoters) (Ma et al. 2015b), cloning difficulty (due to repetitive use of the same promoters and terminators), and challenging to package in a virus vector (due to cargo limits of the vector) (Ali et al. 2015; Scholthof et al. 2017). For the record, a construct can express up to 24 gRNAs in plants, although the efficiency is reduced (Stuttmann et al. 2021). For the successful development of multiplex CRISPR editing, stoichiometry of gRNA:Cas complex is important, as the editing efficiency is sensitive to the concentration of the gRNA:Cas complex (Stuttmann et al. 2021; Zhang et al. 2021). From here, the term CRISPR/Cas9 genome editing will be used synonymously with genome editing.

23.4 Plant Genome Editing

23.4.1 Procedure

Plant genome editing is achieved with the following steps: (i) selection of appropriate nuclease for the target sequence; (ii) construction of genome editing vectors; (iii) validation of the activity of the vectors using protoplasts; (iv) delivery of genome editing reagents into plant cells; (v) regeneration of genome-edited cells into plantlets via tissue culture; and (vi) screening and genotyping the resulted genome-edited plants (Gao 2021) (see Figure 23.8(a), Gao 2021).

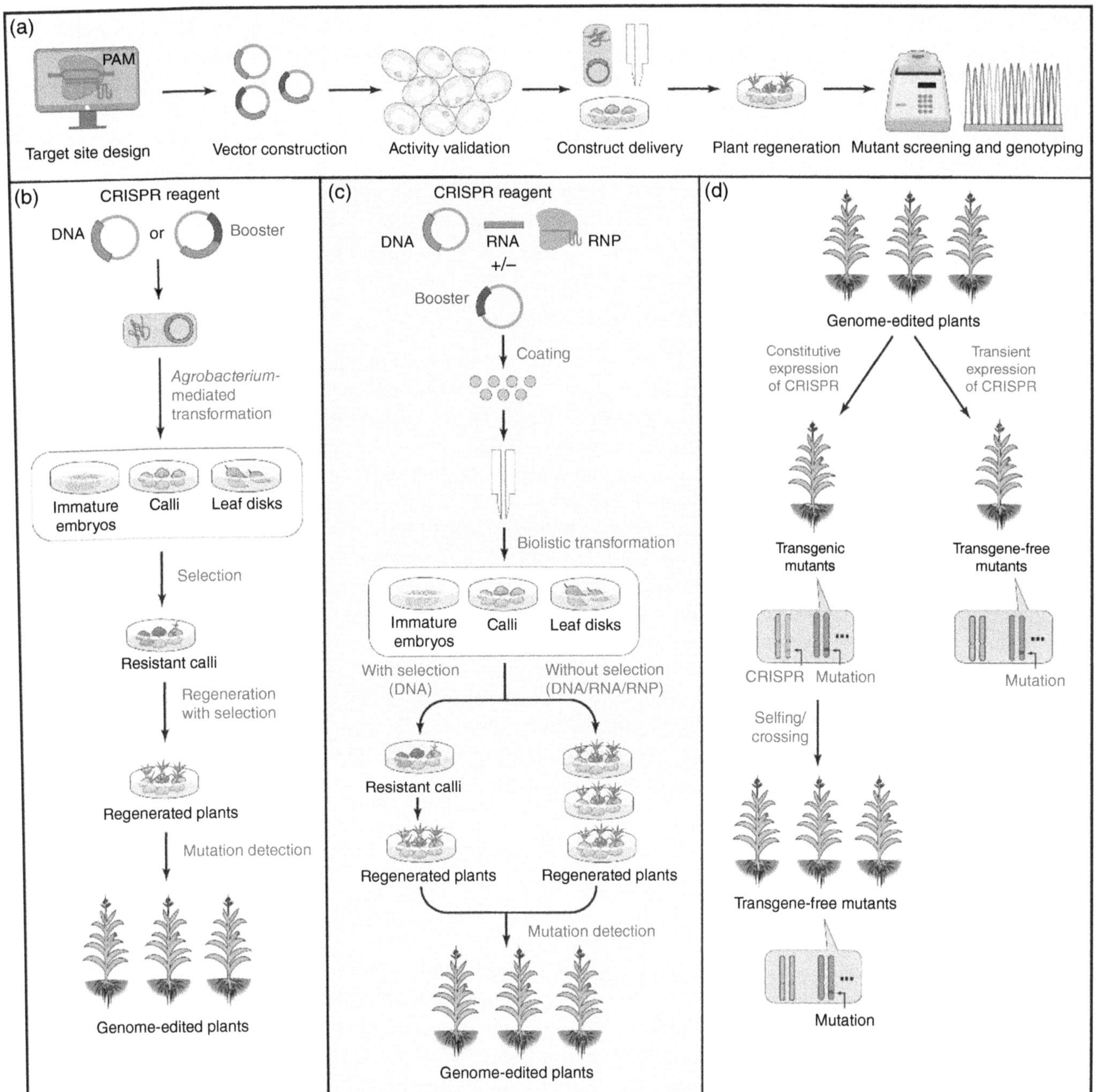

Figure 23.8 General procedure for plant genome editing. (a) Schematic illustration of the six major steps in plant genome editing. (b) Genome-edited plants generated by Agrobacterium-mediated delivery of CRISPR DNA. (c) Conventional and transient expression methods for particle bombardment-mediated genome editing by delivery of CRISPR DNA, RNA, or RNP. (d) Two strategies used to obtain transgene-free mutants.

Usually, genome editing reagents are transformed either by using a gene gun or Agrobacterium into the plant cells, but it is ideal if it is genotype independent, tissue culture free, and directly applicable to tissue system (Gao 2021). Other available methods include transformation mediated by polyethylene glycol (PEG), viral vectors, and nanoparticles. The Agrobacterium as well as the particle bombardment methods are shown in the Figure 23.8(b) and (c) (see Gao 2021; Liang et al. 2017; Ran et al. 2017; Svitashev et al. 2016; Zhang et al. 2016). Most recently, plant RNA viruses like the tobacco mosaic virus (Cody et al. 2017), tobacco rattle virus (Ali et al. 2015, 2018), ssDNA cabbage leaf curl virus (Yin et al. 2015) can deliver sgRNA in plants with up to 80% efficiency (Zhu et al. 2020). Tobacco rattle virus RNA2 is given sgRNA coupled with RNA mobile elements to prevent mutation brought on by the virus vector, and after successful delivery, the mobile elements introduce heritable mutation with efficiencies of up to 100% (Ellison et al. 2020; Zhu et al. 2020). Boosters or developmental regulators that transform plant cells into a somatic totipotent cell for further regeneration in the tissue culture are also needed for obtaining a successfully genetically engineered plant (Altpeter et al. 2016; Atkins and Voytas 2020; Gao 2021). For example, overexpression of WUSCHEL (WUS) and BABY BOOM (BBM) developmental regulators enhances the chances of regeneration in various transformation-recalcitrant genotypes and species (Lowe et al. 2016). After the delivery of the DNA into the target cells, the further tissue culture can be done either by selection of resistant tissue with a selection reagent (the conventional method) or without selection (as the transient DNA expression method), shown in Figure 23.8(c) (Gao 2021; Zhang et al. 2016). After the generation of the mutant plant, in the conventional method, selfing is required to segregate the transgene used for the selection, and for the transient method, no such method is required, thus reducing off target editing events (see Figure 23.8(d)) (Gao 2021; Liang et al. 2017; Svitashev et al. 2016; Zhang et al. 2016).

23.4.2 Plant Improvement Strategies Based on Genome Editing

The door for modifying the traits of plants has been wide open with the advent of genome editing, however limited. The genome editing can bring about various strategies to engineer a prospective trait or to enhance a recessive phenotype through rewiring of various metabolic pathways. From the traditional mendelian breeding techniques, the precise and efficient genome editing of plant is more favored. Thus, genome editing can – direct mutagenesis (creating or transferring the desired traits rather than extensive genetic crossing and genotyping) (Gao 2021; Lassoued et al. 2019); perform multiplex genome editing (it efficiently stacks various desired gene simultaneously creating a milestone for speed breeding technology) (Gao 2021; Hickey et al. 2019); edit quantitative trait loci (QTL) (it has the potential to overcome the limitations of classical genetic analysis of QTLs) (Cooper et al. 2009; Gao 2021; Liu et al. 2020); accelerate *de novo* domestication (multiplex CRISPR can edit the polygenic beneficial traits genes to the wild relatives for domestication) (Gao 2021; Li et al. 2018b; López-Marqués et al. 2020); induce haploid and artificial apomixis (by directed editing or knocking out of certain genes lead to production of defective male gametophytes and a maternal haploid induction, e.g. *MTL/PLA1/NLD* in maize, rice, and wheat) (Gao 2021; Zhu et al. 2020); achieve large-scale screening for trait discovery (genome-wide characterization of the link between genotypes and phenotypes through sgRNA sequencing) (Gaillochet et al. 2020; Gao 2021; Lu et al. 2017; Meng et al. 2017); mediate plant synthetic biology (by redesigning pathways and introducing new genes can protect against various biotic and abiotic stresses) (Gao 2021; Gunn et al. 2020; South et al. 2019); and engineer plant microbiome (to favor plant growth and develop resistance to pathogen) (Arif et al. 2020; Gao 2021).

23.5 Plant Stress

23.5.1 Plant Stress and Their Types

Under normal environmental conditions, plants biochemical and physiological pathways run in coordination, and they grow and reproduce properly. When exposed to unfavorable environmental conditions plants face adverse effects and its growth and reproduction hamper. Stress is any unfavorable environmental condition that affects plants growth and development and reduces the overall production of plants. Stress is of two types: abiotic and biotic. Biotic stress includes attacks by pathogens, bacteria, fungi, viruses, nematodes, insects, and herbivores. Abiotic stress includes salinity, drought, heavy metal, flooding, and cold (refer Figure 23.9) (Wang et al. 2003; Wania et al. 2016).

Over the years, plants develop advanced signaling mechanisms and defense mechanisms, which makes them resistant toward biotic and abiotic stresses and enable them to survive under such harsh conditions (Suzuki et al. 2014). Biotic stress results in plant chlorosis and necrosis, stunted growth, local lesions, reduced fertility, altered metabolic processes, hormonal

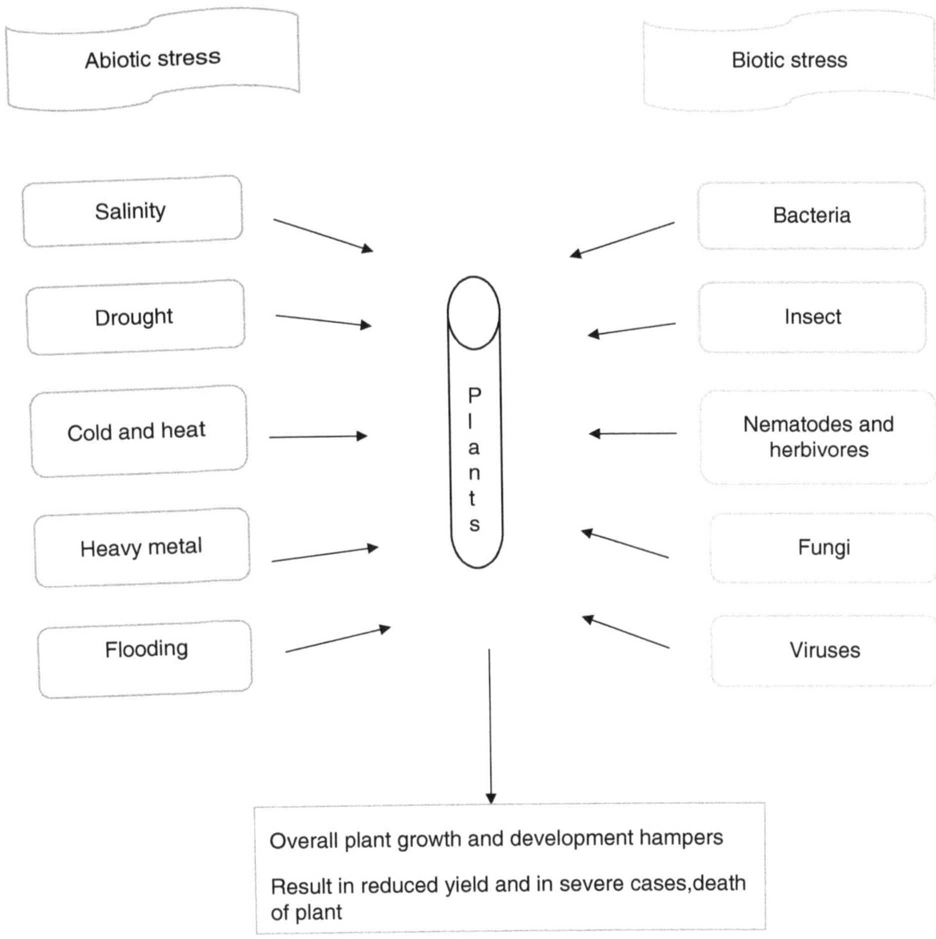

Figure 23.9 Types of plant stress.

imbalance, decreased yield, and some pathogens produce toxins that kill the plant (Gimenez et al. 2018). Plants show defensive mechanisms against these biotic agents. Attack of insects and herbivores triggers herbivore-associated molecular patterns (HAMPs) and mitogen-activated protein kinases (MAPKs) in plants, which activates defense mechanisms to reduce the harmful effects of infection. Salicylic acid pathway and octadecanoid pathway are activated, and down the line they also activate the defense genes (Lincoln et al. 2015). Microorganism-associated molecular patterns (MAMPs) are specific patterns that indicate the attack of viruses and bacteria (Boller and He 2009). Other methods to detect infection are signals that induce pattern recognition immunity (PTI) and effect-triggered immunity (ETI), resulting in a defense mechanism activating pathogen-related genes (PR) and JA defense genes (Tsuda and Katagiri 2010). Abiotic stress causes membrane destabilization and stomatal closure, which results in reduced gas exchange, decrease in carbon dioxide fixation, alteration of transpiration rate, decrease in biochemical reaction rate, occurrence of ion cytotoxicity, inhibition of photosynthesis and respiration, disturbance in the expression of genes and protein production, photosystem damage, change in enzymatic activity, disturbed uptake of water and nutrient, cessation of plant growth and development, and reduced overall plant productivity (refer Figure 23.10). To overcome these negative effects of abiotic stress, plants develop defense mechanisms to improve their physiological, biochemical, and molecular conditions (Ramegowda and Senthil-Kumar 2015).

23.5.2 Plant Remedial Measures Toward Stress

Mechanisms plants adapt to protect themselves from stress includes escape, avoidance, and tolerance/resistance mechanisms. Escape mechanisms result in shortening of life cycle, early flowering and maturity, and short vegetative phase (Levitt 1980). Tolerance/resistance mechanisms work at signaling level. Plants perceive signals after the attack of an abiotic or biotic stress agent down the line; these signals activate the stress-responsive genes, antioxidant defense system, early and

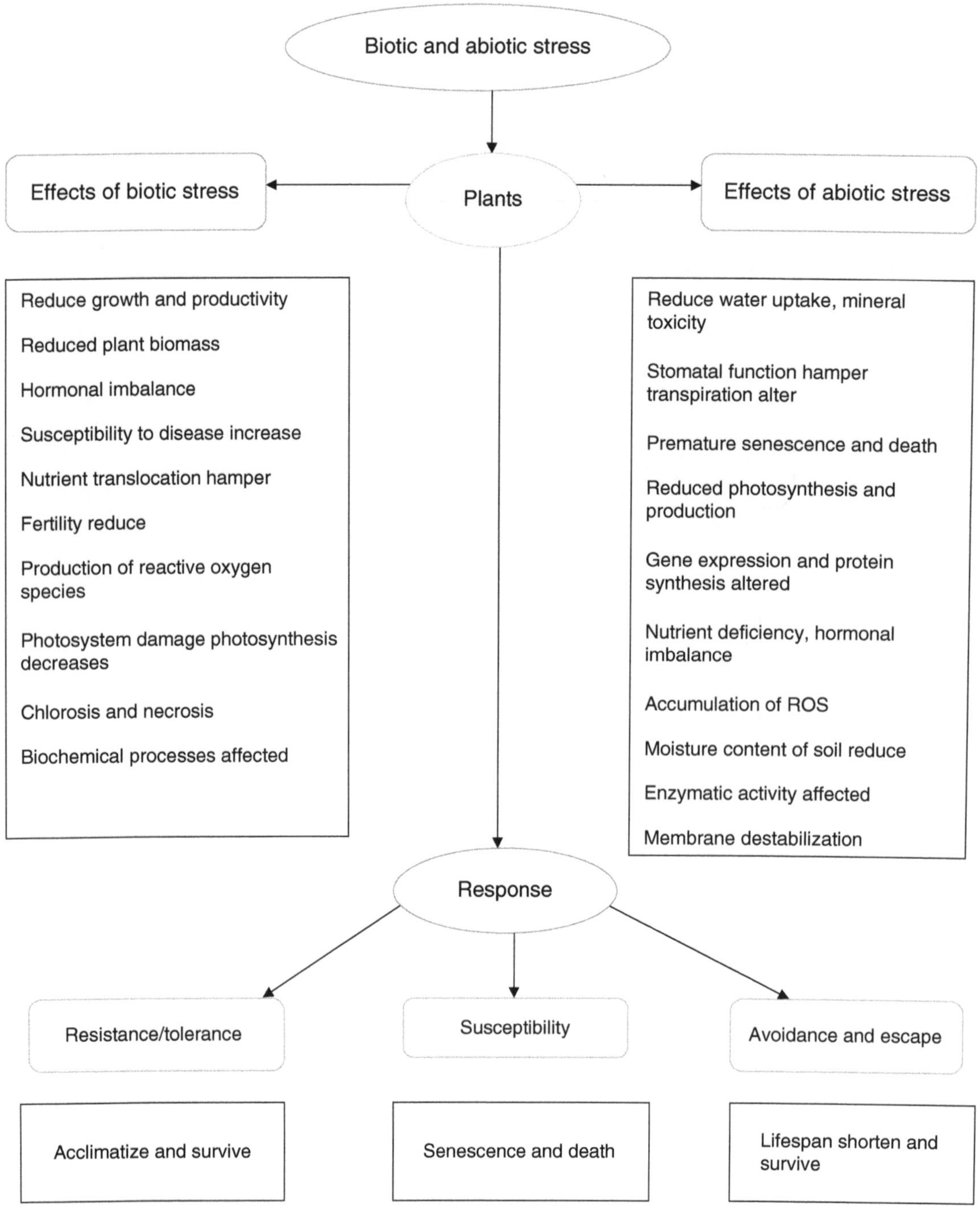

Figure 23.10 Effects of biotic and abiotic stress.

delayed stress genes. Product of these genes stabilizes the overall physiological and biochemical processes resulting in stress tolerance (refer Figure 23.11) (Arif et al. 2020).

23.6 Genome Editing for Plant Stress

Throughout evolution, plant architecture has helped plants sustain themselves against various biotic and abiotic conditions. Phytohormones, phenotypes, and genetic materials all contribute to their survival and continuation. The dynamic world has pressured plant natural homeostasis and tested their limits of enduring stresses. At this point, the remedial

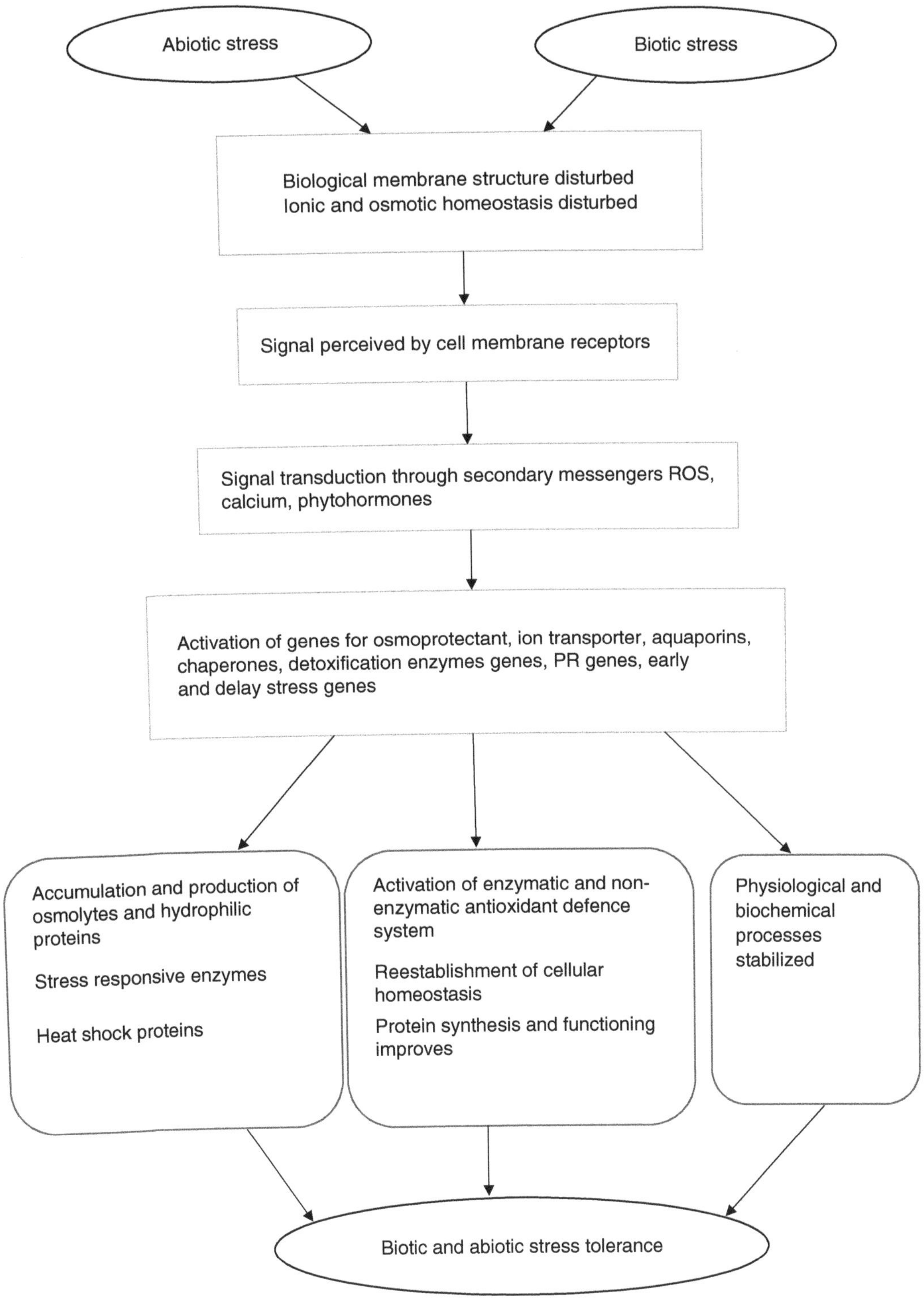

Figure 23.11 Mechanisms of biotic and abiotic stress tolerance.

aspect of genome editing comes in handy in ameliorating plants' stress management. Protection from stress in plants is orchestrated through various plant metabolites responsible for different functions in interactions with other organisms as well as ecological adaptation (Gottwald et al. 2002). If a way to augment plant stress management is discovered, conventional breeding and gene manipulation will take a lot of time, and in the case of trees, decades (Janick 2005). However, genetic engineering is optimized efficiently to the specific need of the plant. Genetically engineered crops are produced either by the use of Agrobacterium or by the direct DNA transfer method (Lobato-Gómez et al. 2021; Song et al. 2019). In the following, some examples of genome editing in light of improving plant stress management will be discussed briefly. The discussion is not all encompassing, but rather an establishment of perspective for the applicability of CRISPR/Cas9 for plant stress management.

23.6.1 Biotic Stress

23.6.1.1 Bacterium

Susceptibility (S) genes help the bacteria in successful invasion and growth, facilitating infection and compatibility through the symptoms of the disease (Boch et al. 2014; Bogdanove et al. 2010; van Schie and Takken 2014). Mutation to such susceptibility genes can induce resistance to the pathogen or even disrupt the general mechanism for other pathogen infections (Blanvillain-Baufume et al. 2017; Boch et al. 2014; Gawehns et al. 2013; Li et al. 2012; Ma et al. 2021; McGrann et al. 2014). Pathogens like *Xanthomonas* secrete transcription activator-like effector (TALE) proteins that manipulate the host susceptibility genes to their advantage (Kay et al. 2007). Disease susceptibility of such pathogen depends on the interaction between the TALE pathogenic gene and the corresponding S gene (Yang et al. 2006).

Xanthomonas citri (Xcc) is a pathogen that causes citrus canker, which has the recognizable signs of hyperplasia (cell over division), hypertrophy (cell enlargement), and necrosis (cell death) as symptoms (Brunings and Gabriel 2003; Gottwald et al. 2002). The TALE, PthA4, when translocated to host cell, suppresses plant immunity and leads to pathogen growth, inducing canker development (Jia et al. 2016). PthA4 acts as a promoter to express the S gene (Lateral organ boundaries 1 [CsLOB1]) that aids the infection (Jia et al. 2016). To have the S gene activated for transcription, PthA4 specifically binds to the effector binding elements (EBE_{PthA4}) in the promoter region (Hu et al. 2014; Jia et al. 2016). The best resistance to Xcc was seen in *Citrus sinensis* Osbeck when the whole EBEPthA4 gene was deleted from both CsLOB1 alleles (Peng et al. 2017). Cas9/CsLOB1sgRNA plasmids were used to generate homozygous targeted mutations in the first regenerated shoots, and AtU6-1 promoter (*Arabidopsis thaliana* L.) was used for driving its expression (Peng et al. 2017). In another experiment, WRKY22 of *C. sinensis* was found to be another S gene, and reducing its expression through CRISPR/Cas9-mediated editing reduced susceptibility (Wang et al. 2019). Similarly, a broad-spectrum bacterial blight resistance was also achieved by disrupting EBEs of the S genes of rice (OsSWEET11 and OsSWEET14) (Xu et al. 2019).

23.6.1.2 Virus

Geminiviruses are DNA viruses that have ssDNA genome and replicate either by rolling-circle mechanism through dsDNA replicative form or by recombination-dependent mechanism (Gutierrez 1999; Hanley-Bowdoin et al. 2013). They are known to cause serious losses and economic consequences in many crops (Mansoor et al. 2003; Moffat 1999). Another plant virus, potyvirus which is a positive sense, RNA (1ssRNA) damages many plants (Karasev and Gray 2013). It uses its **viral protein genome-linked** (VPg) protein to manipulate the host translation complex to translate its own protein (Carrington et al. 1989; Pyott et al. 2016).

In a Cas9 overexpressing *Nicotiana benthamiana*, sgRNAs targeting sequences of geminivirus virus (tomato yellow leaf curl virus [TYLCV]) are delivered through tobacco rattle virus (TRV) (Ali et al. 2015a, b). When the plant was inoculated after the delivery with the TYLCV, the amount of TYLCV titer was significantly lower in the sgRNA infiltrated plants than the controls (Ali et al. 2015a). This indicates the applicability of CRISPR/Cas9 mediated interference in creating virus resistance in plants. In another experiment using another geminivirus, beet severe curly top virus (BSCTV), sgRNA–Cas9 overexpressing plants (*Arabidopsis* and *N. benthamiana*) were highly resistance to the virus (Ji et al. 2015). Potyvirus turnip mosaic virus (TuMV) VPg in *Arabidopsis* requires the host eIF(iso)4E, an isoform of eIF4E (eukaryotic initiation factor) for its RNA translation (Wittmann et al. 1997). The targeted point mutations by the use of CRISPR/Cas9 at the eIF(iso)4E locus confers a complete resistance to TuMV (Pyott et al. 2016).

23.6.1.3 Fungus

Blumeria graminis f. sp. *tritici* (Bgt), the pathogen for powdery mildew, devastates wheat (*Triticum aestivum* L.) (Singh et al. 2016). Hexaploidy, a large genome of 17 Gb, and more frequencies of repetitive sequences in *T. aestivum* are some challenges for forward and reverse genetic analysis (Slade et al. 2005). RNAi was used to silence the enhanced disease resistance1 (EDR1), which was reported to have a negative role against *Arabidopsis* resistance against powdery mildew (Frye et al. 2001; Zhang et al. 2017). The virus-induced gene silencing of TaEDR1 in the mutant enhanced resistance against mildew, but this was not enough to produce heritable modification (Unniyampurath et al. 2016; Zhang et al. 2017).

CRISPR/Cas9 system with sgRNAT-EDR1 was used to introduce homozygous frameshift mutations in the three homoeologs of TaEDR1 through particle bombardment in immature embryos (Zhang et al. 2017). It was confirmed that the mutants have much higher resistance and reduced susceptibility to Bgt infection (Zhang et al. 2017).

In the pathogenesis of *Rhizoctonia solani* in maize, CRISPR/Cas9 knockout technique is used to identify the involvement of cinnamyl alcohol dehydrogenase, CAD (ZmCAD) in the disease resistance pathway (Li et al. 2019). Lignin acts as a physical barrier and elicits PAMP- and DAMP-triggered immunity (Malinovsky et al. 2014; Underwood 2012). F-box protein, ZmFBL41, mediates banded leaf and sheath blight resistance in maize, and FBL41 failure to recruit CAD degradation leads to more lignin synthesis, increasing the resistance (Li et al. 2019).

23.6.1.4 Insect

Insect pests pose a threat to the well-being of plants. Some plants have evolved to manage this biotic stress by making themselves unpalatable to consume. And with the intervention of genetic engineering, plants producing insecticidal proteins are also produced so far. The crystal protein (CRY) gene, the vegetative insecticidal protein (VIP) gene, the protease inhibitor (PI) gene, and insect resistance (IR) gene are used to genetically engineer plants against insect pests (Kumar et al. 2020; Saikhedkar et al. 2019; Zhang et al. 2022). This approach raises issue due to the use of foreign genetic material (Domingo and Bordonaba 2011; Gao 2021; Pozebon et al. 2020; Raman 2017; Schiemann et al. 2019).

Helicoverpa armigera and *Spodoptera litura* are leaf-chewing insect pests in soyabean (*Glycine max*) (Haile et al. 2021; Otuka et al. 2020). In the QTL-M, Glyma.07g110300 marker (encoding a UDP-glycosyltransferase) plays an important role in *H. armigera* and *S. litura* resistance (Ortega et al. 2017). CRISPR/Cas9 was used to perform GmUGT gene knockout mutants and was observed to increase resistance to the pests in the mutants (Zhang et al. 2022). The resistance to the pest may be attributed to the altered production of various flavonoids metabolites (7,4-dihydroxyflavone, eupatilin, isoformononetin, ononin, and daidzein) in the *G. max* due to the loss of function of the GmUGT (Zhang et al. 2022).

23.6.2 Abiotic Stress

23.6.2.1 Chemicals

Without doubt, the growing population has become largely dependent on chemicals. This really increases the plant stresses due to herbicides, heavy metals, pollutants etc. The novel method of genome editing through CRISPR/Cas9 has made it possible to address this issue by producing herbicide-tolerant plant, salt and reactive oxygen species (ROS) tolerant plant, and heavy metal tolerant plant.

CRISPR-mediated base editors were used to manipulate the full-length intronless coding of the acetolactate synthase (ALS) of rice i.e., OsALS1 (Kuang et al. 2020). ALS is targeted by commercial herbicides, and altering the homolog OsALS1 confers tolerance to the herbicide, which was tested by using bispyribac-sodium screen (Garcia et al. 2017; Kuang et al. 2020; Xia et al. 2017). Cas9n base editors along with a sgRNA facilitate the production of artificial evolution of the endogenous OsALS1 gene in rice cells for developing herbicide-tolerant rice germplasms (Kuang et al. 2020). Base editing CRISPR/Cas9 has been successfully used to engineer herbicide-tolerant rice and watermelon (Kuang et al. 2020; Shimatani et al. 2017; Tian et al. 2018).

When there is an excess of ROS in plants, oxidative stress occurs, and it has a damaging effect on the plants (Demidchik 2015; Yousuf et al. 2012). ROS is involved in many biological signaling and processes in plants (Foyer et al. 2017; Noctor et al. 2014, 2018). Paraquat herbicide induces oxidative stress in plants (Summers 1980). *Arabidopsis* PARAQUAT TOLERANCE 3 (AtPQT3) being E3 ligase, interacts with PRMT4b (methyltransferase) and degrades it. PRMT4b, however, is involved in enhancing stress resistance through its epigenetic histone modification pathway (Luo et al. 2016). Previously, AtPQT3 knockout was demonstrated to enhance resistance to abiotic stress (Luo et al. 2016). Alfatih et al. targeted the rice homolog OsPQT3 through CRISPR/Cas9 mediated editing to generate null mutations in the gene (2020). The mutants of the OsPQT3 were found to have significant tolerance toward paraquat as well as salt stress (NaCl) (Alfatih et al. 2020). Also, CRISPR/Cas9 knockout of OsARM1, an arsenate-responsive genes identified through global transcriptome analysis, improved tolerance to As(III) (Wang et al. 2017; Yu et al. 2012).

23.6.2.2 Environmental

To keep pace with the global warming and erratic climatic changes, it is advantageous for plants to utilize their mechanisms of coping up with the extremes. The phytohormone abscisic acid (ABA) during drought stress facilitates the closure of stomatal guard cells (Himmelbach et al. 2003). ABA-induced transcription repressors (AITRs) have negative regulation on ABA signaling and abiotic stress responses (Tian et al. 2017). The six genes of AITRs in *A. thaliana* are targeted simultaneously using CRISPR/Cas9 technique, thus generating *aitr2 aitr3 aitr4 aitr5 aitr6* (*aitr23456*) quintuple and *aitr1 aitr2 aitr3 aitr4 aitr5 aitr6* (*aitr123456*) sextuple mutants (Chen et al. 2021; Tian et al. 2017). The researchers, by knocking out the entire genes, successfully increased drought and salinity tolerance, without compromising the overall fitness of the plant (Chen et al. 2021).

Another way of modifying plants toward a responsive drought tolerance is through activating positive regulator of drought stress response, ABA-responsive element binding factor 1 (AREB1/ABF2) (Barbosa et al. 2013; Fujita 2005; Nakashima et al. 2014; Oh 2005; Roca Paixão et al. 2019; Todaka et al. 2015; Yoshida et al. 2010). Here, dCas9 fused with HAT (Histone acetyltransferase from *A. thaliana*) through epigenetic histone core acetylation makes the DNA more available for transcription of the AREB1 gene (Eberharter 2002; Roca Paixão et al. 2019; Schneider et al. 2013). This epigenome editing generated phenotypes suitable for drought resistance (Roca Paixão et al. 2019).

23.7 Elimination of CRISPR/Cas from the System After Genetic Editing

From an ethical stand, it is crucial for transgenes to be removed from the plant system after they have achieved their purpose (He et al. 2018). Generation of transgene-free CRISPR-edited plant with heritable stable traits is a big issue (Gao and Zhao 2014). This is important because of the possibility that the editing construct could harm the environment, or the transgene product (pollen/seed) might not be acceptable for government approval (Callaway 2018; Huang et al. 2016). Also, the CRISPR/Cas construct, if present, might further continue to edit other targets (Gao and Zhao 2014). Different approaches can be employed to obtain a transgene-free CRISPR/Cas edited plant (Chen et al. 2018; Gao et al. 2016; He et al. 2017, 2018, 2019; Liu et al. 2019; Luo et al. 2015; Woo et al. 2015).

The first-generation plants (T_0) of CRISPR/Cas targeted genome editing are usually heterozygous of the transgene locus; however, after selfing, the transgene and edited gene segregate according to Mendelian laws (He and Zhao 2020; Jeon et al. 2000). This may produce transgene-free T_1 plants up to 25%, even though this strategy is difficult in the case of linked transgene and edited targeted gene (He and Zhao 2020). Drug (bentazon) sensitivity using a selection screen is a well-known approach for removing transgenic plants (He and Zhao 2020; Lu et al. 2017). A fluorescence marker (mCherry) can be used to spot the transgene based on the expression and the intensity of the fluorescence protein linked with the CRISPR components (transgene) in *Arabidopsis* (Gao et al. 2016; He et al. 2017; He and Zhao 2020; Yu and Zhao 2019). Thus, the transgenic T_1 seeds can be easily identified, and selection of the transgene-free seed can be done (Yu and Zhao 2019).

23.8 Prospects and Limitations

Owing to many flavors of CRISPR/Cas genome editing like deletion, insertion, point mutation, and editing of the sequence in the targeted genome, it has become one of the most widely used methods in genetic engineering, and its popularity and versatility are only going to grow more and more in coming days. Many precise modifications in plants' genome have also been proved valuable for the development of disease resistance and stress resistance. Especially, in this dynamic world where changes are inevitable, and unavoidable, such windows of opportunities opened recently by this technique are inexpensive. It can construct synthetic transcription factors to upregulate or downregulate specific pathways that can augment plant productivity and survival (Jusiak et al. 2016). Transfer of Nod factors to the legumes will eliminate the need of the use of inorganic nitrogen fertilizers (Biswas et al. 2021). Domestication of wild relative species of crops in a short time using CRISPR is another way of increasing global food security (Zhu et al. 2020). An average of 33% of all the fruits and vegetable crops produced globally go to waste due to their short shelf-life, which leads to postharvest loss and waste (FAO 2011). Through the CRISPR engineering of desirable traits, it can be reduced considerably (Shipman et al. 2021). Moreover, it is another possibility to develop a plant factory for the production of specific metabolites or proteins, like the production of Monoclonal Antibodies in animal (Bortesi and Fischer 2015).

The efficiency of gene editing by the CRISPR/Cas system is significantly variable in plants (Bortesi and Fischer 2015). The dependency in the motif specificity of PAM and the length of the gRNA for the generation of DSBs limits versatility of the CRISPR/Cas9 system (Lozano-Juste and Cutler 2014). Off-target is a concern that needs to be explicitly addressed in plant system (Lozano-Juste and Cutler 2014). Another main challenge in plant breeding is the generation of transgene-free progeny (He and Zhao 2020). Lastly, the large size of SpCas9 is a problem in delivery using viral-based vector in plants (Hassan et al. 2021).

Nevertheless, we have come a long way from figuring out the seven traits of peas by Mendel, to precisely generate a genetic sequence through the utilization of enzymes, RNA, and other discoveries on the road to genome editing. It should, however, be appreciated that we are more and more conscious about life, and its value that we innovate ways to help the living around us through their struggles in evolution.

References

Alfatih, A., Wu, J., Jan, S.U. et al. (2020). Loss of rice PARAQUAT TOLERANCE 3 confers enhanced resistance to abiotic stresses and increases grain yield in field. *Plant Cell Environ.* 43 (11): 2743–2754. https://doi.org/10.1111/pce.13856.

Ali, Z., Abulfaraj, A., Idris, A. et al. (2015a). CRISPR/Cas9-mediated viral interference in plants. *Genome Biol.* 16: 238. https://doi.org/10.1186/s13059-015-0799-6.

Ali, Z., Abul-Faraj, A., Li, L. et al. (2015b). Efficient virus-mediated genome editing in plants using the CRISPR/Cas9 system. *Mol. Plant.* 8: 1288–1291. https://doi.org/10.1016/j.molp.2015.02.011.

Ali, Z., Eid, A., Ali, S., and Mahfouz, M.M. (2018). Pea early browning virus-mediated genome editing via the CRISPR/Cas9 system in *Nicotiana benthamiana* and *Arabidopsis*. *Virus Res.* 244: 333–337.

Altpeter, F., Springer, N.M., Bartley, L.E. et al. (2016). Advancing crop transformation in the era of genome editing. *Plant Cell* 28: 1510–1520.

Anders, C., Niewoehner, O., Duerst, A., and Jinek, M. (2014). Structural basis of PAM-dependent target DNA recognition by the Cas9 endonuclease. *Nature* 513: 569–573. https://doi.org/10.1038/nature13579.

Anton, T., Bultmann, S., Leonhardt, H., and Markaki, Y. (2014). Visualization of specific DNA sequences in living mouse embryonic stem cells with a programmable fluorescent CRISPR/Cas system. *Nucleus* 5: 163–172. https://doi.org/10.4161/nucl.28488.

Anzalone, A.V., Randolph, P.B., Davis, J.R. et al. (2019). Search-and-replace genome editing without double-strand breaks or donor DNA. *Nature* 576: 149–157.

Arif, I., Batool, M., and Schenk, P.M. (2020). Plant microbiome engineering: expected benefits for improved crop growth and resilience. *Trends Biotechnol.* 38: 1385–1396.

Atkins, P.A. and Voytas, D.F. (2020). Overcoming bottlenecks in plant gene editing. *Curr. Opin. Plant Biol.* 54: 79–84.

Barbosa, E.G.G., Leite, J.P., Marin, S.R.R. et al. (2013). Overexpression of the ABA-dependent AREB1 transcription factor from *Arabidopsis thaliana* improves soybean tolerance to water deficit. *Plant Molecular Biology Reporter* 31: 719–730.

Bétermier, M., Bertrand, P., and Lopez, B.S. (2014). Is non-homologous end-joining really an inherently error-prone process? *PLoS Genet.* 10 (1): e1004086. https://doi.org/10.1371/journal.pgen.1004086.

Bibikova, M., Carroll, D., Segal, D.J. et al. (2001). Stimulation of homologous recombination through targeted cleavage by chimeric nucleases. *Mol. Cell. Biol.* 21: 289–297.

Bibikova, M., Golic, M., Golic, K.G., and Carroll, D. (2002). Targeted chromosomal cleavage and mutagenesis in Drosophila using zinc-finger nucleases. *Genetics* 161: 1169–1175.

Bikard, D., Jiang, W., Samai, P. et al. (2013). Programmable repression and activation of bacterial gene expression using an engineered CRISPR-Cas system. *Nucleic Acids Res.* 41: 7429–7437. https://doi.org/10.1093/nar/gkt520.

Biswas, D., Saha, S.C., and Dey, A. (2021). CRISPR-Cas genome-editing tool in plant abiotic stress-tolerance. *Plant Gene* 26: 100286.

Blanvillain-Baufume, S., Reschke, M., Sole, M. et al. (2017). Targeted promoter editing for rice resistance to *Xanthomonas oryzae* pv. oryzae reveals differential activities for SWEET14-inducing TAL effectors. *Plant Biotechnol. J.* 15: 306–317.

Blin, K., Pedersen, L.E., Weber, T., and Lee, S.Y. (2016). CRISPy-web: an online resource to design sgRNAs for CRISPR applications. *Synth. Syst. Biotechnol.* 1: 118–121.

Boch, J., Scholze, H., Schornack, S. et al. (2009). Breaking the code of DNA binding specificity of TAL-type III effectors. *Science* 326: 1509–1512.

Boch, J., Bonas, U., and Lahaye, T. (2014). TAL effectors–pathogen strategies and plant resistance engineering. *New Phytol.* 204: 823–832.

Bogdanove, A.J., Schornack, S., and Lahaye, T. (2010). TAL effectors: finding plant genes for disease and defense. *Curr. Opin. Plant Biol.* 13: 394–401.

Boller, T. and He, S.Y. (2009). Innate immunity in plants: an arms race between pattern recognition receptors in plants and effectors in microbial pathogens. *Science* 324: 742–744.

Bolotin, A., Quinquis, B., Sorokin, A., and Ehrlich, S.D. (2005). Clustered regularly interspaced short palindrome repeats (CRISPRs) have spacers of extrachromosomal origin. *Microbiology* 151: 2551–2561. https://doi.org/10.1099/mic.0.28048-0.

Bortesi, L. and Fischer, R. (2015). The CRISPR/Cas9 system for plant genome editing and beyond. *Biotechnol. Adv.* 33 (1): 41–52. https://doi.org/10.1016/j.biotechadv.2014.12.006.

Brocken, D.J.W., Tark-Dame, M., and Dame, R.T. (2018). dCas9: A versatile tool for epigenome editing. *Curr. Issues Mol. Biol.* 26: 15–32. https://doi.org/10.21775/cimb.026.015.

Brouns, S.J., Jore, M.M., Lundgren, M. et al. (2008). Small CRISPR RNAs guide antiviral defense in prokaryotes. *Science* 321: 960–964.

Brunings, A.M. and Gabriel, D.W. (2003). *Xanthomonas citri*: breaking the surface. *Mol. Plant Pathol.* 4: 141–157.

Callaway, E. (2018). CRISPR plants now subject to tough GM laws in European Union. *Nature* 560: 16. https://doi.org/10.1038/d41586-018-05814-6.

Capecchi, M.R. (1989). Altering the genome by homologous recombination. *Science* 244: 1288–1292.

Carrington, J.C., Freed, D.D., and Sanders, T.C. (1989). Autocatalytic processing of the potyvirus helper component proteinase in *Escherichia coli* and *in vitro* Cl-l. *J. Virol.* 63 (10): 4459–4463.

Castel, B., Yang, Y., and Jones, J.D.G. (2019). Optimization of T-DNA architecture for Cas9-mediated mutagenesis in *Arabidopsis*. *PLoS One* 14: e0204778.

Chen, B., Gilbert, L.A., Cimini, B.A. et al. (2013). Dynamic imaging of genomic loci in living human cells by an optimized CRISPR/Cas system. *Cell* 155: 1479–1491. https://doi.org/10.1016/j.cell.2013.12.001.

Chen, L., Li, W., Katin-Grazzini, L. et al. (2018). A method for the production and expedient screening of CRISPR/Cas9-mediated non-transgenic mutant plants. *Hortic Res.* 5: 13.

Chen, S., Zhang, N., Zhou, G. et al. (2021). Knockout of the entire family of AITR genes in *Arabidopsis* leads to enhanced drought and salinity tolerance without fitness costs. *BMC Plant Biol.* 21 (1): 137. https://doi.org/10.1186/s12870-021-02907-9.

Cheng, A.W., Wang, H., Yang, H. et al. (2013). Multiplexed activation of endogenous genes by CRISPR-on, an RNA-guided transcriptional activator system. *Cell Res.* 23: 1163–1171. https://doi.org/10.1038/cr.2013.122.

Chevalier, B.S., Kortemme, T., Chadsey, M.S. et al. (2002). Design, activity, and structure of a highly specific artificial endonuclease. *Mol. Cell* 10: 895–905. https://doi.org/10.1016/S1097-2765(02)00690-1.

Christian, M., Cermak, T., Doyle, E.L. et al. (2010). Targeting DNA double-strand breaks with TAL effector nucleases. *Genetics* 186: 757–761.

Chylinski, K., Le Rhun, A., and Charpentier, E. (2013). The tracrRNA and Cas9 families of type II CRISPR-Cas immunity systems. *RNA Biol.* 10: 726–737.

Chylinski, K., Makarova, K.S., Charpentier, E., and Koonin, E.V. (2014). Classification and evolution of type II CRISPR-Cas systems. *Nucleic Acids Res.* 11: 6091.

Cody, W.B., Scholthof, H.B., and Mirkov, T.E. (2017). Multiplexed gene editing and protein overexpression using a tobacco mosaic virus viral vector. *Plant Physiol.* 175: 23–35.

Cong, L., Ran, F.A., Cox, D. et al. (2013). Multiplex genome engineering using CRISPR/Cas systems. *Science* 339: 819–823.

Cooper, M., van Eeuwijk, F.A., Hammer, G.L. et al. (2009). Modeling QTL for complex traits: detection and context for plant breeding. *Curr. Opin. Plant Biol.* 12: 231–240.

Deltcheva, E., Chylinski, K., Sharma, C.M. et al. (2011). CRISPR RNA maturation by trans-encoded small RNA and host factor RNase III. *Nature* 471: 602–607.

Demidchik, V. (2015). Mechanisms of oxidative stress in plants: from classical chemistry to cell biology. *Environ. Exp. Bot.* 109: 212–228.

Domingo, J.L. and Bordonaba, J.G. (2011). A literature review on the safety assessment of genetically modified plants. *Environ. Int.* 37: 734–742. https://doi.org/10.1016/j.envint.2011.01.003.

Doudna, J.A. and Charpentier, E. (2014). Genome editing. The new frontier of genome engineering with CRISPR-Cas9. *Science* 346 (6213): 1258096. https://doi.org/10.1126/science.1258096.

Eberharter, A. (2002). Histone acetylation: a switch between repressive and permissive chromatin: second in review series on chromatin dynamics. *EMBO Rep.* 3: 224–229.

Ellison, E.E., Nagalakshmi, U., Gamo, M.E. et al. (2020). Multiplexed heritable gene editing using RNA viruses and mobile single guide RNAs. *Nat. Plants* 6: 620–624.

FAO (2011). Global food losses and food waste – extent, causes and prevention. http://www.fao.org/docrep/014/mb060e/mb060e00.pdf.

Feng, C., Su, H., Bai, H. et al. (2018). High-efficiency genome editing using a dmc1 promoter-controlled CRISPR/Cas9 system in maize. *Plant Biotechnol. J.* 16: 1848–1857.

Fonfara, I., Le Rhun, A., Chylinski, K. et al. (2014). Phylogeny of Cas9 determines functional exchangeability of dual-RNA and Cas9 among orthologous type II CRISPR-Cas systems. *Nucleic Acids Res.* 42: 257.

Foyer, C.H., Ruban, A.V., and Noctor, G. (2017). Viewing oxidative stress through the lens of oxidative signalling rather than damage. *Biochem. J.* 474: 877–883.

Frye, C.A., Tang, D., and Innes, R.W. (2001). Negative regulation of defense responses in plants by a conserved MAPKK kinase. *Proc. Natl. Acad. Sci. U.S.A.* 98: 373–378.

Fu, Y., Foden, J.A., Khayter, C. et al. (2013). High-frequency off-target mutagenesis induced by CRISPR-Cas nucleases in human cells. *Nat. Biotechnol.* 31: 822–826.

Fujita, Y. (2005). AREB1 is a transcription activator of novel ABRE-dependent ABA signaling that enhances drought stress tolerance in *Arabidopsis*. *Plant Cell Online* 17: 3470–3488.

Gaillochet, C., Develtere, W., and Jacobs, T.B. (2020). CRISPR screens in plants: approaches, guidelines, and future prospects. *Plant Cell* 33 (4): 794–813.

Gao, C. (2021). Genome engineering for crop improvement and future agriculture. *Cell* 184: 1621–1635.

Gao, Y. and Zhao, Y. (2014). Specific and heritable gene editing in *Arabidopsis*. *Proc. Natl. Acad. Sci. U.S.A.* 111: 4357–4358. https://doi.org/10.1073/pnas.1402295111.

Gao, X., Chen, J., Dai, X. et al. (2016). An effective strategy for reliably isolating heritable and Cas9-free *Arabidopsis* mutants generated by CRISPR/Cas9-mediated genome editing. *Plant Physiol.* 171: 1794–1800. https://doi.org/10.1104/pp.16.00663.

Garcia, M.D., Nouwens, A., Lonhienne, T.G., and Guddat, L.W. (2017). Comprehensive understanding of acetohydroxyacid synthase inhibition by different herbicide families. *Proc. Natl. Acad. Sci. U.S.A.* 114: E1091–E1100.

Gasiunas, G., Barrangou, R., Horvath, P., and Siksnys, V. (2012). Cas9-crRNA ribonucleoprotein complex mediates specific DNA cleavage for adaptive immunity in bacteria. *Proc. Natl. Acad. Sci. U.S.A.* 109: E2579–E2586. https://doi.org/10.1073/pnas.1208507109.

Gaudelli, N.M., Komor, A.C., Rees, H.A. et al. (2017). Programmable base editing of A•T to G•C in genomic DNA without DNA cleavage. *Nature* 551: 464–471.

Gawehns, F., Cornelissen, B.J., and Takken, F.L. (2013). The potential of effector target genes in breeding for plant innate immunity. *Microb. Biotechnol.* 6: 223–229.

Gerashchenkov, G.A., Rozhnova, N.A., Kuluev, B.R. et al. (2020). Design of guide RNA for CRISPR/Cas plant genome editing. *Mol. Biol. (N.Y.)* 54: 24–42.

Gilbert, L.A., Larson, M.H., Morsut, L. et al. (2013). CRISPR-mediated modular RNA-guided regulation of transcription in eukaryotes. *Cell* 154: 442–451. https://doi.org/10.1016/j.cell.2013.06.044.

Gimenez, E., Salinas, M., and Manzano-Agugliaro, F. (2018). Worldwide research on plant defense against biotic stresses as improvement for sustainable agriculture. *Sustainability* 10: 391.

Gottwald, T.R., Graham, J.H., and Schubert, T.S. (2002). Citrus canker: the pathogen and its impact. *Plant Health Prog.* 10: 1–34.

Gunn, L.H., Martin Avila, E., Birch, R., and Whitney, S.M. (2020). The dependency of red Rubisco on its cognate activase for enhancing plant photosynthesis and growth. *Proc. Natl. Acad. Sci. U.S.A.* 117: 25890–25896.

Gutierrez, C. (1999). Geminivirus DNA, replication. *Cell. Mol. Life Sci.* 56 (3–4): 313–329.

Haft, D.H., Selengut, J., Mongodin, E.F., and Nelson, K.E. (2005). A guild of 45 CRISPR-associated (Cas) protein families and multiple CRISPR/Cas subtypes exist in prokaryotic genomes. *PLoS Comput. Biol.* 1: e60.

Haile, F., Nowatzki, T., Storer, N., and Davis, J. (2021). Overview of pest status, potential risk, and management considerations of *Helicoverpa armigera* (Lepidoptera: noctuidae) for U.S. soybean production. *J. Integr. Pest. Manag.* 12: 1–10.

Hale, C.R., Zhao, P., Olson, S. et al. (2009). RNA-guided RNA cleavage by a CRISPR RNA-Cas protein complex. *Cell* 139: 945–956.

Hanley-Bowdoin, L., Bejarano, E.R., Robertson, D., and Mansoor, S. (2013). Geminiviruses: masters at redirecting and reprogramming plant processes. *Nat. Rev. Microbiol.* 11: 777–788.

Hanna, R.E. and Doench, J.G. (2020). Design and analysis of CRISPR-Cas experiments. *Nat. Biotechnol.* 38: 813–823.

Hassan, M.M., Zhang, Y., Yuan, G. et al. (2021). Construct design for CRISPR/Cas-based genome editing in plants. *Trends Plant Sci.* 26 (11): 1133–1152. https://doi.org/10.1016/j.tplants.2021.06.015.

He, Y. and Zhao, Y. (2020). Technological breakthroughs in generating transgene-free and genetically stable CRISPR-edited plants. *aBIOTECH* 1 (1): 88–96. https://doi.org/10.1007/s42994-019-00013-x.

He, Y., Wang, R., Dai, X., and Zhao, Y. (2017). On improving CRISPR for editing plant genes: ribozyme-mediated guide RNA production and fluorescence-based technology for isolating transgene-free mutants generated by CRISPR. *Prog. Mol. Biol. Transl. Sci.* 149: 151–166. https://doi.org/10.1016/bs.pmbts.2017.03.012.

He, Y., Zhu, M., Wang, L. et al. (2018). Programmed self-elimination of the CRISPR/Cas9 construct greatly accelerates the isolation of edited and transgene-free rice plants. *Mol. Plant.* 11 (9): 1210–1213. https://doi.org/10.1016/j.molp.2018.05.005.

He, Y., Zhu, M., Wang, L. et al. (2018). Programmed self-elimination of the CRISPR/Cas9 construct greatly accelerates the isolation of edited and transgene-free. *Rice Plants Mol. Plant* 11: 1210–1213. https://doi.org/10.1016/j.molp.2018.05.005.

He, Y., Zhu, M., Wang, L. et al. (2019). Improvements of TKC technology accelerate isolation of transgene-free CRISPR/Cas9-edited rice plants. *Rice Sci.* 26: 109–117. https://doi.org/10.1016/j.rsci.2018.11.001.

Hickey, L.T., N Hafeez, A., Robinson, H. et al. (2019). Breeding crops to feed 10 billion. *Nat. Biotechnol.* 37: 744–754.

Hilton, I.B., D'Ippolito, A.M., Vockley, C.M. et al. (2015). Epigenome editing by a CRISPR-Cas9-based acetyltransferase activates genes from promoters and enhancers. *Nat. Biotechnol.* 33: 510–517. https://doi.org/10.1038/nbt.3199.

Himmelbach, A., Yang, Y., and Grill, E. (2003). Relay and control of abscisic acid signaling. *Curr. Opin. Plant Biol.* 6: 470–479.

Hsu, P.D., Scott, D.A., Weinstein, J.A. et al. (2013). DNA targeting specificity of RNA-guided Cas9 nucleases. *Nat. Biotechnol.* 31: 827–832.

Hsu, P.D., Lander, E.S., and Zhang, F. (2014). Development and applications of CRISPR-Cas9 for genome engineering. *Cell.* 157 (6): 1262–1278. https://doi.org/10.1016/j.cell.2014.05.010.

Hu, Y., Zhang, J., Jia, H. et al. (2014). Lateral organ boundaries 1 is a disease susceptibility gene for citrus bacterial canker disease. *Proc. Natl. Acad. Sci. U.S.A.* 111: E521–E529.

Hu, X., Meng, X., Liu, Q. et al. (2018). Increasing the efficiency of CRISPR-Cas9-VQR precise genome editing in rice. *Plant Biotechnol. J.* 16: 292–297.

Hua, K., Tao, X., Liang, W. et al. (2020). Simplified adenine base editors improve adenine base editing efficiency in rice. *Plant Biotechnol. J.* 18: 770–778.

Huang, S., Weigel, D., Beachy, R.N., and Li, J. (2016). A proposed regulatory framework for genome-edited crops. *Nat Genet* 48: 109–111. https://doi.org/10.1038/ng.3484.

Ishino, Y., Shinagawa, H., Makino, K. et al. (1987). Nucleotide sequence of the iap gene, responsible for alkaline phosphatase isozyme conversion in *Escherichia coli*, and identification of the gene product. *J. Bacteriol.* 169: 5429–5433.

Jacquier, A. and Dujon, B. (1985). An intron-encoded protein is active in a gene conversion process that spreads an intron into a mitochondrial gene. *Cell* 41: 383–394. https://doi.org/10.1016/S0092-8674(85)80011-8.

Janick, J. (ed.) (2005). The origins of fruits, fruit growing, and fruit breeding. In: *Plant Breeding Reviews*, vol. 25. Springer.

Jeon, J.S., Lee, S., Jung, K.H. et al. (2000). T-DNA insertional mutagenesis for functional genomics in rice. *Plant J.* 22: 561–570.

Ji, X., Zhang, H., Zhang, Y. et al. (2015). Establishing a CRISPR-Cas-like immune system conferring DNA virus resistance in plants. *Nat. Plants* 1: 15144. https://doi.org/10.1038/nplants.2015.144.

Jia, H., Orbovic, V., Jones, J.B., and Wang, N. (2016). Modification of the PthA4 effector binding elements in Type I CsLOB1 promoter using Cas9/sgRNA to produce transgenic Duncan grapefruit alleviating XccΔpthA4:dCsLOB1.3 infection. *Plant Biotechnol J.* 14 (5): 1291–1301. https://doi.org/10.1111/pbi.12495.

Jiang, F. and Doudna, J.A. (2017). CRISPR-Cas9 structures and mechanisms. *Annu. Rev. Biophys.* 46: 505–529.

Jin, S., Fei, H., Zhu, Z. et al. (2020). Rationally designed APOBEC3B cytosine base editors with improved specificity. *Mol. Cell* 79: 1–13.

Jinek, M., Chylinski, K., Fonfara, I. et al. (2012). A programmable dual-RNA–guided DNA endonuclease in adaptive bacterial immunity. *Science* 337: 816–821. https://doi.org/10.1126/science.1225829.

Jinek, M., Jiang, F., Taylor, D.W. et al. (2014). Structures of Cas9 endonucleases reveal RNA-mediated conformational activation. *Science* 343: 1247997.

Johnson, R.A., Gurevich, V., Filler, S. et al. (2015). Comparative assessments of CRISPR-Cas nucleases' cleavage efficiency in planta. *Plant Mol. Biol.* 87: 143–156.

Jusiak, B., Cleto, S., Perez-Piñera, P., and Lu, T.K. (2016). Engineering synthetic gene circuits in living cells with CRISPR technology. *Trends Biotechnol.* 34: 535–547. https://doi.org/10.1016/j.tibtech.2015.12.014.

Kang, B.C., Yun, J.-Y., Kim, S.-T. et al. (2018). Precision genome engineering through adenine base editing in plants. *Nat. Plants* 4: 427–431.

Karasev, A.V. and Gray, S.M. (2013). Continuous and emerging challenges of potato virus Y in potato. *Annu. Rev. Phytopathol.* 51: 571–586. https://doi.org/10.1146/annurev-phyto-082712-1023.

Kay, S., Hahn, S., Marois, E. et al. (2007). A bacterial effector acts as a plant transcription factor and induces a cell size regulator. *Science* 318: 648–651.

Kearns, N.A., Pham, H., Tabak, B. et al. (2015). Functional annotation of native enhancers with a Cas9-histone demethylase fusion. *Nat. Methods* 12: 401–403. https://doi.org/10.1038/nmeth.3325.

Kim, Y.G., Cha, J., and Chandrasegaran, S. (1996). Hybrid restriction enzymes: zinc finger fusions to Fok I cleavage domain. *Proc. Natl. Acad. Sci. U.S.A.* 93: 1156–1160. https://doi.org/10.1073/pnas.93.3.1156.

Kishi-Kaboshi, M., Aida, R., and Sasaki, K. (2017). Generation of gene-edited *Chrysanthemum morifolium* using multicopy transgenes as targets and markers. *Plant Cell Physiol.* 58: 216–226.

Komor, A.C., Kim, Y.B., Packer, M.S. et al. (2016). Programmable editing of a target base in genomic DNA without double-stranded DNA cleavage. *Nature* 533: 420–424.

Kuang, Y., Li, S., Ren, B. et al. (2020). Base-editing-mediated artificial evolution of OsALS1 in planta to develop novel herbicide-tolerant rice germplasms. *Mol. Plant.* 13 (4): 565–572. https://doi.org/10.1016/j.molp.2020.01.010.

Kumar, K., Gambhir, G., Dass, A. et al. (2020). Genetically modified crops: current status and future prospects. *Planta* 251: 91. https://doi.org/10.1007/s00425-020-03372-8.

Lassoued, R., Macall, D.M., Hesseln, H. et al. (2019). Benefits of genome-edited crops: expert opinion. *Transgenic Res.* 28: 247–256.

Levitt, J. (ed.) (1980). Chilling, freezing, and high temperature stresses. In: *Responses of Plants to Environmental Stress*, vol. 2, 23–445. New York, NY: Academic Press.

Li, T., Liu, B., Spalding, M.H. et al. (2012). High-efficiency TALEN-based gene editing produces disease-resistant rice. *Nat. Biotechnol.* 30: 390–392.

Li, J., McCormack, M., Zhang, D. et al. (2013). Multiplex and homologous recombination mediated genome editing in *Arabidopsis* and *Nicotiana benthamiana* using guide RNA and Cas9. *Nat. Biotechnol.* 31: 688–691.

Li, C., Zong, Y., Wang, Y. et al. (2018a). Expanded base editing in rice and wheat using a Cas9-adenosine deaminase fusion. *Genome Biol.* 19: 59.

Li, T., Yang, X., Yu, Y. et al. (2018b). Domestication of wild tomato is accelerated by genome editing. *Nat. Biotechnol.* 36: 1160–1163.

Li, N., Lin, B., Wang, H. et al. (2019). Natural variation in ZmFBL41 confers banded leaf and sheath blight resistance in maize. *Nat. Genet.* 51 (10): 1540–1548. https://doi.org/10.1038/s41588-019-0503-y.

Li, C., Zhang, R., Meng, X. et al. (2020). Targeted, random mutagenesis of plant genes with dual cytosine and adenine base editors. *Nat. Biotechnol.* 38: 875–882.

Liang, Z., Chen, K., Li, T. et al. (2017). Efficient DNA-free genome editing of bread wheat using CRISPR/Cas9 ribonucleoprotein complexes. *Nat. Commun.* 8: 14261.

Lieber, M.R. (2010). The mechanism of double-strand DNA break repair by the nonhomologous DNA end-joining pathway. *Annu. Rev. Biochem.* 79: 181–211. https://doi.org/10.1146/annurev.biochem.052308.093131.

Lin, Q., Zong, Y., Xue, C. et al. (2020). Prime genome editing in rice and wheat. *Nat. Biotechnol.* 38: 582–585.

Lincoln, T., Eduardo, Z., Max, M.I., and Angus, M. (2015). *Plant Physiology and Development*, 706. USA: Sinauer Associations, Inc.

Liu, Y., Zeng, J., Yuan, C. et al. (2019). Cas9-PF, an early flowering and visual selection marker system, enhances the frequency of editing event occurrence and expedites the isolation of genome-edited and transgene-free plants. *Plant Biotechnol. J.* 17: 1191–1193. https://doi.org/10.1111/pbi.13118.

Liu, H.J., Jian, L., Xu, J. et al. (2020). High-throughput CRISPR/Cas9 mutagenesis streamlines trait gene identification in maize. *Plant Cell* 32: 1397–1413.

Lobato-Gómez, M., Hewitt, S., Capell, T. et al. (2021). Transgenic and genome-edited fruits: background, constraints, benefits, and commercial opportunities. *Hortic Res.* 8 (1): 166. https://doi.org/10.1038/s41438-021-00601-3.

López-Marqués, R.L., Nørrevang, A.F., Ache, P. et al. (2020). Prospects for the accelerated improvement of the resilient crop quinoa. *J. Exp. Bot.* 71: 5333–5347.

Lowder, L.G., Zhang, D., Baltes, N.J. et al. (2015). A CRISPR/Cas9 toolbox for multiplexed plant genome editing and transcriptional regulation. *Plant Physiol.* 169: 971–985.

Lowe, K., Wu, E., Wang, N. et al. (2016). Morphogenic regulators BABY BOOM and WUSCHEL improve monocot transformation. *Plant Cell* 28: 1998–2015.

Lozano-Juste, J. and Cutler, S.R. (2014 May). Plant genome engineering in full bloom. *Trends Plant Sci.* 19 (5): 284–287. https://doi.org/10.1016/j.tplants.2014.02.014.

Lu, Y., Ye, X., Guo, R. et al. (2017). Genome-wide targeted mutagenesis in rice using the CRISPR/Cas9 system. *Mol. Plant* 10: 1242–1245.

Lu, H.P., Liu, S.-M., Xu, S.-L. et al. (2017). CRISPR-S: an active interference element for a rapid and inexpensive selection of genome-edited, transgene free rice plants. *Plant Biotechnol. J.* 15: 1371–1373. https://doi.org/10.1111/pbi.12788.

Luo, S., Li, J., Stoddard, T.J. et al. (2015). Non-transgenic plant genome editing using purified sequence-specific nucleases. *Mol. Plant* 8: 1425–1427. https://doi.org/10.1016/j.molp.2015.05.012.

Luo, C., Cai, X.-T., Du, J. et al. (2016). PARAQUAT TOLERANCE3 is an E3 ligase that switches off activated oxidative response by targeting histone-modifying PROTEIN METHYLTRANSFERASE4b. *PLoS Genet.* 12 (9): e1006332.

Ma, H., Naseri, A., Reyes-Gutierrez, P. et al. (2015a). Multicolor CRISPR labeling of chromosomal loci in human cells. *Proc. Natl. Acad. Sci. U.S.A.* 112: 3002–3007. https://doi.org/10.1073/pnas.1420024112.

Ma, X., Zhang, Q., Zhu, Q. et al. (2015b). A robust CRISPR/Cas9 system for convenient, high-efficiency multiplex genome editing in monocot and dicot plants. *Mol. Plant* 8: 1274–1284.

Ma, H., Li, J., Ma, L. et al. (2021). Pathogen-inducible OsMPKK10.2-OsMPK6 cascade phosphorylates the Raf-like kinase OsEDR1 and inhibits its scaffold function to promote rice disease resistance. *Mol Plant.* 14 (4): 620–632. https://doi.org/10.1016/j.molp.2021.01.008.

Makarova, K.S., Grishin, N.V., Shabalina, S.A. et al. (2006). A putative RNA-interference-based immune system in prokaryotes: computational analysis of the predicted enzymatic machinery, functional analogies with eukaryotic RNAi, and hypothetical mechanisms of action. *Biol. Direct* 1: 7. https://doi.org/10.1186/1745-6150-1-7.

Makarova, K.S., Aravind, L., Wolf, Y.I., and Koonin, E.V. (2011a). Unification of Cas protein families and a simple scenario for the origin and evolution of CRISPR-Cas systems. *Biol. Direct* 6: 38.

Makarova, K.S., Haft, D.H., Barrangou, R. et al. (2011b). Evolution and classification of the CRISPR-Cas systems. *Nat. Rev. Microbiol.* 9: 467–477.

Makarova, K.S., Wolf, Y.I., and Koonin, E.V. (2013). Comparative genomics of defense systems in archaea and bacteria. *Nucleic Acids Res.* 41: 4360–4377.

Mali, P., Yang, L., Esvelt, K.M. et al. (2013a). RNA-guided human genome engineering via Cas9. *Science* 339: 823–826.

Mali, P., Aach, J., Stranges, P.B. et al. (2013b). Cas9 transcriptional activators for target specificity screening and paired nickases for cooperative genome engineering. *Nat. Biotechnol.* 31: 833–838.

Malinovsky, F.G., Fangel, J.U., and Willats, W.G. (2014). The role of the cell wall in plant immunity. *Front. Plant Sci.* 5: 178.

Mansoor, S., Briddon, R.W., Zafar, Y., and Stanley, J. (2003). Geminivirus disease complexes: an emerging threat. *Trends Plant Sci.* 8: 128–134.

McGrann, G.R., Stavrinides, A., Russell, J. et al. (2014). A trade off between mlo resistance to powdery mildew and increased susceptibility of barley to a newly important disease, Ramularia leaf spot. *J. Exp. Bot.* 65: 1025–1037.

Meng, X., Yu, H., Zhang, Y. et al. (2017). Construction of a genome-wide mutant library in rice using CRISPR/Cas9. *Mol. Plant* 10: 1238–1241.

Mikami, M., Toki, S., and Endo, M. (2015a). Parameters affecting frequency of CRISPR/Cas9 mediated targeted mutagenesis in rice. *Plant Cell Rep.* 34: 1807–1815.

Mikami, M., Toki, S., and Endo, M. (2015b). Comparison of CRISPR/Cas9 expression constructs for efficient targeted mutagenesis in rice. *Plant Mol. Biol.* 88: 561–572.

Miller, J.C., Holmes, M.C., Wang, J. et al. (2007). An improved zinc-finger nuclease architecture for highly specific genome editing. *Nat. Biotechnol.* 25: 778–785.

Miller, J.C., Tan, S., Qiao, G. et al. (2011). A TALE nuclease architecture for efficient genome editing. *Nat. Biotechnol.* 29: 14.

Moffat, A.S. (1999). Geminiviruses emerge as serious crop threat. *Science* 286: 1835.

Mojica, F.J., Díez-Villaseñor, C., Soria, E., and Juez, G. (2000). Biological significance of a family of regularly spaced repeats in the genomes of archaea, bacteria and mitochondria. *Mol. Microbiol.* 36: 244–246. https://doi.org/10.1046/j.1365-2958.2000.01838.x.

Molla, K.A. and Yang, Y. (2019). CRISPR/Cas-mediated base editing: technical considerations and practical applications. *Trends Biotechnol.* 37 (10): 1121–1142. https://doi.org/10.1016/j.tibtech.2019.03.008.

Moscou, M.J. and Bogdanove, A.J. (2009). A simple cipher governs DNA recognition by TAL effectors. *Science* 326: 1501.

Nakashima, K., Yamaguchi-Shinozaki, K., and Shinozaki, K. (2014). The transcriptional regulatory network in the drought response and its crosstalk in abiotic stress responses including drought, cold, and heat. *Front. Plant Sci.* 5: 170.

Nishimasu, H., Ran, F.A., Hsu, P.D. et al. (2014). Crystal structure of Cas9 in complex with guide RNA and target DNA. *Cell* 156: 935–949.

Noctor, G., Mhamdi, A., and Foyer, C.H. (2014). The roles of reactive oxygen metabolism in drought: not so cut and dried. *Plant Physiol.* 164 (4): 1636–1648. https://doi.org/10.1104/pp.113.233478.

Noctor, G., Reichheld, J.P., and Foyer, C.H. (2018). ROS-related redox regulation and signaling in plants. *Semin. Cell Dev. Biol.* 80: 3–12. https://doi.org/10.1016/j.semcdb.2017.07.013.

Oh, S.-J. (2005). Arabidopsis CBF3/DREB1A and ABF3 in transgenic rice increased tolerance to abiotic stress without stunting growth. *Plant Physiol.* 138: 341–351.

van der Oost, J., Westra, E.R., Jackson, R.N., and Wiedenheft, B. (2014). Unravelling the structural and mechanistic basis of CRISPR–Cas systems. *Nat. Rev. Microbiol.* 12: 479–492.

Ordon, J., Bressan, M., Kretschmer, C. et al. (2020). Optimized Cas9 expression systems for highly efficient *Arabidopsis* genome editing facilitate isolation of complex alleles in a single generation. *Funct. Integr. Genomics* 20: 151–162.

Ortega, M.A., Lail, L.A., Wood, E.D. et al. (2017). Registration of two soybean germplasm lines containing leaf-chewing insect resistance QTLs from PI 229358 and PI 227687 introgressed into 'Benning'. *J. Plant Regist.* 11: 185–191. https://doi.org/10.3198/jpr2016.04.0019crg.

Osakabe, Y., Watanabe, T., Sugano, S.S. et al. (2016). Optimization of CRISPR/Cas9 genome editing to modify abiotic stress responses in plants. *Sci. Rep.* 6: 26685.

Otuka, A.M., Matsumura, M., and Tokuda, M. (2020). Dispersal of the common cutworm, *Spodoptera litura*, monitored by searchlight trap and relationship with occurrence of soybean leaf damage. *Insects* 11: 427. https://doi.org/10.3390/insects11070427.

Paszkowski, J., Baur, M., Bogucki, A., and Potrykus, I. (1988). Gene targeting in plants. *EMBO J.* 7 (13): 4021–4026.

Pattanayak, V., Lin, S., Guilinger, J.P. et al. (2013). High-throughput profiling of off-target DNA cleavage reveals RNA programmed Cas9 nuclease specificity. *Nat. Biotechnol.* 31: 839–843.

Peng, A., Chen, S., Lei, T. et al. (2017). Engineering canker-resistant plants through CRISPR/Cas9-targeted editing of the susceptibility gene CsLOB1 promoter in citrus. *Plant Biotechnol. J.* 15 (12): 1509–1519. https://doi.org/10.1111/pbi.12733.

Popp, M.W. and Maquat, L.E. (2016). Leveraging rules of nonsense-mediated mRNA decay for genome engineering and personalized medicine. *Cell* 165: 1319–1322.

Pozebon, H., Marques, R.P., Padilha, G. et al. (2020). Arthropod invasions versus soybean production in Brazil: a review. *J. Econ. Entomol.* 113: 1591–1608. https://doi.org/10.1093/jee/toaa108.

Pyott, D.E., Sheehan, E., and Molnar, A. (2016). Engineering of CRISPR/Cas9-mediated potyvirus resistance in transgene-free *Arabidopsis* plants. *Mol. Plant Pathol.* 17 (8): 1276–1288. https://doi.org/10.1111/mpp.12417.

Qi, L.S., Larson, M.H., Gilbert, L.A. et al. (2013). Repurposing CRISPR as an RNA-guided platform for sequence-specific control of gene expression. *Cell* 152: 1173–1183. https://doi.org/10.1016/j.cell.2013.02.022.

Raman, R. (2017). The impact of genetically modified (GM) crops in modern agriculture: a review. *GM Crops. Food* 8: 195–208. https://doi.org/10.1080/21645698.2017.1413522.

Ramegowda, V. and Senthil-Kumar, M. (2015). The interactive effects of simultaneous biotic and abiotic stresses on plants: mechanistic understanding from drought and pathogen combination. *J. Plant Physiol.* 176: 47–54.

Ran, Y., Liang, Z., and Gao, C. (2017). Current and future editing reagent delivery systems for plant genome editing. *Sci. China Life Sci.* 60: 490–505.

Ren, B., Yan, F., Kuang, Y. et al. (2018). Improved base editor for efficiently inducing genetic variations in rice with CRISPR/Cas9-guided hyperactive hAID mutant. *Mol. Plant* 11: 623–626.

Roca Paixão, J.F., Gillet, F.X., Ribeiro, T.P. et al. (2019). Improved drought stress tolerance in *Arabidopsis* by CRISPR/dCas9 fusion with a Histone AcetylTransferase. *Sci. Rep.* 9 (1): 8080. https://doi.org/10.1038/s41598-019-44571-y.

Saikhedkar, N.S., Joshi, R.S., Yadav, A.K. et al. (2019). Phyto-inspired cyclic peptides derived from plant Pin-II type protease inhibitor reactive center loops for crop protection from insect pests. *Biochim. Biophys. Acta. Gen. Subj.* 1863: 1254–1262. https://doi.org/10.1016/j.bbagen.2019.05.003.

Sapranauskas, R., Gasiunas, G., Fremaux, C. et al. (2011). The *Streptococcus thermophilus* CRISPR/Cas system provides immunity in *Escherichia coli*. *Nucleic Acids Res.* 39: 9275–9282. https://doi.org/10.1093/nar/gkr606.

van Schie, C.C. and Takken, F.L. (2014). Susceptibility genes 101: how to be a good host. *Annu. Rev. Phytopathol.* 52: 551–581.

Schiemann, J., Dietz-Pfeilstetter, A., Hartung, F. et al. (2019). Risk assessment and regulation of plants modified by modern biotechniques: current status and future challenges. *Annu. Rev. Plant. Biol.* 70: 699–726. https://doi.org/10.1146/annurev-arplant-050718-100025.

Schneider, A., Chatterjee, S., Bousiges, O. et al. (2013). Acetyltransferases (HATs) as targets for neurological therapeutics. *Neurotherapeutics* 10: 568–588.

Shimatani, Z., Kashojiya, S., Takayama, M. et al. (2017). Targeted base editing in rice and tomato using a CRISPR-Cas9 cytidine deaminase fusion. *Nat. Biotechnol.* 35: 441–443.

Shipman, E.N., Yu, J., Zhou, J. et al. (2021). Can gene editing reduce postharvest waste and loss of fruit, vegetables, and ornamentals? *Hortic Res.* 8 (1): 1. https://doi.org/10.1038/s41438-020-00428-4.

Shockey, J. (2020). Gene editing in plants: assessing the variables through a simplified case study. *Plant Mol. Biol.* 103: 75–89.

Singh, R.P., Singh, P.K., Rutkoski, J. et al. (2016). Disease impact on wheat yield potential and prospects of genetic control. *Annu. Rev. Phytopathol.* 54: 303–322.

Slade, A.J., Fuerstenberg, S.I., Loeffler, D. et al. (2005). A reverse genetic, nontransgenic approach to wheat crop improvement by TILLING. *Nat. Biotechnol.* 23: 75–81.

Smith, J., Grizot, S., Arnould, S. et al. (2006). A combinatorial approach to create artificial homing endonucleases cleaving chosen sequences. *Nucleic Acids Res.* 34: e149.

Song, G., Prieto, H., and Orbovic, V. (2019). Agrobacterium-mediated transformation of tree fruit crops: methods, progress, and challenges. *Front. Plant Sci.* 10: 226.

South, P.F., Cavanagh, A.P., Liu, H.W., and Ort, D.R. (2019). Synthetic glycolate metabolism pathways stimulate crop growth and productivity in the field. *Science* 363: eaat9077.

Sternberg, S.H., Redding, S., Jinek, M. et al. (2014). DNA interrogation by the CRISPR RNA-guided endonuclease Cas9. *Nature* 507: 62–67. https://doi.org/10.1038/nature13011.

Stuttmann, J., Barthel, K., Martin, P. et al. (2021). Highly efficient multiplex editing: oneshot generation of 8× *Nicotiana benthamiana* and 12× *Arabidopsis* mutants. *Plant J.* 106: 8–22.

Summers, L.A. (1980). *The Bipyridinium Herbicides*. London: Academic Press.

Suzuki, N., Rivero, R.M., Shulaev, V. et al. (2014). Abiotic and biotic stress combinations. *New Phytol.* 203: 32–43.

Svitashev, S., Schwartz, C., Lenderts, B. et al. (2016). Genome editing in maize directed by CRISPR-Cas9 ribonucleoprotein complexes. *Nat. Commun.* 7: 13274.

Szczelkun, M.D. et al. (2014). Direct observation of R-loop formation by single RNA-guided Cas9 and Cascade effector complexes. *Proc. Natl. Acad. Sci. U.S.A.* 111: 9798–9803. https://doi.org/10.1073/pnas.1402597111.

Tian, H., Chen, S., Yang, W. et al. (2017). A novel family of transcription factors conserved in angiosperms is required for ABA signalling. *Plant Cell Environ.* 40: 2958–2971.

Tian, S., Jiang, L., Cui, X. et al. (2018). Engineering herbicide-resistant watermelon variety through CRISPR/Cas9-mediated base-editing. *Plant Cell Rep.* 37 (9): 1353–1356. https://doi.org/10.1007/s00299-018-2299-0.

Todaka, D., Shinozaki, K., and Yamaguchi-Shinozaki, K. (2015). Recent advances in the dissection of drought-stress regulatory networks and strategies for development of drought-tolerant transgenic rice plants. *Front. Plant Sci.* 6: 84.

Tsuda, K. and Katagiri, F. (2010). Comparing signaling mechanisms engaged in pattern-triggered and effector-triggered immunity. *Curr. Opin. Plant Biol.* 13: 459–465.

Underwood, W. (2012). The plant cell wall: a dynamic barrier against pathogen invasion. *Front. Plant Sci.* 3: 85.

Unniyampurath, U., Pilankatta, R., and Krishnan, M.N. (2016). RNA interference in the age of CRISPR: will CRISPR interfere with RNAi? *Int. J. Mol. Sci.* 17: 291–305.

Urnov, F.D., Miller, J.C., Lee, Y.L. et al. (2005). Highly efficient endogenous human gene correction using designed zinc-finger nucleases. *Nature* 435: 646–651.

Wang, W., Vinocur, B., and Altman, A. (2003). Plant responses to drought, salinity and extreme temperatures: towards genetic engineering for stress tolerance. *Planta* 28: 1–14.

Wang, Z.-P., Xing, H.-L., Dong, L. et al. (2015). Egg cell-specific promoter-controlled CRISPR/Cas9 efficiently generates homozygous mutants for multiple target genes in *Arabidopsis* in a single generation. *Genome Biol.* 16: 144.

Wang, F.Z., Chen, M.X., Yu, L.J. et al. (2017). OsARM1, an R2R3 MYB transcription factor, is involved in regulation of the response to arsenic stress in rice. *Front. Plant Sci.* 8: 1868. https://doi.org/10.3389/fpls.2017.01868.

Wang, L., Chen, S., Peng, A. et al. (2019). CRISPR/Cas9-mediated editing of CsWRKY22 reduces susceptibility to *Xanthomonas citri* subsp. *citri* in Wanjincheng orange (*Citrus sinensis* (L.) Osbeck). *Plant Biotechnol. Rep.* 13: 501–510.

Wang, S., Zong, Y., Lin, Q. et al. (2020). Precise, predictable multi-nucleotide deletions in rice and wheat using APOBEC-Cas9. *Nat. Biotechnol.* 38: 1460–1465.

Wania, S.H., Kumar, V., Shriram, V., and Sah, S.K. (2016). Phytohormones and their metabolic engineering for abiotic stress tolerance in crop plants. *Crop J.* 4: 162–176.

Wittmann, S., Chatel, H., and Fortin, M.G. (1997). Interaction of the viral protein genome linked of turnip mosaic potyvirus with the translational eukaryotic initiation factor (iso) 4E of *Arabidopsis thaliana* using the yeast two-hybrid system. *Virology* 92 (234): 84–92.

Woo, J.W., Kim, J., Kwon, S.I. et al. (2015). DNA-free genome editing in plants with preassembled CRISPR-Cas9 ribonucleoproteins. *Nat. Biotechnol.* 33: 1162–1164.

Xia, L., Zou, D., Sang, J. et al. (2017). Rice expression database (RED): an integrated RNA-Seq-derived gene expression database for rice. *J. Genet. Genomics.* 44 (5): 235–241. https://doi.org/10.1016/j.jgg.2017.05.003.

Xie, K., Minkenberg, B., and Yang, Y. (2015). Boosting CRISPR/Cas9 multiplex editing capability with the endogenous tRNA-processing system. *Proc. Natl. Acad. Sci. U.S.A.* 112: 3570–3575.

Xu, Z., Xu, X., Gong, Q. et al. (2019). Engineering broad-spectrum bacterial blight resistance by simultaneously disrupting variable TALE-binding elements of multiple susceptibility genes in rice. *Mol. Plant.* 12 (11): 1434–1446. https://doi.org/10.1016/j.molp.2019.08.006.

Xu, R., Li, J., Liu, X. et al. (2020a). Development of plant prime-editing systems for precise genome editing. *Plant Commun.* 1: 100043.

Xu, W., Zhang, C., Yang, Y. et al. (2020b). Versatile nucleotides substitution in plant using an improved prime editing system. *Mol. Plant* 13: 675–678.

Yamamoto, A., Kimura, Y., Sawa, S. et al. (2019). Developing heritable mutations in *Arabidopsis thaliana* using a modified CRISPR/Cas9 toolkit comprising PAM-altered Cas9 variants and gRNAs. *Plant Cell Physiol.* 60: 2255–2262.

Yang, B., Sugio, A., and White, F.F. (2006). Os8N3 is a host disease-susceptibility gene for bacterial blight of rice. *Proc. Natl Acad. Sci. U.S.A.* 103: 10503–10508.

Yin, K., Han, T., Liu, G. et al. (2015). A geminivirus-based guide RNA delivery system for CRISPR/Cas9 mediated plant genome editing. *Sci. Rep.* 5: 14926.

Yoshida, T., Fujita, Y., Sayama, H. et al. (2010). AREB1, AREB2, and ABF3 are master transcription factors that cooperatively regulate ABRE-dependent ABA signaling involved in drought stress tolerance and require ABA for full activation. *Plant J.* 61: 672–685.

Yousuf, P.Y., Hakeem, K.U.R., Chandna, R., and Ahmad, P. (2012). Role of glutathione reductase in plant abiotic stress. In: *Abiotic Stress Responses in Plants* (ed. P. Ahmad and M.N.V. Prasad), 149–158. New York, NY: Springer.

Yu, H. and Zhao, Y. (2019). Fluorescence marker-assisted isolation of Cas9-free and CRISPR-edited *Arabidopsis* plants. *Methods Mol. Biol.* 1917: 147–154.

Yu, L.J., Luo, Y.F., Liao, B. et al. (2012). Comparative transcriptome analysis of transporters, phytohormone and lipid metabolism pathways in response to arsenic stress in rice (*Oryza sativa*). *New Phytol.* 195: 97–112. https://doi.org/10.1111/j.1469-8137.2012.04154.x.

Zhang, Y., Liang, Z., Zong, Y. et al. (2016). Efficient and transgene-free genome editing in wheat through transient expression of CRISPR/Cas9 DNA or RNA. *Nat. Commun.* 7: 12617.

Zhang, Y., Bai, Y., Wu, G. et al. (2017). Simultaneous modification of three homoeologs of TaEDR1 by genome editing enhances powdery mildew resistance in wheat. *Plant J.* 91 (4): 714–724. https://doi.org/10.1111/tpj.13599.

Zhang, Y., Ren, Q., Tang, X. et al. (2021). Expanding the scope of plant genome engineering with Cas12a orthologs and highly multiplexable editing systems. *Nat. Commun.* 12: 1944.

Zhang, Y., Guo, W., Chen, L. et al. (2022). CRISPR/Cas9-mediated targeted mutagenesis of GmUGT enhanced soybean resistance against leaf-chewing insects through flavonoids biosynthesis. *Front. Plant Sci.* 13: 802716. https://doi.org/10.3389/fpls.2022.802716.

Zhu, H., Li, C., and Gao, C. (2020). Applications of CRISPR-Cas in agriculture and plant biotechnology. *Nat. Rev. Mol. Cell Biol.* 21: 661–677.

Zong, Y., Wang, Y., Li, C. et al. (2017). Precise base editing in rice, wheat and maize with a Cas9-cytidine deaminase fusion. *Nat. Biotechnol.* 35: 438–440.

Zong, Y., Song, Q., Li, C. et al. (2018). Efficient C-to-T base editing in plants using a fusion of nCas9 and human APOBEC3A. *Nat. Biotechnol.* 36: 950–953.

24

Ethylene Mediates Plant-Beneficial Fungi Interaction That Leads to Increased Nutrient Uptake, Improved Physiological Attributes, and Enhanced Plant Tolerance Under Salinity Stress

Priya Yadav[1], *Mohammad Wahid Ansari*[1], *Narendra Tuteja*[2], *and Ratnum K. Wattal*[1]

[1] *Department of Botany, Zakir Husain Delhi College, University of Delhi, New Delhi, India*
[2] *Plant Molecular Biology Group, International Centre for Genetic Engineering and Biotechnology, New Delhi, India*

24.1 Introduction

Global climate change majorly affects agricultural yield. Rising population of the world creates constraints to cope with the food needs. There is a need to improve plant productivity in a way that it can cope with the changing climate (Dahal et al. 2019). Various abiotic factors include drought, heat, cold, salinity. Rise in temperature and irregular rainfall result in reduced available water in the soil, which creates drought condition as the water level decreases, soil salt increases, and salinity stress arises (Wang and Frei 2011). Salinity stress is a major abiotic stress that harms the plants globally (Munns et al. 2020). During salinity stress, water uptake by plants is greatly affected as osmotic pressure gets disrupted. According to the present scenario, 20% of cultivable land face salinity stress, and this percentage is rising continuously (Otlewska et al. 2020). High salinity stress leads to over accumulation of ions, which results in oxidative damage and also affects physiological processes like photosynthesis, respiration, stomatal movement (Annunziata et al. 2017). Stomatal movement gets affected due to salinity stress, evapotranspiration gets reduced as stomata undergo closure, and reactive oxygen species (ROS) gets accumulated. Production and accumulation of ROS lead to cell death as they affect biomolecules, e.g. oxidation of protein, DNA (Moller et al. 2007). To overcome the harmful effects of ROS accumulation, plants have some enzymatic and non-enzymatic components. These compounds include superoxide dismutase (SOD), glutathione reductase (GR), glutathione transferase (GST), ascorbate peroxidase (APX), catalase (CAT), ascorbic acid (AsA), glutathione (GSH), carotenoids, and flavonoids (Parida and Das 2005). Ethylene is one of the phytohormones that regulate plant growth and development. Ethylene shows a triple response: inhibition of hypocotyl and root elongation, exaggerated tightening of apical hook, and swelling of hypocotyl (Guzman and Ecker 1990). Along with its role in plant growth and development, ethylene is also involved in abiotic and biotic stress responses. An increased endogenous level or exogenous treatment of ethylene was found beneficial in mitigating salinity stress. It is found that ethylene biosynthesis or signaling inhibited plants become more sensitive toward salinity stress (Gharbi et al. 2017). Plant microbes play an important role in mitigating the harmful impacts of stress on the growth and production of plants (Kaushal and Wani 2016). These microbes produce hormones that trigger endogenous phytohormones in plants. Also, the growth and release of nutrients are improved. These microbes elevate the synthesis of antioxidants and osmolytes, which reduce the osmotic damage (Barnawal et al. 2016). Fungi show superior results in alleviating abiotics' stress conditions. *Piriformospora indica* enhances antioxidant enzymes and plant growth to decrease stress levels in plants (Baltruschat et al. 2008). This chapter provides an insight into the role of fungi and ethylene in improving production under stress conditions. Also, it provides information on mechanism and signaling to understand stress tolerance.

24.2 Plant Response Towards Salinity Stress

Plants show salinity stress response in four different forms: morphological, physiological, biochemical, and molecular responses. Salinity stress influences the growth and development of plants. It affects plants at every developmental stage, right from seed germination up to flowering and reproduction. Salinity stress results in reduced water and nutrient uptake;

Global Climate Change and Plant Stress Management, First Edition. Edited by Mohammad Wahid Ansari, Anil Kumar Singh, and Narendra Tuteja.

Table 24.1 Various plant responses under salinity stress.

Morphological response	Physiological response	Biochemical response	Molecular response
• Reduced biomass • Reduced plant growth • Reduced plant height • Leaf rolling and burning	• Reduced water content • Reduced photosynthesis • Reduced seed germination • Stomatal closure • Leaf gas exchange increased	• Accumulation of toxic ions • Accumulation of ROS • Altered metabolism • Osmolytes biosynthesis	• Expression of stress regulatory genes • Altered protein synthesis

increased Na^+ and Cl^- ions disturb the osmotic and ionic balance, which interfere with the functions of stomata, membrane potential, and some enzymes (Wang et al. 2003). High salinity stress causes cell injury, inhibits cell division, leaf transpiration, and photosynthesis and leads to early senescence (Munns 2006) (Table 24.1).

Plants perceive signals through cellular signaling cascades during stress conditions. Abiotic stress signals are perceived by sensing mechanisms in plants. But the exact mechanism of how plants perceive these signals is still not understood (Lamers et al. 2020). There are abiotic stress sensors that sense the stress. For salinity stress, there are two sensory mechanisms: one is the osmosensor, which senses the osmotic stress caused by salinity, and the other is for sensing ionic stress (Munns and Tester 2008). Salt stress has two phases: early osmotic phase and late ionic phase, both later result in oxidative stress due to overproduction and accumulation of ROS (Yang et al. 2019).

The first phase of salt stress is osmotic stress, which results in increased turgor pressure. Plants cope with this problem by taking in an increased number of inorganic ions (Shabala and Lew 2002). Osmotic stress also leads to the closure of stomata as xylem pressure decreases due to low external water potential (Wegner et al. 2011). The second phase of salt stress is "ionic stress" or "ionic toxicity"; this is related to overaccumulation of Na^+ and Cl^-. Accumulation of excess Na^+ results in sodium toxicity, which affects the activity of many enzymes (Cheeseman 2013). High Cl^- level or chloride toxicity will result in dysfunction of many physiological processes (Cheeseman 2013). Oxidative stress arises due to accumulation of ROS, H_2O_2 (hydrogen peroxide), O_2^- (superoxide), and OH^- (hydroxide). Low concentration of ROS acts as a signal molecule, but at high concentration, it will result in oxidative stress. The increased number of ROS starts reacting with organic molecules. This reaction results in enzyme inhibition, DNA and RNA damage, protein oxidation (Tuteja 2007). ROS activates ROS-sensitive ion channels, which in turn interrupt ionic homeostasis (Pei et al. 2000). ROS also influences PCD (programmed cell death), cell cycle, pathogen defense, growth, and development. Plants have ROS scavengers in the form of enzymatic and non-enzymatic antioxidant defense systems, which protect the plant against the harmful effects of oxidative stress (Figure 24.1).

24.3 Plant–Fungal Interaction and the Mechanism of Plant Growth Promotion by Fungi

Plant–fungal interaction covers a range from pathogenicity to mutual symbiosis. The interaction may be beneficial in the form of mutual symbiosis or harmful in the form of pathogens that cause infection in the plants (Dudeja et al. 2012). There are specific endophytic fungi that interact with plants and promote growth of the plant. These fungi produce growth-promoting metabolites (Zhao et al. 2010). Almost all plants were colonized by fungi, and these plants and fungi coevolved over the years (Rai and Agarkar 2016) (Tables 24.2 and 24.3).

Endophytic fungi show direct and indirect benefits in stress tolerance. Direct benefits include acquisition of nutrients and production of phytohormones. This increase in nutrient uptake and phytohormone production directly increases the plant's biomass and improves its yield (Bamisile et al. 2018).

24.3.1 Nutrient Acquisition and Phytohormones Production

Endophytic fungi are able to improve nutrient uptake, such as potassium, nitrogen, phosphorus, magnesium, zinc, copper, and iron, and their supply to the plant. Also, they produce phytohormones, auxins, gibberellins, and cytokinins (Khan et al. 2015; Rana et al. 2020). Harmful effects of chemical fertilizers can be reduced by using microbes that solubilize soil nutrients and make them available for plant uptake and transport. Many studies

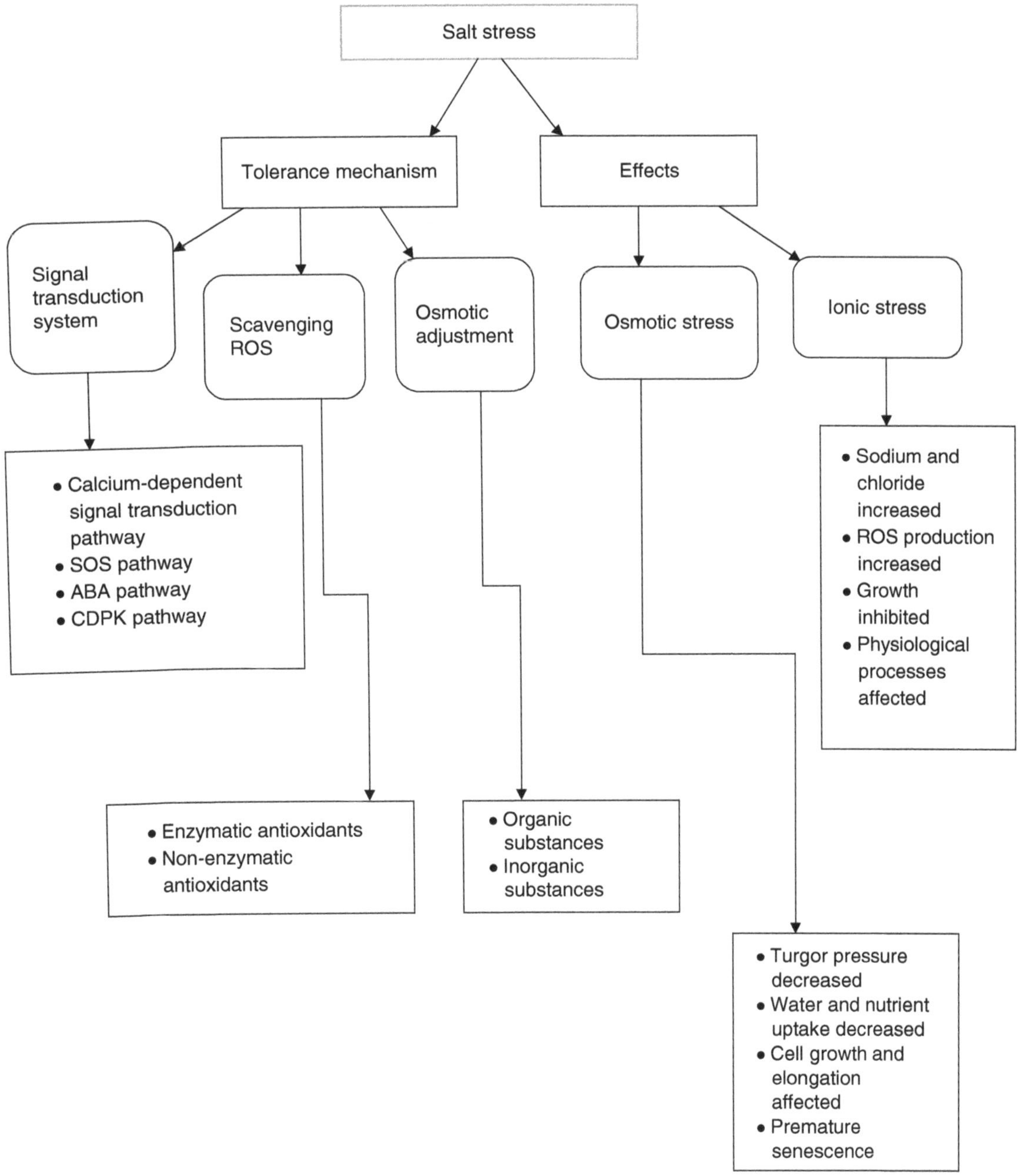

Figure 24.1 Salinity stress affects and tolerance mechanism operating in plants.

Table 24.2 Endophytic fungi-mediated salinity stress tolerance in plants.

Fungi	Plant	Mechanism	References
Piriformospora indica	*Arabidopsis*, tomato, barley	Maintains Na^+ and K^+ homeostasis, increases tolerance genes and transcription factor expression increase, production of osmoprotectant	Arshad et al. (2017), Abdelaziz et al. (2019), and Waller et al. (2005)
Aspergillus flavus CHS-1	*Glycine max*	Stimulates antioxidant system	Asaf et al. (2018)
Trichoderma harzianum	Tomato	Alters expression of 1243 genes in roots	De Palma et al. (2019)
Fusarium sp.	Rice	Stabilizes chlorophyll index and improves photosynthesis	Sampangi-Ramaiah et al. (2020)
Alternaria sp.	Tomato	Increases photosynthesis and biomass and improves water usage	Bogner et al. (2016)

Table 24.3 Beneficial role of endophytic fungi in salinity stress tolerance.

Fungi	
Direct benefits	**Indirect benefits**
• Nutrient acquisition • Phytohormones production	• Activation of systemic resistance • Production of siderophores • Production of antibiotics and secondary metabolites • Protection to biotic and abiotic stress

have been conducted since years that report that endophytic fungi are able to transfer nutrient to plants, e.g. *P. indica, Metarhizium virgatum, Trichoderma asperellum, Aspergillus sydowii. Aspergillus sydowii* enhanced the phosphorus uptake in *Zea mays* (Baron et al. 2018). *Trichoderma asperellum* improves phosphorus uptake in onions, as reported by (Ortega-Garcia et al. 2015). Arbuscular mycorrhizal fungi (AMF) also improve the rate of phosphorus uptake and its transportation (Hiruma et al. 2018). *Aureobasidium pullulans, Barnettozyma californica, Dothideomycetes* sp., and *Torulaspora* sp. make Zn available and promote plant growth (Fu et al. 2016). These fungi produce organic acid, which solubilizes Zn and makes it available to plant (Fomina et al. 2004; Sutjaritvorakul et al. 2017).

IAA (auxin) and GA (gibberellin) produced by *Phoma glomerata* and *Paecilomyces formosus* stain LHL10 reported by Khan et al. (2012) and Waqas et al. (2012). Production of GA from *Penicillium citrinum* IR-3-3 was reported by (Khan et al. 2008). *Fusarium oxysporum is* found in mutual relationship with the roots of maize and produces secondary metabolites and IAA, which promote plant growth and development (Mehmood et al. 2018). *Piriformospora indica* modulated through ethylene promoted growth in *Arabidopsis*. Ethylene helps *P. Indica* colonize with plants as reported by Ansari et al. (2013). In recent studies, it was found that fungi not only acquire nutrients or produce phytohormones but also promote growth and increase plant biomass and chlorophyll content (Bader et al. 2020).

Indirect benefits include the production of siderophores, the production of antibiotics and secondary metabolites, the activation of systemic resistance, and the protection from biotic and abiotic stress.

24.3.2 Activation of Systemic Resistance

Endophytic fungi activate systemic resistance (SR) in plants against pathogens. Plants that show mutual relation with endophytic fungi have improved self-defense systems and higher growth (Berg 2009; Busby et al. 2016). Induce systemic response, which leads to production of signaling molecules (ethylene, salicylic acid), flux of ions, production of ROS, and expression of defense-related genes (Chadha et al. 2015). Pathogen attack is recognized by pathogen recognition receptor (PRR) and (pathogen associated molecular pattern (PAMPs) triggered immunity defense response. This defense response triggers a signaling cascade that involves activation of stress hormones salicylic acid (SA), jasmonic acid, and ethylene in plants (Shigenaga et al. 2017).

24.3.3 Production of Siderophores

Siderophores are produced by some microorganisms including endophytic fungi, and these siderophores chelate iron (Chowdappa et al. 2020). *Trichoderma koningii* ST-KKU$_1$, *Macrophomina phaseolina* SS$_1$L$_{10}$ and *M. phaseolina* SS$_1$R$_{10}$ produce siderophores, as reported by Suebrasri et al. (2020). Siderophores' ability to bind with ferric ions solves the problem of iron uptake by plants. Siderophore directly or indirectly improves plant growth and also helps in plant defense against pathogens (Yadav et al. 2017). Siderophores not only promote plant productivity but are also used as biocontrol, as reported by Devi et al. (2020).

24.3.4 Production of Antibiotics and Secondary Metabolites

Endophytic fungi produce not only phytohormones but also antibiotics and secondary metabolites. These antibiotics and secondary metabolites act against pathogens and help plants improve self-defense (Card et al. 2016; Lugtenberg et al. 2016). Secondary metabolites include alkaloids, steroids, terpenoids, flavonoids, quinols, phenols, chlorinated compounds, and volatile organic compounds. Antibiotics include antiviral and antimicrobial are being used to treat diverse type of infections (Card et al. 2016; Latz et al. 2018). There are reports that show *Trichoderma* sp., *Penicillium* sp., *Cladosporium* sp., *Aspergillus* sp., etc., produce antibiotics and show antimicrobial activity. *Trichoderma viridae* show antimicrobial activity, as reported by Awad et al. (2018). *Penicillium daleae* Ef4 was isolated from *Enteromorpha flexuosa Linn. It* produces antibiotic components and shows antimicrobial activity, as reported by Iswarya and Ramesh (2019). They identified that this fungus was beneficial against *Staphylococcus aureus* and *Pseudomonas aeruginosa*. Endophytic fungi are notable sources that produce antibiotic and secondary metabolites and help plants survive under stress conditions.

24.3.5 Protection to Biotic and Abiotic Stress

Endophytic fungi play an important role in fighting biotic and abiotic stress conditions. Endophytic fungi help plants adapt according to stress conditions through different mechanisms (Aly et al. 2011; Khan et al. 2015). Abiotic stress affects plants in many ways; one main problem is leakage of electrolytes through the membrane. Endophytic fungi prevent this electrolyte leakage by changing the lipid composition of membranes. Endophytic fungi help plants increase production of osmolytes, antioxidant enzymes that reduce salinity stress level in plants (khan et al. 2015; Yan et al. 2019). During biotic stress, endophytic fungi activate the ISR and acquired systemic resistance (ASR) mechanisms in plants; both of these mechanisms improve plant growth and protect plants from pathogens. Also, endophytic fungi produce some metabolites that act against many pests and pathogens in plants (Chadha et al. 2015; chitnis et al. 2020).

24.4 Fungi and Ethylene Production and Its Effects

Ethylene has an effect on root development and root-fungal symbiosis (Kaska et al. 1999). But the exact mechanism is not known (Rupp et al. 1989). Camehl et al. (2010) reported that ethylene plays an important role during symbiosis between *Arabidopsis* and *P. indica*. They also reported that impaired ethylene signaling results in more colonization of *P. indica* with *Arabidopsis* roots. Ethylene is also produced by fungi like other phytohormones. One of the pathways of ethylene biosynthesis has been well described in the *Botrytis cinerea* KMBA (α-keto y-methylthio butyric acid) pathway (Chague et al. 2002). Another pathway of ethylene biosynthesis involves specific enzyme activity and is described in *Penicillium digitatum* or *F. oxysporum*. The enzyme involved in the ethylene production was EFE (Ethylene Forming Enzyme) that utilizes α-ketoglutarate (AKG) and arginine (L-Arg) as substrates in ethylene formation (Chou and Yang 1973; Hottiger and Boller 1991). Both these processes of ethylene biosynthesis are different from conventional methods, which involve three steps: conversion of methionine into *S*-adenosyl methionine (SAM). Enzyme involved in this step is Ado-met synthase. Further, SAM converted into 1-aminocyclopropane 1-carboxylic acid (ACC) with the help of ACC synthase enzyme. The last step involves conversion of ACC into ethylene with the help of ACC oxidase enzyme. This process is regulated by an ACS enzyme (Thomann et al. 2009; Yoon 2015). *Rhizophagus irregularis* germinating spores produce ethylene through the KMBA (α-keto y-methylthio butyric acid) pathway reported by Pons et al. (2020). Ethylene in general proves as a negative regulator for mutualistic root symbiosis. But for *P. indica*, it proves as a positive regulator. Ethylene supports *P. indica* root colonization in barley and *Arabidopsis* (Khatabi et al. 2012). Ethylene is produced by arbuscular mycorrhizal fungi, but the exact effects of ethylene are difficult to find. But it was found that ethylene acts as a negative regulator and prevents excess colonization of AM fungi with plants (Torres de Los Santos et al. 2011). Also, it was found that ethylene concentration plays an important role in AM fungi colonization inhibition (Khatabi et al. 2012). Ethylene production increases under salt stress, and the growth of the plant gets reduced. As the activity of ACS (ACC synthase) and ACO (ACC oxidase) increase, the concentration of ethylene increases during salinity stress. Endophytic fungi *Phomopsis liquidambaris* produce ACCD (1-Aminocyclopropane-1-carboxylate Deaminase) and concentration of ACS and ACO ethylene biosynthesis enzymes reduced. ACCD works very effectively by converting ACC into ammonia and α-keto butyrate. Hence, the concentration of ethylene reduced and the salt tolerance increased. *Phomopsis liquidambaris* promotes plant growth, increases chlorophyll *a* content, increases biomass during salinity stress using ACCD activity reported by Siddikee et al. (2021) (Figure 24.2).

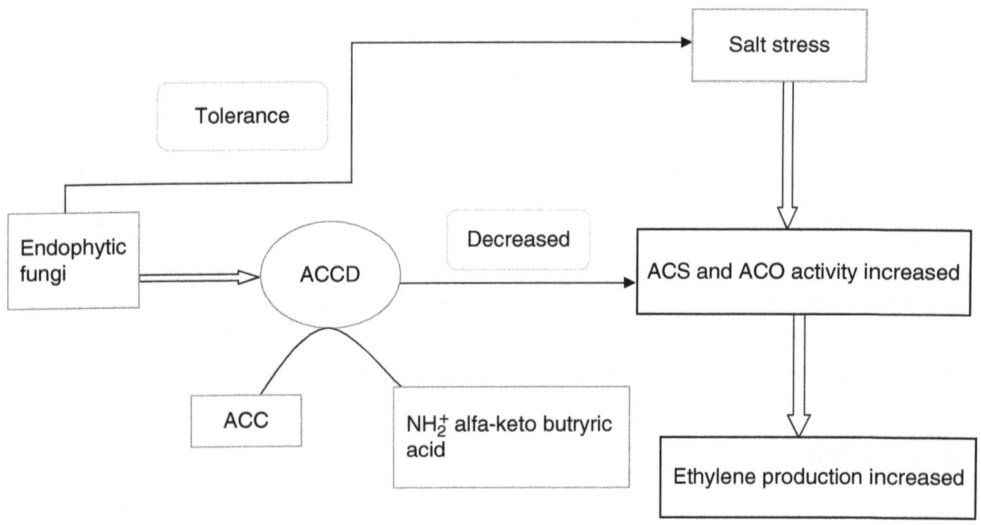

Figure 24.2 Regulation of salinity stress by endophytic fungi.

24.5 Role and Mechanism of Ethylene in Salinity Stress Tolerance

Ethylene acts as a positive regulator as well as a negative regulator under salinity stress. It was discovered that practically all ethylene biosynthesis genes' expression changed and responded to salinity stress. These genes involved ethylene receptor genes (ETR1, ETR2, and EIN4), ethylene biosynthesis genes (ACS1, ACS12, ACO1 and ACO3), ethylene signaling pathway genes (ERF1, ERF2, EIN3, and MKK1-MKK2-MPK4/6 kinases), and feedback mechanism gene (CTR1) (Peng et al. 2014b). It was reported that in cotton plants, ethylene signaling genes get upregulated during salinity stress (Peng et al. 2014a). Inhibition of ethylene action in *Arabidopsis* was found beneficial to cope up salinity stress (Peng et al. 2014b). In *Arabidopsis* during salt stress, ETO1 (ethylene overproduction 1) overexpresses, which increases the formation of ROS and also helps in Na^+/K^+ homeostasis, resulting in the maintenance of plant growth and development (Li et al. 2014). But in the case of wheat, overproduction of ethylene during salinity stress due to ACO1 does not help plants remain sensitive toward salinity stress (Chen et al. 2014). *Solanum chilense* shows increase in salinity stress sensitivity if ethylene biosynthesis and signaling get inhibited probably because ethylene helps *S. chilense* to cope with salinity stress by maintaining osmotic adjustment, stomatal conductance, etc. (Gharbi et al. 2017), whereas in the case of rice, it acts as a negative regulator because it was found that exogenous application of ethylene increases the salinity stress in rice and reduces the growth and development (Muller and Munné-Bosch 2015; Yang et al. 2015). Exogenous application of ethylene was found to be beneficial to induce seed germination. Seed germination is hampered due to excess of H_2O_2 during salinity stress. Ethylene regulates H_2O_2 level and ROS homeostasis, which help in seed germination and plant growth (Lin et al. 2013). Ethylene signaling genes respond differently under salinity stress during seed germination. It was found that ETR2 regulates seed germination in a positive way, while ETR1 and ETR4 inhibit seed germination (Wilson et al. 2014). Ethylene inhibits seed germination in many plants, e.g. *Cucumin sativus*. Ethylene promotes soil-salinity tolerance via improved Na/K homeostasis mediated by RBOHF-dependent regulation of Na accumulation and RBOHF-independent regulation of K accumulation (Jiang et al. 2013). Pretreated *Capsicum annuum* seeds with SA show seed germination because ethylene gets suppressed, and low level of ethylene promotes seed germination (Ahmed et al. 2020). As salinity stress causes ionic and osmotic stress, which also increases the ROS production, disturbs ion homeostasis that negatively affects photosynthesis, hampers gas exchange, and results in stomatal closure (Cheeseman 2013). Due to ionic imbalance, sodium concentration increases, and it will affect the proton motive force that results in impaired photosynthesis (Van Zelm et al. 2020). Ethylene regulates sodium potassium concentration and ROS production, hence helping in salinity stress tolerance, maintaining photosynthesis and plant growth (Tao et al. 2015). Ethylene produced through the KMBA pathway by *Enterobacter* sp. SA187 elevates salinity stress tolerance in *Arabidopsis* and *alfalfa* reported by Splivallo et al. (2009) and Nouri et al. (2015). All these studies show that ethylene plays an important role in maintaining ion homeostasis, ROS production, photosynthesis, and plant growth during salinity stress.

24.6 Conclusion

This chapter summarizes some of the research studies that highlight the role of fungi and ethylene in salinity stress tolerance in plants. Besides technology, the use of fungi proves beneficial because they provide direct or indirect benefits to the plants during stress conditions. Phytohormones and metabolites produced by endophytic fungi play a critical role in plant growth and development during stress conditions. Future studies need to focus on the molecular signaling mechanisms of endophytic fungi and host cells during salinity stress tolerance. Cross-talk of ethylene with other phytohormones has been found beneficial during stress conditions, but the exact and detailed mechanism needs to be explored.

References

Abdelaziz, M.E., Abdelsattar, M., Abdeldaym, E.A. et al. (2019). *Piriformospora indica* alters Na^+/K^+ homeostasis, antioxidant enzymes and LeNHX1 expression of greenhouse tomato grown under salt stress. *Sci. Hortic.* 256: 108532.

Ahmed, W., Imran, M., Yaseen, M. et al. (2020). Role of salicylic acid in regulating ethylene and physiological characteristics for alleviating salinity stress on germination, growth and yield of sweet pepper. *PeerJ.* 8: e8475.

Aly, A.H., Debbab, A., and Proksch, P. (2011). Fungal endophytes: unique plant inhabitants with great promises. *Appl. Microbiol. Biotechnol.* 90 (6): 1829–1845.

Annunziata, M.G., Ciarmiello, L.F., Woodrow, P. et al. (2017). Durum wheat roots adapt to salinity remodeling the cellular content of nitrogen metabolites and sucrose. *Front. Plant Sci.* 7: 2035.

Ansari, M.W., Trivedi, D.K., Sahoo, R.K. et al. (2013). A critical review on fungi mediated plant responses with special emphasis to *Piriformospora indica* on improved production and protection of crops. *Plant Physiol. Biochem.* 70: 403–410.

Arshad, M., Gruber, M.Y., Wall, K., and Hannoufa, A. (2017). An insight into microRNA156 role in salinity stress responses of alfalfa. *Front. Plant Sci.* 8: 356.

Asaf, S., Hamayun, M., Khan, A.L. et al. (2018). Salt tolerance of *Glycine max* L. induced by endophytic fungus *Aspergillus flavus* CSH1, via regulating its endogenous hormones and antioxidative system. *Plant Physiol. Biochem.* 128: 13–23.

Awad, N.E., Kassem, H.A., Hamed, M.A. et al. (2018). Isolation and characterization of the bioactive metabolites from the soil derived fungus *Trichoderma viride*. *Mycology* 9 (1): 70–80.

Bader, A.N., Salerno, G.L., Covacevich, F., and Consolo, V.F. (2020). Native *Trichoderma harzianum* strains from Argentina produce indole-3 acetic acid and phosphorus solubilization, promote growth and control wilt disease on tomato (*Solanum lycopersicum* L.). *J. King Saud Univ. Sci.* 32 (1): 867–873.

Baltruschat, H., Fodor, J., Harrach, B.D. et al. (2008). Salt tolerance of barley induced by the root endophyte *Piriformospora indica* is associated with a strong increase in antioxidants. *New Phytol.* 180 (2): 501–510.

Bamisile, B.S., Dash, C.K., Akutse, K.S. et al. (2018). Fungal endophytes: beyond herbivore management. *Front. Microbiol.* 9: 544.

Barnawal, D., Bharti, N., Tripathi, A. et al. (2016). ACC-deaminase-producing endophyte *Brachybacterium paraconglomeratum* strain SMR20 ameliorates Chlorophytum salinity stress via altering phytohormone generation. *J. Plant Growth Regul.* 35 (2): 553–564.

Baron, N.C., Costa, N.T.A., Mochi, D.A., and Rigobelo, E.C. (2018). First report of *Aspergillus sydowii* and *Aspergillus brasiliensis* as phosphorus solubilizers in maize. *Ann. Microbiol.* 68 (12): 863–870.

Berg, G. (2009). Plant–microbe interactions promoting plant growth and health: perspectives for controlled use of microorganisms in agriculture. *Appl. Microbiol. Biotechnol.* 84 (1): 11–18.

Bogner, C.W., Kariuki, G.M., Elashry, A. et al. (2016). Fungal root endophytes of tomato from Kenya and their nematode biocontrol potential. *Mycol. Prog.* 15: 1–17.

Busby, P.E., Ridout, M., and Newcombe, G. (2016). Fungal endophytes: modifiers of plant disease. *Plant Mol. Biol.* 90 (6): 645–655.

Camehl, I., Sherameti, I., Venus, Y. et al. (2010). Ethylene signalling and ethylene-targeted transcription factors are required to balance beneficial and nonbeneficial traits in the symbiosis between the endophytic fungus *Piriformospora indica* and *Arabidopsis thaliana*. *New Phytol.* 185 (4): 1062–1073.

Card, S., Johnson, L., Teasdale, S., and Caradus, J. (2016). Deciphering endophyte behaviour: the link between endophyte biology and efficacious biological control agents. *FEMS Microbiol. Ecol.* 92 (8): fiw114.

Chadha, N., Mishra, M., Rajpal, K. et al. (2015). An ecological role of fungal endophytes to ameliorate plants under biotic stress. *Arch. Microbiol.* 197 (7): 869–881.

Chagué, V., Elad, Y., Barakat, R. et al. (2002). Ethylene biosynthesis in *Botrytis cinerea*. *FEMS Microbiol. Ecol.* 40 (2): 143–149.

Cheeseman, J.M. (2013). The integration of activity in saline environments: problems and perspectives. *Funct. Plant Biol.* 40 (9): 759–774.

Chen, D., Ma, X., Li, C. et al. (2014). A wheat aminocyclopropane-1-carboxylate oxidase gene, TaACO1, negatively regulates salinity stress in *Arabidopsis thaliana*. *Plant Cell Rep.* 33 (11): 1815–1827.

Chitnis, V.R., Suryanarayanan, T.S., Nataraja, K.N. et al. (2020). Fungal endophyte-mediated crop improvement: the way ahead. *Front. Plant Sci.* 11: 561007.

Chou, T.W. and Yang, S.F. (1973). The biogenesis of ethylene in *Penicillium digitatum*. *Arch. Biochem. Biophys.* 157 (1): 73–82.

Chowdappa, S., Jagannath, S., Konappa, N. et al. (2020). Detection and characterization of antibacterial siderophores secreted by endophytic fungi from *Cymbidium aloifolium*. *Biomolecules* 10 (10): 1412.

Dahal, K., Li, X.Q., Tai, H. et al. (2019). Improving potato stress tolerance and tuber yield under a climate change scenario–a current overview. *Front. Plant Sci.* 10: 563.

De Palma, M., Salzano, M., Villano, C. et al. (2019). Transcriptome reprogramming, epigenetic modifications and alternative splicing orchestrate the tomato root response to the beneficial fungus *Trichoderma harzianum*. *Hortic. Res.* 6: 5.

Dudeja, S.S., Giri, R., Saini, R., Suneja-Madan, P. and Kothe, E., 2012. Interaction of endophytic microbes with legumes. *J. Basic Microbiol.* 52(3), pp.248–260.

Devi, R., Kaur, T., Kour, D. et al. (2020). Beneficial fungal communities from different habitats and their roles in plant growth promotion and soil health. *Microb. Biosyst.* 5 (1): 21–47.

Fomina, M., Alexander, I.J., Hillier, S., and Gadd, G.M. (2004). Zinc phosphate and pyromorphite solubilization by soil plant-symbiotic fungi. *Geomicrobiol. J.* 21 (5): 351–366.

Fu, S.F., Sun, P.F., Lu, H.Y. et al. (2016). Plant growth-promoting traits of yeasts isolated from the phyllosphere and rhizosphere of *Drosera spatulata* Lab. *Fungal Biol.* 120 (3): 433–448.

Gharbi, E., Martínez, J.P., Benahmed, H. et al. (2017). Inhibition of ethylene synthesis reduces salt-tolerance in tomato wild relative species *Solanum chilense*. *J. Plant Physiol.* 210: 24–37.

Guzman, P. and Ecker, J.R. (1990). Exploiting the triple response of Arabidopsis to identify ethylene-related mutants. *Plant Cell* 2 (6): 513–523.

Hiruma, K., Kobae, Y., and Toju, H. (2018). Beneficial associations between Brassicaceae plants and fungal endophytes under nutrient-limiting conditions: evolutionary origins and host–symbiont molecular mechanisms. *Curr. Opin. Plant Biol.* 44: 145–154.

Hottiger, T. and Boller, T. (1991). Ethylene biosynthesis in *Fusarium oxysporum* f. sp. tulipae proceeds from glutamate/2-oxoglutarate and requires oxygen and ferrous ions *in vivo*. *Arch. Microbiol.* 157 (1): 18–22.

Iswarya, B. and Ramesh, K.V. (2019). Promising antimicrobial compounds from an endophytic fungus *Penicillium daleae* EF 4 isolated from the seaweed *Enteromorpha flexuosa* Linn. *Res. Rev.* 9: 26–39.

Jiang, C., Belfield, E.J., Cao, Y. et al. (2013). An Arabidopsis soil-salinity–tolerance mutation confers ethylene-mediated enhancement of sodium/potassium homeostasis. *Plant Cell* 25 (9): 3535–3552.

Kaska, D.D., Myllylä, R., and Cooper, J.B. (1999). Auxin transport inhibitors act through ethylene to regulate dichotomous branching of lateral root meristems in pine. *New Phytol.* 142 (1): 49–57.

Kaushal, M. and Wani, S.P. (2016). Rhizobacterial-plant interactions: strategies ensuring plant growth promotion under drought and salinity stress. *Agric. Ecosyst. Environ.* 231: 68–78.

Khan, S.A., Hamayun, M., Yoon, H. et al. (2008). Plant growth promotion and *Penicillium citrinum*. *BMC Microbiol.* 8 (1): 1–10.

Khan, A.L., Hamayun, M., Kang, S.M. et al. (2012). Endophytic fungal association via gibberellins and indole acetic acid can improve plant growth under abiotic stress: an example of *Paecilomyces formosus* LHL10. *BMC Microbiol.* 12 (1): 1–14.

Khan, A.L., Hussain, J., Al-Harrasi, A. et al. (2015). Endophytic fungi: resource for gibberellins and crop abiotic stress resistance. *Crit. Rev. BioTechnol.* 35 (1): 62–74.

Khatabi, B., Molitor, A., Lindermayr, C. et al. (2012). Ethylene supports colonization of plant roots by the mutualistic fungus *Piriformospora indica*. *PLoS One* 7 (4): e35502.

Lamers, J., Van Der Meer, T., and Testerink, C. (2020). How plants sense and respond to stressful environments. *Plant Physiol.* 182 (4): 1624–1635.

Latz, M.A., Jensen, B., Collinge, D.B., and Jørgensen, H.J. (2018). Endophytic fungi as biocontrol agents: elucidating mechanisms in disease suppression. *Plant Ecol. Divers.* 11 (5–6): 555–567.

Li, J., Jia, H., and Wang, J. (2014). cGMP and ethylene are involved in maintaining ion homeostasis under salt stress in Arabidopsis roots. *Plant Cell Rep.* 33 (3): 447–459.

Lin, Y., Yang, L., Paul, M. et al. (2013). Ethylene promotes germination of Arabidopsis seed under salinity by decreasing reactive oxygen species: evidence for the involvement of nitric oxide simulated by sodium nitroprusside. *Plant Physiol. Biochem.* 73: 211–218.

Lugtenberg, B.J., Caradus, J.R., and Johnson, L.J. (2016). Fungal endophytes for sustainable crop production. *FEMS Microbiol. Ecol.* 92 (12): fiw194.

Mehmood, A., Khan, N., Irshad, M. et al. (2018). IAA producing endopytic fungus *Fusariun oxysporum* wlw colonize maize roots and promoted maize growth under hydroponic condition. *Eur. J. Exp. Biol* 8: 24.

Møller, I.M., Jensen, P.E., and Hansson, A. (2007). Oxidative modifications to cellular components in plants. *Annu. Rev. Plant Biol.* 58: 459–481.

Müller, M. and Munné-Bosch, S. (2015). Ethylene response factors: a key regulatory hub in hormone and stress signaling. *Plant Physiol.* 169 (1): 32–41.

Munns, R.S. (2006). Growth and phytohormones. In: *Salinity: Environment—Plants—Molecules*, 271–290. Dordrecht: Springer.

Munns, R. and Tester, M. (2008). Mechanisms of salinity tolerance. *Annu. Rev. Plant Biol.* 59: 651.

Munns, R., Passioura, J.B., Colmer, T.D., and Byrt, C.S. (2020). Osmotic adjustment and energy limitations to plant growth in saline soil. *New Phytol.* 225 (3): 1091–1096.

Nouri, M.Z., Moumeni, A., and Komatsu, S. (2015). Abiotic stresses: insight into gene regulation and protein expression in photosynthetic pathways of plants. *Int. J. Mol. Sci.* 16 (9): 20392–20416.

Ortega-Garcia, J.G., Montes-BelOnt, R., Rodriguez-Monroy, M. et al. (2015). Effect of *Trichoderma asperellum* applications and mineral fertilization on growth promotion and the content of phenolic compounds and flavonoids in onions. *Sci. Hortic.* 195: 8–16.

Otlewska, A., Migliore, M., Dybka-Stępień, K. et al. (2020). When salt meddles between plant, soil, and microorganisms. *Front. Plant Sci.* 11: 1429.

Parida, A.K. and Das, A.B. (2005). Salt tolerance and salinity effects on plants: a review. *Ecotoxicol. Environ. Saf.* 60 (3): 324–349.

Pei, Z.M., Murata, Y., Benning, G. et al. (2000). Calcium channels activated by hydrogen peroxide mediate abscisic acid signalling in guard cells. *Nature* 406 (6797): 731–734.

Peng, Z., He, S., Gong, W. et al. (2014a). Comprehensive analysis of differentially expressed genes and transcriptional regulation induced by salt stress in two contrasting cotton genotypes. *BMC Genomics* 15 (1): 1–28.

Peng, J., Li, Z., Wen, X. et al. (2014b). Salt-induced stabilization of EIN3/EIL1 confers salinity tolerance by deterring ROS accumulation in Arabidopsis. *PLoS Genet.* 10 (10): e1004664.

Pons, S., Fournier, S., Chervin, C. et al. (2020). Phytohormone production by the arbuscular mycorrhizal fungus *Rhizophagus irregularis*. *PLoS One* 15 (10): e0240886.

Rai, M. and Agarkar, G. (2016). Plant-fungal interactions: what triggers the fungi to switch among lifestyles? *Crit. Rev. Microbiol.* 42 (3): 428438.

Rana, K.L., Kour, D., Kaur, T. et al. (2020). Endophytic microbes: biodiversity, plant growth-promoting mechanisms and potential applications for agricultural sustainability. *Antonie Van Leeuwenhoek* 113 (8): 1075–1107.

Rupp, L.A., Mudge, K.W., and Negm, F.B. (1989). Involvement of ethylene in ectomycorrhiza formation and dichotomous branching of roots of mugo pine seedlings. *Can. J. Bot.* 67 (2): 477–482.

Sampangi-Ramaiah, M.H., Dey, P., Jambagi, S. et al. (2020). An endophyte from salt-adapted Pokkali rice confers salt-tolerance to a salt-sensitive rice variety and targets a unique pattern of genes in its new host. *Sci. Rep.* 10 (1): 1–14.

Shabala, S.N. and Lew, R.R. (2002). Turgor regulation in osmotically stressed Arabidopsis epidermal root cells. Direct support for the role of inorganic ion uptake as revealed by concurrent flux and cell turgor measurements. *Plant Physiol.* 129 (1): 290–299.

Shigenaga, A.M., Berens, M.L., Tsuda, K., and Argueso, C.T. (2017). Towards engineering of hormonal crosstalk in plant immunity. *Curr. Opin. Plant Biol.* 38: 164–172.

Siddikee, M.A., Zereen, M.I., Wu, M. et al. (2021). *Phomopsis liquidambaris* reduces ethylene biosynthesis in rice under salt stress via inhibiting the activity of 1-aminocyclopropane-1-carboxylate deaminase. *Arch. Microbiol.* 203 (10): 6215–6229.

Splivallo, R., Fischer, U., Göbel, C. et al. (2009). Truffles regulate plant root morphogenesis via the production of auxin and ethylene. *Plant Physiol.* 150 (4): 2018–2029.

Sutjaritvorakul, T., Chutipaijit, S., and Sihanonth, P. (2017). Solubilization and bioprecipitation of zinc oxide nanoparticles by fungi isolated from zinc sulfide mineral ores. *Mater. Today Proc.* 4 (5): 6562–6566.

Suebrasri, T., Harada, H., Jogloy, S. et al. (2020). Auxin-producing fungal endophytes promote growth of sunchoke. *Rhizosphere* 16: 100271.

Tao, J.J., Chen, H.W., Ma, B. et al. (2015). The role of ethylene in plants under salinity stress. *Front. Plant Sci.* 6: 1059.

Thomann, A., Lechner, E., Hansen, M. et al. (2009). Arabidopsis CULLIN3 genes regulate primary root growth and patterning by ethylene-dependent and-independent mechanisms. *PLoS Genet.* 5 (1): e1000328.

Torres de Los Santos, R., Vierheilig, H., Ocampo, J.A., and GarcíaGarrido, J.M. (2011). Altered pattern of arbuscular mycorrhizal formation in tomato ethylene mutants. *Plant Signal. Behav.* 6 (5): 755–758.

Tuteja, N. (2007). Mechanisms of high salinity tolerance in plants. *Methods Enzymol.* 428: 419–438.

Van Zelm, E., Zhang, Y., and Testerink, C. (2020). Salt tolerance mechanisms of plants. *Annu. Rev. Plant Biol.* 71: 403–433.

Waller, F., Achatz, B., Baltruschat, H. et al. (2005). The endophytic fungus *Piriformospora indica* reprograms barley to salt-stress tolerance, disease resistance, and higher yield. *Proc. Natl. Acad. Sci. U. S. A.* 102 (38): 13386–13391.

Wang, Y. and Frei, M. (2011). Stressed food–the impact of abiotic environmental stresses on crop quality. *Agric. Ecosyst. Environ.* 141 (3-4): 271–286.

Wang, W., Vinocur, B., and Altman, A. (2003). Plant responses to drought, salinity and extreme temperatures: towards genetic engineering for stress tolerance. *Planta* 218 (1): 1–14.

Waqas, M., Khan, A.L., Kamran, M. et al. (2012). Endophytic fungi produce gibberellins and indoleacetic acid and promotes host-plant growth during stress. *Molecules* 17 (9): 10754–10773.

Wegner, L.H., Stefano, G., Shabala, L. et al. (2011). Sequential depolarization of root cortical and stelar cells induced by an acute salt shock–implications for Na^+ and K^+ transport into xylem vessels. *Plant Cell Environ.* 34 (5): 859–869.

Wilson, R.L., Kim, H., Bakshi, A., and Binder, B.M. (2014). The ethylene receptors ETHYLENE RESPONSE1 and ETHYLENE RESPONSE2 have contrasting roles in seed germination of Arabidopsis during salt stress. *Plant Physiol.* 165 (3): 1353–1366.

Yadav, A.N., Kumar, R., Kumar, S. et al. (2017). Beneficial microbiomes: biodiversity and potential biotechnological applications for sustainable agriculture and human health. *J. Appl. Biol. Biotechnol.* 5 (6): 4–7.

Yan, L., Zhu, J., Zhao, X. et al. (2019). Beneficial effects of endophytic fungi colonization on plants. *Appl. Microbiol. Biotechnol.* 103 (8): 3327–3340.

Yang, C., Ma, B., He, S.J. et al. (2015). MAOHUZI6/ETHYLENE INSENSITIVE3-LIKE1 and ETHYLENE INSENSITIVE3-LIKE2 regulate ethylene response of roots and coleoptiles and negatively affect salt tolerance in rice. *Plant Physiol.* 169 (1): 148–165.

Yoon, G.M. (2015). New insights into the protein turnover regulation in ethylene biosynthesis. *Mol. Cells* 38 (7): 597.

Zhao, J., Zhou, L., Wang, J. et al. (2010). Endophytic fungi for producing bioactive compounds originally from their host plants. *Curr. Res. Technol. Educ. Trop. Appl. Microbiol. Microb. Biotechnol.* 1: 567–576.

25

Role of Chemical Additives in Plant Salinity Stress Mitigation

Priya Yadav[1], *Mohammad Wahid Ansari*[1], *and Narendra Tuteja*[2]

[1] *Department of Botany, Zakir Husain Delhi College (University of Delhi), New Delhi, India*
[2] *Plant Molecular Biology Group, International Centre for Genetic Engineering and Biotechnology, New Delhi, India*

25.1 Introduction

Agricultural production is greatly affected by environmental stress conditions. One of the main environmental stresses that cause major reductions in cultivable land area and cause agricultural losses by reducing plant growth is salinity stress (Parida and Das 2005). It was found that soil salinity increased by 100 mha from 1986 to 2016 globally, and it is increasing by 2 mha year^{-1}, approximately (Ivushkin et al. 2019). Soil is considered saline when the electrical conductance (EC) exceeds 4 dsm^{-1} at 25 °C in the root zone. Main and the most abundant ion during salinity stress is sodium (Na^+); other ions include calcium (Ca^{2+}), magnesium (Mg^{2+}), chloride(Cl^-), and sulfates ($SO_4{}^{2-}$). Access to these ions affects plant growth (Gupta et al. 2021). Oster and Jayawardane (1998) reported that salinity stress not only affects soil physical properties but also affects the physiological and biochemical function of plants through the negative influence of osmotic and ionic components. Salinity stress causes ion toxicity, which results in high osmotic potential, and this high osmotic potential reduces the water absorption from soil (Machado and Serralheiro 2017). Many plants have tolerance mechanisms that get activated when plants get exposed to salinity stress. This tolerance mechanism makes plants acclimatize under stressful conditions. Mechanism involves changes in the development process, activation of defense and repair systems, synthesis of osmolytes, altered metabolism, and redirection of physiological processes (Acosta-Motos et al. 2017). Some of the studies reported that to understand mechanisms of alleviation of stress impact at tissue and organ levels, it is necessary to know signal processes at these levels to activate adaptive processes. Improving salt tolerance at the microlevel is important to make plants acclimatize (Manchanda and Garg 2008). Impact of salt stress depends on the concentration of salt, sensitivity of crop, and genetic makeup of plants (Munns 2005). Plants have mechanisms to reduce the effects of oxidative stress caused by salinity by reducing the reactive oxygen species (ROS) accumulation (Dalton et al. 1993). Plants show high antioxidant enzyme activity during salt stress. These enzymes reduce or detoxify ROS. For example, ascorbate peroxide (APX) regulates H_2O_2 level (Meloni et al. 2003; Matamoros et al. 2006). Plant priming is one of the most successful methods to enhance plants tolerance toward environmental stress. Plant priming using chemical additives against abiotic stress is called chemical priming. Mode of application plays an important role in the efficiency of chemical priming. Primed plants show more tolerance against environmental stress as compared to non-primed plants. Chemicals used as priming agents are sodium nitroprusside, melatonin, polyamines, hydrogen peroxide, and sodium hydrosulfide (Savvides et al. 2016). This chapter highlights the importance of chemical priming under salinity stress conditions and the application and mode of action used during chemical priming. Recent advances in salinity stress tolerance are discussed.

Global Climate Change and Plant Stress Management, First Edition. Edited by Mohammad Wahid Ansari, Anil Kumar Singh, and Narendra Tuteja.

25.2 Types of Chemical Additives and Their Source

Chemical additives are of two types: natural and synthetic.

Natural chemical compounds that acts as priming agents include:

- **Polyamines**: polyamines are nitrogen-containing low-molecular-weight compounds with two or more amino groups (Xu et al. 2009). In plants, polyamines are found in free as well as in conjugated forms. Free form polyamines are putrescine (put), spermidine (spd), and spermine (spm) (Mustafavi et al. 2018). Conjugated polyamines are present in covalently conjugated (CC-PA) and non-covalently conjugated (NCC-PA) (Gholami et al. 2013). Polyamines regulate plant growth and development as well as provide stress tolerance to plants (Alcázar et al. 2010).
- **Reactive species,** say ROS, reactive nitrogen species (RNS), reactive sulfur species (RSS), act as priming agents during stress conditions. These reactive species are obtained from normal metabolism inside plants. They are formed as bioproducts during metabolism pathways. ROS includes mainly H_2O_2 (hydrogen peroxide), RNS includes NO (nitric oxide) (Tanou et al. 2009; Christou et al. 2014). Nitric oxide (NO) is a gaseous signaling molecule involved in many physiological processes. NO also regulates plant growth and development (Beligni and Lamattina 2001). Involvement of NO in defense system of plant during stress conditions has been reported (Fancy et al. 2017). NO found effective in alleviating salinity stress in many plants (Zhang et al. 2006). It was found that ROS and RNS also act as signal transduction molecules in plants during physiological and biochemical processes (Molassiotis and Fotopoulos 2011).
- **Melatonin** is an idolic compound and discovered as a plant hormone by Dubbels et al. (1995). Melatonin biosynthesis in plants has similarity with IAA biosynthesis. Tryptophan acts as a precursor for melatonin synthesis (Murch and Saxena 2002). Melatonin synthesized inside mitochondria and chloroplast in plants (Tan et al. 2013). Melatonin shows an important role in plants vegetative as well as reproductive development (Byeon and Back 2014; Hardeland 2016). Melatonin protects photosystem proteins from damage (Zhou et al. 2016). Melatonin regulates expression of GA and ABA biosynthesis genes (Zhang et al. 2014; Li et al. 2015). It was found that the endogenous level of melatonin is affected by light (Byeon et al. 2012). Arnao and Hernández-Ruiz (2013) reported that melatonin level decreases during the day in *Vitis vinifera* and increases at night.
- **Hormones:** almost all phytohormones play a defensive role during environmental stress conditions in one or other way, e.g. salicylic acid (SA) and ethylene (ET).

  Salicylic acid is a phenolic compound in which almost all plants synthesize SA through shikimate and isochorismate pathways (Boukraa et al. 2013). Salicylic acid helps in many physiological processes in plants (El-Beltagi et al. 2017). Also, SA plays an important role in salinity stress mitigation (khan et al. 2015). Yuan and Lin (2008) reported that moderate concentrations of SA found beneficial in mitigating stress, if plants face high or very low SA concentration become more susceptible toward stress.

  Ethylene is a gaseous plant hormone that easily diffuses from one cell to other, regulates plant growth and development, and shows triple response: exaggeration of apical hook, inhibition of stems and root elongation, and swelling of hypocotyl (Guzman and Ecker 1990). Ethylene is mainly synthesized in older plant parts, and more synthesis occurs during fruit ripening and leaf senescence (Burg and Burg 1967). Ethylene plays an important role in counteracting salinity stress by detoxifying ROS and reducing its effect, maintaining stomatal conductance, and managing oxidative stress (Wang et al. 2016a).
- **Ascorbic acid (AsA)** is a water-soluble antioxidant found in plants (Smirnoff 2000). Ascorbic acid also acts as a cofactor for some enzymes (Davey et al. 2000). In the plants leaves, AsA is present in higher concentration compared to other parts (Smirnoff et al. 2004). Muller-Moule (2003, 2004) reported the antioxidant behavior of AsA; they found that in *Arabidopsis* mutants with AsA were more sensitive toward UV-B light. Endogenous level of AsA found beneficial to reduce harmful effects of oxidative stress. Exogenous application of AsA increases the endogenous level of AsA (Chen and Gallie 2004).
- **Trehalose** is a non-reducing disaccharide sugar that acts as an osmoprotectant and stabilizing molecule (Jain and Roy, 2009). Trehalose also acts as a signaling molecule in dehydration stress (Schluepmann et al. 2003). Trehalose synthesis occurs in plants with the help of two molecules: uridine-diphospho-glucose and glucose-6-phosphate (Blazquez et al. 1998). Trehalose is found in low levels inside plants, but during stress conditions, its level rises. It helps in adaptation during salinity stress by maintaining osmotic potential (Shahbaz et al. 2017).

Table 25.1 Exogenous application of priming agent mitigates salt stress in plants.

Chemical compound	Plant name	Effects	References
Salicylic acid	*Zea mays*	Maintains sodium–potassium ions ratio; regulates the activity of SOD, CAT, POD; also, maintains the proline and sugar content	Tufail et al. (2013) and Fahad and Bano (2012)
Salicylic acid	*Glycine max*	Regulates SOD activity and maintains ascorbate level that acts as osmoprotectant; maintains sodium to potassium ratio	Ardebili et al. (2014)
Salicylic acid	*Hordeum vulgare*	Maintains sodium to potassium ratio and MDA content	Fayez and Bazaid (2014)
Polyamine spd	*Zoysia japonica steud*	Decreases the hydrogen peroxide level and MAD content and increases polyamine biosynthesis	Li et al. (2016)
Polyamine spd	*Cucumis sativus*	Increases the activity of SOD, CAT, and POD and decreases the oxidative damage	Wu et al. (2018)
Nitric oxide	Soybean	Reduces hydrogen peroxide level and activates antioxidant defense system	Egbichi et al. (2013)
Nitric oxide	Rice	Increases biomass, upregulates expression of OsGR1, OsP5CS2, and OsHIPP38	Adamu et al. (2018)
Melatonin	*C. sativus*	Increases chlorophyll content, antioxidant activity, and ascorbate and reduces MDA, hydrogen peroxide content	Wang et al. (2016a)
Melatonin	*Citrullus lanatus*	Increases stomatal conductance and photosynthesis, reduces glutathione, MDA, and superoxide concentration	Li et al. (2017)
Melatonin	*Malus hupehensis* L.	Increases dry weight, chlorophyll content, and antioxidant activity, maintains sodium to potassium ratio	Li et al. (2012)
Ethylene	*Arabidopsis*	Enhances seed germination	Wilson et al. (2014)
Ascorbic acid	*Pisum*	Increases cell division, cell enlargement, and antioxidant system	Cabo et al. (1996)
Trehalose	*Catharanthus roseus*	Increases soluble sugars, free amino acids, leaf gas exchange, and ionic flow	Chang et al. (2014)
Kresoxim-methyl	*Medicago truncatula*	Reduces salinity stress effects by modulating biochemical and molecular processes	Filippou et al. (2016)

A synthetic chemical compound that acts as a priming agent includes:

- **Strobilurin** is fungicidal in nature; it blocks electron transport chain work against fungi (Sauter 2007) also improves plant growth parameters (Diaz-Espejo et al. 2012).
- **Kresoxim-Methyl** is a fungicide and has been found helpful in mitigating salinity stress. It is formed by the chemical modification of strobilurin A (Filippou et al. 2016).
- **NOSH – aspirin (NBS-1120)** is a chemical compound that acts as priming agent against stress. It donates NO and hydrogen sulfide, and these components protect plants against harmful effects of stress and alleviate stress conditions (Antoniou et al. 2014).

These are some natural and synthetic chemical compounds that act as priming agents during stress conditions (Table 25.1).

25.3 Application and Mechanism of Action

It was found in different studies that the pretreated plants with priming agents showed enhanced tolerance against the stress condition as compared to non-treated plants. This treatment of priming agents was given at different developmental stages: at seed germination and at vegetative or reproductive stage. Stress shows less harmful effects on the growth and physiology of primed plants. Exact mode of action of these priming agents is still under investigation. But some evidence was found related to the activation of tolerance mechanisms in plants against various stress conditions (Savvides et al. 2016).

25.4 NO (Nitric Oxide) in Salt Stress Tolerance

Many studies confirm the role of NO in salinity stress tolerance. Exogenous application of No increases the APX activity which converts excess H_2O_2 into H_2O and monodehydroascorbate. As a result, the level and harmful effects of H_2O_2 inside the cell are reduced (Shigeoka et al. 2002). Increased level of ROS and H_2O_2 leads to damage in plants NO help in H_2O_2 scavenging, regulates ROS accumulation, and restores redox balance of plants (Liu et al. 2007). Proline, being an osmoprotectant, helps in reducing salinity stress, but greater accumulation of proline is not found beneficial in salinity stress tolerance. NO induce excess proline degradation and help plant in stress tolerance (Lopez-Carrion et al. 2008). Exogenous application of nitric oxide is found beneficial in stress tolerance as it improves growth and physiological parameters, increases biomass, and plays a role in the tolerance mechanism (Molassiotis et al. 2010).

25.5 Melatonin in Salt Stress Tolerance

Melatonin shows protective effects against salt stress and acts as an antioxidant against ROS and RNS (Moustafa-Farag et al. 2019). Melatonin endogenous level increased by exogenous application as melatonin is permeable through plasma membrane and it inhibits oxidative stress. It increases the TaSNAT transcript, which encodes melatonin biosynthesis enzymes in wheat plants (Ke et al. 2018). Melatonin also increases the levels of antioxidant enzymes catalase (CAT), glutathione reductase (GR), superoxide dismutase (SOD), glutathione peroxidase (GPX), and ascorbate oxidase (APX) by upregulating the genes for their synthesis. Hence, it promotes the antioxidant capacity of plants (Wei et al. 2015). Melatonin upregulates ABA catabolism genes (e.g. CsCYP707A1 and CsCYP707A2) and GA (gibberellin) biosynthesis (e.g. GA20ox and GA3ox). This increase in GA level and decrease in ABA induce seed germination in plants (Zhang et al. 2014). Exogenous application of MT (melatonin) also increases the endogenous level of NO, which in turn helps in salinity stress tolerance (Zhao et al. 2018). Also, it regulates sodium potassium concentration and increases expression of MdNHX1 and MdAKT1 ion channels; they help maintain ion homeostasis (Yu et al. 2018) Figure 25.1. Along with the melatonin, its related metabolites and intermediates also show positive effects against salt stress (Kang et al. 2011; Hardeland 2017). Exogenous MT found beneficial in improving photosynthetic rate, and stomatal functions also upregulates expression of photosynthetic reaction genes (Ye et al. 2016).

25.6 Polyamines in Salt Stress Tolerance

Polyamines show salinity stress tolerance level, and the extent of its effect depends on plant species and stress intensity (Chen et al. 2019). Spermine is found as an important indicator of salinity stress tolerance (Li and He 2012). Exogenous application of polyamine improves plant growth, photosynthesis rate, and ROS metabolism (Baniasadia et al. 2018). Sun et al. (2018) reported that polyamines in combination with ABA help in salinity tolerance in grapes. Synthesis of polyamines spm, spd, and put are interlinked. Exogenous spd increases the endogenous levels of spd and spm and decreases put. Polyamines increase antioxidative enzymes and ROS scavenging during stress tolerance (Wu et al. 2018). It was reported that exogenously applied put (putrisine) increases the endogenous level of put and will activate genes that encode amine oxidases in rice. Also, it increases ethylene production. Hence, it alleviates effects of salt stress (Quinet et al. 2010). Exogenous application of spd activates antioxidative enzymes, prolines and decreases MDA content, which in turn helps in countering stress (Duan et al. 2008). Polyamines can be applied at different developmental stages of plants. Presoaked rice seeds with spd show better tolerance against salt stress. These seeds germinate at a good pace and show better growth and improved photosynthesis and biomass (Chunthaburee et al. 2014). Spd decreases the ROS production and MDA content in tomato and also increases ascorbate-glutathione cycle components (Zhang et al. 2016). Polyamines are found to activate different pathways, which play a role in alleviating salinity stress. Jiang et al. (2020) reported that exogenous spd increases antioxidant enzymes, improves photosynthesis and growth, and also balances redox homeostasis in rice. All the above findings support the role of polyamines toward alleviating salinity stress in plants.

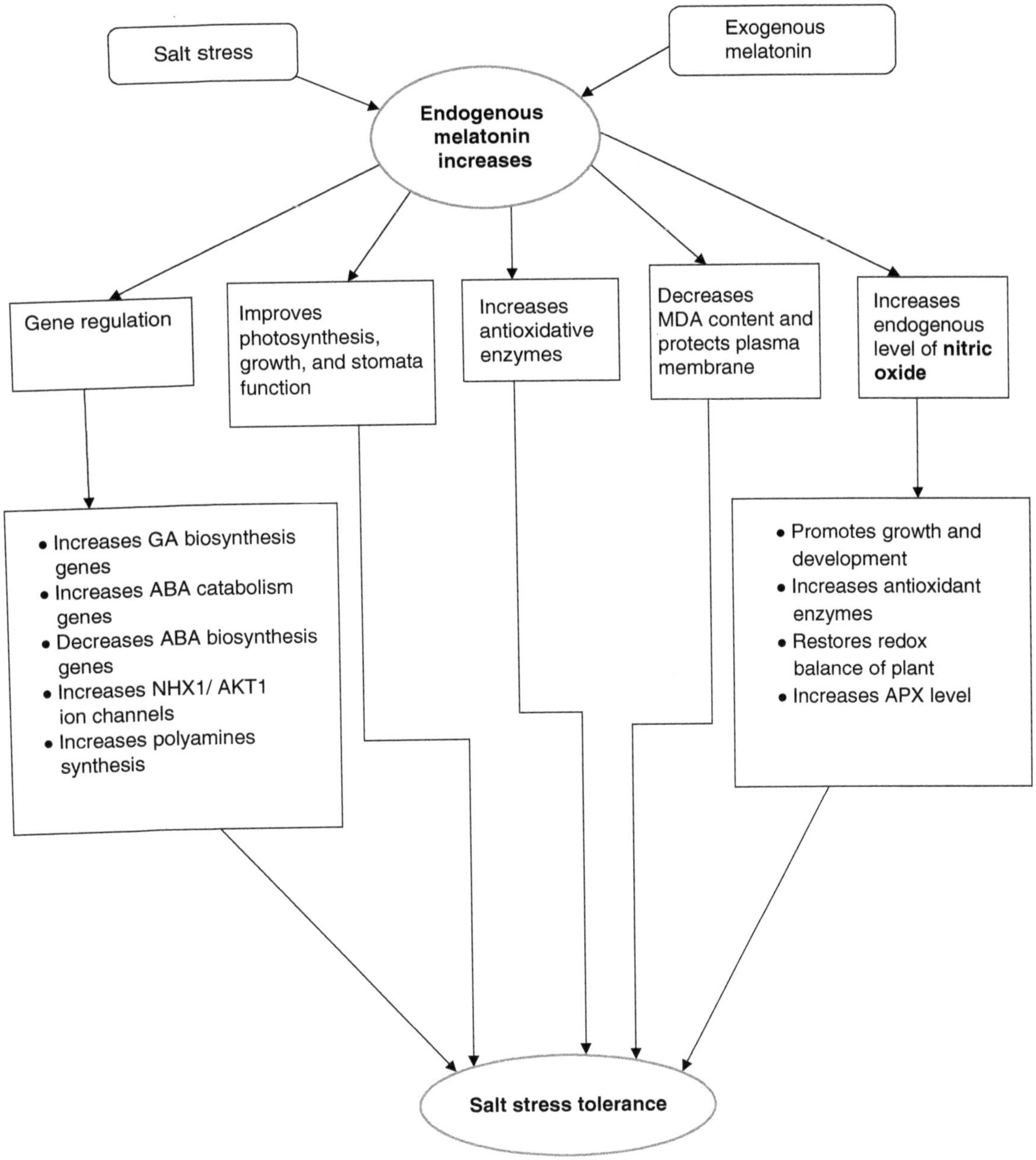

Figure 25.1 Mechanism of melatonin in salinity stress tolerance.

25.7 Salicylic Acid (SA) in Salt Stress Tolerance

SA found beneficial in mitigating salinity stress. It was found that presoaked seeds with SA shows enhanced germination, more photosynthesis, increased activity of APX and GPX, and decreased activity of POX and SOD; these changes positively regulate salinity stress (Hamada 2001; Mutlu et al. 2009). Khan et al. (2014) reported that a foliar spray of SA suppressed the deleterious effects of stress. SA induces glycine betaine accumulation, enhances activity of the antioxidant system, promotes photosynthesis and growth, increases methionine, and reduces ethylene production. Similar results were obtained when INA (2,6,dichloro-isonicotine acid) applied it is a SA analog also AVG (aminoethoxyvinylglycine) ethylene biosynthesis inhibitor also result in decreased ethylene, increase methionine and glycinebetaine. SA application on wheat increases growth, increases the endogenous level of ABA and IAA but does not influence cytokine content, accumulates antioxidant enzymes, and reduces harmful effects of salinity stress (Sakhabutdinova et al. 2003). SA application in maize

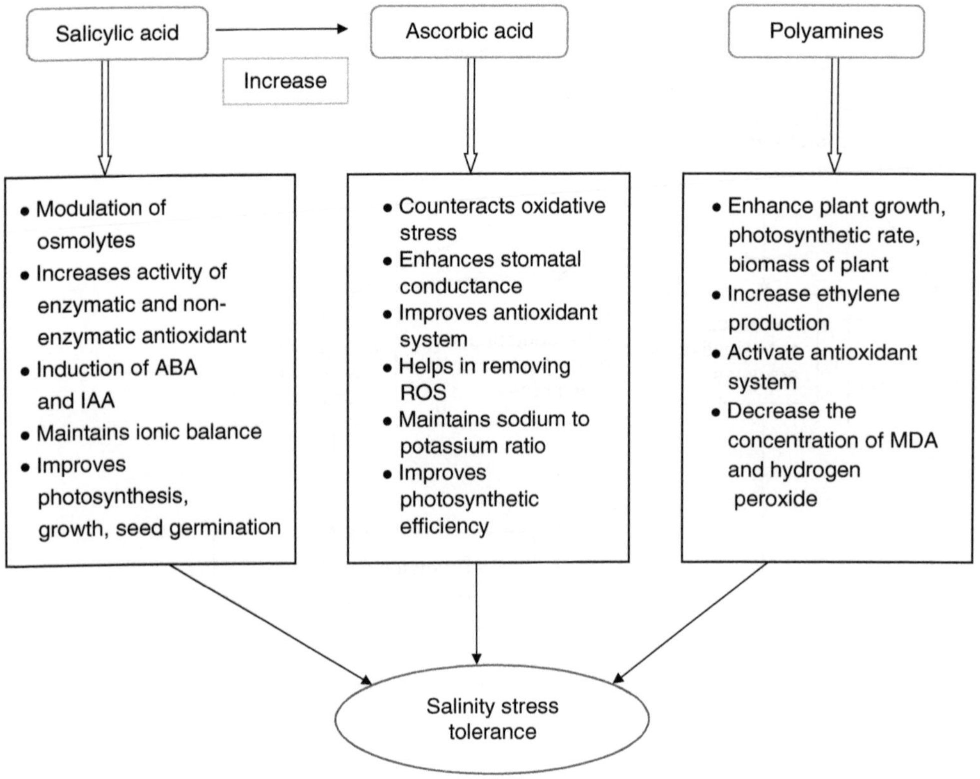

Figure 25.2 Mechanism of priming agents in salinity stress tolerance.

improves biomass, photosynthesis, stomatal conductance, and chlorophyll content and maintains sodium to potassium ratio by increasing potassium and calcium and decreasing sodium (Tufail et al. 2013) Figure 25.2 SA along with sulfur (S) also shows positive effects against salt stress reported by Nazar et al. (2011) in mung bean plants. SA and S improve antioxidant enzyme activity; NR (nitrate reductase) and NiR (nitrite reductase) activities also decrease ROS and reduce oxidative stress impact.

25.8 Ethylene in Salinity Stress Tolerance

Increase in endogenous ethylene or exogenous application of ethylene found beneficial in salinity stress tolerance. Some reports suggest ethylene acts as a negative regulator for salt stress response, like in rice, where the ethylene inhibitor 1-methylcyclopropene decreases the ethylene concentration and improves physiological and biochemical processes that result in salinity stress tolerance (Hussain et al. 2019), whereas some show positive role of ethylene in salinity stress tolerance, like in maize, where ethylene modulates polyamines biosynthesis and catabolism and hydrogen peroxide production (Freitas et al. 2018), and in *Arabidopsis*, ethylene maintains sodium to potassium ratio and reduces salinity stress deleterious effects (Yang et al. 2014). Melatonin increases ethylene production through MYB108A-mediated ethylene biosynthesis in grapevine, and this increased concentration of ethylene helps alleviate salinity stress (Xu et al. 2019). Ethylene manages seed germination during salinity stress ETR2 positively enhances seed germination in *Arabidopsis* (Wilson et al. 2014). Ethylene and NO interactions also positively increase seed germination by decreasing the hydrogen peroxide level in the seed and maintaining ROS homeostasis. Exogenous ethylene treatment also plays a role in decreasing hydrogen peroxide levels and improving seed germination (Lin et al. 2013). Brassinosteroid (BR) and ethylene act synergistically in salinity stress tolerance: BR increases ethylene level and detoxifies excess ROS (Jiroutova et al. 2018), whereas ethylene and gibberellin act antagonistically and activate salinity stress tolerance defense system (Khan et al. 2020).

25.9 Trehalose in Salinity Stress Tolerance

Trehalose maintains ROS, ions balance, and programmed cell death PCD in *Arabidopsis and* rice (Yang et al. 2014, Wang et al 2016b). Mostofa et al. (2015) pretreated plants with trehalose show salinity stress tolerance, improve AsA, GSH, and chlorophyll content, and decreased ROS and MDA. non-enzymatic antioxidant and enzymatic antioxidant systems are also activated. In the rice plant, trehalose has been found to be beneficial in maintaining the sodium to potassium ratio; it decreases SOD and POX and increases the activity of APX. These changes are helpful to the tolerance mechanism (Nounjan et al. 2012).

25.10 Kresoxim-Methyl (KM) in Salinity Stress Tolerance

Salinity stress lowered stomatal conductance, NR activity, proline content treatment with KM enhanced stomatal conductance, NR activity, proline content also modulates biochemical and molecular processes and reduce the effects of salinity stress in *Medicago truncatula* (Filippou et al. 2016).

25.11 Conclusion

Some of the priming agents like NO, melatonin, and polyamines also act as stress signaling molecules. The mode of action of these priming agents required extensive research in the future to better understand the mechanism. Some priming agents show effects against combined abiotic stress, e.g. melatonin is effective against salinity stress as well as against heat, drought, and heavy metal stress. Further detailed investigation needs to be carried out on the combined effect of these agents. One priming agent can manage more than one stress and a single stress can be managed by more than one priming agent. More efforts should be made to investigate the detailed action pathways of these chemical priming agents.

References

Acosta-Motos, J.R., Ortuño, M.F., Bernal-Vicente, A. et al. (2017). Plant responses to salt stress: adaptive mechanisms. *Agronomy* 7 (1): 18.

Adamu, T.A., Mun, B.G., Lee, S.U. et al. (2018). Exogenously applied nitric oxide enhances salt tolerance in rice (Oryza sativa L.) at seedling stage. *Agronomy* 8 (12): 276.

Alcázar, R., Altabella, T., Marco, F. et al. (2010). Polyamines: molecules with regulatory functions in plant abiotic stress tolerance. *Planta* 231: 1237–1249.

Antoniou, C., Chatzimichail, G., Kashfi, K., and Fotopoulos, V. (2014). P77: Exploring the potential of NOSH-aspirin as a plant priming agent against abiotic stress factors. *Nitric Oxide* 39: S39.

Ardebili, N.O., Saadatmand, S., Niknam, V., and Khavari-Nejad, R.A. (2014). The alleviating effects of selenium and salicylic acid in salinity exposed soybean. *Acta Physiol. Plant.* 36 (12): 3199–3205.

Arnao, M.B. and Hernández-Ruiz, J. (2013). Growth conditions influence the melatonin content of tomato plants. *Food Chem.* 138 (2-3): 1212–1214.

Baniasadi, F., Saffari, V.R., and Moud, A.A.M. (2018). Physiological and growth responses of *Calendula officinalis* L. plants to the interaction effects of polyamines and salt stress. *Sci. Hortic.* 234: 312–317.

Beligni, M.V. and Lamattina, L. (2001). Nitric oxide in plants: the history is just beginning. *Plant Cell Environ.* 24 (3): 267–278.

Blazquez, M.A., Santos, E., Flores, C.L. et al. (1998). Isolation and molecular characterization of the Arabidopsis TPS1 gene, encoding trehalose-6-phosphate synthase. *Plant J.* 13 (5): 685–689.

Boukraâ, D., Benabdelli, K., Belabid, L., and Bennabi, F. (2013). Effect of salinity on chickpea seed germination pre-treated with salicylic acid. *Sci. J. Biol. Sci.* 2 (4): 86–93.

Burg, S.P. and Burg, E.A. (1967). Molecular requirements for the biological activity of ethylene. *Plant Physiol.* 42 (1): 144–152.

Byeon, Y. and Back, K. (2014). An increase in melatonin in transgenic rice causes pleiotropic phenotypes, including enhanced seedling growth, delayed flowering, and low grain yield. *J. Pineal Res.* 56 (4): 408–414.

Byeon, Y., Park, S., Kim, Y.S. et al. (2012). Light-regulated melatonin biosynthesis in rice during the senescence process in detached leaves. *J. Pineal Res.* 53 (1): 107–111.

Cabo, R.C., Gonza-lez-Reyes, J.A., Cordoba, F., and Navas, P. (1996). Rooting has-tended in onions by ascorbate and ascorbate free radical. *J. Plant Growth Regul.* 15: 53–56.

Chang, B., Yang, L., Cong, W. et al. (2014). The improved resistance to high salinity induced by trehalose is associated with ionic regulation and osmotic adjustment in *Catharanthus roseus*. *Plant Physiol. Biochem.* 77: 140–148.

Chen, Z. and Gallie, D.R. (2004). The ascorbic acid redox state controls guard cell signaling and stomatal movement. *Plant Cell* 16 (5): 1143–1162.

Chen, D., Shao, Q., Yin, L. et al. (2019). Polyamine function in plants: metabolism, regulation on development, and roles in abiotic stress responses. *Front. Plant Sci.* 9: 1945.

Christou, A., Filippou, P., Manganaris, G.A., and Fotopoulos, V. (2014). Sodium hydrosulfide induces systemic thermotolerance to strawberry plants through transcriptional regulation of heat shock proteins and aquaporin. *BMC Plant Biol.* 14 (1): 1–11.

Chunthaburee, S., Sanitchon, J., Pattanagul, W., and Theerakulpisut, P. (2014). Alleviation of salt stress in seedlings of black glutinous rice by seed priming with spermidine and gibberellic acid. *Notulae Botanicae Horti Agrobotanici Cluj-Napoca* 42 (2): 405–413.

Dalton, D.A., Baird, L.M., Langeberg, L. et al. (1993). Subcellular localization of oxygen defense enzymes in soybean (*Glycine max* [L.] Merr.) root nodules. *Plant Physiol.* 102 (2): 481–489.

Davey, M.W., Montagu, M.V., Inze, D. et al. (2000). Plant L-ascorbic acid: chemistry, function, metabolism, bioavailability and effects of processing. *J. Sci. Food Agric.* 80 (7): 825–860.

Diaz-Espejo, A., Cuevas, M.V., Ribas-Carbo, M. et al. (2012). The effect of strobilurins on leaf gas exchange, water use efficiency and ABA content in grapevine under field conditions. *J. Plant Physiol.* 169 (4): 379–386.

Duan, J., Li, J., Guo, S., and Kang, Y. (2008). Exogenous spermidine affects polyamine metabolism in salinity-stressed *Cucumis sativus* roots and enhances short-term salinity tolerance. *J. Plant Physiol.* 165 (15): 1620–1635.

Dubbels, R., Reiter, R.J., Klenke, E. et al. (1995). Melatonin in edible plants identified by radioimmunoassay and by high performance liquid chromatography-mass spectrometry. *J. Pineal Res.* 18 (1): 28–31.

Egbichi, I., Keyster, M., Jacobs, A. et al. (2013). Modulation of antioxidant enzyme activities and metabolites ratios by nitric oxide in short-term salt stressed soybean root nodules. *South Afr. J. Botany* 88: 326–333.

El-Beltagi, H.S., Ahmed, S.H., Namich, A.A.M., and Abdel-Sattar, R.R. (2017). Effect of salicylic acid and potassium citrate on cotton plant under salt stress. *Fresen. Environ. Bull.* 26 (1A): 1091–1100.

Fahad, S. and Bano, A. (2012). Effect of salicylic acid on physiological and biochemical characterization of maize grown in saline area. *Pak. J. Bot.* 44 (4): 1433–1438.

Fancy, N.N., Bahlmann, A.K., and Loake, G.J. (2017). Nitric oxide function in plant abiotic stress. *Plant Cell Environ.* 40 (4): 462–472.

Fayez, K.A. and Bazaid, S.A. (2014). Improving drought and salinity tolerance in barley by application of salicylic acid and potassium nitrate. *J. Saudi Soc. Agric. Sci.* 13 (1): 45–55.

Filippou, P., Antoniou, C., Obata, T. et al. (2016). Kresoxim-methyl primes *Medicago truncatula* plants against abiotic stress factors via altered reactive oxygen and nitrogen species signalling leading to downstream transcriptional and metabolic readjustment. *J. Exp. Bot.* 67 (5): 1259–1274.

Freitas, V.S., de Souza Miranda, R., Costa, J.H. et al. (2018). Ethylene triggers salt tolerance in maize genotypes by modulating polyamine catabolism enzymes associated with H_2O_2 production. *Environ. Exp. Bot.* 145: 75–86.

Gholami, M., Fakhari, A.R., and Ghanati, F. (2013). Selective regulation of nicotine and polyamines biosynthesis in tobacco cells by enantiomers of ornithine. *Chirality* 25 (1): 22–27.

Gupta, S., Schillaci, M., Walker, R. et al. (2021). Alleviation of salinity stress in plants by endophytic plant-fungal symbiosis: current knowledge, perspectives and future directions. *Plant Soil* 461 (1): 219–244.

Guzman, P. and Ecker, J.R. (1990). Exploiting the triple response of Arabidopsis to identify ethylene-related mutants. *Plant Cell* 2 (6): 513–523.

Hamada, A.M. (2001). *Salicylic Acid Versus Salinity-Drought-Induced Stress on Wheat Seedlings*. Czech Republic: RostlinnaVyroba-UZPI.

Hardeland, R. (2016). Melatonin in plants–diversity of levels and multiplicity of functions. *Front. Plant Sci.* 7: 198.

Hardeland, R. (2017). Taxon-and site-specific melatonin catabolism. *Molecules* 22 (11): 2015.

Hussain, S., Bai, Z., Huang, J. et al. (2019). 1-Methylcyclopropene modulates physiological, biochemical, and antioxidant responses of rice to different salt stress levels. *Front. Plant Sci.* 10: 124.

Ivushkin, K., Bartholomeus, H., Bregt, A.K. et al. (2019). Global mapping of soil salinity change. *Remote Sensing Environ.* 231: 111260.

Jain, N.K. and Roy, I. (2009). Effect of trehalose on protein structure. *Protein Sci.* 18 (1): 24–36.

Jiang, D.X., Chu, X., Li, M. et al. (2020). Exogenous spermidine enhances salt-stressed rice photosynthetic performance by stabilizing structure and function of chloroplast and thylakoid membranes. *Photosynthetica* 58 (1): 61–71.

Jiroutova, P., Oklestkova, J., and Strnad, M. (2018). Crosstalk between brassinosteroids and ethylene during plant growth and under abiotic stress conditions. *Int. J. Mol. Sci.* 19 (10): 3283.

Kang, K., Kong, K., Park, S. et al. (2011). Molecular cloning of a plant N-acetylserotonin methyltransferase and its expression characteristics in rice. *J. Pineal Res.* 50 (3): 304–309.

Ke, Q., Ye, J., Wang, B. et al. (2018). Melatonin mitigates salt stress in wheat seedlings by modulating polyamine metabolism. *Front. Plant Sci.* 9: 914.

Khan, M.I.R., Asgher, M., and Khan, N.A. (2014). Alleviation of salt-induced photosynthesis and growth inhibition by salicylic acid involves glycinebetaine and ethylene in mungbean (*Vigna radiata* L.). *Plant Physiol. Biochem.* 80: 67–74.

Khan, M.I.R., Fatma, M., Per, T.S. et al. (2015). Salicylic acid-induced abiotic stress tolerance and underlying mechanisms in plants. *Front. Plant Sci.* 6: 462.

Khan, N., Bano, A., Ali, S., and Babar, M. (2020). Crosstalk amongst phytohormones from planta and PGPR under biotic and abiotic stresses. *Plant Growth Regul.* 90 (2): 189–203.

Li, Y. and He, J. (2012). Advance in metabolism and response to stress of polyamines in plant. *Acta Agric. Boreali. Sin.* 27: 240–245.

Li, C., Wang, P., Wei, Z. et al. (2012). The mitigation effects of exogenous melatonin on salinity-induced stress in *Malus hupehensis*. *J. Pineal Res.* 53 (3): 298–306.

Li, C., Tan, D.X., Liang, D. et al. (2015). Melatonin mediates the regulation of ABA metabolism, free-radical scavenging, and stomatal behaviour in two Malus species under drought stress. *J. Exp. Bot.* 66 (3): 669–680.

Li, S., Jin, H., and Zhang, Q. (2016). The effect of exogenous spermidine concentration on polyamine metabolism and salt tolerance in Zoysia grass (*Zoysia japonica Steud*) subjected to short-term salinity stress. *Front. Plant Sci.* 7: 1221.

Li, H., Chang, J., Chen, H. et al. (2017). Exogenous melatonin confers salt stress tolerance to watermelon by improving photosynthesis and redox homeostasis. *Front. Plant Sci.* 8: 295.

Lin, Y., Yang, L., Paul, M. et al. (2013). Ethylene promotes germination of Arabidopsis seed under salinity by decreasing reactive oxygen species: evidence for the involvement of nitric oxide simulated by sodium nitroprusside. *Plant Physiol. Biochem.* 73: 211–218.

Liu, Y., Wu, R., Wan, Q. et al. (2007). Glucose-6-phosphate dehydrogenase plays a pivotal role in nitric oxide-involved defense against oxidative stress under salt stress in red kidney bean roots. *Plant Cell Physiol.* 48 (3): 511–522.

Lopez-Carrion, A.I., Castellano, R., Rosales, M.A. et al. (2008). Role of nitric oxide under saline stress: implications on proline metabolism. *Biol. Plant.* 52 (3): 587–591.

Machado, R.M.A. and Serralheiro, R.P. (2017). Soil salinity: effect on vegetable crop growth. Management practices to prevent and mitigate soil salinization. *Horticulturae* 3 (2): 30.

Manchanda, G. and Garg, N. (2008). Salinity and its effects on the functional biology of legumes. *Acta Physiol. Plant.* 30 (5): 595–618.

Matamoros, M.A., Loscos, J., Coronado, M.J. et al. (2006). Biosynthesis of ascorbic acid in legume root nodules. *Plant Physiol.* 141: 1068–1077.

Meloni, D.A., Oliva, M.A., Martinez, C.A., and Cambraia, J. (2003). Photosynthesis and activity of superoxide dismutase, peroxidase and glutathione reductase in cotton under salt stress. *Environ. Exp. Bot.* 49: 69–76.

Molassiotis, A. and Fotopoulos, V. (2011). Oxidative and nitrosative signaling in plants: two branches in the same tree? *Plant Signal. Behav.* 6 (2): 210–214.

Molassiotis, A., Tanou, G., and Diamantidis, G. (2010). NO says more than 'YES' to salt tolerance: salt priming and systemic nitric oxide signaling in plants. *Plant Signal. Behav.* 5: 209–212.

Mostofa, M.G., Hossain, M.A., and Fujita, M. (2015). Trehalosepretreatment induces salt tolerance in rice (*Oryza sativa* L.) seedlings: oxidative damage and co-induction of antioxidant defense and glyoxalase systems. *Protoplasma* 252 (2): 461–475.

Moustafa-Farag, M., Almoneafy, A., Mahmoud, A. et al. (2019). Melatonin and its protective role against biotic stress impacts on plants. *Biomolecules* 10 (1): 54.

Muller-Moulé, P., Havaux, M., and Niyogi, K.K. (2003). Zeaxanthin deficiency enhances the high light sensitivity of an ascorbate-deficient mutant of Arabidopsis. *Plant Physiol.* 133 (2): 748–760.

Muller-Moulé, P., Golan, T., and Niyogi, K.K. (2004). Ascorbate-deficient mutants of Arabidopsis grow in high light despite chronic photooxidative stress. *Plant Physiol.* 134 (3): 1163–1172.

Munns, R. (2005). Genes and salt tolerance: bringing them together. *New Phytol.* 167 (3): 645–663.

Murch, S.J. and Saxena, P.K. (2002). Melatonin: a potential regulator of plant growth and development? *Vitro Cell Dev. Biol. Plant* 38 (6): 531–536.

Mustafavi, S.H., NaghdiBadi, H., Sękara, A. et al. (2018). Polyamines and their possible mechanisms involved in plant physiological processes and elicitation of secondary metabolites. *Acta Physiol. Plant.* 40 (6): 1–19.

Mutlu, S.A.L.I.H., Atici, Ö., and Nalbantoglu, B. (2009). Effects of salicylic acid and salinity on apoplastic antioxidant enzymes in two wheat cultivars differing in salt tolerance. *Biol. Plant.* 53 (2): 334–338.

Nazar, R., Iqbal, N., Syeed, S., and Khan, N.A. (2011). Salicylic acid alleviates decreases in photosynthesis under salt stress by enhancing nitrogen and sulfur assimilation and antioxidant metabolism differentially in two mungbean cultivars. *J. Plant Physiol.* 168 (8): 807–815.

Nounjan, N., Nghia, P.T., and Theerakulpisut, P. (2012). Exogenous proline and trehalose promote recovery of rice seedlings from salt-stress and differentially modulate antioxidant enzymes and expression of related genes. *J. Plant Physiol.* 169 (6): 596–604.

Oster, J. and Jayawardane, N.S. (1998). Agricultural management of sodic soils. In: *Sodic Soils: Distribution, Properties, Management and Environmental Consequences* (ed. M.E. Sumner and R. Naidu), 125–147. New York: Oxford University Press.

Parida, A.K. and Das, A.B. (2005). Salt tolerance and salinity effects on plants: a review. *Ecotoxicol. and Environ. Safety* 60 (3): 324–349.

Quinet, M., Ndayiragije, A., Lefevre, I. et al. (2010). Putrescine differently influences the effect of salt stress on polyamine metabolism and ethylene synthesis in rice cultivars differing in salt resistance. *J. Exp. Bot.* 61 (10): 2719–2733.

Sakhabutdinova, A.R., Fatkhutdinova, D.R., Bezrukova, M.V., and Shakirova, F.M. (2003). Salicylic acid prevents the damaging action of stress factors on wheat plants. *Bulg. J. Plant Physiol.* 21: 314–319.

Sauter, H. (2007). Strobilurins and other complex III inhibitors. *Modern Crop Protect. Comp.* 2: 457–495.

Savvides, A., Ali, S., Tester, M., and Fotopoulos, V. (2016). Chemical priming of plants against multiple abiotic stresses: mission possible? *Trends Plant Sci.* 21 (4): 329–340.

Schluepmann, H., Pellny, T., van Dijken, A. et al. (2003). Trehalose 6-phosphate is indispensable for carbohydrate utilization and growth in *Arabidopsis thaliana*. *Proc. Nat. Acad. Sci.* 100 (11): 6849–6854.

Shahbaz, M., Abid, A., Masood, A., and Waraich, E.A. (2017). Foliar-applied trehalose modulates growth, mineral nutrition, photosynthetic ability, and oxidative defense system of rice (*Oryza sativa* L.) under saline stress. *J. Plant Nutrition* 40 (4): 584–599.

Shigeoka, S., Ishikawa, T., Tamoi, M. et al. (2002). Regulation and function of ascorbate peroxidase isoenzymes. *J. Exp. Bot.* 53: 1305–1319.

Smirnoff, N. (2000). Ascorbate biosynthesis and function in photoprotection. *Philos. Trans. R. Soc. Lond. B Biol. Sci.* 355 (1402): 1455–1464.

Smirnoff, N., Running, J.A., and Gatzek, S. (2004). Ascorbate biosynthesis a diversity of pathways. In: *Vitamin C* (ed. H. Asard, J. May, and N. Smirnoff), 8–33. Taylor & Francis.

Sun, L., Yu, S., and Zhao, F. (2018). Effects of salt stress on polyamines and hormone metabolism in Grape seedlings. *Xinjiang Agric. Sci.* 55: 66–73.

Tan, D.X., Manchester, L.C., Liu, X. et al. (2013). Mitochondria and chloroplasts as the original sites of melatonin synthesis: a hypothesis related to melatonin's primary function and evolution in eukaryotes. *J. Pineal Res.* 54 (2): 127–138.

Tanou, G., Molassiotis, A., and Diamantidis, G. (2009). Hydrogen peroxide-and nitric oxide-induced systemic antioxidant prime-like activity under NaCl-stress and stress-free conditions in citrus plants. *J. Plant Physiol.* 166 (17): 1904–1913.

Tufail, A., Arfan, M., Gurmani, A.R. et al. (2013). Salicylic acid induced salinity tolerance in maize (*Zea mays*). *Pak. J. Bot.* 45 (S1): 75–82.

Wang, L.Y., Liu, J.L., Wang, W.X., and Sun, Y. (2016a). Exogenous melatonin improves growth and photosynthetic capacity of cucumber under salinity-induced stress. *Photosynthetica* 54 (1): 19–27.

Wang, Y., Wu, J., Kim, S.G. et al. (2016b). Magnaportheoryzae-secreted protein MSP1 induces cell death and elicits defense responses in rice. *Mol. Plant Microbe Inter.* 29 (4): 299–312.

Wei, W., Li, Q.T., Chu, Y.N. et al. (2015). Melatonin enhances plant growth and abiotic stress tolerance in soybean plants. *J. Exp. Bot.* 66 (3): 695–707.

Wilson, R.L., Kim, H., Bakshi, A., and Binder, B.M. (2014). The ethylene receptors ETHYLENE RESPONSE1 and ETHYLENE RESPONSE2 have contrasting roles in seed germination of Arabidopsis during salt stress. *Plant Physiol.* 165 (3): 1353–1366.

Wu, J., Shu, S., Li, C. et al. (2018). Spermidine-mediated hydrogen peroxide signaling enhances the antioxidant capacity of salt-stressed cucumber roots. *Plant Physiol. Biochem.* 128: 152–162.

Xu, C., Wu, X., and Zhang, H. (2009). Impact of D-Arg on drought resistance and endogenous polyamines in mycorrhizal *Pinus massoniana*. *J. Nanjing Forestry Univ.* 52 (04): 19.

Xu, L., Xiang, G., Sun, Q. et al. (2019). Melatonin enhances salt tolerance by promoting MYB108A-mediated ethylene biosynthesis in grapevines. *Hortic. Res.* 6: 114.

Yang, L., Zhao, X., Zhu, H. et al. (2014). Exogenous trehalose largely alleviates ionic unbalance, ROS burst, and PCD occurrence induced by high salinity in Arabidopsis seedlings. *Front. Plant Sci.* 5: 1–11.

Ye, J., Wang, S., Deng, X. et al. (2016). Melatonin increased maize (*Zea mays* L.) seedling drought tolerance by alleviating drought-induced photosynthetic inhibition and oxidative damage. *Acta Physiol. Plant.* 38 (2): 1–13.

Yu, Y., Wang, A., Li, X. et al. (2018). Melatonin-stimulated triacylglycerol breakdown and energy turnover under salinity stress contributes to the maintenance of plasma membrane H^+–ATPase activity and K^+/Na^+ homeostasis in sweet potato. *Front. Plant Sci.* 9: 256.

Yuan, S. and Lin, H.H. (2008). Role of salicylic acid in abiotic stress tolerance. *Z Naturforsch C J. Biosci.* 63 (6): 313–320.

Zhang, Y., Wang, L., Liu, Y. et al. (2006). Nitric oxide enhances salt tolerance in maize seedlings through increasing activities of proton-pump and Na^+/H^+ antiport in the tonoplast. *Planta* 224: 545–555.

Zhang, H.J., Zhang, N.A., Yang, R.C. et al. (2014). Melatonin promotes seed germination under high salinity by regulating antioxidant systems, ABA and GA 4 interaction in cucumber (*Cucumis sativus* L.). *J. Pineal Res.* 57 (3): 269–279.

Zhang, Z., Chang, X.X., Zhang, L. et al. (2016). Spermidine application enhances tomato seedling tolerance to salinity-alkalinity stress by modifying chloroplast antioxidant systems. *Russ. J. Plant Physiol.* 63 (4): 461–468.

Zhao, G., Zhao, Y., Yu, X. et al. (2018). Nitric oxide is required for melatonin-enhanced tolerance against salinity stress in rapeseed (*Brassica napus* L.) seedlings. *Int. J. Mol. Sci.* 19 (7): 1912.

Zhou, X.T., Zhao, H.L., Cao, K. et al. (2016). Beneficial roles of melatonin on redox regulation of photosynthetic electron transport and synthesis of D1 protein in tomato seedlings under salt stress. *Front. Plant Sci.* 7: 1823.

26

Role of Secondary Metabolites in Stress Management Under Changing Climate Conditions

Priya Yadav[1] *and Zahid Hameed Siddiqui*[2,3]

[1] *Department of Botany, Zakir Husain Delhi College, University of Delhi, New Delhi, India*
[2] *Department of Biology, Faculty of Science, University of Tabuk, Tabuk, Saudi Arabia*
[3] *Genomic and Biotechnology Unit, Department of Biology, Faculty of Science, University of Tabuk, Tabuk, Saudi Arabia*

26.1 Introduction

Changing climate conditions create a stressful environment for all living entities. Plants face different abiotic stresses (including drought, salinity, temperature, and heavy metals) due to climate change and global warming; other than this, plants also face biotic stresses (including pathogens, herbivores, and insects). These stresses create a serious threat to global food security for the growing population (Verma and Deepti 2016). As per the reports from recent years, these stresses were found to be the main reason for the reduction in crop yield and production of up to 50% (Mantri et al. 2012). Over the period of time, plants develop mechanisms to mitigate these stresses; one of the mechanisms is modulation of secondary metabolites (Hopkins et al. 2017). Secondary metabolites start accumulating when plants are exposed to stressful conditions and play a role in defense signaling (Eid et al. 2015). Accumulation and synthesis of these metabolites depend on the plant species, the environmental conditions, in which they are growing, and the kind of stress (Radušienė et al. 2012). Secondary metabolites of plants are small molecules produced as byproducts of primary metabolism. Almost 200,000 secondary metabolites have been discovered in plants, including phenolics, glucosinolates, cyanogenic glucosides, alkaloids, quinones, peptides, polyacetylenes, terpenes, and amines (Willis 2017). These metabolites are of different chemical structure and perform different functions within the plant; unlike primary metabolites, they are not involved in normal growth and development of the plant but rather involved in specific functions like defense responses, signaling, attracting pollinating insects, UV-protection, adaptation, and interaction with the environment (Kessler and Kalske 2018). Secondary metabolites are different from primary metabolites in their structure and functions, but they are interlinked because secondary metabolites are synthesized as by-products of primary metabolic pathways (War et al. 2020). Secondary metabolites' basic functions include inhibiting the growth of fungi, viruses, and bacteria (biotic), repelling insect herbivores (abiotic), providing defense against abiotic stress, inhibiting the germination of other competing plants, storing nitrogen, facilitating inter-species communication, and attracting pollinators. Advancements in technology explain their role to some extent, but their complete importance is not known (Hussain et al. 2012). This chapter provides information on secondary metabolite types, biosynthesis, and recent advances related to their role in protecting plants from biotic and abiotic stresses. It also provides insight on the effective use of secondary metabolites to develop ecofriendly, cost-effective, and sustainable strategies to enhance crop yield and overall production under stressful conditions.

26.1.1 Types of Plant Secondary Metabolites

In recent years, people have elucidated the role of secondary metabolites in the defense system of plants, and their biosynthesis and applications have been investigated using recent analytical tools and techniques (Wink 2003). Secondary metabolites are mainly categorized into three different groups: phenolics, terpenoids, and nitrogen-containing compounds.

Global Climate Change and Plant Stress Management, First Edition. Edited by Mohammad Wahid Ansari, Anil Kumar Singh, and Narendra Tuteja.

26.1.1.1 Phenolics

Phenolic compounds are the most pronounced secondary metabolites, which structurally contain a hydroxyl functional group (phenyl group) on an aromatic ring. Phenolic compounds include simple phenols (e.g. catechols and hydroxybenzoic acid derivatives), lignins, condensed tannins, catechol melanins, flavonoids, and stilbenes. These phenolic compounds play an important role not only in defense but also in accelerating pollination, antioxidant, antifungal, and antibacterial activities. Peroxidases and polyphenol oxidases convert phenolics into toxic metabolites that cause disturbance in the growth and development of many herbivores, hence the need to protect plants from herbivores (Bhonwong et al. 2009) (Figure 26.1).

26.1.1.2 Terpenoids

Terpenoids/terpenes are one of the largest and most diverse groups of secondary metabolites. They are lipophilic in nature and hydrophobic. Structurally, main body of terpenoids consists of five-carbon isopentyl units, called 2-methyl-1,3-butadiene, or isoprene. Terpenoids contain additional functional group (ketone group or heterocyclic and hydroxyl rings), whereas terpenes contain only isoprene unit. Almost all terpenes perform special function as they are plant secondary metabolites, but there are some terpenes (brassinosteroids, gibberellins) that perform the primary function (growth and development) of plants. Terpenes comprise largest group of secondary metabolites and include 25,000 compounds. Terpenes protect plants from herbivore, act as attractants for pollinators, and provide feeding deterrence (Aharoni et al. 2005) (Figure 26.2).

26.1.1.3 Nitrogen-Containing Secondary Metabolites

It includes alkaloids, cyanogenic glycosides, and glucosinolates.

Alkaloids are grouped into three categories:

- true alkaloids (e.g. morphine, nicotine, atropine, and quinine);
- proto-alkaloids (e.g. hordenine, mescaline, and yohimbine); and
- Pseudo-alkaloids (e.g. solanidine, caffeine, and capsaicin).

True and proto-alkaloids are obtained from amino acid, but pseudo-alkaloids are not derived from amino acids (Aniszewski 2007). Alkaloids are widely distributed and reported in primitive genera of plants, gymnosperms, and most

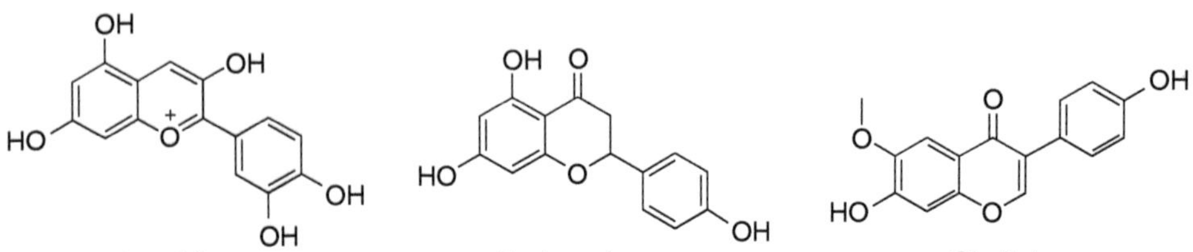

Cyanidin (anthocyanidins) | Naringenin (flavanones) | Glycitein (isoflavones)

Figure 26.1 Examples of different flavonoids: a major phenolic group of secondary metabolites.

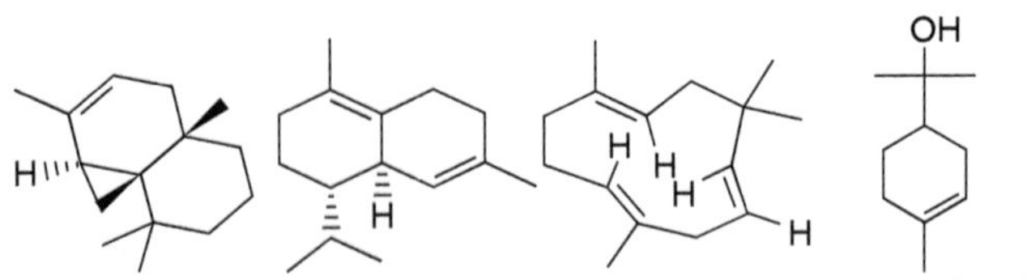

Thujopsene | δ-Cadinene | α-Humulene | α-Terpineol

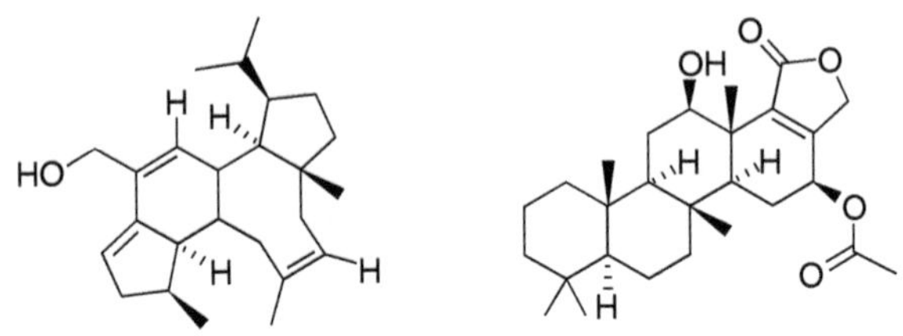

Nitiol | Sesterstatin 7

Figure 26.2 Some of the members of terpenoid plant secondary metabolites.

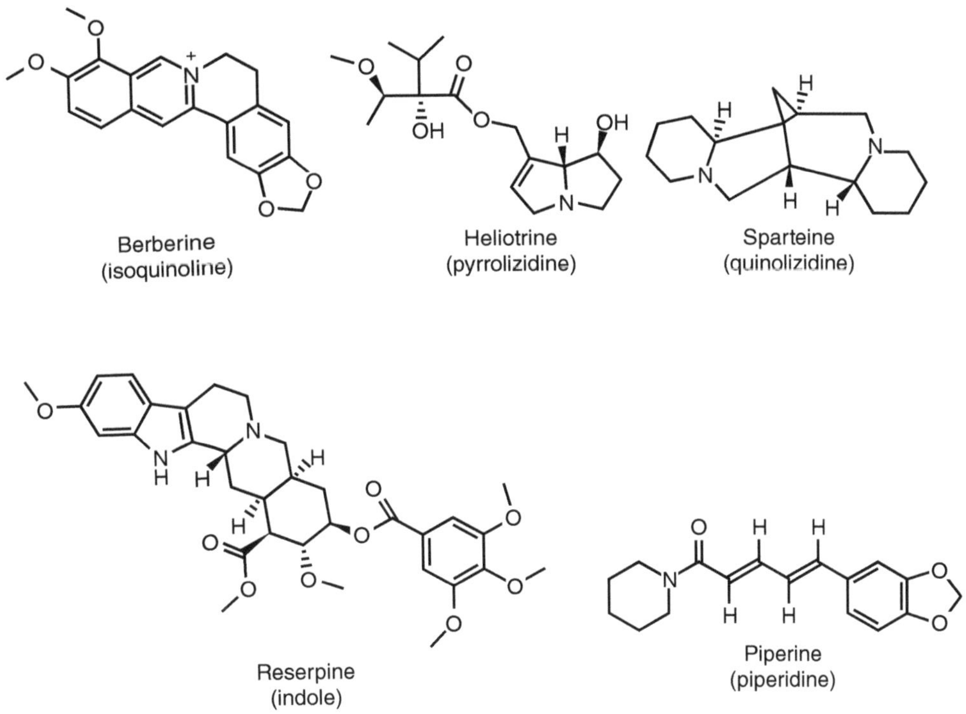

Figure 26.3 Different types of alkaloids (N-containing compounds) produced in plants and their chemical structure.

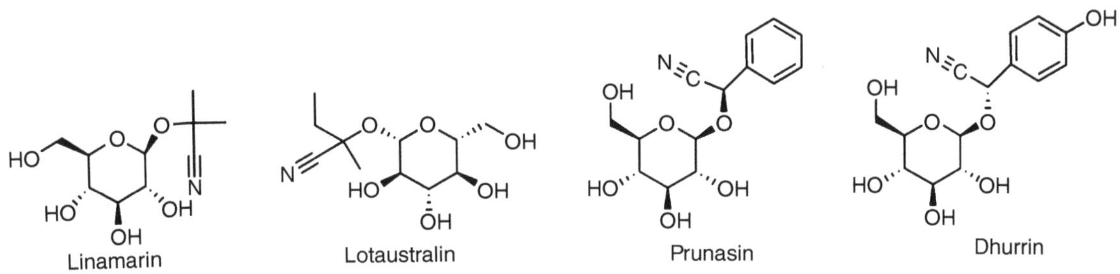

Figure 26.4 Some of the cynogenic glycosides in plants.

abundant in angiosperms. It was found that plant families produce one type of alkaloids, but some families have broad spectrum of alkaloids (e.g. Solanaceae and Rutaceae) (Nowacki and Waller 1973). Alkaloids biosynthesis and production rate are influenced by environmental factors like temperature, light, soil nutrients, and moisture (McCulley et al. 2014) (Figure 26.3).

Cynogenic glycosides and glucosinolates are N-containing groups derived from amino acids. Glucosinolates are found in ferns, gymnosperms, and angiosperms, whereas cynogenic glycosides are found only in the order Carppanals and in one of the genes (Drypetes) of Euphorbiaceae (Halkier and Gershenzon 2006). Both glucosinolates and cynogenic glycosides accumulate in younger leaves and reproductive parts of plants, but some plants like *Prunus turneriana* under low light accumulate cynogenic glycosides in their older leaves (Gleadow and Woodrow 2000; Gleadow and Møller 2014) (Figure 26.4).

26.2 Biosynthesis of Plant Secondary Metabolites

Plants synthesize 100,000 different secondary metabolites under unfavorable conditions. Primary metabolism switches to secondary metabolism and produces secondary metabolites by different metabolic pathways in plants. The rate of synthesis and quality of metabolites are influenced by environmental factors (Meena et al. 2017). Production of these metabolites is tissue-, organ-, or species-specific in nature, and most of them are synthesized in cytoplasm (Yang et al. 2018). Specific transcription factors are involved in the activation of genes for secondary metabolite synthesis. Different metabolites are produced in different organs; for example, coniine, furanocoumarins, caffeine, and terpenes (e.g. monoterpenes, diterpenes, carotenoids, and phytol) are synthesized in the chloroplast, whereas the endoplasmic reticulum (ER) or the cytosol is the site for sesquiterpenes, sterols, and dolichols synthesis (Wink and Schimmer 2010). Most of the secondary metabolites are synthesized in biosynthetic pathways of secondary metabolites are interlinked with primary metabolic pathways, or their precursors come from the primary metabolic pathway. Nitrogen-containing compounds are synthesized from aliphatic amino acids or from intermediates of TCA cycle (Figure 26.5).

Phenolic compounds are synthesized by shikimic acid pathway in plants, whereas in the case of fungi and bacteria, phenolic compounds are produced via malonic acid pathway (Cheynier et al. 2013). These compounds are originated from phenylalanine and, therefore, are also called as phenylpropanoids (Figure 26.6). In plants, terpenoid biosynthesis is carried out under two pathways: (i) mevalonate pathway, (ii) 2-*C*-methylerythritol 4-phosphate (MEP) pathway also called as non-mevalonate pathway. Precursor of MEP pathway (pyruvate + glyceraldehyde-3-phosphate) and precursor of mevalonate pathway (acetyl CoA) are components of glycolysis and tricarboxylic acid cycle (TCA), respectively. These precursors undergo multiple step process and produce isopentenyl pyrophosphate (IDP) and dimethyl allyl pyrophosphate (DMAPP). These two compounds further produce different terpenoids (Figure 26.7). It was reported that enzymes

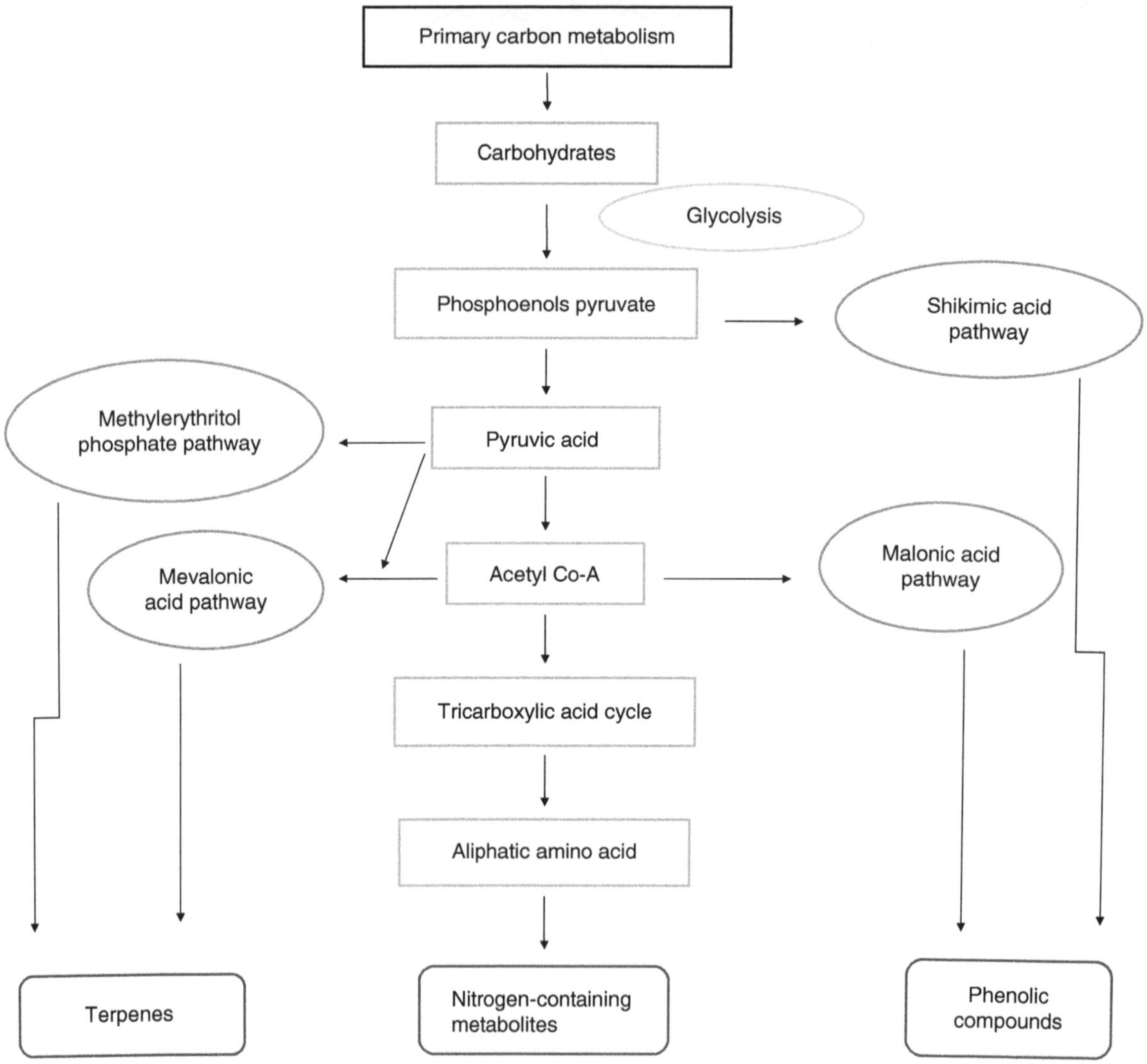

Figure 26.5 Basic outline for biosynthetic pathways of secondary metabolites and their interconnection with primary metabolic pathway. Nitrogen-containing compounds are synthesized from continuing primary carbon metabolic pathway, and precursors for terpenes and phenolic compounds are components of primary metabolic pathway.

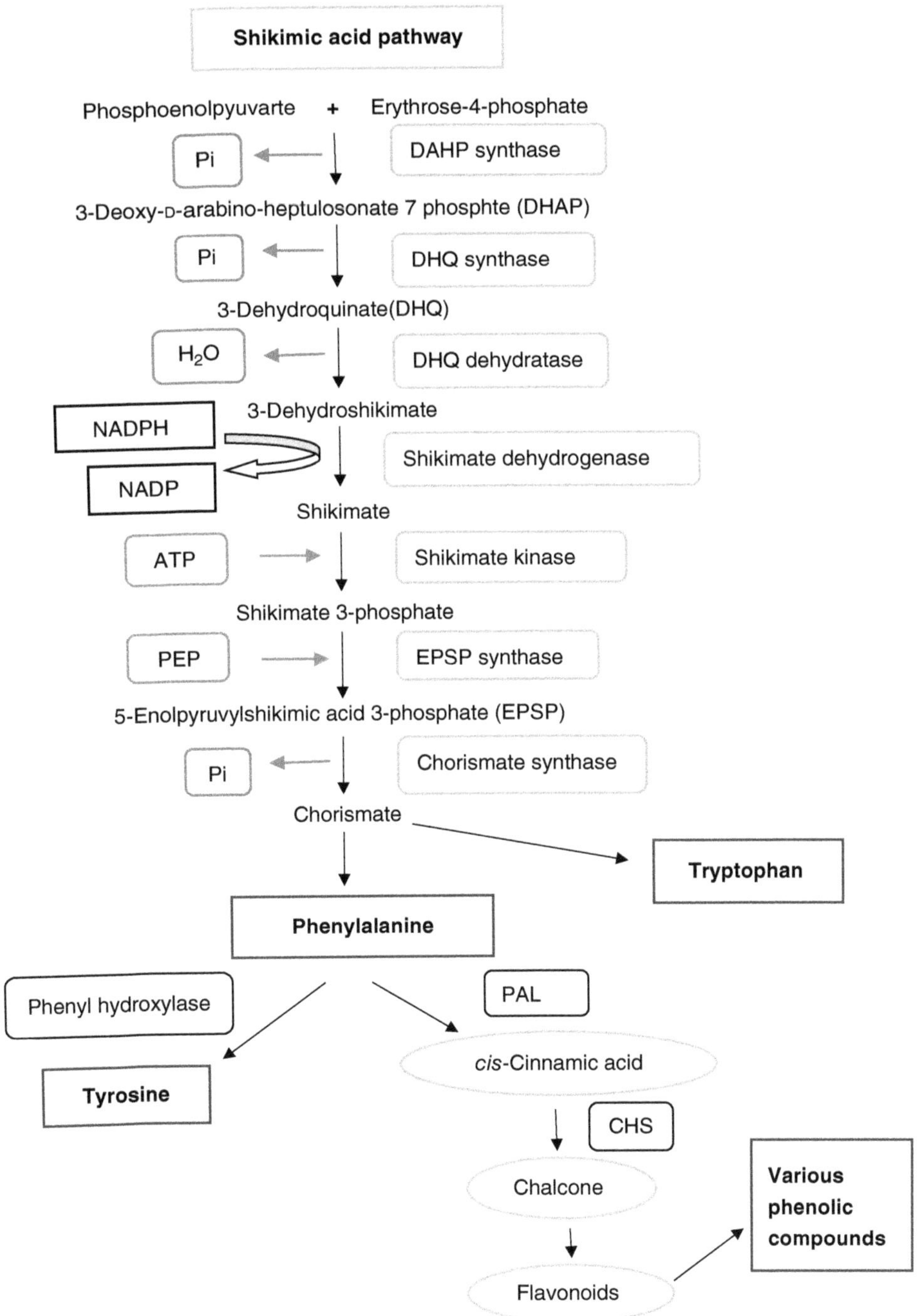

Figure 26.6 Phenolic compounds biosynthetic pathway in plants (shikimic acid pathway). Through this pathway, first phenolic compound formed is phenylalanine, and it acts as a precursor for other phenolic compounds. **ATP**, adenosine triphosphate; **CHS**, chalcone synthase; **PAL**, phenylalanine ammonia lyase; **PEP**, phosphoenolpyruvate; **Pi**, inorganic phosphate; **NADPH**, nicotinamide adenine dinucleotide phosphate.

involved in terpenes biosynthesis have their expression elevated under biotic and abiotic stress conditions (Sankari et al. 2019). Water-soluble secondary metabolite accumulation takes place in vacuoles, whereas lipophilic secondary metabolite accumulates in oil cells, trichomes, raisin ducts, or cuticles; their accumulation sometimes increases more than 10% of the dry weight of plant (Wink and Schimmer 2010). Studies on the distribution of secondary metabolites demonstrate that they are neither ubiquitous in occurrence nor follow any evolutionary patterns (Wink 2003). A detailed understanding of the biosynthesis of plant secondary metabolites is still a subject of further research.

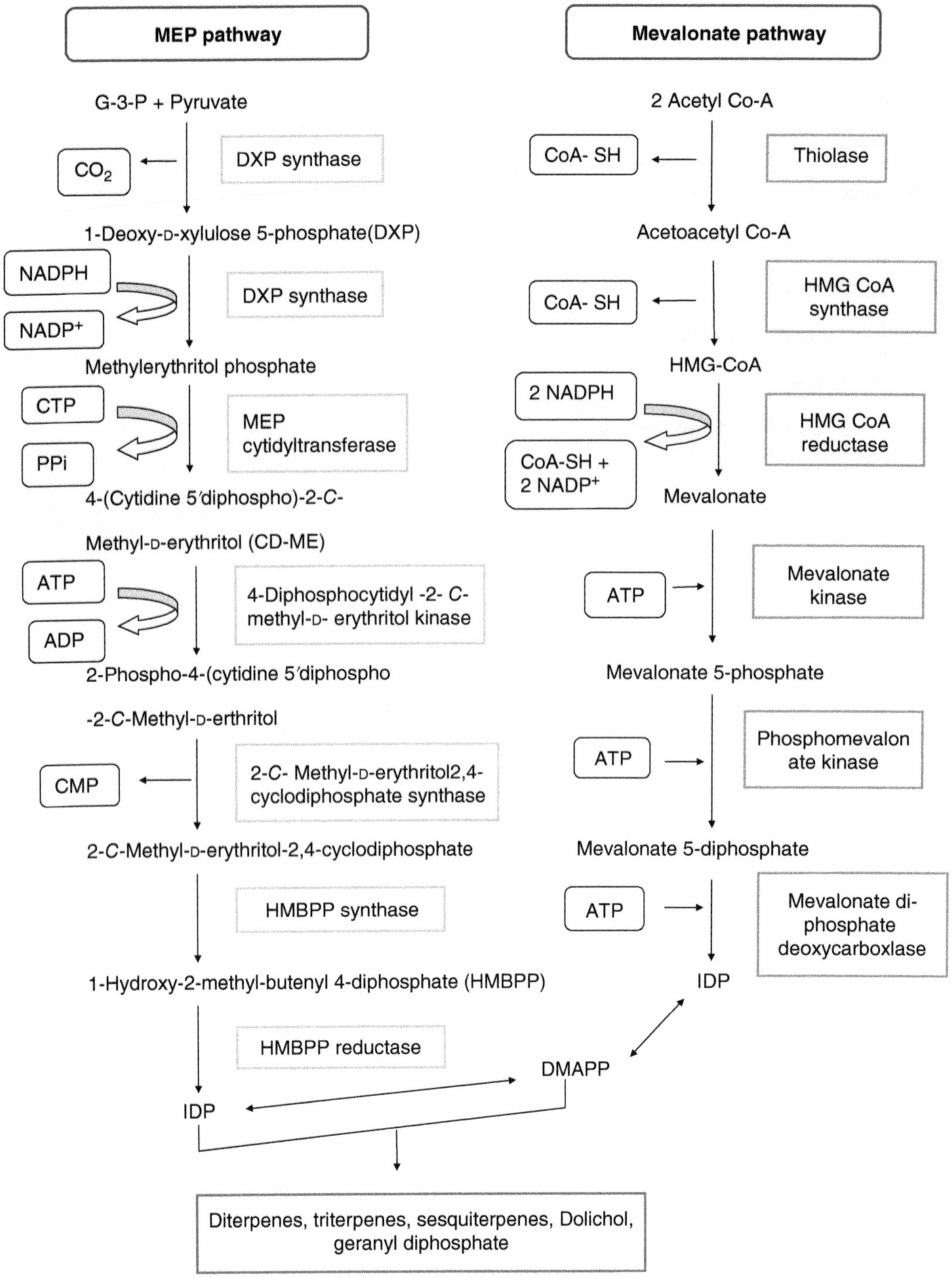

Figure 26.7 The figure shows two pathways (MEP pathway and mevalonate pathway) for the biosynthesis of terpenoids in plants. G-3-P, glyceraldehyde-3-phosphate; HMBPP-1-hydroxy-2-methyl-butenyl 4-diphosphate; MEP, Methylerythritol phosphate; DXP, 1-deoxy-d-xylulose 5-phosphate, HMG, 3-hydroxy-3-methylglutaryl; ATP, adenosine triphosphate; NADPH, nicotinamide adenine dinucleotide phosphate; HMAPP, dimethylallyl diphosphate; IDP, isopentenyl diphosphate.

26.2.1 Role of Secondary Metabolites in Mitigating Abiotic Stress

Plants are exposed to various stresses (drought, heavy metal, temperature, salinity, etc.) during their life span. It was reported by many studies that secondary metabolites play a crucial role in stress management; mechanisms are not fully studied, but it was seen that under stressed condition, the production of secondary metabolites increases.

26.2.2 Secondary Metabolites in Drought Stress Mitigation

Drought stress causes changes in essential physiological and biochemical processes including changes in plant secondary metabolite production. During drought stress, water potential and turgor pressure decrease due to high transpiration rate and less available water for the plant. Altered synthesis of secondary metabolites induces drought resistance in plants through various mechanisms (Verma and Shukla 2015). Secondary metabolites mitigate stress in many ways: they enhance proline biosynthesis, which helps in osmotic stress management (Qu et al. 2019), decrease lipid peroxidation (Treml and Smejkal 2016), promote glycolysis and TCA cycle, and promote ROS scavenging system (Yang et al. 2018). Under drought stress, endogenous level of secondary metabolites increases. Verma and Shukla (2015) reported that in the case of *Artemisa*, high levels of artemisinin, betulinic acid, quercetin, and rutin were observed. *Matteuccia struthiopteris* (L.) Todar. and *Athyrium multidentatum* (Doll.) show increased levels of (proanthocyanidins, flavonoids, and total phenols) content, and these secondary metabolites maintain the osmotic level and enzyme activity (ascorbate peroxidase, catalase, and superoxide dismutase) under drought stress (Wang et al. 2019).

26.2.2.1 Phenolic compounds and drought stress

Phenolic compounds are aromatic compounds that include phenols, lignins, tannins stilbenes, coumarins flavonoids, and catechol melanins. During stress, lignin helps in plant cell wall strengthening and flavonoids activate the ROS scavenging system. Phenols themselves scavenge ROS and protect the plant from oxidative damage (Jamwal et al. 2018). *Echinacea purpurea* and *Labisia pumila* show increases in total phenol content, which helps the plant increase root dry weight; root cichoric acid also promotes PAL activity, photosynthesis rate, and quantum yield (Gray et al. 2003; Jaafar et al. 2012). An increase in quercetin, rutin, 1,5-dihydroxyxanthon, isouliginosin B in *Hypericum brasiliense* promotes antioxidant activity and protects plant from damage (de Abreu and Mazzafera 2005). With the help of metabolomics and proteomics studies, it was revealed that increasing levels of phenol and flavonoids helps in drought resistance in *Arabidopsis* (Nakabayashi et al. 2014). In case of rice, upregulation of flavonoids biosynthesis-related genes *OsDfr* and *OsAns* transcripts were found by the influence of factor OsC1-MYB under drought conditions (Ithal and Reddy 2004). Increased amounts of isoprenoids, phenols, and alkaloids enhance the yield of medicinal plants (Kleinwachter and Selmar 2014). In wheat leaves and shrubs from Argentina, biosynthesis and accumulation of flavonoids were seen throughout their life span, but there was a significant increase in flavonoids production and accumulation reported during drought stress (Ma et al. 2014; Varela et al. 2016). Lignin production and accumulation show effective results under drought stress, and lignin-based hydrogels effectively mitigate drought stress in maize. Lignin-based hydrogels help in enhancing water content; increasing biomass, phosphorus, and proline; and decreasing electrolyte leakage (Mazloom et al. 2020). Phenolic compound accumulation modifies biochemical processes under stressful conditions, hence its utilization as a novel strategy for drought stress mitigation and plant survival.

26.2.2.2 Terpenoids in drought stress tolerance

Terpenoids can be volatile or non-volatile (abscisic acid (ABA), chlorophylls, and carotenoids). Under drought stress, plants produce and accumulate more terpenes and their derivates (Turtola et al. 2003; Ormeno et al. 2007). Apple plant under low humidity induces the production of terpenes such as α-pinene, β-pinene, camphene, β-caryophyllene, (*E*,*E*)-α-farnesene, and limonene (Vallat et al. 2005). Diterpenes were found beneficial for drought stress tolerance in *Salvia* plants. These diterpenes induce the ROS scavenging systems, which reduce the oxidative stress caused by drought (Munne-Bosch et al. 2001). In *Cistus creticus*, some genes named *1-deoxy-*D*-xylulose-5-phosphate synthase* (*CcDXS*), *3-hydroxy-3-methyl-glutaryl-CoA reductase* (*CcHMGR*), *geranylgeranyl diphosphate synthases (CcGGDPS1)*, *1-deoxy-*D*-xylulose-5-phosphate reductoisomerase* (*CcDXR*), and *CcGGDPS2* involved in terpenoids biosynthesis were found to be upregulated under drought stress (Ma et al. 2014). Rapeseed plant increase the production of monoterpenes, (*E*,*E*)-α-farnesene (a sesquiterpene) and camphor (a Terpenoid) under drought stress also some plants emit volatile terpenes or isoprenoids during drought stress to alter the nearby plants against stress (Loreto et al. 2014).

N-containing compounds in drought stress tolerance include phytoalexins, non-protein amino acids, cyanogenic glucosides, indole alkaloids, and glucosinolates. These compounds are directly involved in drought tolerance. N-containing compounds help increase photosynthesis and leaf water status, and these compounds alleviate oxidative damage on the cellular membrane under drought stress in rice (Yang et al. 2007). In sweet potato, ß-aminobutyric acid production helps in drought stress tolerance (Sós-Hegedűs et al. 2014), whereas citrulline content stimulates oxidative stress mitigation in the case of watermelon (Akashi et al. 2001). Cyanogenic glucosides have limited information about their contribution to drought stress tolerance. In some forage plants, cyanogenic glucosides act as antioxidants and help under water deficit condition, whereas antioxidants like thionine and glutathione are involved in detoxifying ROS activity and control the oxidative damage caused by drought stress (Sun et al. 2018). Non-protein amino acid, ornithine, citrulline, β-alanine, ß-aminobutyric acid are some compounds that act as antioxidants and osmoprotectant and alleviate drought stress (Vranova et al. 2011). N-containing compounds have a role in stress mitigation, but they have limitation due to specificity regarding their presence in different plants.

26.2.3 Secondary Metabolites in Mitigating Salinity Stress

Salinity stress is a major constraint to crop productivity; 20% of irrigated land is affected by soil salt. Increased salinity limits the growth of plant, causes osmotic and ionic stress, and disturbs physiological and biochemical processes. Osmotic stress results in low soil water potential reducing water uptake, while ionic stress results in plant cell membrane disturbance and interferes with various metabolic processes (Greenway and Munns 1980; Bernstein 2019). Plants develop various defense mechanisms to cope with salinity stress, including ion compartmentalization and exclusion of toxic ions, prevent salt accumulation (at the whole plant or cellular level), avoid toxic effects, initiating antioxidant pathways, increase production of compatible organic solute including sucrose, proline, glycine-betaine, and sorbitol, change morphological, physiological and biochemical processes (Al Hassan et al. 2015). Flavonoids (secondary metabolites) found helpful in mitigating salinity stress as they act as antioxidant and the detoxify effect of ROS (Di Ferdinando et al. 2012). Anthocyanin accumulation increased during salinity stress and detoxified accumulated salt (Kaliamoortiiy and Rao 1994). In the case of soybean, studies demonstrated that the use of genistein (an isoflavonoid) increases the plant growth (Miransari and Smith 2007). Expression of genes synthesizing enzymes (flavonol synthase, chalcone synthase, and phenylalanine ammonia lyase) responsible for flavonoid synthesis is upregulated under salt stress in *Solanum nigrum* (Ben Abdallah et al. 2016). An increase in tannic acid, flavonoids, gossypol, proteins, and soluble sugars was observed in cotton and sufflower under salt stress (Gengmao et al. 2015; Wang et al. 2015). Secondary metabolites show significant role in salt stress mitigation, yet more detailed study is needed for a better understanding.

26.2.4 Secondary Metabolites as UV Scavengers

UV radiation are short wavelength, high energy radiation that causes damage to plants by breaking chemical bonds. UV radiations are of three types based on the wavelength of radiation: 315–400 nm (UV-A), 280–315 nm (UV-B), and less 280 nm (UV-C). The shorter the wavelength, the more the energy. UV-C has the highest energy and can ionize certain molecules, whereas UV-B affects metabolic molecular processes like photosynthesis, cell division, starch production, also promoting cellular damage, increasing disease susceptibility and hampering the overall growth of plants (Kovacs and Keresztes 2002).

Flavonoids (secondary metabolite) show protective role against UV radiation inside plant cell. Flavonoids, mainly anthocyanin, filter the UV-B radiation, and these flavonoids detoxify the ROS produced by the photochemical reaction done by UV radiation (Chen et al. 2015). A study carried out in tomato demonstrated that high radiation increases the production of protective compounds (phenylpropanoid), which absorb the UV-B and increase the UV receptors and chlorophyll content (Mariz-Ponte et al. 2018). In the case of barley, a significant increase in flavonoid content was seen when plants were exposed to high radiation; these flavonoids (saponarin and lutonarin) help the plants absorb the radiation and mitigate the harmful effects of high radiation (Schmitz-Hoerner and Weissenböck 2003).

26.3 Heavy Metal Stress and Secondary Metabolites

Metal stress is a type of abiotic stress that causes oxidative damage and hampers the metabolic processes of plants. Accumulation of these metals above threshold concentration is toxic for plants; one of the reasons for bioaccumulation and biomagnification of these metals is industrial pollution. Different metal remediation strategies, such as bioaccumulation, rhizofiltration, bio-sorption, volatilization, biotransformation, and bioextraction, are used to remove toxic metals (Verma and Sharma 2017). In contrast, the most economical and environmentally beneficial method of metal treatment is phytoremediation. Internal compartmentalization, metal sequestration, intracellular chelation by phytochelatins and metallothioneins, synthesis of osmoprotectants, and activation of the antioxidant system are some of the intrinsic mechanisms that plants have evolved to alleviate metal stress (Dalvi and Bhalerao 2013; Sharma et al. 2016), recently role of secondary metabolites (isoprenoids, carotenes, and flavonoids) in metal stress tolerance is increasing (Khare et al. 2020).

26.3.1.1 Phenolic compounds and metal stress

Phenolic compounds have redox properties that directly scavenge active oxygen species and help in balancing ROS concentration (Okem et al. 2015). In the case of *Belamcanda chinensis*, total flavonoids (Iristectorigenin A, tectorigenin and tectoridin) increase proline production; enhanced SOD and GPX activities cause copper toxicity tolerance (Zhu et al. 2020). Cadmium stress is reduced by increasing levels of alkaloids, phenolic compounds, and flavonoids, as these compounds ameliorate oxidative stresses caused by cadmium in *Catharanthus roseus* (Abnosi et al. 2015). Márquez-García et al. (2012) reported that the amount of cinnamic acid derivatives, rutin, and epicatechin was increased when cadmium was present below the threshold concentration, helping *Erica andevalensis* survive in the metal-polluted sites. Endogenous levels of anthocyanins and sinapoyl esters increase and activate enzymatic and non-enzymatic antioxidants system to protect the plant from Cu stress in *Brassica oleracea* (Posmyk et al. 2009). Under Pb (lead) stress, *Prosopsis farcta* shows increase in ferulic acid, daidzein, vitexin, and phenolic acids, which enhance the activity of CAT, POD, GPX, and PAL and confer tolerance to Pb (Zafari et al. 2016). According to the studies, phenols reduce uptake of metal ions and detoxify them by activating different defense mechanisms.

Plant produces various secondary metabolites that modify the basic metabolic process of plant during metal stress. Activation of antioxidant system is a strong approach to mitigate oxidative stress caused by metal stress. Antioxidant system comprises enzymatic and non-enzymatic components. Non-enzymatic components include phenolic compounds like anthocyanins, coumarin, flavonoids, lignins, and tannins, as well as ascorbate, glutathione, tocopherol, and carotenoids (Figure 26.8). Most of them are secondary metabolites that help in scavenging ROS and confer metal stress tolerance (Maleki et al. 2017).

Secondary metabolites help in metal chelation. Chelation is an intracellular mechanism to maintain the free metal ion concentration by sequestering the metal ions to vacuole or to the other compartments from the cytosol of the plant cell. Compounds that help in chelation include histidine, phytochelatins, glutathione, metallothioneins, organic acids, and nicotinamide. Compounds like phenols, alkaloids and saponins also act as chelators and help in metal sequestering directly

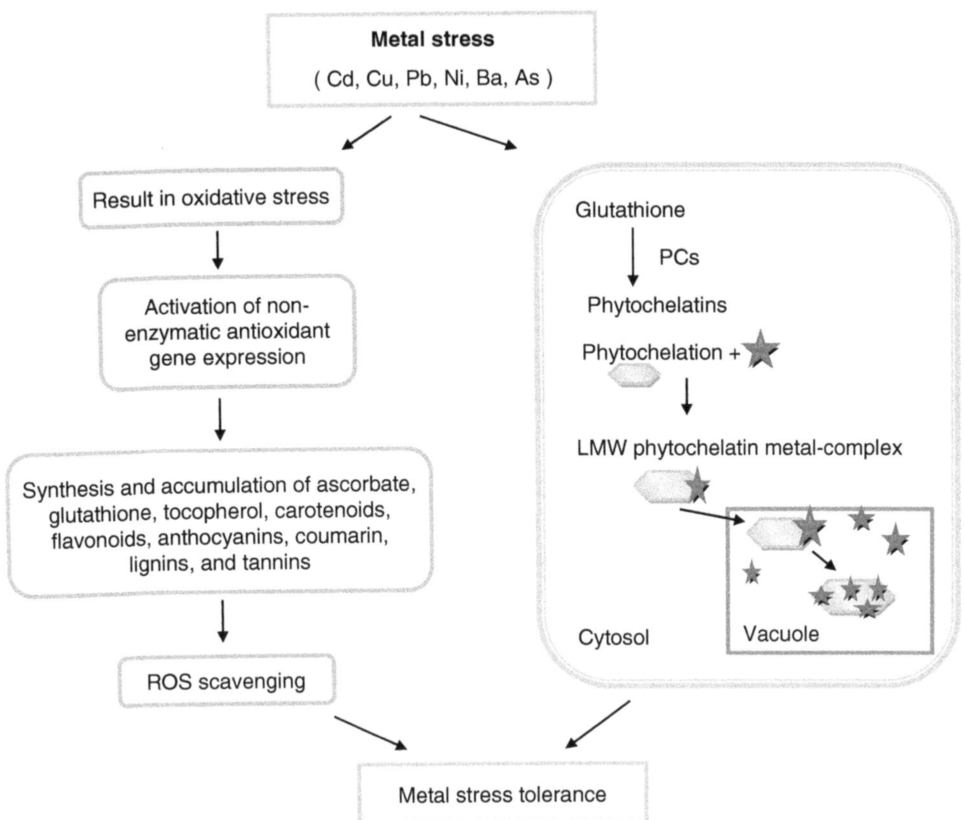

Figure 26.8 Heavy metal stress tolerance by non-enzymatic antioxidant system, which comprises members of secondary metabolites and metal sequestration with the help of phytochelatins. **PCs**, phytochelatin synthase; **LMW**, low molecular weight; , high molecular weight phytochelatin metal-complex; , heavy metal.

or in chelation complex formation (Anjum et al. 2015; Nobahar et al. 2021). Studies reported involvement of secondary metabolites as chelators, e.g. in vitro chelation of Pb, Cr, and Hg carried out with the help of polyphenols in *Nymphaea* (Lavid et al. 2001). Alkaloids in *Cinchona* form chelation complex with Fe, Pb, Cu, and Co and help in metal stress tolerance (Boratyński et al. 2019). Anthocyanins can also chelate with metals due to the presence of adjacent hydroxyl groups. Cyanidin forms a complex with zinc (cyanidin-Zn) in *Zea mays* when exposed to Zn stress and confers the Zn stress tolerance (Janeeshma et al. 2021). Reports indicate the role of glutathione in metal stress tolerance as it detoxifies ROS and helps in chelation of metal within the plant cell. 3-Glutamyl cysteine dipeptidyl transpeptidase called phytochelatin synthase (PCS) synthesize phytochelatins from GSH. These phytochelatins form complex with free metal ions in cytosol and move to vacuole (Vatamaniuk et al. 2004), metal sequestration is mediated by phytochelatins as represented in Figure 26.8. Transgenic *Arabidopsis thaliana* plants are sensitive to Cd with low GSH level as compared to plants with high GSH (Xiang et al. 2001). This study confirms the role of GSH in detoxification and metal sequestration.

26.3.1 Role of Secondary Metabolites in Biotic Stress Mitigation

Due to their sessile nature, plants are vulnerable to a variety of herbivores, nematodes, and diseases (fungus, bacteria, and viruses). Different strategies have been developed by plants to fend off these diseases' effects and attacks. Most of these pathogens are specific to host plant; they show incompatible interaction with non-host plants and the plants that are resistant to that particular host; however, they cause serious damage to the compatible host plants that are susceptible (Koes et al. 1994). Plants show three lines of defense against foreign invaders. These lines of defense are physical barriers, chemical barriers, and specific adaptive responses. Physical barriers/structural barriers are cutin, suberin, wax coating, trichomes, and epidermal cells, which prevent penetration of pathogens. In chemical defense, plant produces secondary nonspecific chemicals like phytoalexins, glucosinolates, saponins, and cyanogenic glycosides, which counteract the pathogen. Specific adaptive defense work through signaling pathways like mitogen-activated protein kinase (MAPK) and hormonal signaling, which begin with the recognition of foreign invader and result in the activation of defense responsive genes. Recognition of the foreign invader is done by two ways: the plant either recognizes the injury caused by invader or recognizes the secretion (elicitors) of invader (Bonaventure 2014).

26.3.1.1 Terpenoids and Biotic Stress

Terpenoids are primary components of essential oils, and these essential oils protect plant from insect and pathogens attack. *Chrysanthemum* plants produce some monoterpenes ester, like pyrethrins, which acts as neurotoxin for insects and protects the plants from insects (Rosell et al. 2008). Whereas pine trees produce alpha and beta pines (monoterpenoids), which repel insects from their significant odor. Some diterpenoids also play a role in plant protection like Azadirachtin from *Azadirachta indica* and Citronella from *Cymbopogon citratus* (lemon grass), act as strong repellants for insects, *Gossypium hirsutum* produce gossypol acts against bacterial and fungal pathogens (Freeman and Beattie 2008).

26.3.1.2 Phenolic Compounds and Biotic Stress

Flavonoids (phytoalexins) are the largest and most-described class of phenols in plant defense. These toxins alter pathogens cellular structure and metabolism. In response to the invasion of particular pathogen, plants produce different phytoalexins to combat the effect of the pathogen. Flavonoids as insect repellent, *Medicago sativa* produces medicarpin; tomatoes and potatoes produce rishitin and camalexin, produced by *A. thaliana* (Freeman and Beattie 2008). Isoglabratephrin (þ)-glabratephrin, tephroapollin-F, and lanceolatin-A are some flavonoids that show toxic effect against *Sitophilus oryzae, Rhyzopertha dominica*, and *Tribolium* insects, respectively (Nenaah 2014). In the case of clover plant, genistein and formononetin (isoflavonoids) disturb the harmful activity of *Hylastinus obscures* (Quiroz et al. 2017).

Flavonoids against nematodes: Glyceollin (an isoflavonoid) acts against *Meloidogyne incognita* and *Heterodera glycines*. Soybean glyceollin inhibits nematode oxygen uptake; studies show a significant increase in glyceollin concentration as infection increases, whereas no glyceollin is produced in the control sample. (Kaplan et al. 1980; Huang and Barker 1991).

Flavonoids against bacteria: Coumestrol, phytoalexin (isoflavonoid), are produced in response to the *Pseudomonas* spp. infection in bean plants (Lyon and Wood 1975), whereas Phaseollin isoflavan and kievitone inhibit the effects of *Xanthomonas* and *Achromobacter* species (Wyman and Van Etten 1978).

Flavonoids as protectant against pathogenic fungi: *Phytophthora citrophthora* fungi cause disease in citrus fruits; accumulation of certain flavonoids (heptamethoxyflavone, nobiletin, sinensetin, and tangeretin) found effective against this fungal infection and reduces the harmful effect (del Río et al. 2004).

26.3.1.3 Nitrogen-Containing Compound and Biotic Stress

Nitrogen-containing compounds (alkaloids, saponins, cyanogenic glycosides, and glucosinolates) are the secondary metabolites present in healthy plants in active or inactive (cyanogenic glycosides and glucosinolates) forms and activated when pathogens attack or damage the plant cell; their activation involves the role of enzymes produced in response to cell damage (Mansfield 1983). Caffeine is a type of alkaloid produced by *Coffea arabica*, *Camellia sinensis*, and *Theobroma cacao* plant. It shows antifungal activity also inhibits the effect of insect. Caffeine not only protects plant from pathogen attack but also shows allelopathy (inhibit the germination of other plants in its vicinity) (Freeman and Beattie 2008).

Saponin in biotic stress mitigation: Based on the structure, saponins are of three types: a triterpenoid saponin, a steroid saponin, and a steroidal glycoalkaloid saponin. Triterpenoid saponin is mainly present in dicots and in some monocots, whereas, steroid saponin is mostly present in monocots and in some dicots. Steroidal glycoalkaloid saponin is mainly found in Solanaceae and Liliaceae family (Hostettmann and Manton 1995). Oats have two types of saponins, the triterpenoid saponin (avenacins) and the steroidal saponin (avenacosides), and these saponins perform antifungal activity to protect the plants from pathogenic fungi (Osbourn 1996).

Cyanogenic glycosides and biotic stress: Cyanogenic glycosides on break down produce hydrogen cyanide (HCN) that acts as respiratory poison for most of the organisms and provides defense against pathogens and herbivores (Davis 1991). Glucosinolates are mainly found in Brassicaceae. Thioglucosidases enzymes act on glucosinolates and produce cyanide gas. Glucosinolates mainly act as feeding deterrents and protect plants from herbivores (Mithen and Magrath 1992).

26.4 Counteradaptation of Insects Against Secondary Metabolites

Insect herbivores use different strategies to adapt against secondary metabolites, like mutation of target sites, detoxifying plant toxins or storing them in adipose tissue, improve activity of digestive enzyme, develop antioxidant defense, and special excretion of toxins. Insects store some toxic secondary metabolites and later use them to protect their eggs from predators (Afroz et al. 2021). Particular proteins, olfactory receptors (ORs), gustatory receptors (GRs), and odorant-binding proteins (OBPs) aid in toxin identification and adaption, with GRs genes primarily detecting secondary compounds (Leal 2013). Insects develop a specific feeding habit; they usually search for those plants species that have less amount of secondary metabolites; they fasten up their digestive system and excrete out the toxin; and they have some microorganisms in their intestine, which help in toxin degradation (Pennisi 2017). In order to develop pest management strategies, one should know the details of insect counteradaptation as insect counteradaptation strategies are complex and need in-depth investigation.

26.5 Sustainable Crop Protection and Secondary Metabolites

To develop as cost-effective and ecofriendly approach for pest management and crop protection, secondary metabolites are good substitute for chemical pesticides; being natural compounds, secondary metabolites are biodegradable, less toxic to humans, and non-pollutant, hence they did not show any negative effect on the environment (Alam et al. 2016). Crop varieties with enhanced secondary metabolites production do not need chemically synthesized insecticide for pest management, e.g. nicotine produced in tobacco plant acts as herbivore repellent and also helps in attracting the predator *Geocoris pallens* when attacked by *Manduca sexta* (Zhou et al. 2017). Pyrethrins and azadirachtin are important plant secondary metabolites used as insect repellents and also used in chemically synthesized insecticides that are less toxic to humans and the environment (Isman and Grieneisen 2014).

26.6 Conclusion

Biotic and abiotic stresses are main culprit in reducing crop yield. These stresses not only affect the normal physiological and biochemical processes but also, in some conditions, cause cellular toxicity. In addition to physiological and biochemical processes, sometimes these stresses affect the plants at a molecular level as well. Hence, plants evolved some defense system at morphological, physiological, and biochemical levels to cope with the destructive effects of stresses. Secondary metabolites play an indispensable role against plant biotic and abiotic stresses and maintain the crop growth. Manipulating biosynthetic pathways of these secondary metabolites can be a good approach for mitigating stress as reports suggested that the overexpression or increased synthesis of secondary metabolites is beneficial in combating stress. Herbivores develop counteradaptation strategies against the plant defense; hence, there is a need to quantify and identify unidentified secondary metabolites using a metabolomic approach. To fulfil the aim of sustainable crop production even under stressful environment, we need a more detailed investigation and better strategies because plant faces significant decrease in crop production despite sophisticated internal defense mechanism and adaptation at morphological and physiological levels. Hence further research is needed to understand the fine details of the signaling mechanism for defense and come up with better strategies for mitigating the harmful effects.

References

Abnosi, M., Amirjani, M., Mahdiyeh, M., and Moradipoor, H. (2015). Biochemical and cellular response of *Catharanthus roseus* callus cells to cadmium toxicity. *J. Genet. Resour.* 1 (2): 101–114.

de Abreu, I.N. and Mazzafera, P. (2005). Effect of water and temperature stress on the content of active constituents of *Hypericum brasiliense* Choisy. *Plant Physiol. Biochem.* 43: 241–248.

Afroz, M., Rahman, M., and Amin, R. (2021). Insect plant interaction with reference to secondary metabolites: a review. *Agric. Rev.* 42 (4): 427–433.

Aharoni, A., Jongsma, M.A., and Bouwmeester, H.J. (2005). Volatile science? Metabolic engineering of terpenoids in plants. *Trends Plant Sci.* 10 (12): 594–602.

Akashi, K., Miyake, C., and Yokota, A. (2001). Citrulline, a novel compatible solute in drought tolerant wild watermelon leaves, is an efficient hydroxyl radical scavenger. *FEBS Lett.* 508: 438–442.

Al Hassan, M., Fuertes, M.M., Sánchez, F.J.R. et al. (2015). Effects of salt and water stress on plant growth and on accumulation of osmolytes and antioxidant compounds in cherry tomato. *Notulae Botanicae Horti Agrobotanici Cluj-Napoca* 43 (1): 1–11.

Alam, M.Z., Haque, M., Islam, M. et al. (2016). Comparative study of integrated pest management and farmers practices on sustainable environment in the rice ecosystem. *Int. J. Zool.* 2016: 1–12.

Aniszewski, T. (2007). *Alkaloids-Secrets of Life: Aklaloid Chemistry, Biological Significance, Applications and Ecological Role.* Elsevier.

Anjum, N.A., Hasanuzzaman, M., Hossain, M.A. et al. (2015). Jacks of metal/metalloid chelation trade in plants—an overview. *Front. Plant Sci.* 6: 192.

Ben Abdallah, S., Aung, B., Amyot, L. et al. (2016). Salt stress (NaCl) affects plant growth and branch pathways of carotenoid and flavonoid biosyntheses in *Solanum nigrum*. *Acta Physiol. Plant.* 38 (3): 1–13.

Bernstein, N. (2019). Plants and salt: plant response and adaptations to salinity. In: *Model Ecosystems in Extreme Environments* (ed. J. Seckbach and P.H. Rampelotto), 101–112. Academic Press.

Bhonwong, A., Stout, M.J., Attajarusit, J., and Tantasawat, P. (2009). Defensive role of tomato polyphenol oxidases against cotton bollworm (*Helicoverpa armigera*) and beet armyworm (*Spodoptera exigua*). *J. Chem. Ecol.* 35 (1): 28–38.

Bonaventure, G. (2014). Plants recognize herbivorous insects by complex signalling networks. *Annu. Plant Rev. Insect Plant Interact.* 47: 1–35.

Boratyński, P.J., Zielińska-Błajet, M., and Skarżewski, J. (2019). Cinchona alkaloids—derivatives and applications. *Alkaloids Chem. Biol.* 82: 29–145.

Chen, H., Gao, H., Fang, X. et al. (2015). Effects of allyl isothiocyanate treatment on postharvest quality and the activities of antioxidant enzymes of mulberry fruit. *Postharvest Biol. Technol.* 108: 61–67.

Cheynier, V., Comte, G., Davies, K.M. et al. (2013). Plant phenolics: recent advances on their biosynthesis, genetics, and ecophysiology. *Plant Physiol. Biochem.* 72: 1–20.

Dalvi, A.A. and Bhalerao, S.A. (2013). Response of plants towards heavy metal toxicity: an overview of avoidance, tolerance and uptake mechanism. *Ann. Plant Sci.* 2 (9): 362–368.

Davis, R.H. (1991). Glucosinolates. In: *Toxic Substances in Crop Plants* (ed. J.P. DMello, C.M. Duffus, and J.H. Duffus), 202–225. Cambridge, UK: Royal Society of Chemistry.

Del Rio, J.A., Gómez, P., Baidez, A.G. et al. (2004). Changes in the levels of polymethoxyflavones and flavanones as part of the defense mechanism of *Citrus sinensis* (Cv. Valencia Late) fruits against *Phytophthora citrophthora*. *J. Agric. Food Chem.* 52 (7): 1913–1917.

Eid, S.Y., El-Readi, M.Z., Fatani, S.H. et al. (2015). Natural products modulate the multifactorial multidrug resistance of cancer. *Pharmacol. Pharm.* 6 (03): 146.

Ferdinando, M.D., Brunetti, C., Fini, A., and Tattini, M. (2012). Flavonoids as antioxidants in plants under abiotic stresses. In: *Abiotic Stress Responses in Plants* (ed. P. Ahmad and M.N.V. Prasad), 159–179. Springer Science.

Freeman, B.C. and Beattie, G.A. (2008). *An Overview of Plant Defenses against Pathogens and Herbivores*, The Plant Health Instructor. St. Paul: *American Phytopathological Society*.

Gengmao, Z., Yu, H., Xing, S. et al. (2015). Salinity stress increases secondary metabolites and enzyme activity in safflower. *Ind. Crops Prod.* 64: 175–181.

Gleadow, R.M. and Møller, B.L. (2014). Cyanogenic glycosides: synthesis, physiology, and phenotypic plasticity. *Annu. Rev. Plant Biol.* 65 (1): 155–185.

Gleadow, R.M. and Woodrow, I.E. (2000). Temporal and spatial variation in cyanogenic glycosides in *Eucalyptus cladocalyx*. *Tree Physiol.* 20 (9): 591–598.

Gray, D.E., Pallardy, S.G., Garrett, H., and Rottinghaus, G.E. (2003). Acute drought stress and plant age effects on alkamide and phenolic acid content in purple conefower roots. *Planta Med.* 69: 50–55.

Greenway, H. and Munns, R. (1980). Mechanisms of salt tolerance in nonhalophytes. *Annu. Rev. Plant Physiol.* 31 (1): 149–190.

Halkier, B.A. and Gershenzon, J. (2006). Biology and biochemistry of glucosinolates. *Annu. Rev. Plant Biol.* 57 (1): 303–333.

Hopkins, D.P., Cameron, D.D., and Butlin, R.K. (2017). The chemical signatures underlying host plant discrimination by aphids. *Sci. Rep.* 7: 8498.

Hostettmann, K.A. and Manton, A. (1995). *Saponins. Chemistry and Pharmacology of Natural Products*. Cambridge, UK: Cambridge University Press.

Huang, J.S. and Barker, K.R. (1991). Glyceollin I in soybean-cyst nematode interactions: spatial and temporal distribution in roots of resistant and susceptible soybeans. *Plant Physiol.* 96 (4): 1302–1307.

Hussain, M.S., Fareed, S., Saba Ansari, M. et al. (2012). Current approaches toward production of secondary plant metabolites. *J. Pharm. Bioallied Sci.* 4: 10.

Isman, M.B. and Grieneisen, M.L. (2014). Botanical insecticide research: many publications, limited useful data. *Trends Plant Sci.* 19 (3): 140–145.

Ithal, N. and Reddy, A.R. (2004). Rice favonoid pathway genes, OsDfr and OsAns, are induced by dehydration, high salt and ABA, and contain stress responsive promoter elements that interact with the transcription activator, OsC1-MYB. *Plant Sci.* 166: 1505–1513.

Jaafar, H.Z., Ibrahim, M.H., and MohamadFakri, N.F. (2012). Impact of soil field water capacity on secondary metabolites, phenylalanine ammonia-lyase (PAL), maliondialdehyde (MDA) and photosynthetic responses of Malaysian Kacip Fatimah (*Labisia pumila* Benth). *Molecules* 17: 7305–7322.

Jamwal, K., Bhattacharya, S., and Puri, S. (2018). Plant growth regulator mediated consequences of secondary metabolites in medicinal plants. *J. Appl. Res. Med. Aromat. Plant* 9: 26–38.

Janeeshma, E., Rajan, V.K., and Puthur, J.T. (2021). Spectral variations associated with anthocyanin accumulation; an apt tool to evaluate zinc stress in *Zea mays* L. *Chem. Ecol.* 37 (1): 32–49.

Kaliamoortiiy, S. and Rao, A. (1994). Effect of salinity on anthocyanin accumulation in the root of maize. *Science* 248: 1637–1638.

Kaplan, D.T., Keen, N.T., and Thomason, I.J. (1980). Studies on the mode of action of glyceollin in soybean incompatibility to the root knot nematode, *Meloidogyne incognita*. *Physiol. Plant Pathol.* 16 (3): 319–325.

Kessler, A. and Kalske, A. (2018). Plant secondary metabolite diversity and species interactions. *Annu. Rev. Ecol. Evol. Syst.* 49: 115–138.

Khare, S., Singh, N.B., Singh, A. et al. (2020). Plant secondary metabolites synthesis and their regulations under biotic and abiotic constraints. *J. Plant Biol.* 63 (3): 203–216.

Kleinwachter, M. and Selmar, D. (2014). Influencing the product quality by applying drought stress during the cultivation of medicinal plants. In: *Physiological Mechanisms and Adaptation Strategies in Plants Under Changing Environment* (ed. P. Ahmad and M.R. Wani), 57–73. Springer.

Koes, R.E., Quattrocchio, F., and Mol, J.N. (1994). The flavonoid biosynthetic pathway in plants: function and evolution. *BioEssays* 16 (2): 123–132.

Kovacs, E. and Keresztes, A. (2002). Effect of gamma and UV-B/C radiation on plant cells. *Micron* 33 (2): 199–210.

Lavid, N., Schwartz, A., Yarden, O., and Tel-Or, E. (2001). The involvement of polyphenols and peroxidase activities in heavy-metal accumulation by epidermal glands of the waterlily (Nymphaeaceae). *Planta* 212 (3): 323–331.

Leal, W.S. (2013). Odorant reception in insects: roles of receptors, binding proteins, and degrading enzymes. *Annu. Rev. Entomol.* 58 (1): 373–391.

Loreto, F., Pollastri, S., Fineschi, S., and Velikova, V. (2014). Volatile isoprenoids and their importance for protection against environmental constraints in the Mediterranean area. *Environ. Exp. Bot.* 103: 99–106.

Lyon, F.M. and Wood, R.K.S. (1975). Production of phaseollin, coumestrol and related compounds in bean leaves inoculated with *Pseudo monas* spp. *Physiol. Plant Pathol.* 6 (2): 117–124.

Ma, D., Sun, D., Wang, C. et al. (2014). Expression of flavonoid biosynthesis genes and accumulation of flavonoid in wheat leaves in response to drought stress. *Plant Physiol. Biochem.* 80: 60–66.

Maleki, M., Ghorbanpour, M., and Kariman, K. (2017). Physiological and antioxidative responses of medicinal plants exposed to heavy metals stress. *Plant Gene* 11: 247–254.

Mansfield, J.W. (1983). Antimicrobial compounds. In: *Biochemical Plant Pathology* (ed. J.A. Callow), 237–265. Chichester, UK: Wiley.

Mantri, N., Patade, V., Penna, S. et al. (2012). Abiotic stress responses in plants: present and future. In: *Abiotic Stress Responses in Plants* (ed. P. Ahmad and M.N.V. Prasad), 1–19. New York: Springer.

Mariz-Ponte, N., Mendes, R.J., Sario, S. et al. (2018). Tomato plants use non-enzymatic antioxidant pathways to cope with moderate UV-A/B irradiation: a contribution to the use of UV-A/B in horticulture. *J. Plant Physiol.* 221: 32–42.

Márquez-García, B., Fernández-Recamales, M., and Córdoba, F. (2012). Effects of cadmium on phenolic composition and antioxidant activities of *Erica andevalensis*. *J. Bot.* 2012: 936950.

Mazloom, N., Khorassani, R., Zohury, G.H. et al. (2020). Lignin-based hydrogel alleviates drought stress in maize. *Environ. Exp. Bot.* 175: 104055.

McCulley, R.L., Bush, L.P., Carlisle, A.E. et al. (2014). Warming reduces tall fescue abundance but stimulates toxic alkaloid concentrations in transition zone pastures of the US. *Front. Chem.* 2: 88.

Meena, K.K., Sorty, A.M., Bitla, U.M. et al. (2017). Abiotic stress responses and microbe-mediated mitigation in plants: the omics strategies. *Front. Plant Sci.* 8: 172.

Miransari, M. and Smith, D.L. (2007). Overcoming the stressful effects of salinity and acidity on soybean nodulation and yields using signal molecule genistein under field conditions. *J. Plant Nutr.* 30 (12): 1967–1992.

Mithen, R. and Magrath, R. (1992). Glucosinolates and resistance to *Leptosphaeria maculans* in wild and cultivated Brassica species. *Plant Breed.* 108: 60–68.

Munné-Bosch, S., Mueller, M., Schwarz, K., and Alegre, L. (2001). Diterpenes and antioxidative protection in drought-stressed *Salvia offcinalis* plants. *J. Plant Physiol.* 158: 1431–1437.

Nakabayashi, R., Yonekura-Sakakibara, K., Urano, K. et al. (2014). Enhancement of oxidative and drought tolerance in Arabidopsis by over accumulation of antioxidant flavonoids. *Plant J.* 77: 367–379.

Nenaah, G.E. (2014). Toxic and antifeedant activities of prenylated flavonoids isolated from *Tephrosia apollinea* L. against three major coleopteran pests of stored grains with reference to their structure–activity relationship. *Nat. Prod. Res.* 28 (24): 2245–2252.

Nobahar, A., Carlier, J.D., Miguel, M.G., and Costa, M.C. (2021). A review of plant metabolites with metal interaction capacity: a green approach for industrial applications. *BioMetals* 34 (4): 761–793.

Nowacki, E. and Waller, G.R. (1973). A study on the origin and protective function of alkaloids in plants: 1. *Protoparce sexta*, an insect which is tolerant to a broad spectrum of alkaloids. *Flora* 162 (1-2): 108–117.

Okem, A., Stirk, W.A., Street, R.A. et al. (2015). Effects of Cd and Al stress on secondary metabolites, antioxidant and antibacterial activity of *Hypoxis hemerocallidea* Fisch. & CA Mey. *Plant Physiol. Biochem.* 97: 147–155.

Ormeno, E., Mevy, J., Vila, B. et al. (2007). Water deficit stress induces different monoterpene and sesquiterpene emission changes in Mediterranean species. Relationship between terpene emissions and plant water potential. *Chemosphere* 67: 276–284.

Osbourn, A.E. (1996). Saponins and plant defence-A soap story. *Trends Plant Sci.* 1: 4–9.

Pennisi, E. (2017). How do gut microbes help herbivores? Counting the ways. *Science* 355: 6322.

Posmyk, M.M., Kontek, R., and Janas, K.M. (2009). Antioxidant enzymes activity and phenolic compounds content in red cabbage seedlings exposed to copper stress. *Ecotoxicol. Environ. Saf.* 72 (2): 596–602.

Qu, X., Wang, H., Chen, M. et al. (2019). Drought stress–induced physiological and metabolic changes in leaves of two oil tea cultivars. *J. Am. Soc. Hortic. Sci.* 144: 439–447.

Quiroz, A., Mendez, L., Mutis, A. et al. (2017). Antifeedant activity of red clover root isoflavonoids on *Hylastinus obscurus*. *J. Soil Sci. Plant Nutr.* 17 (1): 231–239.

Radušienė, J., Karpavičienė, B., and Stanius, Ž. (2012). Effect of external and internal factors on secondary metabolites accumulation in St. John's worth. *Bot. Lithuanica* 18 (2): 101–108.

Rosell, G., Quero, C., Coll, J., and Guerrero, A. (2008). Biorational insecticides in pest management. *J. Pestic. Sci.* 33 (2): 103–121.

Sankari, M., Hridya, H., Sneha, P. et al. (2019). Implication of salt stress induces changes in pigment production, antioxidant enzyme activity, and qRT-PCR expression of genes involved in the biosynthetic pathway of *Bixa orellana* L. *Funct. Integr. Genomics* 19 (4): 565–574.

Schmitz-Hoerner, R. and Weissenböck, G. (2003). Contribution of phenolic compounds to the UV-B screening capacity of developing barley primary leaves in relation to DNA damage and repair under elevated UV-B levels. *Phytochemistry* 64 (1): 243–255.

Sharma, R., Bhardwaj, R., Handa, N. et al. (2016). Responses of phytochelatins and metallothioneins in alleviation of heavy metal stress in plants: an overview. In: *Plant Metal Interaction: Emerging Remediation Techniques* (ed. P. Ahmad), 263–283. Elsevier.

Sós-Hegedűs, A., Juhász, Z., Poór, P. et al. (2014). Soil drench treatment with ß-aminobutyric acid increases drought tolerance of potato. *PLoS One* 9: e114297.

Sun, Z., Zhang, K., Chen, C. et al. (2018). Biosynthesis and regulation of cyanogenic glycoside production in forage plants. *Appl. Microbiol. Biotechnol.* 102: 9–16.

Treml, J. and Smejkal, K. (2016). Flavonoids as potent scavengers of hydroxyl radicals. *Compr. Rev. Food Sci. Food Saf.* 15: 720–738.

Turtola, S., Manninen, A.-M., Rikala, R., and Kainulainen, P. (2003). Drought stress alters the concentration of wood terpenoids in scots pine and Norway spruce seedlings. *J. Chem. Ecol.* 29: 1981–1995.

Vallat, A., Gu, H., and Dorn, S. (2005). How rainfall, relative humidity and temperature influence volatile emissions from apple trees *in situ*. *Phytochemistry* 66: 1540–1550.

Varela, M.C., Arslan, I., Reginato, M.A. et al. (2016). Phenolic compounds as indicators of drought resistance in shrubs from *Patagonian shrublands* (Argentina). *Plant Physiol. Biochem.* 104: 81–91.

Vatamaniuk, O.K., Mari, S., Lang, A. et al. (2004). Phytochelatin synthase, a dipeptidyltransferase that undergoes multisite acylation with γ-glutamylcysteine during catalysis: stoichiometric and site-directed mutagenic analysis of *Arabidopsis thaliana* PCS1-catalyzed phytochelatin synthesis. *J. Biol. Chem.* 279 (21): 22449–22460.

Verma, A.K. and Deepti, S. (2016). Abiotic stress and crop improvement: current scenario. *Adv. Plants Agric. Res.* 4 (4): 345–346.

Verma, N. and Sharma, R. (2017). Bioremediation of toxic heavy metals: a patent review. *Recent Pat. Biotechnol.* 11 (3): 171–187.

Verma, N. and Shukla, S. (2015). Impact of various factors responsible for fluctuation in plant secondary metabolites. *J. Appl. Res. Med. Aromat. Plant* 2: 105–113.

Vranova, V., Rejsek, K., Skene, K.R., and Formanek, P. (2011). Non-protein amino acids: plant, soil and ecosystem interactions. *Plant Soil* 342: 31–48.

Wang, Q., Eneji, A.E., Kong, X. et al. (2015). Salt stress effects on secondary metabolites of cotton in relation to gene expression responsible for aphid development. *PLoS One* 10 (6): e0129541.

Wang, Y., Gao, S., He, X. et al. (2019). Growth, secondary metabolites and enzyme activity responses of two edible fern species to drought stress and rehydration in Northeast China. *Agronomy* 9: 137.

War, A.R., Buhroo, A.A., Hussain, B. et al. (2020). Plant defense and insect adaptation with reference to secondary metabolites. In: *Co-Evolution of Secondary Metabolites* (ed. J.M. Mérillon and K.G. Ramawat), 795–822. Cham: Springer.

Willis, K.J. (Ed.) *State of the World's Plants*; Report Royal Botanic Gardens: London, 2017; https://www.kew.org/about-us/press-media/state-of-the-worlds-plants-2017 (accessed 15 March 2023).

Wink, M. (2003). Evolution of secondary metabolites from an ecological and molecular phylogenetic perspective. *Phytochemistry* 64 (1): 3–19.

Wink, M. and Schimmer, O. (2010). Molecular modes of action of defensive secondary metabolites. In: *Annual Plant Reviews Volume 39: Functions and Biotechnology of Plant Secondary Metabolites*, vol. 39 (ed. M. Wink), 21–161. Wiley.

Wyman, J.G. and Van Etten, H.D. (1978). Antibacterial activity of selected isoflavonoids. *Phytopathology* 68: 583–589.

Xiang, C., Werner, B.L., Christensen, E.L.M., and Oliver, D.J. (2001). The biological functions of glutathione revisited in Arabidopsis transgenic plants with altered glutathione levels. *Plant Physiol.* 126 (2): 564–574.

Yang, J., Zhang, J., Liu, K. et al. (2007). Involvement of polyamines in the drought resistance of rice. *J. Exp. Bot.* 58: 1545–1555.

Yang, L., Wen, K.-S., Ruan, X. et al. (2018). Response of plant secondary metabolites to environmental factors. *Molecules* 23: 762.

Zafari, S., Sharifi, M., Chashmi, N.A., and Mur, L.A. (2016). Modulation of Pb-induced stress in *Prosopis* shoots through an interconnected network of signaling molecules, phenolic compounds and amino acids. *Plant Physiol. Biochem.* 99: 11–20.

Zhou, W., Kügler, A., McGale, E. et al. (2017). Tissue-specific emission of (E)-α-bergamotene helps resolve the dilemma when pollinators are also herbivores. *Curr. Biol.* 27 (9): 1336–1341.

Zhu, Y., Chen, Y., Zhang, X. et al. (2020). Copper stress-induced changes in biomass accumulation, antioxidant activity and flavonoid contents in *Belamcanda chinensis* calli. *Plant Cell Tiss. Organ Cult.* 142: 299–311.

27

Osmolytes

Efficient Oxidative Stress-Busters in Plants

Naser A. Anjum[1], Palaniswamy Thangavel[2], Faisal Rasheed[1], Asim Masood[1], Hadi Pirasteh-Anosheh[3,4], and Nafees A. Khan[1]

[1] *Department of Botany, Aligarh Muslim University, Aligarh, India*
[2] *Department of Environmental Science, Periyar University, Salem, India*
[3] *Department of Agronomy Research, National Salinity Research Center, Agricultural Research, Education and Extension Organization (AREEO), Yazd, Iran*
[4] *Natural Resources Department, Fars Agricultural and Natural Resources Research and Education Center, AREEO, Shiraz, Iran*

27.1 Introduction

27.1.1 Plant Health, Stress Factors, and Oxidative Stress and Its Markers

In their life span, plants inevitably encounter various abiotic stresses. As a common response, elevated or toxic metal-metalloids, drought, temperature extremes, and soil acidity/saline soils widely cause restricted water availability and cellular turgor loss and, eventually, impact plant health and productivity (Anjum et al. 2014a; Batista-Silva et al. 2019). An increased production of reactive oxygen species (ROS) and eventual damages to proteins, lipids, and nucleic acids are consequences of the stomatal closure, altered nutrient, and cellular redox homeostasis followed by inhibited CO_2 assimilation (Batista-Silva et al. 2019).

In plants growing under optimal conditions, aerobic metabolism in chloroplasts, mitochondria, and peroxisomes yields as by-products varied ROS, the hydroxyl radical (OH•), hydrogen peroxide (H_2O_2), superoxide ($O_2^{\bullet-}$), and singlet oxygen ($^1O_2$). Interestingly, the availability of several antioxidants in these organelles keeps a tight control over the cellular ROS and tends to maintain their level at optimum, thereby achieving cell redox homeostasis (Mittler 2002; Mittler et al. 2004; Gill and Tuteja 2010). Physiological optimal levels of ROS contribute to signaling mechanisms and are thereby involved in growth, development, and stress responses (Møller and Sweetlove 2010; Mittler et al. 2011; Anjum et al. 2022). In fact, a fine homeostasis of redox conditions is maintained in the cells with the help of the synchronous action of different components of the plant's antioxidant defense system. However, as an overlapping response, most of the mentioned abiotic stresses cause oxidative stress by causing an imbalance between oxidants (ROS) and antioxidants. On the one hand, the list of oxidative markers in stressed plants includes the status of ROS ($O_2^{\bullet-}$, OH•, and H_2O_2) and the ROS-led protein and membrane lipid products, reactive/protein carbonyls, and thiobarbituric acid reactive substances (TBARS)/malondialdehyde (MDA) (Anjum et al. 2010; 2014a, b; Gill and Tuteja 2010) (Figure 27.1). On the other hand, efficient antioxidant metabolism and eventual the cellular redox homeostasis are accomplished by antioxidant enzymes, such as superoxide dismutase (SOD), catalase (CAT), monodehydroascorbate reductase (MDHAR), glutathione sulfo-transferase (GST), guaiacol peroxidase (GPX), glutathione reductase (GR), ascorbate peroxidase (APX), dehydroascorbate reductase (DHAR), and non-enzymatic antioxidants, such as ascorbate (AsA), carotenoids, glutathione (GSH), and phenolicstocopherols (Anjum et al. 2010, 2014a, b; Gill and Tuteja 2010; Gill et al. 2014).

27.1.2 Modulators of Oxidative Stress Markers and Antioxidant Metabolism

As mentioned earlier, the cellular status of major oxidative markers, namely $O_2^{\bullet-}$, OH•, and H_2O_2, and reactive/protein carbonyls (indication of oxidation of proteins) and TBARS/MDA (lipid peroxidation indicators) is elevated by the type and extent of different stress factors and also by the age, organ, and genotype of the test plants (Anjum et al. 2010; 2014a, b; Gill

Global Climate Change and Plant Stress Management, First Edition. Edited by Mohammad Wahid Ansari, Anil Kumar Singh, and Narendra Tuteja.

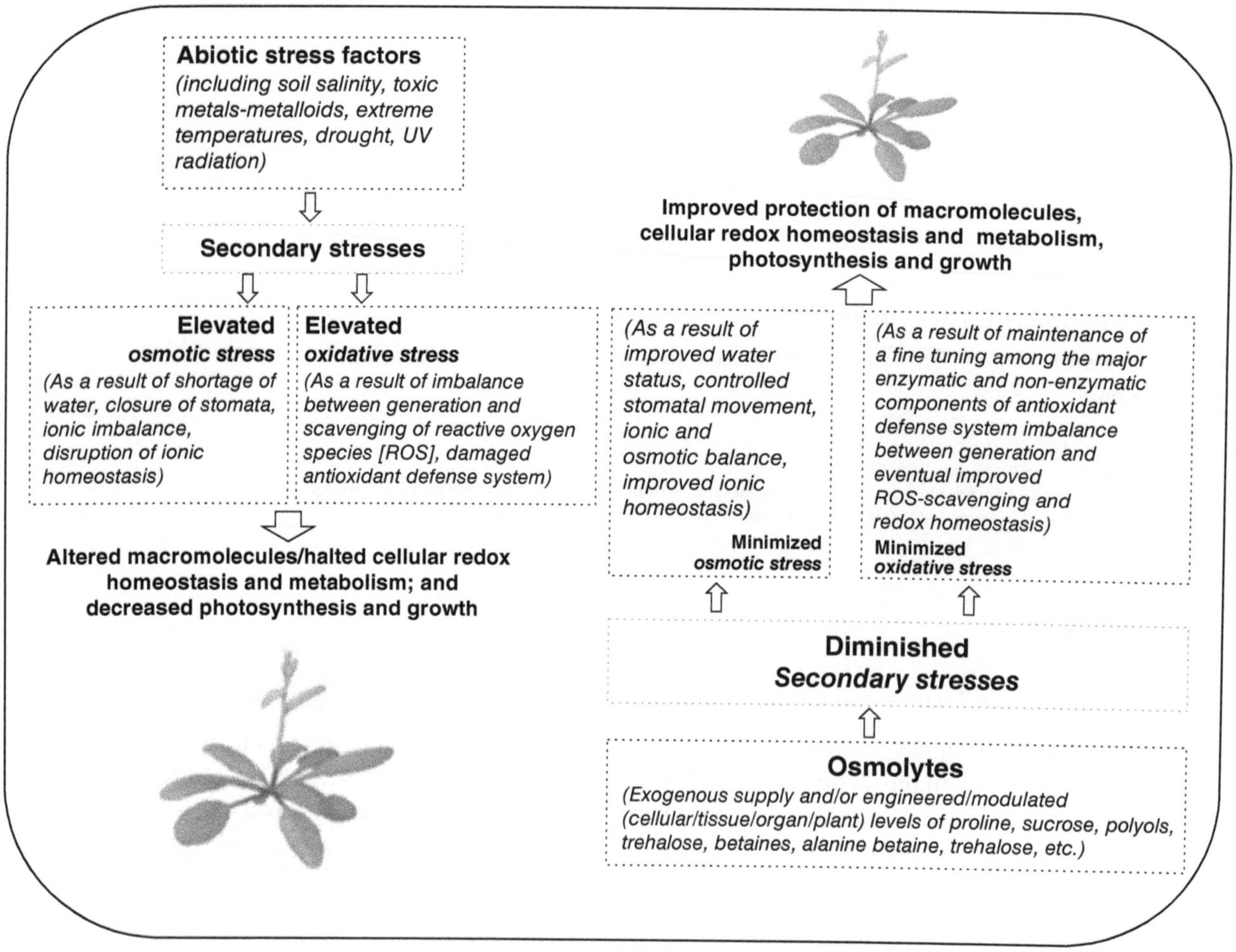

Figure 27.1 Schematic representation of the major abiotic stress-impacts and the major mechanisms underlying osmolytes-mediated improvements the main plant health parameters.

and Tuteja 2010). Notably, the plant's inherent capacity for neutralizing the extent of oxidative stress markers and/or ROS-accrued potential anomalies can be further strengthened by exogenous application of phytohormones (Ryu and Cho 2015; Ahanger et al. 2020; Jahan et al. 2021; Rasheed et al. 2021), mineral plant nutrients (Astolfi and Zuchi 2013; Iqbal et al. 2015; Fahad et al. 2016; Coskun et al. 2016), antioxidant metabolites (Hasanuzzaman et al. 2017, 2020), and/or via spiking the seeds with a range of priming agents (Carvalho et al. 2011; Jisha et al. 2013; Hossain et al. 2015; Paparella et al. 2015; El-Serafy et al. 2021; Johnson and Puthur 2021; Ashraf et al. 2022).

27.2 Osmolytes – An Overview

Plants exposed to drought stress, high levels of salinity, and/or toxic or elevated metals usually exhibit osmotic stress through impaired plant water relations and oxidative stress caused by uncontrolled generation of varied ROS as a common response (Benavides et al. 2005; Anjum et al. 2010, 2014a). Notably, at the physiological level, osmotic stress caused by impaired plant–water relations is overcome via osmotic adjustment as a result of varyingly synthesizing different osmolytes (osmoprotectants) (Sharma et al. 2019). Osmolytes are highly soluble, low-molecular-weight organic compounds, also known as compatible solutes, since even elevated cellular levels of most of these compounds are non-toxic and also do not interfere with normal metabolic reactions (Slama et al. 2015). Sugars (glucose, sucrose, fructans, trehalose, raffinose, and fructose), amino acids (pipecolic acid, γ-aminobutyric acids, ectoine, Pro, etc.), polyols or sugar alcohols (sorbitol, mannitol, inositol, and glycerol), betaines (glycine betaine (GB), β-alanine betaine, choline-*O*-sulfate), and tertiary sulfonium compound dimethylsulfoniopropionate (DMSP) are among the major osmolytes widely known for their important roles in stress tolerance in plants (Anjum et al. 2014b; Sharma et al. 2019). Osmolytes can be classified in three chemical classes, such as betaines and related compounds (such as *N*-methylated amino acid derivatives, DMSP, choline-*O*-sulfate), amino acids (such as proline and ecotine), polyols (such as myo-inositol, pinitol); and non-reducing sugars (such as trehalose). Though DMSP, choline-*O*-sulfate, trehalose and D-ononiltol are not common in plants, the plants under varied stresses exhibit a differential accumulation of proline, GB, and mannitol. On the other hand, ecotine is found only in bacteria (Majumder et al. 2010; Bhardwaj et al. 2014). The biosynthesis of osmolytes is outside thescope of this chapter.

27.2.1 Role of Major Osmolytes in Protection of Plants Against Oxidative Stress

Instances of the important roles of the major osmolytes in minimization of the cellular status of varied ROS and their anomalies in abiotically stressed plants are briefly discussed hereunder, taking into account relevant recent literature (Figure 27.1).

27.2.1.1 Betaines and Related Compounds

Glycine betaine (*N*,*N*,*N*-trimethyl glycine; GB), β-alanine betaine, proline betaine, choline-*O*-sulphate, hydroxyproline betaine, and pipecolate betaine are the quaternary ammonium compounds accumulated in plants (Ashraf and Harris 2004). Synthesized mainly from choline, GB is the major quaternary ammonium compound most abundantly occurring in plants exposed to dehydration due to salinity, drought, heat, and cold stress (Guo et al. 2009; Lokhande and Suprasanna 2012; Anjum et al. 2014b; Slama et al. 2015). Involvement of endogenous and exogenously applied GB-mediated ROS-detoxification and eventual protection of cellular macromolecules has been suggested as mechanisms responsible for GB-induced abiotic stress tolerance in plants (Chen and Murata 2008; Giri 2011; Bhardwaj et al. 2014). In drought-exposed transgenic plants, GB-supply improved ROS-detoxification and helped in reducing ROS-accumulation (Chen and Murata 2008). In chilling-exposed hawthorn fruit, GB-treatment enhanced endogenous GB (and also proline accumulation), whereas GB-treatment led to decreases in H_2O_2 accumulation via enhancing SOD, CAT, and APX activity (Razavi et al. 2018). The expression of ROS-scavenging enzymes in heat-exposed tobacco plants (Yang et al. 2005) and the suppression of ROS-accumulation in chilling-stressed *Arabidopsis* (Einset et al. 2007) were reported with exogenously and/or accumulated GB. In another instance, GB-supply led to significant decrease in ROS-accumulation in chilling exposed chickpea (Nayyar et al. 2005). In transgenic wheat (versus wild-type wheat), the exhibition of improved cold tolerance involved decreased ROS-production and membrane lipid peroxidation, and enhanced activity of ROS-scavenging enzymes was observed with GB-supply (Zhang et al. 2010). In another instance, exogenously applied GB protected mungean seedlings against salt-induced oxidative stress by reducing lipid peroxidation and H_2O_2 level and by enhancing antioxidant defense (Hossain and Fujita 2010). In earlier studies, GB-accumulation and/or its exogenous application triggered off many genes involved in modulation of antioxidants such as CAT and peroxidases (POXs) (Park et al. 2006; Einset et al. 2007; Banu et al. 2009). Choline-*O*-sulfate stands second to none among the major sulphonium compounds, and its accumulation has been reported especially in the members of the halophytic genus *Limonium* and other *Plumbaginaceae* under salinity stress, which provokes the synthesis of choline-*O*-sulfate by inducing choline sulfotransferase (Gul et al. 2022).

27.2.1.2 Proline

Accumulation of amino acids has been observed in plants exposed to abiotic stress (Rhodes et al. 1986; Lugan et al. 2010). The free and a water-soluble amino acid proline occurs widely in higher plants and is accumulated therein under stressful environments. Directly and/or indirectly proline plays key roles in plant defense system against oxidative stress in plants growing under abiotic stress (Ejaz et al. 2020). Proline performs the scavenging varied ROS and also acts as a molecular chaperone stabilizing the structure of proteins in plants exposed to major abiotic stresses (Szabados and Savoure 2010; Krasensky and Jonak 2012; Siddique et al. 2018). Serving as a sink for excess reductants, the accumulated proline is known to provide the NAD^+ and $NADP^+$ required in the processes of respiration and photosynthesis (Kavi Kishor et al. 2005; Slama et al. 2015). In addition to maintaining the cellular redox homeostasis, proline is also known to administer energy for growth, influence programmed cell death, act as antioxidant defense and signaling molecule, and also behave as a metal chelator (Szabados and Savoure 2010; Hayat et al. 2012). Proline-mediated protection against NaCl-induced cell death involved both the decreased level of ROS and its consequences (such as lipid peroxidation) and improved membrane integrity via increasing antioxidant gene (Banu et al. 2009). Proline-mediated plant stress tolerance also involved proline-led increase in the activity of major antioxidant enzymes including CAT, POX, SOD, APX, MDHAR, and dehydroascorbate reductase (DHAR) and eventual enhanced ROS-scavenging and minimized lipid peroxidation (Hoque et al. 2007; Islam et al. 2009). Exogenous proline treatment enhanced accumulation and reestablishment of the redox homeostasis, and eventual improved Cd-tolerance were also reported (Wang et al. 2022). Maintained cellular homeostasis, water uptake, osmotic adjustment, redox balance, and eventual mitigation of oxidative damage were widely reported with the overproduction of proline in plant cells. Therefore, proline has been aptly considered as a multifaceted signaling molecule in plant responses to abiotic stress (Ghosh et al. 2022).

There occurs a close relation between proline and glutathione (GSH), an important water-soluble multifunctional non-enzymatic antioxidant (Anjum et al. 2014a). This is because proline and GSH share common precursor, glutamate. Hence, switching of accumulation of proline or GSH is argued in stressed plants in order to avert stress-caused potential consequences (Anjum et al. 2014a; Ejaz et al. 2020). Additionally, proline was reported to modulate the activity of glutathione reductase (GR), an enzyme involved in GSH-regeneration. Research reports also confirmed the involvement of proline in the maintenance of GSH-pool via enhancement in GR activity, thereby maintaining higher GSH/GSSH ratios and consequently redox homeostasis (Aggarwal et al. 2011; Yan et al. 2011; Hasanuzzaman et al. 2014; Wang et al. 2022). Proline poses ROS-scavenging potential and also reduces ROS-induced K^+ efflux (Cuin and Shabala 2007; Hayat et al. 2012). In addition, exogenously applied proline was reported to significantly enhance the activity of the major antioxidant enzymes of AsA-GSH cycle including CAT, POX, APX, monohydro ascorbate reductase, dihydro ascorbate reductase, and SOD in stressed plants (Hoque et al. 2007; Kaul et al. 2008; Hong et al. 2000). Proline was argued to improve the cellular level of SOD, the first line of defense in stressed plants (Ben Ahmed et al. 2010; Gill et al. 2015). Accumulation of varied ROS and the membrane lipid peroxidation were minimized with applied proline in plants exposed to metals such as chromium (Yu et al. 2017) and Cd (Wang et al. 2022). Interestingly, proline can also modulate the cellular status of phytochelatins (PCs), known to accumulate in metal-stressed plants, where PCs form PCs-metals via thiol (–SH) and stop their intracellular circulation as free ions (de Knecht et al. 1994; Serrano et al. 2015; Wei et al. 2021; Huang et al. 2022).

27.2.1.3 γ-Aminobutyric Acid (Gamma Amino Butyric Acid)

Gamma amino butyric acid (GABA), a non-protein amino acid, has been widely reported for its involvement in various metabolic processes and protection of plants against abiotic stress caused oxidative damage as a result of its contribution to increasing leaf turgor, osmolytes, membrane stability, and stimulation of major antioxidants (enzymes and non-enzymes/metabolites); and significant decrease in ROS-accumulation and scavenging and membrane lipid peroxidation (Barbosa et al. 2010; Renault et al. 2010; Krishnan et al. 2013; Aghdam et al. 2016; Li et al. 2016; Vijayakumari and Puthur 2016; Sita and Kumar 2020). In heat-stressed mungbean seedlings, GABA application decreased multi-fold the content of H_2O_2 and MDA and elevated the activity of SOD, CAT, APX, and GR, as well as the contents of AsA and GSH, and thereby improved the test plant health (Priya et al. 2019).

GABA is equally important in plants exposed to metals such as aluminum, arsenic, cadmium, and chromium (Song et al. 2010; Mahmud et al. 2017; Kumar et al. 2019; Seifikalhor et al. 2020). When supplied to aluminum-exposed barley seedlings, GABA mitigated the oxidative injury caused by aluminum toxicity mainly by activating the antioxidant defense system and decreasing the elevated levels of ROS-caused protein carbonylation and membrane lipid peroxidation (Song et al. 2010). GABA-supply also protected the tested maize against cadmium-caused oxidative stress by significantly decreasing the major cellular ROS and also the content of malondealdehyde (MDA, a membrane lipid peroxidation marker) (Seifikalhor et al. 2020). In chromium-exposed *Brassica juncea*, Mahmud et al. (2017) reported involvement of GABA-mediated modulation of the antioxidant defense (and glyoxalase systems) in GABA-assisted conferment of the test plant to chromium. In As (III) (25 μM)-rice seedlings, GABA-supply significantly reduced the H_2O_2 and TBARS levels, and thereby recovered the growth parameters against As (III) (Kumar et al. 2019).

GABA-supply mediated chilling tolerance in peach fruit involved maintenance of higher energy charge and membrane integrity and permeability; and enhancements in the activity of CAT and SOD, glutathione peroxidase (GPX), glutathione *S*-transferase (GST), APX, DHAR and MDHAR; and also increased ATP content to sustain higher energy charge maintains membrane integrity and permeability (Yang et al. 2011). In the roots of salt-exposed *Caragana intermedia*, GABA-mediated decreased H_2O_2 accumulation involved the regulation of H_2O_2-generation related genes (Shi et al. 2010). GABA-induced tolerance of pepper plants to polyethylene glycol-caused drought stress involved increased activity of SOD and GPX, significantly decreased MDA content, and eventual lowered lipid peroxidation (Vijayakumari and Puthur 2016).

27.2.1.4 Polyols

Hydroxyl groups available in polyols (sugar alcohols; e.g. mannitol, inositol, glycerol, and ribitol) form hydration spheres around macromolecules and thereby prohibit their metabolic inactivation under osmotic stress (Sharma et al. 2014). Notably, polyols-mediated control of abiotic stress consequences has been reported to involve the ROS-scavenging and an eventual decrease in lipid peroxidation and cell damage (Ghosh et al. 2021). In particular, myo-inositol, D-pinitol, and D-ononitol significantly scavenge ·OH radicles. Myo-inositol is another important osmolyte, via myo-inositol oxygenase, myo-inositol has been reported to modulate the content of ascorbic acid in plants (Lisko et al. 2013). In the leaves of CAT-deficient *Arabidopsis*, myo-inositol reverted H_2O_2-triggered salicylic acid induced cell death (Chaouch and Noctor 2010). The role of myo-inositol in ROS-scavenging may also be via raffinose-biosynthesis (Valluru and Van den Ende 2011). Regarding pinitol, D-pinitol (3-*O*-methyl-D-chiroinositol) has been a major solute in the plant families Pinaceae, Leguminosae, and Caryophyllaceae (Nguyen and Lamant 1988).

27.2.1.5 Sugars

Sugars are the major source of carbon and energy and are known to perform regulatory functions, influence all phases of the plant life cycle, and also act as crucial players in plants under oxidative challenge (Bolouri-Moghaddam et al. 2010; Keunen et al. 2013; Jeandet et al. 2022). In particular, plant stress responses largely involve three major types of water-soluble carbohydrates namely disaccharides (sucrose, trehalose), raffinose family oligosaccharides, and fructans (Keunen et al. 2013). Notably, a close link between raffinose family oligosaccharides and fructans was argued in ROS signaling pathways (Bolouri-Moghaddam et al. 2010). Sugars (such as disaccharides, galactinol, RFOs, fructans, and sugar alcohols) have been widely considered as antioxidants in plants (Nishizawa et al. 2008; Hernandez-Marin and Martínez 2012; Peshev et al. 2013). °OH scavenging ability of galactinol and raffinose in vitro was reported comparable with GSH (Nishizawa et al. 2008). Involvement of sugars in the protection of chloroplasts and stabilization of photosynthesis has been reported in plants under stress conditions (Nishizawa et al. 2008; Schneider and Keller 2009; reviewed by Keunen et al. 2013). Sugars have also been reported to prevent ROS-formation in plant mitochondria (Bolouri-Moghaddam et al. 2010; reviewed by Keunen et al. 2013). Sugars were found to enhance the dismutation of superoxides as a result of increasing the activity of SOD in yellow lupine (Formela-Luboińska et al. 2020). Earlier, sugar variation-mediated modifications in the expression of genes associated with SOD, heat shock proteins, and GST were also reported in plants under abiotic stresses (Koch 1996; Price et al. 2004). Literature is full on the significant roles of sugars (glucose, sucrose, fructose, raffinose, trehalose, and fructans) as compatible solutes or osmoprotectants, antioxidants, and also as innate immunity in plants exposed to various abiotic stresses (Moghaddam and Ende 2012; Kido et al., 2013; Formela-Luboińska et al. 2020).

27.2.1.5.1 Trehalose Trehalose, a non-reducing glucose disaccharide, gets accumulated in plants (in particular some resurrection plants) when exposed to drought, salinity, temperature variations, and heavy metal stresses and varyingly confers stress tolerance in plants (Sharma et al. 2014). Trehalose-supply protected maize seedlings against salinity and P-deficiency as a result of decreased contents of ROS and MDA and enhanced activity of SOD, CAT, APX, GPX, GR, MDHAR, and DHAR enzymes (Rohman et al. 2019). In exploring the physiological role of trehalose in enhancing salinity tolerance of wheat plant, Sadak (2019) reported the involvement of trehalose (foliar spray)-mediated reduction in H_2O_2 and enhanced accumulation of antioxidant compounds (phenolics) and compatible osmolytes, and also the improved membrane stability. The highlighted findings of Sadak (2019) coincided with the findings of Nounjan and Theerakulpisut (2012), Ma et al. (2013), Liu et al. (2013), and Sadak et al. (2019), where the authors reported decreased oxidative stress of different plants in different abiotic stresses, mainly as a result of trehalose-mediated improved ROS-scavenging and membrane stabilizing. In cold-stressed tomato plants, Liu et al. (2020) observed exogenous trehalose-mediated enhanced cold-stress tolerance, which the authors argued was a result of regulated antioxidant defense system. In addition, the supply of trehalose was also found to be an exogenous inducer of H_2O_2 (and also nitric oxide), which in turn was argued to regulate antioxidase gene expression and activities, reduced membrane lipid peroxidation and cellular injury, and eventually lead to enhanced cold-stress tolerance. In a recent study on salt stressed tomato, trehalose application regulated ROS-metabolism, osmolyte synthesis, and also trehalose metabolic pathways (Yang et al. 2022). Similar to the results obtained by Yang et al. (2022), some other studies observed the involvement of trehalose in modulation of antioxidant enzyme activity and enhancement of the detoxification effect of ROS, reducing cellular damage and death and alleviating stress-induced inhibition (Aldesuquy and Ghanem 2015; Zhao et al. 2019; Liu et al. 2020). Earlier, trehalose-supply was argued to activate the ROS-scavenging by increasing the activity of SOD, CAT, POX, and ascorbic acid (Shafiq et al. 2015; Akram et al. 2016). In another report on Tre-*deficient Saccharomyces cerevisiae* mutant cells, Benaroudj et al. (2001) observed a positive correlation between trehalose content and ROS resistance.

27.2.1.5.2 Sucrose As one of the most abundant sugars in plants, sucrose is involved in ROS-quenching system. Ascorbic acid is among the most abundant antioxidant, whereas anthocyanins are involved in scavenging different types of reactive oxygen and nitrogen species. Sucrose-mediated accumulation of ascorbic acid and anthocyanins has been reported (Guo et al. 2011; Ejaz et al. 2020). Solfanelli et al. (2006) also reported sucrose-supply led increase in accumulation of anthocyanins. Sucrose-supply was reported to alleviate ROS stress in abiotic stressed plants (Loreti et al. 2005; Sulmon et al. 2007; Rosenwasser et al. 2011). Lowest levels of $O_2^{\bullet-}$, H_2O_2, and MDA levels were observed with sucrose-pretreatment to chilling-exposed cucumber seedlings (Cao et al. 2014).

27.3 Conclusion and Perspectives

Exposure of plants/crops to varied stress factors is inevitable. Plant tolerance to varied stresses may be conferred by controlling cellular osmotic balance, mainly synthesizing osmolytes (and also ion homeostasis), and activating antioxidant defense system components. This chapter attempted to discuss the significant roles (and potential mechanisms) of the major osmolytes in minimizing abiotic stress-caused oxidative stress. Notably, the surroundings and cytoplasm of cells are osmotically adjusted with tiny but big-impact molecules called osmoprotectants or compatible solutes. These solutes stabilize stress inducers by forming hydrophilic complexes. Osmolytes help cells to retain water and size without disturbing normal cellular functions, stabilize proteins and cellular structures, maintain cell turgor by osmotic adjustment, and also promote redox metabolism to remove excess levels of ROS and re-establish the cellular redox balance. Being a vital osmoticum, proline enhanced the plant tolerance to environmental stress. The synthesis and accumulation of amino acids (such as proline, GB and GABA) in abiotically stressed plants have provoked the exploration of candidate genes for genetic manipulation of these amino acids in order to achieve improved abiotic stress tolerance in plants. Much of the information is focused at the physiological and biochemical insights into their dynamics in different stressed plants. However, there exist broad scopes for engineering the major osmolytes in order to further enhance their unique potential for stress-impact mitigation. Approaches incorporating omics have helped in the identification and characterization (and underlying mechanisms) of osmoprotectants. Notably, the huge crop losses to the agricultural sector as a varied abiotic stress factors can be overcome through getting insights into expression of osmolyte synthetic genes and their potential interaction outcomes with major enzymatic and non-enzymatic antioxidants.

The inherent capacity of plants/crops (especially important agronomical and/or horticultural ones) for the synthesis of major osmolytes can be augmented with exogenously supplied osmolytes, which eventually would improve their stress tolerance capacity. Molecular-genetic efforts should be made to analyze the role of the major osmolytes associated with their perception to stress factors causing osmotic and/or oxidative stress as secondary stress; and also to enlighten the relation of osmolytes with other stress metabolites and phytohormone, and cumulative role cellular redox homeostasis under individual and/or combined stresses. If done, the research outcomes would be rewarding in the face of climate change and impending recurrence of stress factors targeting mainly especially important agronomical and/or horticultural plants/crops.

References

Aggarwal, M., Sharma, S., Kaur, N. et al. (2011). Exogenous proline application reduces phytotoxic effects of selenium by minimising oxidative stress and improves growth in bean (*Phaseolus vulgaris* L.) seedlings. *Biol. Trace Elem. Res.* 140: 354–367.

Aghdam, M.S., Naderi, R., Sarcheshmeh, M.A.A., and Babalar, M. (2016). Amelioration of postharvest chilling injury in anthurium cut flowers by γ-aminobutyric acid (GABA) treatments. *Postharvest Biol. Technol.* 110: 70–76.

Ahanger, M.A., Mir, R.A., Alyemeni, M.N., and Ahmad, P. (2020). Combined effects of brassinosteroid and kinetin mitigates salinity stress in tomato through the modulation of antioxidant and osmolyte metabolism. *Plant Physiol. Biochem.* 147: 31–42.

Akram, N.A., Waseem, M., Ameen, R., and Ashraf, M. (2016). Trehalose pretreatment induces drought tolerance in radish (*Raphanus sativus* L.) plants: some key physio-biochemical traits. *Acta Physiol. Plant* 38 (1): 1–10.

Aldesuquy, H. and Ghanem, H. (2015). Exogenous salicylic acid and trehalose ameliorate short term drought stress in wheat cultivars by up-regulating membrane characteristics and antioxidant defense system. *J. Hortic.* 2: 139.

Anjum, N.A., Umar, S., and Chan, M.T. (2010). *Ascorbate-Glutathione Pathway and Stress Tolerance in Plants*. Dordrecht: Springer.

Anjum, N.A., Aref, I.M., Duarte, A.C. et al. (2014a). Glutathione and proline can coordinately make plants withstand the joint attack of metal(loid) and salinity stresses. *Front. Plant Sci.* 5: 662.

Anjum, N.A., Gill, S.S., and Gill, R. (2014b). *Plant Adaptation to Environmental Change: Significance of Amino Acids and their Derivatives*, 1e. Wallingford, UK: CABI https://doi.org/10.1079/9781780642734.0000.

Anjum, N.A., Gill, S.S., Corpas, F.J. et al. (2022). Editorial: recent insights into the double role of hydrogen peroxide in plants. *Front. Plant Sci.* 13: 843274.

Ashraf, M. and Harris, P.J.C. (2004). Potential biochemical indicators of salinity tolerance in plants. *Plant Sci.* 166: 3–16.

Ashraf, M.A., Rasheed, R., Hussain, I. et al. (2022). Effect of different seed priming agents on chromium accumulation, oxidative defense, glyoxalase system and mineral nutrition in canola (*Brassica napus* L.) cultivars. *Environ. Pollut.* 309: 119769.

Astolfi, S. and Zuchi, S. (2013). Adequate S supply protects barley plants from adverse effects of salinity stress by increasing thiol contents. *Acta Physiol. Plant* 35: 175–181.

Banu, N.A., Hoque, A., Watanabe-Sugimoto, M. et al. (2009). Proline and glycinebetaine induce antioxidant defense gene expression and suppress cell death in cultured tobacco cells under salt stress. *J. Plant Physiol.* 66: 146–156.

Barbosa, J.M., Singh, N.K., Cherry, J.H., and Locy, R.D. (2010). Nitrate uptake and utilization is modulated by exogenous γ-aminobutyric acid in *Arabidopsis thaliana* seedlings. *Plant Physiol. Biochem.* 48: 443–450.

Batista-Silva, W., Heinemann, B., Rugen, N. et al. (2019). The role of amino acid metabolism during abiotic stress release. *Plant Cell Environ.* 42: 1630–1644.

Ben Ahmed, C., Ben Rouina, B., Sensoy, S. et al. (2010). Exogenous proline effects on photosynthetic performance and antioxidant defense system of young olive tree. *J. Agric. Food Chem.* 58: 4216–4222.

Benaroudj, N., Lee, H., and Goldberg, A.L. (2001). Trehalose accumulation during cellular stress protects cells and cellular proteins from damage by oxygen radicals. *J. Biol. Chem.* 276: 24261–24267.

Benavides, M.P., Gallego, S.M., and Tomaro, M.L. (2005). Cadmium toxicity in plants. *Braz. J. Plant Physiol.* 17: 21–34.

Bhardwaj, R.B., Indu Sharma, I.S., Resham Sharma, R.S., and Poonam, P. (2014). Betaines and related osmoprotectants-significance in metabolic engineering of plant stress resistance. In: *Plant Adaptation to Environmental Change: Significance of Amino Acids and Their Derivatives*, 1e (ed. N.A. Anjum, S.S. Gill, and R. Gill), 281–299. Wallingford, UK: CABI.

Bolouri-Moghaddam, M.R., Le Roy, K., Xiang, L. et al. (2010). Sugar signalling and antioxidant network connections in plant cells. *FEBS J.* 277: 2022–2037.

Cao, Y.Y., Yang, M.T., Li, X. et al. (2014). Exogenous sucrose increases chilling tolerance in cucumber seedlings by modulating antioxidant enzyme activity and regulating proline and soluble sugar contents. *Sci. Hortic.* 179: 67–77.

Carvalho, R.F., Piotto, F.A., Schmidt, D. et al. (2011). Seed priming with hormones does not alleviate induced oxidative stress in maize seedlings subjected to salt stress. *Sci. Agric.* 68: 598–602.

Chaouch, S. and Noctor, G. (2010). Myo-inositol abolishes salicylic acid-dependent cell death and pathogen defence responses triggered by peroxisomal hydrogen peroxide. *New Phytol.* 188: 711–718.

Chen, T.H.H. and Murata, N. (2008). Glycinebetaine: an effective protectant against abiotic stress in plants. *Trends Plant Sci.* 13: 499–505.

Coskun, D., Britto, D.T., Huynh, W.Q., and Kronzucker, H.J. (2016). The role of silicon in higher plants under salinity and drought stress. *Front. Plant Sci.* 7: 1072.

Cuin, T.A. and Shabala, S. (2007). Compatible solutes reduce ROS-induced potassium efflux in *Arabidopsis* roots. *Plant Cell Environ.* 30: 875–885.

Einset, J., Nielsen, E., Connolly, E.L. et al. (2007). Membrane-trafficking RabA4c involved in the effect of glycine betaine on recovery from chilling stress in *Arabidopsis*. *Plant Physiol.* 130: 511–518.

Ejaz, S., Fahad, S., Anjum, M.A. et al. (2020). Role of osmolytes in the mechanisms of antioxidant defense of plants. In: *Sustainable Agriculture Reviews 39* (ed. E. Lichtfouse), 95–117. Cham: Springer.

El-Serafy, R.S., El-Sheshtawy, A.N., Atteya, A.K. et al. (2021). Seed priming with silicon as a potential to increase salt stress tolerance in *Lathyrus odoratus*. *Plants* 10: 2140.

Fahad, S., Hussain, S., Saud, S. et al. (2016). Exogenously applied plant growth regulators affect heat-stressed rice pollens. *J. Agron. Crop Sci.* 202: 139–150.

Formela-Luboińska, M., Chadzinikolau, T., Drzewiecka, K. et al. (2020). The role of sugars in the regulation of the level of endogenous signaling molecules during defense response of yellow lupine to *Fusarium oxysporum*. *Int. J. Mol. Sci.* 21: 4133.

Ghosh, U.K., Islam, M.N., Siddiqui, M.N., and Khan, M.A. (2021). Understanding the roles of osmolytes for acclimatizing plants to changing environment: a review of potential mechanism. *Plant Signal. Behav.* 16: 1913306.

Ghosh, U.K., Islam, M.N., Siddiqui, M.N. et al. (2022). Proline, a multifaceted signalling molecule in plant responses to abiotic stress: understanding the physiological mechanisms. *Plant Biol.* 24: 227–239.

Gill, S.S. and Tuteja, N. (2010). Reactive oxygen species and antioxidant machinery in abiotic stress tolerance in crop plants. *Plant Physiol. Biochem.* 48: 909–930.

Gill, S.S., Gill, R., and Anjum, N.A. (2014). Target osmoprotectants for abiotic stress tolerance in crop plants—glycine betaine and proline. In: *Plant Adaptation to Environmental Change: Significance of Amino Acids and Their Derivatives*, 1e (ed. N.A. Anjum, S.S. Gill, and R. Gill), 97–108. Wallingford, CT: CAB International.

Gill, S.S., Anjum, N.A., Gill, R. et al. (2015). Superoxide dismutase—mentor of abiotic stress tolerance in crop plants. *Environ. Sci. Pollut. Res.* 22: 10375–10394.

Giri, J. (2011). Glycinebetaine and abiotic stress tolerance in plants. *Plant Signal. Behav.* 6: 1746–1751.

Gul, Z., Tang, Z.-H., Arif, M., and Ye, Z. (2022). An insight into abiotic stress and influx tolerance mechanisms in plants to cope in saline environments. *Biology* 11: 597.

Guo, P., Baum, M., Grando, S. et al. (2009). Differentially expressed genes between drought-tolerant and drought-sensitive barley genotypes in response to drought stress during the reproductive stage. *J. Exp. Bot.* 60: 3531–3544.

Guo, R., Yuan, G., and Wang, Q. (2011). Effect of sucrose and mannitol on the accumulation of health-promoting compounds and the activity of metabolic enzymes in *Broccoli sprouts*. *Sci. Hortic.* 128: 159–165.

Hasanuzzaman, M., Alam, M.M., Rahman, A. et al. (2014). Exogenous proline and glycine betaine mediated upregulation of antioxidant defense and glyoxalase systems provides better protection against salt-induced oxidative stress in two rice (*Oryza sativa* L.) varieties. *Biomed. Res. Int.* 2014: 757219.

Hasanuzzaman, M., Nahar, K., Anee, T.I., and Fujita, M. (2017). Glutathione in plants: biosynthesis and physiological role in environmental stress tolerance. *Physiol. Mol. Biol. Plants* 23: 249–268.

Hasanuzzaman, M., Bhuyan, M.H., Zulfiqar, F. et al. (2020). Reactive oxygen species and antioxidant defense in plants under abiotic stress: revisiting the crucial role of a universal defense regulator. *Antioxidants* 9: 681.

Hayat, S., Hayat, Q., Alyemeni, M.N. et al. (2012). Role of proline under changing environments: a review. *Plant Signal. Behav.* 7: 1456–1466.

Hernandez-Marin, E. and Martínez, A. (2012). Carbohydrates and their free radical scavenging capability: a theoretical study. *J. Phy. Chem. B.* 116: 9668–9675.

Hong, Z., Lakkineni, K., Zhang, Z., and Verma, D.P. (2000). Removal of feedback inhibition of delta(1)-pyrroline-5-carboxylate synthetase results in increased proline accumulation and protection of plants from osmotic stress. *Plant Physiol.* 122: 1129–1136.

Hoque, M.A., Okuma, E., Banu, M.N.A. et al. (2007). Exogenous proline mitigates the detrimental effects of salt stress more than exogenous betaine by increasing antioxidant enzyme activities. *J. Plant Physiol.* 164: 553–561.

Hossain, M.A. and Fujita, M. (2010). Evidence for a role of exogenous glycinebetaine and proline in antioxidant defense and methylglyoxal detoxification systems in mung bean seedlings under salt stress. *Physiol. Mol. Biol. Plants* 16: 19–29.

Hossain, M.A., Bhattacharjee, S., Armin, S.M. et al. (2015). Hydrogen peroxide priming modulates abiotic oxidative stress tolerance: insights from ROS detoxification and scavenging. *Front. Plant Sci.* 6: 420.

Huang, Y., Chen, J., Sun, Y. et al. (2022). Mechanisms of calcium sulfate in alleviating cadmium toxicity and accumulation in pak choi seedlings. *Sci. Total Environ.* 805: 150115.

Iqbal, N., Umar, S., and Khan, N.A. (2015). Nitrogen availability regulates proline and ethylene production and alleviates salinity stress in mustard (*Brassica juncea*). *J. Plant Physiol.* 178: 84–91.

Islam, M.M., Hoque, M.A., Okuma, E. et al. (2009). Exogenous proline and glycinebetaine increase antioxidant enzyme activities and confer tolerance to cadmium stress in cultured tobacco cells. *J. Plant Physiol.* 166: 1587–1597.

Jahan, B., Rasheed, F., Sehar, Z. et al. (2021). Coordinated role of nitric oxide, ethylene, nitrogen, and sulfur in plant salt stress tolerance. *Stresses* 1 (3): 181–199.

Jeandet, P., Formela-Luboińska, M., Labudda, M., and Morkunas, I. (2022). The role of sugars in plant responses to stress and their regulatory function during development. *Int. J. Mol. Sci.* 23: 5161.

Jisha, K.C., Vijayakumari, K., and Puthur, J.T. (2013). Seed priming for abiotic stress tolerance: an overview. *Acta Physiol. Plant.* 35: 1381–1396.

Johnson, R. and Puthur, J.T. (2021). Seed priming as a cost effective technique for developing plants with cross tolerance to salinity stress. *Plant Physiol. Biochem.* 162: 247–257.

Kaul, S., Sharma, S.S., and Mehta, I.K. (2008). Free radical scavenging potential of L-proline: evidence from in vitro assays. *Amino Acids* 34: 315–320.

Kavi Kishor, P.B., Sangam, S., Amrutha, R.N. et al. (2005). Regulation of proline biosynthesis, degradation, uptake and transport in higher plants: its implications in plant growth and abiotic stress tolerance. *Curr. Sci.* 88: 424–438.

Keunen, E., Peshev, D., Vangronsveld, J. et al. (2013). Plant sugars are crucial players in the oxidative challenge during abiotic stress: extending the traditional concept. *Plant Cell Environ.* 36: 1242–1255.

Kido, E.A., Ferreira Neto, J.R., Silva, R.L. et al. (2013). Expression dynamics and genome distribution of osmoprotectants in soybean: identifying important components to face abiotic stress. *BMC Bioinform.* 14: S7.

de Knecht, J.A., Van Dillen, M., Koevoets, P. et al. (1994). Phytochelatins in cadmium-sensitive and cadmium-tolerant *Silene vulgaris* (chain length distribution and sulfide incorporation). *Plant Physiol.* 104: 255–261.

Koch, K.E. (1996). Carbohydrate-modulated gene expression in plants. *Annu. Rev. Plant Physiol. Plant Mol. Biol.* 47: 509–540.

Krasensky, J. and Jonak, C. (2012). Drought, salt, and temperature stress-induced metabolic rearrangements and regulatory networks. *J. Exp. Bot.* 63: 1593–1608.

Krishnan, S., Laskowski, K., Shukla, V., and Merewitz, E.B. (2013). Mitigation of drought stress damage by exogenous application of a non-protein amino acid γ–aminobutyric acid on perennial ryegrass. *J. Am. Soc. Hortic. Sci.* 138: 358–366.

Kumar, N., Gautam, A., Dubey, A.K. et al. (2019). GABA mediated reduction of arsenite toxicity in rice seedling through modulation of fatty acids, stress responsive amino acids and polyamines biosynthesis. *Ecotoxicol. Environ. Saf.* 173: 15–27.

Li, M.F., Guo, S.J., Yang, X.H. et al. (2016). Exogenous gamma-aminobutyric acid increases salt tolerance of wheat by improving photosynthesis and enhancing activities of antioxidant enzymes. *Biol. Plant.* 60: 123–131.

Lisko, K.A., Torres, R., Harris, R.S. et al. (2013). Elevating vitamin C content via overexpression of myo-inositol oxygenase and L-gulono-1, 4-lactone oxidase in *Arabidopsis* leads to enhanced biomass and tolerance to abiotic stresses. *in vitro Cell Dev. Biol. Plant* 49: 643–655.

Liu, X., Hung, C., Chen, Q. et al. (2013). Alleviates effects of exogenous trehalose on oxidative damage metabolism in *Pleurotus pulmonaris* under heat stress. *Acta Hortic. Sin.* 40: 1501–1508.

Liu, T., Ye, X., Li, M. et al. (2020). H_2O_2 and NO are involved in trehalose-regulated oxidative stress tolerance in cold-stressed tomato plants. *Environ. Exp. Bot.* 171: 103961.

Lokhande, V.H. and Suprasanna, P. (2012). Prospects of halophytes in understanding and managing abiotic stress tolerance. In: *Environmental Adaptations and Stress Tolerance of Plants in the Era of Climate Change* (ed. P. Ahmad and M.N.V. Prasad), 29–56. New York: Springer.

Loreti, E., Poggi, A., Novi, G. et al. (2005). A genome-wide analysis of the effects of sucrose on gene expression in *Arabidopsis* seedlings under anoxia. *Plant Physiol.* 137: 1130–1138.

Lugan, R., Niogret, M.F., Leport, L. et al. (2010). Metabolome and water homeostasis analysis of *Thellungiella salsuginea* suggests that dehydration tolerance is a key response to osmotic stress in this halophyte. *Plant J.* 64: 215–229.

Ma, C., Wang, Z., Kong, B., and Lin, T. (2013). Exogenous trehalose differentially modulate antioxidant defense system in wheat callus during water deficit and subsequent recovery. *Plant Growth Regul.* 70: 275–285.

Mahmud, J.A., Hasanuzzaman, M., Nahar, K. et al. (2017). γ-aminobutyric acid (GABA) confers chromium stress tolerance in *Brassica juncea* L. by modulating the antioxidant defense and glyoxalase systems. *Ecotoxicology* 26: 675–690.

Majumder, A.L., Sengupta, S., and Goswami, L. (2010). Osmolyte regulation in abiotic stress. In: *Abiotic Stress Adaptation in Plants: Physiological, Molecular and Genomic Foundation* (ed. A. Pareek, S.K. Sopory, H.J. Bohnert, and Govindjee), 349–370. Dordrecht: Springer.

Mittler, R. (2002). Oxidative stress, antioxidants and stress tolerance. *Trends Plant Sci.* 7: 405–410.

Mittler, R., Vanderauwera, S., Gollery, M., and Van Breusegem, F. (2004). Reactive oxygen gene network of plants. *Trends Plant Sci.* 9: 490–498.

Mittler, R., Vanderauwera, S., Suzuki, N. et al. (2011). ROS signaling: the new wave? *Trends Plant Sci.* 6: 300–309.

Moghaddam, M.R.B. and Ende, W.V.D. (2012). Sugars and plant innate immunity. *J. Exp. Bot.* 63: 3989–3998.

Møller, I.M. and Sweetlove, L.J. (2010). ROS signalling–specificity is required. *Trends Plant Sci.* 15: 370–374.

Nayyar, H., Chander, K., Kumar, S., and Bains, T. (2005). Glycine betaine mitigates cold stress damage in chickpea. *Agron. Sustain. Dev.* 25: 381–388.

Nguyen, A. and Lamant, A. (1988). Pinitol and myo-inositol accumulation in water-stressed seedlings of maritime pine. *Phytochemistry* 27: 3423–3427.

Nishizawa, A., Yabuta, Y., and Shigeoka, S. (2008). Galactinol and raffinose constitute a novel function to protect plants from oxidative damage. *Plant Physiol.* 147: 1251–1263.

Nounjan, N. and Theerakulpisut, P. (2012). Effects of exogenous proline and trehalose on physiological responses in rice seedlings during salt-stress and after recovery. *Plant Soil Environ.* 58: 309–315.

Paparella, S., Araújo, S.S., Rossi, G. et al. (2015). Seed priming: state of the art and new perspectives. *Plant Cell Rep.* 34: 1281–1293.

Park, E.J., Jeknic, Z., and Chen, T.H. (2006). Exogenous application of glycinebetaine increases chilling tolerance in tomato plants. *Plant Cell Physiol.* 47: 706–714.

Peshev, D., Vergauwen, R., Moglia, A. et al. (2013). Towards understanding vacuolar antioxidant mechanisms: a role for fructans? *J. Exp. Bot.* 64: 1025–1038.

Price, J., Laxmi, A., St. Martin, S.K., and Jang, J.C. (2004). Global transcription profiling reveals multiple sugar signal transduction mechanisms in *Arabidopsis*. *Plant Cell* 16: 2128–2150.

Priya, M., Sharma, L., Kaur, R. et al. (2019). GABA (γ-aminobutyric acid), as a thermo-protectant, to improve the reproductive function of heat-stressed mungbean plants. *Sci. Rep.* 9: 7788.

Rasheed, F., Sehar, Z., Fatma, M. et al. (2021). Involvement of ethylene in reversal of salt stress by salicylic acid in the presence of sulfur in mustard (*Brassica juncea* L.). *J. Plant Growth Regul.* 41: 1–18.

Razavi, F., Mahmoudi, R., Rabiei, V. et al. (2018). Glycine betaine treatment attenuates chilling injury and maintains nutritional quality of hawthorn fruit during storage at low temperature. *Sci. Hortic.* 233: 188–194.

Renault, H., Roussel, V., El Amrani, A. et al. (2010). The *Arabidopsis* pop2-1 mutant reveals the involvement of GABA transaminase in salt stress tolerance. *BMC Plant Biol.* 10: 20.

Rhodes, D., Handa, S., and Bressan, R.A. (1986). Metabolic changes associated with adaptation of plant cells to water stress. *Plant Physiol.* 82: 890–903.

Rohman, M.M., Islam, M.R., Monsur, M.B. et al. (2019). Trehalose protects maize plants from salt stress and phosphorus deficiency. *Plants* 8: 568.

Rosenwasser, S., Rot, I., Sollner, E. et al. (2011). Organelles contribute differentially to reactive oxygen species-related events during extended darkness. *Plant Physiol.* 156: 185–201.

Ryu, H. and Cho, Y.G. (2015). Plant hormones in salt stress tolerance. *J. Plant Biol.* 58: 147–155.

Sadak, M.S., El-Bassiouny, H.M.S., and Dawood, M.G. (2019). Role of trehalose on antioxidant defense system and some osmolytes of quinoa plants under water deficit. *Bull. Natl. Res. Cent.* 43: 5.

Schneider, T. and Keller, F. (2009). Raffinose in chloroplasts is synthesized in the cytosol and transported across the chloroplast envelope. *Plant Cell Physiol.* 50: 2174–2182.

Seifikalhor, M., Aliniaeifard, S., Bernard, F. et al. (2020). γ-Aminobutyric acid confers cadmium tolerance in maize plants by concerted regulation of polyamine metabolism and antioxidant defense systems. *Sci. Rep.* 10: 3356.

Serrano, N., Díaz-Cruz, J.M., Ariño, C., and Esteban, M. (2015). Recent contributions to the study of phytochelatins with an analytical approach. *Trends Anal. Chem.* 73: 129–145.

Shafiq, S., Akram, N.A., and Ashraf, M. (2015). Does exogenously-applied trehalose alter oxidative defense system in the edible part of radish (*Raphanus sativus* L.) under water-deficit conditions? *Sci. Hortic.* 185: 68–75.

Sharma, R., Bhardwaj, R., Thukral, A.K. et al. (2014). Osmolyte dynamics: new strategies for crop tolerance to abiotic stress signals. In: *Emerging Technologies and Management of Crop Stress Tolerance Vol. 2 A Sustainable Approach* (ed. P. Ahmad and S. Rasool), 405–430. Academic Press.

Sharma, A., Shahzad, B., Kumar, V. et al. (2019). Phytohormones regulate accumulation of osmolytes under abiotic stress. *Biomolecules* 9: 285.

Shi, S.Q., Shi, Z., Jiang, Z.P. et al. (2010). Effects of exogenous GABA on gene expression of *Caragana intermedia* roots under NaCl stress: regulatory roles for H_2O_2 and ethylene production. *Plant Cell Environ.* 33: 149–162.

Siddique, A., Kandpal, G., and Kumar, P. (2018). Proline accumulation and its defensive role under diverse stress condition in plants: an overview. *J. Pure Appl. Microbiol.* 12: 1655–1659.

Sita, K. and Kumar, V. (2020). Role of Gamma Amino Butyric Acid (GABA) against abiotic stress tolerance in legumes: a review. *Plant Physiol. Rep.* 25: 654–663.

Slama, I., Abdelly, C., Bouchereau, A. et al. (2015). Diversity, distribution and roles of osmoprotective compounds accumulated in halophytes under abiotic stress. *Ann. Bot.* 115: 433–447.

Solfanelli, C., Poggi, A., Loreti, E. et al. (2006). Sucrose-specific induction of the anthocyanin biosynthetic pathway in *Arabidopsis*. *Plant Physiol.* 140: 637–646.

Song, H., Xu, X., Wang, H. et al. (2010). Exogenous γ-aminobutyric acid alleviates oxidative damage caused by aluminium and proton stresses on barley seedlings. *J. Sci. Food Agric.* 90: 1410–1416.

Sulmon, C., Gouesbet, G., Binet, F. et al. (2007). Sucrose amendment enhances phytoaccumulation of the herbicide atrazine in *Arabidopsis thaliana*. *Environ. Pollut.* 145: 507–515.

Szabados, L. and Savouré, A. (2010). Proline: a multifunctional amino acid. *Trends Plant Sci.* 15: 89–97.

Valluru, R. and Van den Ende, W. (2011). Myo-inositol and beyond – emerging networks under stress. *Plant Sci.* 181: 387–400.

Vijayakumari, K. and Puthur, J.T. (2016). γ-Aminobutyric acid (GABA) priming enhances the osmotic stress tolerance in *Piper nigrum* Linn. plants subjected to PEG-induced stress. *Plant Growth Regul.* 78: 57–67.

Wang, Y., Tan, P., Chang, L. et al. (2022). Exogenous proline mitigates toxic effects of cadmium via the decrease of cadmium accumulation and reestablishment of redox homeostasis in *Brassica juncea*. *BMC Plant Biol.* 22: 182.

Wei, W., Peng, H., Xie, Y. et al. (2021). The role of silicon in cadmium alleviation by rice root cell wall retention and vacuole compartmentalization under different durations of Cd exposure. *Ecotoxicol. Environ. Saf.* 226: 112810.

Yan, Z.M., Guo, S.R., Shu, S. et al. (2011). Effects of proline on photosynthesis, root reactive oxygen species (ROS) metabolism in two melon cultivars (*Cucumis melo* L.) under NaCl stress. *Afr. J. Biotechnol.* 10: 18381–18390.

Yang, X., Liang, Z., and Lu, C. (2005). Genetic engineering of the biosynthesis of glycinebetaine enhances photosynthesis against high temperature stress in transgenic tobacco plants. *Plant Physiol.* 138: 2299–2309.

Yang, A., Cao, S., Yang, Z. et al. (2011). γ-Aminobutyric acid treatment reduces chilling injury and activates the defence response of peach fruit. *Food Chem.* 129: 1619–1622.

Yang, Y., Yao, Y., Li, J. et al. (2022). Trehalose alleviated salt stress in tomato by regulating ROS metabolism, photosynthesis, osmolyte synthesis, and Trehalose etabolic pathways. *Front. Plant Sci.* 13: 772948.

Yu, X.Z., Lin, Y.J., Fan, W.J., and Lu, M.R. (2017). The role of exogenous proline in amelioration of lipid peroxidation in rice seedlings exposed to Cr (VI). *Int. Biodeter. Biodegr.* 123: 106–112.

Zhang, X.Y., Liang, C., Wang, G.P. et al. (2010). The protection of wheat plasma membrane under cold stress by glycine betaine overproduction. *Biol. Plant.* 54: 83–88.

Zhao, D.Q., Li, T.T., Hao, Z.J. et al. (2019). Exogenous trehalose confers high temperature stress tolerance to herbaceous peony by enhancing antioxidant systems, activating photosynthesis, and protecting cell structure. *Cell Stress Chaperones* 24: 247–257.

Index

Global Climate Change and Plant Stress Management, First Edition. Edited by Mohammad Wahid Ansari, Anil Kumar Singh, and Narendra Tuteja.

b

h

m

www.ingramcontent.com/pod-product-compliance
Ingram Content Group UK Ltd.
Pitfield, Milton Keynes, MK11 3LW, UK
UKHW050952290926
15470UKWH00005B/1780

9 781119 858522